Medicines from Animal Cell Culture

Medicines from Animal Cell Culture

Editors

Glyn Stacey

National Institute for Biological Standards and Control
South Mimms, UK

John Davis

Bio-Products Laboratory
Elstree, UK

John Wiley & Sons, Ltd

Telephone (+44) 1243 779777

Email (for orders and customer service enquiries): cs-books@wiley.co.uk
Visit our Home Page on www.wileyeurope.com or www.wiley.com

Reprinted with corrections November 2007

Other Wiley Editorial Offices

John Wiley & Sons Inc., 111 River Street, Hoboken, NJ 07030, USA

Jossey-Bass, 989 Market Street, San Francisco, CA 94103-1741, USA

Wiley-VCH Verlag GmbH, Boschstr. 12, D-69469 Weinheim, Germany

John Wiley & Sons Australia Ltd, 42 McDougall Street, Milton, Queensland 4064, Australia

John Wiley & Sons (Asia) Pte Ltd, 2 Clementi Loop #02-01, Jin Xing Distripark, Singapore 129809

John Wiley & Sons Canada Ltd, 6045 Freemont Blvd, Mississauga, ONT, Canada, L5R 4J3

Wiley also publishes its books in a variety of electronic formats. Some content that appears in print may not be available in electronic books.

Anniversary Logo Design: Richard J. Pacifico

Library of Congress Cataloging-in-Publication Data

Medicines from animal cell culture / [edited by] Glyn Stacey, John Davis.
p. ; cm.
Includes bibliographical references.
ISBN: 978-0-470-85094-7
1. Animal cell biotechnology. 2. Pharmaceutical biotechnology. 3. Stem cells–Transplantation. 4. Recombinant proteins–Therapeutic use. I. Stacey, G. (Glyn) II. Davis, John, 1952-
[DNLM: 1. Cell Culture Techniques–methods. 2. Biotechnology–methods. 3. Drug Compounding–methods. QS 525 M489 2007]
TP248.27.A53M43 2007
615′.19–dc22 2006029344

British Library Cataloguing in Publication Data

A catalogue record for this book is available from the British Library

ISBN: 978-0-470-85094-7 (H/B)

Typeset in 9.5/11.5pt Times Roman by Thomson Digital
Printed and bound in Great Britain by Antony Rowe Ltd, Chippenham, Wiltshire
This book is printed on acid-free paper responsibly manufactured from sustainable forestry in which at least two trees are planted for each one used for paper production.

Cover photograph taken by Howard Brundrett in a Novartis facility.

Contents

Contributors

Kym Baker
Lonza Biologics plc
228 Bath Road
Slough
Berkshire SL1 4DY
UK

Stephen G Ball
UK Centre for Tissue Engineering
University of Manchester
Michael Smith Building
Oxford Road
Manchester M13 9PT
UK

Louise Barnes
Faculty of Life Sciences
University of Manchester
Simon Building
Brunswick Street
Manchester
M13 9PL
UK

Jean Bender
Genentech, Inc.
1 DNA Way
South San Francisco, CA 94080
USA

Genoveva Berná
Andalusian Center of Molecular Biology and Regenerative Medicine
Avda. Americo Vespucio s/n
41092 Seville
Spain

John Birch
Lonza Biologics plc
228 Bath Road
Slough
Berkshire SL1 4DY
UK

Ann E Canfield
UK Centre for Tissue Engineering
University of Manchester
Michael Smith Building
Oxford Road
Manchester M13 9PT
UK

Nigel Chesterton
Validation Department
Bio-Products Laboratory
Dagger Lane
Elstree
Hertfordshire WD6 3BX
UK

Lily Chu
Mail Stop R80Y-115
Merck & Co. Inc.
P O Box 2000
Rahway, NJ 07065
USA

Angus G Dalgleish
Department of Oncology
St George's Hospital Medical School
Cranmer Terrace
London, SW17 0RE
UK

John Davis
Research & Development Department
Bio-Products Laboratory
Dagger Lane
Elstree
Herts WD6 3BX
UK

Johan Descamps
Glaxo SmithKline Biologicals
89 rue de l'Institut
1330 Rixensart
Belgium

Michele Duchene
Glaxo SmithKline Biologicals
89 rue de l'Institut
1330 Rixensart
Belgium

Richard Festen
2261 Market Street #250
San Francisco, CA 94114
USA

Stephen Flatman
Lonza Biologics plc
228 Bath Road
Slough
Berkshire SL1 4DY
UK

Roland Fleck
Division of Cell Biology and Imaging
National Institute for Biological Standards and Control
South Mimms
Hertfordshire EN6 3QG
UK

Barry Fuller
University Department of Surgery
Royal Free & UCL Medical School
Pond Street
London NW3 2QG
UK

Pascal Grabarek
Novartis Pharma SAS
Center of Biotechnology
8 rue de l'Industrie
B P 355
68333 Huningue
France

Alex Green
Pharmaceutical Equipment Validation (PEV) Ltd.
Pinewood
Chineham Business Park
Basingstoke
Hampshire RG24 8AL
UK

J. Bryan Griffiths
5 Bourne Gardens
Porton Down
Wiltshire SP4 0NU
UK

Roland Guenther
Global Biotech CMC
Novartis Pharma AG
CH-4002 Basel
Switzerland

Timothy E Hardingham
UK Centre for Tissue Engineering
University of Manchester
Michael Smith Building
Oxford Road
Manchester M13 9PT
UK

Paul Hill
Chiron Corporation
4560 Horton Street
Emeryville, CA 94608-2916
USA

David W Jayme
Department of Biochemistry and Physical Sciences
Brigham Young University – Hawaii
55-220 Kulanui Street
Box 1967
Laie, HI 96762
USA

Simon A Jeffs
GU Medicine Sub-Section
Infectious Diseases Section
Faculty of Medicine
Imperial College
4th Floor Medical School Building
Praed Street
London W2 1PG
UK

J Jones
Institute of Bioengineering
Avda. de la Universidad s/n
03202 Elche
Spain

Cay M Kielty
UK Centre for Tissue Engineering
University of Manchester

Michael Smith Building
Oxford Road
Manchester M13 9PT
UK

Chris Mannix
18a Chiswick End
Meldreth
Royston SG8 6LZ
UK

Franz Martin
Andalusian Center of Molecular Biology and Regenerative Medicine
Avda. Americo Vespucio s/n
41092 Seville
Spain

Paul Matejtschuk
National Institute for Biological Standards and Control
South Mimms
Hertfordshire EN6 3QG
UK

Anthony Meager
National Institute for Biological Standards and Control
South Mimms
Hertfordshire EN6 3QG
UK

Barbara Orton
QA Department
Bio-Products Laboratory
Dagger Lane
Elstree
Herts WD6 3BX
UK

Peter Phillips
National Institute for Biological Standards and Control
South Mimms
Hertfordshire EN6 3QG
UK

Isabelle Pierard
Glaxo SmithKline Biologicals
89 rue de l'Institut
1330 Rixensart
Belgium

Kirsty E Ratcliffe
UK Centre for Tissue Engineering
University of Manchester
Michael Smith Building
Oxford Road
Manchester M13 9PT
UK

Peter L Roberts
Research and Development Department
Bio-Products Laboratory
Dagger Lane
Elstree
Hertfordshire WD6 3BX
UK

David K Robinson
Mail Stop R80Y-115
Merck & Co., Inc.
PO Box 2000
Rahway NJ 07065
USA

Dorothea Sesardic
National Institute for Biological Standards and Control
South Mimms
Hertfordshire EN6 3QG
UK

Geoffrey Sharpe
Quality Assurance Department
Cobra Biomanufacturing Plc
The Science Park
Keele
Staffordshire ST5 5SP
UK

Bernat Soria
Andalusian Center of Molecular Biology and Regenerative Medicine
Avda. Americo Vespucio s/n
41092 Seville
Spain

Glyn Stacey
Division of Cell Biology and Imaging
National Institute for Biological Standards and Control

South Mimms
Hertfordshire EN6 3 QG
UK

Thibaud S Stoll
Novartis Pharma AG
Forum 3-2.100
CH - 4002 Basel
Switzerland

Edward Tarelli
Department of Haematology
St George's Hospital Medical School
Cranmer Terrace
London SW17 0RE
UK

Simon R Tew
UK Centre for Tissue Engineering
University of Manchester
Michael Smith Building
Oxford Road
Manchester M13 9PT
UK

Lincoln Tsang
Partner Arnold & Porter LLP
Tower 42 Level 40
The International Financial Centre
25 Old Broad Street
London EC2N IHQ
UK

Neil J Turner
UK Centre for Tissue Engineering
University of Manchester
Michael Smith Building
Oxford Road
Manchester M13 9PT
UK

P Vaca
Institute of Bioengineering
Avda. de la Universidad s/n
03202 Elche
Spain

Stephen P Vranch
Jacobs Engineering
17 Addiscombe Road
Croydon
Surrey CR0 6SR
UK

Mike A Whelan
Department of Oncology
St George's Hospital Medical School
Cranmer Terrace
London SW17 0RE
UK

Paul Whitehead
ELGA LabWater
Lane End
High Wycombe
Buckinghamshire HP1 3JH
UK

Mark Wilson
Downstream and Formulation Development
Xenova Ltd
Milton Road
Cambridge CB4 0WG
UK

Preface

Over the years the field of animal cell culture has made numerous contributions to the development of new biomedical products. This book covers the full range of these products from natural biomolecules expressed by cells to genetically engineered molecular products, and also therapeutic cells for implantation in humans. All of these are critically dependent on the quality of the components required for successful culture of animal cells, such as the culture medium, growth supplements, and the majority component of cell culture media, water. There are also key technical and regulatory requirements focused on translating laboratory-scale techniques to provide the quality and quantity of material required to manufacture biomedical products from cell culture processes.

The path from a new research development to marketed product in the biotechnology field is long and expensive, even for a relatively simple biomolecule such as a growth factor or monoclonal antibody. Careful product-specific risk assessment is vital and, as cell therapy, tissue engineering and gene therapy mature into a diverse range of new biological medicines, new issues are arising to challenge those developing these new and exciting products.

This book is an attempt to capture the broad range of issues facing basic scientists, biotechnologists, manufacturers and regulators, who are trying to overcome the problems of understanding and delivering new biological medicines. It provides, from experts in the field, an appreciation of developments in the various products that can be derived from animal cells, including the use of the cells themselves as vaccines and regenerative therapies. The extensive list of chapters covers topics from the culture of the cell to the regulatory requirement for products used internationally, and begins with a review of the whole field from one of the best-known innovators in the field of animal cell technology. The book is intended to provide a valuable general reference for graduates, professional scientists, managers and regulators with an interest in animal cell technology and biological medicines in general. We hope it will give readers a full perspective on the potential and reality of the ever-expanding range of medicines that can be produced from animal and human cells.

Glyn Stacey
John Davis

List of Abbreviations

2-AA	2-aminoacridone
50×	liquid ingredients concentrated by fifty-fold
AAV	adeno-associated virus
ABS	adult bovine serum, from animals 12–72 months old
AcMNPV	*Autographa californica* multiple nucleopolyhedrosis virus
Adv	adenoviral vectors
AEI	N-acetylethyleneimine, chemical inactivant for virus
AGT™	Advanced Granulation Technology
ALS	amyotrophic lateral sclerosis
API	active pharmaceutical ingredient
Asn	L-asparagine (H_2N-CH (CH_2CONH_2)-CO_2H)
AU	absorbance units
BHK	Baby Hamster Kidney (cell line)
BHV	bovine herpes virus
BL# (-LS)	Biosafety Level # (- large scale)
BLA	Biologics License Application
BLV	bovine leukemia virus
BMR	batch manufacturing record
BPL	beta-propiolactone, chemical inactivant for virus
BRSV	bovine respiratory syncitial virus
BSE	bovine spongiform encephalopathy
BTV	bluetongue virus
BV	budded virus
BVD or BVDV	bovine viral diarrhoea virus
CBER	FDA Center for Biologics Evaluation and Research
CDER	FDA Center for Drug Evaluation and Research
CEA	carcinoembryonic antigen
CFR	Code of Federal Regulations (FDA)
cGMP	current Good Manufacturing Practice
CHEF	contour-clamped homogeneous electric field
CHMP	Committee for Medicinal Products for Human use
CHO	Chinese hamster ovary (cell line)
CIM	computer-integrated manufacturing
CIP	clean-in-place
CJD	Creutzfeldt–Jakob disease
CLR	cationic lipid reagent
CMVie	cytomegalovirus immediate early promoter
CO2	carbon dioxide
CoA	certificate of analysis
COO	certificate of origin
COP	cleaning-off-place
COSHH	Control Of Substance Hazardous to Health
CPA	cryoprotective agents

CPE	cytopathic effect
CPMP	Committee for Proprietary Medicinal Products
CPV	canine parvovirus
CRAd	conditionally replication-competent adenovirus
CS	calf serum, from animals less than 12 months old
CSO	contract service organisation
CTL	cytotoxic T-lymphocyte
Cys	L-cysteine (H_2N-CH (CH_2SH)-CO_2H)
DAPI	4′,6-diamidino-2-phenylindole (Fluorescent stain for double-stranded DNA)
DC	dendritic cell
DCS	distributed control system
DEAE	diethyl aminoethyl
DF	diafiltration
DH	Department of Health (UK)
dhfr or DHFR	dihydrofolate reductase
DMEM	Dulbecco's Modified Eagle's Medium
DMF	Drug Master File
DMSO	dimethyl sulphoxide
DNA	deoxyribonucleic acid
DO or dO2	dissolved oxygen
DOE	design of experiments
dpc	days post-coitum
DQ	Design Qualification
DSC	differential scanning calorimetry
dsDNA	double stranded deoxyribonucleic acid
dsRNA	double stranded ribonucleic acid
DTA	differential thermal analysis
DTH	delayed-type hypersensitivity
EBA	expanded bed adsorption
EBNA	Epstein-Barr virus Nuclear Antigen
EBs	embryoid bodies
EC cells	embryonal carcinoma cells
ECGS	endothelial cell growth supplement
ECM	extracellular matrix
ECS	extracapillary space (of a hollow-fibre bioreactor)
EDQM	European Directorate for the Quality of Medicines
EG cells	embryonic germ cells
EGF	epidermal growth factor
EIA	early intermediate activator
ELISA	enzyme-linked immunosorbent assay
EMC	encephalomyocarditis virus
EMEA	European Medicines Agency
EMEM	Eagle's Minimal Essential Medium
EPO	erythropoietin
ePTFE	expanded polytetrafluorethylene
ER	endoplasmic reticulum
ERP	enterprise resource planning (computer system)
ES cell(s)	embryonic stem cell(s)
ESI	electrospray ionization
EU	European Union

FAB	fast atom bombardment
FACE	fluorophore-assisted carbohydrate electrophoresis
FACS	fluorescence-activated cell sorting
FAT	factory acceptance test(ing)
FBS	foetal bovine serum
FCS	foetal calf serum
FDA	Food and Drug Administration (USA)
FDS	Functional Design Specification
FeLV	feline leukemia virus
FGF	fibroblast growth factor
FIA	flow-injection analysis
FISH	fluorescence *in situ* hybridisation
FMD	foot and mouth disease
FS	Functional Specifications
FTIR	Fourier transform infrared spectroscopy
Fuc	L-fucose
GABA	gamma-aminobutyric acid
GAG	glycosaminoglycan
Gal	D-galactose
GalNAc	N-acetyl-D-galactosamine
GCCP	Good Cell Culture Practice
GC–MS	gas chromatography–mass spectrometry
GCP	Good Clinical Practice
G-CSF	granulocyte-colony-stimulating factor
GDNF	glial-derived neurotrophic factor
Glc	D-glucose
GlcNAc	N-acetyl-D-glucosamine
GLP	Good Laboratory Practice
GLSP	Good Large-Scale Practice (biosafety level)
GM-CSF	granulocyte/macrophage colony stimulating factor
GMO	genetically modified organism
GMP	Good Manufacturing Practice
GPCRs	G protein-coupled receptors
GS	glutamine synthetase
HAP	hamster antibody production (test)
HAV	hepatitis A virus
HBSS	Hanks' Balanced Salt Solution
HBV	hepatitis B virus
hCMV	human cytomegalovirus
HCP	host cell protein
HCT/P	human cellular and tissue-based products
HCV	hepatitis C virus
HDS	Hardware Design Specification
HEK	Human Embryonic Kidney
HEK-293	Human Embryonic Kidney 293 (cell line)
HEMA	2-hydroxyethyl methacrylate
HEPA	high efficiency particulate air
HETP	height equivalent to a theoretical plate
HFEA	Human Fertilisation and Embryology Authority (UK)
HIC	hydrophobic interaction chromatography

HIV	human immunodeficiency virus
HIV-1	human immunodeficiency virus Type 1
HLA	human leukocyte antigen
HMI	human-machine interface
HMSO	Her Majesty's Stationery Office
HPAEC–PAD	high pH anion exchange chromatography–pulsed amperometric detection
HPLC	high performance liquid chromatography
HPV	human papilloma virus
HS cells	human stem cells
HSP	heat shock protein
HSV	herpes simplex virus
HSVEC	human saphenous vein endothelial cells
HTLV-1	human T-lymphotropic virus
HTS	high throughput screening
HTST	high-temperature short-time
HUVEC	human umbilical vein endothelial cells
HVAC	heating, ventilating and air conditioning
IBR or IBRV	infectious bovine rhinotracheitis virus
ICH	International Conference on Harmonization
ICM	inner cell mass
IEF	isoelectric focusing
IFN	interferon
IFPMA	International Federation of Pharmaceutical Manufacturers and Associations
IGF	insulin-like growth factor
IgG	Immunoglobulin G
IL	interleukin
IL-1	interleukin-1
IMAC	immobilized metal affinity chromatography
IND	investigative new drug
IPC	in-process control
IPG	immobilised pH gradient
IPTG	isopropylthiogalactoside
IPV	inactivated polio vaccine
IQ	Installation Qualification
IQA	Institute of Quality Assurance
IRES	internal ribosomal entry site
iu	infectious units
IVD	*in vitro* diagnostic
LAF	laminar air flow cabinet
LC–ESMS	liquid chromatography–electrospray mass spectrometry
LC–MS	liquid chromatography–mass spectroscopy
LCMV	lymphocytic choriomeningitis virus
LCR	locus control region
LDC	limiting dilution cloning
LDH	lactate dehydrogenase
LIF	leukaemia inhibitory factor
LN	liquid nitrogen
LOD	limit of detection
LV	lentiviral vectors
LWDS	liquid waste decontamination system

MALDI–MS	matrix-assisted laser desorption ionisation – mass spectroscopy
MALDI–TOF MS	matrix-assisted laser desorption ionisation – time of flight mass spectrometry
Man	D-mannose
MAP	mouse antibody production (test)
MAPC	multipotent adult progenitor cells
MCA	Medicines Control Agency (now MHRA)
MCB	master cell bank
MCS	multiple cloning site
MDBK	Madin-Darby bovine kidney (cell line)
MF	microfiltration
MHC	Major Histocompatibility Complex
MHC	myosin heavy chain
MHC-I	Major Histocompatibilty Complex Class I
MHRA	Medicines and Healthcare products Regulatory Agency
MIR	mid-infrared
MLV	murine leukaemia virus
MOI	multiplicity of infection
mRNA	messenger RNA
MRP	material requirements planning (computer system)
MRP II	manufacturing and resource planning (computer system)
MS	Mechanical Specification
MSC	mesenchymal stem cell
MSC	microbiological safety cabinet
MSX	methionine sulphoximine
MTT	3-(4,5-dimethylthiazol-2-yl)-2,5-diphenyltetrazolium bromide
MTX	methotrexate
MVA	modified vaccinia Ankara
MVM	minute virus of mice
MVSS	master virus seed stock
MW	molecular weight
MWCB	manufacturer's working cell bank
MWCO	molecular weight cut-off (of an ultrafiltration membrane)
NASA	(US) National Aeronautics and Space Administration
NBS	newborn bovine serum, from animals less than 10 days old
Neu5Ac	N-acetylneuraminic acid
Neu5Gc	N-glycolylneuraminic acid
NFF	normal flow filters
NGF	nerve growth factor
NIR	near infrared
NK	natural killer
NMR	nuclear magnetic resonance
NYVAC	New York vaccinia
OD280	optical density (measured at 280 nm)
ODV	occlusion-derived virus
OECD	Organisation for Economic Co-operation and Development
OQ	Operational Qualification
ori	origin of replication
OV	oncoviral vectors
PAR	proven acceptable range
p.i.	post infection

PCR	polymerase chain reaction
PCS	process control system
PCV	packed cell volume
PDA	Parenteral Drug Association
PDGF	platelet-derived growth factor
PDGF-BB	platelet-derived growth factor BB
PDL	population doubling level
PEG	polyethylene glycol
PERT	product-enhanced reverse transcriptase
PFM	protein-free medium
PGA	polyglycolic acid
PGC	primordial germ cells
pI	isoelectric point
PI3	parainfluenza virus type 3
PID	proportional integral derivative (controller)
PK	pharmacokinetics
PLC	programmable logic controller
Poly(A)	polyadenosine
PPD	purified protein derivative
PPV	porcine parvovirus
PQ	Performance Qualification
PrP	prion protein
PrPC	cellular form of prion protein
PrPSc	abnormal (scrapie) form of prion protein
PRV	pseudorabies virus
PS	porcine serum
PSA	prostate-specific antigen
PSMA	prostate-specific membrane antigen
PVDF	poly(vinylidene difluoride)
QA	Quality Assurance
QC	Quality Control
RAP	rat antibody production (test)
RE	restriction endonuclease
rHPC	recombinant human protein C
RNA	ribonucleic acid
RP HPLC	reverse phase HPLC
RSD	relative standard deviation
RSV	Rous sarcoma virus
RSV-LTR	Rous sarcoma virus long terminal repeat promoter
RT	reverse transcriptase
RTD	resistance temperature device (temperature probe)
RT-PCR	reverse transcription-polymerase chain reaction
S-	Serum-Free
S+	Serum containing
SAT	site acceptance test(ing)
SC	stem cell
SCID	severe combined immunodeficiency
SDS	Software Design Specification; also sodium dodecyl sulphate
SEM	scanning electron micrograph(y)
Ser	L-serine (H_2N-CH (CH_2OH)-CO_2H)

SF	serum-free
SFM	serum-free medium
SHIV	simian/human chimaeric immunodeficiency virus
shRNAs	short hairpin RNAs
s-ICAM	soluble intercellular adhesion molecules
SIP	(steam) sterilization-in-place
siRNAs	small interfering RNAs
SIV	simian immunodeficiency virus
SM	smooth muscle
SMC	smooth muscle cell
SOPs	Standard Operating Procedures
ssDNA	single stranded deoxyribonucleic acid
SSEA	stage-specific embryonic antigen
SSM	serum-supplemented medium
ssRNA	single stranded ribonucleic acid
SV40	simian virus 40
TAA	tumour-associated antigen
TBE	tick-borne encephalitis
TCID	tissue culture infectious doses
TEM	transmission electron micrograph(y)
TFF	tangential flow filtration
Tg or Tg‘	glass transition temperature
TGA	thermogravimetric analysis
TGF	transforming growth factor
TGF-β	transforming growth factor – β
Thr	L-threonine (H_2N-CH (CH_3CHOH)-CO_2H)
TMP	transmembrane pressure
TNBP	tri-n-butyl phosphate
TNF	tumour necrosis factor
TOC	total organic carbon
tpa or tPA	tissue plasminogen activator
tRNA	transfer ribonucleic acid
TS cell	trophoblast stem cell
TSA	tumour-specific antigen
TSE	transmissible spongiform encephalopathy
TTV	transplant-transmitted virus
UCOEs	ubiquitous/universal chromatin opening elements
UF	ultrafiltration
URS	User Requirement Specification
UTR	untranslated region
vCJD	variant Creutzfeldt–Jakob disease
VEGF	vascular endothelial growth factor
VERO	an African green monkey kidney cell line
VVM	vessel volumes per minute
vWF	von Willebrand factor
WFI	water for injection (as defined in the US or EU Pharmacopoeia)
WHO	World Health Organisation
WLF	Williams Landel Ferry (Kinetics)

1 The Development of Animal Cell Products: History and Overview

B Griffiths

1.1 INTRODUCTION

A review of the development and application of products manufactured by animal cells in culture is in essence a review of animal cell biotechnology. Given the definition of biotechnology as: 'The integration of natural sciences and engineering in order to achieve the application of organisms, cells, parts thereof and molecular analogues for products and services' (based on Houwink 1989) the obvious beginning of animal cell biotechnology dates to 1954. This was when the first product from the *in vitro* cultivation of animal cells, polio vaccine, was manufactured and administered to the human population. The vaccine was an inactivated form produced in primary cultures of monkey kidney cells (Salk & Gori 1954). Credit for the enabling work that led to this product must go to Enders and his coworkers, who were the first to grow viruses in cell culture, i.e. *in vitro*, as opposed to whole organisms or organ culture (Enders *et al.* 1949). The monkey kidney was chosen because the cells gave good yields of virus and were large organs, providing about 5000 million cells per kidney. This product had an enormous impact, largely eliminating polio from North America and Europe, and effectively launched animal cells as a tool for major manufacturing industry. One of the reasons for this impact can be explained with reference to viral vaccine development and the need for safe and clean substrates for virus propagation.

1.2 HISTORY OF VIRAL VACCINES

The first vaccine (Table 1.1), Jenner's smallpox (1798), was produced on the skin of living animals and was a very 'dirty' preparation. The next vaccine, rabies (1885) produced in spinal cord preparations, was equally contaminated with host proteins and caused severe anaphylactic shock and other side effects. The need for cleaner and safer vaccines led to the use of embryonated chicken eggs (yellow fever, 1935; influenza, 1936) and although an improvement, these preparations were still often contaminated with microorganisms. Thus the use of cultured primary cells was seen as a great breakthrough in terms of microbiological quality and purity (i.e. low levels of extraneous contaminating protein). However, subsequent research showed that monkey kidney cells were host to a wide range of intrinsic viruses such as a collection of simian viruses (SV), herpes B virus, etc. Some of these, like SV40, were known to be transforming viruses and thus concerns were felt over the possibility of introducing tumorigenic material with the vaccine. In fact a batch of polio vaccine contaminated with SV40 was given to vaccinees, causing considerable anxiety, but a follow-up over subsequent years has thankfully shown no increased incidence of cancer in this group over that in the population as a whole. This, together with incidents of insufficient inactivation,

Medicines from Animal Cell Culture Edited by G. Stacey and J. Davis

Table 1.1 History of viral vaccines.

Key Development	Period	Examples
First viral vaccines	Before 1900	Smallpox (1798) from animal skin Rabies (1885) from spinal cord
Safer primary cell substrates	1930s–1950s	Yellow Fever (1935) from chick eggs Influenza (1936) from chick eggs Polio (1954) from primary monkey cells
Controlled cell substrates	1960s	Measles (1963) Mumps (1967) Rubella (1969)
Improved scale-up technology	1970s	Microcarrier culture
Recombinant solution	1980s onwards	Hepatitis (1986) HIV, herpes, CMV, etc.

slowed down the development of new vaccines considerably. Human vaccine manufacture from animal cells only accelerated when the human diploid cell, WI-38, was introduced (Hayflick & Moorhead 1961). This cell strain was shown to be free of all known intrinsic viruses, behaved as a normal (non-tumourigenic) cell in culture, had reproducible growth characteristics, and aged normally before dying out after 50–60 population doublings. Another great advantage was that a batch of cells could be grown up, quality controlled, and hundreds of replicate ampoules banked in liquid nitrogen. Each ampoule could be used for just one vaccine batch and all ampoules would behave in an identical manner. The concept of 'cell banking' is a key milestone in the development of animal cell biotechnology, and one only has to look to the 'HeLa scandal' to know why.

1.3 THE HeLa SCANDAL

During the 1940s and 1950s all attempts at establishing human cell lines failed until 1952, when the HeLa cell line was derived by Gey and coworkers (1952) from a cervical cancer. Everyone was waiting for such an event in order to initiate studies of human cancer, and the cell line was distributed to hundreds of laboratories worldwide. There then followed a proliferation of reports of new cell lines established from all sorts of human tumours. However, with the advent of karyotype analysis, suspicions were raised about the authenticity of many of these lines when human lines were found to have mouse, or monkey, chromosomes and *vice versa*. Gartler's isoenzyme test conclusively showed that 18 cell lines he tested were not as described, but were all HeLa. Despite this evidence there was too much vested interest in published work and cancer grants for the scientific community to accept this and Gartler's work was rebutted. Nelson-Rees, who was curator of a cell culture collection, recognized the truth, and eventually persuaded government agencies to act after publishing a list of all cell lines that were in fact HeLa in *Science* (Nelson-Rees & Flandermeyer 1976).

The truth of the matter was that bad laboratory discipline and lack of SOPs, plus lack of standardization/characterization tests, led to extensive cross-contamination of cell cultures and the fast growing HeLa cell took over. The cost in wasted research was huge in time and money (estimated at over \$20 million), and vaccines were produced in HeLa instead of Henle intestine cells (adenovirus) and monkey cells (polio vaccine) (Gold 1986). The message was clear: if animal cell technology was to advance it was essential to work from cell banks with well-characterized and authenticated material, and to have a battery of quality control procedures to ensure reproducibility and validity of results and products (Table 1.2).

Table 1.2 Landmarks in standardization.

Year	Landmarks
1962	American type culture collection and reference cells initiated. WI-38 (Hayflick & Moorhead 1961).
1963	Master cell bank concept (CCC sub-committee).
1964	Standardized powdered growth medium.
1965	Isoenzyme analysis.
1967	Karyological techniques (Giemsa banding of metaphase spreads).
1974	Publication by Nelson-Rees in *Science* on cell cross-contamination (Nelson-Rees, Flandermeyer & Hawthorne 1974).
1976	Second *Science* publication by Nelson-Rees; cell lines that were actually HeLa contaminants identified (Nelson-Rees & Flandermeyer 1976).
1981	Third *Science* publication by Nelson-Rees (Nelson-Rees, Daniels & Flandermeyer 1981): laboratories 'named and shamed.' Antibody assays.
1990s	Application of DNA fingerprinting. Defined serum-free growth media.

1.4 PERIOD OF VACCINES AND UNCERTAINTY (1954–1975)

The enabling period up to 1954 (Table 1.3) with the establishment of single cell culture (largely thanks to trypsinization), and long-term culture due to the advent of antibiotics, followed by the work of Enders and Salk, gave way to a period when vaccines were the only new products. The impetus for the rush of new vaccines (Table 1.4) in the late 1960s and early 1970s came about due to the establishment of the WI-38, and later the MRC-5, human diploid cell (HDC) strains and to standardization being brought out of the chaos in tissue culture laboratories described above.

Table 1.3 Enabling period for animal cell biotechnology (pre-1954).

1916	Trypsinization (Rous & Jones)
1940s	Antibiotics
1948	L cell line (first transformed and cloned line)
1949	Virus culture *in vitro* (Enders *et al.* 1949)
1953	HeLa cell line (first continuous human line)
1954	Salk polio vaccine

Table 1.4 First period (1954–1975).

	'Viral vaccines, uncertainty and chaos'
1954	Salk Polio Vaccine
1961	WI-38 human diploid cell line (Hayflick & Moorhead 1961)
1960s	FMDV vaccine (large scale – 2×10^8 doses p.a.)
1963	Measles vaccine (WI-38)
1964	Rabies vaccine (WI-38)
1966	Cell line chaos (cross-contamination)
1969	Mumps and rubella vaccines (WI-38)
1970s	HeLa recognized as 'endemic'
1974	Varicella, CMV, TBE vaccines

To summarize the position in 1975, viral vaccines were the only products. In human vaccine production the technology was simple, multiple flasks or roller bottles, dictated by the fact that HDC were anchorage dependent. In the veterinary vaccine field, where the constraints on cell substrates were fewer, a suspension cell (BHK) was being used in large-scale fermenter systems (500 l) based on microbial technology. FMDV vaccine was being produced at 200 million doses p.a., by far the largest single animal cell product. However a range of quality control procedures had been introduced to validate cell lines (karyotype and chromosome banding, isoenzyme analysis), and to ensure freedom from contaminating microorganisms. Also more reproducible media (bulk powders available), cell banking procedures, SOPs and end-product tests were accepted by licensing authorities (such as the FDA) all giving confidence for the further exploitation of animal cells.

1.5 PERIOD OF CONTROL, CONSOLIDATION AND CHEMICAL ENGINEERING (1975–1986)

A problem besetting human vaccine manufacturers was the low productivity of cell cultures, in particular HDC, that were anchorage dependent (therefore needing a large surface area), and tended to grow only as a monolayer giving extremely low cell densities per unit area. Consequently, in this period huge efforts went into developing scaleable systems for monolayer cells with many ingenious chemical engineering methods being applied (Table 1.5). Although it was problem enough for vaccine manufacturers, who used systems employing up to 28 000 replicate roller bottle cultures a week, it was critical for a whole range of potential new products that were secreted from cells in low concentrations (Table 1.6). The diversity of cell culture systems developed (Griffiths 2000) is summarized in Table 1.7, and demonstrates alternative methods of overcoming the limiting factors in scale-up of increasing the surface area:volume ratio, and overcoming nutrient and oxygen shortage and toxic metabolite build up. Microcarrier culture evolved as the dominant technology having the most realistic potential for industrial scale-up.

1.5.1 Microcarrier Technology

Van Wezel (1967) developed the concept of growing cells as monolayers on small spheres (100–500 micron diameter) that could be put into stirred culture systems. It answered the need for a scaleable unit process and provided a huge surface area for growth per unit culture volume

Table 1.5 Second period (1975–1986).

'Control, consolidation and chemical engineering'	
• Bioreactor development and scale-up (multiple to unit processes)	
• Development of genetic products and licensing of non-HDC cells	
1975	Hybridoma (monoclonal antibody) technique
1979	Microcarrier technology established in industry; first recombinant cell line (influenza expression)
1980	Interferon (IFN) from Namalwa cells at 8000 l (Wellcome)
1981	First monoclonal antibody diagnostic kit
1986	αIFN licensed

Table 1.6 Cell productivity and clinical doses required (1983 data).

Product	Productivity (pg/cell/day)	Clinical dose culture volume (l)
Polio vaccine		0.001
FMDV		0.010
Interferon	0.3	0.100
Factor IX	0.25	3
tPA	0.07	10
Monoclonal antibodies (therapeutic)	0.10	50

(25 000 cm^2/l). Developing a suitable sphere to support good cell attachment and growth and withstand the rigours of stirring took a long time, and it was not until the mid-1970s that the first really efficient microcarriers appeared on the market. These were the low surface charge Cytodex (Pharmacia) carriers, and by 1979 the first industrial process based on microcarriers was being used to produce FMDV vaccine. Human vaccines followed in 1982. Currently it is a widely used technique with a diverse range of commercially available microcarriers and has been scaled up to over 1000 l at low density (<5 g/l Cytodex), and to 500 l at high density using spin filters to perfuse cultures with 10–15 g/l Cytodex. The availability of porous microcarriers (Rundstadler *et al.* 1989; Griffiths & Looby 1998) has greatly increased the scope of the method by increasing unit yield and specific productivity due to enhanced perfusion efficiency, and protecting fragile cells from culture turbulence.

Table 1.7 Scale-up diversity.

Flask	225 cm^2
Roller bottle	1500 cm^2
Modified roller bottle (plates, spiral, glass tubes)	
Multi-tray	40 000 cm^2
Cell cube	85 000 cm^2
Tubing	25 000 cm^2
Plates	20 000 cm^2
Unit low population density scale-up	
Stacked plate	250 000 cm^2
Glass sphere	10 000 cm^2/l
Microcarrier	25 000 cm^2/l
High population density scale-up	
Hollow fibres	1 l
Ceramic cartridges	1 l
Membranes	3 l
Encapsulation	40 l
Developments for adherent cells in suspension	
Spin filter	20 × 10^6/ml, 500 l
Porous microcarrier	100 × 10^6/ml, 100 l
Developments for suspension cultures	
Spinner to stirred tank	20 000 l
Airlift	5 000 l

1.5.2 New Products

1.5.2.1 Interferon

Wellcome developed a large-scale (8000 l) suspension cell culture process based on Namalwa cells for the production of alpha interferon (Wellferon) during the 1970 s (Phillips *et al.* 1985). Interferon was believed to be an anticancer agent and was needed in large quantities to carry out appropriate clinical trials. The development of the technology was relatively straightforward, being based on experience gained from producing FMDV vaccine in suspension BHK cells (Radlett *et al.* 1985). However, the Namalwa cell line was cancer a human B-lymphoblastoid cell line, and tremendous efforts had to be made to convince licensing authorities that the final product was safe for human administration. This was achieved and alpha interferon production, licensed in 1981, is one of the landmarks in the evolution of animal cell biotechnology as it opened up the opportunity to use cells other than HDC in pharmaceutical manufacture. There is a wide range of both interferons and interleukins occurring naturally (Griffiths 1991), and to date, alpha and gamma interferons and interleukins 2, 3, 4, 6, 11 and 12 have been manufactured in culture (Clayton 2000a).

1.5.2.2 Antibodies

The production of monoclonal antibodies (MAB) by Kőhler and Milstein (1975) was based on the fusion of a fast growing myeloma cell with an antibody secreting non-transformed lymphoblast. For the first time, preparations of a single specific antibody were possible rather than harvesting mixed soups of antibodies. The potential in diagnostics and as a research tool was enormous. As their potential as biopharmaceuticals and *in vivo* diagnostics was realized so the need for large-scale and licensed production processes increased. Interferon had set the precedent for the use of cancer cell lines and the demand for MABs in the mid-1980s gave a huge impetus to the development of industrial animal-cell biotechnology. It stimulated the innovation of many novel 'turnkey' culture units that could be used by staff relatively inexperienced in cell culture in laboratory-scale production facilities. This enabled a whole range of new start-up biotechnology companies to become established. The production technology was often based on high cell density perfusion devices such as hollow fibre, ceramic cartridge, or porous microcarriers in fluidized beds (Griffiths & Looby 1998). This stimulated many new and unique devices and procedures that have been incorporated into today's manufacturing capability, including serum- and protein-free culture media.

The use of MABs expanded from a low concentration requirement (dose) for diagnostics to large concentration doses for therapeutics (HIV, CMV, cancer, allergic diseases, asthma, arthritis, renal prophylaxis, septic shock, transplantation, and anti-idiotype vaccines). Development of recombinant MABs was largely driven by the need to 'humanize' (Harris 1994) the product to reduce immunological incompatibilities that led to short half-lives in patients and limited treatment to single-doses. The field is expanding with the use of adoptive immunotherapy where the patient's cells are altered and grown *in vitro* and perfused back into the patient. Many novel products have been developed, such as CD4-IgG, which is a combination of the genes coding for the soluble form of the CD4 receptor with the gene sequence for IgG molecules, which results in a soluble receptor for HIV.

To add some perspective to the impact MAB technology has made on cell culture, it has been estimated that a total of 1 g was produced worldwide in 1980, 50 kg in 1988, and today 50 kg is considered a modest output by many individual manufacturing companies. Currently 20–25 % of all new biological medicines are MABs with over six MAB products licensed (e.g. Reopro for angioplasty; Rituxan for non-Hodgkins lymphoma; Herceptin for breast cancer; Zenapax and Simulect for transplant rejection) and huge numbers in Phase III clinical trials.

Thus the position in the evolution of animal cell biotechnology at the end of this second period (1975–1986) was that the first product from a cancer cell had been licensed; MABs had come a long way from Köhler and Milstein's original discovery; recombinant products were in the research laboratory; huge improvements in regulatory control of products had been made incorporating newly developed analytical techniques; and viruses were no longer the only animal cell product, being joined by interferon and MABs.

1.6 PERIOD OF GENETICALLY DERIVED CELLS AND PRODUCTS (1986–1996)

We are now entering a period of activity that will be much more familiar to modern day cell technologists as the subsequent 10 years in the evolution of cell culture saw (Table 1.8):

- the arrival of the genetic age with cell line engineering;
- a huge expansion in the biotechnology industry with animal cell products dominating the field of recombinant products;
- the emphasis change from chemical engineering solutions to cell engineering and cell biology to meet low productivity problems (e.g. recombinant MAB clones producing 800 mg MAB/l were replacing native clones producing 10–50 mg/l).

The fact that cells with specialized *in vivo* functions, such as endocrine cells secreting hormones, could not be grown in cell culture while retaining their specialized properties has always been a great disappointment not only for advancing medical studies but also in using cells to manufacture naturally occurring biologicals. Thus genetic engineering techniques that allowed the gene(s) responsible for production of required biologicals from a highly differentiated (non-cultivable) cell to be inserted into a fast-growing robust cell line (Sanders 1990) opened up enormous possibilities for exploitation by animal cell biotechnology.

The landmark event of this period was undeniably the licensing of tPA (tissue plasminogen activator), the first recombinant animal cell therapeutic product. This pioneering work by Genentech opened up the way to a whole range of new recombinant products of which the next was EPO (erythropoietin).

Table 1.8 Third period (1986–1996) 'genetically modified cells and cell products'.

Key issues

- Rapid expansion of biotechnology industry
- Move from bioreactor engineering to cell modification for productivity
- Defined media with growth factors

Products

1987	monoclonal antibody production
	Orthoclone (1987), Myoscint (1989), Centoxin (1990), Oncoscint (1990), Reopro (1994)
1987	First recombinant product (tPA – activase/actilyse)
1989	EPO (Epogen/ Procrit/ Epoetin)
	hGH (Saizen) HBsAg (GenHevac)
1991	IFN (Roferon)
1992	Factor VIII (recombinate)
	Centoxin

1.6.1 Tissue Plasminogen Activator

Tissue plasminogen activator (tPA) enzyme dissolves blood clots and is therefore used for the treatment of myocardial infarction and thrombolytic occlusions. Alternative products that were available, such as urokinase and streptokinase, were less specific and could cause general internal bleeding and other side effects. A means of producing tPA had been sought for many years but production levels from endothelial cells were too low to form a production process. Even a rich *in vivo* source such as the human uterus only yielded 1 mg tPA/5 kg uterus (0.01 mg purified tPA/uterus) (Griffiths & Electricwalla 1987). Some tumour cell lines, such as the Bowes melanoma, secrete tPA at a higher rate (0.1 mg/l) (Cartwright 1992), but this was still considered uneconomical for production, and at the time was considered unsafe, coming from a human melanoma. tPA was thus an ideal product for recombinant enhancement as it was a high-activity/low-concentration product with a huge clinical demand. Genetic engineering not only allowed the product to be produced in a relatively safe cell line (CHO), but was used to amplify productivity from the low native secretion rates discussed above (Kluft *et al.* 1983) to 50 mg/10^9 cells/day. The product developed so successfully by Genentech (Lubiniecki *et al.* 1989) was licensed as Activase/Actilyse and was produced in a 10000-l fermenter-based process.

1.6.2 Erythropoietin

Erythropoietin (EPO), the second recombinant product from animal cells to be licensed, is a hormone produced by the kidney that controls the maturation of erythroid (red blood) cells, with clinical applications in anaemia caused by, for example, chronic renal failure. It was pioneered by Amgen and is produced in a CHO cell line transfected with the pSSVL-gHu Epo plasmid, and productivity is enhanced by gene amplification (Eridani 1990; Patent Application 1986). A roller bottle process was used for an interesting reason. The aim of any start-up biotechnology company is not only to develop a new product to market but also to get to the market first in order to realize their investment and to get a cash flow before the investment capital runs out. The quickest way to the market is to use a simple and well-tested process, such as roller bottles, which requires less development time and less testing for a licence. The product was licensed in 1989 and at that time the roller culture technique could be augmented with robotic operations and high-efficiency roller bottles. The philosophy was that fine-tuning of the production process could come later, once licensed and on the market.

1.6.3 Cell Biology

Up to the 1990s, physiology studies to increase productivity had been very empirical – altering medium constituents or environmental parameters such as oxygen, CO_2, redox, pH etc., or reducing the ammonia and acidity levels in the culture and just observing the response. Undoubtedly improvements to the culture media and environment were made but these were of very low orders of magnitude and were not enough to alter the costs of the product very much. The change in approach to basic cell biology and genetics proved to be far more rewarding. Just one example is that of apoptosis. It was recognized that cell death, as well as being due to stress or hostile culture conditions resulting in lysis, was also genetically regulated (apoptosis). By 1993, the identification of the genes involved followed by ways of effecting their control, was seen as significant in increasing culture efficiency. We now have pro-apoptotic drugs such as Aptosyn (Cell Pathways, Inc.) which inhibits cyclic-GMP phosphodiesterase, that selectively induces apoptosis in cancer cells. Conversely, inhibiting apoptosis may provide a means of extending the productive life of cells in a production process, and thus increase productivity in both batch and continuous-perfusion cultures.

Other areas of cell biology that have been investigated successfully to improve both product quantity and quality are in the understanding and control of protein regulation, transcription, and post-translational processes. The importance of this is huge given the fact that animal cells are

Table 1.9 Productivity comparisons.

	Bacterium (e.g. *E. coli*)	Yeast (e.g. *S. cerevisiae*)	Animal cell (e.g. CHO)
Volume (μm^3)	0.5	50	500
Growth rate (mean gen. time h)	0.3	0.5	18
Productivity (g/l/day)	65	50	1
Culturability	+++	++	+
Medium	Cheap		Expensive
Scale-up	Simple		Difficult
Product quality	+	++	+++
Glycosylation	–	+	+++

more difficult to handle and less productive than bacteria and yeasts (Table 1.9). Animal cells are the preferred choice for manufacturing medicinal products because they secrete complex molecules in a biologically active configuration, i.e. the product has the three-dimensional and post-translational structure to be both active and immunologically inert on administration to humans. The importance of the product being correctly glycosylated is paramount (Table 1.10).

Thus by the end of this third period (1986–1996) cell biology and genetic engineering practices had opened up a whole new range of products previously impossible to produce because of low productivity or the active cell being uncultivable. The way was open for the current period where the emphasis is changing from using products secreted by the cell to using the cell itself.

1.7 PERIOD OF GENETIC MEDICINES (1996–CURRENT)

Progress into new product areas is now possible because (Table 1.11):

- culture media are now defined; assured and repeatable quality;
- chemical engineering facets of the process are well understood and developed with a range of proven bioreactors to meet foreseen needs, although the area of biosensor technology and scale-down still remains important;

Table 1.10 Glycosylation of animal cell products.

Protein	*N*-Linked	*O*-Linked	Comment
tPA	Complex, high mannose	Yes	High mannose form/mannose essential to protein kinase
EPO	Complex	Yes	Required for protein kinase
HGH	No	No	Unusual for secreted protein
G-CSF	No	Yes	For resistance to agglutination and heat denaturation
FVIII	Complex		
DNase	Complex, high mannose	No	Mannose is phosphorylated
Cerebrosidase	Complex	No	Targeted to macrophages by high mannose complex
FSH	Complex	No	For *in vivo* activity

Data from A.Lubiniecki, personal communication

Table 1.11 Fourth period (1996–current) 'genetic medicines'.

Issues
• Engineering specific Cells rather than Processes
• Cell as the Product
• Scale-Down NOT Scale-Up!
Technologies
Gene therapy
ADA-SCID
Directed at cancer, genetic diseases, and HIV
Cell and tissue engineering
Keratinocytes for burns
Neurotrophic factor (spinal cord implant)
Encapsulated transplant cells
(pancreatic islet, chromaffin)
Artificial organs (external)
Liver, Kidney
Stem cell therapy
Drugs, e.g. CAMs (Cylexin, Caladin, Intergratin)
Toxicology, pharmacology, testing

- the regulatory side is now well understood with a philosophy of increasing focus on the end-product utilizing an expanding array of modern analytical techniques

The knowledge and process structure is now in place so that a specific medical need can be targeted. Examples are given below.

1.7.1 Cell Therapy

The replacement, repair or enhancement of damaged or functionally inadequate tissues and organs is the aim of cell therapy. This can be achieved by transplantation of cells to a target organism and tissue – an early example being the injection of normal foetal cells into the brain of Parkinson's disease patients. There is now a specific product for this purpose – Neurocell PD (cell therapy). Another approach is to implant selected/engineered cells to secrete a missing gene product (e.g. for severe combined immunodeficiency). The market for cell therapy is huge. In the tissue repair field 8 million repair surgeries were carried out in 2000 in the USA, and skin, bone and, particularly, cartilage repair surgery alone cost $232 million.

Examples:

- *burns* – These cost $100 million annually in the USA. Sheets of keratinocyte are produced for burns patients (e.g. Dermagraft). Other products are OrCell, which stimulates repair and a re-generation of tissue, and CCS (composite cultured skin), which is a dermal and epidermal layer supported in a bovine type I porous collagen matrix.
- *encapsulation* – the problem with cell implantation is the immune rejection of the beneficial cells. To avoid rejection, active cells are encapsulated in a porous membrane or fibre (e.g. BHK cells secreting neurotrophic factors to combat neurogenerative disease can be implanted in the spinal cord; pancreatic islet cells for diabetes; chromaffin cells for chronic pain)
- *artificial organs* – for the liver and kidney. So far only externally linked artificial organs have been successful but for renal dialysis they have the advantage over conventional dialysis in that

Table 1.12 Targets for cell therapy. Adapted from 'Turning living cells into tomorrow's pills' (Geron Corporation).

Parkinson's disease	Foetal dopamine cells
Burns patients	Keratinocytes and fibroblasts
Brain and spinal cord	Neurotrophin-secreting cells
Diabetes	Pancreatic islet cells
Pain	Chromaffin cells
Duchenne's MD	Myoblasts
Liver disease	Parenchymal keratinocytes
Cartilage damage	Chondrocytes
Cardiovascular disease	Endothelial cells
Cancer	Haemopoietic cells, bone marrow, adoptive cell therapy
Age-related macular degeneration	Retinal pigmented epithelium
Huntington's disease	Foetal neurones

the encapsulated kidney cells will perform metabolic transformation processes thus returning essential nutrients as well as just removing toxic products.

A wide range of potential cell therapies (Gage, 1998) is listed in Table 1.12.
These techniques give rise to a new technological challenge – that of scale-down of bioreactors. For implantation small bioreactors supporting in excess of 10^9 cells in a small, semipermeable unit that will keep cells viable and active for long periods of time are needed. The future is expected to be dominated by stem cell therapy, i.e. multipotent stem cells that have the capability of differentiating into any cell type, tissue or organ once the controlling factors of differentiation are defined (see Chapter 29).

1.7.2 Gene Therapy

There is tremendous potential for treating a wide range of genetic deficiencies, cancer and viral infections but progress has been extremely slow. It was in 1990 that the first application went into clinical trial (T-lymphocyte directed gene therapy of ADA-SCID). Over 500 clinical products are under investigation or in trial but are beset by many problems, including development of safe and efficient gene delivery systems (Clayton 2000b; Chapter 9).

The principal targets are (Anderson 1998):

- cancer – breast, colon, lung, neuroblastoma, melanoma, ovarian, renal;
- genetic diseases – cystic fibrosis, Gaucher's, SCID;
- viral – HIV:

Even when successful, treatment may still be ineffective in many diseases because the offending gene is still present (e.g. Hartington's sickle cell anaemia). A problem is that a single gene deletion or malfunction may need several replacement genes to effect repair of the malfunctioning gene.

Gene therapy techniques are:

- *ex vivo* – remove the cells from the body, treat, and return to the patient (mainly applicable to blood cells);
- *in situ* – inject a vector carrying the functional gene into the affected tissues (e.g. infusion of adenoviral vectors into the trachea and bronchi of cystic fibrosis patients);

- *in vivo* – a vector is injected into the blood stream (method is still largely a goal rather than a routine application);
- *genoplasty* – the introduction of short, sequence-specific, oligonucleotide fragments to stimulate normal DNA sequence and trick the cell into endogenous repair mechanisms.

The main research effort is currently on engineering viruses.

The era of modern medicines has achieved a great deal but we are still currently in this period because there is so much more to be achieved, particularly in overcoming the basic problems in gene therapy. At some stage the critical technical breakthroughs will be achieved that will see a dramatic range of diseases being treated by these novel technologies.

1.8 CONCLUSION

One has to have the hindsight of many years (in this case around 50) to realize what huge progress has been made in animal cell biotechnology from its basic beginning in 1954.
The landmarks have been:

1. polio vaccine (1954) – the beginning of animal cell technology with the first cell culture derived product;
2. human diploid cell lines (1963), cell banking and cell line authentication – an end to unreliable and contaminated cell lines and the basis for successful licensing of processes;
3. interferon from Namalwa cells – the first product from a cancer cell line;
4. monoclonal antibody technology (1980s) – a huge range of products and applications bringing about a huge increase in bioreactors and start-up biotechnology companies;

Table 1.13 Animal cell products – licensed, under trial and potential.

Product range	Target diseases
Vaccines: native, recombinant and DNA	Viral infections, arthritis, multiple sclerosis, cancer
Immunoregulators: interferons, interleukins	Cancer, HIV, transplantation, tissue regeneration,
Blood clotting factors: Factor VII, VIII, IX	Haemophilia A and B
Hormones: hGH, FSH (Gonal-F)	Dwarfism, fertility, contraception
Antibodies (monoclonal)	Diagnostics *in vitro* and *in vivo*, cancer, Vascular remodelling
Tumour necrosis factors	Cancer
Colony stimulating factors	Neutropenia, sepsis, infectious disease
Growth factors	Multiple sclerosis, ulcers, diabetes, tissue repair
Gene therapy	See Table 1.12
CAMs (cell adhesion molecules)	Cancer, atherosclerosis, infections
Others	
tPA	Myocardial infarction, thrombolytic occlusion
Erythropoietins	Anaemia
Dismutases	Oxygen toxicity
Soluble receptors	Asthma, arthritis, septic shock
Antisense	Viral (e.g.HIV), cancer, inflammatory disease
Stem cells	Tissue engineering, cell therapy, cancer

5. tissue plasminogen activator (1987) – the first recombinant product from animal cells giving rise to new products (from previously uncultivable cells without genetic modification, or of products previously too pathogenic to be produced safely in a production process), higher productivity and quality of product;
6. cell engineering for treatment of burns (1991) opening up new fields of genetic medicine.

Against this list of landmarks, the cell culture process itself has shown remarkable progress. In 1954 multiple flasks, followed by roller bottles were used and this developed into 10–20 000l unit processes based on stirred or airlift bioreactors with innovative adaptations such as spin filters for perfusion. Unit cell density has increased from $1\text{–}2 \times 10^6$ cells/ml to over 10^8/ml and long term (50–150 days) perfusion processes giving a high daily yield of product are commonplace. It is interesting to reflect that after 45 years of trying every means possible to scale-up, a current requirement is for scale-down for implantable bioreactors.

Products have been highlighted that have had a particular significance in the development of animal cell technology. These are by no means the only products produced with animal cells and a full list is given in Table 1.13.

The subject is still evolving fast and is no longer the province of just chemical engineers, but now includes a range of disciplines from the biochemist and geneticist through the various engineers to medical practitioners. Also the cell is now becoming the principal product rather than just being a vehicle or factory for producing proteins.

REFERENCES

Anderson WF (1998) *Nature (Supplement)*; 392: 25–30.

Cartwright T (1992) In *Animal Cell Biotechnology*. Eds Spier RE, Griffiths JB. Academic Press, London; Vol 5: 217–246.

Clayton T (2000a) In *Encyclopedia of Cell Technology*. Ed. Spier RE. John Wiley & Sons, Inc., New York; Vol. 1: 423–441.

Clayton T (2000b) In *Encyclopedia of Cell Technology*. Ed Spier RE. John Wiley & Sons Inc., New York; Vol. 1: 441–457.

Enders JF, Weller TH, Robbins FC (1949) *Science*, 109: 85–87.

Eridani S (1990) In *Animal Cell Biotechnology*. Eds Spier RE, Griffiths JB. Academic Press, London; Vol 4: 475–490.

Gage FH (1998) *Nature (Supplement)*; 392: 18–24.

Gey GO, Coffman WD. Kubicek MT (1952) *Cancer Res.*; 12: 364–365.

Gold M (1986) *A Conspiracy of cells*. State University of New York Press, New York.

Griffiths JB (1991) In *Mammalian Cell Biotechnology – A Practical Approach*. Ed Butler M. IRL Press, Oxford; 207–235.

Griffiths JB (2000) In *Animal Cell Culture*. Ed. Masters JW. Oxford University Press, Oxford; 19–68.

Griffiths JB, Electricwalla A (1987) *Adv. Biochem. Eng. Biotechnol.*; 34: 147–166.

Griffiths JB, Looby D (1998) In *Cell and Tissue Culture: Laboratory Procedures in Biotechnology*. Eds Doyle A, Grifiths JB. John Wiley & Sons Ltd, Chichester; 268–281.

Harris WJ (1994) In *Animal Cell Biotechnology*. Eds Spier RE, Griffiths JB. Academic Press, London; Vol 6: 259–280.

Hayflick L, Moorhead PS (1961) *Exp. Cell Res.* 25: 585–621.

Houwink EH (1989) *Biotechnology – Controlled use of Biological Information*. Kluwer Academic Publishers, Dordrecht.

Kluft C, van Wezel AL, van der Velden CAM, Emeis JJ, Verheijen JH, Wijngaards G (1983) *Adv. Biotechnol. Proc.*; 2: 97–110.

Köhler G, Milstein C (1975) *Nature*; 256: 495–497.

Lubiniecki A, Arathoon R, Polastri G, *et al.* (1989) In *Advances in Animal Cell Biology and Technology for Bioprocesses.* Eds Spier R, Griffiths JB, Stephenne J, Crooy PJ. Butterworth & Co Ltd., Oxford; 442–451.

Nelson-Rees WA, Daniels DW, Flandermeyer RR (1981) Science; 212: 446–452.

Nelson-Rees WA, Flandermeyer RR (1976) Science; 191: 96–98.

Nelson-Rees WA, Flandermeyer RR, Hawthorne PK (1974) Science; 184: 1093–1096.

Patent Application WO 85/02610 (1986) Kirin-Amgen Inc.

Phillips AW, Ball GD, Fantes KH, Finter NB, Johnston MD (1985) In *Large-scale Mammalian Cell Culture.* Eds Feder J, Tolbert WR. Academic Press Inc., New York; 87–95.

Radlett PJ, Pay TWF, Garland AJM (1985) *Dev. Biol. Stand.* 60: 163–170.

Rous P, Jones FS (1916) *J. Exp. Med.*; 23: 549–555.

Runstadler PW Jr, Tung AS, Hayman EG, Ray NG, Sample JVG, DeLucia DE (1989) In *Large Scale Mammalian Cell Culture Technology.* Ed. Lubiniecki AS. Marcel Dekker, New York; Vol. 3: 363–381.

Salk JE, Gori JB (1954) *Am. J. Publ. Health*; 44: 563.

Sanders PG (1990) In *Animal Cell Biotechnology.* Eds Spier RE, Griffiths JB. Academic Press, London; Vol. 4: 16–70.

van Wezel AL (1967) *Nature*; 216: 64–65.

Fundamental Elements of Cell Growth Media

2 Water Purity and Regulations

P Whitehead

2.1 INTRODUCTION

Since time immemorial man has purified water. Initially his main concern was making the water fit and palatable to drink. In *The Deipnosophists* dating from 168 BC, Athenaeus of Naucratis describes how the Egyptians purified jars of river water by a combination of exposure to sunlight and air, straining and allowing to settle overnight. Similar techniques are seen in engravings in Egyptian tombs dating from the fifteenth century BC (Purchas 1981). However, it was in the nineteenth and twentieth centuries that drinking-water purification technologies developed on a large scale, for example with the introduction of compulsory filtration of drinking water in London in 1852, and the addition of chlorine to control bacteria levels in water in the UK in 1904 (Scott 2000).

Distillation has long been the method of purifying water for scientific use. A still was standard equipment for alchemists in the Middle Ages (Saunders 1981). It was not until the twentieth century that it was displaced as the major water purification process following the invention of a whole battery of alternative technologies, such as ion-exchange softening in 1905 (Gans 1905), cation and anion exchange in 1935 (Adams 1935), and reverse osmosis in the 1960s (Loeb 1967; Schultz 1966). These and other technologies have been developed and refined to meet the ever more stringent demands for highly purified water of the microelectronics and pharmaceutical industries (see, for example, ASTM 99) and ultra-trace analytical techniques such as ICP-MS and gradient HPLC.

Purified water is a key component in cell culture work and related preparative and analytical activities. The water purity is critical. This water can be provided by a variety of means and the approach chosen is often a function of the other activities on-site. The purity specified and technologies used are based on a combination of the technical requirements of the work and the selection of a suitable standard specification that corresponds with these requirements.

This chapter will provide some background information on impurities present in water and their origins, the standards to be met and the means to achieve the required purity.

2.2 IMPURITIES IN WATER SOURCES

Purified water is usually produced by the multi-stage treatment of a potable water supply. Potable water is sourced from a combination of surface water, river water and underground aquifer. Impurities originally present can be divided into dissolved ionic matter, organic compounds, particulates, colloids, and a range of bacteria and other life forms. Dissolved salts are leached into the water from rocks or soil – calcium, sodium, bicarbonate, chloride and sulphate are the most common ions found. Organic compounds in the feed-water are both naturally occurring and man-made. The former are mainly a complex mixture of fulvic and humic acids and tannins derived from the decomposition of leaves and grasses. In addition there are bacteria, other living creatures, and their

Medicines from Animal Cell Culture Edited by G. Stacey and J. Davis

Table 2.1 Typical mains water impurities and target values for cell culture work.

Parameter	Mains water	Water for cell culture	% reduction
Conductivity (μS/cm)	50 to 900	0.2	99.95
Calcium (mg/l)	20 to 150	<0.01	>99.99
Sodium (mg/l)	20 to 150	<0.01	>99.99
Iron (mg/l)	0.01 to 0.1	<0.001	>98
Bicarbonate (mg/l)	30 to 300	<0.01	>99.99
Chloride (mg/l)	10 to 150	<0.01	>99.99
Sulphate (mg/l)	1 to 100	<0.01	>99.98
TOC (mg/l)	0.2 to 5	0.1	96
Free chlorine (mg/l)	0.1 to 0.5	<0.01	>97
Bacteria (CFU/100 ml)	100 to 1000	<10	>98
Endotoxin (IU/ml)	1 to 10	<0.1	>98
Turbidity	0.1 to 2	<0.01	>99

by-products. Industrial, agricultural and domestic wastes contribute detergents, solvents and oils along with fertilizers, herbicides and pesticides.

As the water is treated to make it suitable for domestic or industrial use, many of the impurities are removed, among them heavy metals and pesticides, but others are introduced, for example plasticizers from plastic pipes and tanks. Other compounds are produced by reactions with the chlorine or ozone used to control bacterial levels.

The main alternatives to potable water as a water source are boreholes that take water directly from an aquifer. The cost savings advantage of this approach can be seriously offset by the need for extensive extra water treatment to bring the water to a sufficient standard for purification. Dissolved iron and carbon dioxide can be particularly problematic in such waters. Where the purified water is to be used for pharmacopoeia applications there is a further requirement that the source water is of an equivalent purity to potable water (USP 2006; EP 2006).

Depending on local sources the impurities in mains water will vary widely. Typical ranges are shown in Table 2.1 along with typical impurity targets for water for cell culture.

2.3 SIGNIFICANCE OF IMPURITIES IN WATER FOR USE IN CELL CULTURE

To avoid interference with the various processes involved in cell culture it is, above all, essential to minimize the presence of biologically active species, for example, endotoxins, bacteria and nucleases. In addition, levels of ionic contaminants, especially multivalent ions and heavy metals, and organic contaminants must be kept low. For ancillary operations, such as initial rinsing of equipment and some media preparation, less tight specifications are acceptable (see Finter *et al.* 1990).

The potential adverse effects of endotoxins on cell culture have been widely reported (see Case Gould 1984). Dawson (1998) produced an interesting review, highlighting the wide range of effects and great variation in sensitivity of even parent and daughter cell lines. A series of papers describing the interactions of endotoxins with cells are included in Levin *et al.* (1993). Nagano *et al.* (1999) reported the beneficial results obtained by eliminating endotoxins from water for preparing protein-free media for the culture of bovine embryos. Endotoxins are well known to have deleterious effects on *in-vitro* fertilization, see for example Fukuda *et al.* (1986) and Dumoulin *et al.* (1991). Cotton (1994) and Weber (1995) both reported that the presence of endotoxin in plasmid preparations lowers the transfection efficiency of endotoxin-sensitive cell lines. Weiss and Goldwasser (1981) observed that the biological effects attributed to erythropoietin were, at least in part, due to the use of material contaminated with bacterial endotoxin. Epstein (1990) found that less than 20 ng/ml of *E.coli* endotoxin had no detectable effect on the cell types tested and used 1 ng/ml as an acceptable limit for all his cell culture media. However, standardized measurements of bacterial endotoxin use the limulus amoebocyte lysate assay (Novitsky 1984, USP 2006) that determines endotoxin levels in international units for which the acceptable maximum limit for water for injection (WFI) is 0.25 IU/ml, equivalent to about 0.05 ng/ml. However, even this standardized test for endotoxin will not detect the non-classified endotoxins of Gram-positive organisms. These are detected only by the use of human peripheral blood leucocytes (Gaines Das *et al.* 2004).

Pseudomonas species are the bacteria most widely found in purified water. Martino *et al.* (1996) have reviewed a number of papers reporting infections caused by *Pseudomonas* species. As described in Whitehead (1999), salts present can form deposits that can act as centres for bacterial growth. Dissolved organic compounds, also act as a source of nutrients for bacteria. Organic compounds present in the water can also cause a variety of problems with trace HPLC and GC analyses including poor detection limits and reproducibility, and contamination of separation media and detectors. Concentrations as low as 1 μg/l can be problematic. Examples of the sensitivity of analyses to trace contaminants are given by Anantharaman *et al.* (1994), Whitehead (1998), and Reust and Meyer (1982). Gabler *et al.* (1983) have also described the organic contamination of high purity water on storage.

2.4 STANDARDS

The principal organizations concerned with standards of purified water relevant to pharmaceutical, clinical and molecular biological activities are the pharmacopoeias, CLSI, CAP, ASTM and ISO. The organizations and relevant water grades are summarized in Table 2.2. Of these, the pharmacopoeial standards are the ones most widely applied to cell culture production activities.

Many countries produce and apply their own pharmacopoeial standards. For the different grades of purified water these pharmacopoeia set broadly similar standards with the main differences relating to restrictions on methods of production and testing. In practice, the key standards are those set down by the US and European pharmacopoeias (USP 2006; EP 2006; see web site addresses at the end of the chapter). The two established grades that are relevant to cell culture work are 'purified water' and 'water for injection' (WFI). In 2002 the European Pharmacopoeia (EP) introduced a new grade of 'highly purified water'. This grade is identical in terms of purity

Table 2.2 Pure water standards.

Organisation	Standards
American Society for Testing and Materials (ASTM)	D 1193–99 Reagent water
College of American Pathologists (CAP)	Refer to CLSI
ISO	BS EN ISO 3696:1995 Water for analytical laboratory use
Clinical and Laboratory Standards Institute (CLSI)	Guideline for Preparation and Testing of Reagent Water in Clinical Laboratories 4th edition (C3–A4)
European pharmacopoeia (EP)	Purified water, Highly purified water, Water for injection
US pharmacopeia (USP)	Purified water, Water for injection

Table 2.3 Specifications for Purified Water.

	Purified Water	
Parameter	USP 29 (2006)	EP (5th Edition, 2006)
Electrical conductivity	<1.3 μS/cm at 25°C	Max 5.1 μS/cm at 25°C
TOC	<0.5 mg/l	<0.5 mg/l or oxidisable substances test
Nitrates (mg/l)		Max. 0.2 ppm
Heavy metals (mg/l)		Max 0.1 ppm
Bacteria (Total aerobic count)	<100 CFU/ml	<100 CFU/ml
Production method	Suitable process	Distillation, ion exchange or other suitable method

specification to WFI (EP), except that it does not specifically require the water to be distilled. This also brings it in-line with WFI (USP). These specifications are summarized in Tables 2.3 and 2.4.

2.5 WATER PURIFICATION

Water that is pure enough for use in cell culture is usually produced by the multistage treatment of potable mains water. Mains water tends to be produced locally and contains a wide variation in impurities that have to be removed before the water is fit for purpose. As shown in Table 2.1, it is necessary to substantially reduce the levels of all types of contaminant. The maintenance of low bacterial levels within a water system is a particular challenge and needs to be considered at each stage of the system design.

Water purification will normally consist of several *pre-treatment* steps, the choice of which is largely governed by the nature of the local feedwater, *main purification* in one or more stages and *final treatment* to achieve and maintain the required water purity.

Table 2.4 Specifications for *WFI* and *Highly Purified Water.*

	Water for Injection		Highly Purified Water
Parameter	USP 29 (2006)	EP (5th Edition, 2006)	EP (5th Edition, 2006)
Electrical conductivity	Max 1.3 μS/cm at 25 °C	Max 1.3 μS/cm at 25 °C	Max 1.3 μS/cm at 25 °C
TOC	Max 0.5 mg/l	Max 0.5 mg/l	Max 0.5 mg/l
Nitrates		Max 0.2 ppm	Max 0.2 ppm
Heavy metals		Max 0.1 ppm	Max 0.1 ppm
Bacteria (Total aerobic count)		<10 CFU/100 ml	<10 CFU/100 ml
Endotoxin	Max 0.25 IU/ml	<0.25 IU/ml	<0.25 IU/ml
Production methods	Distillation or equivalent process	Distillation	Suitable process e.g. double-pass RO, UF

2.6 PRETREATMENT

Pretreatment is required to achieve one or more of the following:

- control of fouling by removal of particulates and organic and microbial impurities;
- control of scaling by removal of hardness and metals, and
- removal of microbial control agents.

The principal techniques used are filtration, softening, and treatment with carbon to remove organics and chlorine.

Depth or media filtration, to remove particulates, is often the first step. Multi-sized sand is common on larger plants; on small systems replaceable depth filters can be used. Control of bacterial growth in the media may be required.

Ion-exchange water softening is very commonly used to avoid precipitation of sparingly soluble salts of divalent and trivalent cations such as calcium carbonate and sulphate, later in the purification process. Passage through a bed of cation ion-exchange resin in the sodium form replaces the great majority of other cations with sodium. Carbon dioxide removal by degassing after acidification can also be advantageous. Addition of a sequestering agent to complex the problem metals is an alternative approach.

The initial removal of organic impurities can be carried out by oxidation with ozone, strong base ion-exchange or barrier filtration (possibly with the addition of a flocculent) but the most common approach is the use of activated carbon to adsorb organics and remove chlorine and chloramines. The main disadvantage of carbon is that it can act as a source of nutrients and a large surface area for microbial growth.

2.7 MAIN PURIFICATION TECHNIQUES

A wide range of processes in various configurations can be used to produce pharmacopoeia-grade purified water.

2.7.1 Reverse Osmosis

Reverse osmosis (RO) uses pressure to force water and other low molecular weight molecules through a semipermeable membrane that resists the passage of ions and acts as a barrier to colloids, bacteria, endotoxins and larger organic molecules. It is the principal technique used for removing well over 90 % of all impurities other than dissolved gases. Its characteristics are illustrated in the left hand side of Figure 2.1.

2.7.2 Ion Exchange

Ion exchange (IE) utilizes the long-established process of exchanging impurity anions and cations for hydroxyl and hydrogen ions, respectively, on passage through beds of anion-and cation-exchange resins. The beds of resin can be separate or mixed. Two-bed systems are very effective for removing the bulk of contaminant ions while the mixed-bed units can achieve the highest levels of ionic purity. The latter are typically used as secondary or 'polishing' systems. When the effective resin capacity is exhausted IE systems require regeneration-with acid for cation beds and alkali for anion beds. The alternative is for the cylinders to be replaced and the regeneration carried out off-site in a specialist regeneration station. For small systems the ion exchange resins are contained in disposable packs.

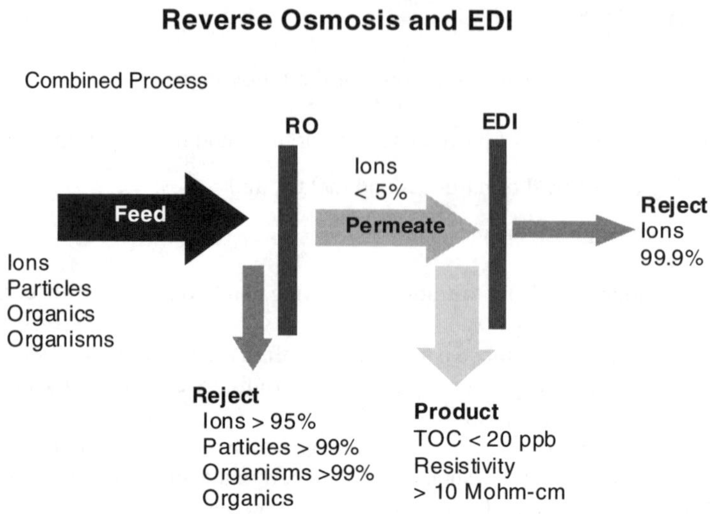

Figure 2.1 Characteristics of Reverse Osmosis and Electrodeionisation (EDI).

2.7.3 Electrodeionization

Electrodeionization (EDI) overcomes the need for regenerant chemical handling and avoids the variation in water purity inherent in conventional ion-exchange cycles. The basic process in EDI involves the removal of ions as they traverse a bed of ion-exchange resin in a 'stack' across which a DC electrical field is applied. Ions are taken up by the resin and move perpendicularly to the water flow in the direction of the electric field. The resin beds are delimited by anion and cation ion-exchange membranes that allow the passage of either anions or cations but not both. A suitable combination of single- or mixed-resin beds facilitates the transfer of impurity ions into waste streams. The current flowing, in effect, maintains the resins in a regenerated form and avoids the need for chemical regeneration. EDI is an increasingly popular approach to ion removal. There are a variety of different designs of EDI systems using various combinations of mixed- and single-resin beds. To reduce ionic load and to avoid contaminating the resins in the electrical stack, EDI uses feedwater pretreated by dechlorination, softening and RO. Figure 2.1 illustrates how the characteristics of EDI complement those of reverse osmosis.

2.7.4 Ultrafiltration

During ultrafiltration water is forced by pressure through a fine filter (typically 2 to 50 nm pore size). The filter prevents the passage of particulates and large molecules including bacteria and endotoxins. Usually a small proportion of the inlet water stream is directed to waste in order to flush the filter surface and minimize build-up of impurities. The major use of ultrafiltration is to provide a barrier to bacteria and large molecules such as endotoxins and RNase.

2.7.5 Microfilters

Microfilters are generally used for microbial retention downstream of potential sources of contamination. They have pore sizes ranging from 0.1 to 0.45 microns and are highly effective. However, they do not prevent the passage of endotoxins and other molecules; if significant concentrations of bacteria collect on the surface, release of additional endotoxins can occur.

2.7.6 Ultraviolet Light

Ultraviolet light at wavelengths from about 240 nm to 300 nm damages DNA in microorganisms and the resultant modifications bring about the destruction of the organism. The exposure of purified water to a sufficiently high dose of UV light can be an extremely effective bactericide and is used as a final treatment step after other purification processes. However, it has little or no effect on endotoxins.

2.7.7 Distillation

Distillation can be used for removing the bulk of impurities in water when fed with suitably pre-treated water. However, due to the high energy requirements for large-scale production, its role is mainly limited to the production of WFI. Distillation is a requirement in the production of WFI for the European Pharmacopoeia and is widely used for WFI for USP also. It can provide very effective (>99.9 %) reduction in endotoxins and bacteria.

In distillation, water is evaporated producing steam and leaving behind dissolved solids and non-volatiles. Volatile and low molecular weight impurities, including endotoxins, entrained with water droplets in the steam, are removed in a separator before the steam is condensed. Multi-effect and vapour-compression stills provide more energy effective alternatives to the basic still. Pre-treatment is needed to prevent scaling and chlorine damage and also to avoid excessive bacterial and endotoxin loads.

Other sources of information on water purification technologies and the latest developments are presented in a list for further reading.

2.8 WATER PURIFICATION SYSTEMS FOR CELL CULTURE

The common practice for cell culture on a production scale is to provide pharmacopoeia-grade *Purified Water* for site applications and use this as a feed to a still to provide WFI. This combination offers considerable regulatory convenience and Bergmann (1990) provides an example.

Purified water may be produced by ion-exchange, distillation, or reverse-osmosis-based systems. In practice, due to the advantages of low running costs, and operational convenience, over 90 % of new systems use primary stage reverse osmosis with final polishing by electrodeionization, ion-exchange or a second reverse osmosis stage (ISPE Guide 2001). An example of this type of system using RO and EDI is shown in Figure 2.2.

As described by Jordain (2002) and Lampard (2002), it combines, on a single 'skid', water softening and micro-filtration pretreatment, reverse osmosis and EDI. UV disinfection and ultra-filtration can be added if required. The stainless steel construction enables hot water sanitization at 85 °C to be used. A simplified flow schematic is given in Figure 2.3. The version shown includes an ultrafilter and is specified to meet the EP 'highly purified water' standard. Without the ultrafilter, performance is well within 'purified water' requirements. If distillation is needed for WFI then the product water can be fed to a still.

On a small scale, for example within laboratories, various other alternatives are possible. Stills can be used, often combined with pretreatment by reverse osmosis or ion-exchange to minimize maintenance. The more frequently used approach is a miniature version of the multiple technology systems described above. This is normally provided in two stages. An initial step involves pretreatment and reverse osmosis, sometimes combined with ion exchange or EDI to fill a reservoir with partially purified water. This water is then 'polished' to achieve its final purity by repeat treatment using a combination of ion exchange, ultraviolet exposure and ultra-filtration. Final, point-of-use filters can provide further protection against bacterial contamination. Such a polishing system is shown schematically in Figure 2.4.

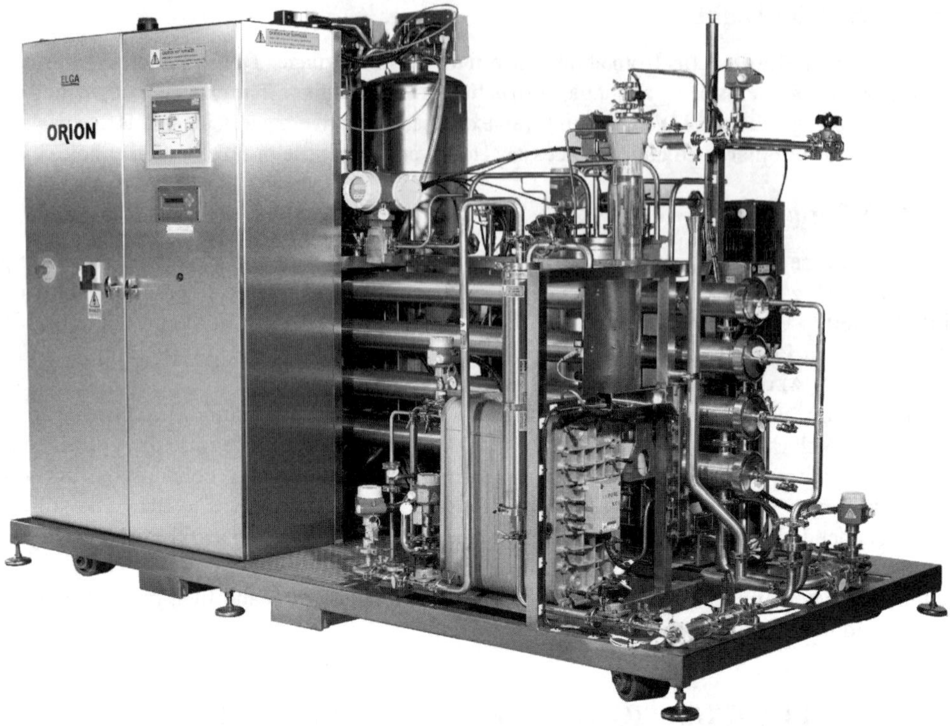

Figure 2.2 Hot-water sanitisable water purification system to produce *Highly Purified Water.*

2.9 DISTRIBUTION SYSTEMS

Having produced water of suitable purity it is vital to provide a means of maintaining that quality. Control over bio-burden and endotoxin is essential. On a large scale this can be achieved most effectively by recirculation at elevated temperatures, typically over 80 °C, to inhibit microbial growth. The water can be cooled locally before use or the loop operated cool during the working day and raised to high temperatures at other times. Such technology is well established for 'Water for Injection' production. Cold recirculating loops are generally less expensive to install but require regular flushing and sanitization, for example with ozone, to maintain water purity.

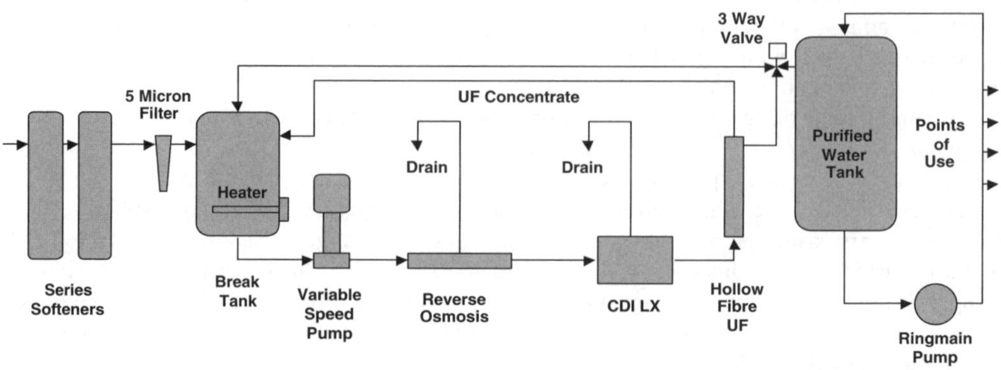

Figure 2.3 Water system for producing *Highly Purified Water* (EP).

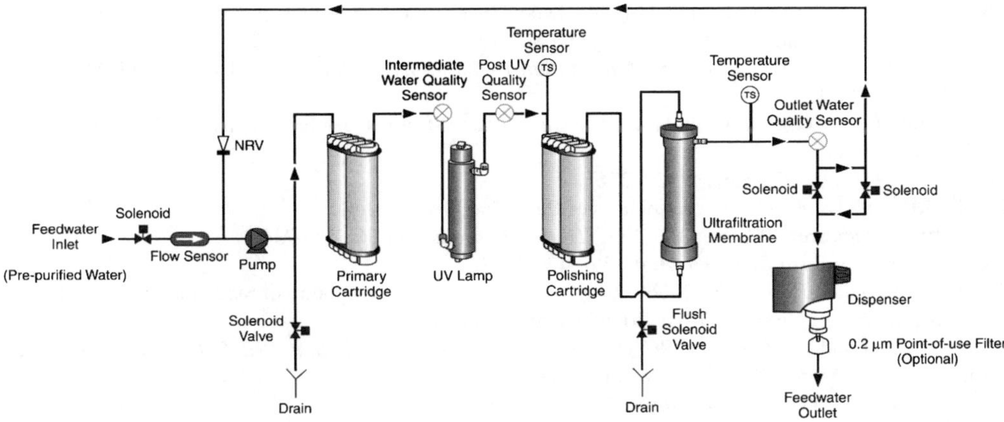

Figure 2.4 Laboratory water system producing ultrapure water suitable for cell culture.

On a laboratory scale good bacterial and endotoxin purity can be maintained using intermittent recirculation through an UV chamber and an ultrafilter combined with periodic chemical sanitization. Such an approach is shown in Figure 2.4. Its advantages are described by Mortimer and Whitehead (2001).

2.10 SYSTEM MONITORING AND VALIDATION

Two essential aspects of any pharmaceutical pure water system are water purity monitoring and validation. The usual parameters monitored on-line are electrical conductivity and total organic carbon (TOC). The requirements are discussed in detail in the pharmacopoeias (USP 2006; EP 2006). To avoid errors in temperature compensation, conductivity measurements without temperature correction are specified. Alternative calibration procedures for the meters are fully specified. For TOC there is a specific requirement to make regular suitability tests on the equipment used. Routine monitoring of the other parameters, including bacteria and endotoxin, when specified, is usually carried out off-line. It is of note that since edition 4.2 (the second supplement to edition 4), EP has specified the use of the low nutrient growth medium R2A agar and incubation at 30 to 35 °C for 5 days for determining total viable bacteria counts. This ensures isolation of a broad range of environmental organisms.

Full system validation is required for pharmaceutical production facilities and is being increasingly sought for laboratory water systems as well. Keer (1995, 1996) and Hill (2002) discuss the former while Mortimer (2002) considers the implications for laboratory water. Section 1231 'water for pharmaceutical purposes' in the General Information of USP (2006) also describes in some detail validation and system design, highlighting bacterial control as the principal challenge. Although system monitoring is essential, validation can be greatly simplified by suitable system design. For example, the use of elevated temperatures for bacterial control is easy to validate by logging temperatures against time, while systems using periodic chemical sanitization can incorporate, for example, conductivity profiling of the sanitization cycles to log contact times.

REFERENCES

Adams BA, Holmes EL (1935) *J. Soc. Chem. Ind.*; T54: 1.

Anantharaman V, Parekh B, Hedge R (1994) *Ultrapure Water*; 11: 30–36.

ASTM (1999) *Standard Guide for Ultrapure Water Used in the Electronics and Semiconductor Industry, D5127 – 99*, American Society for Testing and Materials, Philadelphia, PA, USA.
Bergmann DG (1990) In *Large-scale Mammalian Cell Culture Technology.* Ed Lubiniecki AS. Marcel Dekker, New York.
Case Gould M (1984) *Endotoxin in Vertebrate Cell Culture.* Tissue Culture Association, Gathersburg, MD: 125–136.
Cotton M (1994) *Gene Therapy*; 1: 239–246.
Dawson ME (1998) *LAL Update.* Associates of Cape Cod; Vol. 16: 1–4.
Dumoulin JC, Menheere PP, Evers JL (1991) *Human Reproduction*; 6: 730–734.
Epstein J (1990) *In Vitro Cell. Dev. Biol.*; 26: 1121–1122.
European Pharmacopoeia, (EP) (2006) European Directorate for the Quality of Medicines of the Council of Europe, Strasbourg, France; Fifth edition including supplement 5.5.
Finter NB, Garland AJM Telling RC (1990) In *Large-scale Mammalian Cell Culture Technology.* Ed Lubiniecki AS. Marcel Dekker, New York.
Fukuda A, Noda Y, Yano J (1986) *Sanfu Shinpo*; 38: 39–48.
Gabler R, Hedge R, Hughes D (1983) *J. Liquid Chromatog.*; 6: 2565–2570.
Gaines Das RE, Brugger P, Patel M, Mistry Y, Poole S (2004) *J. Immunol. Methods*; 288: 165–177.
Gans RJ (1905) *Preuss Geol. Landesanstalt*; 26: 179.
Hill R (2002) *Eur. Pharm. Rev.*; 7(1): 55–58.
ISPE (2001) *Pharmaceutical Engineering Guides for New and Renovated Facilities*, Vol. 4: *Water and Steam Systems.*
Jordain PT (2002) *Pharm. Manuf. Pack. Sourcer*; October.
Keer DR (1995) *Ultrapure Water*; 12(9): 24–32.
Keer DR (1996) *Ultrapure Water*; 13(3): 32–42.
Lampard G (2002) *Manuf. Chem.*; July: 41–44.
Levin J, Alving CR, Munford RS, Stutz PL (Eds) (1993) *Bacterial Endotoxin: Recognition and Effector Mechanisms.* Endotoxin Research Series, Vol. 2, Elsevier Science.
Loeb S, Johnson JS (1967) *Chem. Eng. Prog.*; 63: 90.
Martino R, Martinez C, Periclas R *et al.* (1996) *European Journal of Clinical Microbiology and Infectious Diseases*; 15: 610–615.
Mortimer AD, Whitehead P (2001) Pittcon, New Orleans, The Pittsburgh Conference on Analytical Chemistry and Applied Spectroscopy, Pittsburgh, PA, USA; 15235–5503.
Mortimer AD (2002) *LabPlus International*; September: 26–29.
Nagano M, Takahashi Y, Katagiri S (1999) *J. Reprod. Dev.*; 45: 239–242.
Novitsky TJ (1984) Pharm. Engineering 4(2): 21–23.
Purchas DB (1981) In *Handbook of Water Purification.* Ed Lorch W. McGraw-Hill, London; Chapter 5.
Reust JB, Meyer VR (1982) *Analyst*; 107: 673–679.
Saunders (1981) in Handbook of Water Purification. Ed Lorch W. McGraw-Hill, London; Chapter 6.
Schultz J, Newby GA (1966) *GGA Report No. GA-7153.* Gulf General Atomic, San Diego California.
Scott D (2000) In *Advanced Materials for Water Handling: Composites and Thermoplastics*, Elsevier Advanced Technology, Oxford; Chapter 1.
United States Pharmacopoeia (USP) (2006) Rockville, MD 20852, USA; Vol. 29, including Second Supplement.
Weber M (1995) *BioTechniques*; 19: 930–939.
Weiss TL, Goldwasser E (1981) *Biochem J.*; 198: 17–21.
Whitehead P (1998) *Laboratory Solutions*; December, 12–15.
Whitehead P (1999) Healthcare equipment & supplies, June, 21.

Further Reading

Handbook of Water Purification (1981) Ed Lorch W. McGraw-Hill, London.
High Purity Water Preparation (1993) Meltzer TH, Tall Oaks Publishing, Littleton, USA.
ISPE, *Pharmaceutical Engineering Guides for New and Renovated Facilities*, Vol. 4, *Water and Steam Systems*, January 2001.

Water for Pharmaceutical Purposes, General Information, USP 29, USA, 1231. Pharmacopoeial Convention, (2006) Rockville, USA.
Ultrapure Water, Tall Oaks Publishing, Littleton, USA.
Water Treatment Handbook (1991) Degremont, Lavoisier Publishing, Paris; Sixth edition.

Useful Web Sites

American Society for Testing and Materials (ASTM)	www.astm.org
College of Amercian Pathologists (CAP)	www.cap.org
ISO	www.iso.ch
National Committee for Clinical Laboratory Standards (NCCLS)	www.nccls.org
European Pharmacopoeia (EP)	www.pheur.org
US Pharmacopeia (USP)	www.usp.com

3 Development and Optimization of Serum-free and Protein-free Media

D Jayme

3.1 INTRODUCTION

Cultivation of mammalian cells has historically been performed using relatively ill-defined nutrient conditions (Mather 1998; Freshney 2000; Altman & Dittmer 1961). In an attempt to mimic the composition of bodily fluids, tissue explants and monodisperse cells were bathed in an iso-osmotic buffered saline solution augmented by addition of various organic nutrient constituents, and ultimately further supplemented by animal sera at 5–20% (v/v) (Jayme & Blackman 1985; Ham & McKeehan 1979; Barnes *et al.* 1984; Waymouth 1972).

In the decades following these initial cell culture efforts, there has been a dramatic evolution in the parameters and applications that define this field. As noted in Table 3.1, there has been a virtual explosion in the breadth of cell types, cultivation conditions and analytical tools to facilitate optimization and delivery of required nutrients. Concomitant with these trends has been an emergence of applications targeted toward production of biological molecules and engineered cell types for human and veterinary therapeutics that has evoked a rapid progression in the development of nutrient media to support these highly regulated applications (Jayme & Blackman 1985; Jayme & Gruber 1998).

About 15 years ago when I first lectured on development of serum-free media, I crafted a table that defined the various motivations to reduce or eliminate serum (Table 3.2). As a by-product of the cattle industry, animal serum (particularly foetal bovine serum, FBS) experienced extremes of supply and cost pressures that prompted both academic researchers and the emerging biotechnology industry to consider options for serum reduction or to qualify serum-free alternatives.

In addition to these concerns regarding price and availability, there was mounting evidence that serum addition might be problematic for certain cell culture applications (Ham & McKeehan 1979; Barnes *et al.* 1984; Waymouth 1972; Jayme & Gruber 1998). Serum factors typically promoted fibroblast overgrowth in mixed cell populations or failed to provide essential growth factors in adequate abundance to promote epithelial cell growth. Progenitor cells were difficult to maintain in serum-supplemented media without undergoing spontaneous differentiation or apoptosis. Differentiated cells exhibited rapid deterioration of key cellular functions. Serum also contained proteolytic enzymes that degraded cell-secreted products and neutralizing antibodies that reduced viral titres.

Finally, almost as an afterthought, I observed that animal serum might be a source of regulatory concerns, due to the antigenicity of foreign protein elements and the potential for them to introduce adventitious contaminants. At that time, little was known (Spier 1983) regarding viruses

Medicines from Animal Cell Culture Edited by G. Stacey and J. Davis

Table 3.1 Evolution of cell culture applications.

Cell culture parameter	Historical requirement	Current requirement
Range of applicable cell types	Narrow range	Broad range
Serum supplementation	Relatively high	Relatively low or serum-free
Nutritional requirements	Less fastidious	More fastidious
Nutrient optimization method	Less analytical	More analytical
Inoculation density	Narrow range	Broad range
Maximal cell density	10^5–10^6 cells/ml	10^6–10^9 cells/ml
Target culture application	Cell proliferation	Biological production
Regulatory level	Research/*in vitro* diagnostic	Bioproduction/*ex vivo* therapy
Range of bioreactor types	Narrow range	Broad range
Bioreactor controls	Limited manual control	Computer-driven controls

that could pass transplacentally and that would be present even in aseptically collected foetal serum processed through multiple 0.1μm sterilizing filters. Prions were yet to become an issue for biological medicines and transmissible spongiform encephalopathies were not on anyone's radar screen in biotechnology.

Many investigators were reluctant to convert from their established serum-supplemented culture systems because of the anxiety in verifying comparability. When the serum supply and cost returned temporarily to previous levels, many laboratories deferred their prior efforts to develop serum-free media. Those who had ventured to eliminate serum discovered that early commercial attempts to introduce serum-free media resulted in prototypes with diminished stability and biological performance. The early operating assumption had been that addition of selected serum-derived proteins, growth factors and trace metals to traditional basal media would substitute for the broad cell culture functions of serum. Subsequent investigation established that, in addition to serving as a source of growth hormones, the serum additive also served various other roles that also required substitution under serum-free culture conditions to sustain normal cell growth and functionality (Jayme & Blackman 1985).

In our efforts to develop and optimize nutrient formulations for a broad range of cell types and applications, we have found it useful to classify cell culture-based applications as follows (Figure 3.1):

- cells as research and diagnostic tools to investigate normal and aberrant cell function;
- cells as biological factories to produce medicines for human or veterinary therapy; and
- cells as therapeutic products for *ex vivo* therapy or tissue engineering use.

Table 3.2 Motivators to reduce or eliminate serum supplementation.

Product availability	Serum-associated artifacts
• Final product cost and impact on final dosage cost of future product	• Inhibition of proliferation of certain cell types by serum factors
• Raw material cost fluctuation	• Induction of differentiation or apoptosis
• Finite global supply and increased demand	• Proteolytic degradation of product
Downstream processing impact	Regulatory concerns
• Decreased product yield and recovery	• Foreign protein immunogenicity
• Co-purification of serum elements with molecule of interest	• Adventitious agent contamination

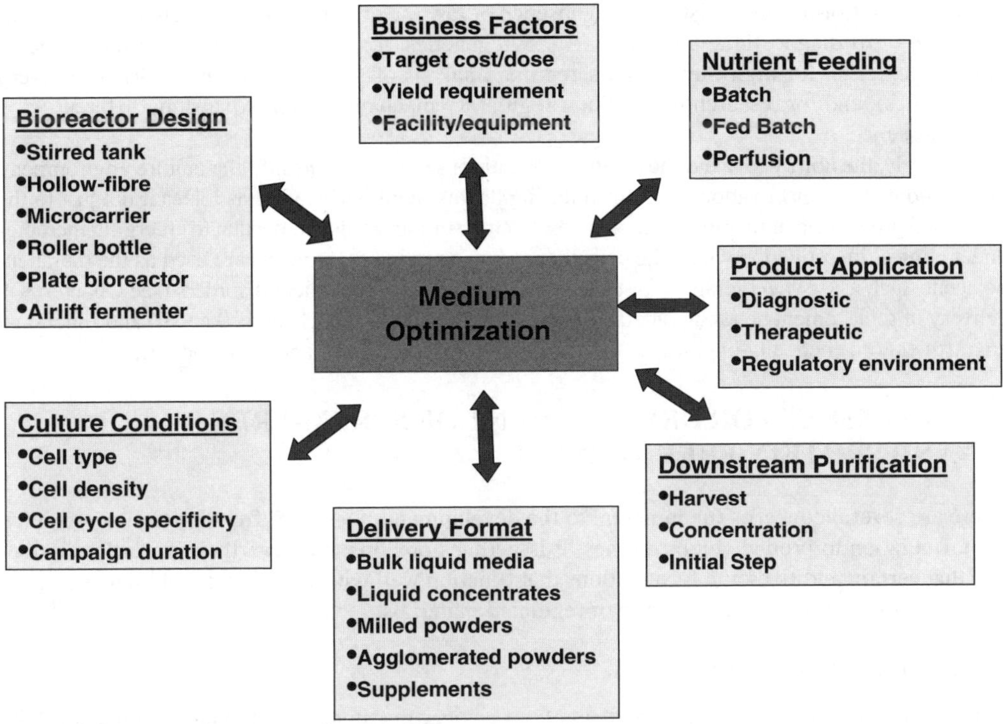

Figure 3.1 Integrated nutrient medium optimization. Effective nutrient medium optimization for biopharmaceutical production applications cannot effectively focus exclusively on biochemical composition. Integration of formulation design optimization within process development through incorporating inputs from bioproduction, bioreactor engineering, downstream purification, regulatory affairs and business perspectives results in technical and economic superiority.

Given the title of this volume, this chapter will focus upon the second of these three categories, i.e. where cultured eukaryotic cells are used as biological factories to manufacture vaccines, interferon, and genetically engineered products (e.g. monoclonal antibodies, recombinant proteins) for human and veterinary therapy. This scope is consciously restrictive, as biomedicines have also been successfully produced in microorganisms and lower eukaryotes, as well as in transgenic animals and plants.

Recombinant proteins produced within bacterial, yeast and even insect cell production systems have typically yielded authentic peptide sequences, but also less complex post-translational modifications (e.g. glycosylation, protein folding, disulfide cross-linkages) that have rendered them less efficacious in prospective therapeutic environments. In parallel with investigation to optimize the nutrient environment for animal cell culture production applications, efforts have been made genetically to engineer lower cell types with 'mammalian-like' post-translational processing activities.

Exploitation of transgenic technologies may prove useful, particularly for production of therapeutic proteins in ton quantities. However, the potential advantages and the technical and regulatory challenges of animal and plant transgenic production systems fall outside of the scope of this book.

While this work will emphasize cell-based bioproduction applications, there also exists considerable overlap with other cell culture focal areas in terms of nutrient medium optimization requirements. Lot-to-lot consistency and absence of raw material contaminants also become critical for *in vitro* diagnostic applications (e.g. cytogenetics, toxicology, irritancy, carcinogenicity, immunogenicity), for high-throughput screening analysis of gene expression for drug discovery (proteomics), and for research into cellular regulatory mechanisms for expansion, differentiation and senescence.

Similarly, the ability to avoid introducing adventitious contaminants into the culture environment, or to validate their inactivation or removal during downstream purification, is relevant not just to the cell-based production of therapeutic molecules. Such assurance is also important to emerging therapies under clinical investigation where the cells themselves or living by-products are used as the therapeutic agent, such as *ex vivo* therapies for cancer and various immunological disorders (see Chapter 30), delivery of replacement genes or vaccines through viral vectors (see Chapters 9 and 6) and functional engraftment of genetically-engineered tissues and neo-organs (see Chapters 28 and 29).

3.2 KEY ISSUES FOR DEVELOPMENT OF SERUM-FREE AND PROTEIN-FREE MEDIUM

There are several concerns fundamental to the development of a serum-free or protein-free nutrient formulation to produce biomedicines. It is perhaps obvious, but nevertheless worth mentioning, that certain additives and formulations that function well within a research laboratory may be impractical for large-scale production of regulated materials.

3.2.1 Fundamental Concerns

Influential factors include the cost and supply of critical raw materials; the biochemical stability of raw materials both independently and within a complex mixture of other biologically active ingredients; the ability to maintain biological activity and potency following sterile processing; robustness and scalability of the nutrient mixture; and compatibility of formulation ingredients and processing components with regulatory mandates for pharmaceutical production (ICH Harmonized Tripartite Guideline Q5E 2003).

3.2.1.1 Formulation design optimization

The elevated cell densities and extended bioreactor campaigns required to improve the space–time utilization efficiency of a manufacturing suite for cost-effective production of human and veterinary biomedicines necessitate both quantitative and qualitative changes in composition of the nutrient medium (Jayme *et al.* 1998, 1999). Constituent levels adequate for short-term incubation or at low cell inoculation density may provide insufficient buffering capacity or metabolic substrates under pilot- or production-scale culture conditions. Additives beneficial to supporting biomass expansion during proliferative phases may prove detrimental to specific productivity and overall yield of the target biomolecule.

By illustration, the early industry practice of elevating glucose concentrations accelerated biomass expansion, but also produced medium acidification that was compensated for by base addition, leading ultimately to increased bioreactor osmolality levels that negatively affected cell viability and specific productivity (Zielke *et al.* 1978). Glutamine elevation resulted in short term conversion from hexose to amino acid as a primary source for metabolic energy, but ultimately resulted in ammonia accumulation within the bioreactor (Fike *et al.* 1993).

3.2.1.2 Integration of upstream and downstream processes

Misalignment of upstream and downstream processes can result in costly delays and yield losses. Development and optimization of the nutrient medium should not proceed independent of subsequent processes that may also be impacted. As suggested above, a nutrient formulation optimized in tissue culture flasks or other small-scale, relatively uncontrolled, environments may be sub-optimal within a bioreactor. Homogeneous bulk-phase bioreactors, such as a stirred tank reactor, will demand different elements of medium optimization from longitudinal flow systems, such as hollow-fibres or plate bioreactors. Other factors, such as regulatory and business issues impacted by the intended end use of the target biomolecule and duration of the production campaign, can also impact medium optimization.

A frequently overlooked element of medium optimization is the impact that medium constituents may exert on downstream processing, particularly the initial capture steps in purification (Jayme *et al.* 1998). By illustration, many serum-free media contained human transferrin as a critical constituent for cellular delivery of iron. Presumed acceptable for biopharmaceutical production due to raw material screening and heat processing, transferrin instantly became problematic when the primary vendor withdrew product from the market due to its implication in a variant form of human Creutzfeld-Jakob disease. Laboratory and commercial attempts to replace transferrin quickly demonstrated that the cytotoxicity of iron salts could be mitigated through chelation by various anionic species. However, some of these chelating agents adversely affected product binding to the ion exchange resins frequently used as the primary step in purification of the harvest supernatant. Similarly, detergents and antifoams that minimize cell disruption due to mechanical shear within the bioreactor may co-elute with the biological product and adversely impact its target application.

3.2.1.3 Robustness

To be commercially relevant as a nutrient medium for biopharmaceutical production, a nutrient medium must exhibit substantial lot-to-lot consistency, must be scalably manufactured to a level adequate to meet full-scale demand, and must be biochemically and functionally stable under intermediate and final storage conditions for a practical duration. Various constituents that exhibit beneficial metabolic or antioxidatitive effects within the laboratory fail to replicate them within a commercial environment, due to performance deterioration that results from inherent instability or processing artefacts, or to adsorptive clearance by sterilizing filters or formulation vessels.

Animal sera were historically quite variable from lot-to-lot, since a commercial batch represented the aggregate contribution of multiple donor animals and various environmental factors (see Chapter 4). However, the blood-derived factors and protein hydrolysates routinely employed in many serum-free and protein-free media also exhibited batch-related variability. Consequently, a chemically defined nutrient formulation, where each biochemical constituent is defined as a robust, low molecular weight compound, offers significant potential benefits to the robustness of both upstream and downstream processes.

An additional factor that impacts medium robustness is exposure to light (Wang 1976; Taylor 1984). Classical nutrient formulations contained phenol red or other biological indicators that, in addition to serving as visual indicators of pH, quenched the effects of incident light and scavenged reactive oxygen species. To eliminate culture artefacts and interference with intended applications, such indicators have frequently been removed from nutrient media designed for biopharmaceutical production applications. However, this omission renders the residual formulation exquisitely sensitive to the destructive effect of exposure to light. Cell-free media may undergo accelerated deterioration, primarily of ring heterocyclic nutrient molecules, when exposed to light. Such effects are further amplified within a cell culture environment due to the elevated incubation temperature and cytotoxic lipid peroxidation effects. It is particularly critical to observe storage guidelines for complete liquid serum-free and protein-free formulations by keeping them refrigerated and minimizing light exposure.

Despite careful efforts at the bench scale to evaluate potential contributors to variability, challenges are frequently encountered at the scale-up phase. Influential factors are manifold and may include genetic instability of the cell line; altered contact surfaces or processes for medium or buffer formulation, cell cultivation and product harvest; altered susceptibility to environmental perturbations; and modified nutrient requirements and consumption profiles.

In our efforts to scale up manufacture of serum-free formulations from laboratory prototypes to production scale, we have encountered various issues that may be useful to consider. Biological raw materials, such as protein hydrolysates, humoral fluid fractions, lipids and conditioned media, should be considered initially during trouble-shooting investigation, due to their intrinsic lot-to-lot variability. However, we have noted various instances where presumptively well-defined biochemical constituents exhibited variable cell culture performance (despite unremarkable incoming chemical characterization) due to residual manufacturing artefacts, such as endotoxin content, ammonia levels from ammonium sulphate precipitation, toxic (or beneficial) trace metals, etc. Small-scale preparation of nutrient media or buffers will typically only use a single batch of these components, and raw materials issues may only emerge with augmented constituent requirements associated with full-scale production volumes.

Other issues relate to selection and proper processing of contact surfaces in formulation vessels, piping and filtration housings. Bench-scale formulation conducted in plastic vessels with plastic tubing and flat filter configurations may yield vastly different results in terms of non-specific adsorption, leachable elements and denaturation dynamics than may ultimately be observed at the production scale with stainless steel tanks, piping and filter housings and cartridge filters. Similar concerns relate to the scale-up of water for production and clean steam generation.

Mechanical and chemical perturbations of the cell culture environment due to scale-up can also affect process robustness. Improved monitoring and control of pH, temperature, dissolved gases and nutrient feeding within a bioreactor may enhance volumetric yield. However, increased impeller shear and gas sparging dynamics, and adjustments to control metabolic medium acidification and carbon dioxide evolution, can negatively impact cell viability and specific productivity. Alterations in oxygen consumption can also exert both qualitative and quantitative changes in nutrient consumption kinetics and relative profiles, as cells shift between competing metabolic pathways to generate energy and to produce intermediary metabolites. Such metabolic transitions may not only impact gross parameters, such as specific and volumetric productivity, but may also affect post-translational product quality parameters, such as glycosylation, disulfide bond formation, secretion and folding.

3.2.1.4 Regulatory perspectives

Typically, biomedical product manufacturers outsource manufacture of nutrient formulations and focus upon their unique core competencies, rather than bear the additional regulatory scrutiny of validating internal media production according to cGMP criteria. Quality assurance is acquired through formal vendor qualification programs and site audits to ensure compliance with regional and international statutes and user requirements. User audits should include a thorough review of all components required for consistent quality of product manufacture, ranging from design criteria for new product introduction, through processes designed to ensure quality, traceability and suitability of raw material constituents, and including all processes associated with product formulation, sterilization, dispensing, storage and delivery (ICH Q7A 2000; see also Chapter 34).

3.2.1.5 Manufacturing format

Historically, nutrient media were formulated as single-strength liquid solutions, supplied in a sterile, ready-to-use format. To add additional stability and minimize costs associated with water shipment, a ball-milled powdered format was commercialized (Young *et al.* 1966). With the advent of serum-free media, the complexity of nutrient formulations resulted in products that were challenging to produce homogeneously or to solubilize from ball-milled powders whilst retaining full biological performance. The biopharmaceutical production industry also expressed concerns over the limited ability to sanitize the ceramic media employed in ball mills, as well as other artefacts associated with powder processing by this method (Jayme *et al.* 2002).

In response to these requirements, several novel formats were introduced, each of which has found significant utility. Liquid medium concentrates (Jayme *et al.* 1992) provide a stable, pre-solubilized kit of nutrient ingredients within a 50× format that may be sterilely reconstituted in batch mode within a formulation tank (Jayme *et al.* 1998) or bioreactor, or continuously processed using a mixing device which is capable of delivering in excess of 30 000 litres of diluted nutrient medium to bulk containers (Jayme *et al.* 1996). The FitzMill system is a stainless steel hammer-mill system that produces homogeneous powdered medium containing thermolabile constituents within a pharmaceutically acceptable manufacturing environment (Jayme & Smith, 2000).

Advanced granulation technology (AGT™) represented a novel application of the common pharmaceutical fluid bed process to generate homogeneous granules of nutrient medium (Jayme *et al.* 2002). AGT granules may be produced for a broad range of catalogue and customized serum-free and protein-free formulations (Fike *et al.* 2001). Further, AGT media exhibit rapid dispersion and dissolution, flowability and stability properties highly compatible with efficient biopharmaceutical processing requirements (Radominski *et al.* 2001; Walowitz *et al.* 2003).

3.2.2 Basic Medium Constituents

The diversity of cell culture functions of the basic medium constituents has been exhaustively reviewed elsewhere (Freshney 2000; Jayme & Blackman 1985; Ham & McKeehan 1979). Such basal media were developed either through approximation of the composition of the native humoral fluids or through comprehensive analysis of the contribution of each prospective medium constituent to cellular proliferation. Accelerated optimization of basal media may now be accomplished through volumetric blending of established formulations (Murakami *et al.* 1984) and through statistical design (Peppers *et al.* 2002). Key medium constituents and their primary roles are noted below.

3.2.2.1 Inorganic salts

Superficial analysis of animal cell culture nutrient formulations reveals that the principal inorganic salt component is sodium chloride, provided to maintain osmotic equilibrium and to energize the co-transport of organic solutes into the cell. Other inorganic cations, e.g. potassium, calcium, magnesium and zinc, participate in metabolic and signaling functions and facilitate attachment and proliferation. Anionic species modulate transmembrane potentials and may serve as precursors for sulphur and nitrogen-containing organic molecules.

3.2.2.2 Amino acids

The naturally occurring amino acids are traditionally supplied at the millimolar level and serve three fundamental roles: (i) precursors for protein/peptide biosynthesis; (ii) metabolic intermediates for synthesis of other biomolecules, and (iii) substrates to generate metabolic energy. Recognizing that similar amino acids compete for common transport and metabolic pathways leads to the fundamental conclusion that balanced delivery is a key objective and that supraphysiological concentrations of a particular solute ultimately may be perceived by the cell as a relative scarcity of competitive solutes.

For example, given that multiple essential neutral amino acids (e.g. leucine, isoleucine, valine, phenylalanine, etc.) compete for a common carrier-mediated transport pathway in most mammalian cells (Oxender & Christensen 1963), compensating exclusively for depletion of a particular solute (e.g. leucine) by *excessive* augmentation would saturate the transport receptor and ultimately be perceived at the cellular level as a relative paucity of competitive nutrient solutes.

3.2.2.3 Vitamins

These organic catalysts facilitate various cellular processes associated with one-carbon metabolism, transamination, carboxylation and decarboxylation, and oxidative phosphorylation. Although the axiom persists that a catalyst is not consumed as a reactant, practical application recognizes that vitamins will degrade over time and require periodic replenishment to maintain optimal performance. Certainly, the concentration of vitamin components needs to be augmented in order to support elevated cell density applications.

3.2.2.4 Carbohydrates

Glucose remains the predominant carbohydrate in mammalian cell culture formulations, although other aldo- and ketohexoses and pentoses may be preferentially utilized to minimize lactate accumulation or to alter protein glycosylation. Sodium pyruvate and inositol are also common constituents to facilitate intermediary metabolism.

3.2.2.5 Water

Ultimately, the primary constituent of a nutrient medium is water. To minimize fluctuations in performance due to variability in water ions, endotoxin, organochemicals and other factors, production water for biopharmaceutical applications should comply with water for injection (WFI) criteria. Typically, WFI-quality water will be processed to remove trace impurities, and in order to minimize biological contaminants it will be maintained at elevated temperature in a circulation loop prior to use. Water quality for cell culture is dealt with in more detail in Chapter 2.

3.2.3 Complex Medium Constituents

These constituents are distinguished from the more basic biochemical ingredients by there often being no compendial monographs that specify purity and because of technical limitations to biochemical or biological performance assay.

3.2.3.1 Proteins, peptides and hydrolysates

With the general trend away from materials of animal origin (Jayme & Smith 2000), native proteins derived from animal blood fractions (e.g. albumin, transferrin, fetuin) or animal tissues (e.g. insulin, epidermal growth factor, fibroblast growth factor) that were common to serum-free media less than a decade ago, have declined in popularity. Where possible, these proteins have been replaced by full-chain or truncated recombinant forms with comparable binding affinity and biological activity.

Protein hydrolysates have been widely used for decades as sources of amino acids and other intermediary metabolites, and of oligopeptides. Historically, these hydrolysates were derived from meat digests or organ infusions, and were obtained through enzymatic cleavage by animal-derived proteases. Regulatory pressures have accelerated a transition to hydrolysates of plant- or yeast-derived proteins obtained either by autolysis or cleavage by bacterial or fungal proteases. Hydrolysates remain among the most variable components of 'protein-free' media, and considerable effort has been focused on replacing hydrolysates with defined components to yield 'chemically defined' formulations, or to refine the raw materials and processes associated with protein hydrolysis to yield a more consistent product (Lobo-Alfonso *et al.* in press).

3.2.3.2 Trace metals

Although supplied as inorganic salts, these elements are grouped separately because the cellular requirement (if any) is in the nanomolar to picomolar range, several orders of magnitude below the inorganic materials mentioned above. Serum supplementation and hydration with lower-quality production water masked the importance of many of these trace metals for decades. Trace metal optimization can be challenging due to the relative insensitivity of analytical methods, difficulty in correlating small changes with biological performance, and difficulty in extrapolating single component changes to a mixture. Many trace metals exhibit clear performance optima, with supra-elevated concentrations proving inhibitory and reduced levels being sub-potent.

3.2.3.3 Lipids

Owing to practical difficulties with maintaining elevated levels of lipid materials stably in aqueous solution, advances in optimization of lipid delivery have been relatively slow. Organic solvent solubilization, Pluronic™ polyol microemulsification and liposome encapsulation have been examined with limited success. We have recently experienced superior stability, lipid delivery, and manufacturing scalability with cyclodextrin solubilization (Gorfien *et al.* 2000). These concentrated lipid additives have facilitated production applications of sterol-requiring cells and augmented the available lipid membrane precursors to support biomass expansion in serum-free culture (Walowitz *et al.* 2003; Gorfien *et al.* 2000) (see also Section 31.5).

3.2.3.4 Dissociating enzymes and attachment factors

A common observation is that end users will specify a cell culture environment entirely free of animal-origin medium constituents, yet persist in dislodging adherent cells from the substratum using porcine pancreatic trypsin. Despite certified vendor processing by gamma irradiation, porcine trypsin represents a significant potential source of parvovirus contamination. We recently investigated and subsequently commercialized a recombinant trypsin-like protease with comparable cell removal kinetics and biological performance to animal origin trypsin, but with superior purity and stability (Nestler *et al.* 2004) (see also Section 31.5).

Like growth factors (noted above), the most widely used attachment factors in cell culture were historically derived from animal blood fractions (e.g. fibronectin, vitronectin). The response to animal origin concerns has been least successful for adherent cells. There exist multiple nutrient formulations that can support production applications of cells already attached in serum-supplemented media to harvest a target protein or vaccine for human or veterinary applications. Modifications to the basal medium may preferentially support cell attachment or synthesis of extracellular matrix elements. Substrata may be modified by charge density or impregnation of synthetic recognition sequences to enhance cell surface protein recognition and initial attachment. However, development of a stable, cost-effective serum-free medium for extended serial passaging of adherent cells without diminished performance has been elusive.

3.2.4 Cell-related Issues

3.2.4.1 Adaptation

The most common concern regarding transition to serum-free or protein-free media is that the cells fail to adapt. In many cases, we have obtained these cells and performed the serum-free adaptation and initial banking internally as a service. This observation suggests that contributing to the difficulty in adapting cells may be a failure at the user laboratory to adhere to specified guidelines and protocols.

Occasionally, one is blessed with a hardy cell line, i.e. one that can be centrifuged, followed by aspiration of serum-containing medium and replacement with serum-free medium, with healthy proliferation ensuing with minimal lag time or cessation of cellular functionality. Such experiences are to be fondly remembered to provide encouragement for the majority of cell lines that require sequential adaptive weaning protocols. Even with the more complex protocols, however, cells may generally be adapted to serum-free culture conditions within a period of 4–6 weeks. Such protocols have been elaborated elsewhere, but in summary:

- *Viability of the cellular inoculum* Cells to be adapted should be obtained from a healthy (>90–95 % viability) culture that was recently passaged and/or provided replenishment with fresh medium.
- *Inoculation density* Recognizing that not all cells will survive the adaptation process and that the target medium is probably sub-optimal, we typically inoculate cells during the adaptation transition process at twice the 'normal' inoculation density. This higher density permits survival of sufficient healthy cells to secrete paracrine conditioning factors that facilitate recovery of the population. Once adaptation has been completed, the normal inoculation density generally may be restored.
- *Passaging frequency* Generally, laboratory personnel have become accustomed to subculturing cells at a particular frequency, given normal inoculation and harvest densities. However, there is clearly a proliferative lag phase during adaptation: if cells are subcultured at the normal frequency, they often fail to survive past one or two subcultures. Instead, cell density should be monitored daily during the adaptive period, subculturing being performed only once the culture has reached mid-to-late log phase. Conversely, if the cell density is allowed to become too high, the cell viability will plummet as cells become apoptotic, and sustained cultivation will become problematic.
- *Physical stress* Serum components provide substantial protection for the cell against mechanical stresses associated with centrifugation, pipetting and culture agitation. Sensitivity to these stressors persists even following adaptation, but during transition to serum-free medium, cells are exquisitely sensitive. Reduction in speed or elimination of centrifugation, modest reduction in impeller velocity or rotational speed of the bioreactor, and minimizing trituration during pipetting may enhance recovery of adapted cells.
- *Gas diffusion* Reducing the barriers to gas diffusion appears to facilitate adaptation. We have experienced greater success with shaker flasks than with tissue culture flasks. Similarly, adaptation has often been facilitated by use of a minimal layer of medium covering the cells.
- *Medium transition* Typically, we have recommended transitioning from serum-supplemented medium (SSM) to serum-free medium (SFM) in smaller sequential steps, each requiring two or three subcultures before proceeding to the next level of adaptation, and always keeping a control culture at the previous level as a back-up source. The SSM source is recommended as a relatively fresh medium (mother liquor) conditioned by cell-secreted factors. A common protocol initiates at 75 % SSM: 25 % SFM. Once cells have adapted to this culture environment, cells are inoculated (2× density) into a 50 % SSM: 50 % SFM mixture. Following adaptation over two or three subcultures in this new combination, cells are transitioned into 25 % SSM: 75 % SFM, and eventually to 100 % SFM.
- *Cell detachment* The difficulty of dissociating cells from the substratum under serum-free culture conditions varies widely with the cell type and its intrinsic ability to deposit extracellular matrix elements, and upon the nature and composition of the substratum. Collagenase and other enzymatic methods may be used to digest matrix elements and to dislodge cells from

collagen-coated matrices. Brief incubation with divalent cation chelators, such as ethylenediaminetetraacetate (EDTA, 1 mM) may be adequate to dissociate loosely adherent monolayers into monodisperse suspensions. EDTA in combination with trypsin may facilitate protease access to basement membrane-associated cellular attachment proteins to accelerate removal of tight-junctional epithelia.

As for most cell culture applications, benefit may derive from optimization of the incubation time and temperature for the detachment process to maximize both cell removal and viable cell recovery. Excessive damage to extracellularly oriented proteins following trypsinization was historically prevented by 'stopping' proteolysis by addition of serum-containing medium, which of course is undesirable for serum-free culture applications. As many serum-free media still contain albumin or other animal-derived proteins as lipid carriers and bulk protectants, addition of bovine or human serum albumin to the dissociation medium may be a simple and cost-effective approach. For animal origin-free cultivation, addition of non-cytotoxic stoichiometric inhibitors of serine proteases, such as soybean-derived trypsin inhibitor, may be a useful alternative. We recently reported (Nestler *et al.* 2004) that with fungal-derived recombinant protease treatment, simple dilution with fresh serum-free medium may be sufficient, presumably due to the absence of residual contaminating proteases such as those found in porcine pancreatic trypsin that may be cytotoxic under serum-free conditions.

3.2.4.2 Cryopreservation and recovery

Another common question: once cells are adapted to serum-free medium and are ready for cryopreservation, is it necessary to add serum to the freezing medium? When cells are recovered from cryostorage, is it necessary to add serum? Answer: No, in both cases. We have successfully cryopreserved a wide variety of cell types in serum-free or protein-free media without adding serum, albumin or any other protein component. Recovery of cryopreserved cells from liquid nitrogen storage has yielded viabilities comparable to serum-supplemented cultures. The same guidelines for the general culture health noted above for adaptation (e.g. high viability, mid-log phase, recent medium replenishment) also apply for stock cultures to be cryopreserved. We also recommend that cultures intended for cryopreservation be passaged for several generations and expanded in antibiotic-free medium to minimize the potential for latent contamination by adventitious agents.

We typically cryopreserve cells in 7–10% dimethylsulfoxide (DMSO) and attempt to minimize their exposure to DMSO at high concentrations or elevated temperatures. To accomplish this objective, we suspend cells in conditioned medium at twice the targeted freezing density. We have previously prepared an equal volume of fresh medium (refrigerated) containing twice the final targeted DMSO concentration. Then, we mix the two fractions and dispense as rapidly as possible into cryovials. Where possible, a controlled rate freezer yields superior recovery results, particularly with high-volume cell banking. If unavailable, a freezing process should be utilized that permits slow transition through crystallization temperature, such as wrapping cryovials in insulating material and storing overnight at −20 °C (or more commonly (using more insulation) −80 °C prior to transferring to liquid nitrogen or ultracold storage freezers, to minimize disruption of cellular and organellar membranes by ice crystals.

The optimal process for recovery of cryopreserved cells remains controversial and may vary with cell type, culture conditions, freezing medium (including cryoprotectant), and other factors. In our experience, cell recovery is best accomplished through rapid thawing of the cryovial and rapid dilution of the contents into pre-warmed fresh medium. The frozen cryovial should be incubated briefly with constant swirling within a 37 °C water bath, exercising caution to avoid introduction of contaminants. The cryovial should be removed from the water bath while some ice crystals remain to preserve lower temperature and to minimize cytotoxicity resulting from cryoprotectant exposure and relative anoxia.

While reducing exposure to elevated DMSO concentrations is important to improve viable cell recovery, we do not recommend centrifugation or vigorous trituration immediately following thawing, as cell membranes remain fragile and prone to disruption by mechanical forces. Rather, we recommend rapid dilution of the cryovial contents into 5–10 volumes of fresh pre-warmed medium. This step simultaneously accelerates warming of the cryopreserved cells and dilutes the DMSO below toxic levels without requiring centrifugation or other mechanical stress. After a brief recovery period (12–24 h), medium containing diluted DMSO is removed and replaced with fresh medium. While this process is effective for many cell types, some are quite sensitive even to diluted concentrations of DMSO and it may be necessary to remove DMSO-containing supernatants rapidly from settled or gently centrifuged cell suspensions prior to inoculation into tissue culture flasks.

Some laboratories have successfully operated with both master cell bank and working cell bank containing cells in serum-supplemented medium, and clear adaptive regimens for cell recovery. We recommend at least that the working cell bank consist of cells adapted to the target serum-free or protein-free medium for biopharmaceutical production.

For a more detailed discussion of cryopreservation see Chapter 21.

3.2.4.3 Medium optimization

Given the accelerated cycle times associated with process development from gene isolation to pilot production of Phase I clinical trial materials, a decision must be made rapidly regarding the preferred cultivation (including optimal nutrient medium) and purification schemes. While proliferative rates are clearly important to the economics of biomass expansion, it is possible to lose valuable time if a scale-down system that truly mimics the pilot- or production-scale bioreactor is not employed, or if nutrient medium optimization screening is based solely upon cell growth, rather than expression and quality characterization of the target biomolecule.

As illustrated by Figure 3.1, multiple factors contribute to determination of the optimal nutrient formulation for biological production applications. Much of the extant literature describing nutrient medium optimization emphasizes determination of the ‘right’ concentrations of metabolic substrates and hormonal factors to stimulate proliferation or bioproduction. However, many of these studies fail to appreciate both the quantitative and qualitative adjustments in nutrient medium composition that may be required as a consequence of the variety of inputs illustrated in the figure. Nutrient concentrations that may be appropriate for batch culture reactions may be inhibitory to growth or production within a fed-batch or continuous perfusion culture system. Constituents acceptable for research or diagnostic applications may be precluded from biopharmaceutical production use due to regulatory considerations associated with potential side effects or adventitious contaminants. Nutrients that potentiate cell density or volumetric productivity may be undesirable because they interfere with critical initial steps of downstream purification. Bioreactor design and culture conditions will similarly affect the biochemical composition and preferred delivery format of nutrient media. Lastly, economic factors associated with raw material cost and stability, product quality and stability, and space/time utilization efficiency within the manufacturing facility may be inconsistent with raw materials that are too expensive, too variable or unavailable in sufficient quantity for production-scale requirements.

3.3 KEY TRENDS

3.3.1 Elimination of Animal Origin Risks

In 1998, I was invited to speak at the Council of Europe as part of a panel investigating the risks and issues associated with utilizing animal sera and derivatives of animal sera for the production of pharmaceutical products (Jayme 1999). Although there were evident technical hurdles that needed to be overcome, the overwhelming recommendation from that conference was that '... Manufacturers of biological, medicinal and veterinary products are encouraged to reduce or eliminate the use of animal-derived products in their manufacturing processes' (Jayme & Smith 2000; Castle & Robertson 1999). Since that date, all regulatory guidance proceeding from Europe and all other jurisdictions participating in harmonizing quality regulations has uniformly been to reduce or eliminate animal-origin constituents. The definition of 'animal origin' has been provisionally expanded beyond primary starting materials to include elements of animal origin that may be used as nutrient additives or as processing components.

To be certain, there exist decades-old production processes that, unless demonstrated unequivocally to be broken, will continue with supplementation by serum or other animal-origin materials. However, it was clear that manufacturing process changes targeted towards alleviation of animal-origin-associated risks would be recommended for expedited consideration. Moreover, new product applications containing constituents and processes that exposed the potential therapeutic product to risk of contamination by animal-origin adventitious agents would require validated reduction of these risks by multiple, orthogonal processes.

While the final chapter clearly remains to be written in the development of serum-free and protein-free media for all cell culture-based applications, it is clear that for many cell types (particularly cells compatible with suspension cultivation), such animal-origin-free formulations are commercially available to alleviate risks from adventitious agents, generally without loss of cell culture performance or cell-specific yield.

3.3.2 Batch versus Fed Batch and Perfusion Culture

Since many early animal cell culture practitioners had transitioned from bacterial fermentation experience, it was logical that early bioproduction protocols mimicked bacterial batch fermentation processes (Kadouri & Spier 1997). The batch production process facilitates materials handling and regulatory definition of what constitutes a 'batch' of product, since all of the raw materials including cells are placed into a single bioreactor and the target product is harvested following an appropriate cultivation period and processed as a single unit. To achieve production-scale capacity, biopharmaceutical manufacturers frequently require multiple production suites containing stainless steel bioreactors of 10–12 000 litre capacity and dedicated purification trains. Given the capital investment of early adopters of these batch processes, and given the preferred compatibility of certain biomedical products to batch processes, batch bioproduction processes are likely to remain highly useful for an extended period.

An ideal production environment would have a relatively small bioreactor that could sustain cells at high viability and productivity for an extended period of time. The continuous delivery of nutrients and removal of waste substances would be highly efficient, and the harvested effluent would be highly concentrated with stable product. Perfusion culture accomplishes some, although not all, of these objectives (Vogel *et al.* 2001). Clearly, the bioreactor scales are downsized by an order of magnitude from stirred tank systems. Many systems have successfully maintained cells at relatively higher viability and productivity for campaign periods lasting for over 100 days, as compared with a batch bioreactor campaign that is typically limited to less than 2 weeks. The efficiency of nutrient utilization, however, varies widely with the user application and the extent of control over bioreactor processes. Consequently, some processes exhibit relatively efficient nutrient consumption and isolate relatively concentrated product from the harvested effluent, while other processes pump excessive volumes of nutrient medium through the bioreactor and require effective capture steps to concentrate dilute product from high harvest volumes. Given the relative complexity of perfusion bioreactor systems, they tend to require a higher level of manual surveillance by trained professionals than might be expected for a batch culture bioreactor. This observation notwithstanding, perfusion bioreactors offer significant capital advantages in terms of the upstream facility footprint and space-time utilization efficiency.

Somewhat intermediate between these two processes is fed-batch culture, which means slightly different things to different people, but basically consists of supplementing a batch bioreactor with nutrient feeds to replenish consumed materials, and may include multiple product harvests. Optimization of nutrients to be included within the basal medium and as part of the feed stream represents a novel, emerging derivative of the field of medium development (Fike *et al.* 1993).

Perhaps you were asking yourself why the author is venturing into the area of bioreactor production protocols in a chapter nominally devoted to medium optimization? There exist significant qualitative and quantitative differences in the optimal medium for cultivating a particular cell type in a batch production environment compared with a fed batch or perfusion culture environment. Virtually all of the nutrient media developed for animal cell culture over the past several decades were designed for batch culture applications. To begin to optimize basal formulations and nutrient feeding regimens for perfusion and for fed batch applications may well involve reexamination of fundamental postulates, and reinvention of approaches for investigating and optimizing nutritional requirements of cell culture bioreactors operating in extended modes. Preliminary efforts for selected cultivation systems have demonstrated that productive lifespan may be extended and specific product yield may be increased significantly with a programmed delivery of a simple, optimized nutrient feed stream (Gorfien *et al.* 2003).

3.4 SUMMARY

Serum-free and protein-free formulations have been developed, consistent with emerging trends to eliminate constituents of animal origin, to accommodate a broad range of cell types, target biomolecules, and modes of bioreactor operation. Transition to optimized serum-free media, as outlined above, will facilitate yield improvement and downstream purification, avoid risk of adventitious agent contamination and expedite regulatory approval of future cell culture-derived biomedical products.

3.5 ACKNOWLEDGEMENTS

The author gratefully acknowledges insightful discussions with Invitrogen colleagues over the past two decades to generate facilitating strategies and technologies for development of effective serum-free and protein-free media. In particular, he notes the helpful perspectives of Shawn Smith, Eric Cornavaca, Mark Stramaglia and David Cady regarding emerging needs of the biopharmaceutical industry, and technical contributions of Steve Gorfien, Dale Gruber and George Jones to this manuscript.

REFERENCES

Altman PL, Dittmer DS (1961) *Blood and Other Body Fluids*. Federation of American Societies for Experimental Biology, Bethesda.

Barnes DW, Sirbasku DA, Sato GH (1984) *Cell Culture Methods for Molecular and Cell Biology*. AR Liss, New York; Vol. 1–4.

Castle P, Robertson JS (1999) 'Summary and Conclusion'. In *Animal Sera, Animal Sera Derivatives and Substitutes Used in the Manufacture of Pharmaceuticals: Viral Safety and Regulatory Aspects*, Dev. Biol. Stand., Karger, Basel; Vol. 99, 191–196.

Fike R, Dadey B, Hassett R, Radominski R, Jayme D, Cady D (2001) *Cytotechnology*; 36: 33–39.

Fike R, Kubiak J, Price P, Jayme D (1993) *BioPharm Int*; 6(8): 49–54.

Freshney RI (2000) *Culture of Animal Cells: A Manual of Basic Technique*, fourth edition. AR Liss, New York.

Gorfien SF, Paul W, Judd D, Tescione L, Jayme DW (2003) *Biopharm Int.* 16(4): 34–38.

Gorfien SF, Paul W, Walowitz J, Keem R, Biddle W, Jayme D (2000) *Biotechnol. Prog.*; 16: 682–687.

Ham RG, McKeehan WL (1979) *Meth. Enzymol.*; 58: 44–93.

ICH Harmonized Tripartite Guideline Q5E (2003) *Comparability of biotechnological/biological products subject to changes in their manufacturing processes.(see www. och.org)*

ICH Harmonized Tripartite Guideline Q7A (2000) *Good manufacturing practice guide for active pharmaceutical ingredients.(see www.ich.org)*

Jayme DW (1991) *Cytotechnology*; 5: 15–30.

Jayme DW (1999) In *Animal Sera, Animal Sera Derivatives and Substitutes Used in the Manufacture of Pharmaceuticals: Viral Safety and Regulatory Aspects*, Dev. Biol. Stand., Karger, Basel; Vol. 99, 181–187.

Jayme DW, Blackman KE (1985) In *Advances in Biotechnological Processes.* Eds Mizrahi A, van Wezel AL. Vol. 5, 1–30.

Jayme DW, Gruber DF (1998), In *Cell Biology: A Laboratory Handbook.* Academic Press, San Diego; Vol 1, 19–26.

Jayme DW, Smith SR (2000) *Cytotechnology*; 33: 27–35.

Jayme DW, DiSorbo DM, Kubiak JM, Fike RM (1992) In *Animal Cell Technology: Basic and Applied Aspects.* Eds Murakami H, Shirahata S and Tachibana H. Kluwer, Dordrecht; Vol 4, 143–148.

Jayme DW, Kubiak JM, Battistoni, Cady DJ (1996) *Cytotechnology*; 22: 255–261.

Jayme D, Kubiak J, Fike R, Rashbaum S, Smith S (1998) In *Animal Cell Technology: Basic and Applied Aspects.* Kluwer, Dordrecht; Vol 9, 223–227.

Jayme DW, Price PJ, Plavsic MZ, Epstein DA (1999) In *Animal Cell Technology: Products from Cells, Cells as Products.* Kluwer, Dordrecht; 459–461.

Jayme D, Fike R, Radominski R, Dadey B, Hassett R, Cady D (2002) In *Animal Cell Technology: Basic and Applied Aspects.* Kluwer, Dordrecht; 12: 155–159.

Kadouri A, Spier RE (1997) *Cytotechnology*; 24: 89–98.

Lobo-Alfonso J, Price P, Jayme D (in press) In *Protein Hydrolysates in Nutrition and Biotechnology.* Ed Pasupuleti VK. Springer, Dordrecht.

Mather JP (1998) In *Animal Cell Culture Methods.* Eds Mather JP, Barnes D. Academic Press, San Diego.

Murakami H, Shimomura T, Nakamura T, Ohashi H, Shinohara K, Omura H (1984) *J. Agricult. Chem. Soc. Japan*; 56: 575–583.

Nestler LL, Evege EK, McLaughlin JA *et al.* (2004) *Quest*; 1: 42–47.

Oxender DL, Christensen HN (1963) *Nature*; 197: 765–767.

Peppers SC, Talley DL, Loke HN, Caple MV (2002) *IBC Eighth International Conference on Antibody Production and Downstream Processing.*

Radominski R, Hassett R, Dadey B, Fike R, Cady D, Jayme D (2001) *BioPharm*; 14(7): 34–39.

Spier RE (1983) In *Advances in Biotechnological Processes.* AR Liss, New York, Vol. 2.

Taylor WG (1984) *In Vitro*; 20: 58–70.

Vogel JH, Prtischet M, Wolfgang J, Wu P, Konstantinov K (2001) In *Animal Cell Technology: From Target to Market.* Kluwer, Dordrecht.

Walowitz JL, Fike RM, Jayme DW (2003) *Biotechnol. Prog.*; 19: 64–68.

Wang RJ (1976) *In Vitro*; 12: 19–22.

Waymouth C (1972) In *Growth, Nutrition, and Metabolism of Cells in Culture.* Eds Rothblat GH, Cristofalo VJ. Academic Press, New York.

Young FB, Sharon WS, Long RB (1966) In *Biochemical Engineering.* Eds Constantinides A, Wieth WR, Subramanian KV. NY Academy of Science USA; 369: 108.

Zielke HR, Ozand PT, Tildon JT, Sevdalian DA, Cornblath M (1978) *J. Cell Physiol.*; 95: 41–48.

Relevant Web Sites

Additional resources	Primary focus	Website
International Conference on Harmonization	Global quality requirements	http://www.ich.org
FDA Points to Consider	US regulatory perspectives	http://www.fda.gov
EU Guidance Documents	European regulatory perspectives	http://www.emea.europa.eu
BioReliance	Contract testing	http://www.bioreliance.com
BioWhittaker (Cambrex)	Media and reagents	http://www.cambrex.com
HyClone (Perbio)	Media and reagents	http://www.hyclone.com
GIBCO (Invitrogen)	Media and reagents	http://www.invitrogen.com
Irvine Scientific	Media and reagents	http://www.irvinesci.com
JRH BioSciences	Media and reagents	http://www.jrhbio.com
Quest (Kerry Bio-Sciences)	Protein hydrolysates	http://pharma-ingredients.questintl.com
Sigma (Sigma-Aldrich)	Media and reagents	http://www.sigmaaldrich.com

4 Understanding Animal Sera: Considerations for Use in the Production of Biological Therapeutics

R Festen

4.1 INTRODUCTION

Animal serum is used worldwide in the production of human and veterinary biologicals derived from cell culture. The addition of serum to basal culture media provides the necessary growth factors, attachment factors, transport proteins, protease inhibitors, lipids, trace elements, hormones and other small molecules required for effective cell growth and protein production. Serum is the universal growth supplement for most cell types and this single addition to basal media can obviate the need for extensive media development.

Although the pharmaceutical and biotechnology industries are working toward removing animal-derived materials from their manufacturing processes, and in some cases from the cell line development and screening process, serum has a number of distinct advantages that can make it an indispensable raw material for cell culture. Serum promotes the attachment of anchorage-dependent cell lines, such as VERO (African green monkey kidney cell line) and MDBK (Madin–Darby bovine kidney cell line) used in roller bottle and microcarrier cultures, and can act as a shear protectant in agitated, suspension culture (Valez *et al.* 1986). Omitting serum from cell culture media has been shown to increase the proteolytic activity of the medium and may result in lower protein quality and reduced cell growth (Kretzmer *et al.* 1994). From a production economics perspective, adding a low cost serum such as adult bovine serum (ABS) to an enriched basal medium can be significantly less expensive than serum-free media formulated with recombinant growth and attachment factors, hormones, and other necessary components.

Being a natural blood product from various animal species, gestational periods and geographical locations, the use of serum in biological production systems is not without certain disadvantages and risks. Risk of cell culture or product contamination with adventitious infectious agents, lot-to-lot variability, geographical animal disease outbreaks and agricultural economics all affect the suitability and availability of animal serum for use in biological manufacture.

It has been shown that more than half of all commercial serum is contaminated with bovine viral diarrhoea virus (BVD) (Bolin *et al.* 1991; Kniazeff *et al.* 1975; Fong *et al.* 1975). Other common bovine viruses such as infectious bovine rhinotracheitis (IBR), bovine respiratory syncitial virus (BRSV), and bluetongue virus (BTV) may also be present. These viruses propagate in cell culture and have been detected in vaccine products (Barkema *et al.* 2001; Harasawa 1995;

Medicines from Animal Cell Culture Edited by G. Stacey and J. Davis

Wilbur *et al.* 1994). Contamination in the manufacturing process or product can lead to product recall, cessation of manufacturing operations and/or adverse events in the patient population.

As serum is a complex and undefined raw material, it is difficult to predict precisely which components will create variability in the growth and production of a particular cell line. Milo *et al.* (1976) showed that serum could vary greatly from supplier to supplier in its ability to promote growth in human lung cells. Each lot of serum must be qualified to growth, production and product quality specifications and this testing may take several months to complete.

From a manufacturing process perspective, supplementing growth medium with serum can increase the protein concentration of the media to many times that of the level of recombinant protein being expressed, and may detrimentally complicate downstream processing.

With sophisticated serum-free and protein-free media available commercially for industrial cell lines, and escalating regulatory requirements for serum usage, it is surprising that the world-wide demand for serum continues to increase and that demand for fetal bovine serum (FBS) may outstrip supply over the next 5–10 years. Most users of animal serum have limited knowledge of the collection, manufacture and composition of the material. This chapter will discuss these topics in depth, along with ways to minimize the risks when using animal serum.

4.2 TYPES OF SERUM

A listing and the descriptions of commonly available sera is shown in Table 4.1. FBS is considered the serum of choice as it has the strongest growth-promoting capacity along with the lowest immunoglubulin (IgG) level. It takes between one and three bovine foetuses to yield a single litre of raw serum. In drought years, ranchers take larger numbers of cattle to market for slaughter to save on feed costs, therefore more material becomes available for processing. Conversely, as herds are expanded, fewer animals are brought to market and this translates into fewer blood collections and increased prices. For these reasons FBS is also the most expensive serum and subject to strong market forces. Other factors can affect worldwide supply and pricing, including large volume purchasing by major biological manufacturers protecting their raw material supply, and disease outbreaks, which can limit collections.

More often than not, pharmaceutical manufacturers neglect to explore other available sera having fewer supply issues and significantly lower cost. The supply of newborn, calf and adult bovine sera range from somewhat limited to nearly limitless. How well any of these sera perform will depend on the cell type, media and culture conditions. Significant quantities of IgG may make these sera undesirable for veterinary vaccines and antibody-based therapeutics, human or animal. Many commercial serum suppliers offer serum substitutes. These run the gamut from serum blended with growth factors and other components, to cocktails containing no serum *per se*, but serum-derived proteins or recombinant proteins along with other constituents such as lipids, hormones and trace elements(see Chapter 3). In some cases, the addition of growth factors to a lower-priced serum such as adult bovine can yield a replacement for foetal bovine serum at a fraction of the cost.

4.3 SERUM COMPONENTS

Serum separated under native gel electrophoresis gives five distinct bands; albumin, α_1, α_2, β- and γ-globulins (Friedlander 2003).

Albumin is the primary protein in serum and comprises approximately 60% of the total protein content. It functions as a carrier of small molecules, a pH buffer and shear protectant, and contributes *in vivo* to the colloid oncotic pressure of the plasma. As albumin carries a negative charge, it binds readily to salts such as Ca^{2+}, Na^{+}, K^{+} and Cu^{2+}, as well as free fatty acids, vitamins and hormones. While not a growth factor, albumin contributes to the overall mitogenic activity of the serum by binding growth-promoting components such as fatty acids and presenting them to the cell at active or mitogenic levels

Table 4.1 Bovine serum comparison chart[a].

Description	Fetal	Newborn	Calf	Iron-fortified calf	Donor calf	Donor adult	Adult
Age	Gestational	< 10 days	< 12 months	4–6 months	3–12 months	1–5 years	1–6 years
Weight	20–70 lb	50–90 lb	350–750 lb	300–400 lb	350–1000 lb	1200–3500 lb	1200–3500 lb
Animal	Dairy and beef cattle	Primarily dairy cattle	Dairy and beef cattle	Dairy bull calves	Beef cattle	Mostly beef cattle	Dairy and beef cattle
Diet (feed)	Varying	Nursing from mother only	Varying, primarily grass and some grain	Special milk formula diet	Controlled diet, hay and grain or grazing	Controlled diet, hay and grain, or grazing	Varying, hay, grain, and grazing
Source	Abattoir	Abattoir	Abattoir	Abattoir	Monitored herd	Controlled herd	Abattoir
Collection method	Closed, cardiac puncture/umbilical cord collection	From jugular or heart, semi-closed or closed collection	From jugular or heart, semi-closed to closed collection	From jugular or heart, semi-closed to closed collection	Semi-closed or closed from jugular	Semi-closed or closed from jugular	From jugular or heart, semi- closed or closed
Supply	Varies (commodity)	Limited	Unlimited	Somewhat limited	Limited	Adjustable to demand	Unlimited
Relative cost	Fluctuates $$$$$+	$–$$	$	$–$$	$$	$$–$$$	$
Total protein g/dl (avg)	3.0–4.5	4.5–6.5	5.8–7.5	5.8–7.5	5.8–7.5	5.8–8.2	5.8–8.2

[a]*The table provides a representation of what is available in the market place, but details may vary from supplier to supplier.*

(Barnes *et al.* 1980; Rockwell *et al.* 1980; Kawamoto *et al.* 1983; Honma *et al.* 1979). Albumin also acts by binding toxic metals and pyrogens from the medium (Ham *et al.* 1979; Iscove & Melchers 1978).

The α_1 and α_2 globulin components contain the major protease inhibitors α_1-antitrypsin and α_2-macroglobulin, respectively. These components provide protection from endogenous cellular proteases released into the medium, which can be a particular problem for product quality in large-scale cultures. They also function to inactivate trypsin used during the subculture of attached cells. β-globulin includes transferrin for iron transport and beta-lipoproteins. In some cases, serum can be separated into β_1 and β_2 components. Gamma globulin contains the major classes of immunoglubulins IgA, IgM and IgG.

Other specific growth factors that may be present are epidermal growth factor (EGF), fibroblast growth factor (FGF), insulin-like growth factor (IGF), platelet-derived growth factor (PDGF) and neural growth factor (NGF). Fetuin, a component specifically found in FBS, plays a role in cellular attachment and speading, and appears to promote cell growth (Dziegielewska *et al.* 1995). Other attachment factors present are fibronectin and laminin.

Table 4.2 provides a listing of various biochemical components of bovine serum and their concentration ranges. It can be seen that, as the animals increase in age, the most notable changes are an increase in total protein, gamma globulin and cholesterol levels. The composition of any lot of animal serum can vary depending upon the sex, age, health, feeding regime, and stress conditions at the time of slaughter, or donor animal collection.

Table 4.2 Biochemical Profile of Various Types of Bovine Sera

Test Parameters	Foetal n=26	Newborn n=13	Calf n=37	Adult n=7
pH	7.1 – 7.5	7.2 – 7.8	7.5 – 8.0	7.6 – 8.0
Osmolality (mOsm/kg)	298 – 325	285 – 302	268 – 313	268 – 287
Endotoxin (EU/ml)	<0.5 – 1.1	3.8 – 176.3	<0.4 – 240	<0.1 – 0.5
Sodium (meq/l)	133 – 142	134 – 143	131 – 144	135 – 142
Potassium (meq/l)	9.7 – 12.7	5.4 – 8.0	5.4 – 7.7	4.5 – 5.9
Chloride (meq/l)	93 – 101	93 – 102	91 – 105	94 – 102
Uric Acid (mg/dl)	1.5 – 3.2	0.1 – 2.1	0.6 – 1.6	0.2 – 0.6
Calcium (mg/dl)	13.5 – 15.0	8.5 – 11.8	9.6 – 11.8	7.8 – 9.4
Phosphorus (mg/dl)	9.1 – 11.6	8.1 – 10.3	7.7 – 11	6.6 – 7.4
Alk. Phosphatase (U/l)	165 – 247	122 –220	137 – 329	45 – 63
LDH (U/l)	409 – 605	667 – 1102	950 – 2388	747 – 980
SGOT/AST (U/l)	22 – 69	27 – 66	47 – 98	16 – 63
SGPT/ALT (U/l)	0 – 8	12 – 20	8 – 23	13 – 20
GGT (U/l)	4 – 7	76 – 425	22 – 47	14 – 26
Cholesterol (mg/dl)	27 – 58	53 – 112	111 – 174	69 – 205
Bilirubin, total (mg/dl)	0.1 – 0.2	0.3 – 0.9	0.1 – 0.4	0.1 – 0.2
Glucose (mg/dl)	83 – 132	63 – 193	86 – 946	17 – 100
Iron, serum (ug/dl)	170 – 194	69 – 100	35 – 144	80 – 169
BUN (mg/dl)	13 – 17	13 – 15	6 – 14	4 – 15
Creatine (mg/dl)	2.6 – 3.5	0.9 – 1.4	0.9 – 1.8	1.1 – 1.6
Triglyceride (mg/dl)	17 – 91	34 – 48	32 – 53	31 – 38
Haemoglobin (mg/dl)	6.0 – 17.9	12.0 – 20.2	6.6 – 23	3.1 – 22.9
Protein, total (g/dl)	3.1 – 3.7	4.8 – 6.3	5.7 – 7.1	5.9 – 8.1
Albumin/Globulin ratio	0.6 – 4.1	0.4 – 0.9	0.4 – 1.1	0.3 – 0.8
Albumin (g/dl)	1.2 – 2.9	1.5 – 2.8	1.9 – 3.7	1.9 – 2.6
Alpha-1 globulin (g/dl)	<0.1 – 0.8	<0.1 – 0.8	<0.1 – 0.1	<0.1
Alpha-2 globulin (g/dl)	0.1 – 1.3	0.5 – 1.3	0.7 – 1.6	0.6 – 1.5
Beta-globulin (g/dl)	0.2 – 1.5	0.8 – 1.2	0.9 – 2.0	0.8 – 1.4
Gamma-globulin (g/dl)	<0.1	0.5 – 2.0	0.6 – 2.7	1.6 – 4.1

4.4 COLLECTION AND PROCESSING OF SERUM

Blood is generally collected from two sources: animals slaughtered for food consumption or from donor herds maintained on controlled farms designed primarily to supply material to the pharmaceutical and biotechnology industries. A schematic overview of the collection and manufacturing steps is provided in Figure 4.1. For FBS, at the time of collection the adult animal is humanely killed, hung on an elevated rail and the gravid uterus is removed to a separate processing area. The

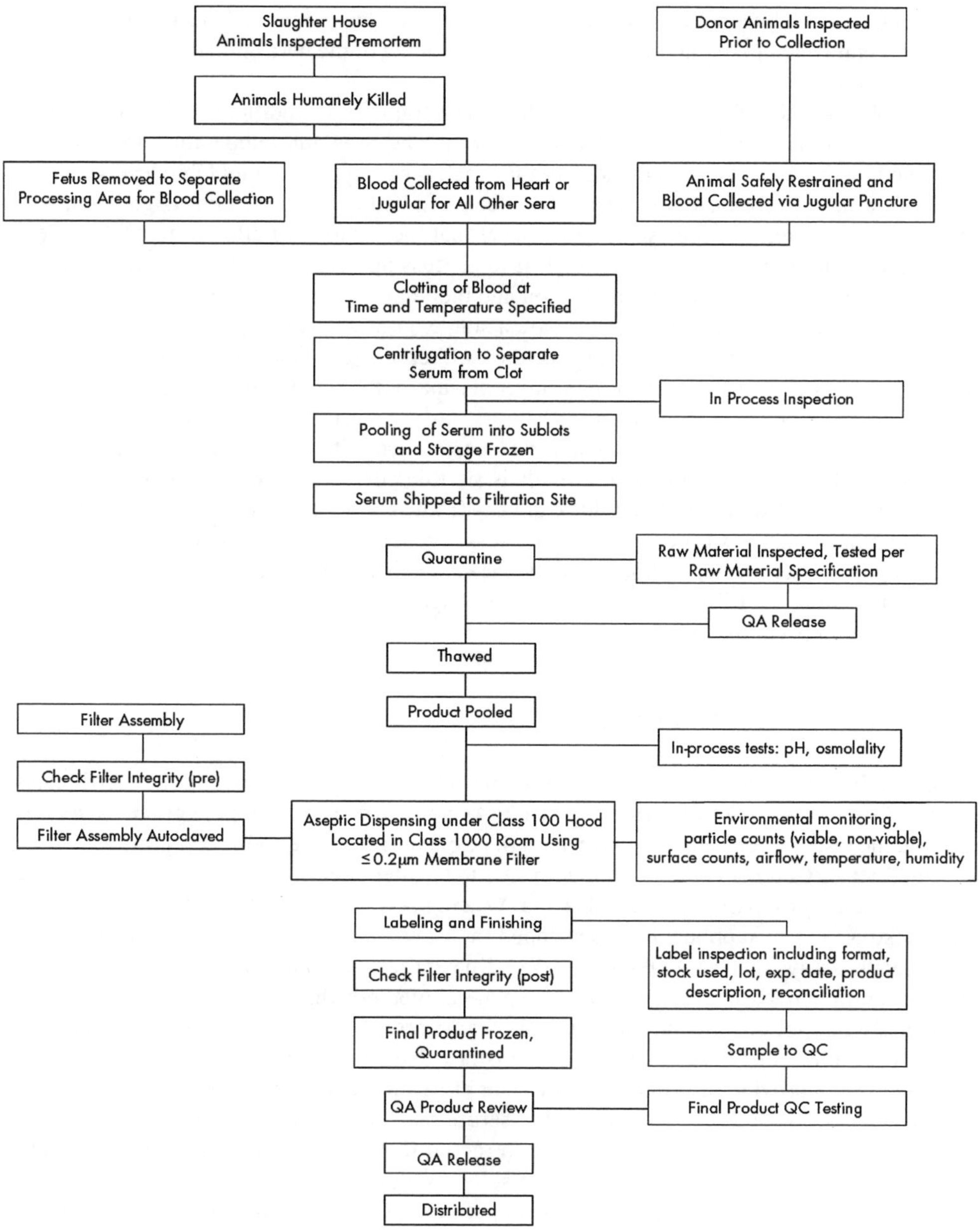

Figure 4.1 Schemetic of the collection and manufacturing steps in the production of animal sera.

foetus can then be separated from the uterus and blood collected aseptically via cardiac puncture or, less frequently, by umbilical cord collection. For newborn (NB), calf (CS) or adult bovine serum (ABS; 12–60 months old) and other adult sera such as porcine serum (PS), the animal is humanely killed and hung. A trocar or canula is then inserted into the heart or jugular vein and the blood collected by gravity into collection devices. These types of sera can normally be traced back to the collection point and date of collection.

Donor animals are those maintained on controlled farms whereby the health and husbandry of individual animals is known. Donor sera can thus be traced back to the specific animal and date of collection. During the collection process, animals are safely restrained, the site of collection is swabbed and a sterile needle inserted into the jugular to collect material. The amount of material collected is controlled to maintain the health and well being of the herd and, ultimately, the quality of the final product.

Once collected, the blood is allowed to clot under controlled conditions and centrifuged. The serum is then pooled into small sub-lots and frozen. The time and temperature for clotting are determined by the typical clotting tendencies for each serum type. Haemoglobin levels in sera are considered a marker for how well serum has been handled and transported. High haemoglobin content (resulting from haemolysis) may indicate that the serum was allowed to sit for extended periods and/or at too high a temperature prior to processing. High endotoxin levels indicate contamination that may be due to poor collection and blood processing techniques.

When enough sub-lots of serum to prepare a single large-volume lot of material are available, most reputable vendors will test samples of the unfiltered sub-lots against set quality specifications. These tests may include endotoxin, mycoplasma, viral screening and bioburden levels. A typical manufacturing scheme is then to thaw the bulk material and filter sterilize it through a series of filters of descending porosity under cleanroom conditions. Typically, the smallest filter is 0.1 μm for FBS. Serum from older animals is significantly higher in protein and lipid content and is most often 0.2 μm filtered, but this can vary from supplier to supplier. The filtered material is collected into a single, sterile, large volume vessel which may exceed 2000 l, depending on the serum supplier. After mixing to homogeneity, the sera can then be dispensed into sterile containers. Once filtration and filling is complete, the packaged material is frozen for long-term storage until final testing is completed.

4.5 FINAL PRODUCT TESTING

Serum suppliers test the final product against written specifications and provide certificates of analysis (CoA) for each lot of serum produced. As no pharmacopoeial monograph exists for serum, specific tests performed and reported on the CoA may follow existing test protocols found in the United States Code of Federal Regulations, or the United States, European or other pharmacopoeias, e.g. tests for sterility and mycoplasma. Where no guidance exists, such as for growth promotion testing, serum suppliers have developed their own testing methodologies or may perform client-specific assays relevant to the client's cell line and/or manufacturing regime. Table 4.3 lists the most common release tests along with the general methods that may be employed.

The desired ranges for the physiochemical parameters, pH, osmolality, appearance and physiochemical profile, are typically based upon historical ranges supplied by the serum producer. While the results provide information that the lot of serum being evaluated is similar to previous lots, it has been this author's experience that serum users rarely use this information as part of their selection criteria for serum, other than to record that testing has been performed, and rely most heavily on the content, performance and safety profile for the product.

In regard to the safety testing of the serum product, of special concern is viral screening for adventitious agents. Testing for viruses in animal sera by serial passage of susceptible cells with

Table 4.3 Example of tests and methods used in releasing final serum products.

Test	Method
Physicochemical characteristics	
Appearance	Visual inspection
pH	Potentiometric
Osmolality	Depression of freezing point
Physicochemical profile[a]	Automated clinical systems, multiple assays
Identity and content	
Identity	Serum electrophoresis band pattern
Species identity	Ouchterlony (double immuno-diffusion assay)
Total protein	Biuret reaction method
Haemoglobin concentration	Colorimetric determination
Bovine IgG content	Enzyme-linked immunosorbent assay
Safety testing	
Sterility	Membrane filtration
Endotoxin	Kinetic chromogenic limulus amebocyte lysate (LAL) assay
Detection of mycoplasma	Hoechst stain (DNA fluorochrome stain) and/or broth/agar cultivation
Adventitious viral agents	Cell culture isolation, fluorescent antibody staining, haemadsorption, cytopathic effects and/or polymerase chain reaction
Virus antibody test	Measurement of neutralizing antibody in serum
Functionality	
Growth performance	Multiple-passage cell growth, clonal growth and/or plating efficiency
Productivity	Viral expression, protein expression

[a]*A list of chemical parameters tested can be found in Table 4.2.*

the test sera is still the industry and government standard (FDA 2001; EMEA2002a; 2003), although other testing such as RT-PCR (reverse transcription polymerase chain reaction) may be employed. Cell lines such as VERO, and another primary or continuous cell line of the same species as the sera being tested, are serially passaged over a period of approximately 3 weeks. At regular intervals, the cell monolayers are observed for cytopathic effects (CPE) and haemagglutination. After the last passage, cells are also screened for specific viruses by fluorescent antibody staining.

Serial passage in cell culture has many disadvantages:

- viruses may be non-cytopathic in cell culture;
- the limit of detection may not be sufficient to detect low level contamination in large serum batches. This is aggravated by the small size of evaluation samples compared with the total lot size. Low viral titres not detected as part of routine viral screening may be amplified during large-scale manufacture.
- The fluorescent antibodies may not detect all viral strains.

Serum that is reported negative on supplier CoA has been proven to contain live virus when assayed by a combination of RT-PCR and serial passage in cell culture (Yanagi *et al.* 1996). In another study, detection of bovine polyoma virus using susceptible cell cultures as a substrate took 4 to 7 weeks, which is approximately twice the length of time of the most common testing regimes (Schuurman *et al.* 1991). It should be assumed, regardless of viral testing results on a CoA, that all commercially available lots of sera that have not been treated with a viral inactivation or removal process contain some unknown level of virus.

PCR (polymerase chain reaction) has been shown to be an extremely sensitive test for detecting viral nucleic acid, but alone it is unable to distinguish between infectious viral contamination and nucleic acid contamination. Combining PCR with culture amplification can positively identify infected lots of serum (Yanagi *et al.* 1996; Schuurman *et al.* 1991). However, much validation of the appropriate probes and methodologies will be required before PCR is accepted as a routine product acceptance testing method, but it may be useful as an in-house screening tool. Validated, more sensitive methods of viral detection may create the quandary that more serum may end up being rejected even though similarly infected lots of serum are most likely used in manufacturing today. With the limitations of serum supply and the need to address safety concerns regarding viral transmission via medicinal products, methods to inactivate or remove contaminating viruses are required. This will be addressed in Section 4.8.

Step	**Information recorded / deposited**	**Provision**	**Archiving**
Collection of raw material, whole blood	Date of collection USDA/AQIS/MAF[a] registration number of abattoir[b]	Raw serum collector	Serum manufacturer
↓			
Processing of whole blood to raw serum	Raw serum sub-lot number Raw serum certificate of origin[c] Shipping documentation	Raw serum collector	Raw serum collector Serum manufacturer
↓			
Manufacture of serum	Origin of batch[d] Raw serum data summary Product certificate of analysis Device history record[e] Health certificate[f]	Serum manufacturer	Serum manufacturer

a United States Department of Agriculture/Australian Quarantine and Inspection Service/ New Zealand Ministry of Agriculture and Forestry

b Documentation on the origin of the batch is retained. Confirmation that the processing of the raw serum occurred on a specific date at a specified controlled facility is documented.

c The certificate of origin documents that the raw serum was collected at USDA/AQIS/MAF approved slaughterhouses.

d The origin of the batch, including the USDA/AQIS/MAF number of the abattoir for collection and date of collection, is documented in the raw material batch records for the serum sub-lots.

e The device history record is the term used for the batch record.

f A health certificate contains a signed affidavit confirming the status of the country of origin, the source and health of the animals and the status of the sites of collection. However, it should be noted that these certificates are not obtained routinely for all batches of material, rather upon specific request.

Figure 4.2 Overview of traceability measures employed.

Some sera are sold as being specifically tested for a particular use, e.g. for insect or hybridoma cells. This may be misleading, as no growth assay can replace growth evaluation in a user's own laboratory with their cells, process and equipment. It is also relevant to point out that the presence of certain components that may be required for a given cell line might not be tested for specifically, such as oestrogen or platelet-derived growth factor.

As part of the CoA, serum origin information is listed along with a declaration of the inspection and health status of the animals, and a declaration of the bovine spongiform encephalopathy (BSE) status of the country of origin. Figure 4.2 provides a schematic presentation of the information deposited and retained by either the raw serum collector or the serum manufacturer to ensure traceability of serum of United States, Australian or New Zealand origin.

When raw serum is received by the serum manufacturer from the raw serum vendor, assurance of the country of origin and that the abattoirs are government registered and approved is provided on a certificate of origin (COO). As stated on the COO, the raw serum vendor retains documentation to confirm the following information on the collection of the whole blood:

- country of origin;
- name of collection facility (abattoir);
- location of collection site, city, state;
- government approved establishment number;
- date of collection and amount collected.

This information is archived by the serum manufacturer as part of the raw material batch record.

4.6 STORAGE AND STABILITY

Storage recommendations vary by supplier, but are typically within the range of −5 to −40°C and lower temperatures. Those down to −80 °C have been recommended for serum packaged in non-glass containers. It should be borne in mind that whilst serum appears to retain its growth promoting characteristics well at a variety of storage temperatures, it cannot be assumed to be completely frozen unless it is kept at low temperatures (e.g.−70 °C). Storage in self-defrosting freezer units should be avoided as multiple freeze/thaws can cause precipitation of salts, lipoproteins, and cold-insoluble globulins, and may reduce the growth-promoting properties of the serum. Stability testing is based upon yearly evaluation of the sera against final product specifications for physiochemical and growth-promoting properties. Most vendors assign an expiration date of 5 years from the date of manufacture.

In our laboratories we have evaluated the growth promoting properties of FBS stored at 4 °C in bottles over a period of 1 year. At predetermined intervals throughout the year, the serum was diluted in cell culture media to either 2 or 10 % and compared to serum that had been stored frozen at −10 to −40°C. The cell lines used to evaluate the serum were VD10 (IgG-secreting hybridoma), CHO K1 and Madin–Darby bovine kidney (MDBK). Cells were sub-cultured in 25-cm^2 tissue culture flasks for three passages and cell counts were compared to control. Our results indicate that the serum maintained its growth promoting properties for VD10 and MDBK cells over the 1-year period, but CHO K1 cells displayed satisfactory growth only until week 24 of the study. In our hands bovine calf serum also appears to be stable for 28–32 weeks at 2–8 °C storage (data not shown). For serum users, these results open the possibility of alternate storage of serum or the ability to purchase complete media, i.e. media plus serum, possibilities that may have not been previously considered due to concern over product stability.

It is likely that storage of serum at 2–8 °C will result in oxidation of lipid and vitamin components, and that serum proteins may precipitate out of solution. Sensitivity to any sort of degradation should be evaluated on a case by case basis.

4.7 SUPPLY AND RISK

We estimate the annual world supply of FBS to be 600 000 l with the US contributing 300 000 l with another 120 000 l from Australia and New Zealand. The remainder of the material is derived from Canada, Central/Southern America and small research-grade volumes from Europe. As mentioned previously, the amount of material available can fluctuate with environmental conditions (drought, flood, natural disasters) and epidemics such as foot and mouth disease (FMD), which would inhibit or preclude collection. Customs authorities have the right to suspend importation of animal-derived materials if these products pose a threat to agricultural health in the importing country. In 1986, the first cases of bovine spongiform encephalopathy (BSE) were reported for the United Kingdom (Wells et al. 1987), and BSE has now also been found in mainland Europe and other countries including Japan, Canada and the United States (OIE 2007). Regulatory guidance prescribes suitable sources for serum employed in biological manufacture (EMEA 2003 ; FDA 2000). Sourcing from appropriate geographical areas where BSE surveillance programs are in place still provides the greatest assurance of freedom from BSE contamination (EMEA 2003). In January 2007, the Office International des Epizooties (OIE) listed 25 countries as reporting or having reported cases of BSE with an additional two countries (Falkland Islands and Oman) reporting cases of BSE from imported animals only (OIE 2007). As these sources are typically unsuitable for serum destined for pharmaceutical manufacture, this limits the worldwide serum supply.

The United States, Australia and New Zealand provide an estimated 70% of all animal sera produced, and more than 90% of all serum used in pharmaceutical manufacture. In May of 2003 a single case of BSE was reported (CFIA 2003) for an animal located in Alberta, Canada. Based on the investigation to date, the cause of the BSE infection is most likely due to the consumption of feed containing meat and bonemeal (MBM) contaminated with the BSE prion prior to the MBM feed ban of August of 1997. Due to strong BSE surveillance programs and stringent measures in Canada, it is likely that the actual number of infected animals in the cattle population is extremely low (CFIA 2003).

The US closed its borders to trade in Canadian live cattle on May 20, 2003 (USDA 2003), but until that time the US accounted for nearly all of Canada's exports. During the years 1998 to 2002, each year approximately 1–1.7 million head of cattle crossed the border into the US Of these, between 100 000 to 500 000 entered into the US cattle population after a 10-day quarantine period and were able to move throughout the US. The remaining animals were sent for slaughter and possible serum collection (AAFC 2003). As the United States does not maintain a countrywide cradle-to-grave tracking system, cattle that cross the border and are integrated into the US cattle population may not be traceable to the country of origin. It must be assumed, with the possible exception of donor herds, that serum from cattle of Canadian ancestry may be collected within the US. Two case of BSE have been reported in the US, one of Canadian origin and the other originating in the United States (OIE, 09, 2007). Further monitoring will determine if the US is facing any greater risk from BSE, and what impact this may have on the suitability of all ruminant source materials for import into other countries (USDA 2003). From a risk management perspective, it is imperative to evaluate, at the least, serum with fewer supply concerns (bovine calf, adult bovine, donor bovine) or serum alternatives and animal protein-free media as part of a comprehensive raw material supply risk management exercise.

4.8 MINIMIZING THE RISK

Several methods exist to remove and/or inactivate microbiological contaminants from serum. These methods can be performed by serum suppliers or at the time of use by the end user. The serum industry standard and the recommendation found in government guidance (EMEA 2002a, b) is to treat serum with gamma radiation, but other methods are available including viral exclusion filtration and chemical inactivation. Methods such as acid treatment or heat inactivation generally work only on limited virus types and/or are deleterious to the final serum product (Hansen *et al.* 1997). While no method can provide 100 % assurance of freedom from adventitious agents, all provide a greater margin of safety assurance when using animal-derived products.

Gamma-irradiation has several key advantages:

- Serum can be irradiated at dosages that reduce the level of tissue culture infectious doses/ml (TCID/ml) of common contaminating viruses by a factor of 10^6 (often referred to as 'six logs') without compromising product integrity.
- Serum can be irradiated in bulk with relative economy and speed.
- Treatment can be performed in the final, sealed, serum container at any time following sterile filtration.
- The process can be validated, and once validated the dosage given can be adjusted to individual requirements.
- This method is accepted by regulatory and customs authorities as an efficacious treatment.
- Gamma irradiation is a clean process that does not leave residues, as is possible with chemical inactivants.

Several major serum suppliers have published data showing that a panel of naturally occurring potential contaminants of serum such as bovine viral diarrhoea, or model viruses chosen in order to represent a spectrum of virus types, i.e. enveloped and non-enveloped viruses, DNA and RNA viruses and relatively large and small viruses (Table 4.4) can be inactivated in excess of 6–7 logs (Hansen *et al.* 1997; Plavsic *et al.* 1999; Purtle *et al.* 2002). Some small viruses such as porcine parvovirus (PPV) and minute virus of mice (MVM) are more difficult to inactivate, but still show approximately 3–4 logs viral reduction when treated at 25–35 kGy (Table 4.5). The data all agree that when serum is irradiated under controlled conditions, the serum will maintain its growth-promoting capacity on a variety of industrially relevant cell lines.

Table 4.4 Virus panel to demonstrate inactivation potential using Gamma-irradiation. Reproduced from Purtle, D *et al.* (2002). Copyright SAFC Biosciences.

Target virus	Family	Characteristics[a]	Model virus
Bovine viral diarrhoea (BVD)	Flaviviridae	ss RNA enveloped	BVD
Parainfluenza Type 3 (PI3)	Paramyxoviridae	ss RNA enveloped	PI3
Infectious bovine rhinotracheitis (IBR)	Herpesviridae	ds DNA enveloped	IBR
Bluetongue (BTV)	Reoviridae	ds RNA enveloped	BTV
Bovine leukemia (BLV)	Retroviridae	ss RNA enveloped	FeLV[b]
Porcine parvovirus (PPV)	Parvoviridae	ss DNA non-enveloped	PPV
Minute virus of mice (MVM)	Parvoviridae	ss DNA non-enveloped	MVM

[a] *Single stranded/double stranded*
[b] *Feline leukemia virus*

Table 4.5 Viral inactivation of model viruses at 25–35 kGy. Reproduced from Purtle, D *et al.* JRH Biosciences. Research Report: Validated Gamma Radiated Serum Products (2002). Copyright SAFC Biosciences.

Organism[a]	0 kGy[b]	25– 35 kGy	Log reduction
BVD	4.32	None detected	⩾4.32
IBR	4.69	None detected	⩾4.69
PI3	7.13	None detected	⩾7.13
MVM	4.6	0.80	3.80
BTV	6.42	3.11	3.31
FeLV	5.87	2.58	3.29

[a]*For full name of viruses refer to Table 4.4.*
[b]*Starting titre* (log_{10}).

Sera that have been subjected to 0.04 µm filtration as opposed to standard 0.1 µm or 0.2 µm filtration can also be purchased. Studies show that reovirus preparations of approximately 70 nm particle diameter can be reduced from 8.3×10^8 virions/ml to undetectable levels (Hyclone Labs, 1989). Further experiments showed that virus-like particles with diameters ranging from 60 to 180 nm can be reduced from 9.8×10^8 to $<2 \times 10^6$ particles/ml. The data indicate that this type of filtration can remove certain quantities of contaminating virus, but that some infectious material may pass through the filters. This would most certainly be true for the smallest of viruses such as PPV or MVM. It may be suitable to use this type of filtration in conjunction with gamma irradiation.

4.9 CHEMICAL INACTIVATION

Historically, beta-propiolactone (BPL) was employed to inactivate serum contaminants chemically. It has been used almost exclusively by veterinary biological companies and is generally considered to be unsuitable for products for human use. Other chemical inactivants such as *N*-acetylethyleneimine (AEI) may also be suitable for inactivating a variety of relevant virus types (Brown *et al.* 1999).

4.10 SUMMARY

The decision to use serum in new manufacturing processes, or to maintain its use in current manufacturing schemes, is a complex one. Input is required from not only the scientific and development staff, but from regulatory affairs, quality, and purchasing departments, to provide insight into their respective areas of expertise. These decisions must be tempered with the realities of production economics and the need to maintain a high level of safety within the patient population. To this end, regulatory agencies have published guidelines on the testing and use of serum in human and veterinary pharmaceuticals (FDA 2000;2001; EMEA 2001a, b, 2003), and the European Pharmacopoeia has issued a monograph on bovine serum (European Pharmacopoeia 2006). It is also necessary to work closely with serum manufacturers in a consultative relationship. These suppliers have a depth of knowledge of the procurement, documentation, manufacturing, testing, and storage of animal sera. In conjunction with their expertise in sera, many also develop their own proprietary serum-free and protein-free media for multiple cell lines and may be able to provide the broadest perspective on how best to achieve the overall requirements of the pharmaceutical and biotechnology manufacturers.

REFERENCES

AAFC (2003) Agriculture and Agri-Food Canada. Livestock Exported to the United States through Ports of Exit. www.agr.gc.ca/misb/aisd/redmeat/02Table4.xls.

Barkema HW *et al.* (2001) *Tijdschr Diergeneeskd*; 126(6): 158–165.

Barnes D, Sato G (1980) *Analytical Biochemistry*; 102: 225–270.

Bolin SR, Mathews PJ, Ridpath JF (1991) *J. Vet. Diagn. Invest.*; 3(3): 199–203.

Brown F *et al.* (1999) In *Animal Sera, Animal Sera Derivatives and Substitutes used in the Manufacture of Pharmaceuticals: Viral Safety and Regulatory Aspects.* Ed Cartwright BF *et al.* Dev. Biol. Stand. Basel, Karger; Vol. 99, 119–130.

CFIA (2003) Canadian Food Inspection Agency *Summary of the Report of the Investigation of Bovine Spongiform Encephalopathy (BSE).* Alberta, Canada. July 2, 2003. www.inspection.gc.ca/english/anima/heasan/disemala/bseeb/evalsume.shtml.

Doyle A and Griffiths JB (eds.) (2000) Cell and Tissue Culture for Medical Research, pp 83, Hoboken, John Wiley & Sons Ltd.

Dziegielewska KM, Brown WM (1995) Molecular Biology Intelligence Unit. Fetuin. R.G. Landes Company. pp. 2–6.

EMEA (2002a) The European Agency for the Evaluation of Medicinal Products *Guideline on the Requirements and Controls Applied to Bovine Serum Used in the Production of Immunological Veterinary Medicinal Products.* January 11, 2002 EMEA/CVMP/743/00 www.emea.eu.int.

EMEA (2002b) The European Agency for the Evaluation of Medicinal Products *Note for Guidance on the Use of Bovine Serum in the Manufacture of Human Biological Medicinal Products* (Draft). November 2002 CPMP/BWP/1793/02 www.emea.eu.int.

EMEA (2003) The European Agency for the Evaluation of Medicinal Products *Note for Guidance on Minimizing the Risk of Transmitting Animal Spongiform Encephalopathy Agents via Human and Veterinary Medicinal Products.* October 2003 EMEA/410/01 revision 2.

European Pharmacopoeia (2006) *Bovine Serum Monograph* 04/2006:2262.

Fong J, Gross PA *et al.* (1975) *J. Clinical Microbiol.*; 1(2): 219–224.

FDA (2000) Food and Drug Administration. *Letter to Manufacturers of Biological Products-Recommendations Regarding Bovine Spongiform Encephalopathy (BSE)* April 2000. www.fda.gov/cber/ltr/bse041900.

FDA(2001) Food and Drug Administration. Code of Federal Regulations, Title 9: *Animals and Animal Products*, Revised January 1, 2001. 578–579.

Friedlander E (2003) www.pathguy.com/lectures/proteins.htm.

Ham RG, McKeehan WL (1979) *Methods Enzymol.*; 58: 44–93.

Hansen G *et al.* (1997) *Art To Science*; 16(2): 1–7.

Harasawa R (1995) *Vaccine*; 13(1): 100–103.

Honma YT *et al.* (1979) *Exp. Cell Res.*; 124: 421–428.

Hyclone Laboratories, Inc. (1989) *Art to Science*; 8(1): 1–2.

Iscove NN, Melchers F (1978) *J. Exp. Med.*; 147: 923–933.

JRH Biosciences Inc. (2000) Technical Bulletin: *Microbe Selection for the SER-TAIN^{TM} Process Validation*, March 2000. www.jrhbio.com.

Kawamoto T *et al.* (1983) *Anal. Biochem.*; 130: 445–453.

Kniazeff AJ *et al.* (1975) *In Vitro*; 11: 400–403.

Kretzmer J *et al.* (1994) In *Animal Cell Technology, Products of Today, Prospects for Tomorrow.* Ed Spier R. Oxford, Butterworth-Heinemann Ltd; 419–421.

Milo GE *et al.* (1976) *In Vitro*; 12(1): 23–30.

OIE (2003) Office International des Epizooties www.oie.int/eng/info/en_esb.htm revision 25/08/2003.

OIE (2007) office International des Epizooties www.oie.int/eng/info/en_esbcarte.htm.

Plavsic MV *et al.* (1999) In *Animal Sera, Animal Sera Derivatives and Substitutes Used in the Manufacture of Pharmaceuticals: Viral Safety and Regulatory Aspects.* Ed & Cartwright BF *et al.* Dev. Biol. Stand. Basel, Karger; Vol 9, 95–109.

Purtle, D *et al.* (2002) JRH Biosciences. Research Report: *Validated Gamma Radiated Serum Products.* Issued March 2002. www.jrhbio.com. Basel, Karger.

Rockwell G, Sato G, McClure D (1980) *J. Cell Phys.*; 102:323.
Schuurman R *et al.* (1991) *J. Gen. Virol.*; 72(11): 2739–2745.
USDA (2003) United States Department of Agriculture *Backgrounder Bovine Spongiform Encephalopathy (BSE)*. US Department of Agriculture and Food and Drug Administration. Release No. bg0166.03 Updated July 10, 2003. www.usda.gov.news/releases/2003/05/bg0166.htm.
Valez D *et al.* (1986) *J. Immunol. Methods*; 86: 45–52.
Wells GAH *et al.* (1987) *Vet. Rec.*; 121: 419–420.
Wilbur LA *et al.* (1994) *J. Am. Vet. Med. Assoc.*; 204 (11): 1762–1765.
Yanagi M *et al.* (1996) *J. Infect. Dis.*; 174 (6): 1324–1327.

List of Useful Web Sites

Agriculture and Agri-Food Canada	www.agr.gc.ca
Canadian Food Inspection Agency	www.inspection.gc.ca
European Medicines Agency	www.emea.europa.eu
Food and Drug Administration	www.fda.gov
JRH Bioseceinces, Inc.	www.jrhbio.com
United States Department of Agriculture	www.usda.gov

Cell Engineering for Recombinant Products

5 Expression of Recombinant Biomedical Products from Continuous Mammalian Cell Lines

SA Jeffs

5.1 INTRODUCTION

Recombinant therapeutic proteins can be expressed in bacterial, plant, yeast, insect and mammalian cells. Animal cell culture is the optimal system for the expression of many therapeutic proteins (particularly glycoproteins) as protein folding, transport and processing follow the same pathways seen in the target organism. Furthermore, the glycosylation pattern of recombinant glycoproteins expressed in mammalian cells more closely resembles that of native glycoprotein than if insect cells are used. This can have important therapeutic implications and is discussed in detail later (see Section 5.2.3).

Essentially, there are three methods whereby biological products can be expressed in mammalian cells:

- by transfection with a modified bacterial plasmid containing a strong mammalian gene expression promoter, polyadenylation signal (see 5.2.1a) and the gene of interest;
- by infection with a replication-competent or replication-incompetent recombinant viral vector engineered for a high level of expression of the desired protein (i.e: adenoviral, retroviral, alphaviral or vaccinia expression systems);
- by infection with a virus of interest, using a whole or modified viral genome (viral protein only).

Figure 5.1 illustrates the components of a mammalian expression system, based on use of a modified bacterial plasmid. The choice made of each component will impact upon the yield and therapeutic efficacy of the recombinant protein, as well as on the time and cost to obtain it, but it is important to recognize that each individual component is linked to another. For example, the choice of expression vector will determine the choice of host cell (or *vice versa*), and the amount of protein required will be related to not only vector and cell line, but also to the culture system employed. This review considers each of these components in turn to enable the correct choices to be made in order to obtain suitable quantities of recombinant protein for a particular application.

Medicines from Animal Cell Culture Edited by G. Stacey and J. Davis

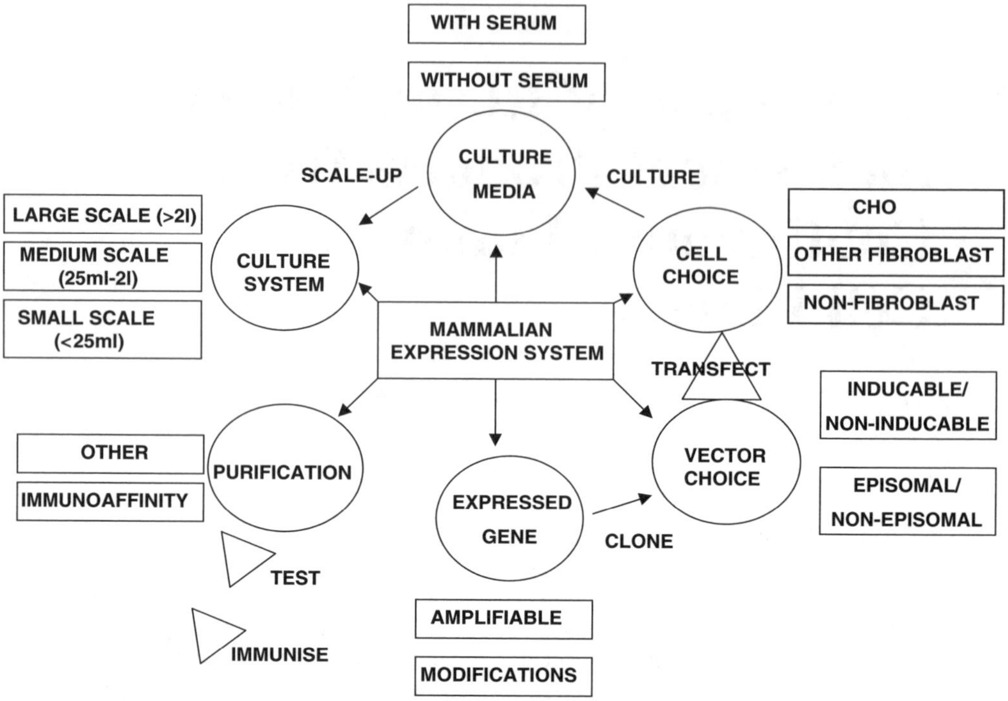

Figure 5.1 Generalized scheme for production of recombinant protein.

5.2 GENE EXPRESSION

5.2.1 Genetic Modification to Expressed Gene

By the use of suitably designed primers, many modifications can be made to the gene of interest. These may include the addition of restriction endonuclease (RE) sites to assist cloning into plasmids, plus linker sequences of the appropriate length to ensure their restriction prior to cloning; purification/detection tags (if not included in the plasmid) and a terminal stop codon. In addition, the expression of human immunodeficiency virus Type 1 (HIV-1) IIIB gp120 and gp160 in both COS and Chinese hamster ovary (CHO) cells may be very poor unless the native signal sequence has been replaced by that from human tissue plasminogen activator (TPA) or herpes simplex virus (HSV) type 1 glycoprotein D (Chapman *et al.* 1991; Lasky *et al.* 1986). Berman *et al.* (1988b) highlighted the importance of the signal sequence in the optimal targeting of the recombinant protein to the cellular secretory machinery and hence to the ambient medium. This not only aids protein purification, but is also less likely to lead to proteolytic degradation or aggregation of the product into poorly soluble high-molecular-weight bodies. If the homologous signal sequence is not effective in a specific expression system, consideration should be given to its replacement. In mammalian cell expression systems, the human interleukin-2 and TPA signal sequences are particularly effective and widely employed (Sasada *et al.* 1988; Chapman *et al.* 1991; Planelles *et al.* 1991). Codon usage effects are a major impediment to the efficient expression of many non-mammalian genes and the systematic replacement ('codon-optimization') of the native gene codons with codons chosen to reflect more closely the codon preference of highly expressed human genes has been shown substantially to increase the expression levels of, for example, both HIV-1 *env* and *gag* genes and to remove the constraints of *cis*- and *trans*-acting elements on their regulation (Haas *et al.* 1996; zur Megede *et al.* 2000). Protein production may also be initiated or enhanced by the removal of certain structural elements or regions, such as fusogenic domains, membrane anchors, multimerization domains, *N*-linked glycans and the cleavage sites between subunits (Fussenegger *et al.* 1999; Liljeqvist & Stahl 1999; Etchevery 1996). However, it is essential that such modifications neither remove nor disable important epitopes required in the final product, and it is recommended that these constructs are characterized for desired functionality at the earliest possible stage of development. Enhanced production of HIV *env* gene expression in mammalian cells was seen following removal of the cytoplasmic and transmembrane regions of gp41 from the full-length precursor polyprotein (gp160) to produce gp140 (Berman *et al.* 1988a; Berman *et al.* 1989; Willey *et al.* 1988; Jeffs *et al.* 2004, 2006), and on the removal of portions of both the *N*- and *C*-termini and V1 and V2 (but not V3) hypervariable loops (Pollard *et al.* 1992; Wyatt *et al.* 1993; Jeffs *et al.* 1996). Enhanced production of a C-clade gp140 from stable CHO-K1 lines was also achieved following ablation of the gp120/41 primary cleavage site, and in addition the proportion of oligomeric over monomeric forms was increased (Jeffs & Viera, personal observations).

5.2.2 Expression Vectors

5.2.2.1 Plasmid vectors

5.2.2.1a Plasmid structure

A wide selection of plasmid-based expression vectors is now available. For many studies of gene expression, small quantities of protein are sufficient and can be obtained using transient assay systems. When larger amounts of protein are required, it is necessary to identify clonal cell lines in which the vector sequences are retained during cellular proliferation ('stable' cell lines). This can be achieved either by episomal plasmid replication or by the integration of the vector into the host cell genome. Cell lines containing foreign deoxyribonucleic acid (DNA) can be identified by the use of a suitable selectable marker gene (Gorman 1985; Chisholm 1995). The choice of vector will depend upon the host cell employed and the amount of protein required. Each vector must contain bacterial 'backbone': sequences that allow replication and maintenance in bacteria, a translation initiation sequence, a promoter/enhancer sequence and a polyadenylation signal. If stable expression is needed, a selectable marker gene is required, while episomal plasmid replication is directed by a variety of viral elements. Optional components may include the presence of an antigen epitope that permits detection/visualization of expressed gene products and a purification tag. A reporter gene, whose expression can give quantitative indications of transfection or transcriptional activity, can be added if required. The read-out from this can be either enzymatic, such as chloramphenicol acetyl transferase, or bioluminescent, such as firefly luciferase or green fluorescent protein (Alam & Cook 1990; Chisholm 1995).

The bacterial backbone always includes an origin of replication (ori) and its associated *cis*-acting control elements, the whole genetic unit being termed a replicon. Replicons derived from the pMB1 (or ColE1) plasmid do not require plasmid-encoded functions for replication, and can function in the absence of ongoing protein synthesis, leading to the accumulation of several thousand copies of the plasmid in the cell (Stadenbauer 1978). One or more antibiotic resistance genes (i.e; *amp*^{r}–ampicillin, *kan*^{r}–kanamycin, *tet*^{r}–tetracycline or *cam*^{r}–chloramphenicol) for selection in *E.coli* must also be present. Inclusion of elements of the *lacZ* operon permits blue/white screening of recombinant plasmids, while a multiple cloning site (MCS) with a wide range of single-cutting RE sites for insertion of the gene to be expressed is essential. The MCS can be designed to permit the cloning of blunt or sticky-ended polymerase chain-reaction-derived cDNA and to allow insertion in a specific orientation. Other genes that may be present include a variety of suppressor transfer ribonucleic acid (tRNA) genes to inhibit bacterial nonsense alleles, a f1 or M13

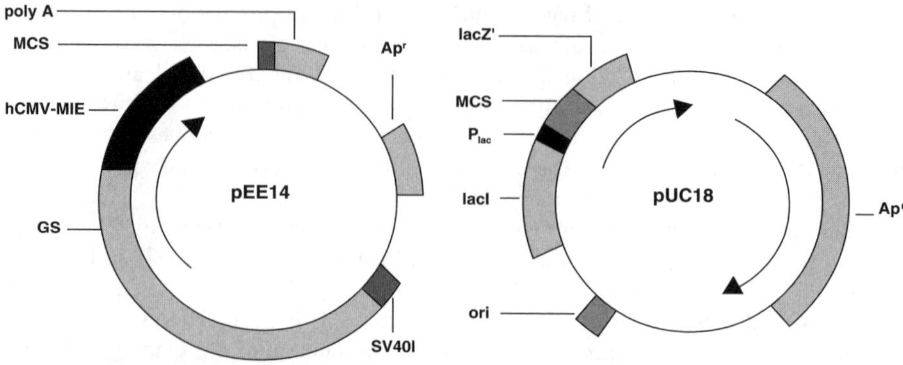

Map of plasmid pEE14, an expression vector for CHO cells. GS= hamster glutamine sythetase minigene (contains a single GS intron expressed from the SV40 late promoter (SV40l); hCMV-MIE= human cytomegalovirus major intermediate early enhancer-promoter; MCS = multiple cloning site; Ap^r = ampicillin resistance gene; poly A = polyadenylation signal

Map of plasmid pUC18, a general bacterial cloning vector. ori = origin of replication; Ap^r = ampicillin resistance gene; lacI =lac repressor gene; lac Z' = lacZ' gene; P_{lac} = lac promoter; MCS = multiple cloning site

Figure 5.2 Example of a general cloning vector (pUC18) and mammalian expression vector (pEE14).

ori to generate single-stranded DNA templates for sequencing or site-directed mutagenesis, and promoter sequences for T3 or T7 bacteriophage DNA-dependent RNA polymerase to permit the generation of RNA transcripts.

To ensure that the inserted gene is optimally expressed, a 'Kozak' translational initiation sequence (CCA/GCATG) is nearly always included before the initiating ATG of cDNA gene expression constructs (Kozak 1986). Polyadenylation of messenger ribonucleic acid (mRNA) is critical for message stability and transport from nucleus to ribosome. Most expression vectors contain either the SV40 early or bovine growth hormone polyadenylation signal to allow for this (Goodwin & Rottman 1992; Kaufman & Sharp 1982). Examples of a basic cloning plasmid (pUC18) and a mammalian expression plasmid (pEE14) are given in Figure 5.2.

5.2.2.1b Selection

Selectable markers are either genes (mostly bacterial), which establish drug resistance in cell culture, or which are dependent on specific mammalian cell genotypes (Table 5.1). The first category of markers is commonly represented in commercial vectors by the bacterial aminoglycoside phosphotransferase gene *aph*, which detoxifies the protein synthesis inhibitor drug G418 (neomycin/geneticin)(Colbere-Garapin *et al.* 1981) hygromycin-B-phosphotransferase (*hph*) which inhibits hygromycin-B (Blochlinger & Diggelmann 1984) and the *ble, blc* and *pac* genes that confer resistance to the antibiotics bleomycin (zeocin), blasticidin and puromycin respectively (Vara *et al.* 1986; Mulsant *et al.* 1988). Many of the second category of genes are 'amplifiable', and will be discussed in relation to strategies that enhance protein expression (Bebbington 1995).

5.2.2.1c Promoters

In many cases, maximal amounts of biologically active, stable recombinant protein are required. The inclusion of specific elements within an expression vector can make the difference between poor and high yields of protein from a given cell line. The first essential choice is that of the promoter/enhancer sequence. When not working with a toxic gene, timing of expression is less important and a constitutively expressing promoter can be used (see below). However, if the gene product is toxic or unstable, expression can be controlled by the use of an inducible promoter

Table 5.1 Genes (and their specific inhibitors) used for selection of clonal cell lines.

Gene	Product	Inhibitor	Reference
Bacterial/fungal selectable markers			
Neo	Aminoglycoside phosphotransferase	G418 (geneticin/neomycin)	Colbere-Garapin *et al.* (1981)
Hph	Hygromycin-B-phosphotransferase	Hygromycin B	Blochlinger and Diggelmann (1984)
Sh ble	High affinity antibiotic binding protein	Bleomycin (zeocin)	Mulsant *et al.* (1988)
Bsd	High affinity antibiotic binding protein	Blasticidin	Mulsant *et al.* (1988)
Pac	Puromycin *N*-acetyltransferase	Puromycin	Vara *et al.* (1986)
trpB	Tryptophan synthetase	Indole	Hartman and Mulligan (1988)
hisD	Histidinol dehydrogenase	Histidinol	
Mammalian cellular genotypes			
Dhfr	Dihydrofolate reductase	Methotrexate	
Gs	Glutamine synthetase	L-Methionine sulfoximine	Bebbington (1995)
Ada	Adenosine deaminase	Deoxycoformycin	
Asns	Asparagine synthetase	Albizzin	
Cad	Aspartate transcarbamylase	*N*-(phosphonacetyl)-L-aspartate	
Polr2K	Metallothionein-I	Heavy metals	
Odc	Ornithine decarboxylase	α-Difluoromethyl ornithine	
mdrI	P-Glycoprotein 170	Adriamycin	
Rnr	Ribonucleotide reductase	Hydroxyurea	
TK	Thymidine kinase (defective)	—	
Xgprt	Xanthine-guanine phosphoribosyl transferase	Mycophenolic acid	

(see 5.2.2.1d). The promoter consists of a DNA sequon where RNA polymerase II transcription is initiated, and is positioned so as to direct transcription in a defined direction. Enhancer motifs augment transcription from a promoter and are essential for the highest levels of transcriptional activation.

The most generic and versatile promoter/enhancer assemblies for mammalian expression are derived from viral systems. The human cytomegalovirus major immediate-early viral promoter/enhancer (hCMV), the Rous sarcoma virus long terminal repeat promoter (RSV-LTR), the simian virus early promoter/enhancer (SV40), and the human elongation factor 1α-subunit promoter are often employed in commercially available vectors (Berg 1981; Gorman *et al.* 1982; Foecking & Hofstetter 1986; Kim *et al.* 1990). hCMV is a stronger promoter than either RSV-LTR or SV40 in most cell lines and is the most commonly used. It is particularly effective in adenovirus-transformed cell lines, such as HEK-293 (Gorman *et al.* 1989). The RSV-LTR is derived from an avian virus, and works best in avian cell lines, while SV40 works well in most cell lines but performs best in lines containing the stably integrated SV40 large T antigen, such as the African green monkey kidney COS cell lines (Berg 1981; Gorman *et al.* 1982).

5.2.2.1d Enhanced expression

Regulated (or inducible) transcription allows the precise expression of the gene of interest and is often used when the protein product is cytotoxic or cytostatic. Simple inducible promoters are based upon such elements as metallothionein or heat-shock promoters, but lack the precision achievable with engineered systems where the regulatory components are stably integrated into the host cell. These have now reached a considerable degree of sophistication and the choice of a suitable vector can allow the low basal expression of a highly toxic protein, a high induced level of expression for maximum production or modulated expression for functional analysis. Current commercial systems allowing high inducible levels of expression are mostly based on bacterial regulated promoters, such as the modified *E.coli lac* operon system available from Stratagene (Wyborski *et al.* 1996); the 'tet-On'/'tet-Off' system based on elements of the Tn-10 encoded tetracycline resistance operon, and the HSV virion protein 16, marketed by Clontech (Gossen & Bujard 1992); and the T-Rex (tetracycline-regulated-expression) Invitrogen system that also employs elements from the *E.coli* tetracycline Tn-10 suppression mechanism to regulate expression from the hCMV promoter (Yao *et al.* 1998). Other promoters are regulated by steroid hormones, such as the ecdysone-inducible system based on a *Drosophila* regulatory mechanism from Invitrogen (No *et al.* 1996). Dual-regulated systems using both tetracycline and streptogramin are under development (Fussenegger 2001; Fussenegger *et al.* 2000). The Invitrogen Geneswitch system uses elements from the yeast Gal4 and adenoviral E1b promoters and remains transcriptionally silent until activated by a regulatory protein consisting of the Gal4 DNA binding domain, a truncated human progesterone receptor ligand binding domain, and the NFκB transcription factor p65. This highly controlled system permits the expression of toxic proteins (Wang *et al.* 1994).

Recent advances include the development of multicistronic expression strategies which permit multigene expression using a single vector to transduce suitable mammalian cell lines. At the most basic level, this allows the expression of product and marker gene from the same plasmid, the most stable coupling being achieved when both genes are expressed from the same promoter. Coordinated expression of the paired genes can be achieved by gene fusions or differential splicing (Kromer *et al.* 1997; Sonenberg 1994; Attal *et al.* 1999), but most multigene systems take advantage of an internal ribosomal entry site (IRES) to direct the translation of the second cistron (Fussenegger *et al.* 1999). Most IRES elements are drawn from poliovirus or encephalomyocarditis virus, IRES-mediated translation efficiency varies greatly between the different IRES elements and also with the cell line used (Fussenegger *et al.* 1999; Borman *et al.* 1997). IRES-based vectors containing three or four cistrons have been developed and applied to a number of novel technologies including the coupled expression of multicomponent/multisubunit proteins; the cloning of high-producer cell lines by the introduction of a selection marker into the last cistron; one-step regulated gene expression systems using positive-feedback circuits; and multigene metabolic engineering of continuous cell lines using sense, antisense or ribosome technology (Fussenegger *et al.* 1998; Burger *et al.* 1999; Lucas *et al.* 1996). Their use as nucleic acid vaccines is also under investigation (Fussenegger *et al.* 1999).

Enhanced production in transient assay systems in many cell lines can be achieved by the episomal replication of the expression plasmid to high copy numbers. This is usually induced by the presence of the SV40 large T antigen or the Epstein–Barr virus origin of replication (*oriP*) and nuclear antigen (EBNA-1). Indeed, the level of expression of several proteins in a HEK-293 EBNA line approached that seen with stable cell lines (Meissner *et al.* 2001) and in a fraction of the time. A complete system based on this 'Mass Transient' technology is now available ('Freestyle 293 Expression System', Invitrogen).

5.2.2.1e Gene amplification

Although the combination of a strong promoter/enhancer, suitable selectable marker and a permissive cell line may lead to increased gene expression from an integrated vector, it is known that the chromosomal location where integration occurs has a profound effect on transcription levels. This so-called 'position effect' can be overcome by increasing vector copy

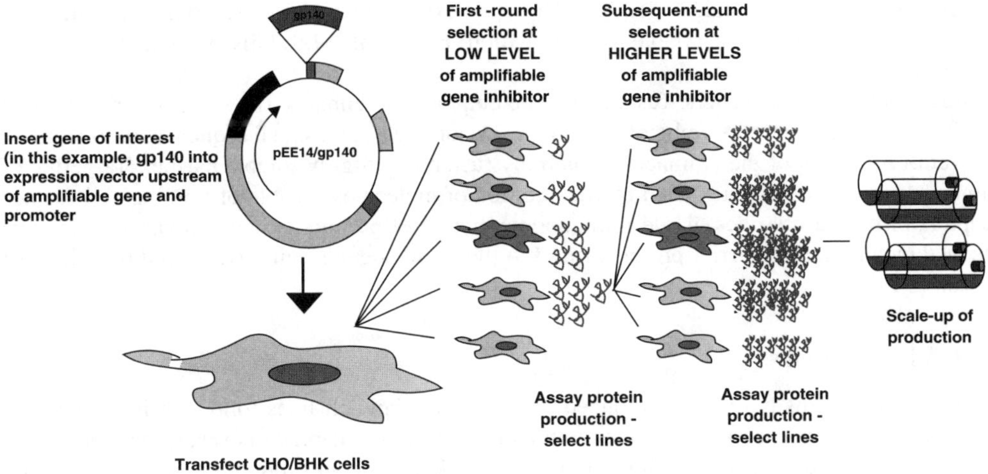

Figure 5.3 Gene amplification.

number, but other approaches exist as described in Chapter 8. This is done by selection for gene amplification.

When cell lines are cultured in the presence of certain toxic drugs, mutant drug-resistant lines can be isolated. These often arise from the overproduction of an essential enzyme the drug inhibits, the overproduction arising from increased mRNA levels ultimately stemming from increased copy number of the enzyme gene. About a dozen genes have been found to amplify in this way and, if cloned and reintroduced into a cell line that lacked endogenous drug resistance activity, the cloned gene was found to be capable of amplification as well (Table 5.1). The significant feature of gene amplification is that regions of chromosomal DNA adjacent to the enzyme gene are also amplified, hence other sequences on an integrated vector containing the enzyme gene will be co-amplified. This provides a means of progressively increasing mRNA transcription by selecting lines in the presence of increasing concentrations of the specific enzyme inhibitor, and has led to a class of vectors with the most efficient expression characteristics currently available (Figure 5.3). The two most commonly used amplifiable systems (gene/selective inhibitor/cell line) are based on dihydrofolate reductase (*dhfr*) + methotrexate in *dhfr*$^-$ CHO lines, and glutamine synthetase (GS) + methionine sulfoximine in CHO-K1 or murine myeloma NS0 cells (Bebbington *et al.* 1992). Amplifiable genes and their inhibitors are listed in Table 5.1. Both systems have been used extensively to express recombinant proteins (Page *et al.* 1991; Rhodes *et al.* 1994; Bebbington 1995; Jeffs *et al.* 1996).

5.2.2.1f Affinity fusion partner technology

Genetic fusions comprise a method whereby desired properties from the fusion partner can be added to the target protein. The commonest use is to add an affinity fusion partner to permit purification by immunoaffinity chromatography. A cleavable link can be engineered between the target protein and the fusion partner. The most widely used approach consists of the addition of six histidine residues to the *N*- or *C*-terminal of the target protein to permit its capture by solid phases incorporating nickel or cobalt ions (see Section 18.5.6). These 'hexahis' tags are very efficient at extracting monomeric, minimally glycosylated proteins, but are less useful for oligomeric glycoproteins (Jeffs personal observations). Other fusion partners commercially available are based on maltose binding protein, calmodulin, cellulose binding domain, glutathione *S*-transferase, HSV glycoprotein D, *S*-protein and T7 gene 10. These are extremely useful when no high affinity protein-specific antibodies are available for purification (Ford *et al.* 1991; Nilsson *et al.* 1996; Hearn & Acosta 2001).

Other applications of genetic fusions may be employed to enhance the secretion of the target protein from the cell into the ambient medium by use of a secretion signal sequence, or to increase the overall solubility of the chimaeric protein by fusion to a highly soluble partner. Targeting of chimaeric proteins can be achieved by the addition of molecules capable of binding cell surface receptors or polysaccharides, while immunopotentiation of protein antigens in vaccines may be achieved by addition of carrier proteins fused to the immunogen subunit (reviewed by Liljeqvist & Stahl 1999).

5.2.3 Viral Vectors

Recombinant proteins have been produced from mammalian cell lines following infection with a chosen virus, or transfection with a recombinant viral vector adapted to overexpress a heterologous protein. The advantage of such vectors is that the heterologous protein is expressed naturally in the context of an infected cell, leading to the production of highly immunogenic molecules. When used as vaccines, they have the advantage of inducing both cellular and humoral responses. The most widely used replication-competent expression systems are derived from poxviruses such as the vaccinia/T7 recombinant virus system and adenovirus. The vaccinia virus has a large genome (180–200 kb) into which a variety of foreign genes can be inserted and expressed without compromising viral replication. The vaccinia virus was originally used to vaccinate against and eventually eradicate smallpox, while a recombinant vaccinia-rabies virus has been successfully used to control rabies in wild foxes (Pastoret & Brochier 1999).

Replication-defective virus vectors have had the genes coding for replication deleted and will replicate to very high titre in recombinant cell lines expressing the deleted genes. Vectors in such cells undergo an abortive infection *in vivo* while expressing the foreign protein gene. Such antigens are, thus, efficiently presented to both the humoral and cell-mediated arms of the immune response and the system is considered safer for immunocompromised hosts. Two attenuated forms of vaccinia recombinants ('modified vaccinia Ankara' (MVA) and 'New York Vaccinia' (NYVAC) (Tartaglia *et al.* 1992; Sutter & Moss 1992), as well as the canarypox vector ALVAC (Perkus *et al.* 1995), have been shown to be both safe and highly immunogenic, and to have potential for the delivery of a number of viral immunogens. Adenoviruses lacking the E1A gene replicate to high titre in HEK-293-E1A cells, and viral antigens can be co-expressed by cloning into the E1A region. Immunization of experimental animals has shown protection against a number of viral antigens, including hepatitis B, measles, HSV and rabies (Lubeck *et al.* 1989; Fooks *et al.* 1995; Gallichan *et al.* 1993; Xiang *et al.* 1996). Similarly, a replication-defective HSV has conferred homologous protection and is now being engineered for use as a vector of foreign viral antigens (Lowry *et al.* 1997). Perhaps the most exciting recent development has been the use of self-replicating vectors based on a number of alphaviruses, including Sindbis, Semliki Forest or Venezuelan equine encephalitis virus, to produce large amounts of protein in both mammalian and insect cells (Agapov *et al.* 1998; Lundstrom *et al.* 2001; Davis *et al.* 1996). These vectors can be used for gene delivery as either naked RNA or as DNA. Recombinant alphavirus RNA or DNA is synthesized from plasmid vectors containing the alphavirus replicon under the control of the SP6 or T7 (RNA) or RNA polymerase II (DNA) promoters. Both systems lead to high levels of protein expression and the efficient induction of both humoral and cellular immune responses in both rodents and non-human primates (Lundstrom *et al.* 2001). Vectors based on DNA are more stable and have lower production costs. Both vector systems allow the use of significantly smaller amounts of DNA than do conventional nucleic acid-based vaccines and, due to the transient nature of the replicons, the risk of DNA integration into the host genome is much reduced (Smerdou & Liljestrom 1999).

5.2.4 Choice of Host Cell

The choice of host cell for protein expression is dependent on a number of related parameters that will impact upon the yield and therapeutic efficacy of the recombinant protein, the time and cost to obtain it, and its possible cytopathic effects on the host cell. This may range from small-scale transient expression in a COS-7 cell line to bulk production of therapeutic agents in industrial-scale bioreactors using suspension-adapted CHO or myeloma cells. The large-scale cultivation of cell lines was driven by two developments. First, the work of Earle and Eagle, who, in 1955, reported a chemically defined medium (EMEM – Eagle's minimum essential medium), which could replace the biological fluids used previously and, second, the development of continuous cell lines that could be subcultured indefinitely and adapted to suspension culture (Kretzmer 2002). Ideally, a recombinant protein to be used as a vaccine should be expressed in the target host (or closely related) cell, but this is rarely possible as most of the host cells do not have the characteristics of continuous cell cultures. In practice, the choice is limited to attached fibroblast-like (e.g. CHO, BHK-21, NIH 3T3, MRC-5) or epithelial-like (HeLa, Vero, HEK293) lines or myeloma and promyelocytic suspension cell lines such as NS0 and U937 respectively. If small (<1 mg) quantities of protein are required, a range of attached cell lines can be considered, but for larger-scale production the use of suspension culture-adapted lines is preferred.

The most important consideration is the ultimate therapeutic or biological functionality of the recombinant protein. There is an abundant literature to support the hypothesis that this can be influenced by the choice of host cell. In general, this is linked to changes in the *N*-linked glycosylation profile of a glycoprotein, although other changes, such as susceptibility to protease activity, have been observed (Hartman *et al.* 1995). For example, the antibody dependent cell-mediated cytotoxicity activity of the therapeutic antibody CAMPATH-1H derived from rat YO myeloma cells was consistently higher than those obtained from murine NSO myeloma or CHO cells (Lifely *et al.* 1995), while candidate HIV gp120/160 vaccines have enhanced immunogenicity when produced in CHO as opposed to insect cell lines (Viscidi *et al.* 1990; Redfield *et al.* 1991; Schwartz *et al.* 1993).

The induction of apoptosis in host cell lines can be a hindrance to the high level expression of many recombinant proteins, in particular viral antigens, and novel strategies will be required to circumvent this. Paradoxically, the induction of apoptosis in the cells of a vaccine recipient may actually be beneficial to vaccine efficacy as there is now considerable evidence that it may enhance antigen uptake and processing by dendritic cells, resulting in presentation by the MHC Class I pathway (Bellone *et al.* 1997; Ronchetti *et al.* 1999). Technologies to overcome apoptosis include genetic engineering of anti-apoptotic pathways to overexpress Bcl-2 and Bcl_{XL} (Kim & Lee 2000; Laken & Leonard 2001; Mastrangelo *et al.* 2000), advances in fed-batch and batch media composition (de Zengotita *et al.* 2000) and media additives such as the growth factor inhibitors rapamycin or suramin (Balcarcel & Stephanopoulos 2001; Zanghi *et al.* 1999). Several studies have indicated that recombinant protein production is closely related to the cell cycle of the host cell. Attempts to exploit these phenomena have used inducible expression of the product gene in concert with a regulatory gene that induces cell-cycle arrest at the G1/S phase border. This leads to as much as a 15-fold increase in specific productivity (Fussenegger *et al.* 1998; Zanghi *et al.* 1999; Balcarcel & Stephanopoulos 2001; Kaufmann *et al.* 2001). Other factors potentially involved in the control of productivity by inducing cell cycle arrest are culture temperature, and manipulation of the tumor suppressor Interferon Regulatory Factor-1 in BHK cell cultures (Geserick *et al.* 2000; Kaufmann *et al.* 2001).

5.2.5 Transfection Technology

5.2.5.1 General considerations

'Transfection' describes the delivery of nucleic acid into cells. There are three parameters to consider. Firstly, the DNA template for transfer must be highly purified. For many years, the method of choice was to use caesium chloride ultracentrifugation, but due to safety considerations and the time consuming nature of this technique, alternative methodologies were devised based upon anion-exchange chromatography and these now provide purified DNA of excellent quality. There are many commercial products available (e.g. Qiagen, Wizard (Promega)). It is recommended that purified DNA is monitored by agarose gel electrophoresis to establish the degree of supercoiling, as expression in a transient assay can be affected by the presence of nicked or relaxed vector. Closed, circular, supercoiled plasmid ('ccc pDNA') is the superior template.

The second parameter of importance is the condition of the cells to be transfected. Optimal culture methodologies for each individual cell line may have to be determined empirically, although some generalizations can be made. Specialized media are available for many mammalian cell lines, and these may be supplemented as required. Antibiotics may be added to prevent bacterial contamination, but this may mask mycoplasma contamination. Testing for mycoplasma is mandatory for any new cell line entering the laboratory, and routine testing at later stages in the development of recombinant cell lines is advisable. Serum, if added, should be from certified sources. If serum-free culture conditions are to be used, most cell lines will need to be gradually weaned from serum-containing to serum-free conditions, and growth parameters monitored throughout. If monolayer cultures of cells are to be transfected, it is essential that they are in the logarithmic phase of growth in the range of 50–80 % confluency and have a viability of >90 %. Cells that have reached confluency prior to subculture may not return to a healthy growth phase before transfection, while an over-dense culture can contain many non-viable cells, thus lowering transfection efficiency. Transfection efficiency may also drop with the age of the cell culture, and it is recommended that lines are not subcultured more than 10–15 times before a fresh culture is established from frozen stocks. Plating densities vary with cell line and transfection methodology employed. In the absence of any specific recommendations, a general guide is to seed $1\text{–}4 \times 10^4$ cells/cm^2 24 hours prior to transfection (Chisholm 1995).

The third parameter, the transfection methodology itself, can use either biochemical/physical methods or an infection process (using a suitable virus (see Section 5.2.3) to transfer nucleic acid into cells. With the exception of the DEAE-dextran technique, all biochemical methodologies are suitable for both transient expression and the selection of stable expression cell lines.

5.2.5.2 Biochemical transfection procedures

5.2.5.2a Calcium and strontium phosphate co-precipitation

When calcium chloride, DNA and phosphate buffer are mixed, extremely small, insoluble particles of calcium phosphate containing condensed DNA are formed that adhere to cell membranes and are internalized by phagocytosis. The method is quick, cheap and simple, works with many adherent and suspension-adapted fibroblast and endothelial cell lines such as CHO, mouse L, HeLa and HEK-293, and can be used to transfect cells both for transient expression and the selection of stable cell lines. Crucial to success are the size and quality of the calcium phosphate particles, DNA quantity (needs to be determined empirically for each vector) and quality (linear DNA is inactive), and the very narrow pH range (6.95–7.2) of the transfection medium and buffers. Modifications to improve efficiency are many, but the use of a $CaCl_2$/*N,N*-bis[2-hydroxyethyl]-2-aminoethanesulfonic acid-buffered saline which has a slightly more acidic pH and allows the DNA-calcium phosphate complex to form more slowly is particularly useful (Chen & Okayama 1987). Transfection efficiencies can be increased in many cell types by additional treatments with glycerol (Frost & Williams 1978), dimethyl sulfoxide (Lopata *et al.* 1984), chloroquine (Luthman & Magnusson 1983) or sodium butyrate (Gorman *et al.* 1983) after the primary exposure of the cells to DNA-calcium phosphate. Some primary epithelial cell lines are very sensitive to calcium ions and some success has been reported by replacing calcium with strontium (Brash *et al.* 1987). However, this technique has been superseded by the use of cationic lipid reagents (see Section 5.2.5.2c).

5.2.5.2b DEAE-dextran and Polybrene

The positively-charged polymers of dextran or Polybrene complex with negatively charged DNA, enabling the complex to bind to the cell surface where it is internalized by osmotic shock in the presence of dimethylsulfoxide (DMSO) or glycerol. Like calcium phosphate-mediated transfection, the method is cheap, quick and simple, works with both adherent and suspension-adapted cells and is particularly effective with COS cells. However, it is not suitable for the selection of stable cell lines, perhaps because the polymer is toxic to many common cell types such as baby hamster kidney (BHK). Crucial to success are the cell number, dextran concentration and method of cell shock (Chisholm 1995; McCutchan & Pagano 1968). DNA quantity is less crucial. As with calcium phosphate-mediated transfection, efficiency is enhanced by the pre-treatment of cells with glycerol, DMSO or chloroquine (Lopata *et al.* 1984; Luthman & Magnusson 1983).

5.2.5.2c Cationic lipid reagents

Cationic lipid reagents (CLRs) form small, unilamellar liposomes in aqueous solutions. The DNA is not encapsulated within the liposomes but, being negatively charged, binds spontaneously to the positively charged liposomes to form reagent complexes. The complex is then internalized and released through endosomes and lysosomes (Felgner *et al.* 1994). A very wide range of CLRs are now commercially available, which can be used reproducibly to transfect most commonly used adherent and suspension eukaryotic (insect and mammalian) cell lines with either DNA or RNA, and this has become the transfection method of choice for most applications (Duzgunes & Felgner 1993). Transfection efficiency is dependent on several parameters that must be optimized for each cell line using a suitable β-galactosidase or chloramphenicol–acetyl-transferase reporter plasmid. These include DNA-to-lipid ratio; incubation time; presence of serum; presence of antibiotics; health of cell culture; cell-plating density; choice of promoter and DNA quality. Optimization protocols are available from CLR suppliers. The manual from Invitrogen is particularly useful.

5.2.5.3 Physical transfection procedures

Electroporation delivers molecules by exposing cells in suspension to a brief electrical pulse of high field strength. This induces a potential difference across the cell membrane and is thought to create temporary pores through which nucleic acids can enter. The method can be used with many cell lines, including some primary cell isolates, but requires specialized equipment, has some safety risks and is often accompanied by high levels of cell death (>50 %). Optimization of the electrical pulse and field strength must be determined for each cell line to prevent membrane damage and cell lysis. A modified technique ('Nucleofector') has been developed by Amaxa Biosystems, based on a combination of specialized buffers and electrical pulses, and which delivers the DNA directly into the cell nucleus by electroporation. Although cell viability after transfection is still low, this method transduced a number of primary and stem cells which were refractile to other methods based on liposome-mediated transfer (Hamm *et al.* 2002).

Mechanical methods of gene delivery include using very fine needles to introduce nucleic acids directly into the cell cytoplasm or nucleus and 'gene gun' technology, a biolistic method of delivering macromolecule-coated or encapsulated microprojectiles (Capecchi 1980; Daniell *et al.* 1990).

5.2.6 The Cell Environment

The basic requirements for the successful maintenance of cultured mammalian cell lines are a sterile environment, a supply of nutrients for growth and biosynthesis of heterologous proteins, a balanced pH (usually between 7.2–7.4), a suitable CO_2 concentration (if bicarbonate buffering systems are being used; usually 5–8 % CO_2), and a constant temperature (nearly always 37 °C).

When cultures are scaled-up and/or maintained for longer periods, other factors such as oxygen limitation, mechanical/shear stress (in suspension or stirred cultures), metabolite toxicity (both from waste products such as lactate and NH_4^+ and secreted recombinant proteins), nutrient (e.g. glutamine) limitation and spatial limitations become important (Birch & Froud 1994; Mather 1998; Chu & Robinson 2001).

5.2.6.1 Making informed choices on culture media

Over the past 30 years various defined media types have been developed and are now available commercially. Basal media, such as minimal essential medium, contain a balanced salt solution, pH buffer and nutrients such as a carbohydrate source (usually glucose and galactose, sometimes maltose or fructose), amino acids, vitamins (especially the B group), fatty acids and lipids, trace elements (e.g. zinc, copper, selenium) and citric acid cycle intermediates. Serum is an important source of vitamins, proteins and peptides, and fatty acids and lipids, and these must be provided if serum-free media are used. Basal media have since been optimized to support the growth of most commonly used cell lines under serum-containing, serum-free or even protein-free conditions. Nearly all growth media are supplemented, and certain cell lines may be either totally dependent on specific additives or only perform optimally when they are present. The most common additives are insulin, transferrin, ethanolamine, β-mercaptoethanol and selenium (Mather 1998). The requirements for media and serum are dicussed further in Chapters 3 and 4.

5.2.6.2 Serum-containing versus serum-free media

The transition from a small-scale (T-flask), anchorage-dependent, serum-containing culture system to a high cell density, serum-free suspension culture can lead to dramatic changes in the growth performance of the cell line and the structural (and potentially therapeutic) characteristics of the recombinant protein. Cell populations undergoing a growth crisis during the adaption process to serum-free conditions may be vulnerable to the outgrowth of undesirable subpopulations of cells that may show low growth rate, poor specific productivity or genetic instability (Ozturk & Palsson 1991; Zang *et al.* 1995). For all these reasons, the growth characteristics of recombinant cell lines, as well as the structural and functional integrity of the expressed protein, must be closely monitored throughout the development process. There is abundant literature within this field, which tends to suggest that the withdrawal of serum affects each protein individually, although production rates tend to be compromised more in insect cell culture systems than in mammalian (particularly CHO) cell lines (Goochee & Monica 1990; Battista *et al.* 1994; Zang *et al.* 1995; Taticek *et al.* 2001). Jeffs *et al.* (2006) compared the productivity of recombinant CHO cell lines expressing HIV-1 monomeric and oligomeric envelope glycoproteins in serum-containing (S+), serum-free (S−) and protein-free (P−) media, and noted maximal expression of oligomeric gp140 in S+ media while monomeric gp120 production was highest in S− media. The amount of gp140 produced in medium-scale cultures (roller bottles) was similar in S+ and S− media, but much reduced in P− media. Purified S+ and S− gp140 possessed identical biological functionality, despite quantitative differences in their *N*-linked glycan profile, as evidenced by CD4 and antibody binding, but P− gp140 was misfolded and non-functional.

5.2.7 Culture Systems

In small-scale culture, most cells are grown in T-flasks ranging from $25\,cm^2$ to $225\,cm^2$. Typical cell yields from a $175\,cm^2$ flask range from 1×10^7 for an attached line, to 1×10^8 for a suspension line, although the actual yield will depend on the cell employed. Although T-flasks are indicated during the development phase, it is difficult to produce larger quantities of cells in this system due to the labour intensity of the system as well as to the associated space demands and cost. There is a wide variety of intermediate solutions to scale-up, that do not require the expert process development required of bioreactors (Mather *et al.* 1997; Griffiths 2001). 'Static' systems include triple flasks and multitray systems ('Cell Factories', Nunc), which are useful for attached cell lines and require no specialized equipment. Roller bottles (with or without expanded surface area) can be used for both attached and suspension-adapted lines, are cheap and disposable but require an automated 'deck' to turn them. Shake flasks and spinner flasks are for suspension culture only, and support high cell densities. However, specialized equipment is required and most vessels need to be cleaned and sterilized before use. The use of bioreactors is beyond the capability of most laboratories and is generally restricted to industrial applications (Kretzmer 2002).

It should be appreciated, however, that the choice of culture system might impact upon the biological functionality of the expressed protein as much as on the host cell or the decision to use serum-free media. Jeffs *et al.* (2006) examined the culture of recombinant CHO cell lines expressing HIV-1 gp140 in four culture systems (T flask, standard and enhanced surface-area roller bottles, and spinner flasks), using both attached and suspension-adapted cell lines. In this case, the cell-specific production rate was found to be highest in attached cells in T flasks, but the greatest yield was obtained from standard roller bottles. Receptor and antibody binding was very similar in all attached cell lines, but gp140 obtained from suspension cultures in roller bottles was misfolded and non-functional. Other studies have shown that maximal expression of a human interleukin-2 variant glycoprotein in BHK-21 cells and human chorionic gonadotropin in CHO cells was obtained with adherent (microcarrier) cultures as opposed to suspension culture (Battista *et al.* 1994; Gawlitzek *et al.* 1995). Unfortunately, the biological functionality of these products from differing culture systems was not compared.

5.3 THERAPEUTIC RECOMBINANT PROTEINS

Having established the therapeutic efficacy of a biological medicine in clinical trials, industrial choices must be made for production in large-scale cell culture. This section briefly reviews large-scale cell culture technology and describes recently approved biotechnology products.

5.3.1 Scaling-up

In general, the technology of choice for large-scale cell culture is the deep tank reactor, with the contents mixed by airlift technology or stirrers. However, for low-volume and speciality applications, such as the manufacture of viral vaccines, gene therapy vectors and certain diagnostic reagents, reactor types may encompass diverse technologies. These include roller bottles, multitray cell factories (such as the Nunc Cell Factory) and hollow fibre perfusion systems (such as the Biovest AcuSyst ™ series of instruments). Anchorage-dependent cell lines are generally more difficult to scale-up, unless microcarrier systems are used, otherwise roller bottles or perfused hollow fibre systems are most commonly employed. Suspension-culture-adapted lines are more straightforward to scale-up in stirred-tank fermenters. These systems also have the advantage of providing more-or-less homogeneous conditions, ensuring optimal monitoring and control of process parameters. Further detailed information on scale-up of mammalian cell cultures is given elsewhere in this book (see Chapter 10).

Most purification protocols will require multiple steps to achieve the desired level of purity, each of which will involve product loss. Thus, although the number of steps may be minimized or combined, there are essentially four sequential stages:

(i) preparation/extraction/clarification of starting material;

(ii) product capture (isolation, concentration and stabilization of product);

(iii) intermediate purification (removal of bulk impurities);

(iv) final purification (to desired level of purity).

Protocols for purification strategies are widely available (Marshak *et al.* 1996; Scopes 2002; Doonan 1996) and are considered elsewhere in this book (Chapter 18). Methods to evaluate product quality may include bioassay for function, *in vitro* toxicology and physical characterization.

5.3.2 FDA Biological License Approvals 1996–2003

Table 5.2 lists Biological License Approvals (BLAs) granted by the United States Food and Drug Administration (FDA) for biotechnology products derived using recombinant DNA technology and mammalian cell lines over the period January 1996 to July 2003 (derived from Chu & Robinson 2001, and updates from the FDA website). There are two categories; Recombinant Therapeutics and Diagnostic Products; products such as inactivated or live viral vaccines and natural products derived from cells or blood are not included. Note that no recombinant viral vaccines derived from mammalian cell lines have yet been approved for use.

It is immediately apparent from Table 5.2 that the majority of approved BLAs have been produced using CHO cells, with the remainder being monoclonal antibodies or antibody fragments produced in murine myeloma or hybridoma cells, and one therapeutic from BHK cells. In general,

Table 5.2 FDA Biological License Approvals 1996–2005.

Product name	Manufacturer	Product type	Cell line
Recombinant therapeutics			
Tenecteplase/TNKase	Genentech	rgp	CHO
Antihemophilic factor/ ReFacto	Genetics Institute	rgp	CHO
Coagulation factor VIIa/ NovoSeven	Novo Nordisk A/S	rgp	BHK
Etanercept/Enbrel	Genentech	IgG_1 fusion with TNF receptor	CHO
Trastuzumab/Herceptin	Genentech	Humanized Mab	CHO
Infliximab/Remicade	Centocor	Humanized Mab	NSO
Palivizumab/Synagis	MedImmune	Mab	NSO
Basiliximab/Simulect	Novartis	Humanized Mab	Recombinant murine myeloma
Daclizumab/Zenapax	Hoffman-LaRoche	Humanized Mab	SP2/0
Rituximab/Rituxan	Genentech	Humanized Mab	CHO
Coagulation factor IX/ Benefix	Genetics Institute	rgp	CHO
Interferon β-1a/Avonex	Genetics Institute	rgp	CHO
Zemaira	Aventis Behring	rgp	CHO
Advate	Baxter Healthcare	rgp	CHO
Diagnostic products			
Nofetumomab/Verluma	Dr Karl Thomae	Murine Fab fragment	'Mammalian cells'
Imciromab pentetate/ Mycoscint	Centocor	Murine Fab fragment	Recombinant murine hybridoma

rgp = recombinant glycoprotein
Mab = monoclonal antibody

recombinant therapeutics, which are mostly represented by recombinant proteins or antibodies, require significant production quantities. Where such information has been provided by the manufacturer, this has been achieved by the large-scale culture of suspension- and serum-free adapted cell lines in stirred-tank bioreactors.

5.4 SUMMARY

As our knowledge and skill in the employment of rDNA technologies grows so does the possibility of new and exciting therapeutic products. The use of animal cells raises a specific set of regulatory issues and challenges for the scale-up of production. It also has the potential to produce products that closely mimic and may even improve upon native biological molecules. Production of therapeutic proteins from recombinant animal host cells is likely to yield an expanding and increasingly sophisticated range of new products for the future.

REFERENCES

Agapov EV, Frolov I, Lindenbach BD, Pragai BM, Schlesinger S, Rice CM (1998) *Proc. Natl. Acad. Sci. USA*; 95(22): 12989–12994.

Alam J, Cook JL (1990) *Anal. Biochem.*; 188(2): 245–254.

Attal J, Theron MC, Puissant C, Houdebine LM (1999) *Gene Expr.*; 8(5-6): 299–309.

Balcarcel RR, Stephanopoulos G (2001) *Biotechnol. Bioeng.*; 76(1): 1–10.

Battista PJ, Tilkins ML, Judd DA, Godwin GP, Gorfien SF (1994) *Am. Biotechnol. Lab.*; 12(5): 64, 66, 68.

Bebbington CR (1995) In *DNA Cloning 4 - Mammalian Systems.* Ed Glover DM. IRL Press, Oxford.

Bebbington CR, Renner G, Thomson S, King D, Abrams D, Yarranton GT (1992) *Biotechnology*; 10(2): 169–175.

Bellone M, Iezzi G, Rovere P *et al.* (1997) *J Immunol.*; 159(11): 5391–5399.

Berg P (1981) *Science*; 213(4505): 296–303.

Berman PW, Nunes WM, Haffar OK (1988a) *J. Virol.*; 62(9): 3135–3142.

Berman PW, Vogt PE, Gregory T, Lasky LA, Kern ER (1988b) *J. Infect. Dis.*; 157(5): 897–902.

Berman PW, Riddle L, Nakamura G *et al.* (1989) *J Virol.*; 63(8): 3489–3498.

Birch JR, Froud SJ (1994) *Biologicals*; 22(2): 127–133.

Blochlinger K, Diggelmann H (1984) *Mol. Cell Biol.*; 4(12): 2929–2931.

Borman AM, Le Mercier P, Girard M, Kean KM (1997) *Nucleic Acids Res.*; 25(5): 925–932.

Brash DE, Reddel RR, Quanrud M, Yang K, Farrell MP, Harris CC (1987) *Mol. Cell Biol.*; 7(5): 2031–2034.

Burger C, Carrondo MJ, Cruz H *et al.* (1999) *Appl. Microbiol. Biotechnol.*; 52(3): 345–353.

Capecchi MR (1980) *Cell*; 22(2 Pt 2): 479–488.

Chapman BS, Thayer RM, Vincent KA, Haigwood NL (1991) *Nucleic Acids Res.*; 19: 3979–3986.

Chisholm V (1995) In *DNA Cloning 4. -Mammalian Systems.* Ed Glover DM IRL Press, Oxford.

Chu L, Robinson DK (2001) *Curr. Opin. Biotechnol.*; 12(2): 180–187.

Colbere-Garapin F, Horodniceanu F, Kourilsky P, Garapin AC (1981) *J Mol. Biol.*; 150: 1–14.

Daniell H, Vivekananda J, Nielsen BL, Ye GN, Tewari KK, Sanford JC (1990) *Proc. Natl. Acad. Sci. USA*; 87: 88–92.

Davis NL, Brown KW, Johnston RE (1996) *J. Virol.*; 70: 3781–3787.

deZengotita VM, Miller WM, Aunins JG, Zhou W (2000) *Biotechnol. Bioeng.*; 69(5): 566–576.

Doonan S (1996) *Methods Mol. Biol.*; 59: 1–16.

Duzgunes N, Felgner PL (1993) *Methods Enzymol.*; 221: 303–306.

Etchevery T (1996) In *Protein Engineering; Principles and Practice.* Eds Cleland JL, Craik CS Wiley-Liss, Inc., New York.

Felgner JH, Kumar R, Sridhar CN *et al.* (1994) *J Biol. Chem.*; 269: 2550–2561.

Foecking MK, Hofstetter H (1986) *Gene*; 45(1): 101–105.

Fooks AR, Schadeck E, Liebert UG *et al.* (1995) *Virology*; 210: 456–465.
Ford CF, Suominen I, Glatz CE (1991) *Protein Expr. Purif.*; 2(2–3): 95–107.
Frost E, Williams J (1978) *Virology*; 91(1): 39–50.
Fussenegger M (2001) *Biotechnol. Prog.*; 17(1): 1–51.
Fussenegger M, Schlatter S, Datwyler D, Mazur X, Bailey JE (1998) *Nat. Biotechnol.*; 16(5): 468–472.
Fussenegger M, Bailey JE, Hauser H, Mueller PP (1999) *Trends Biotechnol.*; 17(1): 35–42.
Fussenegger M, Morris RP, Fux C *et al.* (2000) *Nat. Biotechnol.*; 18(11): 1203–1208.
Gallichan WS, Johnson DC, Graham FL, Rosenthal KL (1993) *J. Infect. Dis.*; 168(3): 622–629.
Gawlitzek M, Hartman S, Valley U, Nimtz M, Wagner R, Conradt HS (1995) *J. Biotechnol.*; 42(2): 117–131.
Geserick C, Bonarius HP, Kongerslev L, Hauser H, Mueller PP (2000) *Biotechnol. Bioeng.*; 69(3): 266–274.
Goochee CF, Monica T (1990) *Biotechnology*; 8(5): 421–427.
Goodwin EC, Rottman FM (1992) *J. Biol. Chem.*; 267: 16330–16334.
Gorman C (1985) In *DNA Cloning: A Practical Approach.* Ed Glover DM. IRL Press, Oxford; Vol. 2, 143–190.
Gorman CM, Merlino GT, Willingham MC, Pastan I. Howard BH (1982) *Proc. Natl. Acad. Sci. USA*; 79: 6777–6781.
Gorman CM, Howard BH, Reeves R (1983) *Nucleic Acids Res.*; 11: 7631–7648.
Gorman CM, Gies D, McCray G, Huang M (1989) *Virology*; 171: 377–385.
Gossen M, Bujard H (1992) *Proc. Natl. Acad. Sci. USA*; 89, 5547–5551.
Griffiths B (2001) *Mol. Biotechnol.*; 17: 225–238.
Haas J, Park EC, Seed B (1996) *Curr. Biol.* 6, 315–24.
Hamm A, Krott N, Breibach I, Blindt R, Bosserhoff AK (2002) *Tissue Eng.*; 8(2): 235–245.
Hartman S, Mulligan R (1988) *Proc. Natl. Acad. Sci. USA*; 85: 8047–8051.
Hearn MT, Acosta D (2001) *J. Mol. Recognit.*; 14(6): 323–369.
Jeffs SA, McKeating J, Lewis S *et al.* (1996) *J. Gen. Virol.*; 77: 1403–1410.
Jeffs S, Goriup S, Kebble B *et al.* (2004) *Vaccine*; 22: 1032–1046.
Jeffs S, Goriup S, Stacey G, Yuen C-T, Holmes H (2006) Appl. Microbiol. Biotechnol.; 72: 279–290.
Kaufman RJ, Sharp PA (1982) *Mol. Cell. Biol.*; 2: 1304–1319.
Kaufmann H, Mazur X, Marone R, Bailey JE, Fussenegger M (2001) *Biotechnol. Bioeng.*; 72: 592–602.
Kim DW, Uetsuki T, Kaziro Y, Yamaguchi N, Sugano S (1990) *Gene*; 91: 217–223.
Kim NS, Lee GM (2000) *Biotechnol. Bioeng.*; 71: 184–193.
Kozak M (1986) *Cell*; 44: 283–292.
Kretzmer G (2002) *Appl. Microbiol. Biotechnol.*; 59: 135–142.
Kromer WJ, Carafoli E, Bailey JE (1997) *Eur. J Biochem.*; 248: 814–819.
Laken HA, Leonard MW (2001) *Curr. Opin. Biotechnol.*; 12: 175–179.
Lasky LA, Groopman JE, Fennie CW *et al.* (1986) *Science*; 233: 209–25.
Lifely MR, Hale C, Boyce S, Keen MJ, Phillips J (1995) *Glycobiology*; 5: 813–822.
Liljeqvist S, Stahl S (1999) *J. Biotechnol.*; 73: 1–33.
Lopata MA, Cleveland DW, Sollner-Webb B (1984) *Nucleic Acids Res.*; 12: 5707–5717.
Lowry PW, Koropchak CM, Choi CY *et al.* (1997) *Antiviral Res.*; 33: 187–200.
Lubeck MD, Davis AR, Chengalvala M *et al.* (1989) *Proc. Natl. Acad. Sci. USA*; 86: 6763–6767.
Lucas BK, Giere LM, DeMarco RA, Shen A, Chisholm V, Crowley CW (1996) *Nucleic Acids Res.*; 24: 1774–1779.
Lundstrom K, Schweitzer C, Rotmann D, Hermann D, Schneider EM, Ehrengruber MU (2001) *FEBS Lett.*; 504(3): 99–103.
Luthman H, Magnusson G (1983) *Nucleic Acids Res.*; 11: 1295–1308.
Marshak DR, Kadonaga JT, Burgess RR, Knuth MW, Brennan WA, Lin S-H (1996) *Strategies for Protein Purification and Characterization: A Laboratory Manual,* Cold Spring Harbor Laboratory Press.
Mastrangelo AJ, Hardwick JM, Zou S, Betenbaugh MJ (2000) *Biotechnol. Bioeng.*; 67: 555–564.
Mather JP (1998) *Methods Cell. Biol.*; 57: 19–30.
Mather JP, Moore A, Shawley R (1997) *Methods Mol. Biol.*; 62: 369–382.
McCutchan JH, Pagano JS (1968) *J. Natl. Cancer Inst.*; 41: 351–357.

Meissner P, Pick H, Kulangara A, Chatellard P, Friedrich K, Wurm FM (2001) *Biotechnol. Bioeng.*; 75: 197–203.
Mulsant P, Gatignol A, Dalens M, Tiraby G (1988) *Somat. Cell Mol. Genet.*; 14: 243–252.
Nilsson J, Larsson M, Stahl S, Nygren PA, Uhlen M (1996) *J. Mol. Recognit.*; 9: 585–594.
No D, Yao TP and Evans RM (1996) *Proc. Natl. Acad. Sci. USA*; 93(8): 3346–3351.
Ozturk SS, Palsson BO (1991) *Biotechnol. Prog.*; 7: 481–494.
Page M, Mills KH, Schild GC *et al.* (1991) *Vaccine*; 9: 47–52.
Pastoret PP, Brochier B (1999) *Vaccine*; 17: 1750–1754.
Perkus ME, Tartaglia J, Paoletti E (1995) *J. Leukoc. Biol.*; 58: 1–13.
Planelles V, Haigwood NL, Marthas ML *et al.* (1991) *AIDS Res. Hum. Retroviruses*; 7: 889–898.
Pollard SR, Rosa MD, Rosa JJ, Wiley DC (1992) *EMBO. J.*; 11: 585–591.
Redfield RR, Birx DL, Ketter N *et al.* (1991) *New Engl. J. Med.*; 324: 1677–1684.
Rhodes AD, Spitali M, Hutchinson G, Rud EW, Stephens PE (1994) *J. Gen. Virol.*; 75: 207–213.
Ronchetti A, Rovere P, Iezzi G *et al.* (1999) *J. Immunol.*; 163, 130–136.
Sasada R, Marumoto R, Igarashi K (1988) *Cell Struct. Funct.*; 13: 129–141.
Schwartz DH, Gorse G, Clements ML *et al.* (1993) *Lancet*; 342: 69–73.
Scopes RK (2002) *Protein Purification: Principles and Practice*: Springer Advanced Texts in Chemistry. Springer-Verlag, New York.
Smerdou C, Liljestrom P (1999) *Curr. Opin. Mol. Ther.*; 1: 244–251.
Sonenberg N (1994) *Curr. Opin. Genet. Dev.*; 4: 310–315.
Stadenbauer WL (1978) *Curr. Topics Microbiol. Immunol.*; 83: 93–99
Sutter G, Moss B (1992) *Proc. Natl. Acad. Sci. USA*; 89: 10847–10851.
Tartaglia J, Perkus ME, Taylor J *et al.* (1992) *Virology*; 188: 217–232.
Taticek RA, Choi C, Phan SE, Palomares LA, Shuler ML (2001) *Biotechnol. Prog.*; 17, 676–684.
Vara JA, Portela A, Ortin J, Jimenez A (1986) *Nucleic Acids Res.*; 14: 4617–4624.
Viscidi, R, Ellerbeck E, Garrison L *et al.* (1990) *AIDS Res. Hum. Retroviruses*; 6: 1251–1256.
Willey RL, Bonifacino JS, Potts BJ, Martin MA, Klausner RD (1988) *Proc. Natl. Acad. Sci. USA*; 85: 9580–9584.
Wyatt R, Sullivan N, Thali M *et al.* (1993) *J. Virol.*; 67, 4557–4565.
Wyborski DL, DuCoeur LC, Short JM (1996) *Environ. Mol. Mutagen.*; 28: 447–458.
Xiang ZQ, Yang Y, Wilson JM, Ertl HC (1996) *Virology*; 219: 220–227.
Yao F, Svensjo T, Winkler T, Lu M, Eriksson C (1998) *Hum. Gene Ther.*; 9: 1939–1950.
Zang M, Trautmann H, Gandor C *et al.* (1995) *Biotechnology*; 13, 389–392.
Zanghi JA, Fussenegger M, Bailey JE (1999) *Biotechnol. Bioeng.*; 64: 108–119.
zur Megede J, Chen MC, Doe B *et al.* (2000) *J. Virol.*; 74: 2628–2635.

Useful Web Site Addresses

Invitrogen Transfection Techniques manual	http://www.invitrogen.com/content/sfs/manuals/transgd.pdf
U.S. Food and Drug Administration web site	www.fda.gov

6 Production of Recombinant Viral Vaccine Antigens

SA Jeffs

6.1 INTRODUCTION

There is little doubt that the advent of worldwide vaccination programmes has led to dramatic reductions in deaths from fatal viral diseases (Plotkin 1993). Most of the vaccines currently employed to control such infections consist of live, attenuated or killed (inactivated) viruses. Live attenuated viruses confer strong, long-lasting protective humoral and cell-mediated immune (CMI) responses. However, a risk of reversion to virulence remains, especially in immunocompromised recipients (particularly children). Killed vaccines cannot replicate but are weaker immunogens, often requiring multiple booster injections and the co-adminstration of adjuvant to enhance immunogenicity. Furthermore, it is difficult to meet the current requirements from regulatory authorities for exact composition and immune competency with live and killed whole-organism vaccines (Liljeqvist & Stahl 1999).

An alternative approach to this type of vaccination is to use purified subunit immunogens either as vaccines or vaccine components. For example, the predominant immunogenic component of most currently licensed trivalent, inactivated, influenza vaccines (TIVs) is the haemagglutinin (HA) polypeptide partially purified from detergent-extracted, inactivated, virions (Kemble & Greenberg 2003), and Phase II trials are underway with purified fusion (F) protein from respiratory syncitial virus (RSV) (Piedra *et al.* 2003). At the time of writing (February 2007), the results of this trial have not yet been reported.

With the advent of recombinant DNA technology, large quantities of purified viral antigens can be produced for use in immunoprophylaxis. Other approaches to immunization include the use of synthetic peptides representing immunodominant epitopes, the use of anti-idiotypic antibodies and nucleic acid (DNA or RNA) vectors, but these are outside the remit of this review. Having identified suitable protective antigens (see Section 6.3), recombinant immunogens can be produced by isolating the gene for the antigen or antigen fragment, cloning it into a suitable expression vector (usually a plasmid or viral vector), then transducing a suitable host cell. By modification of either the gene insert or plasmid, expressed protein can be secreted into the ambient medium. This greatly simplifies purification and reduces the risk of contamination with potentially oncogenic cellular DNA. The purification technique must be gentle enough to maintain the protein in its native state, thus ensuring that the epitopes involved in the induction of neutralizing antibodies are correctly presented.

The major advantages of recombinant subunits may be summarized as:

- pathogen excluded from vaccine production process;
- no risk of contamination with toxic compounds;

Medicines from Animal Cell Culture Edited by G. Stacey and J. Davis

- low risk of reversion to virulent genotype (ensure that viral vector systems are completely disabled);
- no risk from incomplete inactivation of whole-cell vaccines (assuming recombinant host-cell vaccines are not pathogenic);
- recombinant subunit can be optimized for immunogenicity, either by genetic manipulation prior to transfer into host cell, within the host cell by utilizing the cell's biosynthetic capacity or by the addition of adjuvant molecules to the purified protein by chemical methods;
- genetic fusions may be used to obtain chimaeric antigens. This can have many applications, ranging from simplifying purification to immunopotentiation to increasing half-life;
- a very wide range of delivery systems can be employed to tailor the immune response for the specific pathogen against which the vaccine is targeted.

The disadvantages of recombinant subunits include:

- the possibility of proteolytic degradation;
- problems with stability, aggregation and solubility;
- poor presentation of epitopes due to sub-optimal conformation leading to poor or aberrent immunogenicity (particularly with complex glycoproteins);
- variable therapeutic efficiency (often related to glycosylation);
- addition of adjuvants is often required;
- high production costs;
- lengthy development phase;
- possible risk of contamination by prions, endogenous retroviruses and other adventitious organisms, either through the host cell or medium components (particularly bovine serum).

Despite these caveats, recombinant technologies are currently being employed in both the development of vaccines against viral pathogens for which no vaccines currently exist and to improve upon existing vaccines.

6.2 VIRAL VACCINES

Studies with influenza A and paramyxovirus indicated that immunoprophylaxis requires the generation of an immune response to viral proteins that will protect individuals on exposure to the infecting virus. The proteins inducing such a response are termed 'protective antigens' and, in the main, are represented by surface-located glycoproteins in enveloped viruses and capsid proteins in non-enveloped viruses (Epstein *et al.* 1993; Tao *et al.* 2000). Some non-structural proteins have also been found to be protective, such as the NS1 protein of dengue, T antigen of SV40 and Tat and Nef of HIV-1. Currently (2006), 16 viruses are covered by licensed vaccines in the United States. These protect against adenovirus, hepatitis A and B, influenza A and B, Japanese encephalitis virus, measles, mumps, papillomavirus, poliovirus, rabies, rotavirus, rubella, smallpox, varicella and yellow fever.

In general, although MHC Class I-restricted $CD8^+$ cytotoxic T-lymphocytes (CTLs) have a major role in detecting and destroying virus-infected cells, the principal mediator of resistance to infection is by antibody, with the assistance of MHC Class II-restricted $CD4^+$ helper T-cells (T_H). Most of the vaccines listed above work by inducing circulating low-titre antibody against blood-borne or blood-distributed viruses such as hepatitis A and B (HAV/HBV), measles, polio, rubella, smallpox and varicella (VZV). Live vaccines induce both humoral and CTL responses, whereas peptide, inactivated or subunit vaccines are generally poor inducers of cell-mediated responses and, in the case of subunits, may require the addition of adjuvants to induce the required humoral response. At present, only two vaccines, one against hepatitis B and the other against human papillomavirus, are manufactured using recombinant DNA. The remainder of this section summarizes the current efforts being made to produce new recombinant vaccines.

6.2.1 Survey of Recombinant Viral Vaccines Currently under Development

6.2.1.1 Herpes viruses

6.2.1.1a Herpes simplex virus

Genital herpes is caused by the herpes simplex virus type 1 or 2 (HSV-1, HSV-2), or human herpes virus 1 and 2 (HHV-1, HHV-2). Subunit HSV vaccines are based upon two envelope glycoproteins, gB and gD which have been shown to be strongly immunogenic and protective in animal studies (Stanberry 1991). Four separate formulations have been evaluated to date, all derived from CHO expression systems. Three vaccines developed by Chiron contained truncated HSV-2 gD absorbed to alum, gD with a muramyl tripeptide adjuvant, and a bivalent vaccine composed of gD and gB with MF59 adjuvant. Although all these vaccines were immunogenic, the first was only modestly protective, the second caused unacceptable side-effects while the third failed to protect (Corey *et al.* 1999; Langenberg *et al.* 1995; Straus *et al.* 1994; Straus *et al.* 1997). The fourth vaccine (from GlaxoSmithKline) is also based on truncated HSV-2 gD and alum combined with the novel adjuvant 3-de-*O*-acylated monophosphoryl lipid A. This preparation was well tolerated and induced potent humoral and CMI responses in Phase I/II trials (Leroux-Roels *et al.* 1994). A Phase III trial has indicated that although effective at prevention in women who are seronegative for both HSV-1 and -2 (but not those who are seroposition for HSV-1), it has *no* efficacy in men, regardless of their HSV serologic status. This may be due to the enhanced Th1 response noted in women (Stanberry *et al.* 2002). Further work is in progress to determine this.

6.2.1.1b HHV-8

HHV-8 is the most recently isolated human herpes virus, and is closely associated with Kaposi's sarcoma (Whitby & Boshoff 1998). The HHV-8 genome has many similarities to that of Epstein-Barr virus, and, should a vaccine be developed, this similarity may aid its development.

6.2.1.1c Varicella Zoster

Varicella Zoster (VZV or human herpes virus 3 (HHV-3)) causes chickenpox. In adults, complications can develop leading to 20–25 deaths per year in England and Wales (Rawson *et al.* 2001). VZV is currently controlled by a live attenuated vaccine. Animal studies with recombinant VZV glycoproteins B, C, E and I (gB, gC, gE and gI), have confirmed the role of antibody in protection, while the intermediate early protein IE62 appears to be implicated in the generation of VSV-specific CTL responses. A recombinant HSV-1 vector expressing either gE or IE62 induced antibody and CTL responses in mice, supporting the feasibility of a combined HSV/VZV vaccine (Lowry *et al.* 1997).

6.2.1.1d Epstein–Barr virus

An effective vaccine against Epstein–Barr virus (EBV, or human herpes virus 4 (HHV-4)) would not only limit the outgrowth of latently-infected B cells in healthy individuals, but would also block the development of EBV-related cancers such as Burkitt's lymphoma, nasopharyngeal carcinoma and Hodgkin's disease. The virus membrane antigens, gp350 (and its derivative, gp220) and gp85, are the primary candidates for a vaccine. To date, nearly all candidate vaccines have been based on the B95-8 A-type strain gp350, which shares a high degree of structural and functional homology with B-type strains, suggesting that a B95-8-derived vaccine should be equally effective against B-type isolates (Lees *et al.* 1993). Gp350 has been shown to be the principal target of neutralizing antibody responses (Thorley-Lawson & Geilinger 1980), and recombinant gp350, either as a CHO-derived purified subunit or expressed from recombinant vaccinia vectors, has been shown to protect non-human primates (cotton-top tamarins) against EBV-induced B cell lymphomas (Morgan 1992). A recombinant vaccinia expressing latent membrane protein-1 from the BNLF-1 ORF was tested in a small group of human infants. All unvaccinated controls became naturally infected, but only 3/9 vaccinees became infected within 16 months (Gu *et al.* 1995). More recent studies have focused on the use of synthetic polytope vaccines, based on a concatomer of EBV-specific CTL epitopes from gp350 and gp85 (Khanna *et al.* 1999; Macsween & Crawford 2003) and Phase I/II trials are planned. Assuming that such trials are successful, it is envisaged that further trials will assess the effects of such vaccines on EBV-associated cancers. Here, the timescales are much longer, at least 10 years to show an effect against Burkitt's lymphoma and at least 50 years for nasopharyngeal carcinoma (Arrand 1998).

6.2.1.1e Human cytomegalovirus

Human cytomegalovirus (CMV or human herpes virus 5 (HHV-5)) infection persists for life with periodic asymptomatic excretion of virus in body fluids. Up to 100 % of the population of non-industrialized countries are seropositive, while in industrialized countries the figure rises with age to approach 50 % in adulthood. CMV can complicate organ transplants, and is associated with retinitis in about 25 % of advanced HIV cases. However, the key justification for a vaccine is to prevent congenital CMV disease in newborns. Whole live virus vaccines based on the Towne strain tend to become heavily overattenuated and poorly immunogenic (Adler 1995), and alternative strategies using subunit and poxvirus vector vaccines are now under development. The CMV envelope glycoproteins gB and gH contain the majority of CMV epitopes against which neutralizing antibodies are induced. gB contains two major antigenic domains, AD-1 and AD-2, the latter being conserved among 12 clinical strains and contains epitopes for both CTLs and T_H cells (Mitchell *et al.* 2002).

A recombinant gB antigen produced in CHO cells has been evaluated in Phase I trials in both adults and children (Frey *et al.* 1999; Mitchell *et al.* 2002; Pass *et al.* 1999; Wang *et al.* 1996). In adults, 100 % of vaccinees became seropositive and 98 % produced neutralizing antibodies after three immunizations of gB plus the adjuvant MF59. In children, three doses were also required for maximal effect, but antibody titres were six-times higher than those observed in adults. The vaccine was well tolerated in both groups. This is a most encouraging result, given that CMV transmission starts in infants. However, data from transplant patients suggests that CMI responses are critical in the control of CMV infection in this group. Analysis of T-cell responses in healthy CMV seropositive individuals has identified the tegument protein pp65 and the intermediate early protein IE1 as the immunodominant antigens for both $CD4^+$ and $CD8^+$ T-cells (Davignon *et al.* 1995; Wills *et al.* 1996).

Recombinant ALVAC canarypox vectors expressing gB and pp65 have been tested in Phase I trials for their ability to induce both neutralizing antibodies (Nabs) and CTLs. Three doses of ALVAC-CMV-gB induced only low levels of Nabs, but two doses of the viral vector, followed by one of attenuated Towne virus was much more effective (Adler *et al.* 1999). Two doses of ALVAC-CMV-pp65 induced long-lasting CTL responses in all vaccine recipients (Berencsi *et al.* 2001). Finally, a CHO-derived chimaeric IE1-pp65 protein has been shown to stimulate both $CD4^+$ and

$CD8^+$ T-cells from CMV-seropositive individuals *in vivo* (Vaz-Santiago *et al.* 2001). Phase II and III trials with all these vaccines are planned.

6.2.1.2 Oncogenic viruses

On a global basis, cervical carcinoma is the second commonest female cancer (after breast cancer). There is a strong body of evidence to support the link between the presence of high-risk human papillomavirus (HPV) strains such as HPV-16 and -18 and cervical cancer (Phillips & Vousden 1998) and an effective anti-HPV vaccine would have an enormous impact worldwide.

In 2006, the FDA approved the first preventative HPV vaccine, Gardasil (Merck & Co.), while the similar GlaxoSmithKline HPV vaccine, Cervarix, is expected to be licensed in 2007. Both of these vaccines are recombinant virus-like particles (VLPs) composed of the major capsid protein, L1. L1 contains the major immunodominant neutralisation epitopes of the virus and induces high levels of protective neutralising antibodies. These protect girls and women against the two commonest HPV strains (HPV-16 and -18) implicated in cervical cancer, while Gardasil also targets HPV-6 and -11 which cause most cases of genital warts (Lowry & Schiller, 2006).

The products of the E6 and E7 open reading frames (ORFs) of HPV-16 have been implicated in the transformation of cervical epithelia into carcinoma, suggesting that immune responses raised against the E6/7 proteins may form the basis of a therapeutic vaccine to attack established tumours. The French biotechnology company, Transgene, has just completed Phase II trials of TG4001, a vaccine based on the E6 and E7 proteins, plus the cytokine IL2 delivered in an MVA vector (see: http://www.transgene.fr/us/page.php?fam=1&rub=3&iframe=product pipeline/iframe mva hpv il2.htm).

The human T-lymphotropic virus type 1 (HTLV-1) is estimated to infect between 10–20 million people worldwide and causes at least two types of disease – an aggressive T-cell malignancy, adult T-cell leukaemia/lymphoma (ATL) and a variety of chronic inflammatory syndromes, most notably HTLV-1-associated myelopathy. Where HTLV-1 is endemic (in the tropics and subtropics and among certain immigrant groups and intravenous drug users in Europe and North America), these are important causes of mortality and morbidity (Bangham 2000). Protection against HTLV-1 can be conferred to infants through maternal antibodies transferred by breastmilk. Thereafter, infection rates rise with declining levels of maternal antibody (Takahashi *et al.* 1991). Passive transfer experiments in rabbits have confirmed the role of antibody in protection (Sawada *et al.* 1991). The protective epitopes appear to be located on the env glycoprotein, gp46, and rodents can be protected against HTLV-1 infection by immunization with vaccinia-*env,* adenovirus-*env* or avipox-*env* (Bomford *et al.* 1996; Shida *et al.* 1987). CHO-derived HTLV-1 proteins protected pigtailed macaques against challenge with simian T-lymphotropic virus (Dezzutti *et al.* 1990) while cynomolgous macaques can be protected against HTLV-1 challenge by immunization with a recombinant vaccinia-gp46 vaccine which was found to induce both humoral and env-specific CTL activity (Ibuki *et al.* 1997).

Current opinion is that HTLV-1 infection is best controlled in endemic areas by transmission-blocking measures such as screening blood donors and refraining from breast-feeding, but where this is not possible, an env-based vaccine may be of some use. However, at the present time, no such vaccine is undergoing clinical trials.

More than 300 million individuals are carriers of hepatitis B virus (HBV) as identified by expression of surface antigen (HBsAg), resulting in 1 million deaths from the consequences of chronic HBV infection. Individuals who recover from acute HBV infection produce antibodies to HBsAg that confer lifelong protection. The current vaccine contains recombinant HBsAg synthesized in yeast or CHO cells and is safe and effective for all but a subgroup of immuno-compromised patients (the elderly, and infants of infectious mothers). HBV-infected hepatocytes secrete HBsAg as non-infectious, subviral particles in the form of 22nm spheres and tubules, and very similar structures are secreted by CHO cell lines stably transfected with the HBsAg gene. Attempts have been made to extend the effectiveness of the vaccine to the non-responding subgroup. Experiments in mice have indicated that immune responses to the major surface protein and pre-S domain are separate, and epitopes in the latter domain may enhance the response to the major surface protein (Milich 1988). Trials with pre-S2 containing vaccines have not overcome non-responsiveness, but more success (particularly with the elderly) has been noted when the pre-S1 domain is included (Neurath *et al.* 1986; Tron *et al.* 1989; Shouval *et al.* 1994).

Hepatitis C virus (HCV) is thought to be the major causative agent of non-A, non-B hepatitis, which can lead to cirrhosis, liver failure and liver cancer (Flint & McKeating 2000). Currently 170 million people are infected worldwide, and the only available treatment (interferon-α with ribavirin) is expensive and only moderately effective. A vaccine is thus a high priority, but efforts are hampered by the lack of a small animal model and the inability to routinely support HCV replication *in vitro* in cultured cells. Conventional neutralizing assays cannot, therefore, be performed, although there are conflicting reports that vesicular stomatitis virus (VSV)/HCV pseudotyped virus expressing combinations of E1 and E2 glycoproteins can be produced by transfected BHK cells (and possibly primary hepatocyte and human hepatoma cell lines) and be neutralized by homologous antisera (Buonocore *et al.* 2002; Lagging *et al.* 1998; Matsuura *et al.* 2001).

Chronic HCV carriers make antibodies against a variety of viral proteins, particularly the structural nucleocapsid proteins and the surface glycoproteins E1 and E2 (Flint & McKeating 2000; Lesniewski *et al.* 1993) and these antigens are currently the focus of a variety of potential vaccines. An early study showed that surface glycoproteins of HCV purified from recombinant vaccinia-transduced human cells were able to protect chimpanzees from a low challenge dose with the homologous strain (Choo *et al.* 1994). Further chimpanzee studies detected low levels of E2 antibodies, which did not protect the animals from heterologous HCV challenge but did protect them from subsequent chronic infection (Abrignani & Rosa 1998). An important study by Heile *et al.* (2000) examined the optimal form of the E2 antigen from the perspective of its ability to bind to the putative HCV receptor, CD81, and its capacity to generate antibodies that would inhibit the interaction of E2 with CD81. Soluble E2 truncated at amino acid 661 (E2-661) expressed in CHO cells only met the desired criteria when purified from the core-glycosylated intracellular fraction. The complex-glycosylated secreted fraction does not bind CD81 and does not elicit CD81/E2-blocking antibodies. Only glycosylated, monomeric non-aggregated E2 bound CD81, and protein immunization was more immunogenic than E2-encoding DNA vaccination. Identical findings were reported by Flint *et al.* (2000). Replication-deficient recombinant adenoviruses (Ad) efficiently expressed the core, E1 and E2 proteins in 293 cells, and induced high-titre, specific antibodies in mice (Makimura *et al.* 1996). A partially purified recombinant E1E2 fusion protein isolated from Ad/HCV-transfected HeLa cells raised antibodies to the E2, but not E1 protein in mice, while administration of a similar replication-deficient adenovirus expressing core, -E1 and -E2 proteins induced E2-specific CTLs, but not antibodies to E1 or E2. An adenovirus prime/protein boost regime induced both humoral and CMI responses to the E2 protein (Seong *et al.* 1998; 2001). The most interesting system currently under development involves the use of a replication-defective herpes simplex virus type-1 recombinant in which the HSV gH-encoding gene has been replaced by that of HCV E2-661. Vero, CR1 or Hep2 cells infected with the recombinant virus express high levels of correctly folded, non-aggregated (both intracellular and secreted) E2-661 protein which is highly reactive with sera from HCV-infected patients. Mice immunized with the HSV/HCV virus produced high titres of E2 antibodies. By contrast, most of the E2-661 protein produced by transient expression in 293 cells was found to consist of misfolded aggregates (Flint *et al.* 2000). The HSV/HCV system shows considerable promise as a potential vaccine.

6.2.1.3 Haemorrhagic fever viruses

Viral haemorrhagic fevers (VHFs) represent a serious public health problem with recurrent outbreaks worldwide. The most widely distributed virus is dengue, causing 20 million infections per annum, and, despite the availability of an efficient live, attenuated vaccine, the spectre of yellow fever continues to haunt Africa and South America. In many cases, VHF control is primarily by elimination of the mosquito vector, but where the natural host is not known, as in the case of Ebola and Marburg virus, this is not feasible. Other important agents of VHF include Lassa virus and Japanese encephalitis virus (JEV), while a number of other agents (St Louis encephalitis virus, tick-borne encephalitis virus, Rift Valley fever virus, West Nile virus (WNV)) cause periodic localized epidemics. Unfortunately, with the exception of dengue, yellow fever and JEV, these diseases are a low commercial priority for vaccine manufacturers and comparatively little is known about their biology and immunopathology (Baize *et al.* 2001). However, the recent spread of WNV in the United States (3887 cases with 120 deaths by November 2006) has re-focused attention on this group of viruses.

Although tetravalent (DEN 1-4) live attenuated vaccines against dengue have been trialled in adults and children (Kanesa-Thasan *et al.* 2001), alternative control strategies have been slow to emerge. Monkeys have been protected against dengue-2 challenge by immunization with a vaccinia recombinant vector expressing truncated dengue-2 envelope (E) protein (Men *et al.* 2000), but protection against challenge could not be demonstrated in mice vaccinated with recombinant E-protein despite the induction of specific neutralizing antibody (Kelly *et al.* 2000). A more fruitful area of research concerns the use of chimaeric live-attenuated dengue vaccines, assembled using reverse genetics (Halstead & Deen 2002). Following the successful Phase I trial of a chimaeric yellow fever/Japanese encephalitis live attenuated vaccine (Monath *et al.* 2002), a number of similar vaccines have been made by inserting combinations of dengue virus 1-4 premembrane and E genes into the backbone of either the yellow fever virus 17D or attenuated DEN 1-4 (Guirakhoo *et al.* 2001). After promising results in non-human primates, Phase I trials are under way (summarized by Halstead & Deen 2002).

Approximately 100 million doses of yellow fever 17D live attenuated virus are produced annually, and are generally effective and well tolerated. However, incidences of extreme adverse reactions (including death) are not unknown, emphasizing the need for the mechanism of attenuation to be better understood (Marianneau *et al.* 2001).

Arenaviruses, such as Lassa fever, establish chronic infections in rodents leading to zoonotic transmission to humans. Of the VHFs, Lassa fever affects by far the largest number, being endemic in West Africa from Guinea to Nigeria. Although clinically severe, with up to 30% mortality, infection confers life-long immunity, leading to the belief that a vaccine can be developed (Fisher-Hoch & McCormick 2001). Viral clearance appears to be by cell-mediated mechanisms, and the presence of antibody to viral antigens is negatively correlated with survival (ter Meulen *et al.* 2000). The first genetically engineered vaccine candidates appeared in the early 1980s, based on recombinant vaccinia virus expressing the nucleocapsid (NC) or glycoprotein (GP) genes of Lassa virus. Although guinea pigs were protected from challenge with Lassa virus using a vaccinia-NC construct, primates were not, but both primates and rodents were effectively vaccinated with vaccinia-GP. Protection was not correlated with antibody level, suggesting a critical role for CMI responses in viral clearance (Auperin *et al.* 1988; Morrison *et al.* 1989). A broader study of protection in macaques using a range of recombinant vaccinia expressing NC, GP or combinations of these proteins confirmed the role of GP in protection and the dependence on CMI responses (Fisher-Hoch & McCormick 2001). Unfortunately, vaccinia recombinants are not currently tenable as a vaccine in West Africa because of the potential side effects in an HIV-endemic area. In the longer term, a yellow fever/Lassa fever chimaeric live attenuated virus may offer an attractive solution, although the attenuated strains of vaccinia, MVA and NYVAC, are worthy of consideration.

Formalin-inactivated mouse brain-derived Japanese encephalitis virus (JEV) vaccines are widely used in Asia, and a variety of other inactivated, live-attenuated and chimaeric yellow fever/JEV ('ChimeriVax') vaccines are under development (Monath 2002). Konishi *et al.* (2001) have developed a stable CHO cell line to express a secreted form of subviral particle (EPs) containing the envelope glycoprotein (E) and a precursor (prM) of the viral membrane protein. EPs were shown to be protective in mice and to have many of the antigenic and biochemical properties of viral E, and may provide a useful source antigen for JEV vaccines and diagnostics. ChimeriVax technology is also being applied to a yellow fever/West Nile virus vaccine, in which the prM and E structural proteins of yellow fever 17D vaccine virus are replaced with the equivalent West Nile genes (Monath 2001).

A vaccine against Ebola virus may help to contain this lethal infection and protect at-risk groups. Trials in a guinea pig model have shown that protection can be conferred using both nucleoprotein or envelope protein expressed by a Venezuelan equine encephalitis replicon, although the immune correlates of protection remain unclear (Gupta *et al.* 2001; Pushko *et al.* 2000; Wilson & Hart 2001). A prime/boost vaccination strategy using a DNA/envelope prime followed by recombinant adenovirus/envelope boost was 100% protective in a small-scale macaque study, but much more work is required to establish the potential of this, or other, Ebola vaccines for human trials (Sullivan *et al.* 2003). A 2003 trial with this vaccine in humans was unsuccessful (http://www3.niaid.nih.gov/news/newsreleases/2003/ebolahumantrial.htm).

6.2.1.4 Respiratory viruses

The 200-plus serologically distinct viruses transmitted via the respiratory tract are the major cause of community morbidity and hospitalization in the industrialized world, with significant mortality amongst children, the elderly, and those of any age with compromised immune, cardiac or respiratory systems. The majority of respiratory infections are caused by rhinoviruses, but coronaviruses, parainfluenza (PIV), RSV, CMV, AV and influenza viruses A and B also have a major impact (Olszewska *et al.* 2002). The development of respiratory virus vaccines must take into account the unique clinical and immunological character of these infections. As most infections are localized and mucosal, local secreted antibodies and T-cell responses may be sufficient for protection, and the presence of circulating serum antibody may be just an indication of immunological priming. Also, most respiratory viruses are non-lytic, and the disease is mostly due to inflammatory and immune responses to infection. Thus disease can be potentiated in the presence of inappropriate immune priming, which has been noted with formalin-inactivated RSV viruses (Kim *et al.* 1969; Openshaw *et al.* 2001). At present, *no* vaccines are licensed for use against RSV, PIV, CMV or any rhino- or coronavirus.

Protection from RSV infection is correlated with high titres of mucosal IgA, and a number of vaccines based on the G and F surface proteins are under development. Phase II trials are underway with purified fusion (F) protein obtained from RSV-infected Vero cells and are showing some promise (Piedra *et al.* 2003). Four different subunit approaches are currently under development. The first employs a fusion protein comprising the major immunodominant domain of the G protein fused to the streptococcal G protein albumin-binding domain. This effectively protects mice against challenge and has entered Phase I trials (Power *et al.* 2001; Siegrist *et al.* 1999). The second and third strategies use recombinant viral vaccines. One based on MVA expressing RSV-G revealed promising results in mice, but after showing poor immunogenicity in non-human primates, clinical trials were cancelled (Wyatt *et al.* 1999). Another viral vector uses bovine RSV successfully to express human RSV-F and -G proteins (Collins *et al.* 1999) and is immunogenic in mice. Further trials are underway. The fourth, and most novel approach, uses the techniques of reverse genetics, whereby infectious virus is produced entirely from cDNA with mutations introduced so as to attenuate viral pathogenicity while retaining its ability to confer protective immune responses (Murphy & Collins 2002). This technology has been applied to RSV subgroup A strain A2 (Collins *et al.* 1999), PIV3 virus strain JS (Durbin *et al.* 1997) and bovine PIV3 (Schmidt *et al.* 2000). cDNAs from human PIV1 and PIV2 are also now available (Kawano *et al.* 2001; Newman *et al.* 2002). Two candidates, based on 4 attenuating genetic elements in the infectious rA2 RSV strains, have been produced as recombinant viruses in Vero cells and used in Phase I/II trials in both adult and infant cohorts. One candidate was found not to be suitable for use in children, given its level of replication in RSV-seronegative patients. However, the other vaccine was well-tolerated in infants, with detectable IgA responses, and vaccinees were protected against a second, homologous challenge. The next stage is to determine whether the vaccine is protective against wild-type viruses (Karron *et al.* 2005).

Influenza has a notable effect on both the community and individuals. In the USA, up to 20 % of the population may catch 'flu in a given year, and epidemics may cause 20–40 000 deaths. These figures pale into insignificance compared with the great 'flu pandemic of 1918-1919, with millions of deaths worldwide. The epidemiology of 'flu is complicated by the fact that the viruses undergo two forms of antigenic change: *antigenic shift*, which occurs when genes from animal viruses are transferred to a human virus by reassortment in a dual-infected human host, thus creating a new subtype with novel antigenicity and often associated high morbidity and mortality; and *antigenic drift*, caused by point mutations within the surface proteins, which may allow immune evasion and disease with time. Two surface proteins are involved in these effects, haemagglutinin (HA) and neuraminidase (NA). Type A viruses are found in both animal reservoirs (primarily birds and swine) and are susceptible to antigenic shift, whereas the type B virus is exclusively human and cannot undergo reassortment. Vaccines must therefore be based on the circulating A virus HA and NA subtypes, plus a B virus component. Continual monitoring of these parameters is necessary, and chemically inactivated viruses, grown in embryonated chicken eggs, are prepared for each year's 'flu season. For this reason, the possibility of producing a 'universal' 'flu vaccine is enticing. Unfortunately, the most conserved regions of the 'flu genome are less immunogenic and less likely to induce a protective response. Nonetheless, a fusion protein consisting of the minor surface antigen M2 from an influenza A virus plus the core protein of hepatitis B provided a high degree of protection against viral challenge in mice (Neirynck *et al.* 1999), while other trial vaccines based on M2 or NA show strengthened immunogenicity when fused to cytokine genes (Babai *et al.* 2001; Kilbourne *et al.* 1995). An alternative approach may be predicting the antigenic trend of evolving viruses and constructing synthetic strains based on a framework of conserved amino acids (Bush *et al.* 1999). Reverse genetics technology is also being applied, based on the attenuation of the 'flu NS1 protein, which affects viral virulence by modulation of the antiviral IFN response. Preliminary work with mice has shown that IFN-antagonist activity can be down-modulated by truncation of the NS1 gene (Palese & Garcia-Sastre 2002). Prime-boost regimes, based on priming with a DNA-based vaccine followed by a subunit boost may also be worth pursuing, but the immunogenicity of both vaccine components must be optimized to develop worthwhile immunity in the short time-frame available annually (Kemble & Greenberg 2003).

Recently, a serious new infection characterised by severe respiratory system complications (*"Severe Acute Respiratory Syndrome" – SARS*) emerged in China, leading to serious epidemics in the Far East and Toronto, Canada, causing serious concern regarding international spread. Although not conclusively identified as the agent of all SARS outbreaks, the culprit appears to be a coronavirus (Peiris *et al.* 2003), and several isolates have now been fully sequenced (Marra *et al.* 2003; Rota *et al.* 2003). These are termed SARS-associated coronaviruses (SARS-CoV). The data indicate that the virus is only moderately related to other known coronaviruses. Serological and immunopathological evidence suggests that children only show mild disease manifestations, and clearance of the infection is associated with high titres of circulating antibody to the viral membrane proteins spike (S) and envelope (M). This would suggest that a subunit vaccine might be feasible to control the disease in the longer term, although containment and control measures are the most effective measures in the short term. Currently, SARS vaccine development is adopting three approaches: An inactivated SARS-CoV, a full-length S-protein and a fragment containing the major neutralising epitopes of the S protein (Jiang *et al.* 2005). The inactivated SARS-CoV vaccine elicited high titres of S-specific antibodies in immunized mice and rabbits that block receptor binding and virus entry (He *et al.* 2004), and initial results from a Phase I clinical trial in the Guangdong Province of China suggests that this vaccine is safe and induces SARS-CoV-neutralising antibodies (Business Wire, December 6th, 2004)

6.2.1.5 Other viruses

Major efforts are also being directed towards producing improved rabies vaccines. The vaccines currently in use for the immunization of humans and domestic animals are derived from the 'fixed'-type virus of serotype 1/genotype 1, but do not offer protection against other genotypes or bat rabies (Woldehiwet 2002). A vaccinia-rabies recombinant expressing the rabies glycoprotein gene (G) has successfully reduced the incidence of rabies in foxes and other carnivores in Europe (Pastoret & Brochier 1999), increasing the risk of transmission from bat-borne infections (Nadin-Davis *et al.* 2001). Although rabies is generally controlled by immunization of the animal reservoir, because the disease has a long incubation period in humans it is possible to prevent the development of disease by post-exposure vaccination. Currently, such vaccines are live-attenuated preparations, derived from infected animal brain or cell culture but, due to costs of production and the risk of hypersensitivity, new vaccines are desirable. Viral clearance is strongly associated with both humoral and cell-mediated responses, and is best conferred by a live vaccine. Recombinant G subunit vaccines are ineffective (Chappius 1995). Morimoto *et al.* (2001) have produced a range of modified rabies virus G genes to engineer rabies recombinant viruses, which exhibit marked decreases in viral infectivity coupled with higher G protein expression than wild-type viruses. Importantly, they are also non-pathogenic in a murine model, but show the highest level of protection when challenged with homologous virus. Immunogenicity may be improved by the use of chimaeric viruses, including fragments of G protein genes from diverse genotypes. DNA vaccination of mice with chimaeric G protein genes from bat, human and canine lyssavirus, conferred protection to heterologous isolates of lyssavirus and indicates that a combined lyssavirus vaccine may be achievable (Jallet *et al.* 1999).

Rotavirus (RRV) is the most common single cause of severe, dehydrating gastroenteritis worldwide and accounts for 20 % of all diarrhoea deaths in children under 5 years old (de Zoysa & Feachem 1985). At least eight live, attenuated, oral rotavirus vaccines are currently in human trials, but concerns that tetravalent RRV ('Rotashield') may be associated with intussusception in some infants has led to the temporary withdrawal of this particular vaccine (Centre for Disease Control and Protection 1999). However, following additional epidemiological research, this was considered to be an over-reaction to a minor risk, and Rotashield has now been re-licensed (Murphy *et al.* 2003). Alternative vaccine designs based on the use of virus-like particles as subunit vaccines are currently being pursued in animal models, but are at a very early stage of development (Cunliffe *et al.* 2002).

Parvovirus B19 is the only member of the family *Parvoviridae* known to be pathogenic in humans, manifestations of infection ranging from a mild rash to fetal death *in utero* (Heegaard & Brown 2002). An empty B19 VP1/VP2 VLP expressed in baculovirus is potently immunogenic in a number of experimental animals, VP1 inducing a high-titre neutralizing antibody response in human volunteers. Phase I trials produced encouraging results and Phase II trials are now underway (Bansal *et al.* 1993; Bostic *et al.* 1999; Kajigaya *et al.* 1991).

There is growing evidence that SV40 may be involved in the development of certain human cancers and Phase I trials with a recombinant vaccinia expressing modified large T antigen are now underway (Imperiale *et al.* 2001).

Effective live, attenuated virus vaccines are available for the four major viral diseases of childhood, measles (MV), mumps (paramyxovirus – PMV), rubella (RV) and varicella (VZV) and are in worldwide use in various combinations. The benefits of childhood vaccination with the combined measles–mumps–rubella (MMR) vaccine have been clearly demonstrated in the UK and elsewhere and no credible scientific evidence has yet substantiated the claims that MMR vaccine causes Crohn's disease or autism (Miller 2002). Indeed, a tetravalent vaccine including varicella (MMRV) is now proposed for use (Nolan *et al.* 2002). Despite this, novel vaccines are still required. Live attenuated measles vaccines are ineffective in infants aged <6–9 months (Redd *et al.* 1999) leading to significant worldwide mortality in children aged under 12 months, and there are doubts as to the effectiveness of the mumps virus component (Pons *et al.* 2000). A number of live viral vectors expressing the dominant measles haemagglutinin (MV-HA) antigen, which contains both B- and T-cell epitopes, are currently under examination in both rodent and non-human primate models. These include replication-deficient vectors such as ALVAC (Taylor *et al.* 1992), MVA (Stittelaar *et al.* 2000), NYVAC (Kovarik *et al.* 2001) and AV (Fooks *et al.* 1998) and replication-competent vectors such as VSV (Schlereth *et al.* 2000) and PIV3 (Durbin *et al.* 2000). The challenge for such vaccines is to induce adult-like antibodies, Th_1-like and CTL responses in infants. Infant mice respond well to the NYVAC-HA vaccine, but not to ALVAC-HA, despite both encoding the same antigen and showing identical T-cell responses in adult mice (Kovarik *et al.* 2001). Unfortunately, mice cannot be used for challenge studies, and optimal strategies, possibly based on prime-boost regimes will have to be undertaken in macaques before promising modalities are transferred to Phase I/II human trials.

6.3 EXPRESSION OF HIV ENVELOPE PROTEINS – AN HIV/AIDS VACCINE

It is no exaggeration to state that a greater effort has been committed to developing an effective vaccine against HIV than any other infectious agent. It is perhaps fortunate that the emergence of AIDS coincided with the blossoming of recombinant DNA technology and the promise of unlimited supplies of exquisitely characterised antigens, carefully honed to potentiate the components of the human immune system associated with HIV neutralisation or clearance. Unfortunately, despite 20 years of effort, no such vaccine yet exists. The most recent International AIDS Vaccine Initiative (IAVI) database (as of 13th June 2006) shows one Phase III, four Phase II and thirty-two Phase I trials in progress (http://www.iavireport.org/trialsdb/). Indeed, HIV vaccine development has driven the search for an ever-increasing range of vaccination technologies, ranging from initial immunisation with CHO-derived envelope subunits adjuvanted with alum, to complex prime/boost regimes involving combinations of nucleic acid vaccines, recombinant subunits and replicating or non-replicating viral vectors, each expressing one or more HIV genes, and often modified to optimise immunogenicity.

In general, the immunogenicity of a potential vaccine is initially established in rodents, then taken to challenge studies in non-human primates. With the very lengthy development phases, and the logistical difficulties in mounting large-scale human clinical trials, it takes 10–15 years to bring a vaccine from first experiment to initial Phase III trials. Hence, the development phase is always well ahead of the clinical phase, and a severe bottleneck at this point restricts the number of potential vaccines that can be usefully examined. As a consequence, the two Phase III trials employing the VaxGen gp120 bivalent vaccine completed in 2003 were based on 1980s vaccine technology (unmodified recombinant envelope subunits) and the subsequent results showed that this vaccine was, as expected, ineffective (Cohen 2003).

Table 6.1 summarises the expression systems that have been used to date to produce HIV envelope glycoprotein, and reveals the use of bacterial, yeast, insect, mammalian and plant cells expressing gp120/140/160 from a bewildering collecting of expression plasmids, and replicating or non-replicating viral RNA and DNA vectors (Bojak, Deml and Wagner 2002). This review will confine itself to the expression of HIV-1 envelope in mammalian cells, using either plasmid or viral vectors, and focus on those regimens that have successfully entered Phase I trials.

To date (2006), 141 Phase I/II trials of candidate HIV vaccines have been undertaken or are in progress. Most initial trials focused on the envelope glycoprotein based on the premise that this structure is the target for neutralising antibodies. Although the immune correlates of protective immunity for HIV are not known, antibodies are the major components that protect individuals against virus infection. Antibodies can be maintained at high levels in serum and at lower levels at mucosal surfaces and thus could act as an effective barrier to the sexual transmission of HIV. In general, subunit envelope vaccines have been found to be safe and immunogenic in diverse populations, and induced neutralising antibody and specific $CD4^+$ T-cell proliferative responses, but no CTL responses, in nearly all recipients (Graham *et al.* 1996; Stanhope *et al.* 1993). The highest titre antibodies with broadest neutralising ability have tended to be produced by gp120, 140 or 160 vaccines derived from mammalian CHO cells, rather than those produced by insect, yeast or bacterial expression systems (Belshe *et al.* 1994; Dolin *et al.* 1991; Graham *et al.* 1996; Keefer *et al.* 1994; Keefer *et al.* 1996; Kovacs *et al.* 1993). Although gp140 and gp160 can be produced in

Table 6.1 Expression systems for the production of HIV envelope glycoprotein.

Vector	Host cell	References
Prokaryotic plasmids (various)	*E.coli*	(McCune *et al.* 1988)
Plant plasmids (various)	Plants	(Bogers *et al.* 2004)
Eukaryotic Plasmids (various)	Yeast	(Liu, Gao, and Wang 1998)
	Insect	(Wells and Compans 1990)
	Mammalian	(Berman *et al.* 1989)
Viral Vector		
Vaccinia		(Katz and Moss 1997)
NYVAC		(Tartaglia *et al.* 1992)
MVA		(Hanke and McMichael 1999)
ALVAC		(Tartaglia *et al.* 1998)
Adenovirus		(Buge *et al.* 1997)
Adeno-associated Viras		(Excler 2005)
Measles Virus		(Lorin *et al.* 2005)
Rhinovirus		(Smith *et al.* 1998)
Semliki Forest Virus		(Smerdou and Liljestrom 1999)
Sindbis Virus		(Kong *et al.* 2002)
Venezuelan Encephalitis Virus		(Caley *et al.* 1997)
Influenza		(Muster *et al.* 1995)
Polio		(Anderson *et al.* 1997)
Rabies		(Morimoto *et al.* 2001)
Herpes Simplex Virus		(Wang *et al.* 2003)
Mengo Virus		(Altmeyer *et al.* 1994)
Vesicular Stomatitis Virus		(Rose *et al.* 2001)
Yellow Fever Virus		(Monath 2002)

Vaccinia/Vero cell or baculovirus/insect cell expression systems, they have not shown any particular advantage to date over CHO-derived gp120. Challenge studies in non-human primates have indicated that subunit vaccines may be protective (Bruck *et al.* 1994; Girard *et al.* 1991; Letvin *et al.* 1997). However, these early vaccines were nearly all based on American or European B clade isolates of HIV-1 which had been extensively adapted to growth on T-cells (so-called T-cell line adapted (TCLA) strains such as IIIB, LAI, MN or SF-2), and their neutralising ability was restricted to the homologous virus or closely-related TCLA strains. Only restricted activity against syncitia-inducing (SI) primary (field) isolates was observed, and no activity was demonstrated against non-syncitia-inducing (NSI) primary isolates (Mascola *et al.* 1996). A further complication arises from the unique heterogeneity of the viral envelope (and gag) proteins. Nine genetic subtypes or "clades" (A-K) within group M (main), group O (outlier) and group N (non-M or –O), as well as an increasing number of inter-subtype recombinants (circulating recombinant forms – CRFs), have been identified, and the lack of correlation between these phenotypic groupings and neutralisation makes vaccine development ever more complex (Kostrikis *et al.* 1996; Moore *et al.* 1996; Moore, Parren, and Burton 2001). Although a very narrow group of antibodies (bl2, 2G12, 2F5, 4E10) have been isolated which do demonstrate such cross-clade neutralising ability (Buchacher *et al.* 1994; Burton *et al.* 1994; Stiegler *et al.* 2001; Trkola *et al.* 1996), no recombinant envelope-based subunit vaccine has yet been devised that can induce this.

Table 6.2 lists envelope-based vaccines that have entered clinical trials. Nearly all the earlier trials were based on B-clade gp120s, and it is only recently that examples from clades A, C, D and E have been examined. Following the failure of the Phase III VaxGen trials, there is little appetite for trials involving recombinant vaccine alone and only two are currently in progress (EnvPro and UMMS). However, several C clade and modified (loop-deleted) B-clade envelope preparations are part of prime/boost regimens that are ongoing or planned for 2007.

Table 6.2 Envelope-based HIV/AIDS candidate subunit vaccines (all B-clade, except where indicated). *Indicates preparations carried forward to Phase II trials (adapted from Bojak *et al.* 2002).

Antigen	Expression system	Manufacturer	Clinical Trial Phase
LAI gp160	Baculo/insect	MicroGeneSys	I
IIIB gp160	Vaccinia/vero	Immuno-Ag	I
MN gp160	Vaccinia/vero	Immuno-Ag	I
MN/LAI gp160	Vaccinia/BHK-21	Aventis Pasteur	I/II
SF2 env[1]	Yeast	Chiron	I
IIIB gp120	CHO	Genentech	I
W61D gp120	CHO	GSK	I
SF162 gp140	CHO	Chiron	I
CM235, 240, 244	CHO	Chiron	I
gp140 (D)	CHO	St Jude's Children's Research Hospital	I
MN gp120*	CHO	Genentech	II
A, B, C, E gp120s[3]	CHO	Advanced Bioscience Laboratories	I
SF2 gp120*	CHO	Chiron	II
SF2/CM235 (E) gp120*	CHO	Chiron	II
92TH023 (E) gp140, 160	Vaccinia/vero	Aventis Pasteur	I/II
MN/GNE8 gp120[2]	CHO	VaxGen	III
MN/A244 (E) gp120[2]	CHO	VaxGen	III

1 – Deglycosylated env; 2 – AIDSVAX bivalent vaccines; 3 – ABL pentavalent vaccine

Preclinical research is attempting to increase the quality and breadth of neutralizing antibodies, primarily by selecting env proteins from primary, CCR5 coreceptor-utilising isolates and by creating forms that mimic the native envelope complex. New information concerning the interaction between virus and cell, the role of carbohydrate, and novel adjuvenation strategies has led to a raft of new concepts as to how to improve immunogenicity. It is now widely accepted that the monomeric structure of gp120 is a major factor contributing to poor immunogenicity, primarily due to the differential exposure of epitopes in monomeric and oligomeric envelope. gp120-vaccinee sera reacts preferentially with epitopes on denatured gp120, while sera from HIV-infected individuals contains antibodies that react predominantly with conformation-dependent epitopes, including those directed at oligomeric env (Beddows *et al.* 1999; Moore and Ho 1993; Steimer and Haigwood 1991; VanCott *et al.* 1995). Other studies have indicated that binding of Mabs to oligomeric but not monomeric env correlates with neutralisation (Fouts *et al.* 1997; Parren *et al.* 1998; Roben *et al.* 1994; Sattentau and Moore 1995), and by omitting gp41, key epitopes in the membrane proximal region of gp41 are lost, specifically those recognised by the cross-clade neutralising antibodies 2F5 and 4E10 (Zolla-Pazner 2004). Consequently, several investigators have directed their attention to the production of a stable, oligomeric envelope protein that closely mimics the native trimeric envelope complex. Although the propensity for recombinant HIV envelope proteins including elements of the gp41 glycoprotein to undergo spontaneous oligomerisation is well established (Earl *et al.* 1994; Earl, Doms, and Moss 1992; Hallenberger *et al.* 1993; Staropoli *et al.* 2000), the production and purification of a stable trimer has proved difficult. To further complicate matters, although removal of the gp120/41 cleavage site produces proteins that are oligomerized by strong, non-covalent interactions between the gp41 subunits, the observations that such constructs display differential epitope exposure and coreceptor interaction compared to native viral envelope may indicate that they do not truly mimic the native structure (Earl *et al.* 1997; Earl *et al.* 1994; Edinger *et al.* 1999). As a result, a number of diverse approaches have been adopted to obtain a viral envelope mimic. Ultimately, proof of efficacy will come from immunogenicity studies including relevant simian/human chimaeric immunodeficiency virus (SHIV) challenge experiments in vaccinated non-human primates.

The simplest method involves mutation of the primary gp120/41 cleavage site REKR sequon to REKS. This has been shown to produce a mixture of gp140 oligomers using an SFVYBHK expression system (Staropoli *et al.* 2000), a recombinant vaccinia/BS-C-1 expression system (Earl *et al.* 1994) or recombinant CHO cells (Srivastava *et al.* 2002); (Jeffs *et al.* 2003). Stable trimers may be separated by size-exclusion chromatography. Zhang *et al.* (2001) obtained oligomeric gp140 by expressing a fusion protein of B clade ADA gp120 and the external domain of SIVMac32 gp41 in CHO Lec3.2.8.1 cells, which had been previously shown to produce stable trimeric SIV mac32 gp140. In this case, apart from a small amount of aggregated high molecular weight contaminant, the majority of the oligomeric protein was trimeric. All of these constructs have been shown to be correctly folded, as evidenced by receptor and antibody binding studies. Immunogenicity studies in rodents have only been carried out by Jeffs and Srivastava, and, in both cases, the CHO-derived gp140s were shown to induce the production of high titre antibodies (Jeffs *et al.* 2004; Srivastava *et al.* 2002). A more relevant study by Earl and co-workers compared the immunogenicity of recombinant vaccinia-derived, lentillectin purified gp120 and gp140 (cleavage-site deleted) in rabbits. The results indicated that the gp140-induced antibodies had better cross-reactivity in binding assays and displayed higher neutralising titres against TCLA isolates than gp120 antibodies. More importantly, gp140-vaccinated macaques were protected from a SHIV-HXB2 challenge (Earl *et al.* 2001).

Intriguingly, there is some indication that it is not always necessary to remove or ablate the gp120/41 cleavage site to produce oligomeric protein, that the propensity for gp120/41 cleavage to occur when produced in mammalian cells may be particularly associated with clade B envelope and that the pattern of oligomerisation is isolate-dependent. Jeffs *et al.*(2004), have shown that recombinant gp140s from A, C, D, F and O clade isolates produced in CHO cells show minimal cleavage and a characteristic range of oligomerisation patterns, whereas B clade gp140 from the HAN2 and IIIB isolates are substantially processed and predominantly monomeric. Staropoli *et al.* (2000), using an SFV expression system, saw reduced cleavage in primary A and E clade gp140s compared to a B clade example. Similar results were noted with a further D clade gp140 (Daniels & Hickling, personal communication). One oligomeric C clade gp140 derived from the 97CN54 isolate and expressed from stable CHO cells was found to be trimeric in the absence of cleavage site modifications (Jeffs *et al.*, unpublished results). This may suggest that structural factors possibly associated with the hypervariable loops are affecting both the exposure of the cleavage site and the pattern of oligomerisation, and could partially explain why the oligomeric stability of the disulphide bond–stabilised "SOS" gp140 protein (see below) is enhanced by loop deletion (Sanders *et al.* 2000; Schulke *et al.* 2002). Recent work in my laboratory has confirmed that the stability of trimeric gp140 preparations is strongly related to the parent isolate. One C-clade gp140 based on a Zambian isolate has been shown to retain its structure after freeze-drying, freeze/thaw cycles and heating to 60°C (unpublished results).

More inventive protein engineering strategies have also been proposed to obtain stable trimeric envelope protein. Yang and co-workers introduced three modifications to obtain YU2 trimers: mutation of cleavage site, introduction of two cysteine residues into gp41 to enhance subunit cross-linking, and extension of the N-terminal gp41 coiled coil by the C-terminal addition of GCN4 sequences. GCN4 is a transcription factor that increases the propensity of the protein to form trimers (Harbury *et al.* 1993). These modifications resulted in the production of a monomeric gp130, a high MW aggregate and a major trimeric component when expressed in 293-T cells. When purified by size-exclusion chromatography, the trimeric component was shown to be correctly processed by its ability to bind both antibodies and the receptors CD4 and CCR5 (Yang *et al.* 2000). Antibodies from mice immunised with purified trimeric YU2 gp130, in contrast to monomeric YU2 gp120, were shown to efficiently neutralise a number of primary B-clade isolates of HTV-1, but this activity did not extend to non B-clade isolates and titres were generally quite low (Yang, Wyatt, and Sodroski 2001). Replacement of the GCN4 motif with that of the T4 bacteriophage fibritin, resulted in a construct that was more stable to heat and reducing conditions and with an identical antigenic profile (Yang *et al.* 2002). No stable cell lines are yet available to provide sufficient trimer for challenge studies.

Finally, a series of papers by Binley and his co-workers have compared the antigenic and immunogenic properties of a range of JR-FL B clade gp140 envelope constructs which have retain the gp120/41 cleavage site but are stabilised by the provision of a disulphide linkage between the C-terminal region of gp120 and the immunodominant region of gp41. This so-called SOS protein is processed efficiently and has antigenic properties very similar to that of the native complex, but only forms with approximately 50% efficiency in transient 293-T expression systems (Binley *et al.* 2000). When expressed from stable CHO cell lines, co-transfected with furin, SOS gp140 presents as a stable, cleaved monomer whereas deletion of the VIV2 hypervariable loops produced trimeric protein (Schulke *et al.* 2002).

A second strategy involves the exposure of conserved epitopes such as CD4 or coreceptor binding sites which should be common to many isolates, by either eliminating structures which shield these regions, such as carbohydrate glycans or hypervariable loops, or by freezing the envelope/receptor complex in a transitional state ("fusion intermediate"). It is hoped that this will lead to the construction of an immunogen that generates the hoped-for cross-clade neutralising antibodies. Jeffs *et al.* (1996) and Wyatt *et al.* (1993) showed that removal of the V1, V2 and V3 loops from HXB2 and IIIB recombinant gp120 enhanced the exposure of the CD4 binding site and C1 regions, as evidenced by the binding of monoclonal antibodies to these regions. More importantly, antisera from rats immunised with the V1V2V3 loop-deleted IIIB gp120 ("PR12") showed potent neutralising activity against a range of chimaeric HXB2 HTV-1 virions in which the envelope gene had been replaced by examples from clades A through E (Jeffs *et al.* 2002), but no monoclonal antibodies exhibiting such activity could be isolated from either the PR12-immunised rats or by PR12-screening of human monoclonal antibodies expressed by Epstein-Barr virus immortalised B-cell hybridomas derived from non-progressing HIV-infected individuals (Jeffs *et al.* 2001; Jeffs *et al.* 2002; McKeating *et al.* 1996). Barnett *et al.* (2001) showed that a V2 loop-deleted construct of HIV-1 SF162 gp140 generated slightly higher neutralising antibody titre and broader, although weak, neutralising activity against a number of primary HIV-1 isolates as compared to the wild-type gp140 using a DNA prime/protein boost regime in macaques. This is now in Phase I clinical trials. In contrast, Kim *et al.* (2003) showed that although mice primed with recombinant vaccinia expressing full-length or various combinations of loop-deleted gp120 from a dual-tropic isolate, DH12, then boosted with the homologous protein, expressed similar titres of antibody. The highest neutralising activity was conferred by wild-type, followed by –V4 and –V1V2 constructs. Taken together, these results suggest that the immune response can be redirected by protein modification, but that the optimal structure has not yet been determined.

The alternative unmasking strategy involves the engineering of N-linked glycans. This can take three formats: The removal of specific glycan moieties to expose specific epitopes; the addition of glycans to mask immunodominant, but irrelevant, epitopes to induce the generation of antibodies specifically to defined regions; alteration of the glycan structure to enhance immunogenicity. The functions of such glycans are essentially two-fold. The first is to maintain the correct folding and 3-dimensional structure of the viral glycoprotein, which will impact on such parameters as resistance to protease cleavage, CD4, CXCR4/CCR5, Mannose-Binding Lectin, DC-SIGN and DC-SIGN receptor binding (Dirckx *et al.* 1990; Hart *et al.* 2002; Lee *et al.* 1992; Li, Rey-Cuille, and Hu 2001; Lin *et al.* 2003; Morikawa *et al.* 1990; Papandreou and Fenouillet 1998; Polzer *et al.* 2002; Morikawa *et al.* 1991) with subsequent effects on viral infectivity and viral tropism. The second is to modulate the host immune response against the virus. The acquisition of glycans specified by the host cell rather than the viral genome ensures that the carbohydrate portion of the envelope is considered as "self" and is generally not recognised by the human immune system (Olofsson and Hansen 1998). In addition, the presence of N-linked glycans is known to shield a variety of epitopes within both HTV (V1 and V3 domains) and SIV envelope (V1 domain), which, on removal or modification, renders the virus open to neutralisation (Back *et al.* 1994; Chackerian, Rudensey, and Overbaugh 1997; Losman *et al.* 2001; Polzer *et al.* 2002; Reitter, Means, and Desrosiers 1998; Schonning *et al.* 1996).

Field observations on individuals who, although infected with HIV, did not progress to AIDS (so-called Long-Term Non-Progressors – LTNPs) and discordant couples, in which one of the pair, although repeatedly exposed to virus, did not become infected ("Exposed Uninfected" – EU), provided evidence underlining the importance of cytolytic $CD8^+$ T-lymphocytes in controlling HIV infection (Borrow *et al.* 1994; Harrer *et al.* 1994). Furthermore, $CD4^+$ T-cells, in addition to their limited role in lysing HIV-infected cells, are required to generate a Th1 or Th2-biased immunity. Th1 cells are characterised by the secretion of the cytokines interferon-γ and IL-2 upon activation by viral antigens, and initiate cell-mediated responses against intracellular pathogens. In contrast, Th2 cells are involved in secreting cytokines such as IL-4, -5, -6,-10 and -13 which control the activation and differentiation of B-cells into antibody secreting cells, thus promoting humoral immunity. This emphasises the need for an HIV vaccine to induce both the humoral and cell-mediated (CMI) arms of the immune response.

CMI-responses cannot be readily induced by recombinant subunit vaccines, but there is now abundant evidence that DNA vaccines, and either replication-competent or replication-incompetent viral vectors encoding HIV genes, are all potent initiators of CMI. By combining different vaccine delivery systems and/or different antigens, an additive or synergistic response can be achieved that activates both arms of the immune system. This so-called "prime/boost" regime is now the vaccine technology employed in nearly all current clinical trials. Although outside of the remit of this review, one cannot discuss AIDS vaccine development without a brief survey of the role of cellular immune responses. Due to the extreme difficulty in inducing potent, neutralising antibodies by envelope-based immunogens, and notwithstanding the desirability of such phenomena in a potential vaccine, the candidate vaccines in advanced clinical trials are mostly designed to induce CMI responses. Three vectors show particular promise. First are those based on poxvirus vectors, such as the recombinant canarypox vectors manufactured by Aventis or the attenuated MVA and NYVAC vaccinia vectors involved in the Euro Vac trials. The second group are based on recombinant adenovirus (rAd) and the third on adeno-associated viruses (AAV) (Doueck *et al.* 2006). In addition, vaccines based on the alphavirus replicons SFV and VEE are poised to enter Phase I trials (Dong *et al.* 2003; Smerdou and Liljestrom 1999), a VSV-based vaccine has conferred protection to challenge with the highly-pathogenic SHIV89.6P in immunised macaques and rates as a highly-promising candidate (Rose *et al.* 2001) and a recombinant live attenuated measles vaccine vector containing either a CTL polyepitope construct or a B-clade 89.6 gp140 with the V1, V2 and V3 hypervariable loops deleted has been shown to prime effective cytotoxic T-lymphocytes or broadly neutralising antibodies in a murine model (Lorin *et al.* 2006).

As of June 2006, one Phase III trial employing a prime/boost regime, with the ALVAC vector vCP1452 (which contains B clade *env, gag, pol* + CTL epitopes) as the prime, followed by boosting with the AIDSVAX bivalent B/E gp120 subunit vaccine, is underway in Thailand (RV144) and has recruited 8000 participants. Phase II trial IAVI A002 employs gag in an AAV vector, while two further rAd-based trials are designed to examine a mutliclade (A, B, C) DNA *gag, pol, nef, env* prime followed by a rAd5 *gag, pol, env* boost (HVTN 204), and a standalone rAd5 *gag, pol, nef* vector (HVTN502/Merck02) (Doueck *et al.* 2006; www.iavi.org/trialsdb).

Recently, attention has turned to the role of the innate immune response as the first line of defence against invading organisms (Levy, Mackewicz, and Barker 1996). Dendritic cells can enhance Th2-type responses by engaging both natural killer (NK) cells as well as $CD8^+$ CTLs to suppress viral replication (Siegal *et al.* 1999), while non-cytotoxic antiviral responses can be mediated by an as yet unidentified fector produced by another subset of $CD8^+$ cells (Levy, Mackewicz, and Barker 1996). Various components of the complement system can also enhance humoral responses. For example, the addition of human or murine C3d complement to gp120 in a DNA construct has been shown to be effective at enhancing antibody titres in a murine model (Green, Montefiori, and Ross 2003; Ross *et al.*, 2001). Work is underway in my laboratory to optimise this system using various prime/boost modalities with DNA or viral vectors plus purified recombinant protein.

In conclusion, an HIV vaccine is likely to be a complex, multi-component vaccine containing a number of genes or gene fragments encoding antigens or epitopes as well as proteins and peptides designed to induce specific components of the humoral, cell-mediated and innate immune systems. Ideally, it will show reactivity against diverse clades, or at least those clades circulating in the target population, and it will be delivered by a reasoned prime/boost regimen. This is a tall order, and the development of such a vaccine will take a co-ordinated approach. To this end, a number of major initiatives have been established. For example, the European Union "EuroVac" initiative brought together over 20 different academic, research and development, and industrial groups from across Europe to compare the safety and immunogenicity of multiple viral antigens from the *gag, pol, nef, env* genes derived from both European B (BX08) and Chinese C (97CN54) clade isolates using prime/boost regimes in a Phase I trial. The initial studies involved DNA vaccines containing codon-optimised genes as the prime, and recombinant NYVAC viral vectors as the boost (www.eurovac.net). Other multi-site, multi-agency trials are being co-ordinated by the International AIDS Vaccine Initiative (IAVI), The Bill and Melinda Gates Foundation Grand Challenges in Global Health, World Health Organisation and UN/AIDS and many government funded initiatives in the USA, Europe, Africa, South America and Thailand.

In the face of the ever-evolving, global AIDS pandemic, it is gratifying to note that major initiatives to develop an HIV vaccine are securing appropriate funding. We now have a great deal of knowledge concerning the nature of the immune response to infection with HIV and it is to be hoped that the combination of the second- and third-generation vaccine technologies outlined above and the move towards large-scale international collaborations will soon bear fruit with the announcement of the first successful HIV vaccine.

REFERENCES

Abrignani S, Rosa D (1998) *Clin. Diagn. Virol.*; 10: 181–185.

Adams M, Borysiewicz L, Fiander A *et al.* (2001) *Vaccine*; 19: 2549–2556.

Adler SP (1995) *Scand. J. Infect. Dis. Suppl.*; 99: 105–109.

Adler SP, Plotkin SA, Gonczol E *et al.* (1999) *J. Infect. Dis.*; 180: 843–846.

Altmeyer R, Escriou N, Girard M, Palmenberg A, van der Werf S (1994) *Proc. Natl. Acad. Sci. USA*; 91: 9775–9779.

Anderson MJ, Porter DC, Moldoveanu Z, Fletcher TM 3rd, McPherson S, Morrow CD (1997) *AIDS Res. Hum. Retroviruses*; 13: 53–62.

Anon (2003) *Aids Alert*; 18: 41, 43–45.

Arrand JR (1998) In *Viruses and Human Cancer.* Eds Arrand JR, Harper DR. Bios Scientific Publishers, Oxford.

Auperin DD, Esposito JJ, Lange JV *et al.* (1988) *Virus Res.*; 9: 233–248.

Babai I, Barenholz Y, Zakay-Rones Z *et al.* (2001) *Vaccine*; 20: 505–515.

Back NK, Smit L, De Jong JJ *et al.* (1994) *Virology*; 199: 431–438.

Baize S, Marianneau P, Georges-Courbot MC, Deubel V (2001) *Curr. Opin. Infect. Dis.*; 14: 513–518.

Bangham CR (2000) *J. Clin. Pathol.*; 53: 581–586.

Bansal GP, Hatfield JA, Dunn FE *et al.* (1993) *J. Infect. Dis.*; 167: 1034–1044.

Barnett SW, Lu S, Srivastava I *et al.* (2001) *J. Virol.*; 75: 5526–5540.

Beddows S, Lister S, Cheingsong R, Bruck C, Weber J (1999) *J. Virol.*; 73: 1740–1745.

Belshe RB, Graham BS, Keefer MC *et al.* (1994) *JAMA*, 272: 475–480.

Berencsi K, Gyulai Z, Gonczol E *et al.* (2001) *J. Infect. Dis.*; 183: 1171–1179.

Berman PW, Riddle L, Nakamura G *et al.* (1989) *J. Virol.*; 63: 3489–3498.

Binley JM, Sanders RW, Clas B *et al.* (2000) *J. Virol.*; 74: 627–643.

Bogers WMJM, Bergmeier MA, Ma J, Oostermeijer H, Wang Y, Kelly CG, ten Haaft P, Singh M, Heeney JL, Lehner T (2004) *AIDS*; 18: 25–36.

Bojak A, Deml L, Wagner R (2002) *Drug Discov. Today*; 7: 36–46.

Bomford R, Kazanji M, De The G (1996) *AIDS Res. Hum. Retroviruses*; 12: 403–405.

Borrow P, Lewicki H, Hahn BH, Shaw GM, Oldstone MB (1994) *J. Virol.*; 68: 6103–6110.

Bostic JR, Brown KE, Young NS, Koenig S (1999) *J. Infect. Dis.*; 179: 619–626.

Bruck C, Thiriart C, Fabry L *et al.* (1994) *Vaccine*; 12: 1141–1148.

Buchacher A, Predl R, Strutzenberger *et al.* (1994) *AIDS Res. Hum. Retroviruses*; 10: 359–369.

Buge SL, Richardson E, Alipanah S *et al.* (1997) *J. Virol.*; 71: 8531–8541.

Buonocore L, Blight KJ, Rice CM, Rose JK (2002) *J. Virol.*; 76: 6865–6872.

Burton DR, Pyati J, Koduri *et al.* (1994) *Science*; 266: 1024–1027.

Bush RM, Bender CA, Subbarao K, Cox NJ, Fitch WM (1999) *Science*; 286: 1921–1925.

Caley IJ, Betts MR, Irlbeck DM *et al.* (1997) *J. Virol.*; 71: 3031–3038.

Campo MS (1994) In *Human Papillomaviruses and Cervical Cancer: Biology and Immunology.* Eds Stern PL, Stanley MA. Oxford University Press, Oxford.

Chackerian B, Rudensey LM, Overbaugh J (1997) *J. Virol.*; 71: 7719–7727.

Chappius GE (1995) In *Rabies in a Changing World.* Eds Benyon PH, Edney ATG. British Small Animal Veterinary Association, London; 49–59.

Chen LP, Thomas EK, Hu SL, Hellstrom I, Hellstrom KE (1991) *Proc. Natl. Acad. Sci. USA*; 88: 110–114.

Choo QL, Kuo G, Ralston R *et al.* (1994) *Proc. Natl. Acad. Sci. USA*; 91: 1294–1298.

Collins PL, Whitehead SS, Bukreyev A *et al.* (1999) *Adv. Virus Res.*; 54: 423–451.

Corey L, McElrath MJ, Weinhold K *et al.* (1998) *J. Infect. Dis.*; 177: 301–309.
Corey L, Langenberg AG, Ashley R *et al.* (1999) *JAMA*; 282: 331–340.
Cunliffe NA, Bresee JS, Hart CA (2002) *J. Infect.*; 45: 1–9.
Davignon JL, Clement D, Alriquet J, Michelson S, Davrinche C (1995) *Scand. J. Immunol.*; 41: 247–255.
de Zoysa I, Feachem RG (1985) *Bull. World Health Org.*; 63: 569–583.
Dempsey PW, Allison ME, Akkaraju S, Goodnow CC, Fearon DT (1996) *Science*; 271: 348–350.
Dezzutti CS, Frazier DE, Huff LY, Stromberg PC, Olsen RG (1990) *Cancer Res.*; 50: 5687S–5691S.
Dirckx L, Lindemann D, Ette R, Manzoni C, Moritz D, Mous J (1990) *Virus Res.*; 18: 9–20.
Dolin R, Graham BS, Greenberg *et al.* (1991) *Ann. Intern. Med.*; 114: 119–127.
Dong M, Zhang PF, Grieder F *et al.* (2003) *J. Virol.*; 77: 3119–3130.
Doueck DC, Kwong PD, Nabel GL (2006) *Cell*; 124: 677–681.
Durbin AP, Hall SL, Siew JW, Whitehead SS, Collins PL, Murphy BR (1997) *Virology*; 235: 323–332.
Durbin AP, Skiadopoulos MH, McAuliffe JM *et al.* (2000) *J. Virol.*; 74: 6821–6831.
Earl PL, Doms RW, Moss B (1992) *J. Virol.*; 66: 5610–5614.
Earl PL, Broder CC, Long D *et al.* (1994) *J. Virol.*; 68: 3015–3026.
Earl PL, Broder CC, Doms RW, Moss B (1997) *J. Virol.*; 71: 2674–2684.
Earl PL, Sugiura W, Montefiori DC *et al.* (2001) *J. Virol.*; 75: 645–653.
Edinger AL, Blanpain C, Kunstman KJ, Wolinsky SM, Parmentier M, Doms RW (1999) *J. Virol.*; 73: 4062–4073.
Epstein SL, Misplon JA, Lawson CM, Subbarao EK, Connors M, Murphy BR (1993) *J. Immunol.*; 150: 5484–5493.
Excler JL (2005) *Indian J. Med. Res.*; 121(4): 568–581.
Feltkamp MC, Smits HL, Vierboom MP *et al.* (1993) *Eur J. Immunol.*; 23: 2242–2249.
Fisher-Hoch SP, McCormick JB (2001) *Rev. Med. Virol.*; 11: 331–341.
Flint M, McKeating JA (2000) *Rev. Med. Virol.*; 10: 101–117.
Flint M, Dubuisson J, Maidens C *et al.* (2000) *J. Virol.*; 74: 702–709.
Fooks AR, Jeevarajah D, Lee J *et al.* (1998) *J. Gen. Virol.*; 79: 1027–1031.
Fouts TR, Binley JM, Trkola A, Robinson JE, Moore JP (1997) *J. Virol.*; 71: 2779–2285.
Frey SE, Harrison C, Pass RF *et al.* (1999) *J. Infect. Dis.*; 180: 1700–1703.
Girard M, Kieny MP, Pinter *et al.* (1991) *Proc. Natl. Acad. Sci. USA*; 88: 542–546.
Gnann JW Jr., Nelson JA, Oldstone MB (1987) *J. Virol.*; 61: 2639–2641.
Graham BS, Matthews TJ, Belshe RB *et al.* (1993) *J. Infect. Dis.*; 167: 533–537.
Graham BS, Keefer MC, McElrath MJ *et al.* (1996) *Ann. Intern. Med.*; 125: 270–279.
Green TD, Montefiori DC, Ross TM (2003) *J. Virol.*; 77: 2046–2055.
Gu SY, Huang TM, Ruan L *et al.* (1995) *Dev. Biol. Stand.*; 84: 171–177.
Guirakhoo F, Arroyo J, Pugachev KV *et al.* (2001) *J. Virol.*; 75: 7290–7304.
Gupta M, Mahanty S, Bray M, Ahmed R, Rollin PE (2001) *J. Virol.*; 75: 4649–4654.
Hallenberger S, Tucker SP, Owens RJ, Bernstein HB, Compans RW (1993) *Virology*; 193: 510–514.
Halstead SB, Deen J (2002) *Lancet*; 360: 1243–1245.
Hanke T, McMichael A (1999) *Immunol. Lett.*; 66: 177–181.
Harbury PB, Zhang T, Kim PS, Alber T (1993) *Science*; 262: 1401–1407.
Harrer E, Harrer T, Buchbinder S *et al.* (1994) *AIDS Res. Hum. Retroviruses*; 10(Suppl 2): S77–78.
Hart ML, Saifuddin M, Uemura K *et al.* (2002) *AIDS Res. Hum. Retroviruses*; 18: 1311–1317.
He Y, Zhou Y, Siddiqui P, Jiang S (2004) *Biochem. Biophys. Res. Commun.*; 325(2): 445–452.
Heegaard ED, Brown KE (2002) *Clin. Microbiol. Rev.*; 15: 485–505.
Heile JM, Fong YL, Rosa D *et al.* (2000) *J. Virol.*; 74: 6885–6892.
Ibuki K, Funahashi SI, Yamamoto H *et al.* (1997) *J. Gen. Virol.*; 78: 147–152.
Imperiale MJ, Pass HI, Sanda MG (2001) *Semin. Cancer Biol.*; 11: 81–85.
Jallet C, Jacob Y, Bahloul C *et al.* (1999) *J. Virol.*; 73: 225–233.
Jeffs SA, McKeating J, Lewis S *et al.* (1996) *J. Gen. Virol.*; 77: 1403–1410.
Jeffs SA, Gorny MK, Williams C *et al.* (2001) *Immunol. Lett.*; 79: 209–213.
Jeffs SA, Shotton C, Balfe P, McKeating JA (2002) *J. Gen. Virol.*; 83: 2723–2732.
Jeffs S, Goriup S, Kebble B *et al.* (2004) *Vaccine* 22:1032–1046.
Jiang S, He Y, Liu S (2005) *Emerging Inf. Dis.*; 11: 1016–1020.

Kajigaya S, Fujii H, Field A *et al.* (1991) *Proc. Natl. Acad. Sci. USA*; 88: 4646–4650.

Kanesa-Thasan N, Sun W, Kim-Ahn G *et al.* (2001) *Vaccine*; 19: 3179–3188.

Karron RA, Wright PF, Beslhe, RB (2005) *J. Infect. Dis.*; 191: 1093–1104.

Katz E, Moss B (1997) *AIDS Res. Hum. Retroviruses*; 13: 1497–1500.

Kawano M, Kaito M, Kozuka Y *et al.* (2001) *Virology*; 284: 99–112.

Keefer MC, Graham BS, Belshe RB *et al.* (1994) *AIDS Res. Hum. Retroviruses*; 10: 1713–1723.

Keefer MC, Graham BS, McElrath MJ *et al.* (1996) *AIDS Res. Hum. Retroviruses*; 12: 683–693.

Kelly EP, Greene JJ, King AD, Innis BL (2000) *Vaccine*; 18: 2549–2559.

Kemble G, Greenberg H (2003) *Vaccine*; 21: 1789–1795.

Khanna R, Sherritt M, Burrows SR (1999) *J. Immunol.*; 162: 3063–3069.

Kilbourne ED, Couch RB, Kasel JA *et al.* (1995) *Vaccine*; 13: 1799–1803.

Kim HW, Canchola JG, Brandt CD *et al.* (1969) *Am. J. Epidemiol.*; 89: 422–434.

Kim YB, Han DP, Cao C, Cho MW (2003) *Virology*; 305: 124–137.

Kong W, Tian C, Liu B, Yu XF (2002) *J. Virol.*; 76: 11434–11439.

Konishi E, Fujii A, Mason PW (2001) *J. Virol.*; 75: 2204–2212.

Kostrikis LG, Michalopoulou ZH, Cao Y, Moore JP, Ho DD (1996) *AIDS Res. Hum. Retroviruses*; 12: 1667–1669.

Kovacs JA, Vasudevachari MB, Easter M *et al.* (1993) *J. Clin. Invest.*; 92: 919–928.

Kovarik J, Gaillard M, Martinez X *et al.* (2001) *Virology*; 285: 12–20.

Lagging LM, Meyer K, Owens RJ, Ray R (1998) *J. Virol.*; 72: 3539–3546.

Langenberg AG, Burke RL, Adair SF *et al.* (1995) *Ann. Intern. Med.*; 122: 889–898.

Lee WR, Syu WJ, Du B *et al.* (1992) *Proc. Natl. Acad. Sci. USA*; 89: 2213–2217.

Lees JF, Arrand JE, Pepper SD, Stewart JP, Mackett M, Arrand JR (1993) *Virology*; 195: 578–586.

Leroux-Roels G, Van Hecke E, Michielsen W, Voet P, Hauser P, Petre J (1994) *Vaccine*; 12: 812–818.

Lesniewski RR, Boardway KM, Casey JM *et al.* (1993) *J. Med. Virol.*; 40: 150–156.

Letvin NL, Montefiori DC, Yasutomi Y *et al.* (1997) *Proc. Natl. Acad. Sci. USA*; 94: 378–383.

Levy JA, Mackewicz CE, Barker E. (1996) *Immunol. Today*; 17: 217–224.

Li Y, Rey-Cuille MA, Hu SL (2001) *AIDS Res. Hum. Retroviruses*; 17: 1473–1479.

Liljeqvist S, Stahl S (1999) *J. Biotechnol.*; 73: 1–33.

Lin G, Simmons G, Pohlmann S *et al.* (2003) *J. Virol.*; 77: 1337–1346.

Liu WF, Gao D, Wang ZN (1998) *Clin. Diagn. Lab. Immunol.*; 5: 592–594.

Lorin C, Delebecque F, Labrousse V, Da Silva L, Lemonnier F, Brahic M, Tangy F (2005) *Vaccine*; 23: 4463–4472.

Losman B, Bolmstedt A, Schonning K *et al.* (2001) *AIDS Res. Hum. Retroviruses*; 17: 1067–1076.

Lowry PW, Koropchak CM, Choi CY *et al.* (1997) *Antiviral. Res.*; 33: 187–200.

Lowry RW, Schiller JT (2006) *J. Clin. Inv.*; 116(5), 1167–1173.

Macsween KF, Crawford DH (2003) *Lancet Infect. Dis.*; 3: 131–140.

Makimura M, Miyake S, Akino N *et al.* (1996) *Vaccine*; 14: 28–36.

Marianneau P, Georges-Courbot M, Deubel V (2001) *Lancet*; 358: 84–85.

Marra MA, Jones SJ, Astell CR (2003) *Science*; 300: 1399–1404.

Mascola JR, Snyder SW, Weislow OS *et al.* (1996) *J. Infect. Dis.*; 173: 340–348.

Matsuura Y, Tani H, Suzuki K *et al.* (2001) *Virology*; 286: 263–275.

McCune JM, Rabin LB, Feinberg MB *et al.* (1988) *Cell*; 53: 55–67.

McKeating JA, Shotton C, Jeffs S *et al.* (1996) *Immunol. Lett.*; 51: 101–105.

Men R, Wyatt L, Tokimatsu I *et al.* (2000) *Vaccine*; 18: 3113–3122.

Milich DR (1988) *Immunol. Today*; 9: 380–386.

Miller E (2002) *J. Infect.*; 44: 1–6.

Mitchell DK, Holmes SJ, Burke RL, Duliege AM, Adler SP (2002) *Pediatr. Infect. Dis. J.*; 21: 133–138.

Monath TP (2001) *Ann. NY Acad. Sci.*; 951: 1–12.

Monath TP (2002) *Curr. Top. Microbiol. Immunol.*; 267: 105–138.

Monath TP, McCarthy K, Bedford P *et al.* (2002) *Vaccine*; 20: 1004–1018.

Moore JP, Ho DD (1993) *J. Virol.*; 67: 863–875.

Moore JP, Cao Y, Leu J, Qin L, Korber B, Ho DD (1996) *J. Virol.*; 70: 427–444.

Moore JP, Parren PW, Burton DR (2001) *Vaccine*; 10: 563–571.

Morikawa Y, Overton HA, Moore JP *et al.* (1990) *AIDS Res. Hum. Retroviruses*; 6: 765–773.
Morikawa Y, Moore JP, Wilkinson AJ, Jones IM (1991) *Virology*; 180: 853–856.
Morimoto K, McGettigan JP, Foley HD, Hooper DC, Dietzschold B, Schnell MJ (2001) *Vaccine*; 19: 3543–3551.
Morrison HG, Bauer SP, Lange JV, Esposito JJ, McCormick JB, Auperin DD (1989) *Virology*; 171: 179–188.
Murphy BR, Collins PL (2002) *J. Clin. Invest.*; 110: 21–27.
Murphy BR, Morens DM, Simonsen L, *et al.* (2003). *J. Infect. Dis.*; 187: 1301–1308.
Muster T, Ferko B, Klima A *et al.* (1995) *J. Virol.*; 69: 6678–6686.
Muster T, Guinea R, Trkola A *et al.* (1994) *J. Virol.*; 68: 4031–4034.
Nadin-Davis SA, Huang W, Armstrong J *et al.* (2001) *Virus Res.*; 74: 139–156.
Neirynck S, Deroo T, Saelens X, Vanlandschoot P, Jou WM, Fiers W (1999) *Nat. Med.*; 5: 1157–1163.
Neurath AR, Kent SB, Strick N (1986) *J. Gen. Virol.*; 67: 453–461.
Newman JT, Surman SR, Riggs JM *et al.* (2002) *Virus Genes*; 24: 77–92.
Nolan T, McIntyre P, Roberton D, Descamps D (2002) *Vaccine*; 21: 281–289.
Olofsson S, Hansen JE (1998) *Scand. J. Infect. Dis.*; 30: 435–440.
Olszewska W, Zambon M, Openshaw PJ (2002) *Br. Med. Bull.*; 62: 99–111.
Openshaw PJ, Culley FJ, Olszewska W (2001) *Vaccine*; 20(Suppl 1): S27–31.
Oxford JS, Jeffs SA (1996) *Vaccine*; 14: 1712–1717.
Palese P, Garcia-Sastre A (2002) *J. Clin. Invest.*; 110: 9–13.
Papandreou MJ, Fenouillet E (1998) *Virology*; 241: 163–167.
Parren PW, Mondor I, Naniche D *et al.* (1998) *J. Virol.*; 72: 3512–3517.
Pass RF, Duliege AM, Boppana S *et al.* (1999) *J. Infect. Dis.*; 180: 970–975.
Pastoret PP, Brochier B (1999) *Vaccine*; 17: 1750–1754.
Peiris JS, Lai ST, Poon LL *et al.* (2003) *Lancet*; 361: 1319–1325.
Phillips AC, Vousden KH (1998) In *Viruses and Human Cancer.* Eds Arrand JR, Harper DR. Bios Scientific Publishers, Oxford and Washington DC.
Piedra PA, Cron SG, Jewell A *et al.* (2003) *Vaccine*; 21: 2448–2460.
Plotkin SA (1993) *J. Infect. Dis.*; 168: 29–37.
Polzer S, Dittmar MT, Schmitz H, Schreiber M (2002) *Virology*; 304: 70–80.
Pons C, Pelayo T, Pachon I *et al.* (2000) *Euro. Surveill.*; 5: 80–84.
Power UF, Nguyen TN, Rietveld E *et al.* (2001) *J. Infect. Dis.*; 184: 1456–1460.
Pushko P, Bray M, Ludwig GV *et al.* (2000) *Vaccine*; 19: 142–153.
Rawson H, Crampin A, Noah N (2001) *Brit. Med. J.*; 323: 1091–1093.
Redd SC, Markowitz LE, Katz SL (1999) In *Vaccines.* Eds Plotkin SA, Orenstein WA. Saunders, Philadelphia; third edition.
Reitter JN, Means RE, Desrosiers RC (1998) *Nat. Med.*; 4: 679–684.
Roben P, Moore JP, Thali M, Sodroski J, Barbas CF 3rd, Burton DR (1994) *J. Virol.*; 68: 4821–4828.
Rose NF, Marx PA, Luckay A *et al.* (2001) *Cell*; 106: 539–549.
Ross TM, Xu Y, Green TD, Montefiori DC, Robinson HL (2001) *AIDS Res. Hum. Retroviruses*; 17: 829–835.
Rota PA, Oberste MS, Monroe SS *et al.* (2003) *Science*; 300: 1394–1399.
Sanders RW, Schiffner L, Master A *et al.* (2000) *J. Virol.*; 74: 5091–5100.
Sattentau QJ, Moore JP (1995) *J. Exp. Med.*; 182: 185–196.
Sawada T, Iwahara Y, Ishii K, Taguchi H, Hoshino H, Miyoshi I (1991) *J. Infect. Dis.* 164: 1193–1196.
Schlereth B, Germann PG, ter Meulen V, Niewiesk S (2000) *J. Gen. Virol.*; 81: 1321–1325.
Schmidt AC, McAuliffe JM, Huang A *et al.* (2000) *J. Virol.*; 74: 8922–8929.
Schonning K, Jansson B, Olofsson S, Nielsen JO, Hansen JS (1996) *Virology*; 218: 134–140.
Schulke N, Vesanen MS, Sanders RW *et al.* (2002) *J. Virol.*; 76: 7760–7776.
Seong YR, Lee CH, Im DS (1998) *Virus Res.*; 55: 177–185.
Seong YR, Choi S, Lim JS, Lee CH, Lee CK, Im DS (2001) *Vaccine*; 19: 2955–2964.
Shida H, Tochikura T, Sato T *et al.* (1987) *Embo. J.*; 6: 3379–3384.
Shouval D, Ilan Y, Hourvitz A (1994) In *Viral Hepatitis and Liver Disease.* Eds Nishioka K, Suzuki H, Mishiro S, Oda T. Springer-Verlag, Tokyo.

Siegal FP, Kadowaki N, Shodell M *et al.* (1999) *Science*; 284: 1835–1837.
Siegrist CA, Plotnicky-Gilquin H, Cordova M *et al.* (1999) *J. Infect. Dis.*; 179: 1326–1333.
Smerdou C, Liljestrom P (1999) *Curr. Opin. Mol. Ther.*; 1: 244–251.
Smith AD, Geisler SC, Chen AA *et al.* (1998) *J. Virol.*; 72: 651–659.
Srivastava IK, Stamatatos L, Legg H *et al.* (2002) *J. Virol.*; 76: 2835–2847.
Stanberry LR (1991) *Rev. Infect. Dis.*; 13(Suppl 11) S920–923.
Stanberry LR, Spruance SL, Cunningham AL *et al.* (2002) *N. Engl. J. Med.*; 347: 1652–1661.
Stanhope PE, Liu AY, Pavlat W, Pitha PM, Clements ML, Siliciano RF (1993) *J. Immunol.*; 150: 4672–4686.
Staropoli I, Chanel C, Girard M, Altmeyer R (2000) *J. Biol. Chem.*; 275: 35137–35145.
Steimer KS, Haigwood NL (1991) *Biotechnol. Ther.*; 2: 63–89.
Stiegler G, Kunert R, Purtscher M *et al.* (2001) *AIDS Res. Hum. Retroviruses*; 17: 1757–1765.
Stittelaar KJ, Wyatt LS, de Swart RL *et al.* (2000) *J. Virol.*; 74: 4236–4243.
Straus SE, Corey L, Burke RL *et al.* (1994) *Lancet*; 343: 1460–1463.
Straus SE, Wald A, Kost RG *et al.* (1997) *J. Infect. Dis.*; 176: 1129–1134.
Sullivan NJ, Sanchez A, Rollin PE, Yang ZY, Nabel GJ (2000) *Nature*; 408: 605–609.
Sullivan NJ, Sanchez A, Rollin PE, Yang ZY, Nabel GJ (2003) *Nature*; 424: 681–884.
Takahashi K, Takezaki T, Oki T *et al.* (1991) *Int. J. Cancer*; 49: 673–677.
Tao T, Davoodi F, Cho CJ *et al.* (2000) *Vaccine*; 18: 1359–1366.
Tartaglia J, Perkus ME, Taylor J *et al.* (1992) *Virology*; 188: 217–232.
Tartaglia J, Excler JL, El Habib R *et al.* (1998) *AIDS Res. Hum. Retroviruses*; 14(Suppl 3): S291–298.
Taylor J, Weinberg R, Tartaglia J *et al.* (1992) *Virology*; 187: 321–328.
ter Meulen J, Badusche M, Kuhnt *et al.* (2000) *J. Virol.*; 74: 2186–2192.
Thorley-Lawson DA, Geilinger K (1980) *Proc. Natl. Acad. Sci. USA*; 77: 5307–5311.
Trkola A, Purtscher M, Muster T *et al.* (1996) *J. Virol.*; 70: 1100–1108.
Tron F, Degos F, Brechot C *et al.* (1989) *J. Infect. Dis.*; 160: 199–204.
VanCott TC, Bethke FR, Burke DS, Redfield RR, Birx DL (1995) *J. Immunol.*; 155: 4100–4110.
Vaz-Santiago J, Lule J, Rohrlich P *et al.* (2001) *J. Virol.*; 75: 7840–7847.
Wang JB, Adler SP, Hempfling S *et al.* (1996) *J. Infect. Dis.*; 174: 387–392.
Wang X, Wiley RD, Evans TG, Bowers WJ, Federoff HJ, Dewhurst S (2003) *Vaccine*; 21: 2288–2297.
Wells DE, Compans RW (1990) *Virology*; 176: 575–586.
Whitby D, Boshoff C (1998) *Curr. Opin. Oncol.*; 10: 405–412.
Wills MR, Carmichael AJ, Mynard K, Jin X, Weekes MP, Plachter B, Sissons JG (1996) *J. Virol.*; 70: 7569–7579.
Wilson JA, Hart MK (2001) *J. Virol.*; 75: 2660–2664.
Woldehiwet Z (2002) *Res. Vet. Sci.*; 73: 17–25.
Wyatt R, Sullivan N, Thali M *et al.* (1993) *J. Virol.*; 67: 4557–4565.
Wyatt LS, Whitehead SS, Venanzi KA, Murphy BR, Moss B (1999) *Vaccine*; 18: 392–397.
Yang X, Florin L, Farzan M *et al.* (2000) *J. Virol.*; 74: 4746–4754.
Yang X, Wyatt R, Sodroski J (2001) *J. Virol.*; 75: 1165–1171.
Yang X, Lee J, Mahony EM, Kwong PD, Wyatt R, Sodroski J (2002) *J. Virol.*; 76: 4634–4642.
Zhang CW, Chishti Y, Hussey RE, Reinherz EL (2001) *J. Biol. Chem.*; 276: 39577–39585.

7 A Brief Overview of the Baculovirus Expression System in Insect and Mammalian Cells

C Mannix

7.1 INTRODUCTION

The baculovirus expression system has been one of the most successful for production of recombinant proteins over the last decade or more, and has recently been used for additional applications such as metabolic labelling of products (often difficult using traditional animal cell culture because of the complex media), and the transient expression of proteins in a wide range of animal cells (Kost *et al.* 2005). The system consists of two parts: the baculovirus into which the gene of interest is inserted; and the host insect cell lines (or, rarely, larvae) in which the virus will replicate and the product be expressed. The advantages of the system come both from the intrinsic properties of the baculoviruses and from the characteristics of the insect cell lines available. The combination of the two creates a highly flexible system that rapidly gives good yields (average 5–20 mg/l) of many recombinant proteins, including multi-subunit products: the cells can be infected simultaneously with two or more baculoviruses, each containing a protein subunit, in order to express a multi-component protein complex. Insect cells carry out most of the posttranslational processing steps of mammalian cells, including phosphorylation, acylation, acetylation and α-amidation, giving authentically processed product. However, *N*-glycosylation in insect cells usually results in high-mannose polysaccharides rather than the complex glycosylation structures of normal mammalian proteins, and if authentic glycosylation of the product is a requirement then baculovirus may not be the most appropriate system. Secreted proteins are often expressed at much lower levels than are non-secreted products in the usual Sf-9 and Sf-21 insect cell lines, but yields may be improved by use of other insect cultures such as the *Trichoplusia ni* cell line BTI-Tn-B4-1 (see below). These caveats aside, baculovirus is likely to be the first-choice system for rapid production of milligram-to-gram amounts of non-secreted eukaryotic proteins and also for many prokaryotic proteins, especially for initial evaluations, structure-function studies and high-throughput screening. In addition, recent developments have shown that baculovirus is a highly efficient method for gene delivery into mammalian cells, providing a versatile and rapid method of generating cells expressing functional pathways for screening and analysis (see Section 7.5).

7.2 BACKGROUND

Baculoviruses are double-stranded DNA viruses that are specific for arthropod species. The virus most widely used as an expression system is based upon the *Autographa californica* multiple

Medicines from Animal Cell Culture Edited by G. Stacey and J. Davis

nucleopolyhedrosis virus (AcMNPV), which infects a limited number of lepidopteran insect species. During the normal infection cycle in the host, two forms of the virus are produced: an early budded virus (BV) form, in which viral DNA and structural proteins are surrounded by membrane derived from the infected cell; and a late occluded form (occlusion-derived virus, ODV), enveloped viral cores embedded in a crystalline matrix consisting mostly of the virus protein polyhedrin. The budded virus rapidly spreads the infection from cell to cell within the host, resulting ultimately in its liquefaction and the release of occluded virus into the environment. The occluded form protects the released virus, allowing it to survive for long periods in the environment until ingested by another host. In the alkaline environment of the insect gut, the protective protein matrix is removed, and the life cycle is repeated. In insect cell cultures, only the BV form of baculovirus is required, and the polyhedrin gene may be replaced with genes for one or more recombinant protein(s) without any detrimental effect on viral replication. An additional benefit of replacing or deleting polyhedrin is that it effectively makes the virus unable to survive outside the laboratory, an advantage in terms of environmental safety. The system is intrinsically safe to animals, as the AcMNPV virus is unable to replicate in species other than a limited range of insects.

The production of recombinant baculoviruses is now a straightforward and rapid process, thanks to a range of technical developments such as the linearization of baculovirus DNA prior to co-transfection (Kitts & Possee 1993) and the generation of complete recombinant baculovirus genomes in *E.coli* (Luckow *et al.* 1993). Systems such as Bac-to-Bac and Gateway™ from Invitrogen (see website list) allow progression from the gene of interest to initial virus stocks in as little as two weeks. Both systems are based on the site-specific transposition of an expression cassette into a baculovirus genome (bacmid) carried in an *E.coli* host. In the Bac-to-Bac system, a transfer vector containing transposition sites and the new gene is generated by recombination in *E.coli*, isolated and purified, and used to transfect *E.coli* DH10Bac containing the bacmid DNA and the transposition function. The Gateway™ system uses a molecular biology approach to generate the transfer vector, which is then transfected into DH10Bac in the same way. Successful transposition of the transfer vector into the bacmid causes disruption of an indicator *lacZ* gene in the bacmid, allowing identification and isolation of recombinant (white) bacterial colonies in the presence of the inducer IPTG and a chromogenic substrate such as X-gal (see Invitrogen website for detailed information). The complete recombinant baculovirus DNA can then be amplified, isolated and purified, and used to transfect insect cell cultures to give the initial virus preparation. These systems are straightforward to use and highly efficient for the generation of recombinant baculoviruses. For a more detailed discussion of the development of baculovirus expression systems see O'Reilly *et al.* (1994) and Jarvis (1997).

The cell lines most frequently used with AcMNPV are Sf-21 (Vaughn *et al.* 1977) and Sf-9 (Summers & Smith, 1987), derived from *Spodoptera frugiperda*: the clonal line Sf-9 is preferred to the parental Sf-21 cell line. Both have excellent characteristics in terms of growth in suspension and in serum-free media, virus replication, and product expression. Various proprietary serum-free media are available commercially, but the supplemented IPL-41 formulation of Maiorella *et al.* (1988) is completely satisfactory for all work with Sf-9 and Sf-21 cultures. The doubling time (T_d) of the cells in this medium is approximately 18–20 h, and virus titres in the range 10^8–10^9 iu/ml (infectious units per ml) are routinely achieved. Other cell lines may be useful for particular applications, for example the *Trichoplusia ni* cell line BTI-Tn 5B1-4 ('High-5', Wickam & Nemerow, 1993) is frequently used for production of secreted proteins. However virus titres are much lower from this cell line, and virus stocks are routinely produced in Sf-9 for productive infection of High-5 cultures, requiring the maintenance of two cell lines for the use of this system.

Insect cell cultures metabolize glucose (and other carbohydrates) much more efficiently than do mammalian cells, and produce relatively little lactate. The pH of the culture medium (pH 5.8–6.2) is maintained by a phosphate buffer system, and does not vary by more than about 0.2 units during growth or infection (Bedard *et al.* 1993; Neermann & Wagner, 1996). Oxygen is the nutrient most likely to become depleted because of its limited solubility, and it is important to use culture systems such as shake flasks that can provide a relatively high level of aeration (Wang *et al.* 1993). With sufficient oxygen, Sf-9 cells can be grown routinely to a maximum concentration of 3×10^6 cells/ml in supplemented IPL-41.

A. californica and other baculoviruses utilize the viral envelope glycoprotein gp64 to attach efficiently to insect cell membranes and are rapidly (within 30 min) taken up into the cell (Hefferon *et al.* 1999). The viral DNA is transported to the nucleus, and virus replication is initiated. Cell division ceases within a few hours of infection, and host DNA synthesis is halted; the synthesis of baculovirus DNA is detectable from about 8 h post-infection (p.i.), and reaches a maximal rate at 10–12 h p.i. (Ikeda & Kobayashi 1999). gp64 is detectable on the surface of infected cells from about 12 h p.i., and shortly afterwards viral nucleocapsids begin to bud through the cell membrane to form new virus particles. gp64 is required to drive the budding process, and without it very few (and non-infectious) virus particles are formed (Oomens & Blissard, 1999). Synthesis of recombinant proteins driven by the late promotors polyhedrin and p10 begins about 18 h p.i., and generally continues for about 2 days, into the third day p.i. Cell viability usually starts to decline from day 3 p.i. (but for some products may decrease earlier), and by 5–6 days p.i. the majority of cells will have lysed. For intracellular products, cells are usually harvested by centrifugation between two and three days p.i. (the optimum time being determined by a time course evaluation), disrupted by detergent or homogenization, and the recombinant protein purified.

7.3 PRODUCTION AND ASSAY OF VIRUS STOCKS

The successful use of the baculovirus system depends upon the production of a high quality virus stock which is accurately titrated: both are readily achieved. Baculovirus stocks should be prepared by infection of Sf-9 (or other cell) cultures with recombinant virus at a low multiplicity of infection (MOI, 0.01–0.05 infectious particles per cell) in order to minimize production of defective interfering particles (DIPs). Infection at high MOI leads to increased levels of defective particles and a deterioration in performance of the virus stock (Kool *et al.* 1991; Wickham *et al.* 1991). The cell population will increase by about two-fold until the initial baculovirus replication has produced sufficient progeny virus to infect all the cells (approx 24 h p.i.), and cell division ceases. After infection, cell metabolism increases, and the nutrient requirements per cell will be approximately double that of non-infected cells. It is therefore important that cultures be maintained throughout at a maximum of about half the cell concentration achievable during cell growth ($<1.5 \times 10^6$ cells/ml). Baculovirus stocks should be harvested 4, or at the latest 5, days p.i. in order to minimize the accumulation of viral proteases and other degradative enzymes. Cells and debris should be removed by centrifugation (approx. 1000 *g* for 5 min), and the supernatant stored in the dark at 4°C. Foetal bovine serum (FBS) at 2–5 % is normally added to improve stability, and stocks containing FBS can be stored for years without significant reduction in titre. If serum-free virus stock is required, infected cells may be resuspended in serum-free medium once cell multiplication has ceased (1 to 2 days p.i.), the supernatant harvested at day 4 p.i. and stored in the dark at 4 °C. Such preparations should not deteriorate significantly over a short period of time (1–2 months), but should be reassayed immediately before use, to determine titre.

The traditional method for estimation of infectious baculovirus concentrations is the plaque assay: monolayers of susceptible cells are infected with dilutions of the virus, and the number of infectious foci determined visually by counting plaques (areas of dead cells) after 5 or 6 days (Griffiths & Page 1977). The assay requires considerable skill to achieve accurate results, and adds a significant delay to the overall production process. A modified assay, in which early foci are detected by an antibody against the virus protein gp64, allows titres to be estimated after 2–3 days (Kitts & Green 1999). More recently, a technically simple assay based on the inhibition of growth

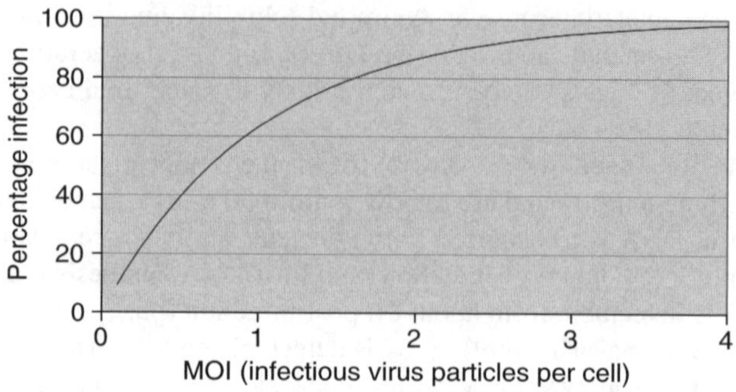

Figure 7.1 Effect of MOI on the proportion of insect cells infected by baculovirus.

of insect cells has been shown to give very rapid (18 h) and accurate results (see Appendix). Since viral infection rapidly (2–3 h p.i.) halts cell division, the degree of growth inhibition compared with controls readily allows estimation of the number of infectious particles added.

7.4 PRODUCTION OF RECOMBINANT PROTEINS AND LABELLED PROTEINS

For expression of recombinant proteins, exponentially growing cultures are infected at a high MOI to ensure synchronous infection of the majority of cells. The relationship between the amount of virus added (MOI) and the number of infected cells is not linear, but is described by the Poisson function (see Figure 7.1): thus an MOI of 1 infectious particle per cell is sufficient to infect only 63 % of the culture, and an MOI of 3 is sufficient to infect 95 %. In practice it is preferable to infect at a nominal MOI of 5 in order to take account of any assay variation. In a productive infection, cell division will cease shortly after virus addition, and cell metabolism and oxygen utilization will increase approximately twofold over the first day post-infection. As for virus production, it is essential to ensure adequate nutrient and oxygen supply by infecting at a reasonable cell concentration (approx. 2×10^6 cells/ml). The efficiency of small-scale systems can be increased by re-suspending cells in fresh medium at higher concentration ($3–3.5 \times 10^6$ cells/ml) before infection, though this is rarely necessary. The optimum time for collection of product should be established by carrying out a time course evaluation. For most proteins, the preferred harvest time is between 48 h and 72 h post-infection, but some products require an earlier collection time because of the product degradation observed later in the infection cycle. Cells and supernatant are separated by centrifugation or filtration for downstream processing. For intracellular products, cells are harvested by centrifugation, washed, and lysed by homogenization, detergent, or other method in a small volume of buffer (generally 1–5 % of the culture volume). Detergents such as NP40 (0.5 % solution in buffer) are a very useful means of disrupting cells, because they give complete lysis of cells but leave nuclei intact. Cell debris and nuclei are removed by centrifugation, and the product is available in concentrated form for assay and purification.

7.4.1 Selenomethionine Labelling

Incorporation of selenomethionine into proteins improves the resolution of X-ray crystallographic studies, and is a valuable technique for progressing structural studies of target proteins. Baculovirus is an ideal system for labelling recombinant products, because it is a transient expression system in which proteins are produced rapidly over a short, defined, time period. Labelled precursors can be added immediately before the start of protein production, minimizing any detrimental effects on the expression system. Selenomethionine is incorporated into proteins in place of methionine, and it is necessary to deplete the levels of methionine in the culture medium to allow efficient labelling. Methionine is present at very high levels relative to its use in many insect media formulations, including IPL-41 (1000 mg/l): insect cell cultures utilize less than 5 % of the available amino acid from IPL-41 during both the growth and infection phases. Cultures should be grown in medium containing low levels of methionine (30 mg/l), infected with the appropriate baculovirus, and incubated for 16 h. At this stage, immediately before the start of recombinant protein synthesis, cells should be collected by centrifugation, resuspended in methionine-free medium containing selenomethionine (50 mg/l), and the incubation continued to day 2 p.i. Using this method, high levels of selenomethionine (>95 %) can be substituted into the recombinant product.

7.4.2 Isotopic Labelling

Advances in NMR stucture determination of proteins have created a demand for isotopically labelled products that can be produced readily using the baculovirus system. Most of the amino acids present in IPL-41 are used only sparingly by Sf-9 cells during growth and infection cycles; many of them are present in the medium in a tenfold excess to usage. One exception is cystine, which can become significantly depleted in cultures at high cell concentration ($>3 \times 10^6$ cells/ml). For labelling purposes, a purified algal amino acid mixture is available as unlabelled or ^{15}N- and ^{13}C/^{15}N-labelled preparations from Cambridge Isotope Laboratories, Inc. (see website list), and used at 4 g/l this can provide the majority of amino acids required (except cystine and small amounts of several amino acids that are missing from the mixture or present at very low levels).

7.4.3 Scale-up of Protein Expression

Expression of recombinant proteins by the baculovirus system is readily scaleable to 100 l or more provided suitable equipment is available (Agathos 1996). The term 'large-scale' refers principally to the methods used rather than to the volume, though in practice large-scale operations are likely to be at or above the 10-l scale. For example, one important element of large-scale culture is the requirement for aseptic inoculation and sampling of vessels *in situ*, where small-scale equivalents would be inoculated and sampled in laminar flow cabinets. Routine large-scale operation would also include *in situ* sterilization of the reactor, associated pipework, and ancillary vessels for media preparation and hold. Because of the reduced air/liquid surface area for a given volume, transfer of oxygen into the culture fluid becomes more difficult, and requires specialized mechanisms to increase the supply of oxygen (such as sparging) and to control the dissolved oxygen concentration (see Chapter 14). However, the parameters that are important for successful scale-up of the baculovirus system are the same as those required at small scale: sterility, supply of adequate nutrients, maintenance of healthy log-phase cultures, minimization of shear forces, and use of high quality baculovirus stocks at suitable MOI. The culture of insect cells and expression of recombinant protein are carried out in essentially the same way in reactors as at laboratory scale, though there are opportunities to increase the cell concentration, and therefore the productivity, of the bioreactor systems. The simplest and most widely used bioreactors for baculovirus expression are animal cell stirred tank reactors fabricated from stainless steel. Complete systems for culture of cells, including all necessary control systems, are available from a large number of manufacturers. Generally, medium for large-scale use will be prepared from a powdered basal mixture and supplements, and sterilized immediately by filtration through a 0.1 (or 0.2) μm filter into a sterile vessel, from where it can be transferred as required to the bioreactor. Liquid transfers are frequently achieved by pressurizing the medium vessel with gas, though pumping is also an option. The bioreactor and other vessels will require temperature control, including the ability to cool vessels for media storage (cooling may also be required to maintain cultures at 27 °C, depending on the environment). Mixing in the bioreactor is required to maintain cells in suspension and to give an even distribution of nutrients, particularly oxygen. In animal cell bioreactors, mixing is usually achieved by using a central impellor, located near the bottom of the vessel. Impellors, such as the four-bladed 45° angled impellor or marine impellor, are designed to give lift in the middle of the reactor (vertical mixing) as well as rotational mixing. Such systems have low shear over the usual range of agitator speeds (50–250 rpm), and are very satisfactory for the serum-free culture of insect cells provided the medium contains Pluronic™ F-68 (0.1–0.2 %).

The most efficient method for supplying adequate amounts of oxygen to stirred tank bioreactors is to sparge air, or air enriched with oxygen, into the culture on demand. Typically, the sterile gas mixture is supplied at a slow rate (0.001–0.005 × reactor volume/min) via a microporous material to the bottom of the reactor, below the impellor. A stream of very small gas bubbles is produced, giving efficient exchange of oxygen with the liquid phase. This process requires a good system for control of dissolved oxygen (dO_2), to prevent oscillation between high and low oxygen concentrations. It is also important to provide a flow of air at higher flow rate over the surface of the culture, to assist removal of carbon dioxide and excess oxygen (thereby improving dO_2 control). Oxygenation can also be improved by increased agitation. Sparging generates additional shear forces (arising from bubble disengagement) but, as previously noted, inclusion of Pluronic™ F-68 gives very good protection against the level of shear forces encountered in a normal sparged reactor (Kioukia *et al.* 1996; Murhammer & Goochee 1988). Antifoam is routinely added to the medium of sparged bioreactors as the build up of foam can lead to the wetting and blocking of filters, and is detrimental to cells.

Dissolved O_2 levels between 30 % and 60 % of air saturation are satisfactory for cell growth and protein production (Schmid 1996). The rate of oxygen utilization of Sf-9 cells is generally accepted to be in the range 4–8 mmol per 10^9 cells per day for non-infected cells: most authors report an increase of between 30 % and 100 % immediately after infection, decreasing at later stages of the process (Schmid 1996). The sparged bioreactor readily provides the amounts of oxygen needed at large scale. Direct scale-up of the baculovirus system from laboratory scale, infecting at concentrations of up to 3×10^6 cells/ml, gives results comparable to laboratory scale on a volumetric basis. In addition, because of the improved transfer of oxygen that can be achieved in sparged culture, it is possible to grow insect cells to much higher concentrations in bioreactors. Provided nutrients such as glucose or glutamine are added to the culture (fed batch process), it is possible to increase the cell concentration at infection to 10^7 cells/ml, with a substantial increase in volumetric productivity (Bedard *et al.* 1994). Although production cultures may be grown to higher cell concentrations, the expression protocol remains the same as in the laboratory-scale process, based on infection at high MOI with the time of harvest determined from small-scale optimization studies.

7.5 BACULOVIRUSES AS EFFICIENT VECTORS FOR TRANSFECTION OF MAMMALIAN CELLS: GENE THERAPY AND CELL CULTURE APPLICATIONS

The use of baculoviruses as a method of gene delivery to mammalian cells was first evaluated for its potential in gene therapy as a safer alternative to adenovirus, retroviruses, and other viral vectors. Hofmann *et al.* (1995) and Boyce and Butcher (1996) reported that genes incorporated into baculovirus under the control of a strong mammalian promoter (Rous sarcoma virus, RSV, promoter and cytomegalovirus immediate early, CMVie, promoter) were efficiently expressed in liver cells (primary hepatocytes and hepatic cell lines), but weakly or not at all in other cell lines tested. The expression level was dependent on the multiplicity of infection, and in resting hepatocytes was stable over more than two weeks. Using a hybrid β-actin/CMVie promoter, Shoji *et al.* (1997) described expression of reporter genes by baculovirus in a wider range of cell lines, and noted that while hepatoma cell lines gave equivalent expression from both this promoter and the standard CMVie promoter, the hybrid promoter gave a more than 10-times higher expression in HeLa cells. Various neuronal cell lines showed efficient transduction *in vitro*, but *in vivo* the cells transfected were predominantly non-neuronal cells (Sarkis *et al.* 2000). A recent report described the construction of a strong neuronal-specific promoter by fusion of the enhancer of the CMVie promoter with the human platelet-derived growth factor (PDGF) β-chain promoter. The hybrid promoter gave expression at high levels similar to the CMVie promoter, but almost exclusively in neuronal cells (Li *et al.* 2004). The expression of transfected genes is thus highly dependent on the choice, or the design, of the promoter system, and this may be manipulated to achieve preferential expression in the target cell or organ. In addition, the surface display of particular protein ligands or antibodies engineered into the major coat protein of baculoviruses offers a potential means for the targeting of specific cell types or tissues (Mottershead *et al.* 2000).

Many human osteosarcoma cell lines have been shown to express very high levels of product after transfection by appropriate baculoviruses (Song *et al.* 2003), and a combination treatment for osteosarcoma (baculovirus-delivered p53 plus chemotherapy) has been proposed (Song & Boyce, 2001). However, some major difficulties would need to be overcome to be able to use baculoviruses as effective gene therapy treatments; it is effectively a transient expression system, and the gene product would be available for only a few weeks, unless the baculovirus is designed to deliver a small self-replicating (or self-integrating) element containing the appropriate gene. Palombo *et al.* (1998) described a baculoviral system containing reporter and selection genes sandwiched between the inverted terminal repeats of adeno-associated virus (AAV) plus the AAV *rep* gene: the construct was able efficiently to transfect human cell lines (293 and diploid MRC-5 cells) and, using the *rep* gene, to integrate both the reporter and selection genes into a specific locus on human chromosome 19. Baculoviruses *in vivo* are rapidly inactivated by the complement system (more than 99.5% inactivation in 30 minutes by human serum, Hofmann & Strauss 1998), and some method of overcoming this will be required. The inactivation can be minimized by complement inhibitors, such as soluble complement receptor 1 (Hofmann *et al.* 1999), and in a later development another inhibitor of complement, decay-accelerating factor, was incorporated into the baculovirus envelope, resulting in viruses that were resistant to inactivation *in vivo* (Hüser *et al.* 2001).

The perception of baculoviruses as a transfection tool was dramatically altered with the publication of a paper demonstrating expression in a very wide range of cells, and amplification of the response (by approximately an order of magnitude in most cell lines evaluated) by inhibitors of histone deacetylase (Condreay *et al.* 1999). It appeared that in many cell lines (hepatoma cells in general being an exception) the baculovirus DNA was successfully transfected, but that transcription was rapidly silenced by deacetylation of the histones associated with the viral DNA. Baculoviruses are frequently more efficient than standard transfection reagents at delivering the target DNA to cells in culture, and expression in the presence of butyrate (a histone deacetylase inhibitor) is generally much higher (Condreay *et al.* 1999). Thus the expression of target genes in this system is subject to epigenetic control, as well as by the choice of promoter for a particular cell line. These findings have led to the widespread use of baculoviruses as transfection vectors in many areas of mammalian cell culture, particularly in the provision of cells for functional assays. The advantages of the system include: the ease of preparation and use of baculoviruses; the ability to express multiple genes simultaneously and at variable levels by using two or more viruses; and its intrinsic safety, both because of the inability of the viruses to replicate in mammalian cells, and their rapid inactivation by complement (in case of inadvertent transfer to laboratory workers). Thus multiple functional assays for G protein-coupled receptors (GPCRs), including expression of both receptor and G proteins, were readily established in the human osteosarcoma cell line

U2-OS by transfection with baculoviruses (Ames *et al.* 2004a), and the system seems likely significantly to advance the discovery of new therapeutic agents that act via GPCRs and other signalling pathways (Ames *et al.* 2004b).

Mammalian cells transfected by baculovirus vectors are also widely used for the study of viruses that are difficult to grow in the laboratory, including hepatitis B (Delaney & Isom, 1998; Delaney *et al.* 2001), hepatitis C (McCormick *et al.* 2002) and defective mutants of human cytomegalovirus (Dwarakanath *et al.* 2001); for the production of virus-like particles (Wang *et al.* 2005), and for the generation of virus antigens as potential vaccines (Aoki *et al.* 1999). The system has been used for the production of fully deleted adenovirus vectors without the use of helper adenoviruses, thus avoiding the problem of contamination with the helper virus (Cheshenko *et al.* 2001). Recombinant adeno-associated virus, a small non-pathogenic DNA virus widely studied for gene therapy applications, has also been produced in mammalian cells by co-infection of two baculoviruses and a helper adenovirus, indicating a simple and efficient method for production and scale-up of these therapeutic vectors.

In the field of RNA interference, Nicholson *et al.* (2005) demonstrated that the baculovirus system can be used to induce RNA interference in several cell lines, with effective knockdown of the corresponding mRNA and protein, thus providing a further option for studies in mammalian cells. Potentially, a baculovirus expressing siRNA or shRNA directed against the appropriate histone deacetylase could, by co-infection at the same time as product baculoviruses, provide a specific and potent method for increasing yields in many mammalian cells without the need to add extraneous inhibitors, such as butyrate, which may have deleterious side effects. Baculoviruses have become not only a rapid method for production of recombinant proteins in insect cells, but also an invaluable reagent for the efficient transfection and manipulation of mammalian cell cultures.

REFERENCES

Agathos SN (1996) *Cytotechnology*; 20:173–189.

Ames R, Nuthulaganti P, Fornwald J, Shabon U, van-der-Keyl H, Elshourbagy N (2004a) *Receptors Channels*; 10: 117–124.

Ames R, Fornwald J, Nuthulaganti P *et al.* (2004b) *Receptors Channels*; 10: 99–107.

Aoki H, Sakoda Y, Jukuroki K, Takada A, Kida H, Fukusho A (1999) *Vet. Microbiol.*; 68: 197–207.

Bedard C, Tom R, Kamen A (1993) *Biotechnology Progress*; 9: 615–624.

Bedard C, Kamen A, Tom R, Massie B (1994) *Cytotechnology*; 15: 129–138.

Boyce FM, Butcher, NLR (1996) *Proc. Nat. Acad. Sci. USA*; 93: 2348–2352.

Cheshenko N, Krougliak N, Eisensmith RC, Krougliak VA (2001) *Gene Ther.*; 8: 846–854.

Condreay JP, Witherspoon SM, Clay, WC, Kost TA (1999) *Proc. Natl Acad. Sci. USA*; 96: 127–132.

Delaney WE 4th, Isom HC (1998) *Hepatology*; 28: 1134–1146.

Delaney WE 4th, Edwards R, Colledge D *et al.* (2001) *Antimicrob. Agents Chemother.*; 45: 1705–1713.

Dwarakanath RS, Clark CL, McElroy AK, Spector DH (2001) *Virology*; 284: 297–307.

Griffiths CM, Page MJ (1997) In *Methods in Molecular Biology.* Eds Pollard JW, Walker JM. Vol. 75: 427–440.

Hefferon KL, Oomens AG, Monsma SA, Finnerty CM, Blissard GW (1999) *Virology*; 258: 455–468.

Hofmann C, Strauss M (1998) *Gene Ther.*; 5: 531–536.

Hofmann C, Sandig V, Jennings G, Rudolph M, Schlag P, Strauss M (1995) *Proc. Natl Acad. Sci. USA*; 92: 10099–10103.

Hofmann C, Hüser A, Lehnert W, Strauss M (1999) *Biol. Chem.*; 380: 393–395.

Hüser A, Rudolph M, Hofmann C (2001) *Nature Biotechnol.*; 19: 451–455.

Ikeda M, Kobayashi M (1999) *Virology*; 258: 176–188.

Jarvis DL (1997) In *The Baculoviruses.* Ed Miller LK. Plenum Press, New York; 389–431.

Kioukia N, Nienow AW, Al-Rubeai M, Emery AN (1996) *Biotechnol. Prog.*; 12: 779–785.

Kitts PA, Green G (1999) *Analyt. Biochem.*;268: 173–178.

Kitts PA, Possee RD (1993) *Biotechniques*; 14: 810–817.
Kool M, Voncken JW, Van Lier FLJ, Tramper J, Vlak JM (1991) *Virology*; 183: 739–746.
Kost TA, Condreay JP, Jarvis DL (2005) *Nature Biotechnol.*; 23: 567–575.
Li Y, Yang Y, Wang S (2004) *Exp. Physiol.*; 90: 39–44.
Luckow VA, Lee SC, Barry GF, Olins PO (1993) *J. Virol.*; 67: 1566–1579.
Maiorella B, Inlow D, Shauger A, Harano D (1988) *Biotechnology*; 6: 1406–1410.
McCormick CJ, Rowlands DJ, Harris M (2002) *J. Gen. Virol.*; 83: 383–394.
Mottershead DG, Alfthan K, Ojala K, Takkinen K, Oker-Blom C (2000) *Biochem. Biophys. Res. Commun.*; 275: 84–90.
Murhammer DW, Goochee CF (1988) *Biotechnology*; 16: 1411–1418.
Neermann J, Wagner R (1996) *J. Cell Physiol.*; 166: 152–169.
Nicholson LJ, Philippe M, Paine AJ, Mann DA, Dolphin CT (2005) *Mol. Ther.*; 11: 638–644.
Oomens AG, Blissard GW (1999) *Virology*; 254: 297–314.
O'Reilly DR, Miller LK, Luckow VA (1994) *Baculovirus Expression Vectors: A Laboratory Manual.* Oxford University Press, Oxford, UK.
Palombo F, Monciotti A, Recchia A, Cortese R, Ciliberto G, La Monica N (1998) *J. Virol.*; 72: 5025–5034.
Sarkis C, Serguera C, Petres S *et al.* (2000) *Proc. Natl. Acad. Sci. USA*; 97: 14638–14643.
Schmid G (1996) *Cytotechnology*; 20: 43–56.
Shoji I, Aizaki H, Tani H *et al.* (1997) *J. Gen. Virol.*; 78: 2657–2664.
Song SU, Boyce FM (2001) *Exp. Mol. Med.*; 33: 46–53.
Song SU, Shin SH, Kim SK *et al.* (2003) *J. Gen. Virol.*; 84: 697–703.
Summers MD, Smith GE (1987) *Texas Agric. Exp. Stn. Bull.* No. 1555.
Vaughn JL, Goodwin RH, Tompkins GJ, McCawley P (1977) *In Vitro*; 13: 213–217.
Wang MY, Kwong S, Bentley WE (1993) *Biotechnol. Progr.*; 7: 355–361.
Wang KC, Wu JC, Chung YC, Ho YC, Chang MD, Hu YC (2005) *Biotechnol. Bioeng.*; 89: 464–473.
Wickham TJ, Davis T, Granados RR, Hammer DA, Schuler ML, Wood HA (1991) *Biotechnol. Lett.*; 13: 483–488.
Wickam TJ, Nemerow (1993) *Biotechnol. Prog.*; 9: 25–30.

Website Addresses

Invitrogen	www.invitrogen.com
Cambridge Isotope Laboratories, Inc.	www.isotope.com

APPENDIX: ESTIMATION OF BACULOVIRUS TITRE BY GROWTH INHIBITION ASSAY (PRESENTED AT THE ANNUAL SCIENTIFIC MEETING OF THE UK BRANCH OF THE EUROPEAN SOCIETY FOR ANIMAL CELL TECHNOLOGY, CAMBRIDGE, UK, JANUARY 2002)

Background

The growth of insect cell cultures such as Sf-9 is inhibited by baculovirus infection, and infection of a single cell is sufficient to prevent its replication (except in cells already committed to division). Therefore the inhibition of growth of test cultures infected at a range of virus dilutions can be used to quantitate the number of infectious particles. Simplistically, infection of a culture containing 2×10^8 cells with 1×10^8 infectious particles would be expected to give approximately 50 % growth inhibition compared with a non-infected control. In practice, the assay requires a slight modification to account for the increasing probability of co-infection as the ratio virus:cells increases. At an average of one infectious particle per cell, only 63 % of the population will be infected, because some cells will be infected with two or more viruses leaving other cells uninfected. For any given multiplicity of infection (MOI, average number of infectious particles per cell), the number of infected cells is described by the Poisson function (Figure 7.1) available on standard spread-sheet programs. The growth inhibition assay is simple, rapid (18 hours), reproducible, and gives results that are in good agreement with those from plaque assays.

Cells and Media

The assay utilizes suspension cultures of Sf-9 cells in logarithmic growth, routinely grown in shake flasks (27 °C, 100–140 rpm) in IPL-41 medium supplemented with 1 % Pluronic™ F-68, 4 g/l TC yeastolate ultrafiltrate, and lipids (Maiorella *et al.* 1988). Other media (e.g. SF900 II) may also be suitable. Cells may be grown routinely in serum-free medium or in medium containing serum, but foetal bovine serum (FBS, 5 %) should be included in the medium used for assay to inhibit cell attachment or clumping.

Assay

(1) Prepare serial fivefold dilutions of the baculovirus in IPL-41 supplemented with 5 % FBS. For the majority of virus preparations, three dilutions, of 5-, 25- and 125-fold, will cover the expected range of titre (5×10^7–5×10^9 infectious particles per ml). Also have available a high-titre virus stock (same or different virus) to allow infection at an MOI of 10 or greater to provide a fully infected culture. Although the assay can be carried out without this control (by assuming that a fully infected culture would increase in cell number by 20 %), this may reduce the accuracy.

(2) Prepare a suspension of Sf-9 cells in medium containing 5 % FBS at a concentration of approximately 1.0×10^6 cells/ml and take a sample for an accurate estimation of cell number. Dispense 10 ml of cell suspension into an appropriate number of 125-ml shake flasks. Keep one flask as an uninfected control, and add 100 µl of each virus dilution to other shake flasks. Add 100–500 µl of the high-titre virus to an additional shake flask to give a fully infected culture (MOI >10). Incubate all flasks at 27 °C on a shaker at 100–140 rpm.

(3) Estimate the cell concentration and viability from the sample (count a minimum of 200 cells by haemocytometer). Calculate the total number of cells per flask.

(4) Incubate for 16–22 h (preferably about 18 h). Estimate the cell concentration and therefore the number of cells in the controls and in each infected flask (count a minimum of 200 cells per sample).

(5) Estimate the increase in cell number from the time of infection compared with the fully infected culture (cell number of each sample or control minus the cell number of the fully infected sample: e.g. 14.0×10^6 cells (sample)–12×10^6 cells (fully infected) = 2.0×10^6 increase in cells compared with fully infected sample).

Some increase in cell number (normally 15–20 %) is seen even with fully infected cultures, presumably due to cells already committed to division. It is therefore important to include the fully infected culture as a control.

Calculation

For the control (non-infected), the increase in total cell number compared with the fully infected control equals 100 % growth (=0 % inhibition).

The relative growth of each sample compared with the control allows calculation of the degree of inhibition, for example, using the following measurements:

starting number of cells	10×10^6 cells (10 ml at 1.0×10^6 cells/ml)
infected control (19 h)	12×10^6 cells
non-infected control (19 h)	20×10^6 cells
sample 1 (100 µl of 1/5) (19 h)	14×10^6 cells

100 % growth is represented by an increase in the non-infected control of 8×10^6 cells compared with the infected control (20×10^6–12×10^6). Sample growth is 2×10^6 cells (14–12) $\times 10^6$, or 25 % of the control, and therefore the inhibition of growth is 75 %.

The degree of inhibition indicates the proportion of cells infected, 75 % in this case, and the MOI (virus particles per cell) can be calculated from the Poisson relationship using the graph in Figure 7.1, or by iterative calculation from the Poisson function. The graph indicates that the MOI needed to infect 75 % of cells is approx 1.4. Therefore the number of virus infectious units (i.u.) in the start sample was 1.4 i.u./cell $\times$ 10 $\times 10^6$ cells = 1.4×10^7 i.u. The titre of sample 1 is calculated from the dilution (fivefold) and volume added (0.1 ml).

$$\text{titre} = 5 \times 1.4 \times 10^7 \text{ particles per 0.1 ml} = 7 \times 10^8 \text{ i.u./ml (infectious units/ml).}$$

The assay can be used for inhibition values between a minimum of 15 % and a maximum of 85 % (MOI of 0.17–1.9), though mid values (30–70 %) are likely to be the most accurate.

8 Stability: Establishing Clones, Genetic Monitoring and Biological Performance

L Barnes

Continuous mammalian cell lines, such as Chinese hamster ovary (CHO), NS0 and baby hamster kidney (BHK) cells, are valuable host cells for the generation of recombinant proteins (see Chapter 1). This is largely because they have the necessary biological machinery to generate recombinant proteins with structures, conformations and post-translational modifications (e.g. glycosylation) that closely reflect those of the native proteins they are intended to mimic (Chapter 24). In addition to cell lines producing recombinant proteins, certain mammalian cells that secrete proteins endogenously, such as hybridomas and Namalwa cells, have also played a role in therapeutic product generation (Johnston 1980; Milstein 1980). These cell lines have allowed the generation of significant quantities of proteins normally only found in low abundance *in vivo*; nevertheless, the necessity for large-scale manufacture of products has led to new demands for scale-up to meet the challenge of obtaining sufficient quantities. The isolation of high-expressing cell lines that exhibit stability of protein production is a vital process for the utilization of mammalian cells in the generation of biological and therapeutic products. Failure to select a cell line that shows stability would lead to problems with process yields, use of money and time, and difficulty with regulatory approval of products, and might ultimately lead to rejection of the product as a therapeutic agent. The challenge of generating high-expressing stable cell lines has been addressed by the cloning of cell lines, and careful selection of the 'best' clone(s), i.e. those with the best combination of productivity and stability, along with the right product quality attributes (protein structure, functional activity, etc.).

8.1 ESTABLISHING CLONES

Cloning has been defined as the generation of a group of organisms from a single parental cell by asexual division of that cell (Oxford Pocket Dictionary 1978). Hence, the word 'clone' is used to define a cell line that has been derived from a single cell (Clarke *et al.* 2002). The generation of a cloned cell line is important in many areas of research and development, but none more so than when generating products for therapeutic use. It is generally considered that 'low-expressing' cells have a lower energy requirement than, and thus may potentially outgrow, higher producers. Accordingly, cloning permits isolation of more productive clones that might otherwise be overgrown by 'low-expressing' clones. Similarly, in the case of hybridomas, cloning can increase the likelihood of producing a pure product by reducing the risk of obtaining a mixture of immunoglobulin

Medicines from Animal Cell Culture Edited by G. Stacey and J. Davis

molecules from a genetically heterogeneous cell population. In addition, homogeneity of both the cell line and product is desirable from a regulatory perspective. Thus cloning is now performed routinely during the early stages of development of hybridomas and genetically modified cell lines. However, there are cases, particularly with hybridoma cells, where recloning at a later stage can be advantageous. After cell lines have been cultured for a period of time, genetic and/or phenotypic drift can occur, and periodic recloning can help to maintain a consistent population of cells. There are now a large number of techniques available for cell cloning. As detailed below, the causes of instability are still largely unknown, so it is not clear whether differences in the methods of cloning can result in different levels of stability in the clones isolated. However, it is important to note that the stresses to which cells are subjected during the cloning procedures may be relevant to the issue of instability.

8.1.1 Cloning Techniques

By far the most widely used method of cloning is limiting dilution cloning (LDC). LDC involves plating cells in multi-well plates at such low population densities that any cell lines that develop have a very high probability of being derived from a single cell (Coller & Coller 1986). However, to maximize the probability of isolating a true clone, it is generally considered necessary to subject cells to at least two rounds of LDC, and at each stage wells are viewed microscopically, so that those containing more than one colony can be rejected. The major advantage of LDC is a saving in terms of cost, as no specialized pieces of equipment are required. However, although the method needs little training time, the cost in terms of man-hours can be excessive if many clones are required, as the technique can be very time consuming. In addition, and despite repeated rounds of LDC, clonality cannot be guaranteed (Lietzke & Unsicker 1985; Underwood & Bean 1988). Due to the problems associated with LDC, many other methods have now been devised which speed up the process of cloning, albeit at a price. A selection of these methods is highlighted below.

The 'spotting' technique is a variation on the LDC method. However, instead of diluting cells to very low concentrations, a single droplet (~1 μl) of a cell suspension containing 500–1000 cells/ml is deposited in each well of a multi-well plate using a Pasteur pipette. Each well is then viewed microscopically to identify wells containing single cells. This technique, like LDC, is simple and easy to perform and, in addition, confirms that the resulting cell lines were derived from a single cell. However, due to the very small volumes in each well there is a tendency for drying to occur before microscopic examination is complete (Clarke *et al.* 2002). The use of cloning rings is another simple and relatively cheap alternative to LDC for the cloning of attached cells. Cells are allowed to grow in standard tissue culture vessels until small colonies are identified. Cloning rings are then placed round colonies to allow cells to be trypsinized and transferred to other vessels for expansion (Clarke *et al.* 2002).

The Quixell system is an example of a capillary-aided micro-manipulation cloning method. This type of system is used to transfer single cells from a source dish to a well using a sealed glass micropipette that collects and ejects the cells by controlling the volume of air within the micropipette (Wewetzer & Seilheimer 1995). Although the cost of purchasing equipment like the Quixell system greatly increases costs compared with LDC, it may eliminate the need to perform repeated rounds of cloning and increases the likelihood of obtaining a clonal population.

Techniques such as cloning in soft agar or methylcellulose have been successfully used for the cloning of suspension cells. The basic principle of these techniques is to limit the spread of a colony by culturing the cells in a high viscosity environment. A single cell is immobilized in agar or methylcellulose and hence the cell population that is derived is localized to the area surrounding the original cell. Colonies immobilized in agar or methylcellulose can then be transferred to other culture vessels for expansion (Davis 1986; Clarke *et al.* 2002).

The gel microdrop assay combines cloning with production analysis. Single cells are immobilized within an agarose-based material. Typically the agarose is conjugated to biotin, and in turn this is bound to a biotinylated capture antibody via a streptavidin bridge. Secreted protein from the immobilized cell binds to the capture antibody and is quantified by addition of a fluorescence-labeled reporter antibody. This also allows the gel microdrop assay to be coupled with fluorescence-activated cell sorting (FACS) and hence high-producing cells can be easily isolated (Gift & Weaver 2000; Hammill *et al.* 2000).

8.2 GENETIC MONITORING OF STABILITY

In biotechnology applications where cells are used to express products, the monitoring of stability is generally taken to mean the monitoring of product formation, either total production per volume of culture or specific productivity per cell (often expressed as pg/cell/day). However, it is also important to monitor the stability of cell phenotype, for example the percent viability and the growth rate of the population, or the gross appearance of the individual cells, as this can also be an indicator of stability. The typical method for testing cell line stability is to culture cells for an extended period of time (e.g. 2–3 months) and to monitor stability indicators throughout this period. Such stability studies are routinely performed on a small scale (e.g. shake flasks) prior to expanding culture volumes to fermenter capacity for production runs.

The most obvious way to monitor stability is at the level of protein production, and one of the most common methods to achieve this is using an enzyme-linked immunosorbent assay (ELISA). However, other methods such as Western blotting and bioassays can also be of value. Bioassays are particularly important when monitoring the functionality of a protein (Fogolin *et al.* 2002). Assessing the stability of mRNA expression can be invaluable both in terms of monitoring and predicting stability of production (Barnes *et al.* 2003). This can be performed using methods such as Northern blotting and quantitative RT-PCR (Borth *et al.* 1999; Raeymaekers 2000). The importance of analysis at the level of DNA has been proven by studies of stability in CHO cells (see Section 8.3.2). Typical DNA analysis would involve copy number analysis by Southern blotting or dot blot assays and also PCR. However, other DNA analysis methods such as contour clamped homogeneous electric field electrophoresis (CHEF) are also of benefit (Cedervall & Radivoyevitch 1996). In addition, chromosomal analysis such as fluorescence *in situ* hybridization (FISH) can also give an indication as to the stability of recombinant cell lines, as it is now well known that the site of incorporation of a recombinant plasmid may influence its expression (see Section 5.2.2.1e) (Ekong & Wolfe 1998).

8.3 THE OCCURRENCE OF INSTABILITY IN INDUSTRIALLY RELEVANT CELL LINES

The mammalian cell lines used within an industrial setting have many advantages, for example high production levels and ease of growth. However, the one disadvantage that they all appear to share is problems with stability of production. This section details the situation with regards to stability of production from a selection of the most frequently used industrial cell lines.

8.3.1 Hybridoma Cell Lines

Hybridoma cell lines have been used for the industrial production of monoclonal antibodies for many years (Bae *et al.* 1995; Kim *et al.* 1996; Lee & Lee 2000). However, hybridomas are notoriously unstable with respect to antibody production. This is reflected in the number of studies that have been performed to investigate instability during prolonged culture of these cell lines (Frame & Hu 1990; Heath *et al.* 1990; Ozturk & Palsson 1990; Lee *et al.* 1991; Chuck & Palsson, 1992; Coco-Martin *et al.* 1992; Merritt & Palsson 1993; Kromenaker & Srienc 1994; Couture & Heath 1995; Kim *et al.* 1996). Although the exact molecular mechanisms that account for instability are controversial, it is thought that the fact that hybridomas are a fusion of two cell types may be a major factor associated with the instability of these cells. Occasionally instability from hybridomas can be transient, with levels of production returning to normal after a period of decreased production, and the extent of instability of different hybridoma cell lines is often variable. However, instability of these cells is a frequent finding and hence cloning to recover cells producing high levels of antibody is a common requirement (Merritt & Palsson 1993; Bae *et al.* 1995).

8.3.2 CHO Cell Lines

As mentioned in Section 5.2.2.1e, Chinese hamster ovary (CHO) cells are commonly used in combination with the dihydrofolate reductase (DHFR) selectable and amplifiable marker, which is amplified using methotrexate (MTX) (Puck *et al.* 1958; Fann *et al.* 2000; Yoshikawa *et al.* 2000a). Despite the fact that they are probably the most commonly used mammalian cell lines within industry, CHO cell lines often suffer from instability. This is highlighted by a variety of reports using this system, including examples for production of antibodies (Kim *et al.* 1998a), tissue plasminogen activator (Fann *et al.* 2000), c-myc (Pallavicini *et al.* 1990), γ-interferon (Cossons *et al.* 1991) and DHFR (Kaufman *et al.* 1981). Instability is usually reported to occur when the selective pressure of the MTX is removed from amplified cell lines. Instability of production has also been observed when the selective pressure is retained, but not to the same extent as when MTX is absent (Kim *et al.* 1998a; Fann *et al.* 2000). By far the most common explanation for instability in DHFR-CHO cells is reduced gene copy number (i.e. gene(s) encoding the recombinant protein are lost from the CHO cell's genome) (Pallavicini *et al.* 1990; Kim *et al.* 1998b; Hammill *et al.* 2000).

8.3.3 NS0 Cell Lines

NS0 is a mouse tumour cell line (myeloma) that has been selected following a series of treatments and cloning so that it no longer produces any endogenous antibody (Galfrè & Milstein 1981; Barnes *et al.* 2000). The historical development of this myeloma cell line is depicted in Figure 8.1 (Stacey 1995; Barnes *et al.* 2000). These cells are most commonly used in combination with the glutamine synthetase (GS) selectable marker. GS also allows amplification through inclusion of methionine sulphoximine (MSX) in the culture medium (Bebbington *et al.* 1992; Barnes *et al.* 2000). NS0 cells are easy to grow and are reported to permit high levels of production. However, until recently, very little information has been published regarding stability of recombinant protein production from NS0 cell lines. Initial reports stated that the system was relatively stable in terms of production. For example, Brown *et al.* (1992) stated that GS-NS0 cells exhibited stable recombinant protein production in the absence or presence of MSX for at least 60 generations (~55 days) in glutamine-free medium. However, more recently reports have shown that instability of recombinant protein production can occur with GS-NS0 cells and that careful selection of cell lines is required to increase the chance of obtaining a cell line that will show stability of production during stability studies and production runs (Barnes

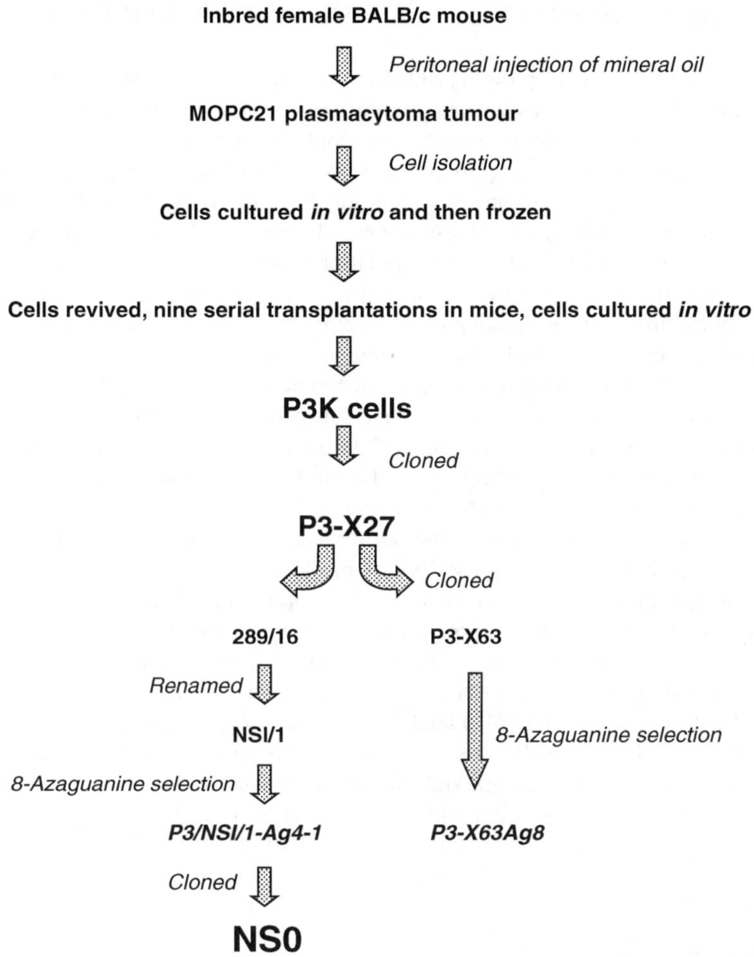

Figure 8.1 Development of the NS0 cell line. (Reproduced from Barnes *et al.* 2000 (Fig 1), with kind permission of Kluwer Academic Publishers)

et al. 2001, 2003; Kearns *et al.* 2003). Evidence has shown that instability from GS-NS0 cells is not related to loss of gene copies, but rather to a decrease in mRNA expression (Barnes *et al.* 2003).

8.3.4 Other Cell Lines

Other cell lines that may play a role in the production of therapeutic proteins also exhibit problems with stability of production. For example, genetic instability has been reported for the murine mammary adenocarcinoma cell line C127I (Stacey *et al.* 1992), and instability during long-term culture has been observed in recombinant BHK-21 cells when selective pressure is removed (Mielke *et al.* 2000). From the examples presented above it is clear that there is no single reason to account for instability. However, the number of reported cases of instability and the wide range of cell lines affected highlights the fact that loss of production from mammalian cell culture is a significant problem.

8.4 MOLECULAR MECHANISMS ASSOCIATED WITH INSTABILITY

As described above, instability of protein production is a substantial problem. The causes of instability are varied and in many cases the exact molecular mechanisms that underlie instability are unknown. The production of proteins by cells is a complex process that is modulated by molecular events at levels ranging from transcription, post-transcriptional processing, translation and post-translational modification, to secretion. Protein production, and hence the stability of production, can be determined by controls at all of these stages. This section details the mechanisms that are known, or are believed, to play a role in the regulation of gene expression, the silencing of genes and the occurrence of instability. However, it is very important to stress that these mechanisms are not mutually exclusive; the regulation of gene expression and the occurrence of instability of production often involves interplay between different mechanisms. Figure 8.2 indicates some of the key molecular sites at which protein production can be regulated.

One of the major, and best documented, causes of instability of production is the loss of the gene(s) encoding a recombinant protein from the host cell genome. As mentioned above, this factor is particularly relevant to instability in the DHFR-CHO system, where loss of productivity is commonly attributed to loss of gene copies (Pallavicini *et al.* 1990; Kim *et al.* 1998b; Hammill *et al.* 2000). The loss of copies can either lead to a decrease in production rate by all the cells in the population, or a decrease/loss of productivity from a subset of the population and possibly the appearance of a non-producing sub-population of cells, that may have a slight growth advantage over the producing population (Lee *et al.* 1991; Chuck & Palsson 1992; Bae *et al.* 1995; Kim *et al.* 1996; Borth *et al.* 1999). However, loss of genes is only one factor that may lead to decreased productivity. Other causes such as gene mutation or changes to the transcriptional capacity of the cell may also lead to decreased protein synthesis (Merritt & Palsson 1993; Kromenaker & Srienc 1994). It is also important to note that environmental factors are thought to have an influence over the stability of recombinant protein production. Both the condition of the initial cell population and serum levels have been reported to show an effect on the appearance of non-producing populations of cells (Chuck & Palsson, 1992; Gaertner & Dhurjati, 1993).

8.4.1 Transcriptional Regulation

The fact that the chromosomal location of recombinant genes within the host cell's genome can affect the expression of recombinant genes and lead to a phenomenon called the 'position effect', has been known for several years (Wilson *et al.* 1990). The 'position effect' or 'position effect variegation' occurs when a recombinant gene is inserted close to an endogenous promoter or enhancer, or alternatively into a region of inactive or silenced genomic DNA (i.e. heterochromatin region). Heterochromatin regions of DNA arise due to the formation of dense packaging of DNA and associated proteins, and the genes present within these highly condensed regions are usually transcriptionally inactive (Babu & Verma 1987; Wallrath 1998). Thus, the expression of any recombinant genes inserted into or close to such a region can be adversely affected (Henikoff 1990; Wilson *et al.* 1990; Wallrath 1998), and this phenomenon has had a significant influence on the design of vectors for stable expression of recombinant proteins.

Much research has been performed in relation to vector construct design, and the basic components (such as promoter elements, introns, polyadenylation sites, bacterial origins of replication) of a successful production vector for expression in mammalian cells are now well defined (see Chapter 5; Makrides 1999). However, in an attempt to increase production levels, and often as a direct result of the need to combat the problem of instability of production, various additional vector elements have been considered. Many of these elements are used to overcome the problem of the 'position effect' during the insertion of recombinant genes into a host cell's genome.

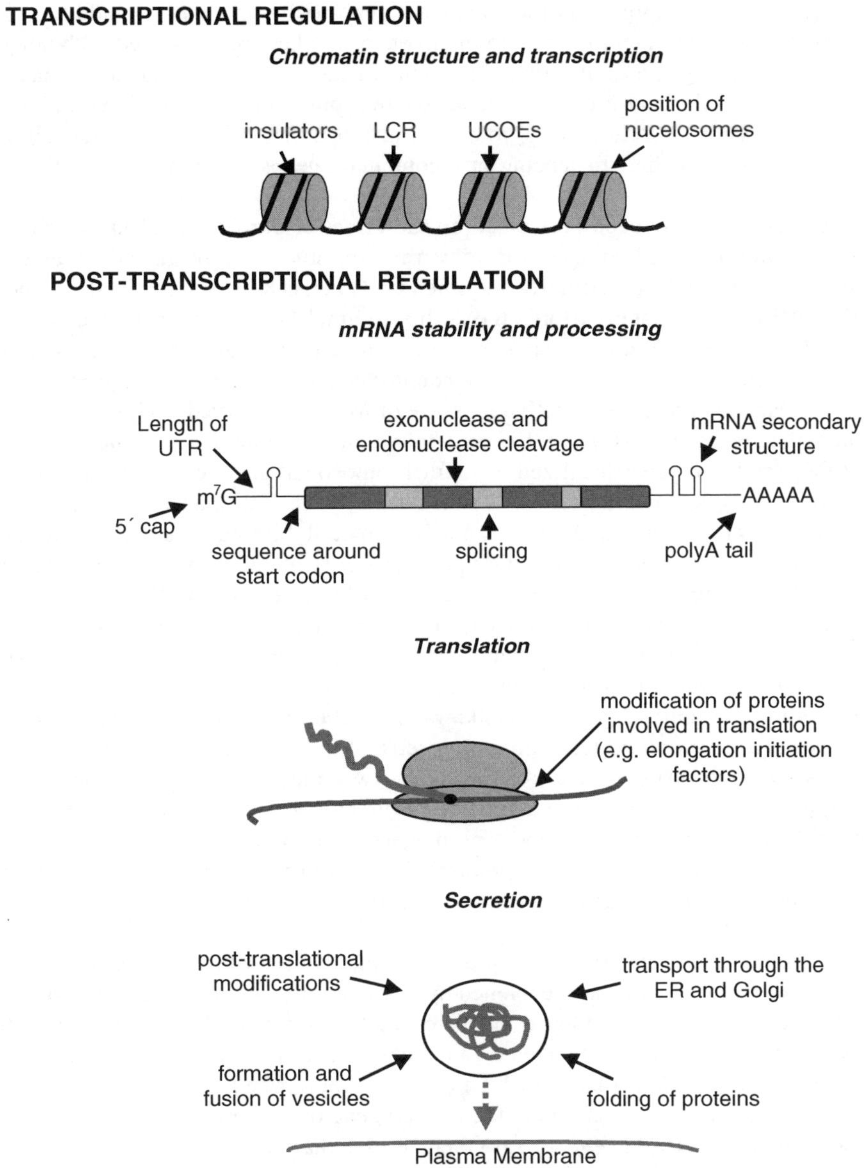

Figure 8.2 Potential sites for regulation of recombinant gene expression in mammalian cells. (Adapted from Barnes *et al.* (2003) with permission of John Wiley & Sons)

Insulator elements are naturally occurring DNA sequences that allow the region between them to act as an independent functional domain not affected by the surrounding genetic environment (Geyer 1997). These elements are often used to flank recombinant genes in order to suppress genomic position effects (Pikaart *et al.* 1998; Kwaks *et al.* 2003). However, to complicate the situation 'anti-insulator' elements have now been found that are thought to allow insulators to be selectively bypassed (Zhou & Levine 1999). Locus control regions (LCRs) are another type of DNA element that can buffer against the position effect (Needham *et al.* 1992; Li *et al.* 1999). LCRs are thought to aid in the opening of chromatin regions and are used to confer position-independent and copy number-dependent expression on recombinant genes (Li *et al.* 1999). In addition, research is also being performed to identify the chromatin elements that are associated with ubiquitously expressed 'housekeeping' genes. It is thought that the inclusion of these elements, termed UCOEs (ubiquitous/universal chromatin-opening elements), in expression vectors will permit integration-independent expression of recombinant proteins. It is hoped that such elements may alleviate the problem of chromatin-mediated silencing of recombinant genes (Benton *et al.* 2002; Antoniou *et al.* 2003).

Many studies have now been performed on the effect of chromosomal location on recombinant gene expression, and some of the most relevant studies from an industrial viewpoint have centred on CHO cells, particularly amplified CHO cells (Kim & Lee 1999). As mentioned before, the DHFR-CHO system affords the ability to amplify recombinant genes by inclusion of MTX in the culture medium. In many cases this can lead to a significant increase in product yield. However, this increase in yield is often counterbalanced by an increase in the occurrence of instability, particularly if the selective pressure of MTX is removed (Pallavicini *et al.* 1990; Kim *et al.* 1998a; Fann *et al.* 2000). It has been found that, without the continued selective pressure of MTX, amplified genes localized on extrachromosomal genetic structures called 'double minutes' are usually lost during mitosis (Wahl *et al.* 1982; Kaufman *et al.* 1983). However, incorporation into chromosomal DNA does not guarantee stability and such genes are often targets for genetic rearrangements, or may disrupt the integrity of endogenous genes upon insertion (Miele *et al.* 1989; Ruiz & Wahl 1990; Kim & Lee 1999). Studies in DHFR-CHO cells during long-term culture in the absence of MTX have indicated that incorporation of amplified recombinant genes close to $(TTAGGG)_n$ sequences (mammalian telomere repeat sequences), which are often found around the telomeric end of amplified arrays in CHO cells, may be associated with the stability of amplified genes (Yoshikawa *et al.* 2000b). A variety of gene amplification mechanisms has been proposed for mammalian cells (Windle & Wahl 1992; Stark 1993). However, there is not necessarily a linear relationship between levels of amplification and levels of expression, as excessive levels of the amplifying drug can be toxic to the cells and, as already discussed, the site of integration is important to expression levels and stability of expression. In amplified CHO cells, gradual stepwise increments in MTX exposure (rather than rapid exposure) have been reported to lead to amplified genes integrating into telomeric regions of chromosomes (Yoshikawa *et al.* 2000b).

The fact that the site of incorporation of a recombinant gene within the host cell's genome is important to its level of expression is exploited by targeted integration. Targeted integration using site-specific recombination is employed by, for example, the Cre/*loxP* and Flp/*FRT* systems. A vector containing a reporter gene flanked by a target sequence (e.g. *loxP* or *FRT*) is inserted randomly into the host cell. Cells that show high levels of reporter gene expression are selected and targeted with a second vector containing the desired gene of interest flanked by the same target sequences. A recombinase (e.g. *Cre*, or *Flp*) then ensures that the reporter gene is removed from the genome and the gene of interest is inserted in its place (Bode *et al.* 2000; Koch *et al.* 2000).

Chromatin and proteins important for the control of transcription can undergo a variety of modifications. In particular two of these modifications are thought to be important for stability of recombinant expression; methylation and acetylation. Various studies of the relationship between these epigenetic mechanisms and stability of production have been made. Methylation typically occurs at cytosine residues that are immediately 5′ to a guanosine residue at, or near to, promoters. It is now widely accepted that methylation of transfected DNA can play a significant role in the regulation of expression. In particular, methylation is known to cause repression of gene expression and it is thought to be a major cause of the maintenance of the silenced heterochromatin regions of DNA. Conversely, hypomethylation of the DNA surrounding gene promoters is often associated with increased transcriptional activity (Razin & Cedar 1991; Paulin *et al.* 1998; Mielke *et al.* 2000). There are examples of methylation leading to decreased expression of transfected genes in mammalian cells, for example instability in BHK cells has been associated with transcriptional shut-off through methylation of regulatory elements of the transfected gene (Mielke *et al.* 2000). Acetylation, the addition of acetyl groups to lysine residues of histones (DNA binding proteins that, together with DNA, form the basic subunit of chromatin, the nucleosome), reduces their affinity for DNA and hence results in unfolding of the chromatin structure. This in turn leads to enhanced accessibility of DNA to proteins involved in transcription (Garcia-Ramirez *et al.* 1995; Grunstein 1997; Hansen *et al.* 1998). The precise mechanism by which acetylation regulates transcription is still elusive (Kouzarides 2000). However, it is now well established that transcriptionally active genes are usually associated with histone acetylation whereas this is not the case for transcriptionally inactive genes and silenced regions of DNA such as heterochromatin (Turner *et al.* 1992; Braunstein *et al.* 1996). In addition, silencers and the transcription factors that bind to them (i.e. repressor proteins) can have a significant role in the regulation of transcription. These proteins can affect, amongst other things, the general assembly of the transcription complex, chromatin structure, intron splicing, cytoplasmic retention of transcription factors and the activity of positive-acting transcription factors (Clark & Docherty 1993; Ogbourne & Antalis 1998).

As well as modifications to the DNA and associated proteins, it is also important to remember that the actual structure of the chromatin can play a role in the regulation of transcription (Wolffe & Guschin 2000). In eukaryotes, DNA is first condensed into units called nucleosomes and then into higher order chromatin structures (Woodcock & Dimitrov 2001). This degree of condensation can hinder the association between the DNA and the transcription complex. Hence, before transcription can be activated chromatin remodelling has to occur (Kingston *et al.* 1996; Felsenfeld 1996; Urnov & Wolffe 2001).

8.4.2 Post-transcriptional Regulation

The structure of mRNA molecules and their precursors can have a profound effect on recombinant protein production levels. The structure can affect the way RNA is translated by affecting its association with translation factors and the ribosome complex. However, the level of mRNA decay is just as important as the rate of synthesis in determining the amount of functional mRNA in a cell, and the structure of mRNA can also affect the rate of degradation by hindering recognition of the RNA by the cellular machinery that defines its half-life. Thus, in considering aspects that might relate to silencing or limiting translation, one has to be aware of the structural elements of mRNA, the proteins that interact with these elements and the way these interacting proteins can be regulated. mRNA stability can be affected by, for example, the 5′ methylguanosine cap structure, the 3′ polyA tail, and internal sequences within the coding region of the transcript itself (Furuichi *et al.* 1977; Bernstein & Ross 1989; Peltz & Jacobson 1992; Ross 1996). Degradation of mRNA can occur by a variety of mechanisms, including shortening of the polyA tail followed by 3′ to 5′ exonuclease activity, or decapping and 5′ to 3′ exonuclease degradation. The shortening of the tail can be promoted by sequence elements such as AU-rich elements (Shaw & Kamen 1986; Chen *et al.* 1994). In addition to exonuclease degradation, endonuclease cleavage at sites within a transcript can also lead to degradation of mRNA (Binder *et al.* 1994).

The binding of certain proteins to mRNA species can also have an effect on stability. For example, AU-rich elements, that are important in the shortening of polyA tails during mRNA degradation, can be regulated by AU-rich element-binding proteins (Burd & Dreyfuss 1994; Chen & Shyu 1995). The polyA tail can be protected from rapid destruction by interaction with polyA-binding proteins (Bernstein *et al.* 1989). Endonuclease cleavage can be regulated by RNA-binding proteins, which, when bound to the cleavage sites, make them inaccessible for endonuclease cleavage (Binder *et al.* 1994). In addition, regulatory proteins can bind to RNA secondary structures and stabilize them (Day & Tuite 1998).

As well as regulation at the level of RNA processing and translation, for the vast majority of recombinant therapeutic proteins there is also the possibility of regulation during the process of secretion. For example, problems with post-translational modifications or protein folding may lead to retention and subsequent degradation of proteins within the endoplasmic reticulum (Cudna & Dickson 2003).

8.5 SUMMARY

There is still a long way to go in deciphering the molecular factors that account for instability of protein production in animal cells. The endogenous processes that control the production of a protein are very complex. The mechanistic events that determine the level and stability of production are complicated further by the problems associated with recombinant gene incorporation into the genome of a host cell. However, the future for stable protein production from animal cells is promising. It is now clear that a multiparametric approach, including monitoring of productivity, mRNA expression and copy number analysis, must be applied when considering procedures for efficient clone selection, and in fact it may be possible to use such parameters as predictive indicators of stability. Early testing of the functionality of products *in vitro* is also critically important. Studies performed on the site of incorporation of recombinant genes into the genome of cells have further increased our knowledge of preferential sites for incorporation. In addition, discoveries such as insulator elements and UCOEs, coupled with the technologies for targeted integration of recombinant genes, have greatly increased the potential for obtaining stable production of recombinant proteins from mammalian cells. These developments are worth pursuing in order to provide therapeutic proteins that more closely mimic their native counterparts. Such technology and expertise may also prove invaluable in cell therapy in the future.

ACKNOWLEDGEMENT

I would like to thank Prof. Alan Dickson for all his help with the proof reading of this manuscript.

REFERENCES

Antoniou M, Harland L, Mustoe T *et al.* (2003) *Genomics*; 82: 269–279.
Babu A, Verma RS (1987) *Int. Rev. Cytology*; 108: 1–60.
Bae SW, Hong HJ, Lee GM (1995) *Biotechnol. Bioeng.*; 47: 243–251.
Barnes LM, Bentley CM, Dickson AJ (2000) *Cytotechnology*; 32: 109–123.
Barnes LM, Bentley CM, Dickson AJ (2001) *Biotechnol. Bioeng.*; 73: 261–270.
Barnes LM, Bentley CM, Dickson AJ (2003) *Biotechnol. Bioeng.*; 81: 631–639.
Bebbington CR, Renner G, Thomson S, King D, Abrams D, Yarranton GT (1992) *Bio/Technology*; 10: 169–175.
Benton T, Chen T, McEntee M *et al.* (2002) *Cytotechnology*; 38: 43–46.
Bernstein P, Ross J (1989) *Trends Biochem. Sci.*; 14: 373–377.
Bernstein P, Peltz SW, Ross J (1989) *Mol. Cell. Biol.*; 9: 659–670.
Binder R, Horowitz JA, Basilion JP, Koeller DM, Klausner RD, Hartford JB (1994) *EMBO J.*; 13: 1969–1980.
Bode J, Schlake T, Iber M *et al.* (2000) *Biol. Chem.*; 381: 801–813.
Borth N, Strutzenberger K, Kunert R, Steinfellner W, Katinger H (1999) *J. Biotechnol.*; 67: 57–66.
Braunstein M, Sobel RE, Allis CD, Turner BM, Broach JR (1996) *Mol. Cell. Biol.*; 16: 4349–4356.
Brown ME, Renner G, Field RP, Hassell T (1992) *Cytotechnology*; 9: 231–236.
Burd CG, Dreyfuss G (1994) *Science*; 265: 615–621.

Cedervall B, Radivoyevitch T (1996) *Electrophoresis*; 17: 1080–1086.
Chen C-YA, Shyu A-B (1995) *Trends Biochem. Sci.*; 20: 465–470.
Chen C-YA, Chen T-M, Shyu A-B (1994) *Mol. Cell. Biol.*; 14: 416–426.
Chuck AS, Palsson BO (1992) *Biotechnol. Bioeng.*; 39: 354–360.
Clark AR, Docherty K (1993) *Biochem. J.*; 296: 521–541.
Clarke J, Porter AJ, Thorpe R, Davis JM (2002) In *Basic Cell Culture: A Practical Approach*. Ed Davis JM. Oxford University Press; Second edition: 275–294.
Coco-Martin JM, Oberink JW, Brunink F, van der Velden-de Groot TAM, Beuvery EC (1992) *Hybridoma*; 11: 653–665.
Coller HA, Coller BS (1986) *Methods Enzymol.*; 121: 412–417.
Cossons NH, Hayter PM, Tuite MF, Jenkins N (1991) In *Production of Biologicals from Animal Cells in Culture*, Eds Spier RE, Meigner B. Butterworth–Heinemann; 309–314.
Couture ML, Heath CA (1995) *Biotechnol. Bioeng.*; 47: 270–275.
Cudna RE, Dickson AJ (2003) *Biotechnol. Bioeng.*; 81: 56–65.
Davis JM (1986) In *Methods in Enzymology*. Eds Langone JJ, Van Vunakis H. Academic Press; Vol 121, 307–322.
Day DA, Tuite MF (1998) *J. Endocrinol.*; 157: 361–371.
Ekong R, Wolfe J (1998) *Curr. Opin. Biotechnol.*; 9: 19–24.
Fann CH, Guirgis F, Chen G, Lao MS, Piret JM (2000) *Biotechnol. Bioeng.*; 69: 204–214.
Felsenfeld G (1996) *Cell*; 86: 13–19.
Fogolin MB, Eberhardt MO, Kratje R, and Etcheverrigaray M (2002) *Elec. J. Biotechnol.*; 5: 243–250.
Frame KK, Hu W-S (1990) *Biotechnol. Bioeng.*; 35: 469–476.
FuruichiY, LaFiandra A, Shatkin AJ (1977) *Nature*; 266: 235–239.
Gaertner JG, Dhurjati P (1993) *Biotechnol. Prog.*; 9: 298–308.
Garcia-Ramirez M, Rocchini C, Ausio J (1995) *J. Biol. Chem.*; 270: 17923–17928.
Galfré G, Milstein C (1981) In *Methods in Enzymology*. Eds Langone JJ, Vunakis HV. Academic Press; Vol. 73, 3–46.
Geyer PK (1997) *Curr. Opin. Genet. Dev.*; 7: 242–248.
Gift EA, Weaver JC (2000) *Cytometry*; 39: 243–249.
Grunstein M (1997) *Nature*; 389: 349–352.
Hammill L, Welles J, Carson GR (2000) *Cytotechnology*; 34: 27–37.
Hansen JC, Tse C, Wolffe AP (1998) *Biochem.*; 37: 17637–17641.
Heath C, Dilwith R, Belfort G (1990) *J. Biotechnol.*; 15: 71–90.
Henikoff S (1990) *Trends Genet.*; 6: 422–426.
Johnston MD (1980) *J. Gen. Virol.*; 50: 191–194.
Kaufman RJ, Brown PC, Schimke RT (1981) *Mol. Cell. Biol.*; 1: 1084–1093.
Kaufman RJ, Sharp PA, Latt SA (1983) *Mol. Cell. Biol.*; 3: 699–711.
Kearns B, Lindsay D, Manahan M, McDowall J, Rendeiro D (2003) *Bioproc. J.*; 2: 52–57.
Kim SJ, Lee GM (1999) *Biotechnol. Bioeng.*; 64: 741–749.
Kim JH, Bae SW, Hong HJ, Lee GM (1996) *Biotechnol. Bioeng.*; 51: 479–487.
Kim NS, Kim SJ, Lee GM (1998a) *Biotechnol. Bioeng.*; 60: 679–688.
Kim, SJ, Kim NS, Ryu CJ, Hong HJ, Lee GM (1998b) *Biotechnol. Bioeng.*; 58: 73–84.
Kingston RE, Bunker CA, Imbalzano AN (1996) *Genes Dev.*; 10: 905–920.
Koch KS, Aoki T, Wang Y *et al.* (2000) *Gene*; 249: 135–144.
Kouzarides T (2000) *EMBO J.*; 19: 1176–1179.
Kromenaker SJ, Srienc F (1994) *Biotechnol. Prog.*; 10: 299–307.
Kwaks THJ, Barnett P, Hemrika W *et al.* (2003) *Nat. Biotechnol.*; 21: 553–558.
Lee GM, Varma A, Palsson BO (1991) *Biotechnol. Prog.*; 7: 72–75.
Lee MS, Lee GM (2000). *Biotechnol. Bioeng.*; 68: 260–268.
Li Q, Harju S, Peterson KR (1999) *Trends Genet.*; 15: 403–408.
Lietzke R, Unsicker K (1985) *J. Immunol. Methods*; 76: 223–228.
Makrides SC (1999) *Protein Express. Purif.*; 17: 183–202.
Merritt SE, Palsson BO (1993) *Biotechnol. Bioeng.*; 42: 247–250.
Miele M, Bonatti S, Menichini P, Ottaggio L, Abbondandolo A (1989) *Mutat. Res.*; 219: 171–178.

Mielke C, Tümmler M, Schübeler D, von Hoegen I, Hauser H (2000) *Gene*; 254: 1–8.
Milstein C (1980) *Sci. Am.*; 243: 56–64.
Needham M, Gooding C, Hudson K, Antoniou M, Grosveld F, Hollis M (1992) *Nucleic Acids Res.*; 20: 997–1003.
Ogbourne S, Antalis TM (1998) *Biochem. J.*; 331: 1–14.
Oxford Pocket Dictionary (1978), Oxford University Press; 148.
Ozturk SS, Palsson BO (1990) *Hybridoma*; 9: 167–175.
Pallavicini MG, DeTeresa PS, Rosette C, Gray JW, Wurm FM (1990) *Mol. Cell. Biol.*; 10: 401–404.
Paulin RP, Ho T, Balzer HJ, Holliday R (1998) *Genet.*; 149: 1081–1088.
Peltz SW, Jacobson A (1992) *Curr. Opin. Cell Biol.*; 4: 979–983.
Pikaart MJ, Recillas-Targa F, Felsenfeld G (1998) *Genes Dev.*; 12: 2852–2862.
Puck TT, Cieciura SJ, Robinson A (1958) *J. Exp. Med.*; 108: 945–959.
Raeymaekers L (2000) *Mol. Biotechnol.*; 15: 115–122.
Razin A, Cedar H (1991) *Microbiol. Rev.*; 55: 451–458.
Ross J (1996) *Trends Genet.*; 12: 171–175.
Ruiz JC, Wahl GM (1990) *Mol. Cell. Biol.*; 10: 3056–3066.
Shaw G, Kamen R (1986) *Cell*; 46: 659–667.
Stacey GN (1995) PhD thesis. Open University.
Stacey GN, Bolton BJ, Morgan D, Clark SA, Doyle A (1992) *Cytotechnology*; 8: 13–20.
Stark GR (1993) *Adv. Cancer Res.*; 61: 87–113.
Turner BM, Birley AJ, Lavender J (1992) *Cell*; 69: 375–384.
Underwood PA, Bean PA (1988). *J. Immunol. Methods*; 107: 119–128.
Urnov FD, Wolffe AP (2001) *Oncogene*; 20: 2991–3006.
Wahl GM, Vitto L, Padgett RA, Stark GR (1982) *Mol. Cell. Biol.*; 2: 308–319.
Wallrath LL (1998) *Curr. Opin. Genet. Dev.*; 8: 147–153.
Wewetzer K, Seilheimer B (1995) *J. Immunol. Methods*; 179: 71–76.
Wilson C, Bellen HJ, Gehring WJ (1990) *Annu. Rev. Cell Biol.*; 6: 679–714.
Windle BE, Wahl GM (1992) *Mutat. Res.*; 276: 199–224.
Wolffe AP, Guschin D (2000) *J. Struct. Biol.*; 129: 102–122.
Woodcock CL, Dimitrov S (2001) *Curr. Opin. Gene. Dev.*; 11: 130–135.
Yoshikawa T, Nakanishi F, Itami S *et al.* (2000a) *Cytotechnology*; 33: 37–46.
Yoshikawa T, Nakanishi F, Ogura Y *et al.* (2000b) *Biotechnol. Prog.*; 16: 710–715.
Zhou J, Levine M (1999) *Cell*; 99: 567–575.

Relevant Web Sites

GS gene expression system	http://www.lonzabiologics.com/group/en/products_services/custommanufacturing/mammalian/geneexpressions/how_gs_works.html
UCOE elements	http://www.cobrabio.com/expression.htm

9 Gene Transfer Vectors for Clinical Applications

A Meager

9.1 INTRODUCTION

Gene therapy, a novel treatment modality for a variety of diseases, the feasibility of which was brought about through technological advances in nucleic acid (NA) manipulation and in methods of NA delivery into cells, was initiated in the final decade of the twentieth century. The possibility of preparing genes with regulatory sequences *in vitro* and incorporating these into DNA plasmids or other 'gene transfer products', commonly referred to as 'vectors', is now routine. Vectors are used for transferring genetic material, either DNA or RNA, into target cells where the incorporated gene (the 'transgene') is expressed to provide the required function for an intended medicinal purpose. For example, for treatment of inherited monogenic disorders such as haemophilia and cystic fibrosis, the symptoms of which are caused by non-functionality of a particular protein due to somatic mutations in the encoding nuclear gene, the introduction of the normal gene as its protein-encoding NA sequence, together with appropriate regulatory control sequences for protein expression, should provide the cell in some measure with the function it lacks. For treatment of cancer, introduction into tumour cells of a gene encoding an enzyme that catalyses conversion of a chemical into a toxic metabolite should provide the means for their killing. In another approach, the introduction of genes encoding biologically active proteins such as cytokines should provide activating signals to stimulate the immune and/or other physiological systems to eliminate infectious, inflammatory or malignant diseases.

It is evident that gene therapy has the potential to treat a range of human diseases. However, that promise has been undermined by several technical and physiological obstacles concerning targeting and efficiency of gene transfer, appropriate regulation of transgene expression, and adverse events caused by undesirable vector-cell interactions and/or immunological responses of the host to vector and/or transgene-expressing cells and/or protein product. To overcome these obstacles and safety concerns and build towards successful treatments for specific diseases, a new scientific area of vector technology has been established. Since gene transfer is to a large degree restricted by the cell membrane, 'vectorologists' have focused on biomaterials and viruses for constructing vectors that efficiently cross this membrane. In this respect, viruses, which have evolved to efficiently package and then transfer their genomes into host cells, have provided the building blocks for the most efficacious vectors currently in use for gene therapy. This chapter focuses on viral vectors developed as biological medicines from tissue culture cells.

The most commonly used viruses as bases for vector development are retroviruses, adenoviruses, small DNA viruses, herpes viruses and poxviruses. As wild-type viruses, these are often human pathogens, but manipulation of their genomes can render them replication deficient and

Medicines from Animal Cell Culture Edited by G. Stacey and J. Davis

thus unable to cause disease. The development and manufacture of replication-deficient viral vectors as gene transfer medicinal products has, however, proved challenging since their production is based on mammalian cell lines, often modified from existing ones by transfection of complementing or 'helper' viral genes to enable vector propagation to take place. Most vectors will have undergone an R&D phase followed by technology transfer to manufacturing facilities and production at large scale. As for all biological medicinal products, compliance with good manufacturing practice (GMP) conditions is mandatory for the production of viral vectors. Regulatory guidance stresses that viral vector products should be well characterized with respect to inherent physical and biological properties, should exhibit consistent composition and long-term physicochemical and biological stability, and should not contain harmful impurities or contaminants, especially adventitious viruses and microorganisms. Nevertheless, viral vectors are complex biological products whose properties are greater than the sum of their individual parts and, as a consequence, quality and safety are relatively difficult to assess. Rigorous preclinical testing by appropriate validated techniques and methodologies to ensure, as far as possible, that such products are fit for their intended purpose is therefore paramount. A comprehensive risk assessment of their use is crucial, while recognizing, however, that certain aspects of their efficacy and safety profile may only become apparent following administration to patients.

Most gene therapy protocols require vectors to exhibit a spectrum of desirable properties targeted to the disease being treated, e.g. highly effective transfection or transduction, appropriate transgene expression level, without manifesting undesirable ones, e.g. high toxicity or immunogenicity, oncogenic activity. Desirable properties are to some extent inherent in vector types, while undesirable properties may be diminished by appropriate procedures. Viruses, which in the natural state are very variable in structure, tropism and pathogenicity, offer a broad 'palette' from which to create viral vectors suited for different medicinal purposes. In the following sections, three distinct categories of viral vectors, in current use for gene therapeutic applications, are covered with emphasis on their design, manufacture and safety in relation to intended purpose.

9.2 ADENO-ASSOCIATED VIRAL VECTORS

One of the most effective ways to deliver genes into cells is to enclose plasmids within viral capsids that efficiently bind to target cells. Certain non-enveloped viruses with relatively simple genomes offer the means for manufacture of 'capsid plus plasmid' viral vectors. Among such viruses, parvoviruses have been those most frequently exploited for this purpose. Adeno-associated viruses (AAV) are small single-stranded DNA parvoviruses of which there are six known human serotypes. AAV-2, the most widely studied serotype, is non-pathogenic with a favourable safety profile (Flotte & Carter 1999). AAV-2 contains a DNA genome of 4.67 kb with only two open reading frames (ORFs) sandwiched between two inverted terminal repeat (ITR) sequences (Srivastasa *et al.* 1983). Viral *cap* encodes capsid proteins, VP-1, VP-2 and VP-3, and *rep*, encodes 'replicase' (Rep) proteins for viral DNA replication (Samulski *et al.* 1989). For wild type AAV-2, there is a highly preferred, Rep-dependent, integration site on human chromosome 19 [19q13.3-qter, known as AAVS1], although such integration has so far only been observed in cultured cells *in vitro* (Surosky *et al.* 1997). AAV-2 is replication deficient, requiring extra functions provided by adenoviruses (or herpes viruses) for virion production (Flotte & Carter 1999). Nevertheless, it can be used to construct vectors by replacement of *rep* and *cap* sequences with a transgene and its promoter, although size of the 'insert' is limited to 4.5 kb (Table 9.1) (Flotte & Carter 1999; Snyder 1999; Russell & Kay 1999; Hitt & Gauldie 2000; Tal 2000; Owens 2002). The viral *rep* and *cap* are incorporated in a separate plasmid to supply Rep and Cap proteins *in trans* (Samulski *et al.* 1989). Originally for AAV-2 vector propagation, a permissive cell line, e.g. HEK 293 or HeLa, was simultaneously transfected with 'transgene' and '*rep/cap*' plasmids and then infected with a wild type adenovirus.

Table 9.1 Comparison of viral vector characteristics.

	Vector					
Characteristic	AAV	OV	LV	AdV	Gutless AdV	CRAd
Capacity	4.5 kb	8 kb	18 kb	8–12 kb	32 kb	(8–12 kb)
Purity/stability	hi/hi	lo/lo	lo/lo	hi/hi	hi/hi	hi/hi
Transduction:						
Dividing cells	+	+	+	+	+	+
Non-dividing cells	+	–	+	+	+	+
Integration	–	+	+	–	–	–
Efficacy	hi	lo	lo	hi	hi	hi
Transgene Expression	m/l	m/l	m/l	s/m	s/m	(s/m)
Immunogenicity	+	–	–	+	–	+
Toxicity	–	–	–	+	+	+
In vivo/Ex vivo	+/+	–/+	+/+	+/–	+/–	+/–

hi, high; lo, low; s, short; m, medium; l, long

However, compared with normal AAV-2 production, this propagation method is inefficient, yielding only relatively low titres of vector (Flotte & Carter 1999; Snyder 1999; Russell & Kay 1999; Hitt & Gauldie 2000; Tal 2000; Owens 2002). Development of stable packaging cell lines containing 'transgene' and '*rep/cap*' constructs is thus seen as necessary to improve vector yields and scale-up (Snyder 1999; Xiao & Samulski 1998; Owens 2002). Although Rep proteins are toxic to tissue culture cells (Yang *et al.* 1994) and an inducible expression system is therefore required (Inoue & Russell 1998), some stable cell lines are currently available for AAV-2 vector manufacture in high yield; however helper adenovirus and low amounts of 'pseudo' wild type AAV-2 generated by recombination between 'transgene' and '*rep/cap*' plasmids are also produced (Wang *et al.* 1998; Salvetti *et al.* 1998). To avoid the co-production of infectious adenovirus, 'helper' adenovirus genes may be supplied in another plasmid (Xiao & Samulski 1998; Owens 2002; Ferrari *et al.* 1996) but, similar to the *rep* gene expression, their expression can also be toxic to vector-producer cells (Snyder 1999). In an alternative approach, a stably '*rep/cap*' transfected HeLa cell line (B50) was firstly infected with an adenovirus defective in E2b (deletion of DNA polymerase and pre-terminal protein) to induce *rep* and *cap* and secondly by a hybrid virus where the AAV vector is cloned in the E1 region of a replication-deficient adenovirus (Gao *et al.* 1998). This system has resulted in high titres of AAV vector free of wild type AAV and adenovirus.

For vector harvest and purification, cells are ruptured by freeze-thawing or other physical means to release the vector particles and these are purified by CsCl density centrifugation (normally not an option for large scale purifications) or chromatographic procedures. If wild-type helper adenovirus is used, the AAV-2 vector harvest also contains adenovirus (see above), which requires removal by physical means, e.g. heat treatment, and extensive purification. Other likely impurities are 'wild type' AAV-2, adenoviral proteins and DNA, residual cellular and plasmid DNAs, and cellular proteins. Advantageously, the small – 18 to 26 nm – AAV-2 particles are very robust, being stable up to 56 °C and resistant to low pH and detergent, solvent, protease or nuclease attack (Arella *et al.* 1990). They are therefore well suited to purification by chromatographic procedures, which can be operated under GMP principles (Snyder 1999; Hauswirth 2000). By exploiting binding of AAV-2 to heparin – cell surface receptors for AAV-2 are heparan sulphate proteoglycans (Summerford & Samulski 1998) – AAV-2 particles can be separated on heparin affinity columns (Summerford & Samulski. 1999). Other chromatographic procedures have also been successfully applied to AAV-2 vector purification (O'Riordan *et al.* 2000).

There are encouraging reports that AAV vectors can be used for *in vivo* gene delivery with efficiencies up to 1 % of infected cells, a higher rate than can be achieved with other vectors (Russell & Hirata 1998; Inoue *et al.* 2001); both dividing and non-dividing cells are transduced (Table 9.1) (Miao *et al.* 2000). AAV-2 vector DNA forms concatemers that persist extrachromosomally for a very long time; site-specific integration cannot occur since the *rep* gene is absent from the vector DNA (Owens 2002; Ponnazhagan *et al.* 1997). However, some vector DNA integration into chromosomal DNA may occur non-specifically with a low risk of insertional mutagenesis (Kearns *et al.* 1996; Ponnazhagan *et al.* 2001; Hiller et al. 2005; McCarty et al. 2004). Some long-term studies in humans and animals of AAV-2 vector administration have indicated that there are no adverse effects, including tumorigenicity and germline transmission (Flotte & Carter 1999; Pearson 2001). Additionally, AAV-2 vectors have low immunogenic and pro-inflammatory activity and, since they lack of any viral coding sequences, transduced cells do not express viral antigens and therefore are not attacked by the immune system (see adenoviral vectors for comparison). However, most humans have immunity to AAV-2 and so AAV-2 vector administration will boost immune responses, making repeat dosing more difficult (Table 9.1) (Chirmule *et al.* 1999). Another disadvantage is the widespread occurrence of heparan sulphate proteoglycans, which act as receptors for AAV-2 and thus make targeting of AAV-2 vectors difficult (Summerford & Samulski 1998). On the other hand, some target cells express low to negligible levels of receptors, e.g. apical surface of lung epithelial cells, so that relatively high doses of vector could be required. Targeting has proved difficult since disruption of the viral capsid to insert 'targeting motifs' usually destroys infectivity, but some progress has recently been reported (Shi *et al.* 2001). Not all AAV serotypes bind heparan sulphate, e.g. AAV-5 (Auricchio *et al.* 2001) and these may offer a means for targeting specific cell types (Wu et al. 2006).

AAV-2 vectors potentially provide very long-term transgene expression with low integration rates and thus a corresponding low risk of insertional mutagenesis. Therefore, they are suitable for gene therapy of monogenic disorders such as cystic fibrosis (Flotte *et al.* 1993) and haemophilia B (Kay *et al.* 2000), and potentially for chronic medical conditions such as atherosclerosis and Parkinson's disease. To date there have been only around 20 clinical trials with AAV-2 vectors, but with the development of vectors based on other AAV serotypes (Wu et al. 2006) and better manufacturing and purification procedures many more trials are sure to follow.

9.2.1 Retroviruses

Although AAV-2 vectors can give long-term transgene expression, it will not be life-long and repeat treatment may not be possible on account of previous immunity (Table 9.1). Moreover, the transgene size that can be accommodated is limited. To overcome these limitations requires larger viral vectors with low immunogenicity and the capability to integrate transgenes for long-term expression. Such characteristics are to be found in retroviruses (Table 9.1).

Retroviruses are usually classified into three sub-families:

(1) *oncoviruses* [also now known as *gamma-retroviruses*, e.g. murine leukaemia viruses (MuLV), mouse mammary tumour virus, human T-lymphotropic viruses];

(2) *lentiviruses* [e.g. human immunodeficiency viruses (HIV), visna retrovirus of sheep, equine infectious anaemia virus (EIAV)];

(3) *spumaviruses* (e.g. simian foamy virus) (Luciw & Leung 1992; Coffin 1996).

After cell entry, oncoviruses and lentiviruses transcribe their RNA genomes by reverse transcription into a DNA copy, called proviral DNA (Luciw & Leung 1992; Coffin 1996; Günzburg & Salmons 1999; Hodgson 2001), which can then be integrated into chromosomal DNA of dividing cells (oncoviruses; Miller *et al.* 1990; Roe *et al.* 1993) or both dividing and non-dividing cells (lentiviruses; Lever 1999; Buchschacher & Wong-Staal 2000; Vigna & Naldini 2000). In contrast, foamy virus replication reverse transcription occurs during virion formation, rather than after cell entry, so infectious particles contain DNA genomes (Moebes *et al.* 1997; Trobridge & Russell 1998).

In the past decade, the majority of retroviral vectors used in gene therapy protocols have been based on simple type C oncoviruses (Yee 2000), in particular murine leukaemia virus (MuLV) (Hitt & Gauldie 2000; Günzburg & Salmons 1999; Danos & Mulligan 1988), but recently there has been much R&D directed towards the development of vectors based on lentiviruses, which have the advantage of transducing non-dividing cells (Lever 1999; Buchschacher & Wong-Staal 2000; Vigna & Naldini 2000). Development of vectors based on spumaviruses has lagged well behind that of oncoviral and lentiviral vectors, although there are now reports of replication-defective vectors derived from human foamy virus (Trobridge & Russell 1998; Vassilopoulis *et al.* 2001).

9.2.2 Oncoviral Vectors

In oncoviral (e.g. MuLV) vectors, all viral genes are removed from the viral genome between the long terminal repeat (LTR) sequences and replaced by a transgene and its regulatory sequences (insert) of up to 6–8 kb in size (Naviaux & Verma 1992). Oncoviral vectors (OV) based on MuLV are produced as follows:

(1) a murine fibroblast cell line is first transfected with a 'helper-retrovirus' genome lacking the packaging signal psi (ψ), which is necessary to pack retroviral RNA into viral envelopes, but containing viral genes required for replication, namely *gag-*, *pol-*, and *env*-genes expressed *in trans* from two separate LTRs integrated in chromosomal DNA (Markowitz *et al.* 1988);

(2) the transfected 'packaging cell' is cloned and grown as a cell line, which produces only empty retroviral particles;

(3) the packaging cells are transfected by a second retroviral (vector) genome, which contains ψ and a primer binding site (PBS) for reverse transcription, but lacks *gag-*, *pol-*, and *env*-genes, which are replaced by the transgene (Günzburg & Salmons 1999). Several variations of this transgene-containing retroviral construct, in which transgene expression is controlled by retroviral LTR or internal promoters, have been developed (Walther & Stein 2000);

(4) following integration, vector genomes are packaged into retroviral envelopes encoded by the integrated 'helper' sequences to yield, after budding from the cell membrane, mature OV particles (Günzburg & Salmons 1999; Yee 2000; Walther & Stein 2000). Although OV particles can infect other (target) cells, make vector proviral DNA and integrate it in chromosomal DNA, no new progeny OV can be made since the target cells lack viral helper sequences (Günzburg & Salmons 1999). The number of vector proviral DNA copies integrated per target cell is dependent on transduction conditions, but may be as high as 30.

A major concern for the safety of OV is that replication-competent retroviruses (RCR) may be produced by recombination events with helper viral sequences either in packaging cells or in target cells (Miller & Buttimore 1986; Markowitz *et al.* 1988; Muenchau *et al.* 1990). Repeated infection by RCR is associated with multiple, apparently random, proviral DNA integrations. However, recent studies (Li *et al.* 2002; Okoh *et al.* 2002) have suggested that the variable structure of chromatin may lead to preferred integration sites, with an increasing risk of insertional mutagenesis. If integrations are upstream of proto-oncogenes, tumour suppressor genes or other growth regulatory genes, they may disrupt gene regulation leading to neoplastic cell transformation and eventually tumours (Boris-Lawrie & Temin 1994). Such integrations may also predispose a cell to other mutagenic stimuli, with the same consequences. An *in vitro* study of retroviral integration in the growth factor-*dependent* human haematopoietic precursor cell line TF-1 showed that growth factor-*independent* mutant cells were formed at a frequency of 2×10^{-7} mutants per integration (Stocking *et al.* 1993). More concerning, murine bone marrow cells transduced with an OV containing a transgene encoding a truncated form of human low affinity nerve growth factor receptor (LNGFR) led to leukaemia development in recipient mice transplanted with transduced cells (Li *et al.* 2002). Other instances of proviral insertions targeting oncogenes or growth factor receptor genes have recently been reported (Meyer *et al.* 2002; Wotton *et al.* 2002). In a single study, replication-competent MuLV was associated with tumour (lymphoma) induction in simians (Donahue *et al.* 1992), suggesting potential pathogenicity in humans. However, until recently, there have been no cases of tumour induction associated with the administration of MuLV-vectors in non-human primates (Cornetta *et al.* 1990) or man (Long *et al.* 1998). Nevertheless, it is a mandatory regulatory requirement to eliminate the generation of replication-competent MuLV during MuLV-based vector propagation since they cannot be eliminated by purification procedures applied to the vector harvest.

The means for eliminating RCR formation during OV propagation lies mainly in the design of OV and its packaging cell line, e.g. by removing any complementary homologous sequences existing between the transgene-containing OV genome and 'helper' sequences. Design improvements have been made by reducing the region of overlap between the *gag-pol* and *env* constructs, by introducing 'codon wobbling' in the construct sequences, which reduces recombination frequency while maintaining the primary protein sequence, by driving *gag-pol* and *env* constructs off independent non-retroviral promoters (Günzburg & Salmons 1999; Morgenstern & Land 1990), and by modification the PBS of the OV genome so that it binds only an 'artificial' tRNA produced in the packaging cell (Günzburg & Salmons 1999); any RCR generated with the modified PBS would be unable to replicate in target cells lacking the 'artificial' tRNA. Alternatively, the OV genome can be modified such that ψ is flanked by direct repeats, which leads to its removal, e.g. using LoxP sites excised by Cre recombinase (Trinh & Morrison 2000). Any RCR formed lack ψ and will therefore be unable to produce infective progeny particles in infected cells (Julias *et al.* 1995; Cholika *et al.* 1996; Russ *et al.* 1996). To prevent vector DNA mobilization, either by co-infection with wild-type RCR or by recombination with endogenous retroviral NA sequences (Purcell *et al.* 1996; Perron & Seigneurin 1999), the OV genome can be modified such that it lacks a complete U3 region in the 3′LTR, which contains the promoter/enhancer, and therefore is self-inactivating (SIN) (Yu *et al.* 1986; Olson *et al.* 1994). Integrated SIN vector proviral DNA, lacking a functional retroviral promoter/enhancer cannot be mobilized by superinfecting wild-type retroviruses, nor can it activate cellular genes through common proviral insertion mechanisms.

One problem in using murine packaging cell lines is that they contain endogenous retrovirus-like particles and sequences, including a high copy number of MuLV-like proviral fragments and about 50 copies of VL30 (virus-like 30) elements (Keshet *et al.* 1980; French & Norton 1997). VL30 are a family of retrotransposons that include sequences that are partially homologous to MuLV *gag* and *pol* genes and terminal regions strongly homologous to MuLV LTRs. Recombination between VL30 and MuLV sequences has been reported to occur (Itin & Keshet 1983). Furthermore, as well as the OV genome, VL30 elements can be efficiently packaged by MuLV Gag proteins into OV particles, and can therefore be inadvertently delivered to target human cells (Scadden *et al.* 1990; Patience *et al.* 1998). Endogenous retroviral sequences may also recombine with OV genome sequences to generate RCR (Chong *et al.* 1994). Therefore, a switch to non-murine, e.g. canine, cells not known to express retrovirus-like particles and sequences, has been suggested as a way of further removing the risk of RCR formation (Smiley *et al.* 1997). Since there is no strong evidence that the human genome contains any integrated proviral fragments with significant homology to the MuLV genome, another possibility is to use modified human cell lines to package OV genomes (Patience *et al.* 1998; Finer *et al.* 1994). An added advantage associated with the use of human packaging cell lines is that, unlike murine cells, they will not post-translationally modify the retroviral or pseudotyping Env with (α 1-3) galactose sugar moiety to which naturally occurring human antibodies bind (Takeuchi *et al.* 1996; Rigg *et al.* 1996). Therefore, OV so produced should not be subject to complement inactivation in human serum (Takeuchi *et al.* 1996; Cosset *et al.* 1995).

Generally, production of high titres of infectious OV has been difficult to achieve; without technical improvements, titres fall between 10^5 to 10^6 focus forming units (ffu)/ml in crude, unpurified supernatants from vector-producing cells. Some improvements in titre have been achieved through alterations to OV design, but sometimes approaches to improve safety resulted in a deleterious effect on titre, e.g. the construction of SIN vectors (Yu *et al.* 1986) (see above). OV particles (and oncoviruses in general) are often fragile and are inactivated at 37 °C (Günzburg & Salmons 1999). One strategy to increase their robustness is to replace or 'pseudotype' the MuLV Env with an Env glycoprotein from another retrovirus, e.g. gibbon ape leukaemia virus (GALV) (Miller *et al.* 1991; Bayle *et al.* 1993) or another family of viruses altogether, e.g. the rhabdovirus vesicular stomatitis virus (VSV), which has a broad range of domestic animals, cattle, horses, swine, as its natural hosts (Emi *et al.* 1991). Additionally, OV so produced have a broad host cell range and infect human cells more efficiently (Friedmann & Yee 1995). For example, pseudotyping with GALV Env has been shown to facilitate infection of CD34+ haematopoietic stem cells (Von Kalle *et al.* 1994; Kiem *et al.*1997; Liu *et al.* 2000). The VSV-Glycoprotein (VSV-G) gene can also be used to produce high-titre pseudotyped, robust, OV (Friedmann & Yee 1995; Von Kalle *et al.* 1994; Kiem *et al.*1997; Liu *et al.* 2000; Burns *et al.* 1993), but production usually requires more complex cellular systems; although VSV-G can be expressed in human cells, e.g. HEK 293, it is toxic and so is usually placed under a tetracycline-inducible system (Chen *et al.* 1996; Ory *et al.* 1996).

Since OV are, due to low titre and potential complement inactivation, poorly effective *in vivo*, their main application has been transduction of human haematopoietic cells *in vitro*. In particular, there has been emphasis on transduction of autologous haematopoietic stem cells (HSC) from bone marrow (BM), peripheral or cord blood for re-engraftment purposes. However, OV can only transduce dividing cells (Table 9.1) and therefore HSC are stimulated to divide by addition of a cocktail of cytokines, including stem cell factor (SCF), interleukin-3 (IL-3), Flt-3 ligand and thrombopoietin (Hacien-Bey *et al.* 2001) followed by multiple rounds of infection with vector. Although this method increases transduction efficiency, there are possible increased risks of insertional mutagenesis. For example, several male children affected by severe combined immunodeficiency syndrome secondary to a genetic deficiency of the common gamma chain (γc) of cytokine receptors, better known as X-SCID, a fatal disorder, have been treated by a French group with autologously transplanted BM cells reconstituted with the normal γc gene *ex vivo* using OV transduction (Cavazzana-Calvo *et al.* 2000, 2001). This treatment has proved successful in that the genetically modified BM cells reconstituted the immune system of most of the X-SCID patients, who as a result were able to live normally with their families at home, but, after 3 years without complications, three children (3/11) have developed T cell leukaemias. In all cases, the cause of leukaemia has been associated with an insertion of the vector proviral DNA in or near the LMO2 gene (Kaiser 2003), which is involved in cell growth regulation and has been associated with childhood acute lymphoblastic leukaemias (ALL) (Rabbitts *et al.* 1997). Thus this is a clear example of insertional mutagenesis associated with an OV and unfortunately may present an unacceptable risk for continuation of this X-SCID gene therapy protocol in new patients (but see Thrasher et al. 2006).

9.2.3 Lentiviral Vectors

OV can only integrate proviral DNA in mitotic, dividing cells (Günzburg & Salmons 1999; Miller *et al.* 1990; Roe *et al.* 1993). This restriction is disadvantageous for tissues and cell populations with normally few dividing cells, e.g. in brain, eye, lung, and pancreas. Although OV are used for *ex vivo* transductions of HSC, cytokine cocktails are required to increase cell division and transduction efficiency, but this process is time-consuming and expensive (see above). In comparison, the proviral DNA complex of lentiviruses has the capacity to traverse the nuclear membrane of non-dividing cells and therefore, following integration, can potentially replicate in these (Bukrinsky & Haffar 1999). Thus, lentiviral vectors (LV) may decrease the need for *ex vivo* cell manipulation and also lead to efficient gene delivery *in vivo* (Buchschacher & Wong-Staal 2000; Vigna & Naldini 2000; Sirven *et al.* 2001). However, lentiviral genomes are more complex than those of the oncovirus group making design of LV a greater challenge (Lever 1999; Buchschacher & Wong-Staal 2000; Frankel & Young 1998). Nevertheless, the very detailed knowledge on lentiviruses, in particular HIV-1 and -2, has made them attractive candidates for vector development.

There are several safety concerns about LV based on HIV-1, principally:

(1) presence of replication-competent lentiviruses (RCL) in the vector preparations;

(2) potential mobilization of vector genomes by wild-type HIV-1, leading to spread of mobilized lentivirions to previously untransduced cells *in vivo* and horizontal transmission with wild type HIV- 1 (Buchschacher & Wong-Staal 2000; Evans & Garcia 2000);

(3) potential insertional mutagenesis.

If wild type HIV-1 infections are considered, where there are multiple proviral integrations in infected T cells and macrophages, there is no evidence of such integrations causing insertional mutagenesis leading to tumour development, e.g. T-cell lymphoma (Kazazian & Moran 1998). However, an apparently rare instance of HIV-1 proviral DNA integration at a site upstream of the *c-fes/fps* proto-oncogene in macrophages being indirectly associated with lymphoma development has been reported (Shiramizu *et al.* 1994; Ng & McGrath 1998). Nevertheless, it is not known whether integrations in other cell types due to the use of VSV-G pseudotyped LV (see below) could cause insertional mutagenesis; the risks of it occurring may be similar to those found for OV (Li *et al.* 2002; Stocking *et al.* 1993). To reduce these safety risks, LV have undergone various developmental stages (Lever 1999; Buchschacher & Wong-Staal 2000; Vigna & Naldini 2000). The more complex replication requirements of lentiviruses in comparison with oncoviruses has so far limited the development of stable packaging cell lines for producing LV (see below). Thus production strategies involving transient transfections of cells, where essential viral proteins are expressed *in trans* from one or more constructs devoid of most viral *cis*-acting sequences, while the latter are linked into an expression cassette for the transgene, have been developed. A transgene size of up to about 18 kb is possible, but large inserts may negatively affect LV titres (Kumar *et al.* 2001). Recombination between constructs, which is most likely to occur during reverse transcription of RNA to DNA, leading to RCL formation, was the main safety problem; steps were therefore taken to minimize or eliminate sequence overlaps between the constructs. In the case of HIV-1, 'virulence genes', e.g. *vpr, vpu, vif,* and *nef,* which in the wild-type virus have some accessory roles in nuclear targeting, integration mechanisms, replication and pathogenesis/virulence (Cullen 1998), were found to be dispensable for vector propagation (Zufferey *et al.* 1997) and therefore eliminated from the constructs (Lever 1999; Buchschacher & Wong-Staal 2000; Vigna & Naldini 2000; Daly & Chernajovsky 2000). This would improve safety as any RCL that did arise would lack the accessory proteins associated with HIV-1 virulence.

Originally, HIV-based LV involved the development of three plasmids, *two* for vector packaging and *one* for the vector genome with incorporated transgene. The first 'packaging' plasmid construct encoded all HIV proteins, except Env, and retained the splice donor site, but the packaging element ψ was eliminated and the 5′LTR substituted by a minimal CMV promoter. The second plasmid contained either a *MuLV Env* or *VSV-G* gene driven by a CMV promoter resulting in pseudotyped vectors, making the generation of a HIV RCL impossible (Vigna & Maldini 2000).

However, vector contamination with an RCL of this type may represent a biohazard because VSV-G, for example, has a broad host cell range (Burns *et al.* 1993). The third plasmid, called the 'transfer vector', contained a CMV promoter-driven transgene and included the packaging signal ψ and the 5′ region of the *gag* gene and the *env*-derived fragment containing the *rev* response element (RRE) between the LTRs. To generate HIV-based LV required the transient transfection of cells, e.g. HeLa cells, simultaneously with all three plasmids (Vigna & Naldini 2000). Further developments have led to the elimination of HIV virulence genes (see above), including *tat*, in the first packaging plasmid, and subsequently splitting the *rev* gene, required for expression of the *gag* and *pol* genes, into a fourth independent plasmid and further reducing the possibility of recombinants between the packaging and transfer vectors (Lever 1999; Buchschacher & Wong-Staal 2000; Vigna & Naldini 2000). Improvements to the transfer vector include substitution of the 5′ LTR by a strong constitutive promoter, e.g. from Rous sarcoma virus (RSV), and introduction of a SIN deletion in part of the U3 region of the promoter in the 3′LTR that includes the TATA box to inactivate its promoter/enhancer activity (Vigna & Naldini 2000). This SIN transfer vector cannot be mobilized by superinfecting HIV-1 wild type viruses (Bukovsky *et al.* 1999) and reduces the risk of proto-oncogene activation by promoter insertion in target cells. However, the production of SIN-containing LV by the four plasmid transient transfection system may compromise vector yields (Vigna & Naldini 2000). One way of reducing the number of plasmids is to remove Rev dependency of *gag-pol* expression. This has recently been achieved by constructing a completely synthetic HIV-1 *gag-pol* gene where the nucleotide sequence has been altered in the majority of codons to retain the primary amino acid sequence, but exploits the favoured codon usage of human cells (Kotspoulou *et al.* 2000). Not only is this artificial sequence more stable than native HIV-1 AU-rich RNAs encoding the same region, but also significantly diminishes the possibility of recombination of *gag-pol* in the packaging vector with *gag* sequences remaining in the transfer vector. Alternatively, ways are being sought to incorporate the Env plasmid construct into 'packaging' cells. However, the pseudotyping VSV-G is toxic to cells and therefore its expression has to be controlled by an inducible promoter, e.g. tetracycline on/off system (Chen *et al.* 1996; Ory *et al.* 1996). One recent study (Farson *et al.* 2001) described the development of a more complex, but stable, packaging line (*Lentikat 2.54*) based on the tetracycline-regulated VSV-G HEK293 (293G) cell line (Ory *et al.* 1996). In this instance, a construct containing HIV-1 *gag-pol, tat,* and *rev* genes was stably introduced into 293G cells under the control of a Tet^0 promoter. In the presence of tetracycline, expression of both VSV-G and Gag p24 was suppressed, but this was released by tetracycline removal and led to significant yields of vector particles, without detectable RCL generation, following transfer vector transfection (Farson *et al.* 2001). Additionally, vector titres, transduction efficiency and transgene expression levels may be positively affected if certain *cis*-acting elements, e.g. the central polypurine tract (cppt) of HIV *pol*, are incorporated in the transfer vector construct (Park & Kay 2001).

Development of LV based on HIV-2 and simian immunodeficiency virus (SIV) has faced similar problems to those encountered for HIV-1-based vectors (Buchschacher & Wong-Staal 2000). A possible advantage of HIV-2-based LV is that HIV-2 is less pathogenic than HIV-1 and, due to its closer relationship with SIV, can be studied in a primate model system. Hybrid LV containing elements from HIV-1 and HIV-2 or from HIV-1 and SIV have been described (Corbeau *et al.* 1998; White *et al.* 1999; Pandya *et al.* 2001) but, although possibly safer than HIV-1-based vectors, it is not known whether these offer any further advantages.

Safety concerns about primate LV has led to a search for suitable non-primate lentiviruses as possible bases for LV development (Buchschacher & Wong-Staal 2000;Vigna & Naldini 2000). Several candidates, including feline immunodeficiency virus (FIV) (Johnston & Power 1999; Stein *et al.* 2001; Curran *et al.* 2000), equine infectious anaemia virus (EIAV) (Mitrophanous *et al.* 1999; Yamada *et al.* 2001), caprine arthritis/encephalitis virus (CAEV) (Mselli-Lakhal *et al.* 1998) and bovine Jembrana disease virus (Metharom *et al.* 2000), which are non-pathogenic for humans, have been identified. The genomes of these non-primate lentiviruses are simpler than HIV-1; EIAV is the simplest known lentivirus (Mitrophanous *et al.* 1999). In the case of FIV the LTR normally needs modification, for example by replacement with a heterologous cytomegalovirus promoter, to enable viral replication in human cells (Buchschacher & Wong-Staal 2000, Johnston & Power 1999). Any LV developed from non-primate lentiviruses will require pseudotyping with an envelope protein, e.g. VSV-G, which confers binding and entry of vector particles into a broad range of cell types. Vector-producing 'systems', similar to those described for HIV-1-based LV, have been developed for FIV and EIAV (Johnston & Power 1999; Mitrophanous *et al.* 1999; Curran *et al.* 2000). However, it is unclear what gains have been made in terms of safety; *in vitro* assays to detect RCLs similar to those being developed for RCLs formed in HIV-1-based LV systems will still need to be developed. Since there is no homology of non-primate lentiviruses with any known human virus, mobilization of non-primate LV appears less likely to occur than with HIV-1-based vectors (Evans & Garcia 2000). Nevertheless, although improbable, a rare mobilizing event might produce a novel pathogenic hybrid lentivirus.

While OV have been used clinically, only one phase I that of LV against HIV infection in AIDS patients has been undertaken (Günzurg & Salmons 1999; Hodgson 2001 (MacGregor 2001; Brann et al 2005)). OV have been used to treat both inherited monogenic diseases and malignancies. Up until recently, they appeared to have a good safety record, but were largely ineffective due to transgene silencing. However, in the one disorder, X-SCID, where they were successful, three recent cases of T-cell leukaemia have highlighted a significant risk of insertional mutagenesis in gene therapy protocols involving retroviral vectors (Rabbitts *et al.* 1997). It may well transpire that there was an interplay of several factors besides the insertional mutagenesis events in or close to the LMO2 gene that ultimately contributed to the development of leukaemia, putting that particular transduction and transplantation procedure in a special high-risk category. However, the risk of insertional mutagenesis must now be considered real for any retroviral vector, OV or LV, mediated gene transfer (Dropulic, 2005) and may lead to their very restricted future use. Progress may now depend on development of retroviral vectors able to target proviral DNA integration to clearly defined 'safe sites' within the genome.

9.2.4 Adenoviral Vectors

Therapy of monogenic diseases requires long-term transgene expression and thus retroviral vectors with their proviral integration capacity seemed the logical choice. However, retroviral vectors can only be produced in low titre and are poorly infectious, making them unsuitable for treatment of malignancies where greater levels of infectivity combined with high levels of transgene expression are desirable in viral vectors (Table 9.1). Vectorologists have therefore concentrated on the adenoviruses (Ads) for development of viral vectors as anticancer agents since Ads can be propagated at high titre, are readily purified, can accommodate large transgenes and can transduce, albeit with transient transgene expression, large numbers of (tumour) target cells *in vivo* (Connelly 1999).

Adenoviruses are large, robust, non-enveloped DNA viruses with a linear genome of 35 kb and thus have potential for development of vectors containing large transgenes for *in vivo* applications (Connelly 1999; Morsy & Caskey 1999). Human Ads have a benign natural history in upper respiratory tract infections and, used as vaccines, have a good safety profile except in immunocompromised individuals (Gaydos & Gaydos 1995; Straus 1984). Since their DNA exists mostly as a 'free' linear DNA molecule in an unstable extrachromosomal state during replication (Horwitz 1990; Shenk 1996), integration in chromosomal DNA and the risk of insertional mutagenesis is very low. Human Ads are not known to be oncogenic (Green *et al.* 1979). Other favourable characteristics include the ability to infect a broad spectrum of both quiescent (non-dividing) and dividing cell types, ready propagation in culture to very high titres, and straightforward purification (Horwitz 1990; Shenk 1996). Therefore, manufacture and QC testing of adenoviral vectors (AdV) may more easily be carried out under

GMP conditions (Lusky 2005; Working et al. 2005). Additionally, the target cell specificity of Ads can be manipulated to achieve targeting *in vivo* (Krasnykh *et al.* 2000) (see below).

Originally, design strategies for constructing AdV were aimed at removing viral genes required for replication and replacing them with a transgene. It has proved relatively easy to prepare replication-defective, deletion mutants of human Ads, e.g. by deleting the 'early' E1 region, which regulates among other things the activation of other early viral genes during the replicative cycle, and replacing it with heterologous DNA of up to about 8.3 kb (Hitt & Gauldie 2000; Connelly 1999; Amalfitano & Parks 2002). Other regions of the adenoviral genome, e.g. E3, may also be deleted and replaced with heterologous DNA. Such AdV have been based mostly on adenovirus type 5 (Ad5) and are designated 'first generation', or AdV1, vectors.

Stable AdV1 vectors may be prepared using conventional recombinant DNA technological methods and appropriate selection methods to isolate vectors by rescue and growth in permissive, E1 complementing, cell lines (Hitt & Gauldie 2000; Connelly 1999; Krasnykh *et al.* 2000; Amalfitano & Parks 2000; Graham *et al.* 1977; Fallaux *et al.* 1998). E1-replacement AdV1 vectors must contain the left end of the viral genome, including the inverted terminal repeat (ITR), packaging signal, transgene 'cassette' and approximately 1 kb of the downstream viral DNA sequence; they may only be propagated in cell lines containing complementing Ad early E1 (E1A and E1B) genes expressed *in trans*. The most widely used is a human embryo kidney (HEK) 293 cell line [now thought to be neuronal in origin (Shaw *et al.* 2002; Toth & Wold 2002)] transformed by sheared Ad5 DNA and containing the left 11 % of the Ad5 genome – including E1A and E1B – integrated in chromosomal DNA (Graham *et al.* 1977). However, Ad5 vectors propagated in HEK 293 are likely to contain low titres of replication-competent Ads (RCA), which are generated by homologous recombination of vector genome with integrated E1 region due to extensive sequence overlap (Lochmüller *et al.* 1994; Louis *et al.* 1997). These RCA have lost the transgene and are similar, but not identical, to wild type Ad5. Removal of homologous sequences between vector and helper should significantly reduce the risk of RCA formation. This has been achieved in the human embryonic retinoblast PER.C6 cell line, which contains the E1 region, as multiple tandem repeats, under the control of a phosphoglycerate kinase (PGK) promoter, but has no flanking Ad sequences with which vector DNA could recombine (Fallaux *et al.* 1998).

Although AdV1 vectors have been shown efficiently to transduce cells *in vitro*, transgene expression may be limited *in vivo* due to host immune responses to Ad proteins, leading to vector clearance and elimination of infected cells expressing viral antigens (Wilson 1995; Christ *et al.* 1997). Since most humans already have immunity to Ads [55 % of adult humans have pre-existing anti-Ad antibodies capable of neutralizing *in vitro* infection by Ad5 (Chirmule *et al.* 1999)], it is expected that secondary responses induced by AdV1 will be correspondingly accelerated and more intense than occur in 'naïves'. In addition, systemic administration of high dose AdV1 vectors is associated with adverse toxicity, including acute-phase response, inflammatory reactions, systemic shock syndrome and potential lethality (Amalfitano & Parks 2002; Brenner 1999; NIH Report 2002). To reduce such adverse responses, further modifications of AdV genomes have been studied, including deletions of early E2a (second generation or AdV2 vectors), or E2a, E3 and E4 (third generation or AdV3 vectors) or all early regions (fourth generation or AdV4 vectors) besides E1 deletion. While reducing immune and inflammatory responses, such modifications require additional complementing Ad genes, e.g. in episomal plasmids or 'amplicons', leading to a more difficult vector propagation procedure (Amalfitano & Parks 2002). Alternatively, chemical modification of Ad capsid, e.g. by coating with polyethylene glycol (PEG) residues to shield from recognition by immunocytes, has been shown to reduce immune responses (O'Riordan *et al.* 1999; Croyle *et al.* 2000).

Recent technical advances have made possible the development of AdV devoid of all Ad genes; these are known as 'gutless' or 'high capacity', AdV (Connelly 1999; Amalfitano & Parks 2002; Kochanek 1999; Maione *et al.* 2000; Balagué *et al.* 2000). Besides reducing potential immune/inflammatory responses, gutless AdV (Table 9.1) can accommodate up to 36 kb of heterologous DNA so that large genes such as the full-length dystrophin gene of 28 kb can be incorporated (Connelly 1999; Kochanek 1999; Maione *et al.* 2000; Balagué *et al.* 2000). However, all Ad genes are required to propagate gutless vector particles and these are supplied by co-transfection of what essentially is a 'helper virus' plasmid (amplicon) (Kochanek 1999). Not only does this complicate gutless AdV production, but low levels of helper virus are also generated (Connelly 1999; Amalfitano & Parks 2002; Kochanek 1999). Thus, while full Ad genome deletion may be desirable, it may not in the end be as practical for vector manufacture as the more limited E1-E4 deletions.

Besides approaches to minimize host responses, AdV are also amenable to modification of capsid proteins, fibre and penton base, to enable specific cell targeting (Chirmule *et al.* 1999; Wickham 2000). Initial cell attachment of most Ads is through binding of the knob domain of fibre protein to Coxsackie/adenovirus receptors (CAR) (Bergelson *et al.* 1997; Roelvink *et al.* 1999; Kirby *et al.* 2000). Secondary attachment and internalization is mediated by penton base interaction with cell surface integrins, $\alpha_v\beta_3$ or $\alpha_v\beta_5$, which act as co-receptors (Nemerow & Stewart 1999). Several human cell types, particularly liver hepatocytes, express both CAR and integrin co-receptors, but expression of CAR by primary fibroblasts, lung epithelial cells, and some neoplastic cells is highly variable and often relatively low, which may limit the usefulness of untargeted adenoviral vectors (Krasnykh *et al.* 2000; Hemmi *et al.* 1998; Walters *et al.* 1999). If the binding domains of fibre knob are ablated (Kirby *et al.* 1999), binding to liver cells is reduced by around 90% and such modification should reduce adverse immune responses and toxicity triggered in the liver by AdV1 vectors. In addition, modifications of the binding specificity by ligation of receptor-binding moieties to the fibre protein, e.g. basic fibroblast growth factor (bFGF), such that AdV1 vectors bind to FGF-receptors at the cell surface, or antibody Fv fragments which direct vectors to the cell surface antigen recognized by the antibody, have been achieved (Krasnykh *et al.* 2000). A major production difficulty for targeted adenoviral vectors lacking tropism for CAR, is that they cannot be propagated in those cell lines, e.g. HEK 293, PER.C6 (Graham *et al.* 1977; Fallaux *et al.* 1998), normally employed for this purpose. Therefore, special cell lines need to be derived to support their propagation and this is usually a far from trivial task (Krasnykh *et al.* 2000; Douglas *et al.* 1999).

A further strategy to obtain tissue specificity involves development of Ads containing specific mutations or deletions, e.g. in early E1b, which, as a result, are conditionally replication competent or 'replication-selective' (Alemany & Zhang 2000; Alemany *et al.* 2000; Curiel 2000; Kirn 2000; Kirn *et al.* 2001; Kruyt & Curiel 2002). These have been designated 'conditionally replication-competent adenoviruses' (CRAds). For example, in wild type Ad infections, E1b 55 kDa protein in association with adenoviral E4orf6 protein binds and inactivates the p53 tumour suppressor protein (p53), thus enabling efficient viral replication. In contrast, an E1b 55 kDa gene-mutated Ad (termed dl1520 or ONYX-015) is replication defective in cells containing normal p53, but is able to replicate in tumour cells containing a dysfunctional p53 leading to cell death (Kirn *et al.* 1998; Barker & Berk 1987). However, recent evidence suggests that replication of dl1520 (ONYX-015) is not absolutely linked to absence of normal p53, i.e. there could be some 'leakiness' and normal cell death (Curiel 2000; Kruyt & Curiel 2002; Hall *et al.* 1998; Vogelstain & Kinzler 2001). 'Oncolytic CRAds' are not typical gene transfer products, since they do not contain a transgene(s). However, a transgene encoding a therapeutic function, e.g. to make tumour cells sensitive to pro-drug toxicity, may be incorporated in CRAds, in which case they may be considered as a special class of AdV (Table 9.1) (Alemany & Zhang 2000; Alemany *et al.* 2000; Curiel 2000; Kirn 2000; Kirn *et al.* 2001; Kruyt & Curiel 2002). For example, insertion of the thymidine kinase (TK) gene leads to killing of CRAd/TK-infected cells on enzymatic conversion of the pro-drug ganciclovir to a cytotoxic product (Wildner *et al.* 1999a, b; Heise *et al.* 1997; Nanda *et al.* 2001). Use of such CRAds in gene therapy of tumours appears attractive since transgene delivery and expression in tumours lasts longer and is found throughout an anatomically more extensive area compared with that effected by replication-defective AdV (Ichikawa & Chiocca 2001). In addition, infection spread can be controlled by pro-drug addition, which would selectively kill CRAd-infected cells, thus increasing safety of treatment. Currently it is not clear whether CRAds can, through recombination or viral genome rearrangements, produce non-selective RCA. If they did, it could present additional problems for safety assurance.

Both AdV and CRAds can be targeted to particular cells by restricting transgene expression of tissue-specific promoters (Connelly 1999; Krasnykh *et al.* 2000). For example, use of muscle creatine kinase (MCK) promoter restricts expression of transgene to muscle cells. Use of prostate specific antigen (PSA) gene promoter sequences to regulate the E1 region, enables CRAd/PSA to replicate specifically in and kill PSA-expressing prostate tumour cells; a selectivity of 10 000-fold over non-PSA-expressing cells has been found (Chen *et al.* 2001). Drug-responsive promoters are also being employed to control transgene expression. For example, rapamycin or the tetracycline '*on*' system, has been used to trigger transgene expression (Urlinger *et al.* 2000; Somia & Verma 2000), while the tetracycline '*off*' system has been used to inactivate expression (Blau & Rossi 1999). However, generally, modifications to increase or change selectivity have reduced the oncolytic efficacy of CRAds to below that of wild type Ads (Kruyt & Curiel 2002).

Although early applications of AdV focused on treatment of monogenic diseases such as cystic fibrosis and ornithine decarbamylase (ODC) deficiency (Connelly 1999; NIH Report 2002), there are now safety concerns about their systemic administration at high dose levels, particularly since a young adult patient, Jesse Gelsinger, died as a direct result of AdV1 therapy of his ODC deficiency (NIH Report 2002). It is known that Ads in high doses can cause inflammatory reactions, cytotoxicity and permanent organ damage (Amalfitano & Parks 2002; Brenner 1999; NIH Report 2002); the ODC deficiency patient received 3.8×10^{13} AdV1 particles and appears to have developed a systemic toxic shock syndrome associated with the release of pro-inflammatory cytokines (NIH Report 2002). However, there have been no further reports of life-threatening toxicities following AdV1 administration, albeit at lower doses, and there remains a high interest in AdV generally, with now much more emphasis on treatment of malignancies. A Phase I clinical trial of intra-arterial p53 AdV1 gene therapy for colorectal tumours metastatic to the liver has demonstrated that doses up to 10^{13} particles can be safely administered (Atencio *et al.* 2001). AdV1 and CRAds have shown some promise in the treatment of head and neck cancers where intratumoral injections are possible (Kirn *et al.* 2001; Kruyt & Curiel 2002). Vector shedding and spread of CRAds beyond the treated patient however remains a safety concern.

9.3 CONCLUSIONS

A range of viral vectors based on known viruses, e.g. AAV-2, OV and LV, and AdV, as illustrated in the previous sections, has been developed for gene therapy protocols of human diseases. Most viral vectors, with one or two exceptions, have so far shown very limited effectiveness in providing therapeutic responses to cure or ameliorate disease, and have in a few cases led to the manifestation of serious adverse events (Rabbitts *et al.* 1997; NIH Report 2002; Thrasher et al. 2006). The case for their continued development and clinical use rests on improvements that need to be made in respect to design, manufacture, transgene delivering capability and targeting expression, and integration to increase their quality, efficacy and safety. It is expected that many advances are still to come, probably mostly in the form of small incremental ones, making it more and more likely that safer vectors and successful treatments will eventually be forthcoming.

REFERENCES

Alemany R, Zhang WW (2000) In *Methods in Molecular Medicine*, Vol. 35: *Gene Therapy: Methods and Protocols*. Eds Walther W, Stein U. Humana Press Inc, Totowa, NJ.

Alemany R, Balague C, Curiel DT (2000) *Nat. Biotech.*; 18: 723–727.

Amalfitano A, Parks RJ (2002) *Curr. Gene Ther.*; 2: 111–133.
Arella M, Garzon S, Bergeron J, Tijssen T (1990) In *Handbook of Parvoviruses*, Ed Tissen P. CRC Press Boca Raton.
Atencio I, Warren R, Venook AP *et al.* (2001) *Molec. Ther.*; 3:S309 (Abstract 871).
Auricchio A, O'Connor E, Hildinger M, Wilson JM (2001) *Mol. Ther.*; 4:372–374.
Balagué C, Zhou J, Dai Y *et al.* (2000) *Blood*; 95: 820–828.
Barker DD, Berk AJ (1987) *Virology*; 156: 107–121.
Bayle JY, Johnson L, St. George J *et al.* (1993) *Hum. Gene Ther.*; 4: 161–170.
Bergelson JM, Cunningham JA, Droguett G *et al.* (1997) *Science*; 275: 1320–1323.
Blau HM, Rossi FMV (1999) *Proc. Natl. Acad. Sci. USA*; 96: 797–799.
Boris-Lawrie K, Temin HM (1994) *Ann. N.Y. Acad. Sci.*; 716: 59–71.
Brann SE, Wong FE, Connole M *et al.* (2005) *Molec. Ther.*; 12: 1157–1167.
Brenner M (1999) *Blood*; 94: 3965–3967.
Buchschacher GL Jr., Wong-Staal F (2000) *Blood*; 95: 2499–2504.
Bukrinsky MI, Haffar OK (1999) *Front. Biosci.*; 4: 772–781.
Bukovsky AA, Song JO, Naldini L (1999) *J. Virology*; 73: 7087–7092.
Burns JC, Friedmann T, Driever W *et al.* (1993) *Proc. Natl. Acad. Sci. USA*; 90: 8033–8037.
Cavazzana-Calvo M, Hacien-Bey S, de Sainte Basile G *et al.* (2000) *Science*; 288: 669–672.
Cavazzana-Calvo M, Hacien-Bey S, Yates F *et al.* (2001) *J. Gene Med.*; 3: 201–206.
Chen ST, Iida A, Guo L *et al.* (1996) *Proc. Natl. Acad. Sci. USA*; 93: 10057–10062.
Chen Y, DeWeese T, Dilley J *et al.* (2001) *Cancer Res.*; 61: 5453–5460.
Chirmule N, Moffett J, Dhagat P *et al.* (1999) *Gene Ther.*; 6: 1574–1583.
Cholika A, Guyot V, Nicholas JF (1996) *J. Virology*; 70: 1792–1798.
Chong H, Starkey W, Vile RG (1994) *J. Virol.*; 72: 2663–2670.
Christ M, Lusky M, Stoeckel M *et al.* (1997) *Immunol. Lett.*; 57: 19–25.
Coffin JM (1996) In *Fundamental Virology.* Eds Fields BN, Knipe DM, Howley PM. Lippincott-Raven Publishers, Philadelphia; third edition.
Connelly S (1999) In *Gene Therapy Technologies, Applications and Regulations: From Laboratory to Clinic.* Ed Meager A. John Wiley & Sons, Chichester.
Corbeau P, Kraus G, Wong-Staal F (1998) *Gene Ther.*; 5: 99–104.
Cornetta K, Moen RC, Culver K *et al.* (1990) *Hum. Gene Ther.*; 1: 14–30.
Cosset F-L, Takeuchi Y, Battini J-C *et al.* (1995) *J. Virology*; 69: 7430–7436.
Croyle MA, Yu QC, Wilson JM (2000) *Hum. Gene Ther.*; 11: 1713–1722.
Cullen BR (1998) *Cell*; 3: 685–692.
Curiel DT (2000) *Clin. Cancer Res.*; 6: 3395–3399.
Curran MA, Kaiser SM, Achacoso PL, Nolan GP (2000) *Molec Ther.*; 1: 31–38.
Daly G, Chernajovsky Y (2000) *Molecular Therapy*; 2: 423–434.
Danos O, Mulligan RC (1988) *Proc. Natl. Acad. Sci. USA*; 85: 6460–6464.
Donahue RE, Kessler SW, Bodine D *et al.* (1992) *J. Exp. Med.*; 176: 1125–1135.
Douglas JT, Miller CR, Kim M *et al.* (1999) *Nat. Biotech.*; 17: 470–475.
Dropulic B (2005) *Curr. Hematol. Rep.*; 4: 300–304.
Emi N, Friedman T, Yee JK (1991) *J. Virology*; 65: 1212–1217.
Evans JT, Garcia JV (2000) *Hum. Gene Ther.*; 11: 2331–2339.
Fallaux FJ, Bout A, van der Velde I *et al.* (1998) *Hum. Gene Ther.*; 9: 1909–1917.
Farson D, Witt R, McGuinness R *et al.* (2001) *Hum. Gene Ther.*; 12: 981–997.
Ferrari F, Samulski T, Shenk T, Samulski RJ (1996.) *J. Virol.*; 70: 3227–3234.
Finer M, Dull T, Qin L *et al.* (1994) *Blood*; 83: 43–50.
Flotte TR, Carter BJ (1999) In *Gene Therapy Technologies, Applications and Regulations: From Laboratory to Clinic.* Ed Meager A. John Wiley & Sons, Chichester.
Flotte TR, Afione SA, Conrad C *et al.* (1993) *Proc. Natl. Acad. Sci. USA*; 90: 10613–10617.
Frankel AD, Young JA (1998) *Ann. Rev. Biochem.*; 67(1): 1–25.
French NS, Norton JD (1997) *Biochim. Biophys. Acta*; 1352: 33–47.
Friedmann T, Yee JK (1995) *Nature Med.*; 1: 275–277.
Gao GP, Qu G, Faust LZ *et al.* (1998) *Hum. Gene Ther.*; 9: 2353–2362.

Gaydos CA, Gaydos JC (1995) *Mil. Med.*; 160: 300–304.
Graham F, Smiley J, Russell WC, Nairn R (1977) *J. Gen. Virol.*; 36: 59–74.
Green M, Wold WSM, Mackey JK, Rigden P (1979) *Proc. Natl. Acad. Sci. USA*; 76: 6606–6610.
Günzburg WH, Salmons B (1999) In *Gene Therapy Technologies, Applications and Regulations: From Laboratory to Clinic*. Ed Meager A. John Wiley & Sons, Chichester.
Hacien-Bey S, Gross F, Nusbaum P *et al.* (2001) *Hum. Gene Ther.*; 12: 291–301.
Hall AR, Dix BR, O'Carroll SJ, Braithwaite AW (1998) *Nat. Med.*; 4: 1068–1072.
Hauswirth WW, Lewin AS, Zolotukhin S, Muzyczka N (2000) *Methods Enzymol.*; 316: 743–761.
Heise C, Sampson-Johannes A, Williams A *et al.* (1997) *Nat. Med.*; 6: 639–644.
Hemmi S, Geertsen R, Mezzacasa A *et al.* (1998) *Hum. Gene Ther.*; 9: 2363–2373.
Hitt MM, J Gauldie (2000) *Curr. Pharmaceut. Design*; 6: 613–632.
Hodgson CP (2001) *Retro-vectors for Gene Therapy*. CD-ROM book. Nature Technology Corporation (Prairie Publishing) Lincoln, Nebraska.
Horwitz MS (1990) In *Virology*. Eds Fields BN, Knipe DM. Raven Press, New York.
Ichikawa T, Chiocca EA (2001) *Cancer Res.*; 61: 5336–5339.
Inoue N, Russell DW (1998) *J. Virol.*; 72: 7024–7031.
Inoue N, Dong R, Hirata RK, Russell DW (2001) *Molec. Ther.*; 3: 526–530.
Itin A, Keshet E (1983) *J. Virology*; 47: 178–184.
Johnston JC, Power CP (1999) *J. Virology*; 73: 2491–2498.
Julias JG, Hash D, Pathnik VK (1995) *J. Virology*; 69: 6839–6846.
Kaiser J (2003) *Science*; 299: 495.
Kay MA, Manno CS, Ragni MV *et al.* (2000) *Nature Genetics*; 24: 257–261.
Kazazian HH Jr., Moran JV (1998) *Nature Genet.*; 19: 19–24.
Kearns WG, Afione SA, Fulmer SW *et al.* (1996) *Gene Ther.*; 3: 748–755.
Keshet E, Shaul Y, Kaminchik J, Aviv H (1980) *Cell*; 20: 431–439.
Kiem HP, Heyward S, Winkler A (1997) *Blood*; 90: 4638–4645.
Kirby I, Davison E, Beavil AJ *et al.* (1999) *J. Virol.*; 73: 9508–9514.
Kirby I, Davison E, Beavil AJ *et al.* (2000) *J. Virol.*; 74: 2804–2813.
Kirn D (2000) *Oncogene*; 19: 6660–6668.
Kirn D, Hermiston T, McCormick F (1998) *Nat. Med.*; 4: 1341–1342.
Kirn D, Martuza RL, Zwiebel J (2001) *Nat. Med.*; 7: 781–787.
Kochanek S (1999) *Hum. Gene Ther.*; 10: 2461–2469.
Kotsopoulou E, Kim VN, Kingsman AJ *et al.* (2000) *J. Virol.*; 74: 4839–4852.
Krasnykh VN, Douglas JT, van Beusechem VW (2000) *Mol. Ther.*; 1: 391–405.
Kruyt FAE, Curiel DT (2002) *Hum. Gene Ther.*; 13: 485–495.
Kumar M, Keller B, Makalou N, Sutton RE (2001) *Hum. Gene Ther.*; 12: 1893–1905.
Lever AML (1999) In *Gene Therapy Technologies, Applications and Regulations: From Laboratory to Clinic*. Ed Meager A. John Wiley & Sons, Chichester.
Li Z, Dullman J, Schiedlmeier B, Schmidt M *et al.* (2002) *Science*; 296: 497.
Liu H, Hung Y, Wissink SD, Verfaillie CM (2000) *Leukemia*; 14: 307–311.
Lochmüller H, Jani A, Huard J *et al.* (1994) *Hum. Gene Ther.*; 5: 1485–1491.
Long Z, Li LP, Grooms T *et al.* (1998) *Hum. Gene Ther.*; 9: 1165–1172.
Louis N, Eveleigh C, Graham F (1997) *Virology*; 233: 423–429.
Luciw PA, Leung NJ (1992) In *The Retroviridae*. Ed Levy JA. Plenum Press, New York.
Lusky M (2005) *Hum. Gene Ther.*; 16: 281–291.
MacGregor (2001) *Hum. Gene Ther.*; 12: 2028–2029.
Maione D, Wiznerowicz M, Delmastro P *et al.* (2000) *Hum. Gene Ther.*; 11: 859–868.
Markowitz D, Goff S, Bank A (1988) *J. Virology.*; 62: 1120–1124.
McCarty DM, Young SM Jr, Samulski RJ (2004) *Annu. Rev. Genet.*; 38: 819–845.
Metharom P, Takyar S, Huiqin Xia H *et al.* (2000) *J. Gene Med.*; 2: 176–185.
Meyer J, Laker C, Janzir N *et al.* (2002) *Growth Factors*; 20: 131–140.
Miao CH, Nakai H, Thompson AR *et al.* (2000) *J. Virol.*; 74: 3793–3803.
Miller AD, Buttimore C (1986) *Mol. Cell. Biol.*; 6: 2895–2902.
Miller AD, Garcia JV, von Suhr N *et al.* (1991) *J. Virology*; 65: 2220–2224.

Miller DG, Adam MA, Miller AD (1990) *Mol. Cell. Biol.*; 10: 4239–4242.
Miller DG, Trobridge GD, Petek LM et al. (2005) *J. Virol.*; 79: 11434–11442.
Mitrophanous K, Yoon S, Rohll J *et al.* (1999) *Gene Ther.*; 6:1808–1818.
Moebes A, Enssle J, Bieniasz PD *et al.* (1997) *J. Virol.*; 71: 7305–7311.
Morgenstern JP, Land H (1990) *Nucleic Acids Res.*; 18: 3587–3596.
Morsy MA, Caskey CT (1999) *Mol. Med. Today*; 5: 18–24.
Mselli-Lakhal L, Favier C, Da Silva T *et al.* (1998) *Arch. Virol.*; 143: 681–695.
Muenchau DD, Freeman SM, Cornetta K, Zwiebel JA, Anderson WF (1990) *Virology*; 176: 262–265.
Nanda D, Vogels R, Havenga M *et al.* (2001) *Cancer Res.*; 61: 8743–8750.
Naviaux RK, Verma IM (1992) *Curr. Opin. Biotechnol.*; 3: 540–547.
Nemerow GR, Stewart PL (1999) *Micro. Mol. Biol. Rev.*; 63: 725–734.
NIH Report (2002) *Hum. Gene Ther.*; 13: 3–13.
Ng VL, McGrath MS (1998) *Immunol. Rev.*; 162: 293–298.
Okoh MP, Kainulainen L, Heiskanen K *et al.* (2002) *Hum. Mutat.*; 20: 480–481.
Olson P, Nelson S, Dornburg R (1994) *J. Virology*; 68: 7060–7066.
Ory DS, Neugeboren BA, Mulligan RC (1996) *Proc. Natl. Acad. Sci. USA*; 93: 11400–11406.
O'Riordan CR, Lachapelle A, Delgado C *et al.* (1999) *Gene Ther.*; 10: 1349–1358.
O'Riordan CR, Lachapelle LA, Vincent KA, Wadsworth SC (2000) *J. Gene Med.*; 2: 444–454.
Owens RA (2002) *Curr. Gene Ther.*; 2: 145–159.
Pandya S, Boris-Lowrie K, Leung NJ *et al.* (2001) *Hum. Gene Ther.*; 12: 847–857.
Park F, Kay MA (2001) *Mol. Ther.*; 4: 164–173.
Patience C, Takeuchi Y, Cosset FL, Weiss RA (1998) *J. Virol.*; 72: 2671–2676.
Pearson H (2001) *Nature*; 413: 9.
Perron H, Seigneurin JM (1999) *Microbes Infect.*; 1: 309–322.
Ponnazhagan S, Erikson D, Kearns WG *et al.* (1997) *Hum. Gene Ther.*; 8: 275–284.
Ponnazhagan S, Curiel DT, Shaw DR, *et al.* (2001) *Cancer Res.*; 61: 6313–6321.
Purcell DFJ, Broscius CM, Vanin EF *et al.* (1996) *J. Virology*; 70: 887–897.
Rabbitts TH, Axelson H, Forster A *et al.* (1997) *Leukemia*; Suppl 3: 271–272.
Rigg RJ, Chen J, Dando JS *et al.* (1996) *Virology*; 218: 290–295.
Roe T, Reynolds TC, Yu G, Brown PO (1993) *EMBO J.*; 12: 2099–2108.
Roelvink PW, Lee GM, Einfeld DA *et al.* (1999) *Science*; 286: 1568–1571.
Russ AP, Friedel C, Grez M, von Melcher H (1996) *J. Virology*; 70: 4927–4932.
Russell DW, Hirata RK (1998) *Nature Genet.*; 18: 325–330.
Russell DW, Kay MA (1999) *Blood*; 94: 864–874.
Salvetti A, Oreve S, Chadeuf G *et al.* (1998). *Hum. Gene Ther.*; 9: 695–706.
Samulski RJ, Chang L-S, Shenk T (1989) *J. Virol.*; 63: 3822–3828.
Scadden DT, Fuller B, Cunningham JM (1990) *J. Virology*; 64: 424–427.
Shaw G, Morse S, Ararat M, Graham FL (2002) *FASEB J.*; 16: 869–871.
Shenk T (1996) In *Virology*. Eds Fields BN, Howley PM. Lippincott Raven Press, Philadelphia; Vol. 1.
Shi W, Arnold GS, Bartlett JS (2001) *Hum. Gene Ther.*; 12: 1697–1711.
Shiramizu B, Herndier BG, McGrath MS (1994) *Cancer Res.*; 54: 2069–2072.
Sirven A, Ravet E, Charneau P *et al.* (2001) *Molec. Ther.*; 3: 438–448.
Smiley WR, Laubert B, Howard BD *et al.* (1997) *Hum. Gene Ther.*; 8: 965–977.
Snyder RO (1999) *J. Gene Medicine*; 1: 166–175.
Somia N, Verma IM (2000) *Nat. Rev. Genet.*; 1: 91–99.
Srivastasa A, Lusby EW, Berns KI (1983) *J. Virol.*; 45: 555–564.
Stein CS, Kang Y, Sauter SL *et al.* (2001) *Molec. Ther.*; 3: 850–856.
Straus SE (1984) In *The Adenoviruses*. Ed Ginsburg HS. Plenum Press, New York.
Stocking C, Bergholz U, Freil J *et al.* (1993) *Growth Factors*; 8: 197–209.
Summerford C, Samulski RJ (1998) *J. Virol.*; 72: 1438–1445.
Summerford C, Samulski RJ (1999) *Nature Med.*; 5: 587–588.
Surosky RT, Urabe M, Goodwin SG *et al.* (1997) *J. Virol.*; 71: 7951–7959.
Takeuchi Y, Porter CD, Strahan KM *et al.* (1996) *Nature*; 379: 85–88.
Tal J (2000) *J. Biomed. Sci.*; 7: 279–291.

Thrasher AJ, Gasper HB, Baun C et al. (2006) *Nature*; 443: E5–E7.
Toth K, Wold WSM (2002) *Mol. Ther.*; 5: 654.
Trinh KR, Morrison SL (2000) *J. Immunol. Methods*; 244: 185–193.
Trobridge GD, Russell DW (1998) *Hum. Gene Ther.*; 9: 2517–2525.
Urlinger S, Baron U, Thellmann M *et al.* (2000) *Proc. Natl. Acad. Sci. USA*; 97: 7963–7968.
Vassilopoulis G, Trobridge G, Josephson NC, Russell DW (2001) *Blood*; 98: 604–609.
Vigna E, Naldini L (2000) *J. Gene Medicine*; 2: 308–316.
Vogelstain B, Kinzler KW (2001) *Nature*; 412: 865–866.
Von Kalle C, Kiem HP, Goele S *et al.* (1994) *Blood*; 84: 2890–2897.
Walters RW, Grunst T, Bergelson JM *et al.* (1999) *J. Biol. Chem.*; 274: 10219–10226.
Walther W, Stein U (2000) *Drugs*; 60: 249–271.
Wang XS, Khuntirat B, Qing K *et al.* (1998) *J. Virol.*; 72: 5472–5480.
White SM, Renda M, Nam NY *et al.* (1999) *J. Virol.*; 73: 2832–2840.
Wickham TJ (2000) *Gene Ther.*; 7: 110–114.
Wildner O, Blaese RM, Morris JC (1999a) *Cancer Res.*; 59: 410–413.
Wildner O, Morris JC, Vahanian NN *et al.* (1999b) *Gene Ther.*; 6: 57–62.
Wilson JM (1995) *J. Clin. Invest.*; 96: 2547–2554.
Working PK, Lin A, Borellini F (2005) *Oncogene*; 24: 7792–7801.
Wotton S, Stewart M, Blyth K *et al.* (2002) *Cancer Res.*; 62: 7181–7185.
Wu Z, Asokan A, Samulski RJ (2006) *Molec. Ther.*; 14: 316–327.
Xiao X, Li J, Samulski RJ (1998) *J. Virol.*; 72: 2224–2232.
Yamada K, Olsen JC, Patel M *et al.* (2001) *Molec. Ther.*; 3: 485–490.
Yang Q, Chen F, Trempe JP (1994) *J. Virol.*; 68: 4847–4856.
Yee JK (2000) In *The Development of Human Gene Therapy*. Ed Friedmann T. Cold Spring Harbor Laboratory Press, Cold Spring Harbor, New York.
Yu SF, von Ruden T Kantoff PW *et al.* (1986) *Proc. Natl. Acad. Sci. USA*; 83: 3194–3198.
Zufferey R, Nagy D, Mandel RJ *et al.* (1997) *Nature Biotechnol.*; 15: 871–875.

Technology and Facilities for Cell Culture Scale-up

10 Systems for Cell Culture Scale-up

J Davis

10.1 INTRODUCTION

The scale-up of animal cell cultures from T-flasks or Petri dishes to full industrial production scale is a huge topic, and no attempt will be made in this chapter to provide a comprehensive treatment. Rather, an outline of the available culture system options will be presented in order to give the reader a jumping-off point for deeper and more specific investigations of his/her particular application.

The goal of scale-up is to satisfy projected market demand, and make the product available at an affordable price by realizing economies of scale. Thus the large number of small batches that would otherwise be required are replaced by a limited number of large batches, resulting in a reduction in the number of procedures, the labour required, and the amount of quality control and other testing. This is not achieved, however, without an increase in risk. The failure – for whatever reason – of one small batch among many may be tolerable, but the loss of a huge single batch could be catastrophic to the producer, and possibly also (in the case of biological medicines) to the patient population. Thus in scale-up there is a real emphasis on reliability, in terms of services, the hardware used, and the process and its control. A balance must also be struck in the size of the unit process, and consequently the number of batches required per year, between:

- economy due to scale (where big is beautiful);
- limitation of losses due to a batch failure (where small is beautiful);
- efficient use of staff time (where the minimum number of staff should be fully occupied all year, but with contingency built in), and
- efficient use of facilities (it may be more economical to run a smaller unit-scale process in an existing building rather than purpose-build a new facility for a larger-scale process).

All of these considerations impinge on the choices to be made in scale-up, but the most important, which must be answered before all others, is: what is the projected market demand? By answering this single question, it may be possible to exclude certain scale-up options, and estimate the final scale required with others; for example, if market demand for a therapeutic monoclonal antibody is projected to be 200 kg/year, then hollow-fibre bioreactor culture, where each machine is unlikely to produce much more than 1 kg/year (see Section 10.3.3.2b), is unlikely to be feasible. Instead, large fermenters are liable to be the system of choice. However, the projected rate of market penetration or growth must also be factored in, and will tend to favour the use of several identical systems, with the number in operation being matched to market demand.

Medicines from Animal Cell Culture Edited by G. Stacey and J. Davis

10.2 PROBLEMS OF SCALE-UP

A variety of issues will need to be addressed in the process of scale up, but some of the most important are as follows:

Oxygen demand: All animal cells require oxygen, but the poor solubility of oxygen in water (around 0.2 mmol/l at 37 °C) can present a significant barrier to the supply of adequate quantities to cells in large-scale culture. At small scale, supplying the oxygen demands of cells in culture is rarely a problem, and the rate of diffusion of oxygen through the few millimetres of medium separating the vessel head-space from the cells is usually sufficient to satisfy the cells' requirements. However, as the culture volume and the concentration of cells increases, supplying adequate amounts of oxygen to the growing cells can become *the* major problem posed by scale-up.

Nutrient and waste product gradients: Cells in culture deplete nutrients from, and secrete waste products into, the surrounding medium. In a static (i.e. non-stirred) culture, a gradient of nutrients and waste products will tend to form around the cell, with the cell sited at the point with the lowest nutrient and highest waste product concentrations. Only diffusion (and possibly a little convective mixing) will dissipate these gradients. The higher the local cell concentration, the steeper these gradients will tend to be. Clearly, for large-scale culture this is inefficient, and mixing mechanisms must be employed. However, these very mechanisms may introduce further problems, including shear and bubbles.

Because they lack a cell wall, animal cells are far more susceptible to the damaging effects of shear than are prokaryotes. However, most mixing mechanisms induce shear, and balancing the damaging effects of shear against the requirements for the mass transport of oxygen to the cells and the minimization of chemical gradients can be a major challenge in scale-up optimization. It has also led to the invention of a variety of different scale-up systems that attempt to solve the problem in different ways.

The introduction of bubbles into a cell suspension can aid both oxygen transport and mixing. The disadvantage is that cells can be damaged upon disengagement of the bubbles at the vessel's liquid/head-space interface. This effect can be controlled by the addition of surface-active agents such as Pluronic F68™ to the medium (Handa-Corrigan 1990).

Logistics: Small culture vessels and their contents can be transported and manipulated by hand, moved from one environment to another (e.g. from an autoclave to a safety cabinet to an incubator), and warmed up or cooled down quickly. As scale increases, transportation and manual manipulation become increasingly difficult, and eventually impossible, at which stage everything must instead be brought to the culture system, and environmental changes effected *in situ*. Thus, for example, steam-in-place sterilization takes over from autoclaving, and other solutions have to be found where tilting of a vessel was required at small scale. Similarly, bringing medium up to the correct incubation temperature may only take minutes at small scale, but in a large fermenter may take many hours. These factors have very important impacts on both the design and operational costs (including down-time) of large-scale systems.

10.3 SYSTEMS

Whilst the various scale-up systems available can be classified in a variety of ways, the most useful for the individual seeking to scale up an existing culture is to categorize them by the nature of the cells for which they are useful. There are then three categories:

- systems suitable for attached or anchorage-dependent cells;
- systems suitable for cells that grow in suspension;
- systems suitable for either of the above.

It should be noted that in some cases the same cells may attach to a substrate or grow in suspension, depending on the conditions under which they are cultured. CHO cells, for example, may attach to a surface if grown in serum-containing medium, yet grow in suspension in protein-free medium. This underlines the importance of ascertaining the essential properties of cells and defining the medium of choice as soon as possible during process development, although in reality it is unlikely that this will have been completed prior to some degree of scale-up. However, all changes become more difficult and more expensive the later they are introduced during development.

A great variety of scale-up systems were introduced in the 1970s and 1980s, but many of these ultimately proved to be of limited use, either for technical or commercial reasons (e.g. the MBR membrane reactor, the Opticell perfused ceramic matrix). These will not be discussed below. Rather, we will confine ourselves to current technologies, plus some mention of those extensively used in the past but which are perhaps now being superseded.

10.3.1 Systems for Attached Cells

10.3.1.1 Roller bottles

The use of rotating bottles for large-scale cell culture was first described in 1933 (Gey 1933), since which time roller systems have been used for culturing a great many different types of attached cell. Clean glass bottles, e.g. 2.5-litre Winchester bottles, were used for many years, but purpose-designed reusable glass bottles or disposable plastic vessels are now available. The inner surface of the bottle or vessel is used as a cylindrical growth surface. Cells are introduced in a limited volume of medium, and the vessel is then placed in a horizontal position on or in an apparatus that will rotate it slowly around an axis parallel to that of the cylindrical growth surface. The cells attach to the vessel wall as it rotates, and are then cyclically immersed in and then removed from the bulk of the medium as the vessel rotates. When not immersed, there is a thin film of medium covering the cells, and the rotation rate must be such that this does not dry out to any significant extent before the cells are re-immersed in the bulk of the medium. Once cell attachment is complete, the amount of medium in the vessel is often increased, typically to around 8–15 % of the vessel's nominal capacity. Rotation rates will depend on the cell line in use and its attachment efficiency. A typical rate would be 0.5–1.0 r.p.m., perhaps with a slower rate (down to 0.1 r.p.m.) used immediately after seeding in order to facilitate cell adhesion. Some workers using cells that attach efficiently have recommended a different approach, using 0.2–0.4 r.p.m. during the attachment phase, then 0.08–0.16 r.p.m. once attachment is complete (Griffiths 1995). Systematic studies have permitted the mixing patterns in roller bottles to be characterized and modeled mathematically (Muzzio *et al.* 1999; Unger *et al.* 2000). From these studies and those of others (Tsao *et al.* 1992), it is clear that for best results optimization of rotation rates should be performed with the vessel/cell/medium combination actually in use. Periodic reversal of the direction of rotation may be useful in facilitating cell adhesion to the vessel walls (Muzzio *et al.* 1999). In addition, the introduction of a vertical rocking motion can improve mixing within the bulk of the medium (Unger *et al.* 2000).

Oxygen transfer is relatively efficient in this system, as the cells spend a large proportion of the time with only a thin film of medium separating them from the gas phase within the vessel. However, in order to avoid oxygen limitation in sealed vessels, it has been recommended that the gas-volume/medium ratio should not be less than 5:1 (Griffiths 1995). Alternatively, vented roller bottles are available. Another approach to improving oxygen transfer has been to attach an external recirculation loop to a roller bottle (Berson *et al.* 2002), but although highly effective this greatly increases the complexity of the system, negating one of the roller bottle's greatest assets: its simplicity.

A single flat-surfaced roller bottle will usually have a surface area of between 490 cm^2 and 1800 cm^2. This area may be increased by using a design with a ribbed culture surface, which can increase the surface area two to 2.5-fold without increasing the external dimensions of the vessel. A further increase in surface area has been achieved without increasing the dimensions of the vessel by adding concentric cylinders (Knight 1977) or plastic spiral films (Griffiths 2001), but such vessels do not seem to be widely available. Whilst longer vessels can be used, further scale-up is usually achieved simply by increasing the number of units used. This makes scale-up from the laboratory to initial production scale and beyond (as the market, or market penetration, increases) relatively simple, as there is no change in the unit process. However, large numbers of units have to be handled, and thus economies of scale can only be realized through automation of the various steps involved in roller bottle handling. The degree of automation can vary, but the largest-scale industrial systems are almost completely automated and have a capacity of tens of thousands of roller vessels. Such industrial systems have been in use in the vaccine industry since at least the mid-1960s (Nardelli & Panina 1976; Panina 1985). The emphasis now seems to be either on the replacement of human operatives with robots in order to reduce labour costs further and decrease the potential for contamination (Kunitake *et al.* 1997) or to convert these processes so that they can be run in fermenters.

A number of modern biological medicines have been produced in large-scale roller bottle facilities, including a number of vaccines as well as recombinant erythropoietin (by Amgen, Johnson & Johnson, and Janssen). However, many of these have now been converted to fermenter processes, either with or without microcarriers (see below).

10.3.1.2 Stacked-plate systems

10.3.1.2a NUNC 'cell factory'/Corning 'CellSTACK®'

These culture systems contain a number of flat culture surfaces stacked in parallel one above the other within a single unit, in order to increase the culture surface area that can be handled in one operation. Unlike roller bottles, they require no agitation and can be thought of for most purposes as large culture flasks. Initial scale-up is by increasing the number of surfaces within a unit, and there may be up to 40 within a single unit (Figure 10.1). Media and other solutions are added through an access port and distributed between the layers by tilting the unit. However, the requirement to move the whole unit, complete with its contents, into different orientations means that the

Figure 10.1 A 40-layer cell factory. Photograph courtesy of NUNC.

Figure 10.2 Part of a rig for handling four 40-layer cell factories simultaneously, shown with only one cell factory in place. Photograph courtesy of NUNC.

largest unit that can realistically be handled by an unaided operator is a 10-stack (10 parallel culture surfaces). Manual rigs are available to facilitate handling of a 40-stack. Beyond this, scale-up is by the use of multiple units, and larger, electrically operated rigs can be used to handle multiple 40-stacks simultaneously (Figure 10.2).

Although the culture environment in these units is very similar to that in a culture flask, there are a number of disadvantages to the system. Unlike in a culture flask or roller bottle, there is no means of direct access to the culture surface, so removal of cells by scraping, for example, is not possible. The units are made of rigid plastic, and are thus susceptible to damage if mishandled or hit by other objects (in contrast to modern plastic roller bottles, which are more resilient). They are also completely dependent on the integrity of the seals between the multiple components of the unit (again in contrast to roller bottles, which are nowadays made as a single, seamless moulding) and these seals are easy to break by simply adding liquids or gases to a unit too quickly. Until relatively recently, it was also difficult (for the above reason) to add a CO_2-enriched atmosphere to the unit, and even more difficult to maintain such an atmosphere in an effective dynamic manner in order to ensure adequate buffering with CO_2/HCO_3^--buffered media as used with many mammalian cells. This problem has now been overcome by NUNC with the introduction of 'active gassed' cell factories that are engineered with a gas distribution system. A CO_2/air mixture is pumped constantly or intermittently at a low flow rate (typically 20 ml per square cm of culture surface per hour) to this distribution system, which ensures even distribution both within and between layers, resulting in efficient buffering and gas exchange without the risk of damaging the unit. Such gas flow rates are easy provided by using a small pump such as would be used in a home aquarium.

'Cell factories' have been available for over 30 years, and have been used with a wide range of cell lines (see Siegl *et al.* 1984; Goetz *et al.* 1992; Bishop *et al.* 1994; Dickinson & Kohwi-Shigematsu 1995; Litjens *et al.* 1997; Nelson *et al.* 1997; Ohtaki *et al.* 1998; Lee *et al.* 1999; Lay *et al.* 2000; Suehiro *et al.* 2000; Loewen *et al.* 2002) and at the industrial scale for vaccine manufacture (Hagen *et al.* 1996). As with roller bottles, scale-up and adjustment to market demand has been relatively simple, being achieved by the use of multiple identical layers, and then multiple identical units. However, economies of scale are again difficult to realize other than by automation.

10.3.1.2b Corning CellCube®

Superficially, the CellCube® is similar in principle to the 'cell factory', having at its heart a culture unit composed of multiple flat surfaces bonded together in parallel. However, unlike the 'cell factory', the CellCube® is a perfusion culture device, with medium being continuously circulated through the system. Also, cells attach to both faces of the flat culture surfaces and, once seeded with cells, these culture surfaces are held in a vertical orientation. There is no air space within the system (as there is between each layer of a 'cell factory'), oxygenation/gas exchange being achieved by the passage of the medium through an external oxygenator during recirculation of the medium through the system. It also requires a controller, and a circulation and medium pump. In fact, the true precursor of this system was not the cell factory, but a vertical-plate system using glass plates separated by PTFE spacers that was first described by Mann (1977) and could be scaled up to a surface area of $10\,m^2$. In the modern CellCube®, by using multiple modules the culture surface area can be increased from $8500\,cm^2$ to $340\,000\,cm^2$ whilst still using a single controller. For exploratory work and initial process development, a simplified version called the 'E-cube' is available.

The CellCube® has not as yet been widely used for the production of GMP-grade material, although GMP-grade retroviral vectors for gene therapy have been produced in the system (Wikstrom *et al.* 2004).

10.3.1.2c Multidisc propagators

Industrial requirements, predominantly in the vaccine industry, for large surface-area single units in which cells could be cultured for weeks and from which multiple harvests of (virus-containing) medium could be removed, resulted in the development of the multidisc propagator (Molin & Heden, 1969; McAleer *et al.* 1975). This normally took the form of a stack of titanium discs, mounted on a spindle with a gap of 2–3 mm between the discs, the whole assembly rotating within a stainless steel vessel partly filled with medium such that the cells were periodically immersed in the medium in a manner analogous to a roller bottle. Typically, an assembly of 100 disks would be contained in a 10-litre vessel (Elliott 1990). However, problems of getting an even distribution of cells on the plates and of confluent sheets sliding off the (vertical) surface of the plates led to a change of configuration, so that the plates were mounted one above the other with their surfaces horizontal and medium being circulated by a pump (Schleicher & Weiss 1968). In this form the systems were scaled up to over 200 litres, but had to be tilted to remove the medium and could only use one of the two surfaces on each plate for cell growth. Thus potential for scale-up was limited. Nevertheless, these vessels in their various forms were used for many years for the production of a variety of vaccines, including those against measles, mumps, rubella and polio (McAleer *et al.* 1975; Elliott 1990). However, use outside the vaccine industry appears to have been limited, and many of the processes that formerly used this system have been converted to fermenter-based systems using microcarriers, in order to overcome the problems already described, to utilize some of the advantages of microcarriers detailed below, and to achieve further increases in the size of the unit process.

10.3.1.3 Microcarriers

Microcarrier beads were conceived as a way of improving the volumetric efficiency (surface area per unit volume) of culture systems for attached cells. Using these tiny beads (typical diameter $200\,\mu m$), surface areas in excess of $30\,cm^2$ per cm^3 of culture medium are easily attainable for use in simple batch culture (van Wezel 1967), and higher values can be employed for more intensive (fed-batch or perfused) cultures. This compares with values of around 3 cm^2/cm^3 for a T-flask or cell factory. However, if one compares surface area with total culture unit volume (a real measure of the amount of incubator/laboratory space required for scale up), then microcarriers in a spinner flask will attain around 10 cm^2/cm^3, whereas a 225 cm^2 T-flask will give c.0.2 cm^2/cm^3, and a 10-stack cell factory c.0.55 cm^2/cm^3.

Originally DEAE-Sephadex A-50 beads were used (van Wezel, 1967), but the density of positive charges on these (equivalent to an exchange capacity of *ca* 4 meq/g) was too high for optimum cell attachment and growth. Reducing the charge density to about 2 meq/g overcame the problem (Levine *et al.* 1977, 1979), and nowadays negatively charged and amphoteric surfaces are also used. A wide range of microcarriers is now commercially available (see Table 10.1), made from a variety of materials. Some have special surface coatings to encourage the attachment of particular cell types, while others are made from materials that can be enzymatically digested in order to release the cells with minimum cellular damage.

A number of other physical properties are also important in defining the utility of microcarrier beads:

- *Density* The beads must be dense enough not to float, but not so dense that they are difficult to keep in suspension. Values between 1.02 and 1.04 g/cm^3 are most frequently used (except in fluidized-bed applications).
- *Transparency* A highly transparent bead material aids microscopic examination of the attached cells.

Table 10.1 Some commercially available microcarriers.

Type	Trade name	Source	Composition
Dextran	Cytodex 1	GE Healthcare	DEAE-substituted crosslinked dextran
	Cytodex 3	GE Healthcare	Collagen-coated dextran
Cellulose	Cytopore 1	GE Healthcare	Macroporous DEAE-coupled cellulose (1.1 meq/g)
	Cytopore 2	GE Healthcare	Macroporous DEAE-coupled cellulose (1.8 meq/g)
Plastic	Cytoline 1	GE Healthcare	Macroporous, silica-weighted polyethylene, density = 1.32 g/cm^3
	Cytoline 2	GE Healthcare	Macroporous, silica-weighted polyethylene, density = 1.03 g/cm^3
	M9650	Sigma-Aldrich	Gelatin-coated plastic beads
	2D MicroHex	NUNC	Flat polystyrene hexagons
	FACT III (F)	Solohill Engineering	Collagen-coated polystyrene
	Collagen (C)	Solohill Engineering	Collagen-coated polystyrene
	Hillex (H)	Solohill Engineering	Trimethylammonium-modified polystyrene
	Hillex II (H)	Solohill Engineering	Trimethylammonium-modified polystyrene
	Glass (G)	Solohill Engineering	Silica glass-coated polystyrene
	Plastic (P)	Solohill Engineering	Polystyrene
	Plastic Plus (PP)	Solohill Engineering	Polystyrene (cationic)
	Pronectin® F (PF)	Solohill Engineering	Recombinant fibronectin-coated polystyrene
Gelatin	CultiSpher-G	Percell Biolytica AB	Crosslinked gelatin
	CultiSpher-S	Percell Biolytica AB	Crosslinked gelatin (increased thermal stability)

Pronectin® F is a registered trade mark of Sanyo Chemical Industries.

- *Porosity* Most microcarrier beads are designed such that cells will grow on their surface. However, some are specifically designed to be macroporous, so that the cells actually grow within the body of the bead. This may have advantages in terms of protecting the cells from bead/bead impacts and high liquid shear environments that can be damaging to cells (Cherry & Papoutsakis 1988), although these mechanisms may only become significant at high agitation rates (Croughan *et al.* 1987, 1988). However, removal of the cells from the beads by trypsinization may be more difficult.
- *Diameter* A large number of beads/cm^3 is required, both to ensure that the suspension is essentially homogeneous, and to obtain the required volumetric efficiency. However, some cells find it difficult to migrate between beads, so each bead should have the ability to carry several hundred cells. Given these constraints, an optimum balance is often obtained with bead diameters in the 150–230 μm range. The size distribution should be as small as possible to ensure an even distribution of cells between the beads, otherwise smaller beads may be colonized at the expense of larger ones (van Wezel 1985).

Clearly, microcarrier beads must not be toxic to the cells, a problem during the early years of microcarrier use that has since been overcome (Giard *et al.* 1977). Similarly, the monomeric material from which they are made, and any other substance (such as surface coatings) liable to leach from the beads, must not be inhibitory to cell growth.

Although most microcarrier beads are spherical, this shape is not essential. Cylindrical cellulose carriers can be used (Reuveny 1990), and NUNC currently market flat, hexagonal polystyrene carriers (see Table 10.1).

One of the advantages of using microcarrier beads is that the culture of attached cells can be carried out in the same type of equipment that is employed for the homogeneous stirred culture of cells growing in suspension (Reuveny 1990), usually with slight modifications to help keep the relatively rapidly sedimenting microcarriers suspended. This has numerous benefits in terms of scale-up and environmental control, and also permits direct sampling and observation of the cells, something that may be difficult if not impossible in some other systems for culturing attached cells. These benefits have led to the widespread use of microcarrier culture, particularly in the vaccine industry where this technology was first commercialized more than 20 years ago (Meignier *et al.* 1980; Montagnon *et al.* 1984).

Macroporous microcarriers have also been utilized in both packed-bed (Looby & Griffiths 1988) and fluidized-bed systems. Formerly, a fluidized-bed system using weighted collagen carriers was available from Verax Corp., but although this is no longer available, a similar unit (the Cytopilot) can now be obtained from GE Healthcare for use with their Cytoline carriers (see Table 10.1). Such systems increase volumetric efficiency (see above) beyond that attainable in a homogeneous suspension system (G.E.Healthcare 2002)

10.3.1.4 Packed-bed systems

Packed-bed systems are used widely in the chemical and water treatment industries to obtain a large surface area in the minimum volume. In cell culture, a variety of packing materials has been investigated for attached cells (Spier 1985), with glass beads having been probably the most widely applied (Burbidge 1980; Robinson *et al.* 1980; Griffiths *et al.* 1982) and having been scaled up to at least 100 litres (Whiteside & Spier 1981). A simplified diagram of such a system is shown in Figure 10.3. However, packing and stability issues (grinding of the cells on the surface of the beads can occur if the bed is sub-optimally packed, or subject to vibration) and a complex relationship between bead diameter (and hence both surface area and inter-bead channel size), cell yield, uniformity of cell distribution, and medium flow rate (and direction) (Griffiths 1990, 2001) have

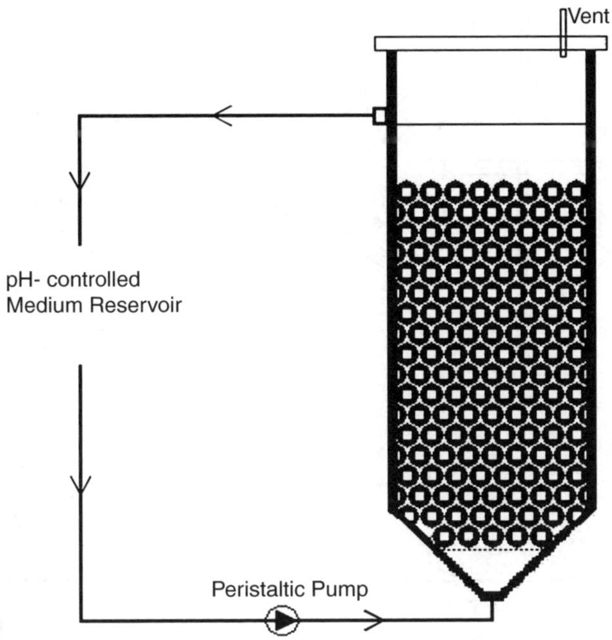

Figure 10.3 Highly simplified diagram of a glass bead packed-bed culture system.

meant that some of the theoretical advantages of the system have been difficult to realize. The system is best suited to the harvesting of a secreted product over a period of time, as it can be difficult to remove cells from the bed efficiently (Robinson *et al.* 1980; Griffiths 2001).

For further discussion of packed beds, see Section 10.3.3.1b.

10.3.2 Systems for Suspension Cells (or attached cells on microcarriers)

10.3.2.1 Spinner flasks

These are cylindrical, agitated vessels, varying in capacity from around 100 ml to 36 litres. They represent an intermediate level of scale-up between flasks or dishes and fermenters, but are far cheaper to purchase and use than a small fermenter of equivalent size. However, far less instrumentation and control is available than on a fermenter, and thus they are not generally of great use as scaled-down models for fermenter process development. Where they may be useful is in the early stages of cell inoculum growth for large-scale systems, for the small-scale production of material, either during the 'proof of principle' or early clinical trial stage of development of a biological medicine, or for the production of diagnostics.

Many different formats are available (two examples are shown in Figure 10.4), with different methods of agitation, different stirrer configurations, different aspect-ratio vessels, different cap types, and different degrees and ease of access to the gas and liquid phases. In general, the options available in this last category increase with the size of the vessel.

10.3.2.2 Shake flasks

Erlenmeyer-style flasks have been used for the culture of mammalian cells since the 1950s. These flasks are secured to a shaker apparatus, which mixes the contents of the flask and keeps the cells in suspension. This culture method is particularly useful for small to moderate

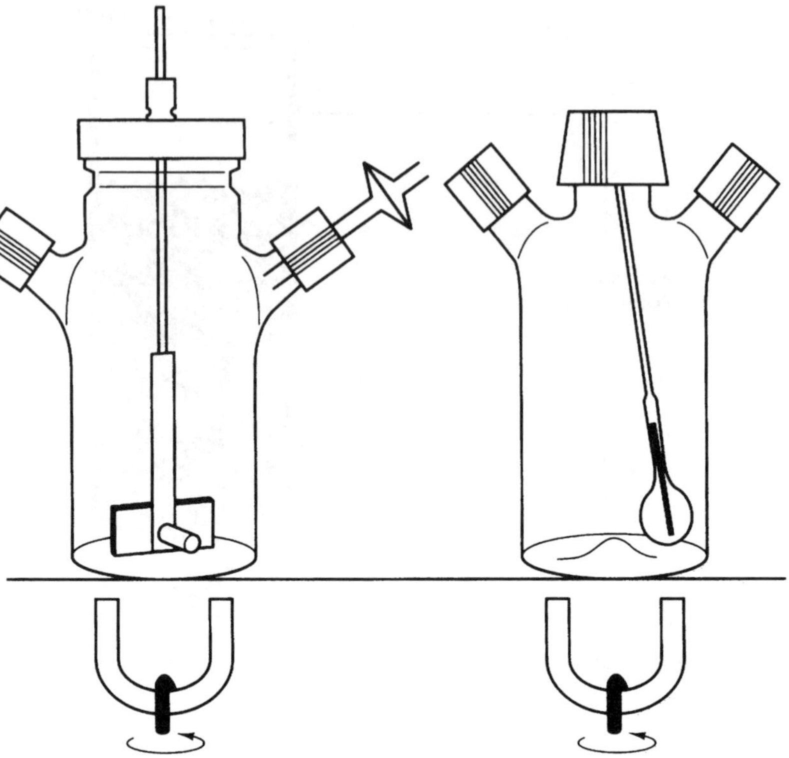

Figure 10.4 Two different types of spinner flasks. Reproduced from Doyle & Griffiths (2000) by permission of John Wiley & Sons Ltd.

volumes of cells having high oxygen requirements, such as insect cells (see Chapter 7). Both reusable and disposable flasks are generally available in sizes from 50 ml to 6 l. Baffles can be added to aid mixing, and vented caps can be used to increase gas exchange. Beyond about 6 litres (i.e. 2 litres of liquid volume) scale-up can only be achieved by the use of multiple flasks.

10.3.2.3 Culture bags

Bags suitable for use as vessels for cell culture have been available for at least 18 years (Schoof *et al.* 1988). Initial applications generally used static, gas-permeable bags in CO_2 incubators, and culture volumes were consequently limited to a few litres at most by the need for oxygen to diffuse through the bag walls. A far more suitable system for use in scale-up was first commercialized by Wave Biotech in 1999, and basically consists of a sterile, single-use, non-gas-permeable pillow-shaped bag that is mounted on a rocking apparatus (see Figure 10.5). The bag is inflated with a CO_2/air mixture suitable for the medium to be used, then the medium is introduced into the bag (and warmed, if necessary) after which cells are added. The bag retains a substantial head-space of CO_2/air, typically equal in volume to that of the medium. Mixing is achieved and gas exchange facilitated by the reciprocating rocking motion of the motorized platform on which the bag is mounted. Shear levels are claimed

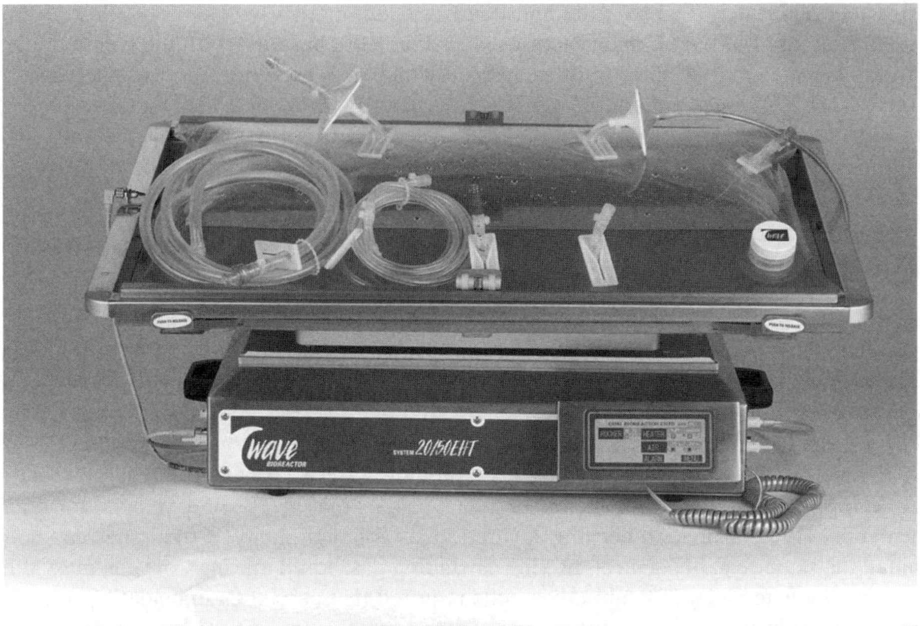

Figure 10.5 Wave bag system. Photograph courtesy of Wave Biotech LLC.

to be very low, as the cells mostly move with the bulk of the liquid, but the rocking must be regulated in order to avoid foaming. The bags come fitted with a fill tube, harvest tube, inlet filter, exhaust filter, sampling port, constant pressure relief valve, and ports for *in situ* pH and dissolved oxygen probes. The system has been used with animal cells both in suspension (Singh 1999; Fries *et al.* 2005) and on microcarriers (Namdev & Lio 2000), as well as plant and prokaryotic cells.

With retention of bag geometry and adjustment of the tilting parameters, linear scale-up is claimed from 100 ml up to 500 litres (www.wavebiotech.com; Pierce & Shabram 2004). Although the makers claim that there is no intrinsic limit to scale-up, other factors such as mechanical and safety problems associated with a large reciprocating mass could become a problem, as could temperature control which is still achieved by heat transfer through the bag wall.

The main advantages of the system, other than its simple principle and low-shear environment, are largely due to the disposable nature of the bag. Thus there is no cleaning or sterilizing (and validation of the cleaning and sterilizing procedures) to be performed by the user, nor is there any need to replace seals or undertake many of the other time-consuming maintenance procedures that are necessary for fermenters.

The manufacturer claims that the system is used by more than ten companies in facilities licensed for the production of human therapeutics. However, in at least some cases the system is used to grow cell inocula for seeding larger, more conventional culture systems rather than for the final full-scale manufacture of product. Nevertheless, up to a certain scale the system has attractive features, and many more systems might be in use but for two factors. The first has been a commercial decision by the manufacturer not to loan out equipment for trial, so some potential users not prepared to invest a substantial sum of money by buying a system 'sight unseen' have been unable to test its efficacy with their cell lines. The second has been problems with the *in situ* sensors (particularly for pH) and the resultant inability to monitor the culture environment accurately in real time and control it automatically. This has allowed the sensor, controller and bioreactor manufacturer Applikon, collaborating with the bag manufacturer Stedim, to bring to market the Appliflex system, which appears similar to the Wave system but has *in situ* pH, DO and temperature sensors interfaced with a controller that allows feedback control of the culture environment. At the time of writing, Appliflex systems are only available for volumes up to 50 litres. A similar system has also been introduced recently by Sartorius, the product of a collaboration with Wave Switzerland.

10.3.2.4 Fermenters

For the industrial-scale production of animal cells and their secreted products, by far the commonest method of culture is submerged culture in stirred-tank (Figure 10.6) or, less commonly, airlift (Figure 10.7) fermenters. This technology has been in use in the brewing and other industries for a great many years, and the principles and engineering involved are well understood (Bailey & Ollis 1986; van't Riet & Tramper 1991; Doran 1995; Asenjo & Merchuk 1994; Stanbury *et al.* 1999; Nielsen *et al.* 2002). Thus early attempts to use fermenters for large-scale mammalian cell culture employed the technology and designs that had been developed for microbial fermentation. However it soon became apparent that, although many of the issues and concerns remained the same [e.g. pH control, mixing (including gas/liquid mixing), oxygen mass transfer] the special characteristics of mammalian cells required that corresponding adaptations be made to fermenter hardware and control strategies. Some examples for a stirred-tank system are given in Table 10.2.

Although the basic issues were the same for airlift fermenters (Figure 10.7), fewer design changes were required, with the issue of cell damage by bubbles being the main problem.

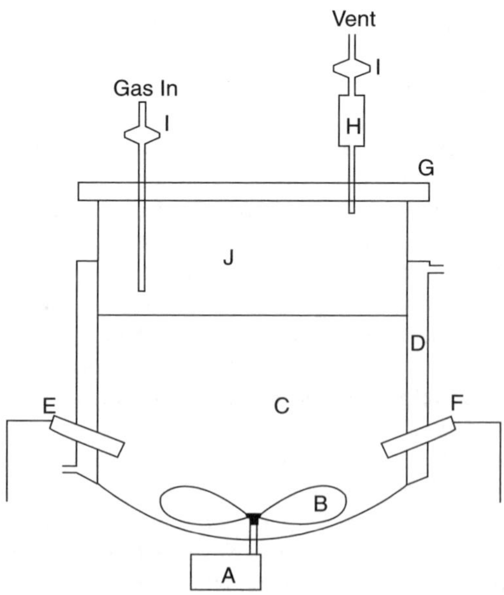

Figure 10.6 Simplified diagram of a stirred tank fermenter. (A) Impeller drive; (B) marine impeller; (C) cell suspension; (D) water jacket; (E) pH probe; (F) DO probe; (G) removable headplate; (H) condenser; (I) gas filter; (J) headspace.

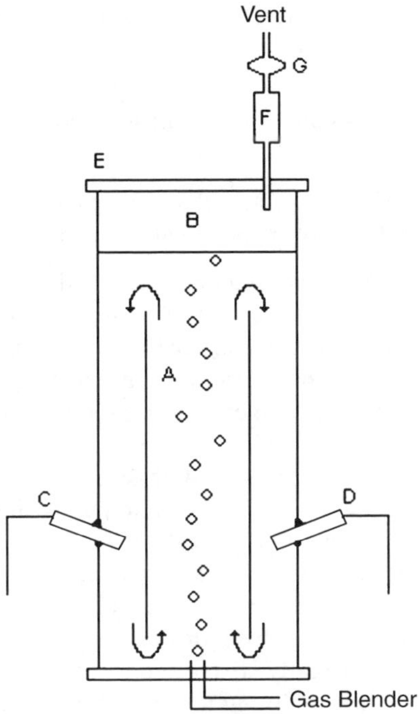

Figure 10.7 Simplified diagram of an airlift fermenter. (A) Cell suspension; (B) headspace; (C) pH probe; (D) DO probe; (E) headplate: (F) condenser; (G) gas filter.

Table 10.2 Some changes in stirred-tank fermenter design/control required to accommodate animal cells rather than microbes.

Microbial characteristic	Animal Cell characteristic	Resultant change
Low shear sensitivity	High shear sensitivity	Change impeller design (e.g. from Rushton turbine to marine impeller) and reduce rotation rate (tip speed); remove baffles
Low sensitivity to damage by bubble disengagement	High sensitivity to damage by bubble disengagement	Increase diameter:height ratio to maximize surface aeration (although effect decreases with culture volume); minimize gas bubble size used in sparging; add surface-active agent (but control foaming)
High medium viscosity	Low medium viscosity	Efficient mixing can be achieved with low-shear impellers rotating at low speeds; curved fermenter bottom improves mixing at low impeller speeds; magnetic drive couplings can be used, overcoming seal problems
High oxygen demands	Lower oxygen demands	Permits lower impeller speeds to be used; reduces the amount of (potentially damaging) sparging required

The stage at which one would change from using, say, a spinner flask to using a fermenter will depend on a number of factors, including:

- *Handling*: Although spinner flasks are available in sizes up to 36 l, manual handling becomes very difficult beyond about 20 l.
- *Control*: pH and DO control of spinner flasks may become difficult, as they cannot usually accommodate *in situ* probes, and are not supplied with feedback controllers.
- *Availability of services*: A fermenter may require some or all of the following services (although a small unit will probably only need a sub-set of these): 3-phase electricity; steam (possibly clean steam); cooling water; hot water; sterile water; drainage; CO_2; O_2; N_2; compressed air.
  If not already available, the expense of installing and running these services may discourage switching if product demand can still be met by using spinner flasks.

Thus the change from spinner flask to fermenter will generally occur when the required culture volumes are in the 10–100 l range, and a useful guide to purchasing fermenters at this scale has been published by Cino and Frey (1996, 1997). At the lower end of this range, glass vessels, and sterilization of vessels in an autoclave, are still an option, but manual handling, and the fragility and poor thermal conductivity of glass, rapidly become an issue. Consequently, all larger fermenters are made of stainless steel, and are sterilized *in situ*.

Scaling up of all fermenters, especially stirred tanks, cannot be proportional. For example, doubling a vessel's dimensions while retaining the same three-dimensional shape will increase its volume eightfold, but the air/liquid interface area of the headspace will only increase fourfold, decreasing the role that headspace gas exchange can play in oxygenation of the culture. Similarly, as a stirred tank increases in diameter it will usually be necessary to increase the diameter of the impeller in order to ensure good mixing. Yet if the impeller rotation rate is kept the same, the shear at the tips will increase, but of course the shear sensitivity of the cells will not change. Many other interacting factors will also vary to different extents. Thus at large scale a good understanding of the cells' physical and metabolic requirements (e.g. shear sensitivity, oxygen requirements) as well as the characteristics of the medium (e.g. density, viscosity, foaming properties) becomes essential in order to define the specification of the fermenter. This data should be supplied to the engineer designing the fermenter in order to help ensure that, when delivered and commissioned, the equipment performs satisfactorily. Currently, the largest stirred tank fermenters in use for animal cell culture have volumes of 20 000 l, whilst the largest airlifts are 5000 l.

The drive for optimum use of available capacity, as well as the desire to minimize both the volume of (expensive) medium to be purchased and the volume of spent, product-containing medium to be processed, has led to the intensification of fermenter processes. Originally, *batch processes* were employed, where a fixed volume of medium was added to the fermenter with the cells, and incubation was continued without addition of further medium until the endpoint of the culture was reached. However this was inefficient, as nutrient depletion tended to limit productivity. In principle this problem could be readily resolved by the addition of fresh medium and/or other nutrient solutions, and has led to the *fed-batch process*, which is now widely implemented in large-scale cell culture. This approach can ultimately become inefficient, however, due to the build up of inhibitory waste products. Overcoming the problem by removing a portion of the culture and replacing it with fresh medium led to the *repeat fed-batch process*. The logical extension of this is continuously to remove spent medium and replace it with fresh, to give a *perfusion process*. Perfusion processes in fermenters can be run either by removing cells from the system along with the medium in so-called chemostat mode (where the rate of removal of medium, and thus cells, has to be carefully balanced against the multiplication rate of the cells in order not to deplete the cell population), or the cell suspension can be passed through a cell-retention device, with the medium harvested while the cells are returned to the fermenter. Where perfusion is used for the production of a secreted product (and most currently available cell-derived medicines are secreted products), the second of these approaches is favoured as it uncouples production rate from cell multiplication rate. Cell retention devices to permit fermenters to be used in this mode can take a number of forms, such as spin-filters within the fermenter, or tangential-flow filters, acoustic filters, gravitational settlers or continuous-flow centrifuges in an external recirculation loop, and each of these is dealt with in detail in Chapter 16.

Further discussion of the ramifications of culture mode on fermenter design, along with numerous other factors influencing fermenter design and construction, can be found in Chapter 14. For an assessment of some of the relative merits of the different culture modes in a real (i.e. uncertain) world, see Lim *et al.* (2006).

10.3.3 Systems for Either Suspension or Attached Cells (not on microcarriers)

These systems can be broadly classified on the basis of whether the medium is mixed within a single compartment, or is intentionally segregated into two compartments separated by a semi-permeable membrane.

10.3.3.1 Single-compartment systems

10.3.3.1a CelliGen®

The CelliGen® is a small to medium-sized, round-bottomed, stirred-tank bioreactor from New Brunswick Scientific that is best characterized by the mixing principle of its cell lift impeller. The rotation of the ports of this specially designed impeller through the upper level of the culture medium creates a negative pressure inside the hollow impeller shaft. This pulls medium up from the bottom of the vessel, through the hollow shaft and out of the impeller ports, thus inducing circulation of medium within the vessel (Figure 10.8). The rate of rotation of the impeller required to induce this flow is relatively low, and thus shear at the impeller tips is kept within acceptable limits. As with other stirred-tanks, it can be run in batch, fed-batch, or perfusion mode.

This bioreactor has been available, in slightly different forms, for more than 20 years, and has attracted a relatively small but loyal user base. It can be used not only as a stirred tank with cells growing in suspension or on microcarrier beads, but also as a perfusion device for use with packed-bed cultures using FibraCel® (see below) or similar bed materials that can both trap or encourage the adherence of suspension cells and support the growth of attached cells. It seems to have found most favour either for the culture of hybridomas (Wang *et al.* 1992; Mercille *et al.* 2000; Kunkel *et al.* 2000) or, with microcarriers or packed beds, for the growth of attached cells for the production of virus particles (Mendonca *et al.* 1994; Gallegos *et al.* 1995; Wang *et al.* 1996; Zhang *et al.* 1997; Hu *et al.* 2000; Merten *et al.* 2001), most notably Gendicine (see below).

The CelliGen® is currently supplied in sizes ranging from 2.2 litres to 14 litres working volume. If required, it can be fitted with conventional impellers, spin filters, etc., so that it can also be used as a normal small stirred tank.

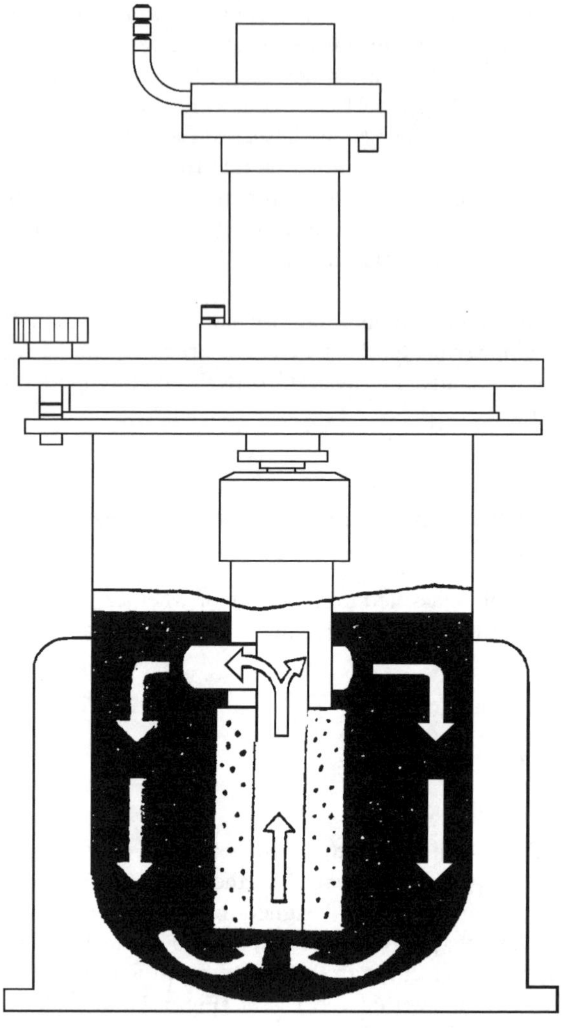

Figure 10.8 Diagram of New Brunswick Scientific's patented low-shear cell lift impeller as used in the CelliGen.

10.3.3.1b Systems for FibraCel-type disks

Fibra-Cel® is a matrix sold by New Brunswick Scientific (NBS) and composed of polyester non-woven fibre and polypropylene (in a 1:1 mixture by volume), ultrasonically bonded together and cut into 6-mm diameter disks. These are then treated to produce a net surface charge. The manufacturer claims that this surface charge facilitates both the attachment of anchorage-dependent cells, and the adherence of suspension cell lines. CESCO Bio of Taiwan produces a similar matrix called BioNOC II®.

These disks are frequently used in packed-bed reactors, although they can also be used in normal stirred vessels such as spinner flasks. Their effective surface area in a packed bed is around $120\,cm^2/cm^3$, much higher than that normally obtained with microcarriers in suspension, and comparable to that in hollow fibre culture units. At small scale, the packed beds can be run

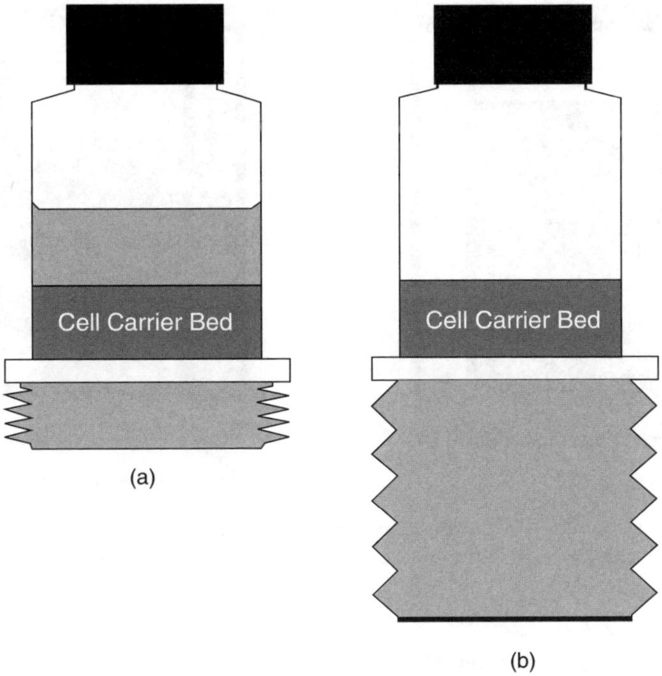

Figure 10.9 Diagram of the principle of action of the BelloCell® culture bottle. (a) As the bellows compress, medium flows through the bed of cell carriers to immerse all the cells. (b) As the bellows expand, medium flows back down the bed, exposing it to the atmosphere.

in systems such as the CelliGen® (see Section 10.3.3.1a) or in purpose-made 'bellows bottles' in the CESCO Bio BelloCell® system (also sold by NBS under the FibraStage™ name – NBS, 2005) In these 'bellows bottles' the packed bed is suspended across the middle of a plastic bottle with a bellows-shaped bottom. Medium is moved up through the bed as the bellows are compressed, and drains out as the bellows expand (Figure 10.9). This cycle repeats indefinitely, achieving mixing of the medium simply by the pumping action of the bellows. A larger-scale system using the same principle is available from CESCO Bio, called the TideCell®, in which the matrix is packed in the inner of two concentric chambers that are interconnected, and air pressure is used to move the medium first into the chamber containing the matrix, then out into the outer chamber (Figure 10.10). Vessels of 5 litres and 25 litres working volume are currently available.

In both the BelloCell® and TideCell® systems, oxygen transfer is achieved by the periodic exposure of the cells to the air, although one might speculate that there are significant differences in the supply of oxygen to cells at the top and bottom of the bed, as the proportion of the cycle time during which the cells are exposed to the air will decrease as one moves down the bed.

It is more difficult to recover cells from such matrices than from microcarriers, and thus they are not recommended for biomass production. However, the packed-bed system has proven useful for the production of secreted or released cell products, and a 14-litre CelliGen Plus® reactor (see Section 10.3.3.1a) with 200 g of FibraCel® disks is employed for the production of the adenoviral vector used in the world's first licensed gene therapy, Shenzhen SiBiono GeneTech's Gendicine (Peng 2004).

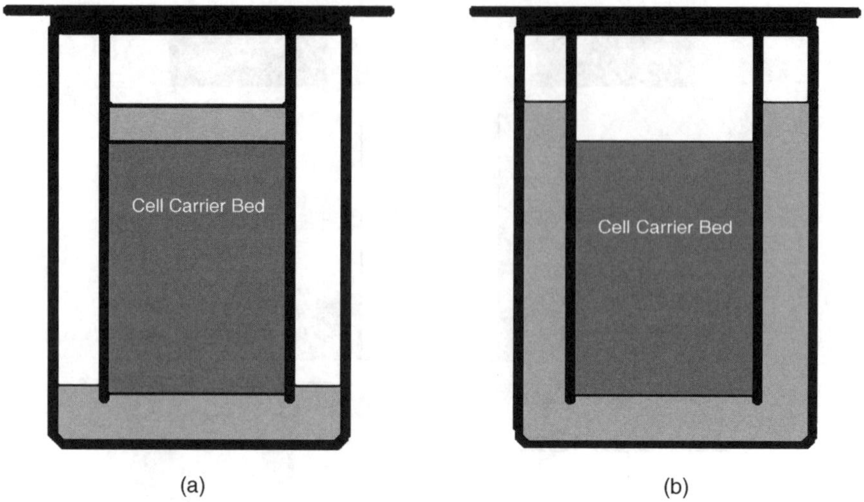

Figure 10.10 The TideCell®, shown with (a) the bed of cell carriers immersed in the culture medium, and (b) the bed of cell carriers exposed to the atmosphere.

10.3.3.2 Dual-compartment systems

10.3.3.2a Dialysis tubing culture systems

These are the simplest type of dual compartment system, being composed of a cell suspension retained within a length or lengths of dialysis tubing sealed at both ends. The tubing is then suspended in a much larger volume of medium within a roller bottle, fermenter or other stirred vessel (see, for example, Pannell & Milstein 1992). The medium and cells are incubated at culture temperature, and high molecular weight secreted products are generated by the cells and retained within the semipermeable dialysis tubing for later harvesting. Such systems are popular for generating quantities of monoclonal antibodies for research use, but are generally home-made, and are not suitable for the generation of material for use in humans.

10.3.3.2b Hollow-fibre systems

Hollow-fibre cell culture systems were first designed with the intention of mimicking the *in vivo* cell environment (Knazek *et al.* 1972). Within tissues, cells live immobilized at high density, and are perfused via capillaries having semipermeable walls. Fluid (blood) circulating through the capillaries delivers oxygen and nutrients to the cells whilst removing CO_2 and other waste products. Hollow-fibre culture systems work in a similar way, with the cells being immobilized on the outside [i.e. in the extracapillary space, (ECS)] of artificial hollow fibres made from semipermeable membranes, while medium is circulated through the lumen of the fibres. This separation of the cells and other contents of the ECS from the majority of the nutrient medium gives the hollow-fibre system a number of advantages, particularly for the production of high molecular weight secreted products such as monoclonal antibodies. In such cases, the molecular weight cut-off (MWCO) of the fibres is chosen such that it is very much smaller than the molecular weight of the product of interest, so that the product is retained in the ECS with the cells whilst nutrients and waste products of low molecular weight equilibrate across the membrane. Thus for the production of an IgG antibody (typical molecular weight 160 kD) an MWCO of 10 kD might be chosen. Only small volumes of high molecular weight nutrients/growth factors then need to be introduced directly into the ECS, and the antibody product can be removed at the same time in a similarly small volume. Some of the advantages of hollow-fibre systems arise directly from the above properties:

- High secreted product concentrations can be attained, typically 20- to 100-times those obtained using the same cells and media in T-flasks (Hanak & Davis 1995). In part this is because the product is contained in only a small fraction (5–10 %) of the total nutrient medium required by the cells, not diluted in the entire volume as would be the case in a homogeneous system.
- Requirements for high molecular weight supplements, such as serum, are low. If needed at all, these need only be added to the small volume of medium in the ECS, and it may be possible to reduce or stop their use without loss of productivity (Tiebout 1990) once the culture is well established, as the cells frequently secrete factors that support their own viability, and these may be retained in the ECS.
- A high ratio of product to medium-derived contaminants is attained. This is a consequence of the first point above, but the effect may be increased further by the second.
- The cells are maintained in a low-shear environment. The cells are separated from the main medium flow in the lumen of the fibres, and oxygenation is performed in a remote gas exchange cartridge, minimizing the cells' exposure to potentially damaging gas bubbles.

Other advantages arise indirectly:

- For equivalent cell numbers, a hollow-fibre system is much smaller than a homogeneous system (e.g. a stirred tank). This is because viable cells are largely retained within the hollow fibre cartridge, meaning that cell densities reach levels of around 10^8/ml in the ECS. Consequently, a bench-top system such as the Biovest AcuSyst-Jr (Figure 10.11) or Maximizer can contain in the region of $1–4 \times 10^{10}$ cells, whereas a production-scale system such as the Biovest Xcell (Figure 10.12) or Xcellerator, despite only being the size of a large fridge/freezer, can maintain in culture $2–4 \times 10^{11}$ cells.

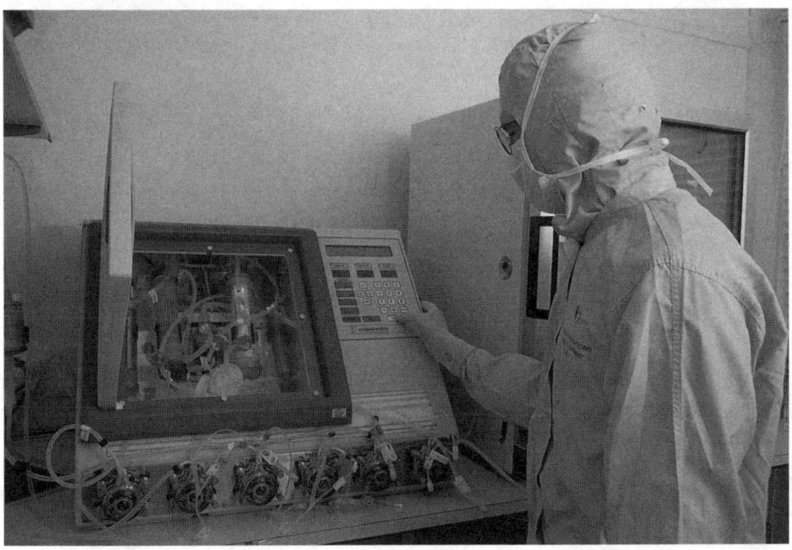

Figure 10.11 Biovest AcuSyst-Jr. bench-scale hollow fibre culture system.

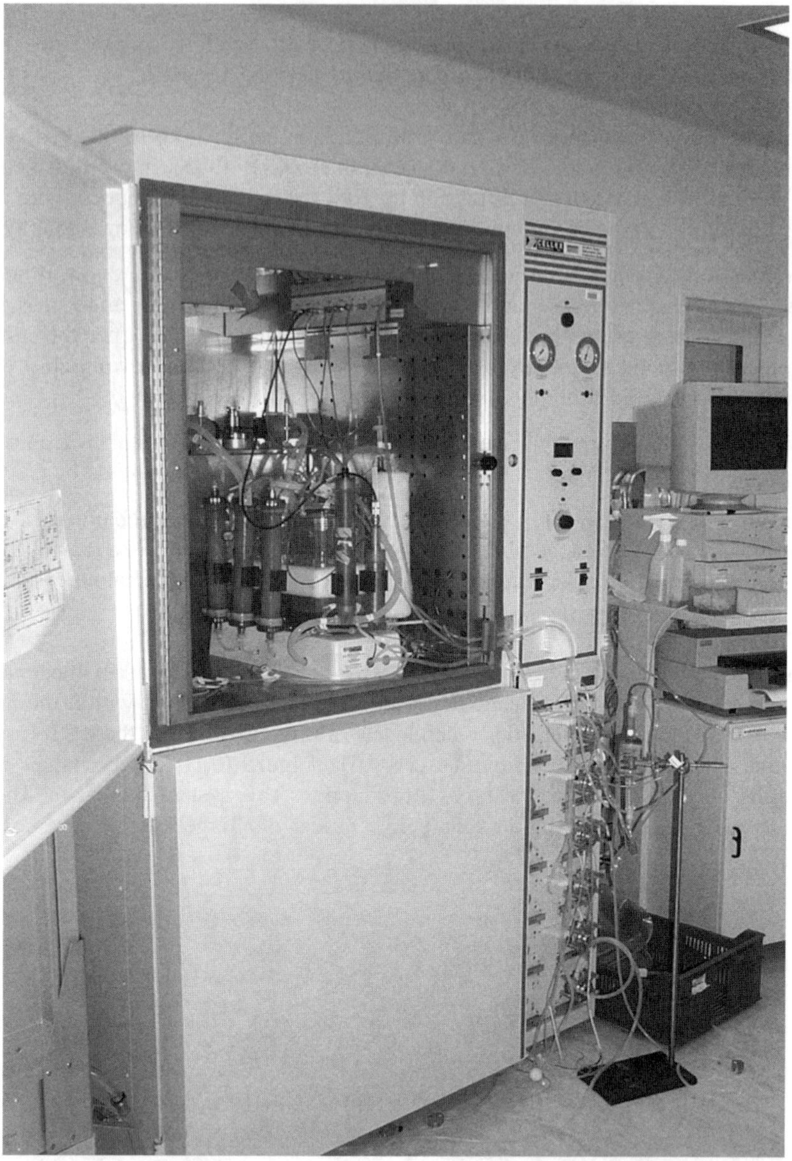

Figure 10.12 Biovest Xcell production-scale hollow fibre culture system.

- Because the units are relatively small, they can be housed within rooms of normal height. This is in contrast to fermenters, where those much larger than 100 litres require additional height to permit removal of the head-plate for cleaning and servicing, and large vessels may take up several storeys of a building specially built or adapted for the purpose. This could have major implications in the case where hollow-fibre systems could be housed in an existing facility but fermenters would require a new building.
- These systems, even at full production scale, are relatively easy and rapid to install and set up, as the only services they require are CO_2 and single-phase electricity, in contrast to a fermenter which may require up to ten different services (see Section 10.3.2.4).

There are, of course, disadvantages to hollow-fibre systems:

- The nature of the supporting material of the fibre walls entraps suspension cells effectively, but is less well suited to the culture of attachment-dependent cells. While such cells can be grown in hollow-fibre systems, they often require larger inocula for the initiation of a successful culture.
- Because of the entrapment of the cells, it is not possible during the course of a culture (which can last many months) to remove a sample of cells for the enumeration of viable cell concentrations and similar purposes. Thus surrogate markers such as lactate production rate must be used, and the relationship of such parameters to cell number will vary not only with the cell line in use, but also with culture conditions such as pH, temperature, and the phase of the culture.
- Hollow-fibre culture is suited to the production of secreted products, but because it is extremely difficult to remove cells from the ECS effectively, it is not suited to the production of biomass.

Scale-up beyond a certain point is limited. Scale-up is achieved not by increasing the size of the hollow-fibre cartridges (as this would merely increase the gradients of oxygen, nutrients and waste products, leading to inefficient colonization of the cartridge), but by the use of multiple cartridges. Whilst up to 20 can be accommodated within a single production-scale machine, further scale-up is achieved through the use of multiple machines. At this point no further advantages of scale-up are available, and more machines mean an essentially linear increase in staffing levels, with batch sizes remaining unaltered. This has limited the application of hollow-fibre systems to markets requiring no more than a few kilograms of secreted protein per year, as a good cell line in a single production-scale machine will probably not produce more than 1–2 kg of product per year. Thus Cytogen's ProstaScint, a radiolabelled monoclonal antibody used for the *in vivo* diagnosis of prostate cancer. is produced in hollow-fibre culture, and was licensed by the FDA in 1997 (and by Health Canada in 2002). However, its unit dose size is only 0.5 mg (see http://www.cytogen.com), very much smaller than most therapeutic monoclonal antibodies where the dose sizes are tens or, more commonly, hundreds of milligrams. To satisfy the world-wide markets for these, tens to hundreds of kilograms of material are required per year, and culture is performed almost exclusively in fermenters. Hollow-fibre systems have, however, proved very popular in the production of monoclonal antibodies for *in vitro* diagnostics, where rather smaller quantities of material are required.

When considering scale up/scale down in hollow-fibre systems, care should be taken (see Section 10.4.1) that the smaller system represents the larger accurately. For example, the AcuSyst-Jr or Maximizer is, in the author's experience, a good scaled-down model for the AcuSyst Xcell. However, attempts to scale-down further would necessitate the use of equipment like the Unisyn C100 that has no in-line pH control, no constant medium replenishment, no constant harvesting from the ECS, and different hollow-fibre membranes. Data obtained from such systems may at best be of limited use, and may even be positively misleading.

10.3.3.2c CELLine flasks

These flasks from Integra Biosciences look similar at first sight to T-flasks, and have a similar footprint, but are actually two-compartment bioreactors. The cells are separated from the bulk of the medium by a semipermeable membrane, as is also the case in hollow-fibre culture systems and the MiniPERM bioreactor (see below). Some of the advantages of a hollow-fibre system, i.e. the higher secreted product concentrations, lower serum requirements, and higher ratio of product-to-medium derived impurities, can be realized at a smaller scale and without the expense of a hollow-fibre system, with the added advantage that cells can be easily removed from the vessel. However, flasks must be kept within a CO_2 incubator, there is not the degree of control available that there is in a hollow-fibre system, and scale-up can only be achieved by the use of multiple units. Thus realistically these flasks can only deliver a limited degree of scale-up from T-flasks.

10.3.3.2d miniPERM vessels

miniPERM vessels can be viewed as a two-compartment roller bottles modified to improve gas exchange. The main, reusable compartment (volume *ca* 550 ml) contains the bulk of the medium and is separated from the disposable cell-containing compartment (volume 5–35 ml) by a semipermeable membrane. The opposite face of the cell compartment is a silicone membrane through which oxygen and carbon dioxide can be exchanged with the atmosphere within an incubator. Mixing within the two compartments should be better than in CELLine flasks, as the vessel is rotated in the same way as a roller bottle (although normally at higher speeds). It is claimed that cell densities beyond 1×10^7/ml (for hybridoma cells) can be attained in the cell compartment (Heraeus 1995). However, as with CELLine flasks, these vessels must be kept within a CO_2 incubator, but also require equipment to rotate them. Scale-up can only be achieved by the use of multiple units, and there is neither the degree of control available that there is with hollow-fibre systems, nor the simplicity inherent in normal roller bottles that facilitates automation. Thus these vessels too are not likely to be used to produce large amounts of material but, like CELLine flasks, have proven popular for producing small quantities of monoclonal antibodies, in place of ascitic fluid production *in vivo* (Bruce *et al.* 2002).

10.3.3.2e Perfused rotary cell culture technology vessels

The Synthecon Rotary Cell Culture Technology systems are based on the rotating wall vessel bioreactor designed by NASA (NASA, 1997). Originally intended for the three-dimensional growth of cells under pseudo-microgravity, certain of their perfused systems combine features of the miniPERM and hollow-fibre systems. Like the miniPERM, the cells are cultured in one compartment of a rotating vessel, and like all dual-compartment systems, the cells are separated from the bulk of the medium by a semipermeable membrane. However like a hollow-fibre system, fresh medium is continuously fed into (and spent medium removed from) the compartment containing the bulk of the medium, and oxygen is supplied to the culture by passing the medium in this compartment through an external oxygenator. These systems are currently less sophisticated than most medium- and larger-scale hollow-fibre systems, and are not yet intended for continuous harvesting of cell-secreted products, thus again placing them somewhere between the miniPERM and hollow-fibre systems. For further details, see the Synthecon website – http://www.synthecon.com/index.htm.

10.3.3.2f Encapsulation

This technology, in a number of different formats including hollow spheres of polylysine (Duff 1985; Rupp 1985) or cellulose (Kloth *et al.* 1995), alginate spheres (Griffiths 1988), and agarose beads (Nilsson *et al.* 1983, 1987), was taken up enthusiastically in the 1980s and early 1990s. The perceived advantages were various. All the methods could potentially protect fragile cells from the shear forces encountered in some culture systems. Hollow spheres had to be semipermeable in order to allow nutrients and oxygen to reach the cells, and to allow waste products out, but the permeability could be easily controlled such that these requirements could be met whilst retaining high molecular weight secreted products such as monoclonal antibodies within the spheres. After harvesting the spheres, these products could then be released at high concentration. (Other encapsulation methods were simpler and allowed the cells access to high molecular weight growth stimulators in the medium, but the particles were not part of a dual-compartment system and did not concentrate secreted products, and the beads produced are perhaps more accurately viewed as macroporous microcarriers.)

A number of problems were encountered with encapsulation technology, most notably difficulties in getting an adequate flux of nutrients and oxygen into, and waste products out of, the beads (and thus of maintaining cell viability), as well as the logistical and technical complexities of applying the technology at large scale (Griffiths 1988). Consequently, the use of this approach became increasingly limited.

Interest has revived in recent years, however, as the technology's potential for use in cell and gene therapy has become apparent. In this case, cells are encapsulated before being introduced into the body of the recipient. The encapsulation process is designed to protect the cells from attack by, or dissemination in, the body into which they have been introduced. As the cells used are usually not those of the host, this prevents immune rejection, or the need to use immunosuppressive drugs. However, the encapsulating material still permits the cells to take up nutrients from the body and secrete the therapeutically active substance that they produce. This approach is proving promising for the treatment of a variety of disorders, such as epilepsy (Guttinger 2005), liver failure (Mai *et al.* 2005), diabetes (Black *et al.* 2006), cancer (Li *et al.* 2006), and haemophilia (Wen *et al.* 2006), and is already in the process of being commercialized (see, for example, http://www.lctglobal.com/).

10.4 OTHER ISSUES

When choosing a culture system for scale-up, other issues must be addressed besides those directly related to its ability to support growth and product generation at the scale required. These largely fall into three categories:

10.4.1 Modelling Potential

The need to model the full-scale production system must be considered during scale-up. It may be that at the early stages of product development, or for the initial steps of cell culture expansion, any similarity to a full-scale system may be immaterial. However, at some stage during development it will become important that the systems used at intermediate scale can be run and controlled in such a way that they mimic the operation of the (proposed) full-scale production unit. This will be essential to predict the performance of the full-scale system, develop and optimize the process, and (possibly) to produce clinical trial material equivalent to that which will be obtained at full scale. Even once full-scale production is in place, these systems will have an important role to play as scaled-down systems for process characterization, support and development studies (Rathore *et al.* 2006).

10.4.2 Cost Implications

These will apply to any large-scale culture system and include:

- *Cost of purchase*
- *Projected lifetime* – This is the projected time in use, or number of batches that can be produced, before the equipment becomes obsolete or uneconomical to maintain.
- *Cost of installation* – This includes providing buildings and all the required services (see Chapter 12 and Chapter 14).
- *Cost of operation* – This includes salaries (i.e. the number, and level of qualification) of the staff required, as well as the cost of running the buildings and services required, the cost of consumables (e.g. media, gases), and other necessary but indirect costs incurred during operation.

- *Cost of servicing* – For example, this includes the cost of seals, filters, maintenance staff, etc.
- *Down time* – When the system cannot be used for product generation, e.g. when it is being cleaned, maintained, prepared and sterilized, it is not generating revenue by making product.

10.4.3 Regulatory Implications

These apply specifically to systems used to generate medicines. All of them have associated cost implications.

- *Materials of manufacture/surface finishes* – These must be of the highest standard (see Chapter 14).
- *Ability to effectively clean/sanitize/sterilize* – Only certain types of fittings and other design features are suitable when generating medicines (again see Chapter 14).
- *Ease of validation* – Validation is expensive (see Chapter 15), so simple systems that can be easily validated require less time and money for this purpose.
- *Amount of user validation required* – If the supplier can provide a ready-made validation package for at least part of the system, then the time required and cost incurred by the user for in house validation will be reduced.
- *Routine/non-routine nature of validation* – With some equipment, the validation requirements are well understood, but with others this is less so. Thus, for example, the validation needed for a stirred-tank fermenter is better defined than for some more innovative systems, which may consequently require more extensive validation studies.
- *Time required for validation* – As well as depending on the other aspects of validation mentioned above, the time required for validation may also depend on the nature of the culture being run in the equipment. Thus batch fermentation should take less time to validate than fed-batch culture, whilst perfusion will take still longer, simply due to the culture periods involved. Thus an incremental approach to culture period may be wise, validating a short period initially to get product to market in a timely fashion, then increasing the validated culture duration subsequently.

10.5 SUMMARY

The choice of system for scaling up animal cell culture for the production of medicines is complex. Many systems are available, and each has its own niche catering for different cell types, products, and product quantities. Define the properties of your cells (e.g. adherent/non-adherent) and projected market demand as early as possible, then investigate the viable options taking into account not only the biological and engineering characteristics of the systems, but the cost and regulatory implications up to and including full-scale licensed production.

REFERENCES

Asenjo J, Merchuk J (1994) *Bioreactor System Design*. Marcel Dekker, Abingdon, UK.

Bailey JE, Ollis D (1986) *Biochemical Engineering Fundamentals*, McGraw-Hill, Columbus, OH, USA; second edition.

Berson RE, Pieczynski WJ, Svihla CK, Manley TR (2002) *Biotechnol. Prog.*; 18: 72–77.

Bishop NE, Hugo DL, Borovec SV, Anderson DA (1994) *J. Virol. Meth.*; 47: 203–216.
Black SP, Constantinidis I, Cui H, Tucker-Burden C, Weber CJ, Safley SA (2006) *Biochem. Biophys. Res. Commun.*; 340: 236–243.
Bruce MP, Boyd V, Duch C, White JR (2002) *J. Immunol. Methods*; 264: 59–68.
Burbidge C (1980) *Dev. Biol. Stand.*; 46: 169–172.
Cino J, Frey S (1996) *BioPharm.*; 9: 52–56.
Cino J, Frey S (1997) *BioPharm.*; 10: 42–48.
Cherry RS, Papoutsakis ET (1988) *Biotechnol. Bioeng.*; 32: 1001–1014.
Croughan MS, Hamel J-F, Wang DIC (1987) *Biotechnol. Bioeng.*; 29: 130–141.
Croughan MS, Hamel J-F, Wang DIC (1988) *Biotechnol. Bioeng.*; 32: 975–982.
Dickinson LA, Kohwi-Shigematsu T (1995) *Mol. Cell. Biol.*; 15: 456–465.
Doran P (1995) *Bioprocess Engineering Principles*, Elsevier, Amsterdam, Netherlands.
Duff RG (1985) *Trends Biotechnol.*; 3: 167–170.
Elliott AY (1990) In *Large-scale Mammalian Cell Culture Technology.* Ed Lubiniecki ES. Marcel Dekker, New York; 207–326.
Fries S, Glazomitsky K, Woods A, Hsu A, Olewinski R, Robinson D, Chartrain M (2005) *BioProcess International*; October, 36–44.
Gallegos GRM, Espinosa LEL, Ramos RL, Kretschmer SR, Aguilar SA (1995) *Arch. Med. Res.*; 26: 59–63.
G.E. Healthcare (2002) http://www6.amershambiosciences.com/aptrix/upp00919.nsf/4a0f132842ea4d354-a25685d0011fa04/a0a6a86bfc6259b4c1256eb400417e4f/$FILE/ATTQWRFP/18114667AB.pdf
Gey G (1933) *Am. J. Cancer*; 17: 752–756.
Giard DJ, Thilly WG, Wang DIC, Levine DW (1977) *Appl. Env. Microbiol.*; 34: 668–672.
Goetz R, Kolbeck R, Lottspeich F, Barde YA (1992) *Eur. J. Biochem.*; 204: 745–749.
Griffiths JB (1988) In *Animal Cell Biotechnology.* Ed Spier RE, Griffiths JB. Academic Press, London; Vol. 3, 179–220.
Griffiths JB (1990) In *Large-scale Mammalian Cell Culture Technology.* Ed Lubiniecki ES. Marcel Dekker, New York; 217–250.
Griffiths JB (1995) In *Cell and Tissue Culture Laboratory Procedures.* Eds Doyle A, Griffiths JB, Newell DG. John Wiley & Sons, Chichester, UK; Module 28A1.
Griffiths JB (2001) *Mol. Biotechnol.*; 17: 223–238.
Griffiths JB, Thornton B, McEntee I (1982) *Dev. Biol. Stand.*; 50: 103–110.
Guttinger M, Fedele D, Koch P, Padrun V, Pralong WF, Brustle O, Boison D (2005) *Epilepsia*; 46: 1162–1169.
Hagen AJ, Aboud RA, DePhillips PA, Oliver CN, Orella CJ, Sitrin RD (1996) *Biotechnol.Appl.Biochem.*; 23: 209–215.
Heraeus (1995) *Instruction Manual: Culture of Hybridoma Cells in the miniPERM Bioreactor.* Heraeus Instruments, Hanau, Germany.
Hanak JAJ, Davis JM (1995) In *Cell and Tissue Culture: Laboratory Procedures.* Eds Doyle A, Griffiths JB, Newell DG. John Wiley & Sons, Chichester; Module 28D:3.
Handa-Corrigan A (1990) In *Animal Cell Biotechnology.* Eds Spier RE Griffiths JB. Academic Press, London; Vol. 4, 123–132.
Hu YC, Kaufman J, Cho MW, Golding H, Shiloach J (2000) *Biotechnol. Prog.*; 16: 744–750.
Kloth C, Riese C, Cho M, Buchholz R, Marx U (1995) In *Animal Cell Technology: Developments Towards the 21st Century.* Eds Beuvery EC, Griffiths JB, Zeijlemaker WP. Kluwer, Dordrecht, Netherlands; 765–769.
Knazek RA, Gullino PM, Kohler PO, Dedrick RL (1972) *Science*; 178: 65–67.
Knight E (1977) *Appl.Environ.Microbiol.*; 33: 666–669.
Kunitake R, Suzuki A, Ichihashi H, Matsuda S, Hirai O, Morimoto K (1997) *J. Biotechnol.*; 52: 289–94.
Kunkel JP, Jan DC, Butler M, Jamieson JC (2000) *Biotechnol. Prog.*; 16: 462–470.
Lay AJ, Jiang XM, Kisker O, Flynn E, Underwood A, Condron R, Hogg PJ (2000) *Nature*, 408: 869–873.
Lee S-Y, Kim SH, Kim VN, Hwang JH, Jin M, Lee J, Kim S (1999) *J.Biomed.Sci.*; 6: 8–17.
Levine DW, Wang DIC, Thilly WG (1977) In *Cell Culture and its Application.* Ed Acton RT, Lynn JD Academic Press, New York; 191–216.
Levine DW, Wang DIC, Thilly WG (1979) *Biotechnol. Bioeng.*; 21: 821–845.
Li AA, Shen F, Zhang T, Cirone P, Potter M, Chang PL (2006) *J. Biomed. Mater. Res. B Appl. Biomater.*; 77: 296–306.

Lim AC, Washbrook J, Titchener-Hooker NJ, Farid SS (2006) *Biotechnol. Bioeng.*; 93: 687–697.
Litjens T, Bielicki J, Anson DS, Friderici K, Jones MZ, Hopwood JJ (1997) *Biochem. J.*; 327: 89–94.
Loewen N, Bahler C, Teo WL, Whitwam T, Peretz M, Xu R, Fautsch MP, Johnson DH, Poeschla EM (2002) *Inv. Ophth. Vis. Sci.*; 43: 3686–3690.
Looby D, Griffiths JB (1988) *Cytotechnology*; 1: 339–346.
Mai G, Huy NT, Morel P *et al.* (2005) *Xenotransplantation.*; 12: 457–464.
Mann GF (1977) *Dev. Biol. Stand.*; 37: 149–152.
McAleer WJ, Bunyak EB, Weibel RE *et al.* (1975) *J. Biol. Stand.*; 3: 381–384.
Meignier B, Mougeot H, Favre H (1980) *Dev. Biol. Stand.*; 46: 249–256.
Mendonca RZ, Vaz-de-Lima LR, Oliviera ML, Pereira CA, Hoshino-Shimizu S (1994) *Braz. J. Med. Biol. Res.*; 27: 1575–1587.
Mercille S, Johnson M, Lanthier S, Kamen AA, Massie B (2000) *Biotechnol. Bioeng.*; 67: 435–450.
Merten O-W, Cruz PE, Rochette C *et al.* (2001) *Biotechnol. Prog.*; 17: 326–335.
Molin O, Heden CG (1969) *Progress in Immunological Standardisation.* Karger, New York; Vol. 4, 106–110.
Montagnon BJ, Fanget B, Vincent-Falquet JC (1984) *Rev. Infect. Dis.*; 6, Suppl. 2: S341–344.
Muzzio FJ, Unger DR, Liu M, Bramble J, Searles J, Fahnestock P (1999) *Biotechnol. Bioeng.*; 63: 185–196.
Namdev P, Lio P (2000) *BioPharm*; 13: 44–50.
Nardelli L, Panina GF (1976) *Develop. Biol. Standard.*; 37: 133–138.
NASA (1997) http://science.msfc.nasa.gov/newhome/br/bioreactor.htm
Nelson K, Bielicki J, Anson DS (1997) *Biochem. J.*; 326: 125–130.
Nielsen J, Villadsen J, Liden G (2002) *Bioreaction Engineering Principles.* Springer, New York, USA; second edition.
Nilsson K, Mosbach K (1987) *Dev. Biol. Stand.*; 66: 183–193.
Nilsson K, Scheirer W, Merten O-W *et al.* (1983) *Nature*; 302: 629–630.
NBS (2005) *Fibrastage™ Disposable, High-yield Cell Culture System.* New Brunswick Scientific, Edison, NJ, USA. See http://www.nbsc.com/files/products/cellculture/fibrastagebrochure.pdf
Ohtaki T, Ogi K, Masuda Y, Mitsuoka K, Fujiyoshi Y, Kitada C, Sawada H, Onda H, Fujino M (1998) *J. Biol. Chem.*; 273: 15464–15473.
Panina GF (1985) In *Animal Cell Biotechnology.* Eds Spier RE, Griffiths JB. Academic Press, London; Vol. 1, 211–242.
Pannell R, Milstein C (1992) *J. Immunol. Methods*; 146: 43–48.
Peng Z (2004) *BioPharm*; 17(5): 42–49.
Pierce LN, Shabram PW (2004) *BioProc. J.*; 3: 1–6.
Rathore A, Krishnan R, Tozer S, Smiley D, Rausch S, Seely J (2006) *Pharm. Tech. Europe*; 18(4): 51–57.
Reuveny S (1990) In *Large-scale Mammalian Cell Culture Technology.* Ed Lubiniecki ES. Marcel Dekker, New York; 271–341.
Robinson JH, Butlin PM, Imrie RC (1980) *Dev. Biol. Stand.*; 49: 173–181.
Rupp RG (1985) In *Large-scale Mammalian Cell Culture.* Eds Feder J, Tolbert WR. Academic Press, Orlando; 19–38.
Schleicher JB, Weiss RE (1968) *Biotechnol. Bioeng.*; 10: 617–624.
Schoof DD, Gramolini BA, Davidson DL, Massaro AF, Wilson RE, Eberlein TJ (1988) *Cancer Res.*; 48: 5007–5010.
Siegl G, deChastonay J, Kronauer G (1984) *J. Virol. Meth.*; 9: 53–67.
Singh V (1999) *Cytotechnol.*; 30: 140–158.
Spier RE (1985) In *Animal Cell Biotechnology.* Eds Spier RE, Griffiths JB. Academic Press, London; Vol. 1, 243–263.
Stanbury PF, Whitaker A, Hall S (1999) *Principles of Fermentation Technology.* Elsevier, London, UK; second edition
Suehiro K, Mizuguchi J, Nishiyama K, Iwanaga S, Farrell DH, Ohtaki S (2000) *J. Biochem. (Tokyo)*; 128: 705–710.
Tiebout RF (1990) *Dev. Biol. Stand.*; 71: 65–71.
Tsao EI, Bohn MA, Ostead DR, Munster MJ (1992) *Ann. N.Y. Acad. Sci.*; 665: 127–136.
Unger DR, Muzzio FJ, Aunins JG, Singhvi R (2000) *Biotechnol. Bioeng.*; 70: 117–130.
van't Riet K, Tramper J (1991) *Basic Bioreactor Design.* Marcel Dekker, Abingdon, UK.
van Wezel AL (1967) *Nature*; 216: 64–65.

van Wezel AL (1985) In *Animal Cell Biotechnology*. Eds Spier RE, Griffiths JB. Academic Press, London; Vol. 1, 265–282.
Wang D, Xiao C, Chen Z, Huang Z (1996) *Chin. J. Biotechnol.*; 12: 119–129.
Wang G, Zhang W, Jacklin C, Freedman D, Eppstein L, Kadouri A (1992) *Cytotechnology*; 9: 41–49.
Wen J, Vargas AG, Ofosu FA, Hortelano G (2006) *J. Gene Med.*; 8: 362–369.
Whiteside JP, Spier RE (1981) *Biotechnol.Bioeng.*; 23: 551–565.
Wikstrom K, Blomberg P, Islam KB (2004) *Biotechnol.Prog.*; 20: 1198–1203.
Zhang L, Zhang Y, Yan C, Yu J (1997) *Appl. Biochem. Biotechnol.*; 62: 291–302.

Useful Web Sites

The American Society of Mechanical Engineers runs a number of courses on bioprocess technology and equipment design each year. See the Education section on	http://www.asme.org/.
Applikon Biotechnology. See	http://www.applikon.com/.
Appliflex single-use bioreactor. See	http://www.singleusebioreactor.com/
Bellco Biotechnology manufactures a number of culture systems for mammalian cells, including stirrer flasks, bioreactors, the FiberCell™ small-scale hollow-fibre system, and the BelloCell® high density culture system. See	http://www.bellcoglass.com/.
The Biotechnical Centre, Riga, Latvia, maintains a listing of many (but not all, despite their claims) of the manufacturers of animal cell bioreactors. This can be found at	http://www.bioreactors.net/eng/producers.html.
Biovest International manufactures a range of hollow-fibre culture equipment, from small-scale investigational units to full-scale production units. See	http://www.biovest.com/.
CESCO Bio markets the BioNOC II® matrix, and the BelloCell® and TideCell®, packed-bed systems. See	http://www.cescobio.com.tw/.
Corning produces the CellCube® and general tissue culture plasticware and glassware. See	http://www.corning.com/.
The Department of Biochemical Engineering, University College, London, provides a number of courses relevant to fermenter and fermentation design. See	http://www.ucl.ac.uk/biochemeng/.
New Brunswick Scientific manufactures fermenters, the Celligen, Fibra-Cel culture discs and the FibraStage packed-bed system. See	http://www.nbsc.com/Main.asp.
NUNC manufactures Cell Factories, microcarriers, and general tissue culture plasticware. See	http://www.nunc.dk/.
Percell Biolytica manufactures microcarriers. See	http://www.percell.se/
Solohill manufactures microcarriers. See	http://www.solohill.com/.
Stedim manufactures single-use bags suitable for various cell culture-associated purposes. See	http://www.stedim.com/.
Synthecon manufactures Rotary Cell Culture Systems. See	http://www.synthecon.com/index.htm.
Wave Biotech LLC. See	http://www.wavebiotech.com/

11 Process Development and Design

DK Robinson and L Chu

11.1 INTRODUCTION

The primary process activities required to bring a therapeutic protein product from research target to a medical treatment are the delivery of appropriate bulk drug product to support preclinical and clinical studies and the development of a robust, reliable, and economically feasible manufacturing process that meets or exceeds regulatory requirements. While this seems straightforward, many barriers must be overcome in order to achieve these goals. We will outline a very general approach for process development and design in the area of mammalian cell culture from the identification of a therapeutic entity to the transfer of the process to manufacturing. Rather than providing detailed descriptions of specific process unit operations, the purpose of this chapter is to highlight issues that need to be considered in order to achieve the primary goals for most mammalian cell culture processes. The typical outline of a task list for a process development group in mammalian cell culture is shown in Figure 11.1. This listing of tasks is not meant to represent sequential events. In fact, many of these tasks are often performed in parallel – such as cell expansion design and purification design. Also in some cases, it would be beneficial to overlap more of these tasks – for example, proponents of statistically designed experiments argue that early range qualification will provide increased understanding of the process. Most of this chapter will focus on bulk process development (from gene to purified entity) with examples from recombinant protein and viral vaccine processes, although we hope that the general concepts will still benefit teams working with other types of cell culture-derived products.

11.2 REGULATORY GUIDANCE

While scientific issues are the primary focus of most research groups, it is necessary to be cognisant of regulatory issues throughout the development process, because development decisions need to be based on both scientific and regulatory requirements. Therefore, we would like to take this opportunity briefly to introduce the regulatory agencies that must be consulted in order to prepare for international approval of a potential product. These include the US Food and Drug Administration (FDA), the European Medicines Agency (EMEA), the World Health Organization (WHO), and the International Conference on Harmonization (ICH). This listing is not comprehensive, but it includes the major regulatory agencies. The FDA is the US federal body that regulates foods and drugs, including biological products. Until recently, biological products were regulated by one of the four bodies that comprise the FDA – Center for Biologics Evaluation and Research (CBER). In 2002–2003, some biological product reviews were transferred to another body of the FDA – Center

Medicines from Animal Cell Culture Edited by G. Stacey and J. Davis

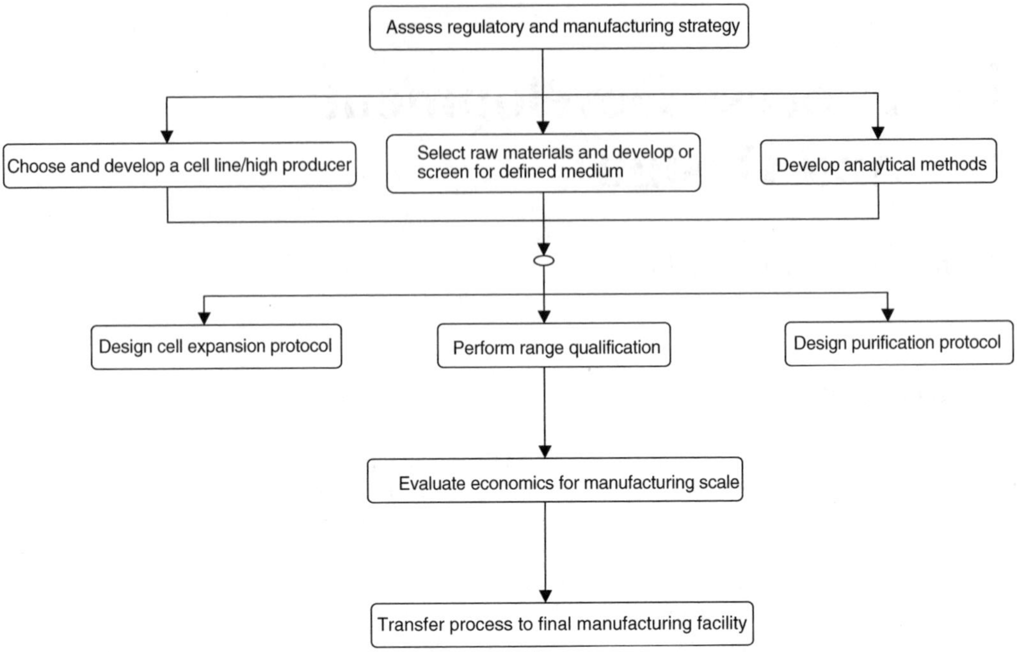

Figure 11.1 Typical task list for a process development group.

for Drug Evaluation and Research (CDER). The biological categories transferred to CDER include 'monoclonal antibodies..., cytokines, growth factors, enzymes, interferons..., proteins extracted from animals or microorganisms other than human blood and blood components and derivatives' where all are intended for therapeutic use. Therefore, manufacturers of biological products will need to interface with different bodies of the FDA depending on their product composition and application. The EMEA is the European agency for the authorization and supervision of medicinal products. This organization is based on 'cooperation between the national competent authorities of the member states and the EMEA'. The Committee for Proprietary Medicinal Products (CPMP) is a scientific subcommittee of the EMEA that issues additional guidances for human medicines. The WHO is a purely advisory international regulatory body. One of the functions delineated in the WHO constitution is 'to develop, establish and promote international standards with respect to food, biological, pharmaceutical and similar products'. The need for a joint regulatory-industry initiative on international harmonization was addressed in 1990 with the formation of ICH. ICH focuses mainly on the 'technical requirements for medicinal products containing new drugs...[with its scope confined to] new drugs and medicines developed in Western Europe, Japan, and the US' as in its mission statement. The guidance documents for these regulatory bodies can be accessed through the websites listed in Table 11.1, and further discussion of the international regulatory framework can be found in Chapter 35. We will highlight where consideration should be given to specific regulatory issues during the evolution of a process in the following sections.

11.3 REGULATORY AND MANUFACTURING STRATEGY

Prior to starting process development, one should review the regulatory and manufacturing strategy, as these considerations may dictate the course of process development. Primary considerations

Table 11.1 Resources for regulatory guidelines.

Regulatory agency	Website
FDA Center for Biologics Evaluation and Research (CBER)	http://www.fda.gov/cber/guidelines.htm
FDA Center for Drug Evaluation and Research (CDER)	http://www.fda.gov/cder/regulatory/default.htm
European Medicines Agency (EMEA)	http://www.emea.europa.eu/index/indexh1.htm
World Health Organization	http://www.who.int/medicines/areas/quality_safety/reulation_legislation/en/
International Conference on Harmonization (ICH)	http://www.ich.org/cache/compo/276–254–1.html.

include the characteristics of the product and the intended patient population, and disease indications for the product. For example, if the product is a glycoprotein whose glycosylation is critical for *in vivo* activity or half-life, then the requirement for such post-translational modification will likely require production as a recombinant gene product in a mammalian cell system. If the product is a live virus paediatric vaccine that will undergo minimal downstream purification, then the requirements for enhanced product safety require an evaluation for production in one of a limited number of mammalian cells (e.g. MRC-5, WI38, Vero) that are more readily accepted by the regulatory agencies for use in this patient segment. If the product is likely to require high doses for efficacy for indications with large patient populations, then process targets should be set so as to allow production at an appropriate volume at production costs that will be acceptable for eventual commercial manufacturing. Potential process changes, including the scale and site of manufacturing, should be considered and coordinated with the clinical development programme to minimize the requirements for clinical bridging between process changes. Similar considerations can help guide the analytical and formulation development efforts as well.

11.4 CELL LINE DEVELOPMENT

Not only is cell line development the first challenge addressed by a process development team, it forms the foundation of process productivity. A process that is developed with a low-producing cell line will be limited in reaching maximum productivity level. Medium improvements and additional processing improvements cannot overcome a poor cell line choice. In addition post-translational modifications such as glycosylation are largely dictated by the host cell line. Finally, the downstream processing steps must be developed to address the potential host cell-related contaminants such as residual DNA, host cell proteins, and potential endogenous adventitious agents.

Fussenegger *et al.* (1999) and Kaufman (1987) provide reviews of the traditional tools and approaches available for development of a cell line that will produce significant levels of a recombinant protein. In light of recent work in the area of genomics and proteomics, Korke *et al.* (2002) list potential techniques that can be used for further analysis of gene expression during cell line development. Assuming that the target gene is already linked to an optimized vector system, the gene must be transfected into a cell line of choice, and subsequently amplified for high-producing clones for further process development. The selection of a cell line can be a difficult decision. The effect of vector systems and choice of cell line on specific productivity is reviewed by Robinson *et al.* (1995). Four cell lines – Chinese hamster ovary (CHO), murine mouse lymphoid (SP2/0, NSO), and rat lymphoid (YB2/0) – have demonstrated the highest expression levels through high-producer selection and amplification. Under optimal conditions, an amplified cell line can reach productivity levels as high as 50–100 pg/cell/day. The most common cell lines for licensed therapeutic proteins are CHO and NSO (Chu & Robinson 2001). Unfortunately, it is difficult to predict *a priori* which expression system will maximize protein expression for a specific recombinant product.

The selection process becomes more complicated for glycosylated proteins. Soluble intercellular adhesion molecule (s-ICAM) production was compared between cell lines (CHO and NS0), and productivity and glycosylation patterns were cell line dependent (Werner *et al.* 1998). Productivity was twofold higher in NS0, but glycosylation in CHO cells was preferable for product stability. Even in a single cell line, it is difficult to predict product glycosylation patterns. Tissue plasminogen activator (tPA), interferon omega, and s-ICAM were each expressed in CHO cells and glycosylation patterns were studied as a function of changing environmental conditions (Werner *et al.* 1998). tPA glycosylation did not change significantly with process times, cultivation methods, or ammonium ion concentrations. Interferon omega and s-ICAM glycosylation were more dependent upon ammonium concentrations and cultivation methods. In order to improve control of glycosylation patterns, researchers have genetically engineered CHO cells to modify their glycosylation capabilities (Weikert *et al.* 1999; Shields *et al.* 2002). Productivity and glycosylation appear to be product-specific, and optimization of both parameters would require cell line screening with each new product candidate. However, in order to meet project timelines, a common approach is to establish one or two cell lines as platform systems based on glycosylation requirements and to redevelop amplification and selection of high producers for each new product candidate.

In addition to technical feasibility, safety issues also need to be considered during cell line development. The history of the cell line, including origin and passage history, should be well documented in order to minimize potential risk of contaminating agents in the final product (see ICH *Guideline Q5d: Derivation and Charactersation of cell substrates used for production of Biotechnological/Biological Products[for website address, see Table 11.1])*. Similarly, a history of the vector-producing cell line should be generated since this is also considered to be a raw material in the process. A risk assessment of the history of the cell line and vector will provide information regarding potential exposure to adventitious agents, which will focus testing efforts. For example, exposure to serum will require bovine virus screens, exposure to porcine trypsin will require porcine virus screens, exposure to humans will require human virus screens, etc. CHO and NS0 cells have both been implicated as containing retroviruses. Electron microscopy of some CHO cell lines indicates the presence of type A and C retroviral particles, although none of these particles has been shown to be functional (Builder *et al.* 1988; Adamson 1998). Retroviral particles (murine leukaemia retroviral family) have also been found in NS0 cultures using transmission electron microscopy and confirmed with a Western blot assay (Taylor *et al.* 2000). The lack of a well documented history will make the design of an adventitious agent-testing program and a transmissible spongiform encephalopathies (TSE) risk assessment more challenging. Unfortunately, these tests are necessary but not sufficient in addressing the presence of contaminating agents. It is impossible to guarantee that all contaminants have been identified and quantified. Therefore, it is necessary to test not only for adventitious agents in the cell line of choice, but it would also be prudent to include purification steps to ensure removal of potential contaminating agents (see Chapter 19).

The cell line also needs to be stable over the course of its manufacturing lifetime in terms of its genetic stability and subsequent product consistency. Berthold (1993) points out that it is important to clarify the definition of genetic stability in this context. Most stable cell lines are actually heterogeneous due to sub-populations that contain a distribution of chromosomal content (Berthold 1993). Consequently, demonstration of genetic stability of the product cell line is defined by the cells' consistent response to the selection marker and generation of a consistent product – 'functional…[rather than] an intrinsic biological property' (Berthold 1993). It is important to conduct an evaluation of genetic stability as part of the selection of a producer cell line, as Barnes *et al.* (2001) observed that the stability of a recombinant NS0 cell line was clone- specific. Another key factor is to ensure appropriate clonal purity of the cell line. Harris *et al.* (1993) described how the lack of such clonal purity may lead potentially to the production of a heterogeneous and inconsistent product. Regulatory guidelines suggest using a clonal isolate to create a two-tiered cell banking system, typically consisting of a master cell bank and a manufacturer's working cell bank to help ensure consistent manufacturing. Data will also be requested to demonstrate the production of a consistent product from cells passaged beyond the generation number that will typically be used for manufacturing (see, for example, ICH Q5B. *Quality of Biotechnologal Products: Analysis of the Expression Construct in Cells Used for Production of r-DNA Derived Protein Products* [for website address, see Table 11.1]). For a typical production process, the cell line must maintain this functional genetic stability for a period of at least 30 generations from the master cell bank.

11.5 RAW MATERIALS AND MEDIUM DEVELOPMENT

Due to increasing regulatory concerns regarding potential viral and prion contaminants in animal-derived raw materials, careful screening of raw materials and designing processes with minimal animal-derived components are prominent goals in process development. Raw materials are defined as any component that is added to, or has contact with, the process. These include medium components, cell lines and their expression systems (see above), purification materials, and even water. There are three strategies to ensure the absence of adventitious agents in raw materials – verification of the raw material source and origin, analytical testing of each raw material of animal origin, and purification steps designed for contaminant removal (Merten 1999; see also Chapter 19). All three strategies are implemented primarily because it is impossible to guarantee the absence of adventitious agents solely through testing. There are strict testing requirements for any raw materials of animal origin that are not sterilized through a generally accepted method. One must demonstrate the absence of mycoplasma, fungi, bacteria, and related adventitious agents (e.g. porcine parvovirus, equine infectious anaemia antibodies, etc.) (9CFR113.53, *Requirements for Ingredients of Animal Origin Used for Production of Biologics* [available at http://a257.g.akamaitech.net/7/257/2422/14mar20010800/edocket.acess.gpo.gov/cfr_2003/9cfr113.53.htm]). Nevertheless, adventitious agents are potentially active at levels equal to or below current analytical capabilities. For example, prions, the causative agents of transmissible spongiform encephalopathies (TSEs), are still not well characterized, and the few assays available are difficult to implement for routine screening of raw materials. An industrial case of large-scale contamination of minute virus of mice (MVM) is another example where raw material testing was insufficient (Garnick 1996). Two separate incidences of MVM contamination occurred in a recombinant protein process and only in-process testing revealed the contaminant. Investigations into the source of contaminant suggest that the MVM was introduced through a raw material at levels undetectable through raw material testing. Therefore, the combination of all three strategies allows one to assess and minimize risk of adventitious agents in the final product.

During the past decade, medium development for recombinant protein processes has evolved towards serum-free medium, specifically animal-free and even protein-free medium, where possible. Serum is undesirable as a raw material since it is animal-sourced and its performance can be inconsistent due to its undefined composition and potential variation from lot to lot and vendor to vendor. In an effort to replace serum, initial serum-free medium formulations replaced serum with animal protein hydrolysates and several animal-derived proteins. More recent medium formulations have successfully replaced animal protein hydrolysates with plant protein hydrolysates or even more defined plant proteins, and some formerly animal-derived proteins of value in cell culture, such as insulin, are now available in recombinant form. Several examples are listed in Table 11.2 (Merten 1999; Jayme 1999; Schenerman *et al.* 1999). Serum-free medium development is beneficial for raw material control and safety issues, but can be challenging in terms of process robustness. Medium formulations that are well defined in composition ("lean medium") are preferred in terms of raw material control, but these formulations are more sensitive to changes in operating conditions

Table 11.2 Replacements for serum in protein-free medium formulations (Data from Merten 1999; Jayme 1999; Schenerman *et al.* 1999).

Serum component	*Protein-free components*
Albumin	Recombinant human serum albumin
Lipids	Plant derived lipids, some conjugated to cyclodextrin to improve solubility
Growth factors	Recombinant growth factors (e.g. EGF, insulin, FGF, IGF, NGF)
Transferrin	Inorganic iron carriers and chelates (e.g. tropolone, ferrous sulphate–EDTA, ferric citrate)
Amino acids	Fermentation-derived amino acids
Serum Functions	*Protein-free (bacterial or fungal) components*
Osmolarity control	Sodium chloride, sucrose, osmoprotective compounds (e.g. proline, glycine, sarcosine, and glycine betaine)
Cell attachment	Arg-His-Asp peptides, heparin, polylysine
Cell growth promotion	Polyamines
Shear force protection	Pluronic F68, polyethylene glycol, dextran, polyvinyl alcohol

including pH, water quality, trace elements, etc. In contrast, serum-containing medium or even complex serum-free medium (e.g. plant protein hydrolysate-containing formulations) can tolerate changes in operating conditions, but some of the raw materials cannot be controlled from lot to lot, which may translate to significant differences in product yield and quality over time.

11.6 ANALYTICAL METHOD DEVELOPMENT

The need for comprehensive and accurate analytical methods is starkly obvious, based on the previous sections regarding process development. One of the advantages of therapeutic proteins is the ability to characterize the final product in detail, which then allows concrete comparisons between process changes. This becomes more difficult when the product consists of more complex protein molecules, in particular for multimeric protein assemblies such as viral vaccines. Analytical methods need to be developed and implemented very early, preferably prior to starting process development work, otherwise critical process decisions will be delayed or in error. Typical product characterization assays for therapeutic proteins include the quantification of concentration and potency, identification and characterization, and glycosylation characterization. In addition, process residuals and adventitious agents must be quantified where possible. Table 11.3 lists some representative analytical methods that have been used in published therapeutic protein processes (Schenerman *et al.* 1999; Moran *et al.* 2000; Garnick 1996). For more detailed coverage of analytical techniques, see Chapter 22–24.

11.7 DESIGN CELL EXPANSION PROTOCOL

Recombinant protein therapies often require significant quantities of product (of the order of 100 mg/dose). In order to produce sufficient quantities of product, it is often necessary to design a cell expansion process that will maximize productivity, maintain reliability, and minimize production costs. The prevailing large-scale model is the stirred tank suspension culture (batch or fed-batch) where the principles of scaling parameters and process control are well understood. Another variation on the traditional stirred tank reactor is a tank designed in a conical shape in order to inoculate directly with smaller volumes and accommodate larger working volumes (Heidemann *et al.* 2002). The primary benefit of this design is reduced equipment changes during the cell expansion process.

Table 11.3 Representative analytical methods for product characterization in therapeutic protein processes. In this case, an antibody is used for illustration. (Data from Schenerman *et al.* 1999; Moran *et al.* 2000; Garnick 1996.)

Purpose	Possible analytical methods
Antibody concentration and potency	Absorbance
	ELISA
	Protein A affinity
	Enzyme immunoassay (binding)
	Complement-mediated cell lysis
Antibody identification and characterization	Isoelectric focusing (pI, size)
	SDS-PAGE (size, presence of degraded Ab)
	Size exclusion chromatography (size, presence of monomers, aggregates or degraded Ab)
	MALDI-TOF mass spectrometry (mass)
	Western blots
	Capillary gel electrophoresis
	Tryptic peptide mapping
	Amino acid analysis
	N-terminal sequencing
	NMR, CD and intrinsic fluorescence spectroscopy (conformation)
Glycosylation characterization	Reverse phase HPLC (oligosaccharide analysis)
	Anion exchange column (oligosaccharide analysis)
	MALDI-TOF mass spectrometry (oligosaccharide analysis)
	Monosaccharide composition analysis
Process residuals	Quantitative PCR (host cell DNA)
	Rabbit pyrogenicity test (endotoxins)
	LAL test (endotoxins)
	EIA (host cell protein, leached Protein A, insulin, other residual medium components)
	Enzymatic (trypsin)
	NMR (Pluronic F-68)
Virus quantification	PCR (retroviruses)
	Tissue culture-based infectivity assay (for example, the 324 K test, developed by Genentech, highly sensitive to MVM as well as other rodent parvoviruses)

Another source of difference in the cell expansion process is maintenance of a continuous seed train in a source bioreactor for multiple production batches rather than using a new vial thaw for each production batch (Moran *et al.* 2000). Some industrial processes have chosen significantly different cell expansion processes, such as perfusion systems (Recombinant Factor VIII, Bayer), roller bottles (EPO, Amgen), and Costar CellCubes (Vaqta®, Merck). These alternative cell expansion processes are more commonly found in vaccine processes where the product dictates the cell line and processes become more specialized. Cell expansion processes are nicely reviewed by Kretzmer (2002), Werner and Noe (1993), and Ramasubramanyan and Venkatasubramanian (1990).

11.8 DESIGN PURIFICATION PROTOCOL

The purification protocol is central to establishing the safety profile of the final product. The choice of specific purification processes should be based not only on product purification, but also with the elimination or reduction of process residuals and adventitious agents in mind. (see Chapters 18 and 19) Preliminary purification schemes may be designed to reduce process residuals in order to meet Phase I clinical timelines, but adventitious agent reduction is a requirement. Common process residuals include host cell protein and DNA, but additional impurities must be considered in order to minimize the potential for an immunogenic response. These include small molecules or proteins from medium components, leached purification ligands (e.g. Protein A), and even aggregates of the product. Finally, adventitious agents must be removed through validated purification steps (see Chapter 19). The typical purification process is composed of a minimum of three chromatography steps – for monoclonal antibody purification these would typically be Protein A affinity column, cation exchange column, and anion exchange column. Protein A affinity columns remove the bulk of the impurities, including host cell proteins, DNA, and endotoxins, by specifically binding antibodies. These columns are significant in cost (30 times the cost of ion exchange columns); therefore, reuse qualification experiments are performed in order to meet cost limitations (Fahrner *et al.* 2001). Residual Protein A is removed by a subsequent cation exchange column, which also removes additional host cell proteins, product aggregates, DNA, and endotoxins. The final polishing step is the anion exchange column for the removal of any remaining DNA and host cell proteins. Both ion exchange columns are also validated for resin reuse, although the degree of reuse is not as efficient as for affinity columns (Fahrner *et al.* 2001). A typical industrial purification process consisting of these three steps can often meet an overall yield of >65 % (Fahrner *et al.* 2001).

Affinity and ion exchange columns may not ensure sufficiently robust viral removal; therefore, it is prudent to include additional viral removal or inactivation steps. These may include extreme pH conditions, heat, organic solvents, chemical detergents, or gamma irradiation for viral inactivation, or filtration for viral removal. By including at least two types of inactivation/removal procedures, there is greater assurance of meeting purity requirements. Validation of viral inactivation/removal can be challenging because analytical limitations do not allow direct measurement of the fold-reduction required by regulatory guidances. As an alternative, viral inactivation/removal steps are validated with spike experiments for individual steps, and the clearance factor for the entire process can be assumed to be the sum of all viral clearance steps. This assumption is reasonable provided each of the clearance steps is based on different principles – e.g. filtration versus inactivation (Berthold & Walter 1994). The choice of model virus spike experiments for recombinant proteins produced in CHO cells was described by Fritsch (1992). Different systems may require different model viruses, and potential viral candidates should be identified for specific systems by direct experimentation. Consideration of the cell line/vector system, additional exposure during processing, and contributions from raw materials are all necessary for a complete viral validation package. This evaluation will help determine the characteristics and load of potential viruses, including virion size and presence of an envelope, and removal/inactivation mechanisms can be designed to address these properties. For example, spike experiments for a CHO cell system with serum-containing medium was designed with four different model viruses – retrovirus, bovine viral diarrhoea virus (BVD), parainfluenza virus, and reovirus. The rationale for these four model viruses was based on the EM identification of retroviruses in the CHO cells, the potential for BVD contamination in serum, and the ability of CHO cells to sustain reovirus and parainfluenza virus based on studies by Weibe (1989). Virus clearance validation experiments are performed with scale-down model systems that maintain critical column parameters including, but not limited to, column height, flow rate, processing time, and pressure. Fritsch (1992) demonstrated that each of these viruses were removed or inactivated differently by different purification steps, and only through systematic experiments can one determine the clearance factors for specific systems while still maintaining product stability and activity. (For more detailed treatment of this whole topic, see Chapter 19.)

11.9 PROCESS RANGE JUSTIFICATION

One key role during development is to provide sufficient data to allow an appropriate selection of the critical process parameters that will be used in process validation and to justify the range of specifications for each of the process parameters, which we refer to here as process range justification. Traditional examples of range justification are described by Schenerman *et al.* (1999) and Moran *et al.* (2000). Schenerman *et al.* (1999) performed studies to examine three critical process parameters – cell line population doubling level, glucose level, and harvest time. By performing experiments beyond the recommended manufacturing ranges, they were able to provide a rationale for the range specifications. Clinical supplies were produced at smaller scale (20–200 l) while manufacturing scale would be much larger (400–10 000 l); therefore, it was necessary to demonstrate comparability between reactor scale and facilities through qualification experiments. Moran et al. (2000) aimed to achieve similar goals by using statistically designed experiments. Their experiments identified critical process parameters and recommended range specifications. Surprisingly, they found that even process temperature could vary from 34–38°C without any significant impact on product quality, although there was the expected impact on cell growth rates. Since they defined the critical quality attribute as product quality, they were able to conclude that their process was robust with respect to their experimental range of process parameters and over a wide production scale (3 l, 80 l, and 7000 l). This example demonstrates the importance of defining critical effects when designing qualification experiments.

It is becoming advantageous to address range specifications earlier in the timeline in order to expedite process development. Shukla *et al.* (2001), Gaertner and Dhurjati (1993a,b) and Ganne and Mignot (1991) demonstrate how early range specification work can improve process design. All these papers implemented the use of statistically designed experiments for their early range specification work. Ganne and Mignot (1991) successfully used this methodology to select and optimize a chemically defined serum-free medium for CHO cells producing recombinant factor VIII:C and confirmed the composition at 1-Litre reactor scale. Gaertner and Dhurjati (1993a,b) used a similar experimental design to gain a better understanding of hybridoma growth and associated antibody production, and they developed a mathematical model based on their experimental results for future reactor design. Shukla *et al.* (2001) used a combination of single variable experiments and statistically designed experiments to provide qualification ranges for metal-affinity chromatography of an Fc fusion protein early in their development process. Therefore, they were able to identify a combination of variable conditions that generated a worst-case scenario with respect to purity, and then design subsequent purification steps to address the potential worst-case scenario.

Many of these publications also demonstrate the practical and viable use of statistical design of experiments (DOE) either early or late in the development process. DOE is the strategy of planning arrays of experiments to determine the effects of multiple experimental factors, particularly how they affect product characteristics collectively. The traditional approach is to examine each factor individually, but the disadvantage of this approach is that one may miss factor interactions, either beneficial or detrimental to process goals, and many more experiments are required to explore such interactions fully. In order to initiate DOE experiments, a list of experimental factors must be identified. Although the number of variables does not impact experimental resources as significantly as do traditional approaches, it is still important to keep the variable list manageable. Initial screening experiments such as Plackett-Burman will require N experiments to evaluate up to N-1 factors. Critical factors (main effects) can be identified with such screening design experiments. The next step is to use factorial design experiments to identify interaction effects between critical factors. In the final stages of development, response surface experiments are designed to identify curvature effects. Therefore, DOE is applicable for all stages of process development, but in our opinion the maximum benefit is reaped when DOE is implemented early in process development, because process design is more easily adjusted based on factor interactions identified by DOE. Additional benefits include the ability to generate more information per experiment than through traditional approaches, resulting in reduced lead time and improved experimental efficiency. Experiments are more organized in data collection and analysis, and the reliability of data can be evaluated based on experimental and analytical variation (Staeheli 1987). Specific examples of the application of DOE to process development problems are described in Staeheli (1987), Shukla (2001), Gaertner and Dhurjati (1993a,b), Ganne and Mignot (1991) and Moran *et al.* (2000).

11.10 ECONOMIC ANALYSIS FOR MANUFACTURING

Early process development concerns usually focus on delivery of adequate quantities of clinical supplies in a timely manner. Given the fact that most product candidates will not proceed past Phase I/II studies, extensive process development efforts are not warranted early in the development life of potential products. Therefore, some process choices are made for reasons of expediency rather than feasibility for a manufacturing facility. As the process progresses through clinical trials with positive results, process groups will begin to consider economic analysis for manufacturing facilities. Included in this analysis is an evaluation of all processing steps with regard to scale-up feasibility and potential cost issues. Also, target productivity values are determined based on market predictions, and this will allow the process group to focus efforts on process steps that require significant development work in order to meet cost and supply requirements. Erythropoietin (EPO) is an example of a high value/low demand drug (total demand in US only 1 kg/year) that was designed using roller bottle technology. This type of process would not be feasible for a typical therapeutic antibody due to the impracticality of scaling up roller bottles for high demand drugs (>10–100 kg/year) (Kretzmer 2002). Consequently, therapeutic antibodies are often produced in stirred tank bioreactors. The choice of scale for stirred tank bioreactors is usually selected to be as large as possible because the capital costs between different size tanks are not significantly different while the operating costs are more favourable with larger tanks. In fact, a tenfold decrease in tank size can cost five times more to operate (per unit of protein) due to the increased number of batches and turn-around costs for CIP, SIP, and medium preparation (Werner 1998). Other factors identified as important process parameters in terms of product cost include productivity, product concentration in the tanks, and bulk harvest purity (Hess 1987; Maiorella 1992).

11.11 PROCESS TRANSFER TO PILOT PLANT AND/OR MANUFACTURING FACILITY

The transfer of a process to a pilot plant or manufacturing facility is often one of the more satisfying accomplishments from a process development perspective. Some believe that a process is only complete when successfully transferred to another group, especially one operating under cGMP conditions. In some companies, the process development groups are responsible only for developing the process and the manufacturing groups are responsible for producing both clinical and marketed products. In these cases, process transfer occurs earlier and perhaps more frequently. In other companies, the process development groups are responsible for both developing the process and manufacturing clinical supplies, and transfer to manufacturing occurs later. In both cases, the transfer process is complicated due to the number of groups involved and the timeline expectations. Some approaches to successful process transfers are presented in Goochee (2002), Dale (2000) and Gerson *et al.* (1998). The common themes in each of these articles are the need for clear communication, clarification of responsibilities and accountability, and defined and timely documentation. In addition, the more robust the process, the greater the likelihood of a successful process transfer. Therefore, many of the earlier process development activities do play a large role in determining potential problems or successes. For example, the optimal process for a manufacturing environment is one that has the capability to absorb process parameter variability with minimal impact on product quality. Therefore, cell expansion and purification steps should be designed for process robustness when possible. This may translate to simpler unit operations because more complicated systems will be more difficult to operate in a plant and still meet regulatory requirements. Perfusion systems are not very common at industrial scale despite the technical advantages, because the systems are quite complex in terms of process validation and cell line stability issues. Similarly, transgenic expression systems are less common at industrial scale due to the complexity of addressing regulatory issues.

Process validation (see Chapter 15) is often the penultimate step in transfer of a process to a final manufacturing facility. Typically, during process validation, a series of manufacturing lots is made that demonstrates the capability to control the process consistently within a predetermined range of critical process parameters for each step that meet a set of critical quality attributes. Regulatory guidance suggests that three to five such consecutively manufactured lots adequately demonstrate manufacturing consistency. These process validation lots should be manufactured using the process and facility intended for licensure. In some cases, these process validation lots may also be evaluated clinically as a measure of clinical consistency.

11.12 INDUSTRIALLY RELEVANT EXAMPLES

Limited published literature exists for industrially relevant processes. However, the few that are available provide a nice overview of the evolution of processes that have been successfully implemented at manufacturing scale.

One of the earlier therapeutic protein processes, the development of a process for tissue plasminogen activator (tPA) production, is described by Builder *et al.* (1988). This process was initially produced in *E. coli*, but inadequate glycosylation and inaccurate three-dimensional structure required a process change to a CHO cell line. Initial clinical supplies were generated from a roller bottle process, but a suspension culture process was implemented early in the development phase. Therefore, only limited bridging work was required. Purification steps were designed to remove or verify the absence of host cell protein, DNA, and viruses by using multiple approaches in order to build redundancy into the process. For example, DNA was removed by more than nine orders of magnitude by validating clarification and ion exchange steps. Product consistency from lot to lot was demonstrated with trypsic peptide mapping, which was necessary for such a large protein.

More recent therapeutic protein processes include the development of Synagis® against respiratory syncitial virus, and an unnamed monoclonal antibody against prostate cancer. These processes were described by Schenerman *et al.* (1999) and Moran *et al.* (2000) respectively. Both of these humanized IgG_1 antibodies were expressed in NS0 cells, and grown in stirred tank bioreactors (in the case of Synagis®, at scales up to 100001). The unnamed antibody was purified by Protein A and ion exchange chromatography, whilst few details were provided on the purification of Synagis®, although it is likely to be similar to other MAb processes. Many analytical methods were used to characterize and compare final product quality. These included matrix-assisted laser desorption ionization – time of flight mass spectrometry (MALDI-TOF mass spectrometry), SDS-PAGE, isoelectric focusing, Western blots, capillary gel electrophoresis, RP-HPLC, anion exchange chromatography, and tryptic peptide mapping. A number of these methods were also used to evaluate and compare different process conditions and the effects upon product quality and productivity. These two examples demonstrate the number of analytical methods that are now routinely used to characterize therapeutic proteins in comparison with early products such as tPA (see also Chapters 23 and 24). The concept of a well-characterized biological allows process groups to address regulatory concerns with concrete evaluation of the impact of process changes upon the final product.

Hagen *et al.* (2000) describes the development of a process for an inactivated hepatitis A vaccine (VAQTA®). This group had the additional challenge of demonstrating comparability for a process that was changed after a pivotal efficacy trial and several safety studies. The upstream process consisted of MRC-5 cell expansion in CellCube bioreactors. Virus was harvested from the CellCube bioreactors and purified through a series of columns in order to remove culture impurities and concentrate the harvest. Serum was still present in this process, therefore additional adventitious agent testing was required for the raw materials and final product. Multiple analytical methods were developed to monitor the process and the final product, which included residual DNA, specific antigenicity, empty capsid content, rate of inactivation, mouse immunogenicity, and SDS-PAGE analysis. All methods demonstrated chemical/physical equivalency, but as a final confirmation, they chose to run bridging clinical studies to demonstrate clinical equivalency. This is another example of the ability to demonstrate product equivalency using the broad range of analytical capabilities now available, even in the case of a complex protein product such as a viral vaccine.

11.13 SUMMARY

The development of a process for the production of therapeutic proteins in mammalian cell culture involves many steps. Key success factors include an early evaluation of the product, regulatory, and manufacturing requirements as this may dictate the direction of development efforts. Specifically, one should consider the selection of the appropriate host cell line for recombinant protein expression, initiate early involvement of the analytical development groups as this is necessary in order properly to gauge the progress of development, and generate and evaluate data supporting the process ranges. Given that the end product of a successful development effort can result in the launch of products that positively impact human health, process development can also be a very satisfying experience for those involved.

REFERENCES

Adamson SR (1998) Dev. Biol. Stand.; 93:89–96. Barnes LM, Bentley CM, Dickson AJ (2001) *Biotechnol. Bioeng.*; 73: 261–270.

Berthold W (1993) *Biologicals*; 21: 95–100.

Berthold W, Walter J (1994) *Biologicals*; 22: 135–150.

Builder SE, van Reis R, Paoni NF, Ogez JR (1988) In *8th International Biotechnology Symposium, Paris 1988: Proceedings*. Ed Durand G. Société Francaise de Microbiologie, Paris; 644–659.

Chu L, Robinson DK (2001) *Curr. Opinion Biotechnol.*; 12: 180–187.

Dale CJ (2000) *BioPharm;October:* 48–56.

Fahrner RL, Knudsen HL, Basey CD *et al.* (2001) *Biotechnol. Gen. Eng. Rev.*; 18: 301–327.

Fritsch E (1992) In *Developments in Biological Standardization*. Eds Brown F, Esber EE, Williams M. Karger, New York; Vol. 76, 239–248.

Fussenegger M, Bailey JE, Hauser H, Mueller PP (1999) *TIBTECH*; 17: 35–42.

Gaertner JG, Dhurjati P (1993a) *Biotechnol. Progr.*; 9: 298–308.

Gaertner JG, Dhurjati P (1993b) *Biotechnol Progr.*; 9: 309–316.

Ganne V, Mignot G (1991) *Cytotechnology*; 6: 233–240.

Garnick RL (1996) In *Developments in Biological Standardization.* Eds Brown F, Esber EE, Williams M. Karger, New York; Vol. 88, 49–56.
Gerson DF, Himes V, Hopfer R *et al.* (1998) *Drug Inf. J.*; 32: 19–26.
Goochee CF (2002) *Cytotechnology*; 38: 63–76.
Hagen A, Aunins J, DePhillips P *et al.* (2000) *Bioproc. Engng.*; 23: 439–449.
Harris RJ, Murnane AA, Utter SL *et al.* (1993) *Bio/technology*; 11: 1293–1297.
Heidemann R, Mered M, Wang DQ *et al.* (2002) *Cytotechnology*; 38: 99–108.
Hess PN (1987) *Arzneim.-Forsch./Drug Res.*; 31: 1210–1215.
Jayme DW (1999) In *Developments in Biological Standards.* Eds Brown F, Cartwright T, Horaud F, Spieser JM. Karger, New York; Vol. 99, 181–187.
Kaufman RJ (1987) *Genet. Engng.*; 9: 155–198.
Korke R, Rink A, Seow TK, Chung MCM, Beattie CW, Hu W-S (2002) *J. Biotechnol.*; 94: 73–92.
Kretzmer G (2002) *Appl. Microbiol. Biotechnol.*; 59: 135–142.
Maiorella B (1992) In *Harnessing Biotechnology for the 21st Century: Proceedings of the Ninth International Biotechnology Symposium and Exposition, Crystal City, Virginia, August 16–21, 1992.* Ed Ladish MR. American Chemical Society, Washington, DC; 26–29.
Merten O-W (1999) In *Developments in Biological Standards.* Eds Brown F, Cartwright T, Horaud F, Spieser JM. Karger, New York; Vol. 99, 167–180.
Moran EB, McGowan ST, McGuire JM *et al.* (2000) *Biotechnol. Bioengng*; 69: 242–255.
Ramasubramanyan K, Venkatasubramanian K (1990) In *Advances in Biochemical Engineering.* Ed Fiechter A. Springer-Verlag, Berlin; Vol. 42, 13–29.
Robinson DK, DiStefano D, Gould SL *et al.* (1995) In *Antibody Engineering.* Eds Wang H, Imanaka T. American Chemical Society, Washington DC; 1–14.
Schenerman MA, Hope JN, Kletke C, Singh JK, Kimura R, Tsao EI, Folena-Wasserman G (1999) *Biologicals*; 27: 203–215.
Shields RL, Lai J, Keck R *et al.* (2002) *J. Biol. Chem.*; 277: 26733–26740.
Shukla AA, Sorge L, Boldman J, Waugh S (2001) *Biotechnol. Appl. Biochem.*; 34: 71–80.
Staeheli J (1987) In *Developments in Biological Standardization.* Eds Spier R, Hennessen WS. Karger, New York; Vol. 66, 143–153.
Taylor FR, Ferrant JL, Foley SF *et al.* (2000) *J. Biotechnol.*; 84: 33–43.
Weibe ME, Becker F, Lazar R *et al.* (1989) In *Developments in Biological Standardization.* Eds Hayflick L, Hennessen W. Karger, New York; Vol. 70, 147.
Weikert S, Papac D, Briggs J *et al.* (1999) *Nature Biotechnol.*; 17: 1116–1121.
Werner RG (1998) *Arzneim.-Forsch./Drug Res.*; 48: 423–426.
Werner RG, Noe W (1993) *Arzneim.-Forsch./Drug Res.*; 43: 1242–1249.
Werner RG, Noe W, Kopp K, Schluter M (1998) *Arzneim.-Forsch./Drug Res.*; 48: 870–880.

Useful Web Sites

Medicines & Healthcare products Regulatory Agency	http://www.mhra.gov.uk/
Food and Drug Administration	http://www.fda.gov/
European Commission	http://pharmacos.eudra.org/
International Conference on Harmonization	http://www.ich.org/
European Medicines Agency	www.emea.europa.eu
United States Department of Agriculture	www.usda.gov

12 Facility Design for Cell Culture Biopharmaceuticals

S Vranch

12.1 PRIMARY AND SECONDARY MANUFACTURE

The manufacture of biopharmaceuticals is split into two parts, namely primary manufacture where the pure drug substance or active pharmaceutical ingredient (API) is made, followed by secondary manufacture, often termed 'fill/finish' where the drug product is made, packaged and labelled. The specification of the dosage form of the drug product will dictate the quality and specification of the API that is made and hence the design of the facility to make it. Most biopharmaceuticals are administered as injections and the dosage forms are small volume parenterals that are either sterile solutions or sterile lyophilized powders in ampoules or vials. However, the field of biological medicines is clearly becoming more complex. Cell therapy and tissue engineering products are just beginning to be developed and the full requirements for production facilities for such products are yet to be determined.

12.2 DESIGN OF PRIMARY MANUFACTURING FACILITIES

The scheme of process development stages outlined in Figure 12.1 is the source of information that the designer uses. Development will have determined the process conditions and the quality-critical parameters, the yield and the optimal sequence of process steps. As well as operating conditions for processing, Development will also have determined conditions for cleaning the plant.

The start of the design process is to collect information from process development and to prepare a user requirement specification (URS) for the facility as this will be used as the basis for the design. The URS should contain the performance criteria by which the facility will be judged. The URS will list the types of cell substrate that will be used and their expected productivity; it will also define the likely markets for the products as these dictate the regulations that apply to the design and operation of the facility.

The design of facilities for the primary manufacture of biopharmaceuticals depends on the scale of operation and on the product to be made. A common feature of biopharmaceutical manufacture is that the drug substance is created very early in the process, either from a natural source from which it is extracted, or by its manufacture in a bioreactor or culture flask. Subsequent processing steps are used to purify and stabilize the drug substance or API so that it can be stored until required for formulation into dosage forms. As many biopharmaceuticals are heat, pH and shear sensitive, these process steps must be designed to maintain the integrity of the API and prevent its exposure to contamination. Good manufacturing practice (GMP) thus starts from the beginning of the process for making a biological API. GMP is mainly concerned with ensuring that the process,

Medicines from Animal Cell Culture Edited by G. Stacey and J. Davis

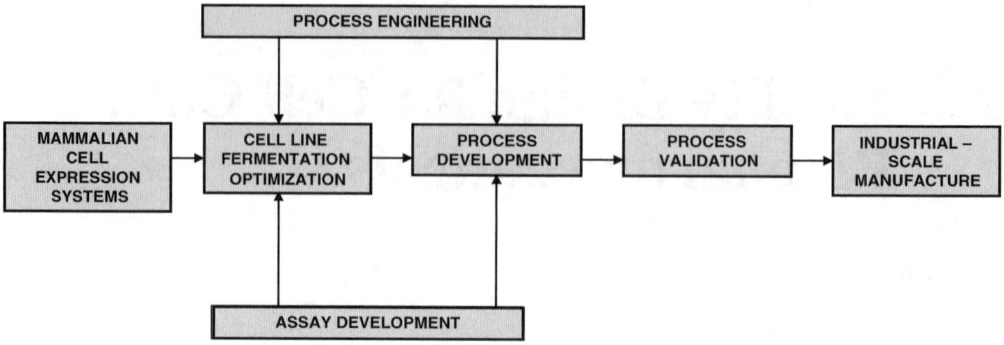

Figure 12.1 Process Development Scheme.

people, procedures and equipment are under control so that the product is consistently made to the required quality (see Chapter 34). The design of the facility itself makes a major contribution to GMP as a good design (CFR21 Part 600 2004; CFR21 Part 210 2004; CFR21 Part 211 2004; EU 1998) will encourage safe and efficient operation.

12.3 MULTI-PRODUCT FACILITIES

There are very few facilities that make products from microbial as well as mammalian cells under the same roof. Even where the same types of cell line are used, there are special challenges for multi-product facilities, the most significant being the need to avoid cross-contamination. A multi-product facility has been defined as one in which more than one product has ever been made. Production of different APIs can be organized on a campaign basis or alternatively more than one product can be made in the building simultaneously. Some rooms and equipment will be dedicated to one product and some will be used by all products, as described in Table 12.1.

Table 12.1 Typical segregation of functions in separate rooms or suites of rooms.

	Segregation	
Function of the room	Common	Product-dedicated
Media preparation	√	
Media storage	√	
Purification buffer preparation	√	
Purification buffer storage	√	
Working cell bank preparation and storage	√	
Inoculum preparation laboratory	√	
Seed bioreactor room	√	√
Production bioreactor room	√	√
Cell separation or harvest room	√	√
Purification		
• Primary purification		√
• Secondary purification		√
Packing purification columns	√	
API bulk filling	√	

Although the capital cost of a facility for the manufacture of a single product will be less than for a multi-product operation, the life of a facility will exceed 10 years and several products are likely to be made over this period. Thus a facility for the manufacture of a bulk biological API will normally be designed to accommodate a group of products. Cell culture-based facilities have special features and the degree of flexibility has to be agreed early in the project. Flexibility must not be at the expense of GMP compliance and there are well-proven techniques to accommodate both, such as the use of well serviced rooms into which modules containing process equipment are added and connected for a particular process.

There are licensed facilities where more than one product is made in different bioreactors in the same room at the same time. This demonstrates the ability to contain these systems and avoid cross-contamination. Purification rooms and even suites remain dedicated to one product at a time.

Facilities used for contract manufacture of drug products are commonly arranged so that a customer can use a dedicated suite that includes bioreactors and purification rooms.

12.4 PROCESSING CONSIDERATIONS

The process is central to the design of the building, and Figure 12.2 shows a process outline.

The facility is primarily built to accommodate the process and each step will require special features. Bioreactors will be taken as an example to illustrate the procedure of integrating the equipment with the facility design.

The type and size of bioreactor dictates the height and dimensions of the processing rooms. The shape of bioreactors was developed for the bulk manufacture of penicillin and the traditional stirred tank has changed little. It is versatile and permits a wide range of operating conditions and volumes, but its performance is not as well characterized as a continuous-flow reactor found in the chemical industry. In the stirred tank, mass and heat transfer are generally achieved with a single impeller and without the need for internal cooling coils, which can give rise to contamination. If the agitator is mounted on top of the bioreactor, then space must be allowed for the removal of the agitator shaft. Many bioreactors are stirred with magnetic drives, thus enhancing containment and safety. A bioreactor volume of 15 000 litres is typical and this size would dictate two working floors with the vessel situated between them. Seed reactor trains are usually located on one floor; the required inoculum ratio (frequently 1/5 to 1/10) is used to establish the volumes of the seed bioreactors.

Air lift bioreactors with an annular draught tube are also used for the growth of animal cells and the environment around the cells can be better controlled (Birch & Arathoon 1990). However

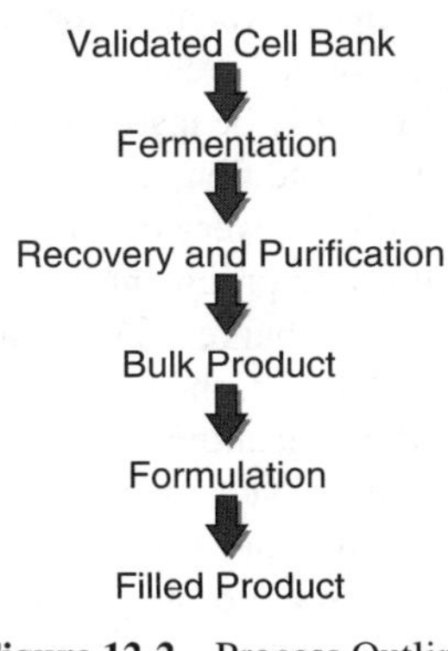

Figure 12.2 Process Outline.

in order to maintain good circulation around the vessel, the height and hence the volume of liquid in the reactor has to be controlled, making the bioreactor virtually a constant volume device. The height has obviously to be accommodated in the design. There are potentially many other types of bioreactor to be accommodated and these are discussed in Chapter 10.

The bioreactor has to contain the culture in order to keep it pure, and if pathogens are being grown, then the vessel must be designed to minimize or to prevent their release. Design measures for cleanliness usually promote containment. For example, an exhaust filter will be fitted on bioreactors for mammalian cell culture in order to maintain culture purity. In normal production operation, the bioreactor is 'closed'. Batching the bioreactor, emptying it, taking samples and adding sterile fluids for pH control are the main interventions and all these can be achieved with a closed system. Air or gas fed to the bioreactor is sterilized by filtration and aseptic connections are made when adding or removing fluids. However, the bioreactor will have to be opened for cleaning, changing instruments and for maintenance. A decision has to be made on the air quality of the room that is needed to ensure that there will be minimal ingress of particles and microbes. It is now considered that where the bioreactors are closed, the room air quality is not critical and increasingly bioreactor rooms are being designed as unclassified.

The same arguments can be used to determine the optimal facility requirements and room classifications for other unit operations, such as centrifugation, chromatography, ultrafiltration and crystallization. The assertion that a system is closed must be challenged and backed up with experimental evidence. Equipment that is dedicated to a particular product should be labelled and securely stored in a clean dry condition until it is required. Such equipment includes the Dewar flasks for storing master and working cell banks, and chromatography columns. Sometimes chromatography work stations are product-dedicated, although this is an expensive option.

When the same equipment is used to make different products, the manufacturing process must be developed and designed to ensure that the equipment can be cleaned between manufacturing campaigns to ensure that there is no possibility of cross-contamination. The success of this policy depends upon the following:

- The cleaning process and its validation must be recognized as an integral part of the process.
- The cleaning process must be designed to be reproducible. Automation can help to achieve this goal. For the same reason, washing machines are preferred to manual cleaning.
- The cleaning process must be capable of being validated.
- Assays for impurities and for any detergents that are used must be developed and validated.

The definition of the acceptable level of cleanliness is the responsibility of the manufacturer and it must be science-based. The acceptance level might be different for cleaning between batches of the same product than for cleaning between batches of different products.

Clean-in-place (CIP) systems should ideally be instrumented and automated to ensure that consistent results are obtained, and that the conditions that appertained during successful cleaning validation are reproduced with the cleaning of every batch (for further details see Chapter 14).

The bioreactors are sterilized and it is a requirement that an axenic culture (a pure culture of a single organism) is maintained to the point of harvest. Downstream processing is performed aseptically to maintain the integrity of the API, and bioburden limits must be pre-defined and based on evidence from process development studies. Cleaning and sanitization techniques will be performed to maintain the bioburden below this pre-defined concentration. Undetectable levels of microbial contamination can adversely affect the quality of a protein. The trend is to use CIP systems that use chemicals followed by rinses with purified water or water for injection (WFI) as the final rinse (see Chapter 14). However, the use of steam-in-place (SIP) is still prevalent in downsteam processing.

SIP installations must be carefully designed so that the steam system does not add cold condensate at the start of the process. Sterilization downstream of the bioreactor is rare, but is recommended for API where large quantities of protein are administered per dose, and the potential for failing the product specification for endotoxin arising from past bacterial contamination is increased. There are several parts of the purification process that cannot be sterilized with steam, such as affinity chromatography matrices, and these will contribute to the bioburden load. SIP is an expensive option; however, it is difficult to qualify a sanitization procedure in comparison with a sterilization one, so the acceptance criteria for contaminants must be scientifically based.

Equipment has to be opened for cleaning, maintenance and insurance inspections, and after these interventions the equipment must be cleaned and sterilized or sanitized using a validated process before it is reused.

The degree of automation and control is a significant cost factor and a key process-related issue. Automation gives more consistent productivity by providing better monitoring and control of the processes, and permits the use of electronic signatures and electronic records. The use of process analytical technology (PAT) where the quality of manufacture is continuously monitored and controlled requires high levels of automation. A URS for the control system is essential and this should include the interfaces with business and data processing systems (see Chapter 13). An early decision on the architecture of the control system and its provider is essential to keep a facility design and construction project on track. The control system interfaces with all other systems and equipment, and the interconnections must be compatible.

12.5 BIOCONTAINMENT

The advantage of recombinant DNA technology is that the use of pathogens for manufacture can usually be avoided by replacing them with safe genetically modified microorganisms (GMO). For regulatory purposes, mammalian cells are generally classed as 'microorganisms' although in some parts of the United States some well characterized mammalian cell lines, such as Chinese Hamster Ovary (CHO) cells, are not considered as microorganisms and are therefore excluded from the regulations for the contained use of GMO. This is not so in Europe, where mammalian cells are included in the contained use regulations. A risk assessment is used to determine the containment measures needed to protect the environment, process operators and other individuals. In Europe this risk assessment is mandatory and there is a prescribed list of measures and procedures that must be addressed during the assessment. Where containment and GMP compliance appear to conflict, such as having rooms at a negative pressure for containment that could draw dirt into the facility, then this conflict must be engineered out so that both conditions are covered. In this example the air that is drawn in from the surroundings would firstly be filtered. As previously noted, where containment is needed to protect the environment, it also ensures that microbes and particles are excluded from the process streams.

12.6 SITE LOCATION AND LAYOUT

At the beginning of a project decisions will be made on:

- site location and permit requirements;
- the optimum number of storeys, for example to allow the location of utilities above and below the manufacturing level as appropriate;
- the list of rooms and their functions;
- the provision of utilities generated within the facility or taken from the site;

- the use of 'grey space' to minimize clean room volume;
- any provision in the design for future expansion;
- the extent of prefabrication and pre-testing of equipment modules.

The location of the facility on a particular site has GMP implications. Surrounding industry and farming should not have a deleterious affect on the products, and a reliable infrastructure is needed.

International GMP regulations have a common requirement that the layout of the facility should encourage operators, visitors, equipment and materials to move or flow in a controlled and logical manner that will avoid mix-ups. For example, the flow of people from an unclassified area to a classified one is via a changing room (or series of rooms) so that the cleanliness of the working area is not compromised. Flow patterns around the facility will help ensure that mistakes and cross-contamination are avoided. Materials and equipment must not be stored in corridors.

Unidirectional flow of people in and out of process rooms is not common in API plants for therapeutic proteins. However the concept of a flow-and-return corridor is still prevalent, gives a straightforward pattern and has frequently been employed in vaccine facilities. The flow corridor supplies materials and clean equipment to the processing rooms and stores, and the 'return' corridor is used to transfer equipment to the washing room and to the waste collection points or decontamination room.

A useful tool called an adjacency diagram is illustrated in Figure 12.3 where adjacencies and key flows between rooms are defined. In this case the facility will be located on an existing site where there are already staff and analytical facilities. The provision of space for future expansion can be included.

Within any area there must be adequate space around equipment for operation, cleaning and maintenance. Processing equipment should be located so as to achieve a logical flow. Good ergonomics is critical, especially if isolators are used, and mock-ups of these are recommended during the design phase. Where solid material flows occur, such as in media preparation, the receiving equipment should be located vertically below, with good access to the chute for cleaning.

The layout is affected by the room air quality classifications or grades. In the USA there are commonly three grades of clean room or area within a room, namely Class 100 in operation, Class 10 000 in operation and Class 100 000 in operation. These grades approximately conform to the EU GMP grades A, B and C respectively (see Section 12.7). It is the custom in Europe for there to be a progression from 'unclassified' to grade D, to grade C and if required to grade B, with a lobby or a corridor at each transition. The gowning should take place in the innermost lobby and the lobby should be the same grade (at rest) as the room it serves. This is not a requirement in the USA where a progression from unclassified to class 100 000 is permitted provided the differential air pressure exceeds 12.5 pascals.

This requirement for lobbies and airlocks can have a major impact on layout. The grade of room should be the minimum required for the duty, then the corresponding lobby can be designed. Sometimes rooms are set at a higher grade than needed in order to save on the number and sequence of lobbies or airlocks. This is not logical and leads to failures to meet the room environmental quality standards and to the ongoing unnecessary cost of environmental monitoring and gowning. However a clean corridor giving access from a gowning room to several clean rooms is acceptable.

A clear policy on gowning is also needed, with the exact details of what happens in each of the gowning rooms specified. This will ensure that gowning rooms are the correct size and that there is sufficient space for storage of clothes and shoes. Hand washing facilities are needed in the outermost part of gowning rooms.

Figure 12.4 shows the typical segregation of rooms.

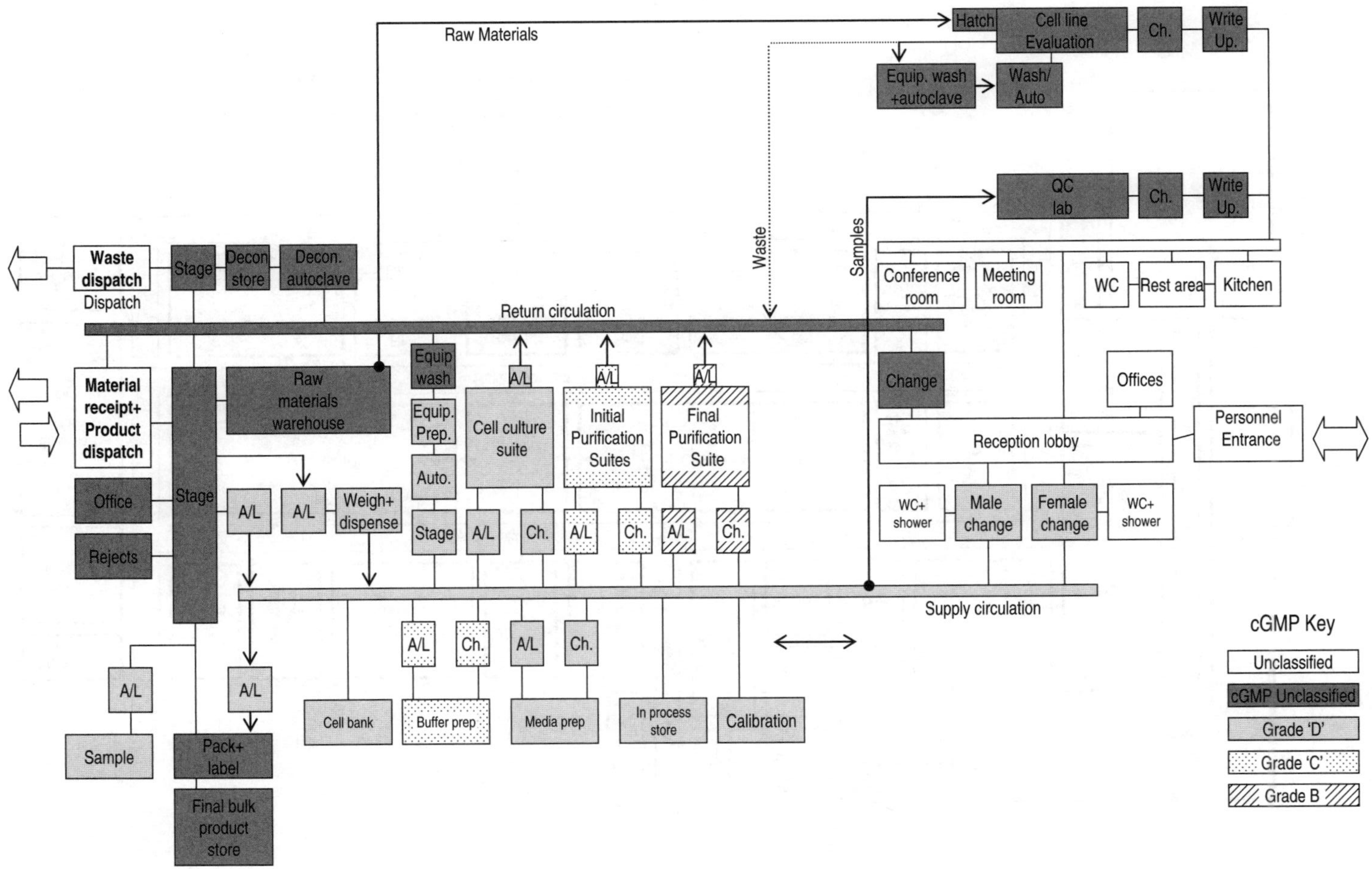

Figure 12.3 Adjacency diagram.

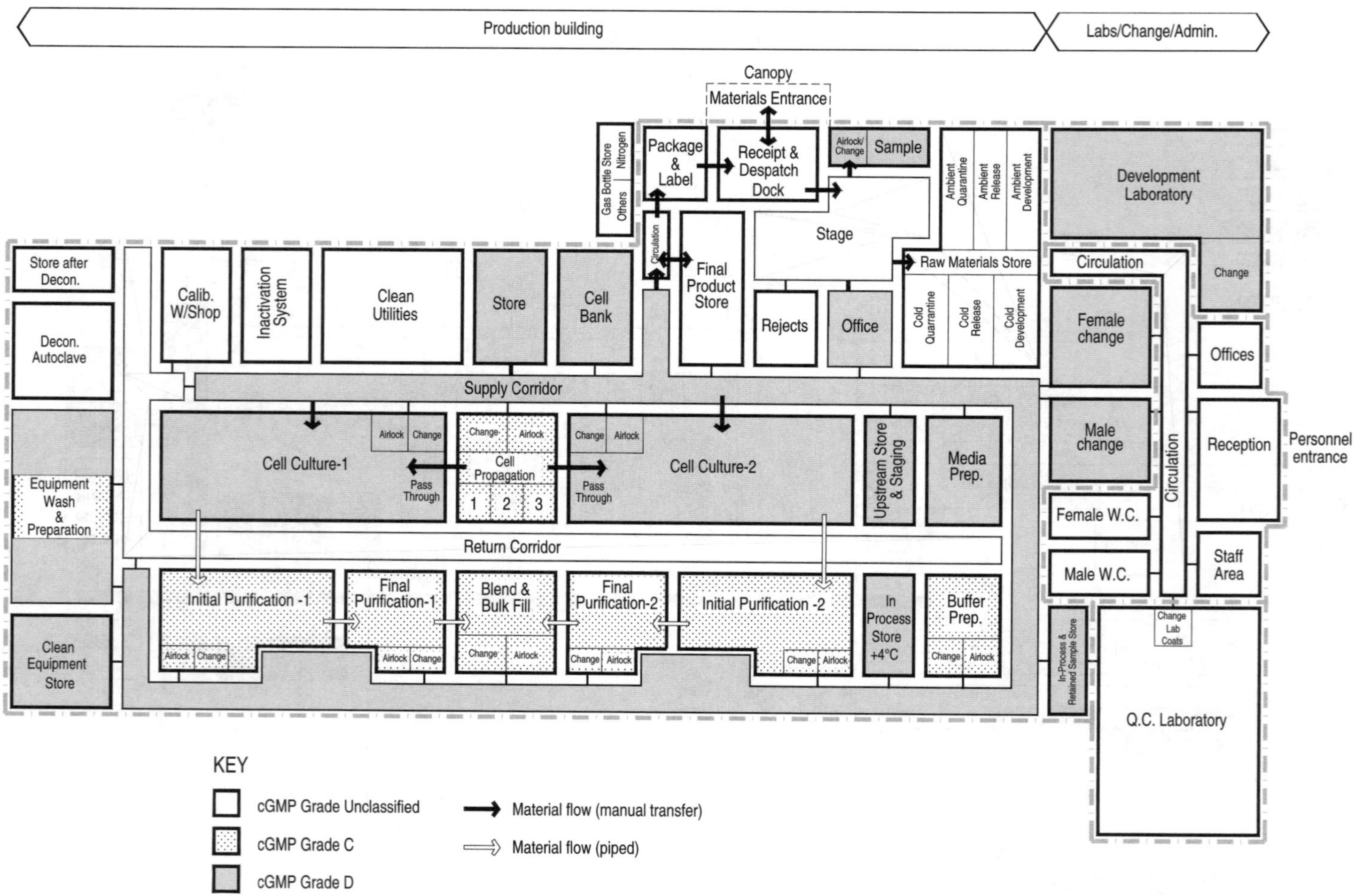

Figure 12.4 Segregation.

Segregation of areas where there are viable and killed microorganisms is essential, and is a key feature of vaccine facilities to avoid the transfer of live organisms into the product.

For mammalian cell culture, the segregation of purification stages before and after a viral inactivation step is a key feature. Almost all mammalian cell lines contain some form of endogenous viruses and, in addition, viruses might be introduced inadvertently during processing (see Chapter 31). The entire purification process must be validated to remove or inactivate harmful viruses (see Chapter 19). A process procedure or step in the purification sequence is identified where viruses are removed to an acceptable level, and at this point the purification process is divided into pre- and post-viral inactivation stages. All purification steps contribute to the removal of viruses and the user must decide which step constitutes the virus removal stage.

There have been some changes in the provision of pre-and post-viral facilities. For example, where processes are closed and the removal and inactivation of viruses has been validated, then the whole train of purification steps can take place in the same room. For open processes, such as solvent/detergent or low pH treatments, it would be expected that the inactivation would be performed in one vessel in room A before being transferred to a clean vessel, usually via a wall port, in room B. There would be separate air handling units for the two areas and personnel would not cross from the pre-inactivation area to the post-activation area without a change of gowning, including footwear. Post-inactivation equipment would be dedicated unless it could be sterilized. Wheeling mobile vessels from pre-inactivation areas into post-inactivation areas would not be allowed unless cleaning is validated.

Many companies have a rule that there can be only one lot of one product in the same room at the same time. So where the entire purification process takes several days, several adjacent purification rooms can be installed and thus the process separated in order to increase productivity, as several lots can be processed at the same time in different rooms.

Where the purification suites have to accommodate different processes, the best approach is to create a well-serviced room or suite, often at grade C, for additional flexibility if the processes might not be closed. Process equipment is then brought from a secure store and connected by hard pipe or fluid transfer panels. At the end of the campaign, the stream can be cleaned and disassembled and a new line created. Utilities can be supplied from service stations within the suites.

Some parts of certain process equipment can be contained in a service area, with only those parts that need access for operation held in the clean room. This method means that there is a minimum of expensive classified clean room volume and also the design permits easy access for maintenance and qualification. This 'grey' side technology has long been used for autoclaves, washing machines and lyophilizers. It is now also used for bioreactors, for example.

The renovation of an existing building to accommodate biopharmaceutical manufacture is challenging. The building might have to be decontaminated. The building's location and shape might not be ideal and compromises are inevitable. It is recommended that the fabric of the existing building is carefully surveyed before it is reused.

12.7 ROOM AIR QUALITY CLASSIFICATION AND AIR CONDITIONING

The quality of the local environment that is needed for reliable production depends on the degree to which the raw materials, API and components are exposed while processing occurs. The air conditioning system has to achieve several objectives:

- Segregation of the environment by ensuring that air and its associated particles cannot flow from one area to another. Examples are the segregation of the areas where there are live and dead cells.

- Providing a suitable environment for processing exposed materials. The definition of clean room environments is described in the EU regulations (EU 1998). The environment should be defined in terms of the desired ranges of temperature, relative humidity and particle concentration. For rooms where there are no process-specific requirements, then human comfort will set the range.
- The removal of toxic or undesirable contaminants and odours.
- The removal of heat from sources such as the equipment and sterilization processes within the room. Air can be recirculated and cooled to save energy costs. The regulations for GMP give no minimal air flow rates but they are typically at least 10 air changes per hour for grade D rooms and 20 air changes per hour for grade C rooms.

Room air-quality classifications have a major effect on facility design, as previously noted. Early in a project a decision must be made on the quality of the room air grades. The different systems of classifying clean rooms are confusing so Table 12.2 gives a comparison showing the approximately equivalent nomenclatures. For both the superseded US FS209E and the current international ISO system it is necessary to state whether the conditions apply to 'at rest' or 'in operation'. The EU nomenclature in the GMP regulations describes both.

The table of comparison is in complete agreement with Table 1 Air Classifications in the FDA *Guidance for Industry, Sterile Products Produced by Aseptic Processing*; the final version was published at the end of September 2006 (FDA 2006). Although it is recognized that the application for bulk drug substances is not covered in this guide, it is the document where room air quality is defined by FDA. Similarly Annex 1 of the EU GMP regulations *Manufacture of Sterile Medicinal Products* is the place where the clean room performance standards are defined. The environmental monitoring requirements for the EU are also given (EU 1998).

In order to achieve the required air quality the room finishes have to be appropriate, and drains carefully designed. For grade C and B rooms, the walls and floors and ceilings must be crevice-free and smooth and impervious. Coving is recommended between walls and floor and ceiling to assist cleaning. Doors and windows should be flush and the use of wood should be avoided. Prefabricated wall systems have the advantage that air conditioning ducting and utilities are incorporated within the walls, avoiding protrusions that are challenges to clean. Pipework should be avoided in clean rooms where possible and be cleanable where it is essential. The specification of materials of construction must support the procedures for clean down, and product changeover. This includes the cleaning of rooms and equipment.

Table 12.3 gives typical grades of room found in bulk biopharmaceutical facilities. Despite their label, unclassified rooms in the facility could have good quality filtered air and controlled temperature. There is no American equivalent of the EU grade D which is only defined in the regulations as the 'at rest' situation. This means that in Europe, the lowest grade of classified room is grade D, whereas in USA it is class 100 000 or IS08 in operation which is equivalent to grade C. This might have lead to the situation where rooms tend to be at a higher class in America for the same required degree of containment in Europe.

Table 12.3 gives an indication where laminar flow cabinets (LAF) are used. A Class II microbiological safety cabinet (MBSC) also provides grade A HEPA filtered air over exposed material, as well as giving some protection to the operators.

Room grades are more stringent in facilities where the processes are open in part, or where many aseptic connections are made. Also, where future needs are not so well defined, a higher grade room may be provided for flexibility. This is again because there might in the future be more open processing and more connections of equipment.

Different air handing units will serve several rooms, and the zoning of these to ensure that there is no cross-contamination is a GMP issue.

Table 12.2 Air quality grade from EU-GMP regulation (revised on 30 May 2003 and enforced from 1 September 2003) compared with ISO 14644 Standard *Clean rooms and Associated Controlled Environments* and with superseded FS209E *Airborne Particulate Cleanliness Classes in Cleanrooms.*

	At rest[b]		In operation[b]	
	Max. permitted number of particles per m³ equal to or above:[a]			
Grade/Class	0.5 µm[d]	5 µm	0.5 µm[d]	5 µm
A	3500	1[e]	3500	1[e]
ISO 5 on operation			3520	29
FS 209E Class 100			3530	—
B[c]	3500	1[e]	350 000	2000
ISO 7 in operation			352 000	2930
FS 209E Class 10 000			353 000	2470
C[c]	350 000	2000	3 500 000	20 000
ISO 8 in operation			3 520 000	29 300
FS 209E Class 100 000			3 350 000	24 700
D[c]	3 500 000	20 000	Not defined[f]	

Notes: (taken verbatim from EU-GMP):

[a]*Particle measurement based on the use of a discrete airborne particle counter to measure the concentration of particles at designated sizes equal to or greater than the threshold stated. A continuous measurement system should be used for monitoring the concentration of particles in the grade A zone, and is recommended for the surrounding grade B areas. For routine testing the total sample volume should not be less than 1 m³ for grade A and B areas.*

[b]*The particulate conditions given in the table for the 'at rest' state should be achieved after a short 'clean up' period of 15–20 minutes (guidance value) in an unmanned state after completion of operations. The particulate conditions for grade A 'in operation' given in the table should be maintained in the zone immediately surrounding the product whenever the product or open container is exposed to the environment.*

[c]*In order to reach the B, C and D grades, the number of air changes should be related to the size of the room and the equipment and personnel present in the room. The air system should be fitted with appropriate filters such as HEPA for Grades A, B and C.*

[d]*The guidance for the maximum number of particles in the 'at rest' and 'in operation' conditions corresponds approximately to the cleanliness classes in the EN/ISO 14644-1 at a particle size of 0.5 µm.*

[e]*These areas are expected to be completely free from particles of size greater than or equal to 5 µm. As it is impossible to demonstrate the absence of particles with any statistical significance the limits are set to 1 particle / m³. During the clean room qualification it should be shown that the areas can be maintained within the defined limits.*

[f]*The requirements and limits will depend on the nature of the operations carried out.*

Recommended limits for microbiological monitoring of clean areas during operation (EU cGMP Annex 1, 5) are as follows (average values):

Grade	Air sample (cfu/m³)	Settle plates (diam. 90mm) (cfu/4 hours)*	Contact plates (diam. 55mm) (cfu/plate)	Glove print 5 fingers (cfu/glove)
A	<1	<1	<1	<1
B	10	5	5	5
C	100	50	25	—
D	200	100	50	—

**Individual settle plates may be exposed for less than 4 hours*

The ventilation systems might be required to accommodate fumigation. Fumigation of rooms for the removal of microorganisms, especially viruses, is generally with formaldehyde or paraformaldehyde in carefully controlled conditions of temperature and relative humidity. The process has long been used in vaccine facilities, for example, where different strains of a virus are used

Table 12.3 Typical room air quality classifications.

Room function	Grade	Features
Media preparation	D	
Media storage	D	
Cell bank storage	NC	
Inoculum preparation	C	Class II MBSC
Bioreactor room	D or NC	
Cell separation room or harvest	D or NC	
Buffer preparation	D or C	
Buffer storage	D or NC	
Purification suite	C or D	
Purification column packing	C	
Filling low or zero bioburden API	C or B	
Filling sterile API	B	Grade A LAF

NC—rooms not classified.

to make the vaccine and fumigation is used to prevent cross-contamination. Fumigation is rarely used in facilities for the manufacture of therapeutic proteins. The use of fumigant imposes special challenges on facility design. The doors and openings have to be sealed during the procedure, which is usually performed overnight. The fabric of the building must not permit leakage that disrupts other areas, and the air conditioning system must be able to isolate the filters, ductwork and rooms to be fumigated. This will affect the number and zoning of air handling units. Limits are set for the rate of discharge of formaldehyde into the environment.

Other fumigants are used such as vapour-phase hydrogen peroxide, which is the best method for sanitizing isolators. It is also used for sanitizing rooms. There are few published papers on the ability of vapour-phase hydrogen peroxide to destroy viruses so it is not yet recommended where the removal of pathogens is the major consideration.

Building management systems (BMS) are used to control the temperature, relative humidity and flow of air. They can be a separate system from the process control system, but it is recommended that quality-critical parameters, such as warm room temperatures, are recorded on the validated PCS. There is no regulatory requirement to qualify a BMS, but it is done increasingly for the best reason, namely to test its consistent reliability. Further details on system management are given in Chapter 13.

12.8 UTILITIES

The utilities that connect directly to the process stream include water of several grades, clean steam, cleaning fluids from CIP systems, compressed air and other gases, and drains. Written specifications are needed for all utilities that are in contact with process surfaces or are in direct contact with the process.

12.8.1 Process Water

The minimum grade of water that is required by the regulators depends on the stage of manufacture. Higher than minimal grades are used to make technology transfer easier between sites, for example the chemical quality of potable water varies widely and the alternative use of deionized or purified water can eliminate these differences. Also there is no regulatory requirement to use purified water or water for injection (WFI) in the media for cell growth. However, in order to have control over the ingredients and avoid differences in water quality in different locations, some companies will use purified water in order to standardize and ease technology transfer.

The specification of the API and the drug product determine the quality of water used for making the API. For sterile and apyrogenic API, water for injection must be used for the final stages of purification and for the final rinses of the equipment used. Where the API is non-sterile but is intended for use in a sterile, parenteral product, then the minimum required grade of water for the final purification stages and for rinsing is purified water with an endotoxin limit of 0.25 EU/ml and control of specified organisms.

Decisions are needed on specifying the correct the grade of water for the different duties required in the facility (see EMEA 2002).

Pure water systems comprise three sections; generation, storage and distribution. Most systems are constructed in stainless steel and the water is circulated to the points of use in either hot or cold loops. Dead legs must be avoided and the system will be routinely tested. Measurements of total organic carbon and conductivity are used for quality control. Once validated, water can be taken directly into the process without having to collect an aliquot and test it. For this reason, regulators are concerned that water systems are designed and operated correctly. Further detail on water quality is given in Chapter 2.

12.8.2 Steam

Clean steam is recommended for sterilization of equipment and for supply to autoclaves. It should also be used to humidify clean rooms. In Europe the quality of clean steam for sterilizing must be routinely tested according to HTM 2010, and EN 285 (HTM 2010 1995; FDA/ISPE 2001). The safe and accessible location of steam sample points is an important design detail.

12.8.3 Compressed Gases

Compressed air should be taken from an oil-free compressor and filtered at the points of use. Compressed gases are sampled before use and automated systems are used to transfer from used to new cylinders. Instrument air has no product contact, but should be filtered to give particulate control in line with the environment into which it is vented.

12.8.4 Waste

Depending on the risk assessment, a biowaste inactivation system may be installed. This waste comprises the bulk liquid waste and condensate, which contains the process cells. A heat inactivation system should be validated using the process host organism.

The design of drainage systems for the facility generally will ensure that there is no back-contamination from the system. Floor drains in clean rooms should be trapped or even avoided if possible.

12.9 SUPPORT FUNCTIONS

The site will often provide many support functions. Laboratories for processing samples for environmental monitoring and in-process control are best incorporated in the biopharmaceutical facility itself. Utility rooms for air conditioning plant and waste treatment must be located in the facility, but offices, maintenance workshop, calibration resources and quality assurance archives can be located elsewhere. It is recommended that personnel areas are restricted to a first aid room, changing rooms, toilets and canteen.

The quality assurance function must be involved in the design of the facility as well as in its operation. The control process and collection and storage of information are fundamental aspects of GMP and the early involvement of QA in the selection of the control system is important (see Chapters 13 and 34).

12.10 SECONDARY MANUFACTURE

The dosage forms for cell-based biopharmaceuticals are most commonly small volume parenterals (SVP) in stoppered vials, ampoules or in pre-filled syringes. Both liquid formulations and solid formulations comprising sterile freeze dried powders are available. Other formulations include large volume parenterals (LVP) and even oral dosage forms. Both SVP and LVP are sterile and administered by injection, and are therefore made to the highest standards of GMP.

In a multi-product secondary manufacturing facility, the range and type of drug products that can share equipment and manufacturing areas is well defined. The complexity of demonstrating and validating effective cleaning between products and the time for proper process development and process cleaning validation might outweigh the cost of additional equipment. A detailed risk analysis is needed where cytotoxic drugs and antibiotics are involved. Separate facilities are used for live and inactivated vaccines.

12.11 DESIGN REVIEWS

A successful project requires an integrated team lead by a project manager. The measure of success is assessed by reference to the User Requirement Specification (URS). This short document defines the purpose of the facility and the key requirements. For example, it will specify the products, the production rate, and the territories where the product is to be sold, which determine the appropriate regulations. The scope of the project will include a list of the functions that will be performed and the timescale for completion, commissioning and qualification. Any known quality-critical factors can be included. The URS will be used as the basis for performance qualification. During the conceptual stage of the design, the URS will give rise to a series of design specifications and philosophies so that early decisions can be made, such as a room list, the room air quality grades and the number of people who will work in the facility.

The facility will be designed for GMP, and the regulations for the manufacture of API have been harmonized and accepted by most countries. Although the FDA regards the document as guidance it is a requirement in the EU and is given as Part II (formerly Annex 18) of the EU regulations *Good Manufacturing Practice for Active Pharmaceutical Ingredients*, **(EU 1998)**.

A series of design reviews is recommended:

- *Safety* A review of safety must be performed during the design phases. The review will cover the safety of the design in relation to the operation of the facility. There will also be a review of the safety measures and procedures that will be employed while the facility is being built. The safety reviews are chaired by an engineer who is independent of the design team, and the use of HAZOP and HAZAN procedures is recommended (Gillett 1997).
- *Constructability* A review is needed by the engineer's construction experts to ensure that the facility can be constructed so that equipment can be safely installed, maintained and replaced in the future.
- *cGMP Compliance* cGMP reviews should be conducted and documented to conform to the requirements of the API Guide and with the EU regulations Annex 15 and Part II (EU 1998).

The purpose is for qualified personnel to review relevant aspects of the proposed design for cGMP compliance in a controlled manner. This will ensure that there is a record that the design and/or the facility meets the cGMP-related parts of the URS and cGMP requirements of the project, and that documentation is produced to demonstrate that the design conforms to these requirements.

A cGMP review is a means to consider formally all aspects of the specification and the design of a facility so that it is best suited to its intended use and is in compliance with the relevant regulations (European Directives, The Rules Governing Medicinal Products in the European Community, US Code of Federal Regulations, etc). The number of cGMP reviews will depend on the size and complexity of the project. cGMP reviews take place in accordance with the project programme. The initial cGMP review should take place towards the end of the front end design/basis-of-design phase, near the end of the preliminary engineering phase, and then typically twice during the detailed design phase. These reviews also contribute to design qualification (DQ), i.e. the documented verification that the proposed design of the facilities, systems and equipment is suitable for its intended use and conforms to the registered process. This includes compliance with the requirements of current GMP and the URS. The cGMP review constitutes a major part of the DQ and DQ is the first step in the qualification and validation process (see Chapter 15).

12.12 TIMESCALES

The timetable for the design, construction, qualification and start-up of a biopharmaceutical facility is typically 5 years from the beginning of design to obtaining a manufacturer's licence. Preliminary engineering design takes about 9 months and detailed design up to a further 12–15 months. Many key items such as bioreactors and disc-stack centrifuges are long-delivery items. Activities can overlap, but construction and installation typically takes 26 months. This time can be reduced by the construction and testing of prefabricated equipment modules and prefabricated buildings that are linked together on site. Qualification can take up to a year for biopharmaceutical facilities. A manufacturer's licence will not be granted until the facility is validated and several batches successfully produced.

12.13 TRENDS IN BIOPHARMACEUTICAL FACILITY DESIGN

The biotechnological revolution is bringing in new pharmaceutical products including those for gene therapy, skin and cartilage from cultured tissue, and novel vaccines. There is also the possibility of tailoring drugs to suit small groups of people. However, there are major challenges in the processing of these materials and making them profitably. By contrast the manufacture of enzymes, monoclonal antibodies and therapeutic proteins is well established.

Processes are being developed to be closed with automated cleaning in place. This means that the room grades are now lower, with fermentation rooms unclassified. The investment in containment is assisted by the trend to use packages or modules containing the equipment and controllers. Packages include centrifuges, bioreactors, ultrafiltration rigs and clean-in-place rigs. These can be tested at the vendor's factory before delivery and thus save time in building the facility.

There are several ways in which the concepts of sustainability are being introduced into a biotech facility. The use of disposable sterile bags removes the need for cleaning vessels and other equipment, thus saving on cleaning chemicals and the subsequent large volumes of high quality rinse water. The insertion of heat exchangers in hot WFI loops can also save water, rather than have dead-ended take-offs with heat exchangers that could put water to waste until the desired temperature is reached. Sometimes it is more economical to make a single grade of water rather than to have two systems. Vapour compression stills can produce WFI economically. Some recent facilities have only a WFI system to deliver both WFI and purified water, thus saving capital and testing costs.

As for processing, there is a revived interest in continuous chromatography and simulated moving bed chromatography. Buffer solutions are being made continuously with savings in capital. The purchase of pre-sterilized media and buffer solutions can save capital cost and reduce facility footprint.

There have been no accidents to date arising from the contained use of genetically modified organisms. Where possible, closed systems are used for productivity reasons and this also enhances safety. The scale of operation of animal cell cultures is up to 20 000 litres and this appears to be an optimal size for large-scale operations.

REFERENCES

Birch JR, Arathoon R (1990) In *Large Scale Mammalian Cell Culture*. Ed Lubiniecki AS. Marcel Dekker, New York.

CFR 21 Part 210 (2004) Title 21, Code of Federal Regulations, Part 210: *Current Good Manufacturing Practice in Manufacturing, Processing, Packing, or Holding of Drugs; General*. US FDA, April 1, 2004.

CFR 21 Part 211 (2004) Title 21, Code of Federal Regulations, Part 211: *Current Good Manufacturing Practice for Finished Pharmaceuticals*. US FDA, April 1, 2004.

CFR 21 Part 600 (2004) (FDA) Title 21, Code of Federal Regulations, Part 600: *Biological Products General*. US FDA, April 1, 2004.

CFR 21 Part 11 (2004) Title 21, Code of Federal Regulations, Part 11: *Electronic Records and Electronic Signatures*. April 1, 2004.

EMEA (2002) (CPMP) *Note for Guidance on Quality of Water for Pharmaceutical Use* (May 2002).

EU (1998) EU Commission of the European Communities. *The Rules Governing Medical Products in the European Community, Good Manufacturing Practice for Medicinal Products*. Volume IV, April 1998, with additional annexes and update of legal references July 2004.

FDA (2006) *Guidance for Industry – Sterile Drug Products Produced by Aseptic Processing – Current Good Manufacturing Practice*. Final, US FDA, September 29, 2006.

FDA/ISPE (2001) *Baseline Pharmaceutical Engineering Guide*, Volume 4: *Water and Steam Systems*. First edition, January 2001.

Gillett JE (1996) *Hazard Study and Risk Assessment in the Pharmaceutical Industry*. Interpharm Press BocaRaton, FL, USA.

HTM2010 (1995) Health Technical Memorandum 2010, *Sterilization*, Part 2: *Design Considerations*. HMSO, London.

List of Useful Websites

Food and Drug Administration	http://www.fda.gov/
European Medicines Agency	www.emea.europa.eu
United States Department of Agriculture	www.usda.gov
Eudralex	http://ec.europa.eu/enterprise/pharmaceuticals/eudralex/homev4.htm

13 Monitoring, Control and Automation in Upstream Processing

TS Stoll and P Grabarek

13.1 INTRODUCTION

Automatic control of a process can be defined as control of the variables without the necessity of manual intervention. The term *automation* has a similar meaning but also refers to control on a higher level, involving a sequence of tasks or operations that are scheduled and performed by a computer system. Control and automation of processes offer several benefits (Beyeler *et al.* 2000). Firstly, optimal conditions for productivity and product quality can be maintained accurately and continuously, even in processes with a high level of complexity. Secondly, recipe-controlled operations also maximise the process reproducibility and minimize run-to-run variations in product quality. Thirdly, safety during production is increased, since the automation system can also continuously check critical parameters and handle some failures according to predefined routines. Even during manual operations, the reliability of personnel can be enhanced by the support of an automation system, via alarms, check lists, etc. Finally, automation can also help to reduce the need for personnel and the duration of operations. For these reasons, control and automation have become key elements of modern pilot- and large-scale biotechnology manufacturing facilities. In upstream processing, automation is actually not only applied to the culture steps *per se* but also to many side operations, such as the supply of medium and gases, the cleaning- and sterilization-in-place of equipment, the transfer of process fluids, etc. (see Chapter 14). In research and development, automatic control is also a very powerful tool for metabolic studies, for the verification of process models, and for the improvement of bioprocesses (Sonnleitner 1997).

The tools used for control and automation can be viewed as being an essential part of the various information systems used in a manufacturing organization or even a whole company. These systems are often classified and organized in a computer-integrated manufacturing (CIM) pyramid, with four or five different functional levels. Each level has the ability to communicate with the level above or beneath in different ways, depending on the degree of automation of the manufacturing organization. Figure 13.1 shows a typical schematic representation of a CIM pyramid.

The field level is discussed in Section 13.2, below, with emphasis on the monitoring of cell cultures; this area has been the object of significant development efforts in the last decade, but is still a weak point in the automation of bioprocesses due to the difficulties in obtaining reliable sensors. The process-control level is discussed in Section 13.3. The higher levels are described in Section 13.4, but only in summary, since they are not specific to animal cell cultivation. This chapter is

Medicines from Animal Cell Culture Edited by G. Stacey and J. Davis

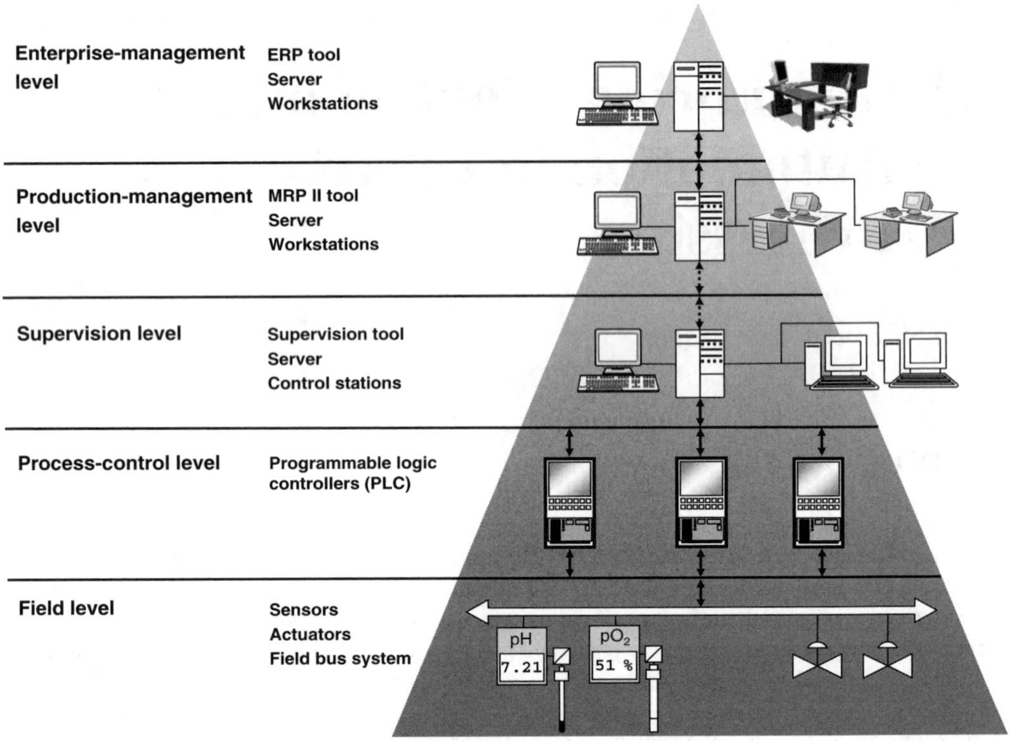

Figure 13.1 Schematic representation of a computer-integrated manufacturing (CIM) pyramid for a manufacturing organization. Only the main elements used at each level are listed and depicted. Double arrows represent, in a simplified way, communication between the levels. Dotted lines indicate communication that is often manual, except in highly automated plants (see Section 13.4.2 for details).

completed by a discussion on the regulatory aspects of automation systems, with emphasis on the requirements set by the US FDA (21 CFR Part 11, 1997).

13.2 FIELD LEVEL

13.2.1 Main Components

13.2.1.1 Sensors

Sensors are one of the main components of the field level. They represent all the devices used to measure the process variables such as temperature, pH, dissolved oxygen, pressure, liquid and gas flowrates, metabolites, etc. *Classical* sensors include, in addition to the sensing function itself, a basic signal-processing function (e.g. amplification) to generate an output signal. The most common output signals are analogue, such as electric current (4–20 mA) or voltage (1–5 V); the signal is then further processed (e.g. converted to a digital one) by external components. *Smart* sensors include advanced signal-processing functions to generate a digital output signal; they are capable of performing various diagnostic, maintenance and time-stamping operations; they can also be configured and calibrated remotely. Traditionally, sensors were directly connected to *indicators*, for local display of the measured value, and to *recorders*, for data storage. In advanced control systems, indicators and recorders may not exist as local or individual devices anymore; instead, they may be integrated into a process-control or plant supervision system (see Sections 13.3.1 and 13.4.1). In this case, the variables are typically displayed in a grouped way on a computer station, and stored in a centralized machine.

For automatic control, sensors should provide measurements *on line*, i.e. in a fully automated way, without the requirement for manual intervention, and in real time.† For this purpose, in the case of cell culture, one can distinguish between *in situ* sensors, i.e. directly mounted in the bioreactor, and *ex situ* ones, i.e. placed on the outside. *In situ* sensors do not require any sample withdrawal and values of the process variable are available 'continuously'. Although they are very practical, few types of such sensors are available, due to the requirements for *in situ* (steam) sterilization and calibration, as well as for stability and reliability over extended periods. *Ex situ* sensors require regular sampling from the bioreactor content and thus provide only a discrete signal; however, if the measurement is fast compared with the time constant of the process, it can be considered as being available in real time, too. Sampling bears several risks, such as line fouling, back-contamination to the bioreactor, and changes in the sample composition during transfer to the sensor. The design of an adequate sampling device is thus critical; examples are discussed elsewhere (van de Merbel 1996). Provided that they are properly automated, monitoring techniques based on *ex situ* sensors can be used for on-line measurement and thus for automatic process control.

A few process variables in a bioreactor are considered as critical, and thus have to be monitored and controlled continuously throughout the culture. Consequently, the corresponding sensors have become standard equipment for a bioreactor. They are discussed in Section 13.2.2. On-line monitoring techniques have also been developed for several other physico-chemical and biological variables (biomass, key metabolites, recombinant proteins), although the measurements have rarely been used for process control at industrial scale; the main reasons are the additional costs and higher risks of failure, which outweigh the relatively modest improvements brought by the automatic control of these variables (see Section 13.3.3). For monitoring only, however, these measurements represent attractive alternatives to corresponding off-line techniques. First, frequent automated measurements can be performed, often very reproducibly, with limited personnel; second, the collected data can allow detailed process analysis and modeling (see Section 13.2.6). These additional techniques are reviewed below (Sections 13.2.3 – 13.2.5). Some key techniques providing data in real time, although from off-line instruments, have also been included in the discussion because they have the potential to be used on line and, if needed, for automatic control provided they are coupled with the proper sampling device. The numerous other off-line analytical techniques used to monitor and characterize cell culture processes and products are outside the scope of this chapter. Only a few are mentioned here, for comparison with corresponding on-line or real-time techniques.

13.2.1.2 Actuators

Actuators are devices that perform some physical action in the process environment, with the goal of controlling one or a set of variables. An actuator consists of three main components: (i) a control-signal processor, that receives and interprets a control command from the process-control level; (ii) a servo-drive or control component, which passes on energy in response to the output of the control-signal processor; (iii) a final control element, such as a valve, pump or motor, which is in physical contact with the process environment. An important milestone

†The expression 'real time' should be used with caution and always with respect to a clearly defined phenomenon or system. Indeed, if one considers the field of animal cell cultivation only, one can find phenomena with time constants ranging anywhere from *ca.* 1 μs to several days (see Sonnleitner, 1999, for a detailed discussion). In this chapter, we will focus on the automatic control of complete cultures at the bioreactor level and thus assume that measurements performed within a couple of minutes are available in real time. This would of course be erroneous if one wanted to study the dynamics of phenomena at the cellular level.

in the automation of bioprocesses was the development of membrane valves in the 1970s, capable of operating under sterile conditions. Actuators for these valves are mostly pneumatically powered and rarely electrically. Only these valves allow the remote control of processes while maintaining sterility, and it is hard to imagine any modern bioprocess without them (Beyeler *et al.* 2000).

13.2.1.3 Field bus system

Traditionally, sensors were individually connected to the automation system in a 'star configuration', and communication at the field level was performed via transmission of an analogue signal. Nowadays, this technique has been replaced by the use of a digital field bus system. This system is a local network with a restricted number of nodes, which can connect some dozens of field devices (sensors and actuators) among themselves, in a serial way, and to computers at the process-control level. Bus systems are suitable for the typical communication requirements at the field level, i.e. the transmission of relatively short data sets within a few milliseconds. Characteristics and advantages of field bus systems are (Steusloff 2002):

- They are compatible with smart sensors and actuators, which require transmission not only of process measurements and control information, but also of their digital status and of control data between themselves and control computers. The additional cabling that used to be required for this purpose can be omitted.
- Cost of cabling is therefore reduced significantly.
- New devices can easily be added to an existing field bus system.
- Conversion of analogue signals to digital ones locally, at the connection to the field bus, prevents interference with the signal on route to the control computer. The redundancy of digital codes also makes it possible to reconstruct erroneous data.

Despite the need and efforts for a standardized field protocol, so that users would not have to worry about the compatibility of devices from different suppliers, several systems coexist currently. The most common ones for industrial applications are the PROFIBUS (PROFIBUS International), used for instance by Siemens, and the FOUNDATION Fieldbus, used for instance by Emerson (ex Fisher-Rosemount). In the R&D environment, the CAN-Bus (available from several manufacturers) is often used (Beyeler *et al.* 2000).

13.2.2 Standard Bioreactor Sensors

13.2.2.1 Introduction

Temperature, pH and pO_2 sensors (or 'probes') are essentially always present on bioreactors for monitoring and control purposes. They are installed *in situ* and are steam-sterilizable in place. They are covered with a housing that provides pressure balance during sterilization and other pressurization operations, as well as protection from contamination. Additionally, pilot- and production-scale bioreactors are commonly equipped with on-line devices to measure the liquid level in, or weight of, the bioreactor and the pressure in the headspace. The use of an *in situ* probe for biomass is also becoming common, at least as a complement to off-line analysis (see Section 13.2.3). Figure 13.2 shows a simplified instrumentation diagram of a pilot- or large-scale bioreactor operated in batch mode, with the standard sensors. Their characteristics are discussed below. Sensors to measure the flow rate of supplied media and gases are not used as independent monitoring devices but are embedded in a combined monitoring and controlling device; they are discussed at the process-control level (Section 13.3.2).

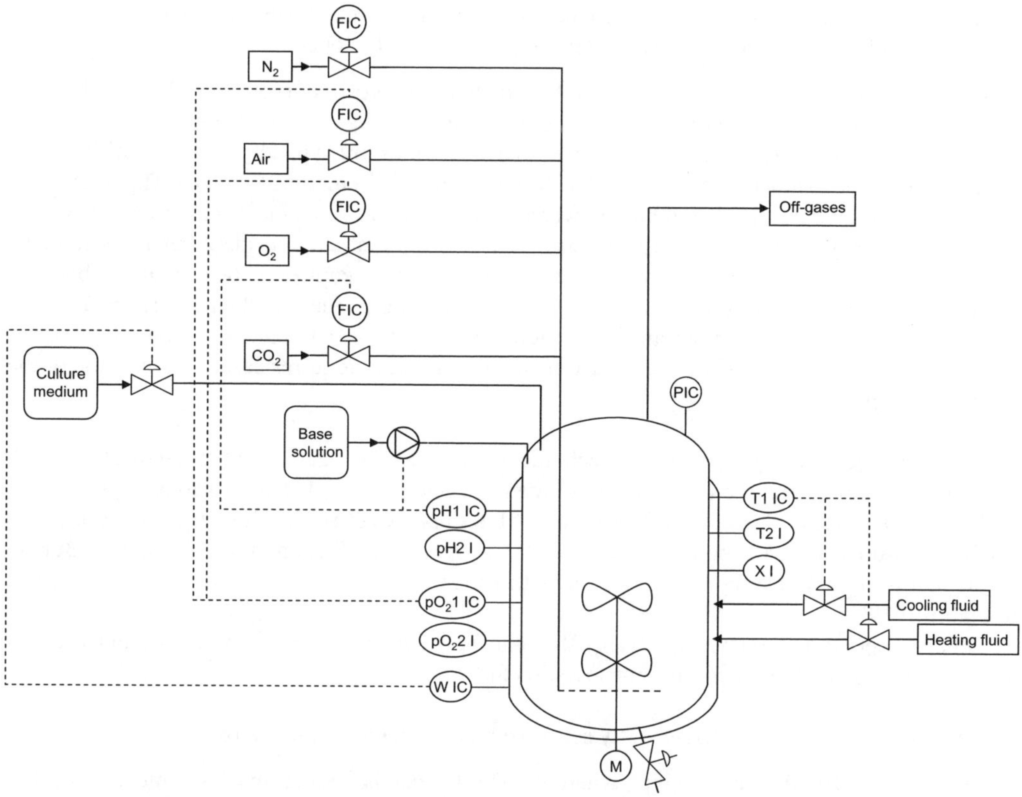

Figure 13.2 Simplified instrumentation diagram of a pilot- or large-scale bioreactor operated in batch mode, with the standard sensors used during cell culture; additional probes (e.g. temperature) and valves used during operations such as cleaning- and sterilization-in-place are not represented. The main control loops (dotted lines) are also shown, in a schematic way (see Section 13.3.2 for details). C: controller; F: flow rate; I: indicator; M: motor; P: pressure; T: temperature; X: biomass; W: liquid weight; '1', '2': indicate probes in duplicate (1 main probe and 1 back-up; see Sections 13.2.2.2–13.2.2.4).

13.2.2.2 Temperature

Animal cells are very sensitive to temperature; this parameter must therefore be monitored and controlled accurately, i.e. within 0.5 °C or less around the setpoint. Temperature probes should be steam-sterilizable in place and stable over several weeks, so that they can be used for *in situ* measurement in fed-batch and perfusion cultures. In pilot- or large-scale bioreactors, duplicate probes are commonly used. Additional temperature probes are also mounted at various points of the tank and associated pipework to monitor sterilization (see Chapter 14, Section 14.4). There are two main types of temperature probe used for bioreactors (Hartnett 1994):

- *Resistance temperature devices* (RTDs): these devices are characterized by a high accuracy and stability. They have a typical time constant of 5 seconds, which is adequate for animal cell cultures. RTDs are based on the measurement of the electrical resistance of a metal, which is known to change with temperature. The most commonly used metal for RTDs is platinum; a classical device is the Pt-100 sensor, referring to a resistance of 100 Ω at 0 °C. Other common metals are nickel, nickel alloys and copper. The non-linear relationship between electrical resistance and temperature requires a careful calibration of the RTD.

- *Thermocouples*: these are considered to be excellent low-cost alternatives to RTDs; their accuracy can actually be equivalent to the latter. While RTDs are preferred for critical measurements in a bioreactor during a culture, thermocouples are typically used for utilities and sterilization. Thermocouples consist of a closed electrical circuit made of two different metals; if there is a temperature difference between the two junctions, an electrical current is generated. One junction is placed in the sensing region and the other, considered as the reference, in an environment where the temperature is either carefully controlled (e.g. in an ice bath) or measured. The temperature at the location of the sensing element is then determined on the basis of the measured current and of the temperature of the reference junction. Here again, the non-linear relationship between current and temperature requires careful calibration of the thermocouple.

Both probe types are equipped with mechanical protection devices called thermowells, which separate the sensor from the process environment. Several standards from various organizations such as the American Society for Testing and Materials (ASTM), the Bureau International des Poids et Mesures (BIPM), the Instrument Society of America (ISA) and the US National Bureau of Standards (NBS) have been defined for both probe types (see, for instance, Kennedy 1983; Mangum 1989).

Other temperature-measuring devices that can be used in upstream processing but are less accurate or less suitable for control purposes, include:

- thermistors, consisting of a thermally sensitive resistor made of a metal oxide;
- gas- or liquid-filled thermometers, where the fluid expands with increasing temperature, thus producing a mechanical movement proportional to temperature;
- bimetal thermometers, made of two metal strips with different coefficients of thermal expansion.

13.2.2.3 pH

pH is routinely measured and controlled in bioreactors, since cells are very sensitive to pH changes. The most common type for bioreactors is a potentiometric steam-sterilizable glass probe filled with liquid or gel electrolyte, combining, in one unit, a reference electrode and a pH-sensing electrode (Hartnett 1994). The electrical signal is temperature-dependent and thus requires a temperature probe for correction or calibration at the working temperature. A three-point calibration is typically performed: one standard for the adjustment of the transmitter output, typically at pH 7, a second one for the slope adjustment, and a third one for a check on linearity. A typical problem with pH probes is the drift of the signal due to fouling of the reference electrode diaphragm with components of the cell culture fluid. Recalibration is thus often required during the culture. This can be performed by adjusting the reading on the basis of an off-line measurement (one-point calibration). An additional common safety measure at pilot- and large-scale production is to use probes in duplicate (Figure 13.2). If the first (main) probe fails, i.e. cannot be recalibrated, for instance because of excessive fouling, it is replaced by the second one to provide the input to the controller. An alternative is to use an interchangeable probe device (for instance the InTrac system of Mettler Toledo), so that the probe can be removed during a culture and either properly recalibrated off line with a two-point method, or serviced. At the end of the culture, chemical and enzymatic cleaning of the diaphragm, and replacement of the electrolyte, is recommended. Sensors based on pH-sensitive dyes and field-effect transistors have recently been developed, but the technology is not yet mature enough for routine use in bioprocessing (Sonnleitner 1999).

13.2.2.4 Dissolved oxygen (pO_2)

The measurement and control of dissolved oxygen during cell cultivation is essential, particularly at high cell density. This measurement is also widely used during the design and scale-up of bioreactors, to determine the efficiency of the gassing system.

The most common sensors are *in situ* sterilizable electrodes, of either *galvanic* (potentiometric) or, more often, *polarographic* (amperometric or Clark) type (Bailey & Ollis 1986). An oxygen-permeable membrane separates the electrode from the cell culture fluid. For both types, the reaction at the cathode (typically platinum) is the reduction of oxygen according to the following reaction:

$$\tfrac{1}{2}\,O_2 + H_2O + 2e^- \xrightarrow{Pt} 2\,OH^-$$

At the anode of the *galvanic* type, the reaction is:

$$Pb \longrightarrow Pb^{2+} + 2\,e^-$$

The resulting current provides a voltage, which is measured.

In the *polarographic* probe, a constant voltage is applied between the cathode and anode; the reaction at the anode is:

$$2\,Ag + 2\,Cl^- \longrightarrow 2\,AgCl + 2\,e^-$$

The resulting current is measured.

In both electrode types, the measured electrical signal is, at steady state, proportional to the oxygen flux to the cathode, which, in turn, is proportional to the oxygen partial pressure in the liquid phase. A two-point calibration is usually performed *in situ* after sterilization but before inoculation, in nitrogen- and then air-saturated culture medium. The main drawback of these electrodes is the drift of the signal, due to the accumulation of hydroxyl or metal ions, the depletion of chloride ions and the external fouling of the membrane surface. Frequent replacement of the membrane and of the internal electrolyte, and cleaning of the electrodes are thus required. As for pH probes, pO_2 probes are typically installed in duplicate in pilot- and large-scale bioreactors. When processes are scaled up, the effect of hydrostatic pressure on the reading of the probe should be taken into account (Marks 2003).

Alternatively, oxygen can be measured with fibre-optic sensors, where an oxygen-sensitive dye is placed at the tip of a fluorescent probe (Marose *et al.* 1999). The very small size and high sensitivity of these probes make them particularly suitable for measuring oxygen at low concentration, in small-scale culture systems, and in biosensors (see Section 13.2.4.2). This technology is, however, still rarely used at pilot and large scale.

13.2.2.5 Dissolved carbon dioxide (dCO_2)

Dissolved carbon dioxide has been shown to affect cell metabolism and protein glycosylation (for instance, see deZengotita *et al.* 1998; Kimura & Miller 1997); the accumulation of CO_2 can become particularly severe in large vessels, due to a relatively low removal rate via the gas phase. Several probes have been developed to measure this variable, although initially not specifically for animal cell cultures (Sonnleitner 1999). A fibre-optic sensor using a fluorescent dye has been tested recently and was found suitable to monitor and control, via N_2 sparging, dissolved CO_2 in a perfusion culture over several weeks (Pattison *et al.* 2000). At present, pilot- and large-scale bioreactors are rarely equipped with such a probe.

13.2.2.6 Pressure

Pilot-and large-scale bioreactors are commonly equipped with a pressure-sensing device in the tank headspace for various applications. After sterilization of the bioreactor, whether during a culture or a period when the tank stays idle before use, a positive pressure is usually applied to minimize the risk of contamination from the environment; additionally, pressure is commonly monitored during sterilization, for a measurement of the degree of steam saturation and for safety reasons. It is also usual to perform a pressure-hold test before sterilization, to detect any leaks in the tank (see Chapter 14, Section 14.4).

Pressure in bioprocessing is normally measured either as a *gauge* pressure, i.e. relative to atmospheric conditions, or as a *differential* pressure, i.e. as a difference between two pressure levels (Hartnett 1994). A measuring device is typically made of a *mechanical sensor* (or *pressure gauge*) and a *converter*, which transforms the mechanical signal into an electrical or a different mechanical one. The whole device is often referred to as a *pressure transducer* or *transmitter.*

The most common mechanical sensors are made of a C-shaped or a helical tube (both called Bourdon tubes), a diaphragm, a capsule or a set of bellows. In each case, a displacement or change in shape is created in response to the applied pressure. When no digital monitoring or automatic control is required, the sensor itself can be used for local readout via a pointing device; alternatively, simple pneumatic pressure converters for a local transmission of the signal can be used. The most common mechanical-to-electrical converter is the *strain gauge*, which is based on the principle that metallic- and semi-conductors subjected to mechanical deformation exhibit a change in their electrical resistance. When the strain gauge is bonded onto the mechanical sensor, the measured electrical resistance of the strain gauge can be correlated with the pressure applied to the sensor. Alternatively capacitance, potentiometric and resonant-wire sensors can be used; in the last of these, the pressure affects the tension of a wire, which in turn changes its resonant frequency; this directly generates a digital signal.

The main requirements of pressure transducers for bioreactors are that they can withstand sterilization temperatures over several cycles and that they are able to provide a reliable measurement at both culture and sterilization temperatures. To maintain sterility, the pressure gauge is normally separated from the bioreactor by a diaphragm.

13.2.2.7 Liquid level and weight

The liquid level in, or weight of, a bioreactor is an essential variable for monitoring and controlling the amount of medium and other nutrient solutions transferred to a bioreactor, both in fed-batch and perfusion cultures. Tanks for the preparation and storage of media and solutions are also often equipped with these devices. Mobile tanks can be placed on a floor scale. For fixed tanks, three types of device are commonly used (Hartnett 1994; Krahe 2002):

- *Differential pressure cells* are probably the most simple and economical devices; they measure the difference between the pressure at the bottom of the liquid and in the tank headspace, and convert the signal into an equivalent height of liquid. The device is made of two pressure sensors. The main characteristic of this measurement is that it is not influenced by the presence of foam and the amount of gas hold-up in the liquid (which may be a drawback, if it is desired to detect excessive foaming during the culture); it can however be affected by pressure fluctuations due to turbulence from agitation and aeration.
- *Conductance* and *capacitance probes* are both electronic sensors, based upon a chemically resistant probe located inside the tank. In a conductance probe, when the liquid level reaches the tip of the probe, an electrical circuit is closed which sends a signal to the reading device. In a capacitance probe, a capacitor is formed between the probe and the vessel;

the measured capacitance is then linearly related to the liquid level. The measurement provided by these probes may be influenced by the presence of foam and the amount of gas hold-up.

With these two *level*-measuring devices, the determination of the corresponding liquid *weight* requires knowledge of the tank cross-sectional area as a function of height.

- *Load cells* are external sensors that are placed at the base of the tank and measure the weight directly. The most common ones are strain-gauge cells, where the load acting on the cells is converted into an electrical signal (see Section 13.2.2.6). In most cases, four gauges are used to obtain maximum sensitivity and temperature compensation. The main problem is avoiding the interference of other external forces on the tank. For this reason, all portions of transfer lines directly connected to the tank should be made of flexible material (e.g. Teflon or silicone) or bent so that they exert only a stable force. Influence of temperature should also be taken into account. Load cells are typically more accurate than the two aforementioned devices, but also more expensive.

13.2.3 Biomass

13.2.3.1 Introduction

Biomass concentration is a key variable in animal cell cultures, and is routinely measured at all stages of process development and manufacturing. It is typically expressed as a mass, volume or number of cells per unit of the bioreactor working volume. The most widely used off-line technique consists in counting the cells in a haemocytometer under a microscope, using a dye such as Trypan blue, in order to distinguish viable from non-viable cells (Freshney 2000) (see also Chapter 32, Table 32.2). Several automated off-line instruments, using the same principle, are now commercially available.

On-line monitoring techniques for biomass can be classified into two main methods: the *direct* method, where a signal, directly related to some physical property of the biomass, is measured, and the *indirect* method, where biomass is derived from the measurement and calculation of other variables, such as metabolic rates or metabolite concentrations. The main techniques of both groups are reviewed below and summarized in Table 13.1 with some of the key references from literature. Additional discussions on these techniques can be found in reviews by Junker *et al.* (1994), Konstantinov *et al.* (1994a), Marose *et al.* (1999), Olsson and Nielsen (1997), and Sonnleitner (1999).

13.2.3.2 Direct method

Probably the most simple and popular technique for the direct method is based on the measurement, in the cell culture fluid, of *light absorption* (turbidity), *light scattering* (nephelometry) or of a combination of both. This is performed via an *in situ* optical probe, usually using visible or near-infrared (NIR) light (750 – 1100 nm). The signal is related to the concentration of solid particles in suspension and can thus be used to measure the total cell concentration, typically over a range of 0.5–20 $\times$ 10^6 cell/ml; usually, the correlation is linear only up to 2–5 $\times$ 10^6 cell/ml; at higher cell densities, a polynomial regression must be applied. Several probes are commercially available (Table 13.1). The main advantages of this technique are the simplicity, low cost and robustness of the probe. The main drawbacks are as follows. First, the signal is not specific to cells and can be affected by interference that changes with time, such as non-cellular solid particles and bubbles in the cell culture fluid, as well as biomass and protein build-up on the surface of the probe. Second, the signal also depends on the cell size and morphology, so that the measured cell concentration may become erroneous if these parameters change during the culture. Finally, no discrimination between viable and non-viable cells is possible. This technique is therefore unsuitable for detecting a decrease in viability.

Table 13.1 List of the main techniques for on-line monitoring of biomass.

Direct method

Measurement principle	Measured variable and correlation with biomass	Application(s)	Range of biomass concentration[1)]	References
Light absorbance and/or scattering	Absorbance of NIR/visible light in the cell culture fluid, correlated with the total cell concentration	Cells in suspension	$0.5-20 \times 10^6$ cell/mL	Akhnoukh *et al.* 1996; Junker *et al.* 1994; Wu *et al.* 1995
	Scattering of NIR/visible light in the cell culture fluid, correlated with the total cell concentration	Cells in suspension	$0.5-20 \times 10^6$ cell/mL	Junker *et al.* 1994; Wu *et al.* 1995
	Combined absorbance and scattering of NIR/visible in the cell culture fluid, correlated with the total cell concentration	Cells in suspension	$0.5-20 \times 10^6$ cell/mL	Junker *et al.* 1994; Wu *et al.* 1995; Zhou and Hu 1994
Dielectric spectroscopy	Dielectric permittivity of the cell culture fluid, proportional to the concentration of cells with intact plasma membranes, the cell radius to the the fourth power and the cell capacitance per unit of membrane area	Cells in suspension	$0.1-3 \times 10^6$ cell/mL	Cannizzaro *et al.* 2003; Cerckel *et al.* 1993; Davey *et al.* 1997; Guan *et al.* 1998; Noll and Biselli 1998; Siano, 1997; Zeiser *et al.* 1999
		Cells on microcarriers	$0.5-5.5 \times 10^6$ cell/mL	Ducommun *et al.* 2002; Noll and Biselli 1998
		Cells in a packed bed	$1-2 \times 10^{11}$ cell/kg	Ducommun *et al.* 2002
Fluorescence	Fluorescence of cell culture fluid, correlated with intracellular NAD(P)H	Cells in suspension	na[2)]	Akhnoukh *et al.* 1996; Junker *et al.* 1994; Siano and Mutharasan 1991
In situ microscopy	Cell counting, via a CCD camera, in an *in situ* flow-through chamber	Cells in suspension	$1-10 \times 10^6$ cell/mL	Joeris *et al.* 2002
Acoustic resonance densitometry	Acoustic resonance of cell culture fluid, correlated with the total cell concentration	Cells in suspension and immobilized	$1-8 \times 10^6$ cell/mL	Kilburn *et al.* 1989
Complex permittivity	Complex permittivity measurement of 3D microporous scaffold, correlated with the total volume of cells inside	Cells in microporous scaffold	$0.4-1.8 \times 10^6$ cells	Bagnaninchi *et al.* 2003
Optical waveguide lightmode spectroscopy	Refractive index of a cell layer, correlated with the number of cells per unit surface area and the cell-surface interaction	Tissue culture	>1000 cells	Hug *et al.* 2001; Hug *et al.* 2002

Table 13.1 *Continued*

Indirect method

Measured/calculated variable(s)	Biomass-related parameter	Application(s)	Range of biomass concentration[1)]	References
Glucose uptake or uptake rate	Concentration of metabolically-active cells	Cells in suspension, on microcarriers and in a packed bed	$1{-}20 \times 10^5$ cell/mL $1{-}8 \times 10^{10}$ cell/kg	Ducommun *et al.* 2001; Ducommun *et al.* 2002; Ozturk *et al.* 1997b; Pelletier *et al.* 1994; Rodrigues *et al.* 1999
Lactate production	Concentration of metabolically-active cells	Cells in a packed bed	$1{-}8 \times 10^{10}$ cell/kg	Ducommun *et al.* 2002; Ozturk *et al.* 1997b
Oxygen uptake or uptake rate	Concentration of metabolically-active cells	Cells in suspension and on microcarriers	$1{-}20 \times 10^5$ cell/mL	Ducommun *et al.* 2001; Kamen *et al.* 1996; Ozturk *et al.* 1997a; Tatiraju *et al.* 1999
Oxygen uptake rate combined with lactate evolution rate	Energy production rate (mol ATP/(m^3 s)) corresponding to the volumetric biomass activity	Cells in suspension[3)]	$1{-}7 \times 10^6$ cell/mL $1{-}20 \times 10^5$ cell/mL	Dorresteijn *et al.* 1996 Eyer and Heinzle 1996
Carbon dioxide evolution rate	Concentration of metabolically active cells	Cells in suspension[3)]	$1{-}10 \times 10^6$ cell/mL	Kamen *et al.* 1996
Recombinant protein production	Concentration of metabolically active cells	Cells in suspension and in a packed bed	$1{-}20 \times 10^5$ cell/mL $1{-}8 \times 10^{10}$ cell/kg	Ducommun *et al.* 2001; Ducommun *et al.* 2002
Redox potential	Viable cell density (via empirical correlation)	Cells in suspension[3)]	$1{-}20 \times 10^5$ cell/mL	Eyer and Heinzle 1996

1) The values give only an order of magnitude of the possible range of measurement. Different methods to determine this range, particularly the detection limit, have been applied by the various authors, and the reported experiments did not necessarily cover the whole possible range of biomass concentration.
2) No direct correlation between biomass concentration and fluorescence was reported in the cited references.
3) This technique could likely be applicable to immobilized cells, too, although it has not been reported in the cited references.

Dielectric spectroscopy is a direct monitoring technique that is used increasingly in animal cell cultures. It is based on the measurement of dielectric permittivity in the bioreactor under an alternating electrical field at a fixed frequency, in the range of 0.1 – 10 MHz. Living cells, i.e. those with an intact plasma membrane, act as small capacitors and thus affect the overall signal. For animal cells in suspension, the overall capacitance, after subtraction of a background signal due to media components, is approximately proportional to the product of the cell concentration, the cell radius to the fourth power, and the cell capacitance per unit of membrane area. Dielectric spectroscopy thus detects only 'viable' cells; solid particles or lysed cells do not get polarized (Noll & Biselli 1998). The viable cell concentration may, however, differ from that determined by a dye-exclusion method, since the segregating principle between viable and non-viable cells is different. Cell densities as low as $1\text{–}2 \times 10^5$ cell/ml can be measured accurately provided that smoothing techniques are applied to the capacitance signal (Guan *et al.* 1998). Careful calibration, using cells and media representative of the culture conditions, must also be performed. The main advantages of dielectric spectroscopy are simplicity and robustness, in addition to the possibility of measuring viable cells. Furthermore, it can also be used to monitor cells growing on macroporous microcarriers and in packed-bed bioreactors (Ducommun *et al.* 2002; Noll & Biselli 1998), for which no other satisfactory on-line technique is available. The main problem with dielectric spectroscopy is that any significant change in media composition, cell size and cell-specific capacitance can affect the signal (Figure 13.3). The effect of cell size, however, can be corrected by an additional (off-line) measurement of this variable (e.g. electronic particle counting, flow cytometry). Alternatively, the influence of cell size and specific capacitance on the signal can provide additional information on the cell physiology, which can be used for process control. For instance Zeiser *et al.* (1999) have shown that the size of insect cells increased significantly after viral infection, such that capacitance measurement was a very promising technique for monitoring and controlling the infection process. Noll and Biselli (1998) observed that in a continuous fluidized-bed reactor with cells growing on macroporous carriers, the cell size decreased significantly during the culture as a result of the limited space available in the pores. The glutamine consumption per unit of capacitance, however, reached a constant value. The capacitance signal could hence be used to adjust the feed rate of glutamine and thus to control accurately the concentration of this species. By measuring capacitance at several frequencies, Cannizzaro *et al.* (2003) obtained additional information on the metabolic state of a culture and estimated the cell size on line.

Measurement of *fluorescence* in the cell culture fluid has also been used in an attempt to monitor biomass; the classical technique actually detects intracellular fluorophores, i.e. mostly NADH and NADPH, at excitation and emission wavelengths of 360 nm and 450 nm respectively. Care must be taken that interfering parameters such as temperature, pH, pO_2 and agitation are kept constant. For determination of biomass from the fluorescence signal, the cellular content of the intracellular fluorophores must of course be constant; this is only the case during particular culture phases or conditions (Olsson & Nielsen 1997). The signal is also strongly affected by media components ('inner-filter effects'), making its interpretation difficult. Fluorescence should thus be viewed as a complementary technique, to use with another biomass measurement method, for an on-line characterization of the cell metabolic state (for instance, Akhnoukh *et al.* 1996; Siano & Mutharasan 1991). The original technique was recently improved by *in situ* scanning, i.e. by measuring fluorescence over a spectrum of excitation and emission wavelengths. This permits the separate quantification of several intra- and extra-cellular fluorophores. By selecting the appropriate fluorophores, a more reliable correlation with biomass may be achieved. A further improvement has been reported, with various microorganisms, where scanning fluorescence and turbidity measurements were performed at the same time with the same probe (Marose *et al.* 1999). The instrument could potentially be used with animal cells, too.

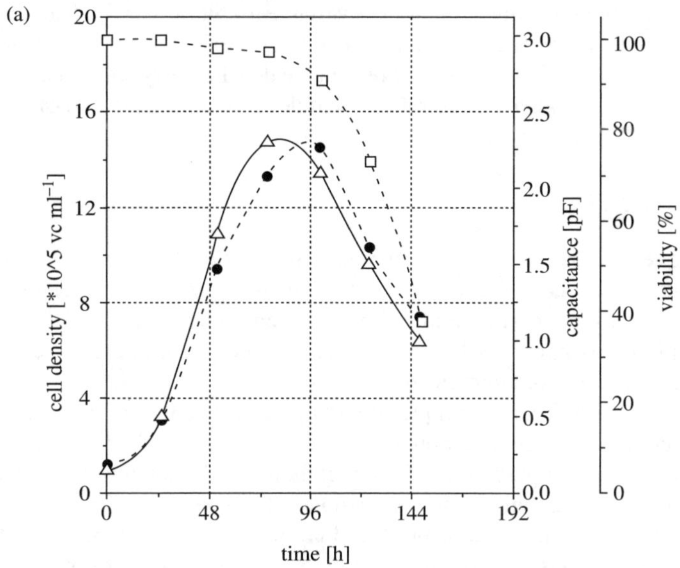

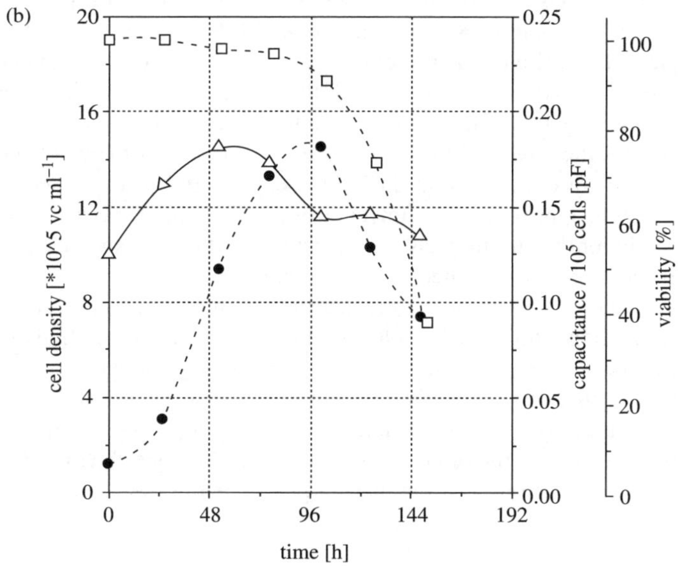

Figure 13.3 Monitoring of biomass during a batch cultivation of hybridoma cells in suspension. (a) Capacitance (measured here off line) (△), cell population density (●) and viability (□); these last two variables were determined by cell counting, using a dye-exclusion method. (b) Cell-specific capacitance (△), cell population density (●) and viability (□). The cell-specific capacitance (i.e. the capacitance per 10^5 viable cells) can be considered as a good approximation of the capacitance per unit of membrane area since the size distribution of viable cells was essentially constant (not shown). At first sight, there seems to be a good correlation between capacitance and cell density (a). However the highest capacitance was reached about 24 h before the highest cell concentration. This can be explained by the variation of the cell-specific capacitance, which reached a maximum during the second half of the exponential growth phase and then decreased (b). (Reproduced with permission from Noll and Biselli (1998).)

There are several other on-line techniques available for biomass, which are, however, used only for very specific applications or are not mature enough yet to be widely used in process monitoring. These include *in situ* microscopy, acoustic resonance densitometry, complex permittivity for cells growing in microporous scaffolds, and refractive index measurements in tissue cultures. Some key references for these techniques are given in Table 13.1.

13.2.3.3 Indirect method

Indirect techniques are based on the on-line measurement of other variables in the culture, from which the biomass or a related estimator is calculated using models of various complexities. This combination of a real sensor with an algorithm is often referred to as a *software sensor* (Chéruy 1997). This approach is particularly useful when a direct method cannot easily be applied, for example with immobilized or aggregated cells.

The most common indirect techniques consist of measuring one or several metabolites (glucose, lactate, glutamine, oxygen, carbon dioxide, etc.), continuously or at high frequency. By calculating the production or uptake rate of the metabolite and knowing the corresponding specific rate (from previous experiments), one can estimate the biomass. Alternatively, the biomass can be determined from the consumed or produced amount of the metabolite and from a previously determined yield of biomass on this metabolite. Several examples are given in Table 13.1, and a general review can be found in Konstantinov *et al.* (1994a). In some of the listed references, only the principle of the technique is presented and the metabolite was actually not measured on line; a proper on-line measurement using one of the techniques discussed below (Section 13.2.4) would, however, make the technique truly on-line for biomass. The main difficulty with indirect techniques is that specific rates and yields are rarely constant over the whole culture duration; in many cases, they are not even constant during the exponential phase. Ducommun *et al.* (2001) proposed an interesting approach to circumvent this problem. In all cases, however, indirect techniques are only valid for the culture phase and conditions where the specific rate or yield has been determined, and cannot be extrapolated outside these limits.

Another type of indirect technique is the determination of a biomass-related variable instead of the biomass itself. Dorresteijn *et al.* (1996), for instance, estimated on line the biomass activity, defined as the volumetric energy production rate, from the oxygen uptake rate and the lactic acid production rate. Although this approach does not give an estimate of the biomass *per se*, it can provide very valuable information on its activity; for control purposes, the biomass activity can actually be more relevant than the biomass amount. This technique can also detect changes in the cell metabolism, such as the onset of the plateau phase, before they become visible through cell counts.

13.2.4 Key Metabolites

13.2.4.1 Introduction

D-Glucose, L-lactate, ammonia and L-glutamine, as well as several other L-amino acids, have long been recognized as being key metabolites in animal cell cultivation and have thus been routinely monitored, first off line, using enzymatic and HPLC methods, and now increasingly using on-line techniques. There are of course numerous other chemical species that play a role in the metabolism of animal cells, such as vitamins, lipids, peptides, various growth factors, etc. These species are rarely, if ever, monitored in large-scale processes, due to technical difficulties and a poor understanding of their actual role. Similarly, on-line analysis of off-gases (O_2, CO_2) is rarely performed, unlike in microbial fermentation (see however some examples in Behrendt *et al.* 1994 and Eyer *et al.* 1995). Therefore the discussion below focuses on the on-line monitoring of the aforementioned key metabolites; the main techniques are reviewed below and summarized in Table 13.2 with some of the key references from literature.

13.2.4.2 Biosensors and flow-injection analysis

The requirement for on-line measurement of metabolites and complex biological species has driven the rapid development, over the past decades, of *biosensors*. A biosensor is a special class of sensor (see Section 13.2.1.1), which combines a *biological sensing element* for the analyte of interest with a *transducer*, to produce a correlated and measurable signal (electrical, optical, etc.). Biosensors are often miniaturized and are typically characterized by a high selectivity and a fast, reversible response, which makes them particularly suitable for high-throughput and automated measurements. A detailed review of the different types and principles can be found elsewhere (Glazier & McCurley 1995; Mulchandani & Bassi 1995; Schügerl 2001).

For the measurement of cell culture metabolites, the most widely used class of biological sensing elements are enzymes, such as oxidoreductases, transferases, hydrolases and lyases, because of their high specificity and their catalytic (amplification) properties (with the exception of ammonia, which is more often detected with non-biological sensors; see Table 13.2). To ensure operational stability and a useful lifespan whilst in use, enzymes are generally immobilized, either by physical entrapment in beads or polymeric films, or by chemical binding to membranes, beads or an electrode. The most common transducers of enzyme-based biosensors are either electrodes (potentiometric and amperometric) or optical probes (often called *optrodes*). Numerous combinations of enzymes and transducer are possible, the choice depending, among other things, on the composition (interfering factors) of the sample to be analysed. This is illustrated below with immobilized oxidases, which are the most widely used enzymes in biosensors, thanks to their availability. For the detection of an analyte, the general reaction, when oxygen is the acceptor, is as follows:

$$XH_2 + O_2 \xrightarrow{\text{oxidase}} X + H_2O_2$$

The concentration of analyte (XH_2) can then be determined in different ways:

(i) Measurement of the hydrogen peroxide produced, by:

- amperometric method, e.g. Pt anode at 0.5 – 0.8 V;
- potentiometric method, e.g. via formation of fluoride from 4-fluoroaniline and detection by a field-effect transistor;
- optical probe, via the chemiluminescent reaction of H_2O_2 with luminol, catalysed by peroxidase.

(ii) Measurement of the oxygen consumed, by:

- amperometric method;
- potentiometric method;
- optical probe, e.g. via a fluorescent dye.

The advantages and disadvantages of the various determination methods are discussed elsewhere (Luong *et al.* 1993; Mulchandani & Bassi 1995).

Although it is desirable, for on-line monitoring, to use biosensors as *in situ* probes, applications are very limited due to several aforementioned technical difficulties (see Section 13.2.1.1). These difficulties are particularly severe in the case of biosensors, since the biological element is not very

Table 13.2 List of the main techniques for on-line monitoring of key metabolites.

Metabolite	Main instrument	Measurement principle Technique/Main Device	Detection principle	Detection limit[1)] (mM)	Assay cycle time[2)] (min)	References
Glucose	***FIA***	Immobilized glucose oxidase[3)]	Amperometric detection of H_2O_2 or of a mediator	0.001–0.1	2–20	Male *et al.* 1997; Meyerhoff *et al.* 1993; Ozturk *et al.* 1997b; Renneberg *et al.* 1991; White *et al.* 1995
			Chemiluminescence detection of H_2O_2	0.001	2–10	Blankenstein *et al.* 1994
			Fluorescence detection of O_2	0.05	2–10	Dremel *et al.* 1992
		Immobilized glucose dehydrogenase	Fluorescence detection of NADH	0.1	1	Spohn *et al.* 1994
	HPLC	Anion-exchange column	Integrated pulsed amperometric (IPAD) or refractive index detection	1	30–60	Favre *et al.* 1990; Kurokawa *et al.* 1994; Larson *et al.* 2002[4)]
	In-situ probe	Immobilized glucose oxidase[3)]	Amperometric measurement via mediators	3	1–3	Bilitewski *et al.* 1993
		Near-infrared spectroscopy		0.1–1	1–3	Arnold *et al.* 2003; Chung *et al.* 1996; Rhiel *et al.* 2004; Riley *et al.* 2001; Yano and Harata 1994
		Mid-infrared spectroscopy		0.5	1	Rhiel *et al.* 2002b
Lactate	***FIA***	Immobilized lactate oxidase[3)]	Amperometric detection of H_2O_2 or of a mediator	0.025	2–10	Ozturk *et al.* 1997b; Renneberg *et al.* 1991
			Chemiluminescence detection of H_2O_2	0.001	2–10	Blankenstein *et al.* 1994
			Fluorescence detection of O_2	0.1	2–10	Dremel *et al.* 1992
		Immobilized lactate dehydrogenase	Fluorescence detection of NADH	0.1	3	Spohn *et al.* 1994
	HPLC	Anion-exchange column	Refractive index detection	1	30	Favre *et al.* 1990
	In-situ probe	Near-infrared spectroscopy		0.2–2	3	Arnold *et al.* 2003; Chung *et al.* 1996; Rhiel *et al.* 2004; Riley *et al.* 2001; Yano and Harata 1994
		Mid-infrared spectroscopy		0.5	1–3	Rhiel *et al.* 2002b

Table 13.2 *Continued*

Glutamine	*FIA*	Immobilized glutaminase + glutamate oxidase[3)]	Amperometric detection of H_2O_2 or of a mediator	0.03–0.1	2–10	Renneberg *et al.* 1991; White *et al.* 1995
			Chemiluminescence detection of H_2O_2	0.001–0.01	2–10	Blankenstein *et al.* 1994; Cattaneo and Luong 1993
		Immobilized glutaminase + glutamate dehydrogenase	Fluorescence detection of NADH	0.1	1	Spohn *et al.* 1994
		Immobilized glutaminase + gas permeable membrane	Spectrophotometric detection of the pH change of an indicator	0.05	2–10	Stoll *et al.* 1996a
			Potentiometric detection of NH_4^+ (ion-selective electrode)	0.1	2–10	Campmajo *et al.* 1994; Meyerhoff *et al.* 1993
	HPLC	Anion-exchange column	Integrated pulsed amperometric (IPAD) or refractive index detection	0.01–1	30–60	Favre *et al.* 1990; Kurokawa *et al.* 1994; Larson *et al.* 2002[4)]
	In-situ probe	Near-infrared spectroscopy		0.05–0.5	1–3	Arnold *et al.* 2003; Chung *et al.* 1996; Rhiel *et al.* 2004; Riley *et al.* 2001; Yano and Harata 1994
Glutamate	*FIA*[5)]					
	HPLC	Anion-exchange column; integrated pulsed amperometric detection (IPAD)		0.01	30–60	Larson *et al.* 2002[4)]
	In-situ probe	Near-infrared spectroscopy		0.6	1–3	Chung *et al.* 1996
Ammonia	*FIA*[6)]	Gas-permeable membrane	Chemical reaction with ammonia + fluorescence detection of resulting product		2–10	Spohn *et al.* 1994
		Immobilized glutamate dehydrogenase + glutamate oxidase	Chemiluminescence detection of H_2O_2	0.001	5–10	Blankenstein *et al.* 1994
		Potentiometric detection of NH_4^+ with ion-selective electrode array		0.01	na	Saez de Viteri and Diamond 1994
	In-situ probe	Near-infrared spectroscopy		0.04–1	1–3	Arnold *et al.* 2003; Chung *et al.* 1996; Riley *et al.* 2001; Yano and Harata 1994

1) The values represent only an order of magnitude, since different methods have been used to determine the detection limit in the cited references. The maximum measurable concentration is not given since it is not considered a critical parameter for these techniques.
2) Cycle time for the injection and analysis of one sample or standard (FIA, HPLC); the frequency during on-line monitoring of a culture was often lower, due to the injection of replicates and of standards for recalibration. For an in-situ probe, the cycle time corresponds to the time for the collection of spectra.
3) See text for additional references on all the detection methods with oxidase-based sensors.
4) Simultaneous detection of glucose and 17 amino acids.
5) See the FIA methods for glutamine which are based on glutamate-oxidase or dehydrogenase.
6) See also the FIA methods for glutamine which are based on glutaminase and detection of the ammonia produced.

stable, and thus is difficult to sterilize and to use without frequent recalibration. An additional barrier is that the optimal physico-chemical environment for biosensors (pH, temperature, ionic strength, etc.) is often different from that of the culture medium, which, additionally, may contain interfering species. In the case of multistep enzymatic reactions, these conditions must actually be modified at each step. Even a relatively simple sensor such as the ammonia-selective electrode (Table 13.2) cannot be used accurately *in situ*, due to strong interference from other ions. As a consequence, *in situ* applications of biosensors are very limited. One successful example with glucose has been reported for bioprocess monitoring (Bilitewski *et al.* 1993).

The second, more practical use of biosensors for on-line monitoring is *ex situ*, in a flow-injection analysis (FIA) system. The FIA technique was first described by Ruzicka and Hansen (1975); principles and applications to bioprocess monitoring are summarized below and reviewed in detail elsewhere (Schügerl 2001; Schügerl *et al.* 1996; van der Pol *et al.* 1996). In brief, a known volume of sample (or standard) is injected into a constant flow of a buffer, then mixed with streams of different reagents or subjected to various physico-chemical treatments, and finally passed through a (bio)sensor. The sensor continuously records the change in absorbance, electrical current, potential or any other physical characteristics of the liquid stream caused by the injection of the sample. This measurement can be correlated with the amount of the species to be analysed in the sample. An essential component of FIA is a sterilizable sampling device that allows the injection of a cell-free sample from the bioreactor (see Section 13.2.1.1). One of the main characteristics of FIA – at least in its original form – is that no complete reaction, equilibrium or even homogeneous mixing is required in the system; this is acceptable provided that the process is well automated and that all operating parameters such as volume, flowrate and mixing time are highly reproducible. Thus, the main advantage of FIA is that relatively fast measurements (typically within a couple of minutes) can be performed despite the complexity of the detection process. Reactions that would not go to completion or not rapidly reach an equilibrium can be employed. Therefore FIA measurements, although not continuous, can be generated at a high frequency and are suitable for on-line monitoring of animal cell cultures. Other advantages offered by FIA for bioprocess monitoring are:

- FIA can incorporate an extremely large variety of chemical and biological reactions, together with physical methods, for which many different types of detection system can be used; thus, a very broad range of chemical species can potentially be measured.
- Small sample volumes are required.
- Dilution by a large factor (e.g. 1000) can be achieved very rapidly (e.g. within 1 min) with high precision ($<1\,\%$) (Sonnleitner 1997; van der Pol *et al.* 1996).
- One instrument can contain an array of biosensors and can thus be used for the 'simultaneous' (actually often sequential) determination of several species with one single injection.
- Frequent automatic recalibrations can be performed easily.

Despite the separation of the sensing and detection elements from the bioreactor content, FIA-based monitoring is often affected by problems of interfering species. Various strategies have been developed to circumvent these effects (Luong *et al.* 1993; Mulchandani & Bassi 1995). In brief, these strategies include:

- The use of an electron acceptor other than oxygen in oxidase-based sensors.
- The removal or degradation of interfering species by use of an ion-exchange column or enzymatic microreactor upstream of the sensor.
- A differential measurement, with two detectors, and a subtraction of the signal due to the interfering species only.

- The protection of the sensor with a selective membrane.
- The careful control of the physico-chemical environment (e.g. pH, pO_2).

D-Glucose, L-lactate, L-glutamine, L-glutamate and ammonia have been measured successfully in different FIA systems over complete culture runs (Table 13.2). Reported detection levels for all these metabolites were in the range 0.001–0.1 mM. This broad range reflects the diversity of techniques and instruments, the different methods applied for determining the detection level, and the different degrees of assay optimization. The maximum measurable concentration is usually not critical, since the injection volume can be easily adapted or a highly reproducible dilution of the sample can be performed prior to detection. Due to the interdependence of glutamine, glutamate and ammonia via the glutaminase- and glutamate oxidase-catalysed reactions, detection methods for glutamine require the elimination – physically, chemically or by calculation – of the contribution of glutamate and ammonia; when properly performed, this operation actually allows the simultaneous quantification of all three species. Blankenstein *et al.* (1994) for instance, reported the simultaneous monitoring of all five aforementioned metabolites in a fluidized-bed reactor over 15 days using a multichannel FIA instrument (Figure 13.4). Although the time for injection and analysis of one species was about 2 minutes, the complete analysis time for all species was 42 minutes, due to injections in triplicate and washing steps before and after each species. Other amino acids can actually be detected using the same principle, for instance with immobilized oxidases. However, as each species requires its own biosensor, the FIA technique tends to becomes very slow and complex for the determination of a large number of species. Commercial instruments are now available with sensors based on immobilized enzymes, to detect the main metabolites in cell cultures (for instance the YSI analyser, YSI, Inc.). These instruments are actually a variant of FIA systems, since measurement is typically performed under equilibrium instead of dynamic conditions. Recently, genetically engineered binding proteins containing a fluorophore have been successfully tested in biosensors to measure D-glucose and L-glutamine off line (Ge *et al.* 2003). The high sensitivity (μM range) and the very small sample volume requirement (<1 μl) make these biosensors attractive alternatives to enzyme-based ones for bioprocess monitoring. The use of biosensors with FIA at pilot- and large-scale is, however, still limited; the modest benefits of on-line monitoring over well-established off-line techniques do not compensate for the additional complexity of the fluid handling system and the lack of robustness of many biosensors.

13.2.4.3 Infrared spectroscopy

Near-infrared (NIR) spectroscopy is another promising technique for detecting metabolites. A sample is exposed to a beam of NIR light through a quartz window and the amount of light absorbed at various frequencies can be correlated to the concentration of some species in the sample. The method offers several advantages. It is non-invasive and can be performed with an *in situ* fibre-optic probe (with all the advantages mentioned in Section 13.2.1.1, particularly the fast response time). The signal is also very stable, and little maintenance is required during operation. Finally, several species can be determined simultaneously with only one sensor (as opposed to one for each species in FIA techniques). Riley *et al.* (2001), for instance, were able to monitor 19 components in serum-containing cell culture media, among which 15 were amino acids, including glutamine. The main difficulty of this technique is that the NIR signal may be affected by several additional (unknown) species in the media; extensive chemometric algorithms, such as partial-least squares regressions, must therefore be applied first, with a set of off-line samples from several previous cultures, to build calibration models (Riley *et al.* 1998; Rhiel *et al.* 2004). Consequently, this technique is particularly suitable for the routine monitoring of cultures that are always operated under similar conditions. However, when new conditions are tested, changes in the cell metabolism may not be detected properly.

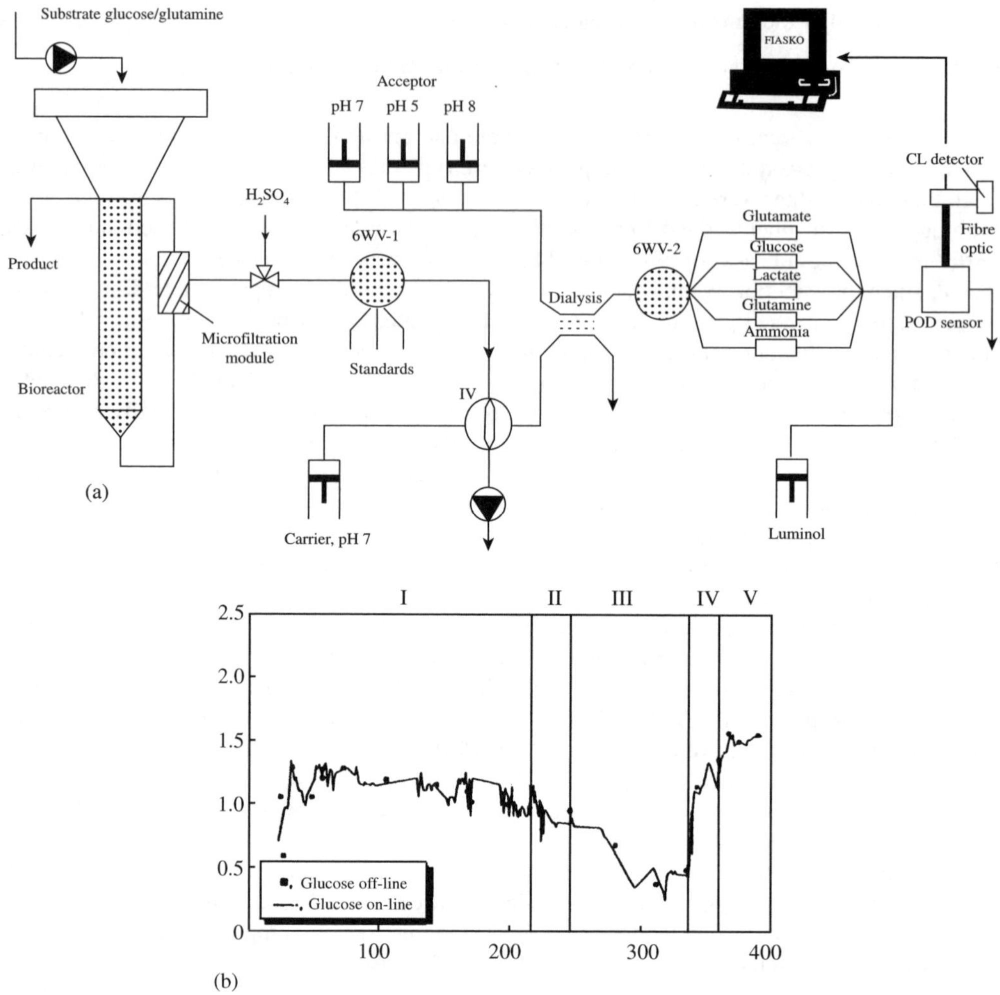

Figure 13.4 Example of a multichannel FIA instrument for on-line monitoring of glucose, lactate, glutamine, glutamate and ammonia in a fluidized-bed bioreactor. (a) Schematic diagram of the FIA instrument. A microfiltered sample from the bioreactor or a standard (selection via a six-way valve (6WV-1)) is transported to a dialysis unit, which is used to remove high molecular weight compounds (proteins) and to dilute the sample further in the stream of an 'acceptor' (buffer solution). A second six-way valve (6WV-2) directs the sample to one of five microreactors with immobilized enzymes. All of them lead to the formation of H_2O_2, which is measured downstream via a chemiluminescent reaction, after addition of a luminol solution and passage through a sensor with immobilized peroxidase (POD). All operations of the FIA are performed automatically. A complete analytical run, with injections in triplicate and washing steps, takes 42 min. IV = injection valve, CL = chemiluminescence (b) Comparison of glucose measurement on line, with the FIA instrument (—), and off-line, with a commercial glucose analyzer (■), during a continuous culture of immobilized hybridoma cells. X-axis: process time (h); Y-axis: glucose concentration (g/l) (Reproduced with permission from Blankenstein *et al.* (1994).)

Mid-infrared spectroscopy (MIR) has recently been developed as a promising variant to NIR for animal cell cultures, since it has a higher sensitivity and selectivity for the metabolites of interest in the so-called 'fingerprint' spectral region (1800 – 800 cm^{-1}). A difficulty, however, is the strong water absorbance in this region. So far, only a few applications for the monitoring of animal cell cultures have been reported (for instance, Rhiel *et al.* 2002b). However, the development of robust calibration models (Rhiel *et al.* 2002a), combined with long-term stability (Rhiel *et al.* 2002b) reported 2.3 years of operation without maintenance, except for daily replenishment of liquid nitrogen for detector cooling) make this technique an attractive basis for future applications.

Compared with FIA, NIR and MIR measurements are less accurate, with a typical detection level of 0.1 – 1 mM for the cell culture key metabolites. However, these techniques are currently undergoing very rapid development such that significant improvements can be expected in the near future. The maximum measurable concentrations are not critical parameters since they are significantly higher than the highest levels found in cell cultures.

13.2.4.4 HPLC and miscellaneous techniques

HPLC, although relatively expensive, is an attractive alternative to biosensor-based techniques, in particular when several species have to be analysed simultaneously. With the proper sampling device, an HPLC instrument can be readily used for on-line monitoring of a bioreactor. Several applications have been reported for glucose, lactate and most of the amino acids present in animal cell cultures, with a complete analysis time in the range of 30–60 min (for instance Larson *et al.* 2002).

Several other (complex) instruments or techniques, typically developed for off-line analysis, can in principle be used to monitor a culture on line or in real time. NMR, for instance can detect a broad range of metabolites in cell cultures including intracellular species, and can be used for metabolic flux analyses (Forbes *et al.* 2000). Very sensitive calorimeters have also been developed, to monitor continuously the heat signal generated by cultivated cells, which is a measurement of their catabolic activity (Kemp & Guan 1997). The complexity and cost of these tools, however, have limited their use to research and development applications.

13.2.5 Recombinant Proteins

13.2.5.1 Introduction

Recombinant proteins and other desired protein products expressed by animal cells have traditionally been quantified off line, using techniques such as ELISA and other similar assays. ELISAs are characterized by a relatively high sensitivity, with a typical detection limit of the order of 1 ng/ml for proteins in culture samples, and a linear dynamic range of 10–220 ng/ml (Baker *et al.* 2002); consequently, sample dilution by several orders of magnitude is required. These assays are thus labour intensive and require several hours to complete. For these reasons, samples are typically analysed in batches, often at the end of the culture. Semi-automated instruments for ELISAs, incorporating sample dilution, have been developed for the high-throughput analysis of off-line samples. These instruments are, however, too expensive and impractical to be used for on-line or real-time monitoring of bioprocesses.

Instead, efforts have been made to develop alternative assays or simplified versions of ELISAs, which can be performed faster, using compact, robust and automated instruments. Significant progress in this direction has been achieved in the past decade. Most of the developed techniques consist of the following two steps, performed in an automated way: (i) a highly specific reaction with the protein to be measured, such as an immuno-based reaction; (ii) the subsequent direct or indirect detection of this reaction, with or without prior washing to eliminate interfering species. Compared with biomass and low molecular weight metabolites, however, few applications of truly on-line monitoring of protein production in cultures have been reported. The main reasons are the higher complexity of detection assays for proteins and the few instances of a real need to monitor these species on line and to control the bioreactor automatically on the basis of this measurement. The main techniques for on-line monitoring of recombinant proteins, as well as those for real-time monitoring, which could be used on line with the proper sampling device, are discussed below and are summarized in Table 13.3 with some of the key references from the literature.

Table 13.3 List of the main techniques for on-line or real-time monitoring of recombinant proteins.

Main instrument	Measurement principle Technique/Main Device	Detection principle	Detection limit[1] (mg/L)	Assay cycle time[2] (min)	References[3]
Techniques with immobilized ligands					
FIA[4]	Biosensor with immobilized antibody or ligand	Optical measurement: surface plasmon resonance	0.001	6	Baker *et al.* 1997
			0.01	8	Jönsson *et al.* 1991
			0.08	5	Karlsson *et al.* 1993
		Optical measurement: waveguiding techniques	0.1	2	Brecht *et al.* 1993
			0.03	15–30	Oroszlan *et al.* 1993
			0.8–10	30	**Polzius *et al.* 1993**
		Optical measurement: resonant mirror	na	na	Cush *et al.* 1993
			1	5	Gill *et al.* 1998
			0.2	2	Holwill *et al.* 1996
		Capacitance measurement	2	22	**Gebbert *et al.* 1994**
	Microcolumn/reactor with immobilized antibody or ligand	Downstream fluorometric measurement of competing labelled-antigen	na	20–25	Degelau *et al.* 1992
			1	10–30	Middendorf *et al.* 1993
		Addition of enzyme-coupled antibody, followed by downstream spectrophotometric measurement of product from enzymatic reaction (= automated sandwich ELISA)	1	11	**Gebbert *et al.* 1994**
		Downstream spectrophotometric measurement of product resulting from enzyme-coupled competing antigen (= automated competitive ELISA)	0.01	6–10	Nilsson *et al.* 1991; Nilsson *et al.* 1992
		Downstream direct fluorometric detection of protein	1	15	**Reinecke and Scheper, 1997**
			5	6	**Stöcklein *et al.* 1991**
		Dual streaming potential electrode system	1	30	Miyabayashi *et al.* 1990

Table 13.3 *Continued*

Main instrument	Measurement principle: Technique/Main Device	Detection principle	Detection limit[1)] (mg/L)	Assay cycle time[2)] (min)	References[3)]
HPLC	Column with Protein-G resin	Downstream fluorometric detection of protein	na	33	**Ozturk *et al.*1995**
	Column with Protein-A resin	Downstream fluorometric detection of protein	1	3–5	**Paliwal *et al.* 1993b**
	Other affinity column	Downstream fluorometric detection of protein	0.005	6	Baker *et al.* 2002
			5	7	Van der Pot ***et al.*** 1997
Liquid-phase immunoassays					
	Mixing chamber with addition of antibody or ligand	Downstream spectrophotometric measurement of turbidimetry	1	2–3	**Fenge *et al.* 1991**
			10	1–5	Middendorf *et al.* 1993
			0.008	60	Baker *et al.* 2002
	Liquid-phase ELISA				
Miscellaneous techniques					
In-situ probe	Near-infrared spectroscopy		0.5	5	Harthun *et al.* 1997
	Fluorescence measurement of protein marker		0.05	na[5)]	Cha *et al.* 1999

1) The value gives only an order of magnitude, since different methods and proteins have been used to determine the detection limit in the cited references. only detection of proteins in cell culture samples have been taken into account. The maximum measurable concentration is not given since it is usually not considered a critical parameter for these techniques.

2) Cycle time for the analysis of one sample or standard. The frequency during on-line monitoring of a culture was often lower, due to the injection of replicates and of standards for recalibration.

3) References with data of truly on-line monitoring of protein concentration in a culture are highlighted in bold.

4) Includes all other types of automated microfluid-handling systems.

5) Measurement was performed with an off-line fluorometer; with an on-line probe, a real-time measurement could be performed.

13.2.5.2 Techniques with immobilized ligands

Several types of biosensor have been developed with antibodies or other ligands immobilized on a flat surface to capture the protein of interest. The formation of the immunocomplex is then detected *in situ*, using optical methods such as surface plasmon resonance (SPR, Jönsson & Malmqvist 1992), waveguiding devices (e. g. interferometry, grating coupling; Brecht *et al.* 1993; Oroszlan *et al.* 1993; Polzius *et al.* 1993) or a combination of both (resonant mirror; Cush *et al.* 1993). Alternatively, the change in capacitance or in other electrical properties can be detected after the formation of the immunocomplex (Gebbert *et al.* 1994). The biosensor can be integrated in an FIA instrument (see Section 13.2.4.2) or a similar microfluid-handling system. The main advantage of these techniques is that no labeled reagent is required. Additionally, the measurement is in general fast, i.e. within a couple of minutes, since the detection can be performed *in situ*, immediately after immunocomplex formation and without removal of the free antigen. However, an incubation step is often introduced (e.g. 20 minutes, Polzius *et al.* 1993) in order to increase the sensitivity of the assay by enhancing the formation of the immunocomplex; a complete regeneration of the sensor surface after measurement also requires rinsing for several minutes. A wide range of assay cycle times (2–60 minutes, see Table 13.3) has been reported, reflecting variations in the affinity constant of the immunocomplexes. Detection limits were in the range of 0.001–10 mg/l, which is suitable for monitoring protein production in most cell culture processes. The maximum measurable concentration is very variable and is not reported in Table 13.3; this parameter is rarely critical with FIA because highly reproducible dilutions can be performed rapidly prior to detection. Commercial instruments based on these techniques are now available (SPR: Biacore™, Biacore AB; resonant mirror: IAsys™, Affinity Biosensors), primarily to be used off line. However, the level of automation and the speed of these assays make them suitable for real-time monitoring of cell cultures and, potentially, for on-line applications. These instruments are actually used not only to quantify proteins, but also to measure the affinity, specificity and kinetics of binding to a ligand; they have thus found numerous applications in life sciences, drug discovery and food analysis.

A second class of techniques is based on a column or a microreactor containing an immobilized antibody or an affinity ligand (e.g. Protein A) to which the analyte binds after injection; various detection methods are then applied downstream, following an elution step. For instance the sample can be injected together with a fixed amount of a fluorophore-labeled antigen, which competes with the free antigen to be analyzed for the binding sites on a column. After elution, the amount of labeled antigen is measured via a fluorescent detector and can be inversely correlated to the amount of free antigen in the original sample (Middendorf *et al.* 1993). Another method consists in injecting, after the sample, an enzyme-coupled antibody (to the antigen of interest) followed by a substrate for the enzyme. After an incubation step, the product of the enzyme reaction is measured downstream, in a similar way to a sandwich ELISA (Gebbert *et al.* 1994). Alternatively, a fixed amount of an enzyme-coupled competing antigen can be injected together or after the sample, by analogy with a competitive ELISA (Nilsson *et al.* 1991, 1992). The protein of interest can also be detected directly, following elution, via a spectrophotometric detector (e.g. Stöcklein *et al.* 1991). These techniques typically use an FIA instrument operated in a stopped-flow mode, i.e. an incubation step is introduced in order to increase the binding of the analyte to the column. The assay is nonetheless performed under non-equilibrium conditions, thanks to the high reproducibility of the instrument. Consequently, the measurement is much faster than in a classical ELISA, where long incubation times are required. The cycle time is here again variable (2–30 min), reflecting differences in the binding affinity of the immunocomplex and in elution conditions. A typical detection limit is about 1 mg/l. Similar techniques based on HPLC instead of FIA have been reported, using a small column with Protein A or

Protein G immobilized on porous beads (e.g. Poros™ media, Applied Biosystems); elution was completed within a few minutes, but column regeneration and equilibration resulted in a total cycle time of about 30 minutes (Ozturk *et al.* 1995; Paliwal *et al.* 1993b; van der Pol *et al.* 1997). Another related technique is based on a differential potentiometric measurement between two electrode columns, one with an immobilized ligand and one with a blank gel (Miyabayashi & Mattiasson 1990).

The major drawback of these techniques is the relatively rapid leaching or degradation of the immobilized ligand. Nilsson *et al.* (1991), for instance, reported a 30 % decrease of the signal after about 10 h. Frequent recalibration of the instrument, typically at each sample injection, is thus required. Commercially available affinity resins for chromatography seem to be an exception, with a reported stability of over 6 months (Baker *et al.* 2002).

13.2.5.3 Liquid-phase immunoassays

Immunocomplexes can also be quantified in solution by turbidimetry or nephelometry measurements. These techniques work best with large proteins that form a turbid solution even with small concentrations of ligands. Commercial automated instruments are now available (e.g. Coulter Array, Beckman). Middendorf *et al.* (1993) reported an FIA application based on this principle for the monitoring of IgG in a hybridoma culture. The immunocomplex was formed in a small mixing coil and the resulting turbidity was measured spectrophotometrically, downstream, in a flow-through cell. Compared with the fluorometric heterogeneous immunoassays mentioned in Section 13.2.5.2, this technique offers the advantages of being simple, very fast and requiring no labeled ligand. The major drawback is the high cost of the reagent, since relatively large amounts are required for each sample (around 150 μg) and no regeneration can be carried out.

'Liquid' ELISAs have also been developed, where an immunocomplex is formed in the liquid phase. A secondary ligand bound to magnetic microbeads is then added, so that the immunocomplex can be temporarily captured at the surface of a detector in a flow-through cell (Baker *et al.* 2002). Unlike in a conventional ELISA, no washing steps are required so that the overall assay time is much shorter. Commercial, semi-automated instruments based on this principle are available (e.g. Origen Analyzer™, Igen). To our knowledge, however, no application to on-line monitoring of bioprocesses has been reported yet.

13.2.5.4 Miscellaneous techniques

Other techniques have also been developed that are not based on the formation of an immunocomplex. An NIR spectroscopy method has been reported, offering the advantage of an *in situ* probe; complex multivariate chemometric methods however are required in order to eliminate interfering signals (Harthun *et al.* 1997). Another promising technique with potential for on-line monitoring is the production of proteins fused with a non-invasive green-fluorescent protein marker. The amount of expressed protein can thus be measured *in situ*, using simple fluorescent scanning; several applications in bacterial fermentation and insect cell cultures, for both upstream and downstream processing, have been reported (for instance Cha *et al.* 1999). In this particular case, cell growth and protein secretion were not affected by the presence of the marker.

Additionally, techniques to assess product *quality* (variants, glycoforms, etc.) in cell culture samples have undergone significant development recently. At present, they are rarely used on line but are now rapid (cycle time of a few minutes) and robust enough so that they can be applied to routine, often real-time monitoring. In addition to the aforementioned immunosensors based upon SPR and resonant mirrors, these techniques include capillary electrophoresis, mass spectrometry coupled with SPR biosensors, and dual-column immunochromatography. They are reviewed in detail elsewhere (Paliwal *et al.* 1993a; Baker *et al.* 2002).

13.2.6 Process Analysis

Apart from automatic control, which is discussed below, one of the key applications of on-line monitoring is to provide data for a detailed analysis and characterization of the process. This can be done at different levels (Ozturk *et al.* 1997a). Concentrations of metabolites and physico-chemical variables, such as temperature, pH and pO_2, represent the first level, i.e. the direct cellular environment. From these measured variables, second-level variables can be derived, i.e. metabolic rates at the bioreactor level, such as glucose and oxygen uptake and lactate production rates. Finally, using on- (or off-) line determination of biomass, cell-specific rates can be calculated, representing a third level of variables. Various techniques to calculate the second- and third-level variable rates are discussed elsewhere (Ducommun *et al.* 2001; Eyer *et al.* 1995).

These different variables can then be used to characterize the process, to assess the run-to-run reproducibility, and to build mathematical models. Applications are reviewed elsewhere (Lübbert & Simutis 1994; Royce 1993; Sanderson *et al.* 1996). Although all of these can be performed in principle with off-line data, on-line monitoring offers the obvious advantage of enabling the collection of a much larger amount of data. This can make the calculation of rates much more accurate and allows a finer detection of dynamic phenomena (for instance Larson *et al.* 2002; van der Pol *et al.* 1996); if needed, metabolic rates, such as oxygen uptake rate, can be determined automatically and in real time.

Furthermore, the FDA has recently issued a guidance document on Process Analytical Technology (PAT), intended to improve product quality through, among others things, innovation in process analysis and control (FDA 2004). The use of in-process real-time measurement is advocated; this will likely stimulate the use of monitoring tools, both on line and off line in close proximity to the process stream.

13.3 PROCESS-CONTROL LEVEL

13.3.1 Main Device

The main device of the process-control level is the programmable logic controller (PLC). A typical PLC consists of a central processing unit (CPU), a memory unit, analogue and digital input/output blocks, as well as a synoptic display and an operating panel that are often combined. All these elements are usually incorporated into a single cabinet designed for rugged industrial environments. One PLC is commonly associated with one process unit (e.g. bioreactor) or a group thereof, for which it is continuously 'available'. A PLC is characterized by its computing capacities, i.e. the number of inputs and outputs and the rate (cycle time) at which these can be handled. Typical cycle-time values are in the range of 50–500 ms.

In addition to performing some operations automatically and to handling certain failures, the key function of a PLC is to control variables at user-defined set-points, via actuators, based upon measurements provided by sensors. The combination of a sensor (with a display and a recorder), a controller and an actuator constitutes a control loop, which is the basic unit in a control system (Hartnett 1994). The most common method for controlling bioprocesses is based upon feedback control with closed loops, using proportional-integral-derivative (PID) or proportional-integral (PI) algorithms (Dunn *et al.* 2003; Gonzalez & Herb 1984). Modern PLCs now have enhanced computing capacities and are often supported by powerful software applications at the supervision level (see Section 13.4.1). This has enabled the use of more advanced control techniques, based on multivariable, adaptive, non-linear and fuzzy logic algorithms as well as on expert systems (Dochain & Perrier 2000; Konstantinov *et al.* 1994b; Lenas *et al.* 1997; Lübbert & Simutis 1994; Schügerl 2001; Shimizu 1993). An example of a control loop for a process variable is illustrated in Figure 13.5. Other examples, for a fermentation plant, are discussed in detail in

(a)

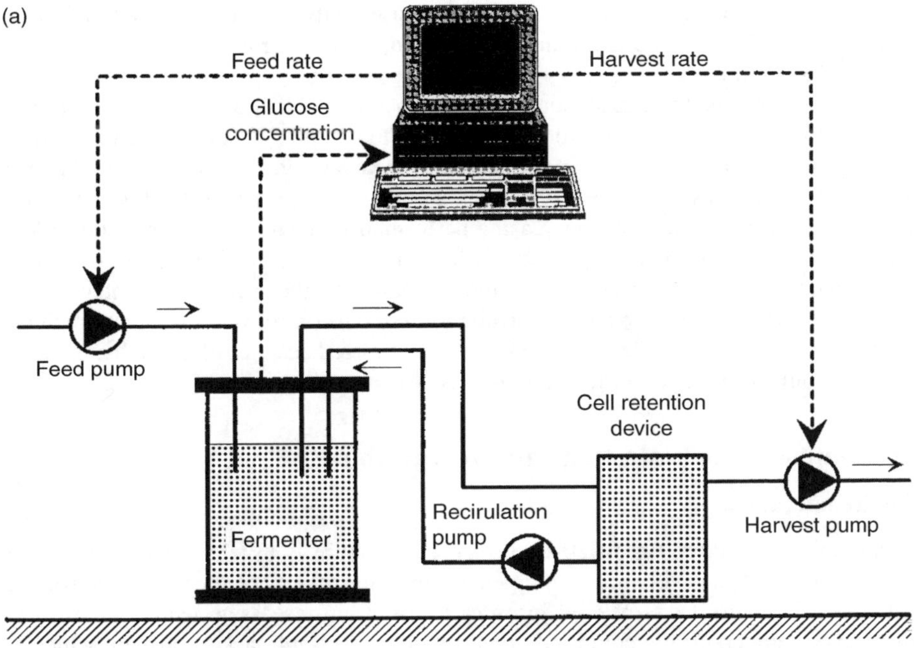

(b)

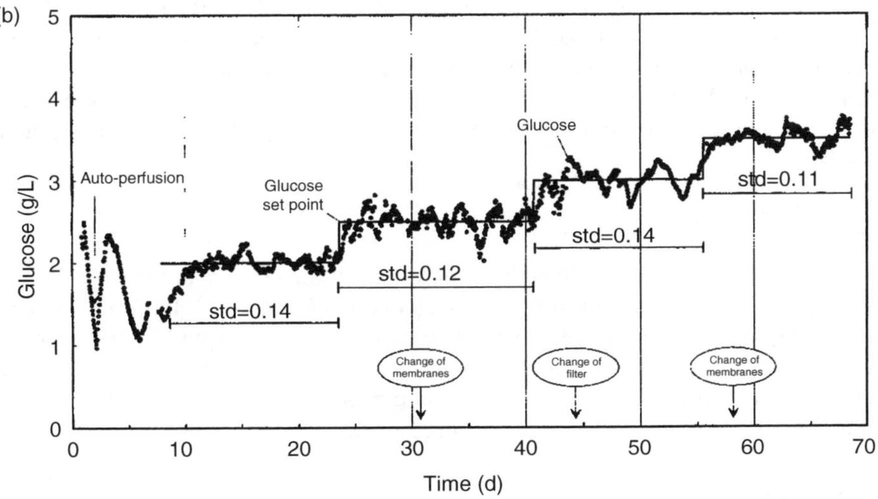

Figure 13.5 Example of automatic control of glucose in a perfusion bioreactor. (a) Scheme of the perfusion bioreactor with the control system. Glucose is monitored on line by an external enzymatic analyser (not shown) and the computer maintains the glucose concentration at the desired set point by adjusting the feed and the harvest rates. (b) Time profile of the glucose concentration measured on line (●) and of the glucose set point (—) in a perfusion CHO cell culture. The set point was increased in a stepwise mode (2.0, 2.5, 3.0 and 3.5 g/l); standard deviations ('std') of measurements are given for each step. (Reproduced from Konstantinov *et al.* (1996), with permission.)

Lam (1992). PLCs typically have a limited memory, used only to store the data required for the control algorithms. Storage and archiving of process data is performed by supervision tools, located on the next CIM level.

Originally, PLCs were built as standalone units, each being dedicated to its own process unit or small set of functions. Communication was limited to a human machine interface (HMI) and to other PLCs to exchange a limited amount of data. Nowadays, PLCs have much higher communication capacities, thanks to field bus systems and other local area networks. Instead of building centralized PLCs with continuously increasing performance, the tendency is now towards a 'distribution of intelligence', i.e. to keep the various control functions spread over several machines, in a decentralized way, and to use a supervision tool to coordinate the operations of the various PLCs (see Section 13.4.1). This leads to a higher availability of the control systems and more flexibility, for instance for validation (see Section 13.5) than a centralized design. Additionally, in the case of a PLC failure, only a small part of the plant will be affected.

13.3.2 Control of the Main Cell Culture Variables

13.3.2.1 Temperature

For bench-scale bioreactors (up to 20–30 litres), temperature is usually controlled via heating only, with a simple electrical blanket wrapped around the tank; since there is no active cooling, a well-tuned PID controller is needed in order to avoid any excessive temperature overshoot. In larger tanks, temperature is commonly controlled via the tank jacket, through which a fluid (usually tap water) is circulated (Krahe 2002). Very large vessels (above 2000 litres) may have several parallel circulation loops. Heating of the jacket fluid can be performed in different ways; for relatively small units (up to about 200 litres), an electrical heater can be mounted on the external part of the loop. For larger tanks, the most common heating method is via an external heat exchanger, using 'black' steam on the other side. Alternatively, steam can be injected directly into the loop. Cooling of the fluid can be performed with cold water, either via direct injection or via a heat exchanger.

13.3.2.2 pH

pH is commonly raised by the pulsed addition of a concentrated base solution (sodium hydroxide or carbonate) via a pump. Since most culture media have a CO_2/HCO_3^- buffering system, stripping CO_2 out of the liquid phase with sparged air can further contribute to raising the pH. This effect can be enhanced via surface aeration. Conversely, pH can be lowered by increasing the amount of CO_2 in the sparged gas mixture. The addition of acid is rarely required as metabolically active cells themselves produce large amounts of lactic acid.

13.3.2.3 pO_2 and dCo_2

Dissolved oxygen is normally controlled via the flowrate and/or composition of the sparged gas mixture. These two parameters can be manipulated in different ways, leading to various control schemes; the scheme can actually be changed automatically in the course of a culture, as demand for oxygen increases. One common method consists of adjusting the flow rate of the sparged gas mixture (often air) at a fixed composition. Alternatively, the gas mixture can be sparged at fixed baseline rate and composition, with additional pulses of air or oxygen as needed. Another solution, if a constant sparging rate is desired, is to adjust the air–oxygen ratio in the gas mixture. In all cases, thermal mass-flow controllers (see Section 13.3.2.4) are widely used for measuring and accurately controlling the various gas flow rates. The headspace pressure in the bioreactor also directly affects the concentration of dissolved oxygen. This parameter however is rarely modified in cell cultures for control purposes. Similarly, adjustment of the stirring rate is not applied (unlike in microbial fermentations) due to the relatively high mechanical fragility of animal cells. Bubble-free oxygenation can also be achieved, by using immersed porous hydrophobic membranes (for instance Schneider *et al.* 1995). This method however is limited to small-scale bioreactors; at large scale, it becomes impractical due to the need for a very large membrane area.

dCO_2 can be controlled via sparged N_2 (Pattison *et al.* 2000); even during periods of intense sparging, pO_2 can be maintained at the setpoint via addition of pure oxygen.

13.3.2.4 Gases and liquids

For the measurement and control of gas flow rates, an inexpensive method is to use rotameters. The reading however, is strongly affected by the gas pressure and temperature; these variables must therefore be carefully controlled or their effect taken into account to correct the reading. For pilot- and large-scale processes, thermal mass-flow meters and controllers, combined in a stand-alone unit, are preferred, due to their higher accuracy. In both cases, the flowmeter operates under non-sterile conditions and must therefore be located upstream of the sterile filters. For the measurement and control of liquid flow rates and amounts in cell culture, rotameters are rarely used; Coriolis-based mass, ultrasonic, and magnetic flowmeters are more common. A key requirement of liquid flowmeters is cleanability; in many cases, such as transfer lines between bioreactors, they must also be sterilizable. Alternatively, a metering pump or a valve controlled by a device continuously measuring the liquid level or weight in the tank (see Section 13.2.2.7) can also be used. More details and selection criteria for the flow controllers mentioned here can be found in Hartnett (1994) and Krahe (2002).

13.3.3 Control of Key Metabolites

13.3.3.1 Fed-batch cultures

In fed-batch cultures the most commonly controlled metabolites are glucose and glutamine. This is mainly because they are the key substrates and major energy sources. When these species are present at high concentrations, most animal cells utilize them in a very inefficient way, referred to as *overflow metabolism*, causing the accumulation of large amounts of lactate and ammonia (see for instance Ljunggren & Häggström 1994). Since these by-products have been recognized as major inhibitory species – at least on the basis of dose-response experiments (for instance, Hasselt *et al.* 1991; Wentz & Schügerl 1992) – a reduction of their accumulation through a careful control of glucose and glutamine at low levels (typically below 0.6 mM and 0.2 mM respectively) may lead to a higher cell density and an extended culture duration. This strategy was first proposed by Glacken *et al.* (1986). Recently, it has also been shown that ammonia can (negatively) affect protein glycosylation (for instance, Gawlitzek *et al.* 1998; Yang & Butler 2000); this finding represents a second incentive to minimize ammonia accumulation during the culture.

Several experiments have been reported where glucose and glutamine were automatically controlled in fed-batch cultures; the best control was achieved when the feed rates of concentrated solutions were adjusted on the basis of the on-line monitoring of these species in the bioreactor (Honda *et al.* 1998; Kurokawa *et al.* 1994, Lenas *et al.* 1997, Siegwart *et al.* 1999). In other studies, feeds were adjusted on the basis of variables related to cell metabolism, such as the oxygen uptake rate or the amount of base solution added to control pH. In these cases, however, glucose and glutamine were not usually controlled with the same accuracy (Eyer *et al.* 1995; Lin *et al.* 2002, Oh *et al.* 1996; Zhou *et al.* 1995, 1997).

As expected, the accumulation of lactate was found to decrease when the glucose set point was reduced. Both a lower consumption of glucose and a lower yield of lactate from glucose contributed to this effect (for instance, Honda *et al.* 1998). A similar observation was made for ammonia when glutamine was controlled at a low level. Interestingly, with BHK cells cultivated at a low fixed glutamine concentration, a reduction of the glucose set point in the range where it is limiting led to a higher consumption of glutamine and a larger accumulation of ammonia (Honda *et al.* 1998). As a consequence, it may not be possible to minimize lactate and ammonia simultaneously by maintaining glucose and glutamine at low (limiting) levels.

Carefully controlling glucose and glutamine at low levels was often found to be beneficial in terms of cell growth and protein productivity, mostly with BHK and hybridoma cells. The optimum was either at the lowest tested glucose and glutamine levels (Honda *et al.* 1998; Kurokawa *et al.* 1994; Zhou *et al.* 1997), or at the combination of glutamine and glucose levels leading to the minimum production of ammonia but not of lactate (Lenas *et al.* 1997). This last example suggests that it may be more important to control ammonia than lactate.

These results, however, do not establish a direct and unique causal effect of ammonia and lactate accumulation on cell growth and productivity. The observed beneficial effects may also be due to the fact that when glucose and glutamine were controlled, they were also supplied in sufficient amounts, unlike in a reference culture (Zhou *et al.* 1997). This is supported by Eyer *et al.* (1995), where a beneficial effect of glutamine feeding was observed, although this amino acid was present during most of the culture time at levels where some overflow metabolism is known to occur.

Simpler ways of controlling glucose and glutamine, based upon off-line measurements or predetermined feeding profiles, also led to a much more efficient utilization of glucose and glutamine (Ljunggren & Häggström 1994), a higher cell density, and a higher concentration of the recombinant protein (Cruz *et al.* 2000) than when these two nutrients were present in excess from the beginning. In other cases, even simple feeds or shots of glucose, glutamine and mixtures of additional nutrients, led to dramatic increases in production yields, even though ammonia and lactate were not necessarily minimized (Bushell *et al.* 1994; Gorfien *et al.* 2003; Xie & Wang 1994). Sauer *et al.* (2000), found that the highest antibody titre was achieved when glucose was controlled at a very high level (4 g/l). This again supports the idea that accumulation of ammonia and lactate are not the sole limiting factors in fed-batch cultures, or at least not in a general way; traditional batch cultures may have often been limited simply by the amount of some nutrients such as glucose and amino acids. Only a few studies have actually been performed where control strategies of various complexities were compared with simpler strategies using the same cell line, in order to quantitate the benefits of automatic control, based upon on-line monitoring. In one report (Zhou *et al.* 1997), higher monoclonal antibody productivity was achieved with independent control of glucose and glutamine than with a glutamine feed simply adjusted to that of glucose. On the other hand, in another experiment, de Tremblay *et al.* (1993) found that a predetermined feeding profile led to a higher final antibody titre than did a feedback control approach. Consequently, although automatic feedback control of metabolites is a powerful technique for metabolic studies and process optimization, it is rarely applied to pilot-and large-scale processes, where simple feeding strategies for nutrients are often preferred. The additional instrumentation, the GMP-compliance costs and the higher risk of failure associated with automatic control are rarely justified by gains in productivity or product quality.

13.3.3.2 Continuous cultures

In continuous cultures, several examples of automatic control of nutrients have been reported. The main purpose was not only to maintain glucose or glutamine at very low concentrations in order to minimize overflow metabolism and accumulation of toxic by-products. Often, the goals were simply to control these nutrients at levels that were thought to be non-limiting for cell growth or viability, while avoiding excessive use of perfused medium (and cell washout in the case of chemostats). Additionally, automatically controlling the concentration of key nutrients, even if the optimal values are not known, is also a way of maintaining the cells under a stable and thus reproducible environment. In principle, if the medium is properly balanced, adjusting the perfusion rate on the basis of only one nutrient level or oxygen uptake rate should ensure that all other non-measured nutrients (e.g. amino acids) are also supplied in sufficient amounts (Kyung *et al.* 1994).

Stoll *et al.* (1996b), for instance, successfully controlled automatically the level of glutamine and ammonia independently in a hollow-fibre reactor, via the medium perfusion rate and the supply of a concentrated solution of glutamine. In other experiments, the level of glucose was controlled via the perfusion rate, either directly (Konstantinov *et al.* 1996, Ozturk *et al.* 1997b) (Figure 13.5) or indirectly, on the basis of the oxygen uptake rate, using a predetermined and constant stoichiometric ratio between oxygen and glucose consumptions (Kyung *et al.* 1994). A more sophisticated approach has been proposed by Konstantinov (1996), where the so-called cell physiological state instead of the cell environment is controlled. Although more complex to implement, this strategy may bring a further optimization of the process.

At present, automatic control of continuous cultures is rarely applied at pilot and large scale, for the reasons mentioned in the previous section. Simpler control strategies are often preferred, such as manual adjustment of the medium perfusion rate once or twice daily, based upon off-line measurements (Dowd *et al.* 2001). At high dilution rates (4–6 bioreactor volumes per day), manually controlled perfusion cultures might become unstable, leading to large oscillations of some process variables (Konstantinov *et al.* 1996). To our knowledge, however, the stabilization of perfusion cultures by closed-loop control has not been assessed by comparative quantitative studies.

13.4 HIGHER CIM LEVELS

13.4.1 Supervision Level

In complex manufacturing facilities, the coordination of PLCs, to automate operations involving several units (e.g. tanks) at the same time, may become extremely complex. At the limit, each PLC would have to mirror the status of all other involved PLCs for a proper synchronization. This challenge has lead to the development of distributed control systems (DCS), where a supervision tool (often termed *supervisory control and data acquisition* (SCADA) application) controls a group of separated PLCs, spread over the various equipment units of the plant. Communication is made possible thanks to a field bus system (see Section 13.2.1.3) and local area networks.

The main functions of the supervision tool are to control the process running on the various process units and to display their status, including all measurements and alarms. A key component of the tool is a human–machine interface, which enables the user to enter individual orders, set points or even complete recipes. Supervision tools are also capable of displaying and generating various graphical and text reports. Data are generally stored temporarily, for a few hours or days, before they are archived; this short period is for safety reasons. For data storage, a reduction algorithm is normally applied, which does not select the data at a fixed frequency but on the basis of a significant-change criterion (Sonnleitner 1999). Figure 13.6 shows a simple example of a DCS for a group of three pilot-scale bioreactors. An example of a more complex DCS, for a fermentation pilot plant, can be found in Lam (1992).

With respect to the architecture of DCSs, one can distinguish two main solutions. The first type is a completely homogeneous system, built by a single supplier. The main advantage is the ease of implementation since communication is standardized, and databases and configuration tools are common to both the PLCs and the supervision tool. Development costs and start-up problems are thus reduced. Furthermore, with a single supplier of hardware and software, fewer people are usually required in the project phase. The main drawback is the fact that the user is linked to a single supplier, such that the evolution of the system may be limited. A second type is a hybrid process control system, made of PLCs and associated field devices on one hand, and

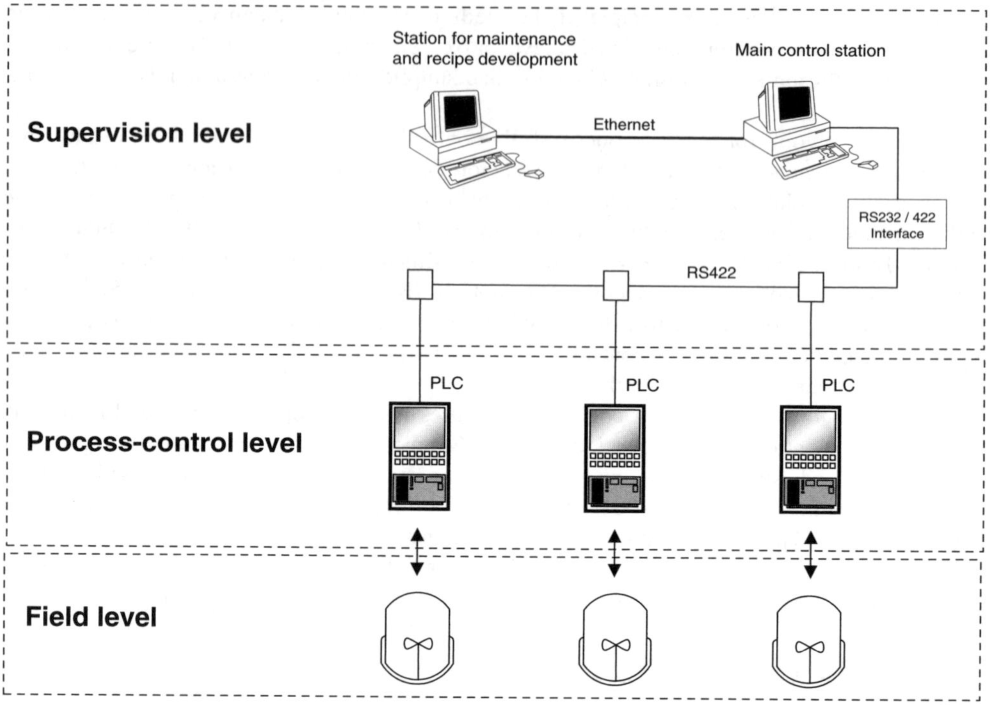

Figure 13.6 Simple example of a DCS for a group of three pilot-scale bioreactors. Each bioreactor is controlled by a separate PLC. In this particular example, PLCs have their own human–machine interface (HMI), with a keyboard for entering data (controller parameters, setpoints, etc.); this is rarely the case in large manufacturing plants with complex DCSs, where there is only one HMI, located at the supervision level. For simplification, only two stations at the supervision level are represented: a main one, for instance in the process room, and a second one, for system maintenance and recipe development.

of a supervision system from another supplier on the other hand. With this type, one can build a customized control system by using the most appropriate devices at each level. Implementation is, however, more complex, takes longer and requires more people with various competencies, since systems and devices from different suppliers must be integrated. The risk of unexpected problems during the project phase is also higher. For the control of complex biotechnology plants, a homogeneous system is thus more common.

13.4.2 Production- and Enterprise-management Levels

Production management represents the next CIM level and corresponds to the coordination of all the production activities for a manufacturing facility. The key component at this level is a manufacturing and resource planning (MRP II) system, whose main element is a material requirements planning (MRP) tool; the primary function of a MRP tool is to transform the demand for products, e.g. from another unit of the company, into a production plan, further broken down into so-called *process* or *shop orders*, and to calculate the amounts and order dates for all raw materials, consumables and intermediates that are required to meet this plan. Additionally, MRP II systems are used for planning, execution and follow-up of process orders as well as archiving of all data, thus providing traceability on all production lots, as required by GMP procedures. Other common functions of MRP II systems include business, sales and operations planning, as well as costing of production campaigns.

Typical additional tools or systems at this level, which may be fully integrated into or simply interfaced with the MRP II system, are:

- A fine-planning tool, to schedule and sequence manufacturing operations at detailed time scales and at finite capacity, i.e. taking into account the equipment resource(s) required at each process step. The tool can thus optimize the utilization of resources common to several operations, such as a CIP unit serving several tanks or a media preparation tank common to several bioreactors; this may be particularly useful when the plant is running close to full capacity.
- A laboratory information management system (LIMS), to handle all IPC and QC samples and the results of analyses.
- A tool for weighing and dispensing raw materials. This tool can supervise all the scales in the plant that are used to weigh raw materials; it can be interfaced with the MRP II system so that inventory levels are automatically updated as soon as raw materials are weighed for manufacturing operations.

Data transfer at this level does not need to occur in real time, as production planning is often performed days in advance; also, collection of information from the execution of process orders is usually done in batch mode, for instance on a daily basis in the case of manual transactions. Computer systems at this level thus work with a time constant of the order of several hours or one day; however, they are characterized by very powerful computing and storage capacities, given the large size and complexity of data they have to handle.

Production-management systems can communicate in different ways with the CIM level underneath, i.e. the supervision level. In plants with a low degree of automation, the two levels may actually not exchange any data automatically. The MRP II and associated systems are operated essentially via manual transactions for the planning, execution and archiving of process orders. In this case, it is common to use, in parallel, paper documents for operating instructions to the plant personnel and for batch records. In highly or fully automated plants, which seems to be the trend for new biotechnology manufacturing plants, all the supervision- and production-management systems may be interfaced to each other or even fully integrated, forming a global distribution and control system, including operating instructions and electronic batch recording, with only one HMI. This implies that the supervision system does not operate only as a function of time but also on a process-order basis. Between these two extreme cases, there are of course all kinds of intermediate systems for plant automation, with different degrees of communication and integration between the supervision and production management levels. Fully automated plants in principle offer all the advantages cited in Section 13.1. The main drawback is the very expensive and time-consuming validation operations (see Section 13.5); the benefits of automation may thus appear only in the long term. Additionally, an automated plant may have less flexibility with respect to process changes or introduction of new processes/products; since several computer systems are integrated, relatively small modifications may require important revalidation operations. A further discussion on the benefits of plant automation up to the production-management level can be found in deSpautz (2000).

The highest level of the CIM pyramid is the enterprise-management level, which corresponds to the global management of a company. Computer systems at this level are particularly useful for medium and large businesses, spread over several sites. They are used for the management of manufacturing, inventory, distribution, personnel, projects, payroll and financial operations. The most common integrated system is the Enterprise Resource Planning (ERP) system; it was originally an extension of a MRP II that has been widened to serve almost all departments within a company. An ERP system is typically based upon packaged modules rather than some proprietary software written for one customer. Modules may, however, be interfaced with an organization's own information systems. One of the most common ERP systems is SAP® (SAP AG).

13.5 REGULATORY ASPECTS OF AUTOMATION SYSTEMS

13.5.1 Main Regulations – 21 CFR Part 11

13.5.1.1 Introduction

Following the increased use of computer and electronic data management systems in the pharmaceutical and biotechnology industry, the US FDA issued a specific regulation in 1997 on these topics (21 CFR Part 11; FDA 1997). 21 CFR Part 11 applies to all industry segments regulated by the FDA, including Good Laboratory, Good Clinical and Good Manufacturing Practices, and has hence been enforced by regulatory centers such as CDER and CBER. The main topic of 21 CFR Part 11 addresses the requirements that electronic records and electronic signatures have to fulfill to become legally equivalent to paper-based records and handwritten signatures. Electronic records are defined by FDA as 'any combination of text, graphics, data, audio, pictorial or other information represented in digital form that is created, modified, maintained, archived, retrieved or distributed by a computer system'. Data that are deleted after the system is switched off, for instance those stored in the random access memory (RAM) of a PLC, are not considered as electronic records. In the case of manufacturing, an example of electronic records is all the quality-relevant electronic data collected during a batch. Electronic signature is defined as a 'computer data compilation of symbols executed, adopted or authorized by an individual to be the legally binding equivalent of the individual's handwritten signature'. Beyond specific requirements on electronic records and electronic signatures, 21 CFR Part 11 requires complete control over computer systems during their whole life cycle, from development till retirement.

After the original Part 11 rule became effective, FDA published a compliance policy guide and several guidance documents. These documents were studied by industry, which then raised concerns about the resulting restriction of technological innovation, the high compliance costs and the lack of clarity on how the regulations should be applied to specific applications (Phoenix & Andrews 2003). Taking these widespread criticisms into account, FDA issued a new guidance document on the scope and application of 21 CFR Part 11 and withdrew all other previous guidance documents (FDA, 2003). The interpretation of this new document and its implications has generated a large number of discussions (Huber & Winter 2004). In brief, the requirements of 21 CFR Part 11 remain the same; the new guidance document, however, introduces the concept of a risk-based assessment of computer systems, to evaluate their potential impact on product quality (Phoenix & Andrews 2003). The result may be that fewer systems will be considered subject to 21 CFR Part 11; records such as audit trails and electronic copies (see Section 13.5.1.2) may not always be required.

In Europe, requirements for computer systems are addressed in Annex 11 of the GMP regulations (European Commission 1998) and in a directive legalizing electronic signatures (European Parliament 2000). The EU regulation is, however, different in many respects from the US version. The discussion below summarizes the main requirements of 21 CFR Part 11, with which all pharmaceutical and biotech companies producing for the US market have to comply. A comprehensive review can be found elsewhere (Bruce 2003; Huber & Winter 2004; Norman 2001; Throm 2002; Winter & Huber 2004a–c).

13.5.1.2 Summary of requirements

- 21 CFR Part 11 makes a distinction between *closed* and *open* computer systems, depending on whether system access is, or is not, controlled by people responsible for the content of the electronic records stored in the system. The first type is much more common and is discussed below. Open systems require additional safeguard measures.
- *System access and authorization*: suitable control mechanisms (both physically and at the software level) must be put in place so that only authorized users can access the system and its data. The authorizations must be appropriately specified, documented and tested. Examples of checks are user ID, password, ID card (badge) and biometric identification (fingerprint, eye scan).
- *Device checks*: appropriate measures must be taken to ensure the validity of the source of data and commands; this can be achieved via checks of the interface or of the status of the input devices (for instance sensors).
- *Audit trail*: the system must be able to generate an audit trail, defined as a record showing who has accessed a computer system and what operations have been performed and when. Any such audit trail must be secure, computer generated and time stamped, and should contain both the original and changed data, plus details of the person responsible for making the change.
- *Training and qualification of personnel*: all the people who work with a computer system, such as users, maintenance personnel and developers, must be suitably trained and formally approved as competent for the tasks they perform.
- *Copies of electronic records*: the computer system handling electronic records must be able to provide full, correct and readable copies (for instance for an inspection), both in electronic and paper formats.
- *Data back-up and archiving*: the data must be accessible in a readable format for the duration of the retention period (often more than 10 years); appropriate back-up procedures for on-line data and archiving are required to ensure data integrity.
- *Electronic signatures*: in principle, these are not mandatory and a handwritten signature on a paper printout of an electronic record represents an acceptable hybrid system. However, when used, electronic signatures must comply with the requirements of 21 CFR Part 11. Electronically signed records must contain the name of the signatory, the date, time and reason (e.g. review, release) of signature. This information must be visible each time the record is viewed or printed. Written directives must exist that assign to each individual the responsibility for the actions associated with the electronic signature. An electronic signature must be assigned to a unique person.
- *Link between signature and electronic record*: whether electronic or handwritten on an electronic record, the signature must be linked to the record in such a way that it cannot be readily separated and misused for other purposes.
- *Components and execution of electronic signatures*: unless it is linked to a biometric form of identification, the signature must contain at least two components such as a user ID and a password. In this case, several checks must be in place, to ensure uniqueness of user ID and password combination, validity, detection of unauthorized use, and adequate actions in case of loss.
- *Validation of computer systems*: computer systems utilizing electronic records and signatures must be validated to ensure that all the aforementioned requirements are met. 21 CFR Part

11 actually does not change the basics of computer system validation, which has long been a regulatory requirement. It does, however, introduce the requirement for the validation of more systems than before, as well as of new software functions, such as the generation of audit trails, including the creation of accurate and complete copies. The main steps for the validation of computer systems are discussed in the next section.

13.5.2 Computer System Validation

13.5.2.1 Life-cycle concept

Because of the complexity of a computer system, it is difficult to rely only on testing to provide a high degree of assurance that the system will always perform as intended. It is acknowledged that quality must be designed, built in and then maintained during both the development and use of the system. This is why a life-cycle concept has been developed for the validation of computer systems, which is now applied to other systems in the plant (Bestetti 2002). The main principle of the life-cycle concept is to control the whole life of the system, from start to finish, following well-defined steps.

The life cycle starts with validation *per se*, which can be broken down into different phases, typically as follows: (i) definition; (ii) design; (iii) development and construction; (iv) qualification. Each phase can be further subdivided into several distinct steps. This approach is usually described in detail in a validation master plan, which must be formally approved by the key stakeholders before validation can actually start. The plan should define, among others, at each step, the required activities, deliverables, responsibilities and standards. The validation phases are often represented in a V-shaped diagram (see Birkin 2002). The definition and design phases are located along the first branch of the V, the development and construction phase at the bottom of the V, whereas the qualification phase is on the second branch. In this way, the various tests of the qualification phase can be represented beside the definition-and design-phase steps to which they relate. Figure 13.7 shows one example of a V-shaped validation diagram.

As discussed in the previous sections of this chapter, modern control systems in manufacturing are made of several layers and are characterized by an increasing 'intelligence' at the field level, with the use of smart devices. This puts a higher challenge on system validation, since each component has to be validated first individually, then as part of subsystems and finally as part of the complete integrated computer system (Thompson 2001).

The life cycle does not end with validation. The validated state of the system needs to be maintained throughout the whole period where the system is operated for its intended use, which actually corresponds to the longest phase of the cycle. The main steps of the life-cycle concept for the validation of plant automation systems are summarized below. More details can be found elsewhere (Bestetti 2002; Glennon 1997; Thompson 2001, Uzzaman 2003). Strategies to increase efficiency of validation are proposed in Budihandojo (2001, 2002).

13.5.2.2 Definition phase

The first step of this phase is a definition of all the needs; these are often formulated by the future user, in the form of a document called the *user requirement specification* (URS). Since the user is often not a computer specialist, the URS is typically written in a non-technical way. It is, however, crucial that the needs are expressed in detail and, if possible, in a quantitative way, so that they can later be verified during the qualification phase. The user should also highlight the criticality, with respect to the process or product, of each requirement. For instance in cell culture, the effects

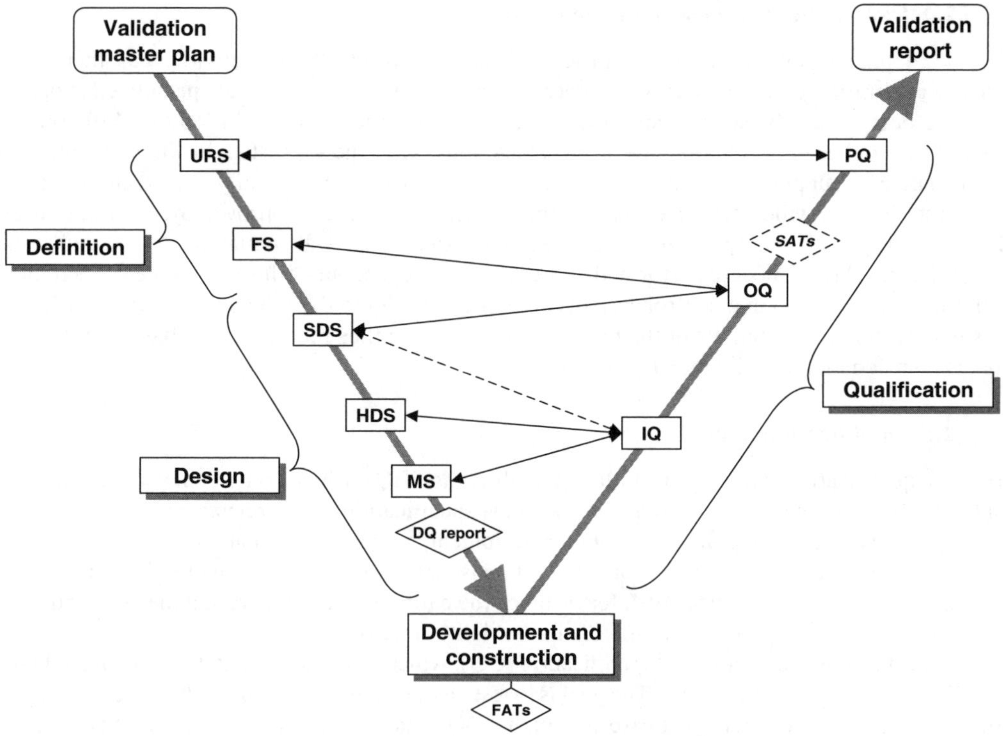

Figure 13.7 Example of a V-shaped validation diagram for a computer system. Double arrows indicate the links between the steps of the definition and design phases, and the corresponding tests of the qualification phase. Diamonds represent milestones. (For the meaning of abbreviations, see main text.)

(technical, regulatory, economic, etc.) of inaccurate control of some parameters or of a temporary failure of the supervision tool should be clearly described. This can greatly help the work of developers as well as providing a base for risk analysis, which is performed later to determine the required level of qualification and testing.

In a second step, a functional analysis is performed to analyse and describe in detail each requirement of the URS and to break it down into elemental functions or functional specifications (FS). The result is usually a much more technical description of the system (hardware and software) to be developed.

13.5.2.3 Design phase

During the design phase, detailed solutions are proposed to cover all FS. The architectures for software and hardware are thus designed, with a description of equipment, tools and application structures, resulting in software design specifications (SDS), hardware design specifications (HDS) and mechanical specifications (MS). The choice between commercial packages and customized ones is typically made at this level. This step is formally completed by a design qualification (DQ), normally in the form of a document that is approved by the key stakeholders.

13.5.2.4 Development and construction phase

The development (software) and construction (mechanical, electrical) activities can then start, usually in parallel. Development is often carried out off site, by an external specialized supplier, who has been carefully selected and audited to assess his know-how and level of GMP compliance. The supplier may actually have been selected just after the URS step, in which case he can participate as a support in formulating the FS. Development ends formally with factory acceptance tests (FATs), which are performed in the presence of, and are approved by, the future user. The goal is to verify that the system is properly built and works well enough to be installed at the manufacturing site. FATs are performed at the supplier site, i.e. out of the environment where they will finally be used, so that control systems are not connected to the production equipment. Tools to simulate inputs and outputs of the process equipment are thus required. Installation of the system is then completed at the manufacturing site.

13.5.2.5 Qualification phase

The first qualification step is the installation qualification (IQ), which consists in verifying that all individual components of the system meet with the specifications. This relates mostly to the hardware and mechanical components (computer equipment, sensors and actuators, connections, etc.), but some SDS, such as the correct software versions, are also checked at this level. Additionally, tests on unit elements are performed, for instance the calibration of sensors and the verification of actuators, including the rotation of motors in the right direction.

The next step is the operational qualification (OQ), which integrates the software with the hardware in the operating environment. The goal is to test on line all the functions of the system individually, both under normal and worst-case conditions. OQ tests are designed such that all SDS and FS can be verified using, for instance, water to simulate culture media. At the end of OQ, the system documentation is usually completed with some user and maintenance manuals. When IQ and OQ are performed by an external contractor, these activities are often grouped into some site acceptance tests (SATs); upon successful completion of these tests, the system is handed over to the future user.

The last step is the performance qualification (PQ), where the system is tested globally, under real conditions. For a process automation system, PQ should be performed by running the manufacturing process, or at least by testing all the functions, in the actual sequence to be used and in real time. Without this proof of usability, the qualification is not complete. While all the previous steps, from the functional analysis, can be subcontracted to external specialists, PQ is normally performed by the future user, who has to verify that the delivered system complies with his own requirements (URS). A validation report is usually written at this point, confirming that all the activities defined in the validation master plan have been completed. Upon approval of the report, the system is released for use.

13.5.2.6 Validation maintenance

To maintain the validation status of the system and to operate it properly, a special organization must be put in place, along with adequate documentation. Personnel must first be trained and approved as competent before operating the system. Procedures to modify the system, both at the hardware and software levels, and to restart or reinstall the system in case of failures (disaster recovery) must be documented. Hardware should be maintained and calibrated regularly; functions should be retested. A maintenance contract with the external supplier is recommended to ensure adequate technical support and upgrades of the system as needed. On-going validation maintenance can help ensure a high degree of compliance while minimizing the overall cost of validation (Thompson 2001). Finally, when the system is retired, a validated migration of data to the new system must be performed.

ACKNOWLEDGEMENT

The authors thank Martin Rhiel and Zoltan Suemeghy for reviewing the manuscript and providing very valuable comments.

REFERENCES

Akhnoukh R, Kretzmer G, Schügerl K (1996) *Enzyme Microb. Technol.*; 18: 220–228.

Bailey JE, Ollis DF (1986) *Biochemical Engineering Fundamentals.* McGraw-Hill Book Company, New York; Second edition: 661–666.

Baker KN, Rendall MH, Patel A *et al.* (2002) *TIBTECH*; 20(4): 149–156.

Behrendt U, Koch S, Gooch DD, Steegmans U, Comer MJ (1994) *Cytotechnol.*; 14: 157–165.

Bestetti C (2002) *Pharmaceutical Eng.*; 22(6): 90–98.

Beyeler W, DaPra E, Schneider K (2000) *Adv. Biochem. Engin./Biotechnol.*; 70: 139–164.

Bilitewski U, Drewes W, Neermann J *et al.* (1993) *J. Biotechnol.*; 31: 257–266.

Birkin S (2002) *Biopharm Europe*; March: 58–64.

Blankenstein G, Spohn U, Preuschoff F, Thömmes J, Kula MR (1994) *Biotechnol. Appl. Biochem.*; 20: 291–307.

Brecht A, Gauglitz G, Polster J (1993) *Biosen. Bioelectr.*; 8: 387–392.

Bruce L (2003) *Pharm. Technol. Europe*; June: 33–36.

Budihandojo R, Bergeson DJ, D'Eramo P *et al.* (2001) *Pharm. Technol. Europe*; December: 27–36.

Budihandojo R, Bergeson DJ, D'Eramo P *et al.* (2002) *Pharm. Technol. Europe*; January: 34–40.

Bushell ME, Bell SL, Scott MF, Spier RE, Wardell JN, Sanders PG (1994) *Biotechnol. Bioeng.*; 44: 1099–1106.

Campmajò C, Cairó JJ, Sanfeliu A, Martínez E, Alegret S, Gòdia F (1994) *Cytotechnol.*; 14: 177–182.

Cannizzaro C, Gügerli R, Marison I, von Stockar U (2003) *Biotechnol. Bioeng.*; 84: 597–610.

Cattaneo MV, Luong JHT (1993) *Biotechnol. Bioeng.*; 41: 659–665.

Cha HJ, Dalal NG, Vakharia VN, Bentley WE (1999) *J. Biotechnol.*; 69: 9–17.

Chéruy A (1997) *J. Biotechnol.*; 52: 193–199.

Chung H, Arnold MA, Rhiel M, Murhammer DW (1996) *Appl. Spectrosc.*; 50: 270–276.

Cruz HJ, Moreira JL, Carrondo MJT (2000) *J. Biotechnol.*; 80: 109–118.

Cush R, Cronin JM, Stewart WJ, Maule CH, Molloy J, Goddard NJ (1993) *Biosens. Bioelectr.*; 8: 347–353.

deSpautz JF (2000) *Pharmaceutical Eng.*; September/October: 44–60.

de Tremblay M, Perrier M, Chavarie C, Archambault J (1993) *Bioprocess Eng.*; 9:13–21.

deZengotita VM, Kimura R, Miller WM (1998) *Cytotechnol.*; 28: 213–227.

Dochain D, Perrier M (2000) In *Bioreaction Engineering – Modeling and Control.* Eds Schügerl K, Beelgardt KH. Springer Verlag, Berlin; 145–166.

Dorresteijn RC, Numan KH, de Gooijer CD, Tramper J, Beuvery EC (1996) *Biotechnol. Bioeng.*; 50: 206–214.

Dowd JE, Kwok KE, Piret JM (2001) *Biotechnol. Bioeng.*; 75: 252–256.

Dremel BAA, Li SY, Schmid RD (1992) *Biosens. Bioelectr.*; 7: 133–139.

Ducommun P, Bolzonella I, Rhiel M, Pugeaud P, von Stockar U, Marison IW (2001) *Biotechnol. Bioeng.*; 72: 515–522.

Ducommun P, Kadouri A, von Stockar U, Marison IW (2002) *Biotechnol. Bioeng.*; 77: 316–323.

Dunn IJ, Heinzle E, Ingham J, Přenosil JE (2003) *Biological Reaction Engineering.* Wiley-VCH, Weinheim.

European Commission (1998) *Good Manufacturing Practice – Medicinal Products for Human and Veterinary Use*, Annex 11: *Computerized Systems.*

European Parliament and Council of the European Union (2000) Directive 1999/93/EC of 13 December 1999 on a community framework for electronic signatures, *Official J. European Communities*; L1: 12–20.

Eyer K, Oeggerli A, Heinzle E (1995) *Biotechnol. Bioeng.*; 45: 54–62.

Favre E, Pugeaud P, Pùringer P (1990) *Biotechnol. Techniques*; 4: 315–320.

FDA (1997) 21 Code of Federal Regulations, Part 11: Electronic records; electronic signatures; final rule. *Federal Register*; 62: 13429–13466.

FDA (2003) Guidance for Industry: Part 11, Electronic records, electronic signatures – Scope and applications. Draft: February 20; final version: September 3.

FDA (2004) Guidance for Industry: PAT – A framework for innovative pharmaceutical development, manufacturing, and quality assurance, September 29.

Forbes NS, Clark DS, Blanch HW (2000) In *Bioreaction Engineering – Modeling and Control*. Eds Schügerl K, Beelgardt KH. Springer Verlag, Berlin; 556–594.

Freshney RI (2000) *Culture of Animal Cells: A Manual of Basic Technique*. Wiley-Liss, New York; Fourth edition: 309–312.

Gawlitzek M, Valley U, Wagner R (1998) *Biotechnol. Bioeng.*; 57: 518–528.

Ge X, Tolosa L, Simpson J, Rao G (2003) *Biotechnol. Bioeng.*; 84: 723–731.

Gebbert A, Alvarez-Icaza M, Peters H, Jäger V, Bilitewski, Schmid RD (1994) *J. Biotechnol.*; 32: 213–220.

Glacken MW, Fleischaker RJ, Sinskey AJ (1986) *Biotechnol. Bioeng.*; 28: 1376–1389.

Glazier SA, McCurley MF (1995) *BioPharm*; June: 38–50.

Glennon B (1997) *J. Biotechnol.*; 59: 53–61.

Gonzalez FJ, Herb SM (1984) *Pharmaceutical Eng.*; September/October: 17–22.

Gorfien SF, Paul W, Judd D, Tescione L, Jayme DW (2003) *BioPharm International*; April: 34–40.

Guan Y, Evans PM, Kemp RB (1998) *Biotechnol. Bioeng.*; 58: 464–477.

Harthun S, Matischak K, Friedl P (1997) *Anal. Biochem.*; 251: 73–78.

Hartnett T (1994) In *Bioprocess Engineering: Systems, Equipment and Facilities*. Eds Lydersen BK, D'Elia NA, Nelson KL. John Wiley & Sons, New York; 413–468.

Hasselt T, Gleave S, Butler M (1991) *Appl. Biochem. Biotechnol.*; 30: 29–41.

Honda H, Lenas P, Watanabe H, Kitade T, Koayashi T (1998) *J. Ferment. Bioeng.*; 85: 532–535.

Huber L, Winter W (2004) *BioPharm International*; February Supplement: 4–9.

Jönsson U, Malmqvist (1992) *Adv. Biosensors*; 2: 291–336.

Junker BH, Reddy J, Gbewonyo K, Greasham R (1994) *Bioprocess Eng.*; 10: 195–207.

Kemp RB, Guan Y (1997) *Thermochim. Acta*; 300: 199–211.

Kennedy RH (1983) *Chem. Eng.*; 90(16): 54–71.

Kilburn D, Fitzpatrick P, Blake-Colman B, Clarke D, Griffiths J (1989) *Biotechnol. Bioeng.*; 33: 1379–1384.

Kimura R, Miller WM (1997) *Biotechnol. Prog.*; 13: 311–317.

Konstantinov KB (1996) *Biotechnol. Bioeng.*; 52: 271–289.

Konstantinov K, Chuppa S, Sajan E, Tsai Y, Yoon S, Golini F (1994a) *TIBTECH*; 12: 324–333.

Konstantinov KB, Zhou W, Golini F, Hu WS (1994b) *Cytotechnol.*; 14: 233–246.

Konstantinov KB, Tsai YS, Moles D, Matanguihan R (1996) *Biotechnol. Prog.*; 12: 100–109.

Krahe M (2002) In *Ullmann's Encyclopedia of Industrial Chemistry*, Wiley-VCH Verlag, Weinheim.

Kurokawa H, Park YS, Iijima S, Kobayashi T (1994) *Biotechnol. Bioeng.*; 44: 95–103.

Kyung YS, Peshwa MV, Gryte DM, Hu WS (1994) *Cytotechnol.*; 14: 183–190.

Lam LH (1992) *Pharmaceutical Eng.*; November/December: 64–76.

Larson TM, Gawlitzek M, Evans H, Albers U, Cacia J (2002) *Biotechnol. Bioeng.*; 77: 553–563.

Lenas P, Kitade T, Watanabe H, Honda H, Kobayashi T (1997) *Cytotechnol.*; 25: 9–15.

Lin J, Takagi M, Qu Y, Yoshida T (2002) *Biochem. Eng. J.*; 11: 205–209.

Ljunggren J, Häggström L (1994) *Biotechnol. Bioeng.*; 44: 808–818.

Lübbert A, Simutis R (1994) *TIBTECH*; 12: 304–311.

Luong JHT, Nguyen AL, Guilbault GG (1993) *Adv. Biochem. Eng./Biotechnol.*; 50: 85–115.

Male KB, Gartu PO, Kamen AA, Luong JHT (1997) *Biotechnol. Bioeng.*; 55: 497–504.

Mangum BW (1989) *Clin. Chem.*; 35: 503–505.

Marks DM (2003) *Cytotechnol.*; 42: 21–33.

Marose S, Lindemann C, Ulber R, Scheper T (1999) *TIBTECH*; 17: 30–34.

Meyerhoff ME, Trojanowicz M, Palsson BO (1993) *Biotechnol. Bioeng.*; 41: 964–969.

Middendorf C, Schulze B, Freitag R, Scheper T, Howaldt M, Hoffmann H (1993) *J. Biotechnol.*; 31: 395–403.

Miyabayashi A, Mattiasson B (1990) *Anal. Biochem.*; 184: 165–171.

Mulchandani A, Bassi AS (1995) *Crit. Rev. Biotechnol.*; 15(2): 105–124.

Nilsson M, Håkanson H, Mattiasson B (1991) *Anal. Chim. Acta*; 249: 163–168.
Nilsson M, Håkanson H, Mattiasson B (1992) *J. Chromatography*; 597: 383–389.
Noll T, Biselli M (1998) *J. Biotechnology*; 63: 187–198.
Norman A (2001) *Pharmaceutical Technol. Europe*; December: 64–69.
Oh GS, Izuishi T, Inoue T, Hu, and Yoshida T (1996) *J. Ferm. Bioeng.*; 81: 329–336.
Olsson L, Nielsen J (1997) *TIBTECH*; 15: 517–522.
Oroszlan P, Thommen C, Wehrli M, Duveneck G, Ehrat M (1993) *Anal. Methods Instrum.*; 1: 43–51.
Ozturk SS, Thrift JC, Blackie JD, Naveh D (1995) *Biotechnol. Bioeng.*;48: 201–206.
Ozturk SS, Blackie J, Wu O *et al.* (1997a) In *Animal Cell Technology*. Eds Carrondo MJT *et al.* Kluwer Academic Publishers, Dordrecht, The Netherlands; 205–214.
Ozturk SS, Thrift JC, Blackie JD, Naveh D (1997b) *Biotechnol. Bioeng.*; 53: 372–378.
Paliwal SK, Nadler TK, Regnier FE (1993a) *TIBTECH*; 11: 95–101.
Paliwal SK, Nadler TK, Wang DIC, Regnier FE (1993b) *Anal. Chem.*; 65: 3363–3367.
Pattison RN, Swamy J, Mendenhall B, Hwang C, Frohlich BT (2000) *Biotechnol. Prog.*; 16: 769–774.
Phoenix K, Andrews J (2003) *Pharm. Eng.*; 23(4): 1–7.
Polzius R, Bier FF, Bilitewski U, Jäger V, Schmid RD (1993) *Biotechnol. Bioeng.*; 42: 1287–1292.
Renneberg R, Trott-Kriegeskorte G, Lietz M *et al.* (1991) *J. Biotechnol.*; 21: 173–186.
Rhiel MH, Amrhein MI, Marison IW, von Stockar U (2002a) *Anal. Chem.*; 74: 5227–5236.
Rhiel M, Ducommun P, Bolzonella I, Marison I, von Stockar U (2002b) *Biotechnol. Bioeng.*; 77: 174–185.
Rhiel MH, Cohen, MB, Arnold MA, Murhammer DW (2004) *Biotechol. Bioeng.*; 86: 852–861.
Riley MR, Arnold MA, Murhammer DW, Walls EL, DelaCruz N (1998) *Biotechnol. Prog.*; 14: 527–533.
Riley MR, Crider HM, Nite ME, Garcia RA, Woo J, Wegge RM (2001) *Biotechnol. Prog.*; 17: 376–378.
Royce PN (1993) *Critical Rev. Biotechnol.*; 13: 117–149.
Ruzicka J, Hansen EH (1975) *Anal. Chim. Acta*; 78: 145–157.
Sáez de Viteri FJ, Diamond D (1994) *Electroanalysis*; 6: 9–16.
Sanderson CS, Phillips PJ, Barford JP (1996) *Cytotechnol.*; 21: 149–153.
Sauer PW, Burky JE, Wesson MC, Sternard HD, Qu L (2000) *Biotechnol. Bioeng.*; 67: 585–597.
Schneider M, Reymond F, Marison IW, von Stockar U (1995) *Enzyme Microb. Technol.*; 17: 839–847.
Schügerl K (2001) *J. Biotechnol.*; 85: 149–173.
Schügerl K, Hitzmann B, Jurgens H, Kullick T, Ulber R, Weigal B (1996) *TIBTECH*; 14: 21–31.
Shimizu K (1993) *Adv. Biochem. Eng./Biotechnol.*; 50: 65–84.
Siano SA, Mutharasan R (1991) *Biotechnol. Bioeng.*; 37: 141–159.
Siegwart P, Côté J, Male K, Luong JHT, Perrier M, Kamen A (1999) *Biotechnol. Prog.*; 15: 608–616.
Sonnleitner B (1997) *J. Biotechnol.*; 52: 175–179.
Sonnleitner B (1999) *Adv. Biochem. Eng./Biotechnol.*; 66: 1–64.
Spohn U, van der Pol J, Eberhardt R, Joksch B, Wandrey C (1994) *Anal. Chim. Acta*; 292: 281–295.
Steusloff H (2002) In *Process Control Engineering, Ullmann's Encyclopedia of Industrial Chemistry*, Chapter 8, Wiley-VCH Verlag, Weinheim.
Stöcklein W, Jäger V, Schmid RD (1991) *Anal. Chim. Acta*; 245: 1–6.
Stoll TS, Ruffieux PA, Schneider M, von Stockar U, Marison IW (1996a) *J. Biotechnol.*; 51: 27–35.
Stoll TS, Ruffieux PA, von Stockar U, Marison IW (1996b) *J. Biotechnol.*; 51: 37–48.
Thompson D (2001) *Pharmaceutical Technol. Europe*; April: 36–43.
Throm S (2002) *Pharmaceutical Technol. Europe*; April: 35–45.
Uzzaman S (2003) *Pharmaceutical Eng.*; 23(3): 52–66.
van de Merbel NC, Lingeman H, Brinkman UAT (1996) *J. Chromatography A*; 725: 13–27.
van der Pol JJ, de Gooijer CD, Biselli M, Wandrey C, Tramper J (1996) *TIBTECH*; 14: 471–477.
van der Pol JJ, Machnik M, Biselli M, Portela-Klein T, de Gooijer CD, Tramper J, Wandrey C (1997) *Cytotechnol.*; 24: 19–30.
Wentz D, Schügerl K (1992) *Enzyme Microb. Technol.*; 14: 68–75.
White SF, Turner APF, Bilitewski U, Bradley J, Schmid RD (1995) *Biosens. Bioelectr.*; 10: 543–551.
Winter W, Huber L (2004a) *BioPharm International*; February: Supplement, 10–16.
Winter W, Huber L (2004b) *BioPharm International*; February: Supplement, 18–20.
Winter W, Huber L (2004c) *BioPharm International*; February: Supplement, 22–25.

Xie L, Wang DIC (1994) *Biotechnol. Bioeng.*; 43: 1175–1189.
Yang M, Butler M (2000) *Biotechnol. Bioeng.*; 68: 370–380.
Yano T, Harata M (1994) *J. Ferm. Bioeng.*; 77: 659–662.
Zeiser A, Bédard C, Voyer R, Jardin B, Tom R, Kamen AA (1999) *Biotechnol. Bioeng.*; 63: 122–126.
Zhou W, Rehm J, Hu WS (1995) *Biotechnol. Bioeng.*; 46: 579–587.
Zhou W, Rehm J, Europa A, Hu WS (1997) *Cytotechnol.*; 24: 99–108.

Relevant Web Sites

FDA (US Food and Drug Administration) Free access to guidance documents of the Center for Biologics Evaluation and Research, the Center for Drug Evaluation and Research and the International Conference on Harmonization (ICH).	www.fda.gov/cber/guidelines.htm www.fda.gov/cder/guidance/guidance.htm
ISA (The Instrumentation, Systems, and Automation Society) Very comprehensive site on automation, including sensors, instruments and computers, not limited to the (bio)pharmaceutical area. Free and charged access to numerous technical references, standards and news.	www.isa.org
ISPE (International Society for Pharmaceutical Engineering) Access with charge to technical and regulatory publications (technical guides, *Pharm. Eng.*, etc) for the (bio)pharmaceutical industry, including computer systems and Process Analytical Technology (PAT). Free access to discussion forums.	www.ispe.org
Dedicated forum on Good Automated Manufacturing Practice (GAMP)	www.ispe.org/gamp/
PDA (Parenteral Drug Association) Access with charge to technical and regulatory publications for the (bio)pharmaceutical industry (technical reports, *PDA J. Pharm. Sci. Technol.*, etc), including computer systems and Process Analytical Technology (PAT). Free access to comments on regulatory documents.	www.pda.org

14 Services and Associated Equipment for Upstream Processing

TS Stoll

14.1 INTRODUCTION

This chapter presents the main systems and equipment for servicing bioreactors, with the focus on pilot and manufacturing installations. The supply of culture media and gases to bioreactors is discussed first, followed by the systems for cleaning and sterilization. The last section addresses some general concepts and requirements for the design of an upstream processing plant. A discussion on the complete design of a facility (room layout, utilities, HVAC) is however beyond the scope of this chapter and can be found in Chapter 12. Similarly, the design of individual components of process equipment such as pipework, pumps and valves is reviewed only briefly. More details can be obtained in general engineering references on this topic such as Adey and Pollan (1994), Dream (1993), and Gonzalez (2001). Some other relevant citations can be found in Section 14.3.2 of this chapter.

14.2 SUPPLY OF CULTURE MEDIA AND GASES

14.2.1 Preparation and Storage of Culture Media and Supplement Solutions

Culture media and supplement solutions are normally prepared on site from dry powder materials, concentrated solutions and high-quality water. Water for injection (WFI), as defined in the US or EU Pharmacopeia, is normally used, although a quality such as USP purified water is also acceptable at this stage of the process (see Chapter 2).

Culture media and supplement solutions in small amounts and up to about 100 litres are typically prepared in mobile open containers, then sterilized and stored in bottles or disposable bags before use. The most common sterilization method is filtration (see Section 14.2.2.1); for some specific supplement solutions, heat sterilization is preferred or required (see Section 14.2.2.4). The batch size is typically chosen to be sufficient for several cell culture runs.

For volumes exceeding 100 litres, media and supplement solutions are usually prepared in a closed stainless steel tank (e.g. 316L grade, Dillon *et al.* 1992), either mobile or fixed, equipped with an agitator, internal baffles and a vent filter; powder components can be added through a manway or via an external recirculation loop. The solution is then usually sterilized by filtration and transferred either directly to the bioreactor or to a container for storage; traditionally, mobile or fixed stainless steel holding tanks have been used for this purpose; recently, however, sterile disposable plastic

Medicines from Animal Cell Culture Edited by G. Stacey and J. Davis

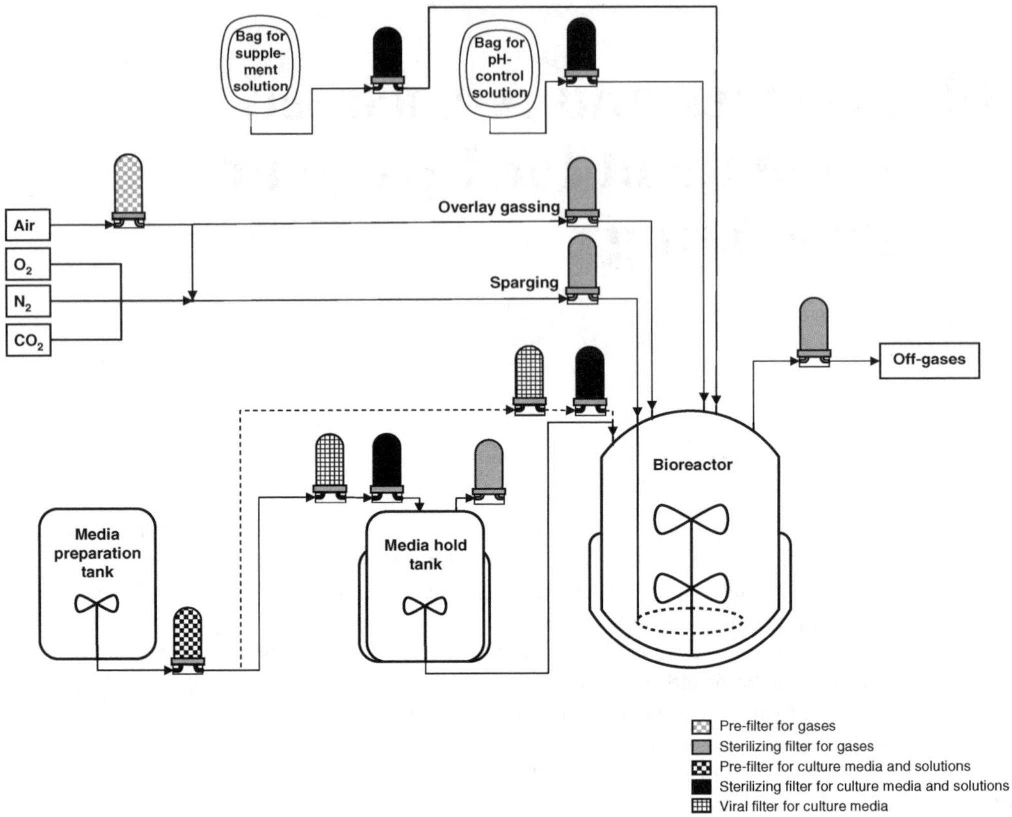

Figure 14.1 Example of sterile and viral filter barriers around a bioreactor. The characteristics of each filter type are discussed in the text (see Section 14.2.4 for gas filters). Dotted lines indicate an alternative flowpath. The size of the filter cartridges is not drawn to scale and their location is only approximate. For clarity, only the main culture medium, one supplement and one pH-control solution are represented. Not shown on the figure are the non-sterilizing vent filter on the media preparation tank, to prevent the release of powder in the room, and the tank(s) and filter(s) used for the preparation and sterilization of the two solutions (typically performed in advance in separate equipment, as discussed in Section 14.2.1).

bags, for instance in ethylene–vinyl acetate or very low-density polyethylene, placed in mobile containers for mechanical support, have become increasingly popular. With mobile containers (tank or bag), the transfer of medium to the bioreactor is performed via flexible tubing, which is connected aseptically using special tube welders; with fixed holding tanks, medium is transferred via hard pipes that have been sterilized in place (see Section 14.4). A schematic flowsheet for the preparation, filtration and storage of media is illustrated in Figure 14.1, together with the main filters.

The choice of storage container – if any – should be based upon careful economic and technical analyses. The advantages of disposable bags over tanks are as follows. First, investment and qualification costs can be reduced. Second, since bags are disposable and can be purchased ready to use, i.e. already γ-sterilized with a mounted sterilizing filter, the capacity of the cleaning and sterilization systems (see Sections 14.3 and 14.4) and of associated utilities (water, plant and clean steam) can be reduced, leading to further savings on both capital and operating costs. Furthermore, costly cleaning and sterilization validation studies can be avoided; this may be particularly advantageous in the case of a multi-product facility, where special care has to be taken with cleaning during product changeover. Additionally, since bag containers can be easily stacked, space and floor area can also be minimized compared with tanks. Finally, disposable bags also increase the flexibility of the plant (at least compared with fixed tanks) since the volumes of prepared media can be adapted easily if the process has to be modified or if a new process has to be introduced. Drawbacks are the absence of efficient mixing and poor heat transfer; bags also lead to additional handling, consumable and disposal costs. Sinclair and Monge (2002) performed a detailed economic comparison of disposable bags with fixed tanks for handling media and buffers in a 2000-l monoclonal antibody process; they claimed that disposable bags could reduce capital investment by about 20 % and cost of goods by about 9 %.

In addition to the aforementioned criteria, a key factor in determining the type of container for storage is, of course, the medium/solution batch size. Whereas, for a few hundred litres, disposable bags, mobile and fixed tanks can all be used, the last of these becomes the best or only practical solution above the 1000-l scale (although disposable bags are now available in sizes up to 3000 litres).

For *perfusion* processes, there is some flexibility with the medium batch size. Provided that the medium is stable, various combinations of prepared amount and frequency can be selected, which will then determine the most suitable container for storing the culture medium. One option is to prepare relatively small amounts (up to a few hundred litres) at a relatively high frequency (e.g. more than once a week); alternatively, larger amounts (e.g. above 1000 litres) can be produced in one batch, sufficient for more than one week of culture, and stored either in a large stainless steel holding tank or in several mobile tanks or bags that are then kept in a cold room. With a larger medium batch size, operational and handling costs are lower; however, investment costs are higher since a larger preparation tank is required, as well as either a storage tank with proper cooling capacity and serviced by cleaning and sterilization systems (see Sections 14.3 and 14.4), or a cold room for mobile containers. In either case, one holding tank or bag is normally dedicated to each bioreactor. However, only one preparation tank is required, which can serve several holding tanks sequentially since the preparation and filtration of the medium can be completed within a few hours.

For *fed-batch* processes, it is common, at least at large scale, to prepare the starting medium 'just in time', i.e. a few hours before it is needed, and then transfer it directly into the bioreactor through a sterilizing filter. By avoiding a storage tank, investment cost is reduced; however, operational costs tend to be higher since each cell culture batch requires the preparation of one batch of medium. Here too, one preparation tank, thanks to its relatively short cycle time, can serve several bioreactors. Feed and supplement solutions are either prepared 'just in time' and directly transferred to the bioreactor, or first stored, depending on the volume and feed rate (bolus *vs* continuous feeding over several days).

14.2.2 Sterilization of Culture Media and Supplement Solutions

14.2.2.1 Sterilization by filtration

Most culture media and supplement solutions for mammalian cell culture are heat-sensitive and must therefore be sterilized by filtration, as described below. Autoclaving and steam sterilization in place are also discussed briefly in Section 14.2.2.4, for the few cases where this method is applied to solutions in mammalian cell culture.

A sterilizing filter is defined as a filter that, when challenged with the microorganism *Brevundimonas diminuta*, formerly called *Pseudomonas diminuta* (ATCC 19146), at a minimum concentration of 10^7 organisms per cm^2 of filter surface, will produce a sterile effluent (FDA 1987). Viruses are not included in the definition. It should be noted that this definition represents a worst case, since no filter is expected to be exposed to this level of challenge during an actual cell culture process. Additional regulatory requirements are that the membrane should be sterilizable by steam or γ-rays, and should not affect the original product attributes. All filter materials in contact with the product or medium should therefore have minimal particle shedding during operation, a low level of extractables, pass USP Class VI toxicity testing and should not adsorb any component to a significant extent (Docksey *et al.* 1999).

Sterilizing filters for liquids are made of a hydrophilic membrane that retains, on its surface, mostly particles that are larger than the pore size, as a result of a sieving (or size exclusion) mechanism. Additionally, particles can be retained by non-specific adsorption and bridging due to accumulation of particles around an open pore (Raju & Cooney 1993). Membrane filters are often called 'absolute'; however, due to the fact that sieving is not the only retention mechanism and that these filters have a pore size distribution that is often unknown, the term 'absolute' is considered improper (Stinavage 2003).

Originally, sterilizing-grade filters were 0.45 μm-rated membranes until *B. diminuta* was identified in the 1960s and found to be small enough to penetrate these membranes at high challenge levels (Jornitz *et al.* 2002a); 0.2 or 0.22 μm-rated membranes then became the standard for sterilizing filters. More recently, several studies have shown that microorganisms such as *B. diminuta* can actually penetrate even these membranes. For example Sundaram *et al.* (2001a, b) found that several bacteria from 'natural' water sources could penetrate 0.2/0.22 μm-rated membranes at challenge levels lower than 10^7 CFU/cm^2. Bacteria with a width smaller than 0.3 μm were actually not limited in their ability to penetrate the filter; as the bacterial width increased above this value, penetration depth into the membranes decreased almost exponentially with size. Interestingly, these penetrating bacteria were on average 20–40 % wider and 40–70 % longer than *B. diminuta* (length 0.88 μm, width 0.31 μm, as determined under the specific experimental conditions). These findings confirm that sieving is not the sole mechanism governing bacterial retention. Most of these bacteria were Gram-negative, oxidase positive, and closely related to *B. diminuta*. They can be considered to be common environmental or ubiquitous organisms. Interestingly, in the latest FDA guidance on sterile products (FDA 2004), the new proposed definition of sterilizing filters does not mention the type and concentration of challenge microorganisms any more; the filter should simply 'remove all microorganisms from a fluid stream, producing a sterile effluent'. For challenge tests, the use of *B. diminuta* at a concentration of at least 10^7 CFU/cm^2 is only suggested.

Similar experiments with 0.1 μm-rated filters revealed significant performance differences between different types, in terms of microbial removal efficiency (Sundaram *et al.* 2001c); actually only those filters that had been qualified with both *B. diminuta* and *Acholeplasma laidlawii* (related to mycoplasma) consistently produced sterile effluents, whereas the other 0.1 μm filters could not prevent the penetration of bacteria. The authors therefore recommended that, apart from knowing the viable bioburden level in the process, one should also try to identify the possible microorganisms present. If there is evidence that the bioburden does not contain any microorganisms that can penetrate the filter more easily than *B. diminuta*, then the use of 0.2/0.22 μm-rated filters may be sufficient. Otherwise, when the presence of mycoplasma or of bacteria with a similar or even slightly larger width than *B. diminuta* cannot be excluded (this is likely to be the case with culture media), tighter, i.e. 0.1 μm-rated, filters should be used. Since there is currently no industry-wide or regulatory standard for rating 0.1 μm filters, those that have been functionally qualified are recommended for instance with, *A. laidlawii* (Meeker *et al.* 1992) or *Hydrogenophaga pseudoflava*, formerly *Pseudomonas pseudoflava* (ATCC 700892) (Sundaram *et al.* 2001d), a bacterium that is slightly longer but narrower than *B. diminuta*. Currently there is, however, no consensus on the benefits of systematically replacing for sterilization all 0.2/0.22 μm-rated filters by 0.1 μm-rated ones (Jornitz *et al.* 2002a).

The most common membrane materials for sterilizing filters for culture media are PVDF, nylon-66 and polyethersulfone; these materials have been modified chemically to become hydrophilic. They typically show steam sterilization resistance up to 134°C. Only a few supplement solutions, such as lipids or vitamins dissolved in ethanol, require a hydrophobic filter in PVDF or PTFE. In this case, special care should be taken to ensure that these chemical compounds do not adsorb significantly to the membrane; saturating the membrane with the solution prior to filtration or flushing the membrane with adequate amounts of ethanol after filtration are recommended.

Sterile filtration is usually performed in a flow-through mode; at small scale, flat disk membranes with a diameter in the range of 13–293 mm (1.3–670 cm^2) in a disposable capsule are used. At intermediate scales, cartridges made of a stack of membrane disks in a pre-assembled disposable capsule are commonly used, providing a filtration area in the range of 100–2000 cm^2. At larger scales, pleated membranes are used; the membranes are supported on a non-woven polyester support, folded to form pleats, wrapped around an inner core, and sealed by two end caps. Support cages are usually placed around the membrane to protect it from mechanical damage (Jornitz *et al.* 2002b). Cartridges are available in a variety of sizes (from 2 to 40 inches nominal length), one

Figure 14.2 Filter cartridges and housings for sterization of gases and liquids. From left to right: 10" cartridge for gases with a collector for condensates; 20" cartridge for liquids (e.g. culture media); 10" cartridge for utilities (e.g. WFI), to be mounted horizontally in line (courtesy of Millipore).

10-inch cartridge providing an effective filtration area of around 0.7 m^2. Cartridges are usually installed in a fixed stainless steel housing (316L grade, Dillon *et al.* 1992) prior to sterilization, as illustrated in Figure 14.2. Nowadays, they can also be purchased pre-assembled and sterilized in a disposable capsule, eliminating the need for cleaning and sterilization of the housing, as well as for the associated validation studies. Filtration can be performed either at a constant differential pressure (typically 1–1.5 bar) or at a constant flow rate. In the first case, the feed solution is pressurized with a compressed gas; in the second case, a pump is connected on the upstream side of the filter. During filtration, pressure or flow rate as well as temperature are commonly measured continuously.

The use of a pre-filter to remove larger particles is recommended, particularly if the medium contains protein hydrolysate or serum, and if 0.1 μm sterilizing filters are used; the required surface area of the sterilizing filter can thus be minimized. Depth filters are typically used for this purpose, thanks to their ability to retain large loads of particles before clogging; they consist of a porous structure of fibres, which form an irregular three-dimensional net. Typical filter materials are polypropylene or cellulose acetate ester fibres, sometimes combined with glass microfibres in a double-layer structure; the membrane is supported by a polypropylene cage. Electrostatic interactions, adsorption, diffusion and impaction of the particles on the filter material are the main retention mechanisms; consequently, particles smaller than the characteristic pore size or dimension of the filter can be captured. These are actually retained not only at the surface (as with membrane filters) but also inside the filter (Raju & Cooney 1993).

14.2.2.2 Removal of viruses

Several cases of viral contamination of bioreactor contents by adventitious viruses have been reported in the past few years. The sources may be raw materials, even process gases, GMP failures by operators, or equipment barrier breaches, and are, in any case, very difficult to identify (Garnick 1998). Viral contamination can often spread to several tanks before it can be detected, unlike bacterial contamination. Liu *et al.* (2000) estimated an overall cost of 6–8 million US$ for a viral contamination in a 10 000-l bioreactor or, assuming an incidence of two contaminations per 1000 batches, an average of 11 000–15 000 US$ per start. There are thus strong incentives to install viral barriers in the upstream process, around the bioreactors, controlling the passage of all liquids and even gases.

Raw materials may be screened for the presence of viruses, although this cannot be applied to all of them, for practical and economical reasons; also, contamination by rodents is typically non-homogeneous and mostly would not be detected. Serum can be γ-irradiated to inactivate potential virus contaminants but this is feasible only at small scale. UV, various chemicals, and low pH, as well as high pressure, have been used successfully to inactivate certain viruses in serum or media. A detailed discussion of serum contamination can be found in Chapter 4. However the two most practical and general viral barriers at large scale, for media that cannot be heat-sterilized, are (Liu *et al.* 2000; Sofer 2003):

(i) High-temperature, short-time (HTST) treatment, i.e. the principle of pasteurization, which inactivates viruses but does not perform real microbial sterilization (a sterile filtration of the medium will still be required). For this purpose, the solution is passed through a circuit consisting of a first heat exchanger, where it is brought to a temperature of, typically, 70–90 °C; it then flows through a holding section for 30 s–2 min, where inactivation takes place, and finally through a second heat exchanger where it is rapidly cooled down. Such a treatment is capital-intensive and implies significant automation and validation costs. Special attention has to be paid to heat-resistant viruses such as parvoviruses.

(ii) Virus-retentive filters, rated at 40 nm or less, in addition to bacterial/mycoplasma-retentive filters (Martin *et al.* 1994). Liu *et al.* (2000) reported the successful use of such a filter, consisting of an ultrafiltration membrane of hydrophilic regenerated cellulose. They observed log reduction values larger than 3 for the 20-nm minute virus of mice, 4.5 for the 28-nm bacteriophage ΦX174, 8.8 for *A. laidlawii*, and 9.2 for *B. diminuta*. The main retention mechanism was size exclusion. The viral filter could be autoclaved and had no detectable effect on the growth promotion properties of the culture medium. This option is particularly suitable at small scale, where HTST equipment may not be justified. Although virus filters are capable of removing bacteria and mycoplasma, it is recommended, if sterility is claimed, to add a 0.1 μm sterilizing filter downstream of the viral barrier. Furthermore, since viral filters are very sensitive to plugging, a pre-filter is required upstream. An example of sterile and viral filter barriers around a bioreactor is shown in Figure 14.1. In this case, a specific viral filter is used only for the culture medium. Although an additional viral barrier for gases could also be installed, as proposed by Liu *et al.* (2000), sterilizing grade hydrophobic filters (0.2 μm) may be adequate since they are also able to retain much smaller particles such as viruses (see Section 14.2.4). The location of the filters is approximate; usually all sterilizing filters for liquids and gases (inlet and outlet) are located on, or very close to, the bioreactor. (For further discussion of virus-retentive filters, see Chapter 19, Section 19.4.2.2).

Additionally, routine testing of product-containing supernatants for the presence of adventitious viruses is recommended, particularly once full production scale is reached, in order to minimize the risk of spreading any viral contamination to downstream processing equipment (Garnick 1996).

14.2.2.3 Validation of filtration

The primary goal of validation of a filtration process is to prove its ability to remove contaminants; this is performed in different steps (Docksey *et al.* 1999; PDA 1998a). First the performance of the filter *per se* should be validated via bacterial-retention testing (usually using *B. diminuta*, as discussed in Section 14.2.2.1) as well as extractable, compatibility and adsorption studies; it is common to have the tests conducted by the filter manufacturer at laboratory scale. Furthermore, the manufacturer should provide documents on its production controls and quality assurance system.

Since the performance of sterilizing filters can be affected by both process conditions (flow rate, differential pressure, temperature, etc.) and the physico-chemical characteristics of the filtered solution, validation should then be performed on site, to show that under these particular conditions, the sterilizing-grade filter meets the requirements of the bacterial challenge test and provides a sterile effluent (PDA 1998a). For this purpose, media fill tests, similar to those performed on sterile pharmaceutical operations (Pfohl & Stärk 2003), are recommended, where the whole upstream process is simulated with cell-free culture medium. A light formulation of the culture medium can be used for this purpose, provided that it is able to promote the growth of the microorganisms recommended for sterility testing (European Pharmacopoeia 2002; United States Pharmacopeia 2003) as well as those identified in the local environment. At the same time, these media fill tests can be used to validate the sterilization conditions of all the culture equipment (see Section 14.4); worst-case values for filtration parameters such as pressure, flow per unit area, filtration time and processed volume should be tested.

During routine operation, regulatory health agencies require that the integrity of sterile filters is tested both before and after filtration (FDA 1987); this is usually performed using so-called *diffusion* or *bubble-point* tests (Martin *et al.* 1994; PDA 1998a; Kuriyel & Zydney 2000), which are non-destructive and much faster than a bacterial challenge (which in any case could not be used in a production environment). In the first type, diffusion of air across the wetted membrane is measured, either directly, at a constant pressure (forward flow test), or from the pressure decay over a specified period of time after pressurization (pressure hold test). A comparison of both measurements can be found in Spanier (2003). In the second type, the pressure at which liquid-filled pores of the membrane are first intruded by a gas is measured. Results of the integrity test can then be correlated with the bacterial-retention properties of the filter. Because of the strong influence of the solution physico-chemical properties on the filter efficiency, it is recommended that integrity-test specifications are used that have been determined with a solution (e.g. culture medium) as close as possible to that of the process. At the same time, these integrity tests can verify proper installation of the filter cartridge and air-tightness of the system. One should, however, keep in mind that a successful test provides only a strong indication that the filter functions properly, with no leak; it does not guarantee the removal of a specific bacterial challenge, since size exclusion is not the only retention mechanism of sterile filters (Jornitz *et al.* 2002a).

For large-area filters, diffusion tests are preferred; for small area, the bubble-point method is recommended, being more accurate. For viral filters, a so-called liquid intrusion test is recommended (although other methods are available–see Chapter 19, Section 19.4.2.2); depth filters cannot be integrity tested (Kuriyel & Zydney 2000). Compact automatic integrity testers are now widely available, directly from most filter manufacturers (Martin *et al.* 1994). Alternatively, at pilot and large scale, the testing recipe can be incorporated into the plant automation system so that the test can be performed *in situ*.

14.2.2.4 Sterilization by heat

For a few liquid supplements used in cell culture, such as concentrated glucose, protein hydrolysate, or antifoam solutions, sterilization by heat instead of filtration is the best or only practical solution. Small amounts that can be stored in bottles are best sterilized in an autoclave. Large amounts are commonly sterilized directly in the stainless steel tank used for their preparation, in a *batch* operation. For this purpose, the solution is heated and then held at the sterilization temperature via steam injection in the tank jacket. In the case of large vessels, additional heating by direct steam injection onto the surface of the solution is recommended. The dilution of the solution due to water condensation must then be taken into account. Special conditions for the sterilization of vessels filled with concentrated nutrient solutions, taking into account their lower vapour pressure and lower activity coefficient, are discussed by Junker *et al.* (1999). An alternative method, although less common, is *continuous* sterilization, similar to an HTST treatment for viral inactivation (see Section 14.2.2.2) although at higher temperature. The advantages over a batch mode are better temperature control (which may or may not be critical), lower steam consumption (typically one fourth or one fifth) and a faster cycle time (Raju & Cooney 1993). The main drawbacks are the higher complexity of the equipment and the requirement for a second tank for post-sterilization storage.

14.2.3 Source of Process Gases

Process gases such as oxygen, carbon dioxide and nitrogen are often produced externally and supplied to the manufacturing site via cylinders or tanks, where they are stored at high pressure, or in a liquid form. Compressed process air, on the other hand, is usually produced on site. In addition to aeration of bioreactors, process air is often used in small amounts for performing transfers under pressure and breaking vacuum after steam sterilization (see Section 14.4.3). For the production of compressed process air, inlet air is first circulated through coarse filters to remove particles and then enters oil-free compressors. Due to adiabatic compression, the air leaving the compressors can easily reach a temperature around 150–200 °C; it is therefore first cooled down to a temperature of typically 40–80 °C via a heat exchanger before being processed through a rotary dryer and stored in a so-called receiver tank, at a typical differential pressure of 2–3 bar, to stabilize flow and pressure variations in the system. Before entering the distribution network, it flows first through a series of filters (e.g. an active charcoal filter and one or two pre-filter(s) (0.2–0.3 μm) made of cellulose, polypropylene or glass fibres). In older facilities, where oil-lubricated compressors are used, a coalescing filter is required to remove oil particles before the pre-filter(s) (Martin *et al.* 1994).

14.2.4 Sterilization of Process Gases

14.2.4.1 Sterilization by filtration

Process gases can contain a mixture of microorganisms such as moulds, yeast, bacteria and possibly phages, and must therefore be sterilized when fed to a bioreactor; exhaust gases must be treated in the same way in order to avoid any back contamination. For this purpose, filtration using hydrophobic membranes is recommended. One of the main challenges of gas filtration is the duration of the process, i.e. from several days for a fed-batch culture to several weeks for perfusion cultures, instead of a few hours or less for the filtration of culture media. Hydrophobic filters are also required on tanks used for storing sterile media and solutions (see Section 14.2.2.1).

Originally, depth filters made, for instance, of glass-wool fibres, were used for the sterilization of gases. Since the early 1970s, they have been replaced by flat or pleated microporous membrane filters, which are much more reliable under both dry and wet conditions, do not release fibres and can be used over more than 50 sterilization cycles (Leahy & Gabler 1984). Typical materials are PVDF and PTFE.

For microorganisms in gases, the main retention mechanism of such filters is adsorption, while sieving plays only a secondary role. It has actually been shown that a 0.22-μm PVDF membrane filter can quantitatively remove particles as small as 0.03 μm in a gas stream (Keating *et al.* 1992). This means that even some viral particles can be removed with such a filter; there is thus no incentive, unlike with liquid filtration, to consider tighter filters, say with a pore size of 0.1 μm. Filters rated at 0.22 μm actually represent the current standard of sterilizing filters for gases. Like hydrophilic filters, hydrophobic filters should be steam sterilizable or autoclavable, have minimal particle shedding during operation, and all the materials in contact with the product or medium should have low extractables, pass USP Class VI toxicity testing, and not adsorb components.

For small-scale applications (tanks up to 20–30 litres), flat disk membranes with typically up to 20 cm^2 of effective filtration area are used in disposable capsules. For larger applications, filter cartridges are used, made of one or two layers of a hydrophobic membrane pleated between two layers of polypropylene support material. These filters can undergo multiple sterilization cycles in either direction of flow or be autoclaved repeatedly. For this reason, they are normally kept for several cell culture batches, unlike hydrophilic filters, which are commonly discarded after use, at least during GMP operations. For tanks up to about 1000 litres, cartridges in disposable capsules (4-inch, equivalent to a filtration area of about 1400 cm^2) are available; for larger volumes, larger cartridges (10- to 40-inch; about 0.7 m^2 for a 10-inch cartridge) are normally installed in a fixed stainless steel housing (Fig. 14.2). If needed, several of them can be combined in one housing.

In the particular case of off-gases, a hydrophobic pre-filter with a typical nominal pore size of 1–2 μm may be required to remove aerosolized particles and liquid droplets (Martin *et al.* 1994). This will extend the service life of the final sterilizing filter. Formation of condensate in the filter (which can lead to filter blinding) can be prevented either by cooling the exhaust line upstream of the filter or by heating the filter housing. In small-scale applications, the condensate can be retained in a simple catch-bottle.

Process gases should be supplied to the bioreactor at a pressure high enough to overcome the pressure within the tank and any other resistance in the exhaust line; the inlet differential pressure is hence normally in the range of 2–3 bar. Air is typically sparged from the lower part of the bioreactor, together with any required oxygen and carbon dioxide, at a typical flowrate of 0.01–0.03 vessel volumes per minute (vvm). Additionally, air can be fed onto the liquid surface at a typical flowrate of 0.02–1 vvm, to maintain some overpressure in the tank and to enhance removal of off-gases. In this case, two filtration cartridges should be used for inlet gases, taking into account the different flowrates of sparged and overlay gases. The filter arrangement for gases is shown in Figure 14.1. The sizing of cartridges should be based upon pressure-drop data provided by the manufacturer; these data depend strongly on the membrane type and filter design. An economic analysis of the energy costs (linked to differential pressure) and filter costs is discussed by Keating *et al.* (1992).

14.2.4.2 Validation of filtration

The same principles of validation discussed for liquid filtration (Section 14.2.2.3) apply to gas filtration. Several particle-retention tests have been developed to validate hydrophobic membrane filters (Keating *et al.* 1992; Rowe *et al.* 1996). Although currently acceptable methods include chemical aerosol and physical retention tests, the most representative challenges of sterilizing filters for gas streams are those based on biological particles. Test conditions should be chosen in order to produce a worst-case challenge of the filter, in the spirit of the FDA guidelines (1987). For this purpose, three types of test are recommended: (i) aerosol-retention tests based upon a very small and shear-resistant virus such as bacteriophage ΦX174 (28 nm in diameter); the test should be performed at relatively high flow rate and humidity (65 % or higher); (ii) an aerosolized bacterial grow-through challenge, at high humidity and over an extended process time (e.g. 21 days), in order to determine if the microorganism can grow through the filter membrane. A test organism such as *B. diminuta* should be used; (iii) a liquid retention test using *B. diminuta* under conditions similar to those applied to sterilizing *hydrophilic* filters (see 14.2.2.1). The purpose of this last test is to verify retention in the event that the filter should inadvertently be wetted out. All these tests should of course be performed at laboratory scale.

In addition to bacterial and viral retention, an important property of hydrophobic filters is their long service life. Consequently, their long-term resistance to a variety of stresses, such as pressure surge, steaming and exposure to air at high temperature should be tested (Keating *et al.* 1992; Rowe *et al.* 1996); the aforementioned retention tests should thus be repeated after exposing the filters to such stresses.

For routine use, a specific integrity test that can be correlated to the filter retention capacity should be performed. The purpose is to verify both the integrity of the membrane and, at large scale, the proper assembly of the cartridge in the housing. Originally, tests using solvents as wetting agents were performed. These have now been replaced by a water-intrusion test, where the housing is filled with water and pressurized, and the amount of water intruding into the membrane over a fixed period of time is then measured. The results have been shown to correlate well with the bacterial retention capabilities of the filter (Rowe *et al.* 1996). As with hydrophilic filters, automatic instruments for integrity testing are now available; alternatively, the testing recipe can be incorporated into the plant automation system so that the test can be performed *in situ*. For reliable data it is, however, essential that the dead volumes of piping and filter housing are minimized.

14.3 CLEANING OF UPSTREAM PROCESSING EQUIPMENT

14.3.1 Introduction

Cleaning of manufacturing equipment has become one of the most important issues in the pharmaceutical and biotech industry; thorough cleaning after each production batch is actually a regulatory requirement. The primary purpose is to reduce or eliminate the possibility of cross-contamination. For cell culture specifically, cleaning is essential to remove any accumulated inhibitory chemical species, thereby ensuring optimum growth performance during the next batch. Cleaning also prepares the equipment for sterilization (see Section 14.4), before which it is important to remove certain contaminants such as chlorides and proteins.

Typical 'soil' types in cell culture equipment are cells and cell debris, proteins, lipids and carbohydrates; in bioreactors these compounds can be present above the normal liquid level as a result of dried foam. In vessels where nutrient solutions are heat-sterilized *in situ*, denatured proteins and caramelized sugars may also be encountered. Bacteria may also be present in the cases of contamination during a culture, or following a long period where some water has remained in an idle vessel (this can of course be avoided by storing the vessel under sterile conditions or by drying it completely at the end of the cleaning cycle). Cleaning should also remove contaminants such as grease and other chemical residues resulting from maintenance or construction operations.

For this purpose, cleaning-in-place (CIP) techniques have been developed and are now widely applied, particularly to large-scale equipment. These techniques consist of cleaning the process equipment with little or no dismantling, using pressurized and high-temperature solutions delivered by a CIP unit or skid; the process is often highly automated. CIP was first developed in the mid-1940s for the dairy industry, and started to be applied to pharmaceutical facilities in the mid-1970s and, later on, to biotechnology facilities (Seiberling 1992). Almost all pieces of equipment such as vessels, piping, pumps, valves, mixers, etc., can now be cleaned in place. By contrast, manual cleaning of equipment, involving some dismantling, is often termed cleaning-off-place (COP).

CIP has several advantages over COP (Adams & Agarwal 1990). First, the fact that little or no dismantling is required leads to a lower risk of injury and no exposure of personnel to any hazardous chemical or biological materials; there is also a lower risk of improper reassembling. Second, thanks to automation, personnel costs are reduced; the cleaning process is also very reproducible and thus reliable, and it is consequently easier to validate. Furthermore, operating costs are often reduced, thanks to more efficient use of water and cleaning solutions during optimized flow-through and recirculation cycles. Finally, CIP is typically faster than manual cleaning, resulting in a shorter downtime of the production equipment.

The discussion below focuses on the design and operation of CIP systems for large-scale upstream processing equipment. COP and cleaning of external equipment surfaces and rooms are not discussed here.

14.3.2 Design of Cleanable and Sterilizable Process Equipment

14.3.2.1 Principles

CIP is only possible if all the process equipment parts including piping have been designed for this purpose. A similar comment can be made regarding the sterilization-in-place (SIP) of equipment, described in Section 14.4. Actually there are several design requirements in common for cleanable and sterilizable process equipment; a key example is drainability, required to ensure that both cleaning and rinse solutions (from CIP) as well as condensed water (from SIP) are removed efficiently. Consequently, both CIP and SIP systems should be designed and constructed together with the process equipment, in an integrated way; this can actually lead to significant savings in equipment cost (piping and valves) (Marks 1999).

The main recommendations for the design of cleanable and sterilizable upstream processing equipment are summarized together below; more details can be found in Adey and Pollan (1994), Gonzalez (2001), Oakley (1994), Purnell (2003), Stewart and Seiberling (1996), Thompson (1994) and Wood (1999). Interesting practical examples of integrated design for CIP and SIP are also illustrated in Marks (1999).

14.3.2.2 Materials

For large-scale applications, pipes and tanks are typically made of austenitic stainless steel (Dillon *et al.* 1992; Gonzalez 2001); AISI 304 type is suitable for most applications where there are no harsh process conditions. Corrosive environments can however cause pitting, which results in surfaces that are difficult to clean and sterilize. Therefore the 316L grade and, in some extreme cases, titanium-stabilized (316Ti) or nickel-enriched (Hastelloy®) steels may be preferred. Other non-ferrous materials (zinc, cadmium, lead, etc.) are generally not acceptable. In general, all surfaces in contact with the product or the cells should be smooth, non-porous and free from pits, cracks and crevices to ensure cleanability and sterilizability. For bioreactors and piping, which must be both sterile and perfectly clean, mechanical polish should be applied, followed by electropolishing, to reach a surface finish of 10–15 μin Ra. For non-sterile applications, a lower surface finish of around 30–50 μin Ra is however acceptable (Gonzalez 2001; Thompson 1994).

Some plastic materials, mainly PTFE, EPDM and PVDF, are suitable for pieces of equipment such as piping, valve diaphragms, seals, gaskets and pump parts. All materials to be sterilized should be able to withstand a temperature of at least 130 °C and a differential pressure of at least 2 bar over several sterilization cycles. This temperature actually puts a heavy burden on the aforementioned pieces such that they have to be replaced regularly. Under high-pressure conditions, plastic tubing should be reinforced with stainless steel. Silicone rubber is acceptable for small-diameter tubing; in this case, however, it is common practice in GMP facilities to discard it after each batch without any cleaning. Glass is also a good material for its cleanability and ease of inspection; it is generally used either for small-scale or highly corrosive applications, but is not recommended for SIP (see Section 14.3.2.3). Polymers to avoid include those with a porous surface, such as natural rubber, and those that tend to leach out additives, such as low-density polyethylene, neoprene and PVC.

14.3.2.3 Vessels

All vessels should drain from the very lowest point in order to remove rinse solutions during CIP, and condensed water during SIP, efficiently. For CIP, this significantly reduces the amount of solution required for rinsing; for SIP, any accumulation of condensed water is undesirable since it would create 'cold' spots, which would not be properly sterilized. Similarly, sensor pockets for pH and pO_2 probes should slope downwards into the vessel to ensure proper drainage. The top of the vessel should permit a person to enter or, on smaller vessels reach into every corner for inspection and manual cleaning if necessary, either via a removable top or a manway in the top dishing. All corners, whether horizontal or vertical, should be rounded with a minimum radius of 1 inch. Although the vessel aspect ratio (height/diameter) should be mainly determined by gassing and stirring considerations, one should be aware that a higher ratio will require higher flow rates for cleaning and rinse solutions. In this case, surfaces are also less accessible to top-mounted spray balls (see Section 14.3.3.2).

Vessels to be sterilized must be able to withstand the required steam pressure with some safety margin; additionally, they must be equipped with the appropriate relief devices for safety. In practice only stainless steel vessels are sterilized in place with steam; glass vessels can explode and should therefore be autoclaved if possible; alternatively, large glass vessels can be sterilized in place with an external steel shield or a plastic film to contain the glass in the event of an explosion. Formation of air pockets should be avoided as they reduce the efficiency of sterilization. Whereas mechanical removal of air via pre-vacuum cycles is common in autoclaves and lyophilizers, it is not in tanks; instead, bleed valves should be installed at all critical locations where air would be likely to be entrapped (Agalloco 1990).

14.3.2.4 Piping

The design of pipework is one of the most critical factors in the cleanability and sterilizability of a plant. As a minimum, for cell culture, the ASTM 270 standards should be applied (Gonzalez 2001). The preferred method of joining pipes is welding. Welds should be ground smooth and polished to the same standard as the other surfaces. For this purpose, automatic orbital welding is recommended wherever possible. Alternatively, where frequent dismantling is required for maintenance and for general flexibility of the process equipment, pipes can be connected via sanitary clamp-type or threaded joints.

Piping bends should have a radius not less than the outside pipe diameter. All pipework should be installed with a fall of approximately 1 % in the direction of a drain point in order to be fully drainable and to prevent the formation of air pockets. Steam should thus be introduced at the highest point(s) and the condensate removed at the lowest point(s).

If possible, dead legs should be avoided or at least not exceed 2–3 pipe diameters, rather than 6, as usually applied in networks of water systems. All dead legs should be sterilizable via a dedicated valve with steam supply and should be fully drainable. Completely vertical ones should, however, be avoided, since entrapped air may prevent the cleaning solution from reaching the upper portion of the fitting.

14.3.2.5 Valves

Diaphragm valves are the most common type in biotechnology because they drain freely and lack internal cracks; in horizontal pipes, diaphragm valves should be installed at an angle of about 15° to ensure self-draining. These valves can also be easily sterilized. For this purpose, steam should circulate through them completely, to a drain or trap on the other side. Compression (or bellows-type) valves of aseptic design can also be used, for instance as harvest valves at the bottom of a bioreactor. The plastic materials used in the valves must be unpigmented and contain no extractables. For hygienic but non-sterile operations, certain types of butterfly, ball and globe valves can also be used. In all cases, crevices must be avoided, for instance by using PTFE lining or proper machining of surfaces.

14.3.2.6 Transfer systems

For efficient plant utilization, it should be possible to perform cleaning and sterilization operations at the same time on adjacent pieces of equipment, without any risk of cross-mixing. This is actually part of the *3-A Accepted Practices for Permanently Installed Sanitary Pipelines and Cleaning Systems* (Stewart & Seiberling 1996). For instance it should be possible to clean a bioreactor together with its transfer line immediately after use, and then to sterilize it in preparation for the next batch, while a culture is running in an adjacent vessel.

A practical solution to prevent cross-mixing is the use of block-and-bleed valve arrangements, which ensure that there are always two valves between incompatible fluids; any leakage through a valve would be diverted to the drain (Thompson 1994). A simple example is illustrated in Figure 14.3.

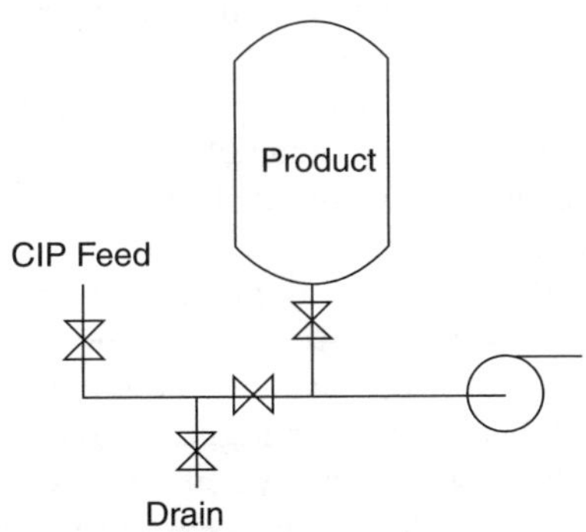

Figure 14.3 Block-and-bleed valve arrangement. This arrangement ensures that there are always two valves between incompatible fluids. Any leakage through the valve is diverted to drain. The system can be made more secure by appropriate use of automated valves (reproduced with permission from Thompson, 1994).

To isolate a sterile area from a non-sterile area efficiently, two valves should be installed where possible, since bacteria can grow between valve membranes. In cases where contamination risks are particularly critical (for economical or biosafety reasons), the use of steam block barriers is recommended, where pressurized steam is fed into the space between two valves (Adey & Pollan 1994).

In large-scale facilities, numerous transfer lines between several bioreactors are needed, all of which must be sterilizable and cleanable; consequently, the number of valves required can increase tremendously, particularly if block-and-bleed arrangements and double valve barriers must be incorporated at each transfer route.

Manual transfer panels represent one practical solution to this challenge and have become very popular in the biotechnology industry (Louie & Williams 2000). A manual transfer panel is composed of a series of stainless steel nozzles or ports welded onto a vertical plate. At the rear of the panel, the ports are connected by hard piping to the inlets and outlets of process vessels and of other process functions. At the front, the panel serves as a 'switchboard': two ports can be connected manually via pivoting elbows or short U- or J-shaped pipes ('jumpers') to select the desired route. Process fluids, as well as utilities such as WFI, CIP supply and return solutions, clean steam, and condensate, can all be circulated via such transfer panels. For sterile transfers, the jumper is installed before sterilizing the whole line; after use, the line is cleaned in place before the jumper is dismantled. Jumpers can be equipped with a safety device called a 'proximity switcher', which communicates the actual position of the jumper to the plant automation system, providing confirmation that the correct route has been selected before the transfer is initiated, thus preventing accidental mistransfers. The design of transfer panels and jumpers is discussed in detail elsewhere (Huang *et al.* 2000; Louie & Williams 2000). Complex transfer panels can even permit a number of simultaneous transfers, yet ensure that there are no cross connections between any two different streams (Seiberling 1992). A simple example, for the selection of the transfer route between one of three source tanks and one of six receiving tanks, is illustrated in Figure 14.4. The use of transfer panels in a large-scale upstream processing plant is also illustrated in Figure 14.11 (Section 14.5.2).

The advantages of manual transfer panels are many. In brief, they provide a true physical isolation of a transfer route, thereby greatly reducing the risk of cross-mixing; they are both highly flexible and adaptable, since ports and piping can be modified without interfering with basic operations. By centralizing process operations, they make efficient use of process space and operator labour. Maintenance is minimal and can be largely confined to non-classified areas behind the panel. These transfer panels do, however, have a few minor limitations. First, setting up a process route is a manual operation and may require operator travel or coordination between different process areas. Second, transfer panels must be fabricated with very narrow tolerances on dimensions in order to ensure a perfect fit of the jumper to the ports.

Ring transfer panels offer an alternative to manual transfer panels, with similar functions. They consist of a compact arrangement of T-, Y-, X- and corner-type diaphragm valves installed in a ring configuration (Wilde 1998). Process fluids as well as utilities can thus be transferred from any inlet valve to any outlet valve without pocketing, and independently of any SIP and CIP cycles taking place on adjacent pieces of equipment. The inlet fluid is split to flow through the entire ring before exiting through the outlet valve (Figure 14.5). The main advantage of this system is the possibility of fully automated operation. The main drawbacks, however, are the higher maintenance cost (particularly due to membrane replacement) and the fact that the system, unlike a manual transfer panel, does not guarantee absolute isolation of a flow path in case of membrane leaks or failure of the control system.

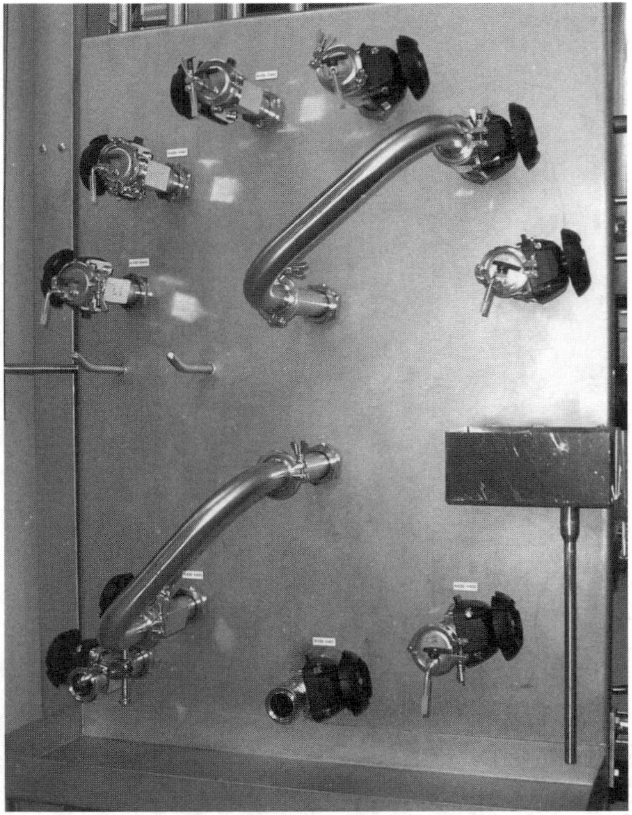

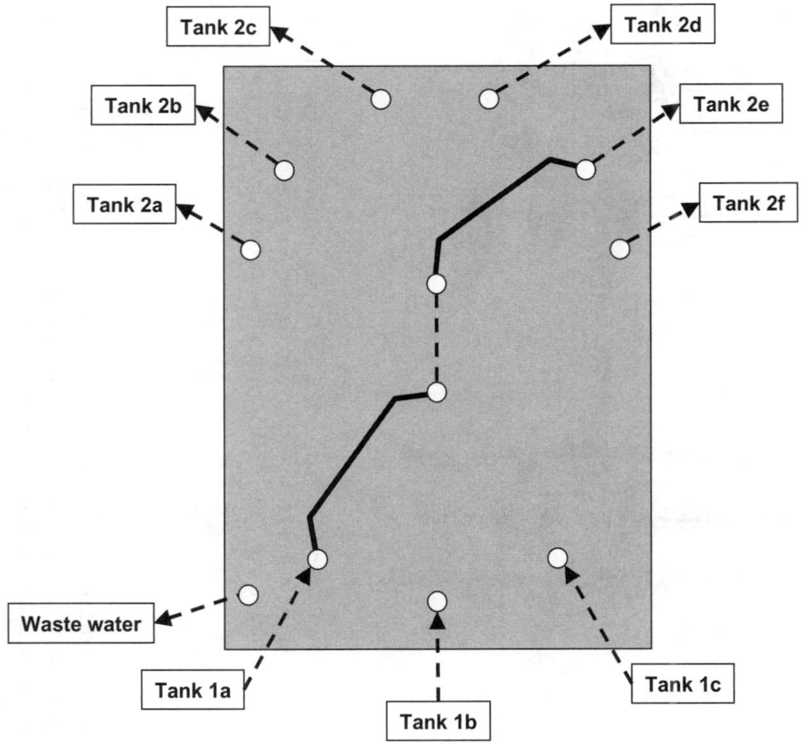

Figure 14.4 (a) Example of a manual transfer panel. (b) Diagram of the manual transfer panel shown in (a). The panel is used for the selection of the transfer route between one of three source tanks (1a–1c) and one of six receiving tanks (2a–2f). Here jumpers are not equipped with proximity switchers; however, each port has a valve that has to be manually opened, after installation of the jumpers, before transfer can take place. When a port is not connected, it is closed by a cap equipped with a small drain; any residual fluid or pressure can thus be released gradually before the cap is removed. The hard pipe connections at the rear of the panel are represented by dotted lines on the diagram (courtesy of Novartis).

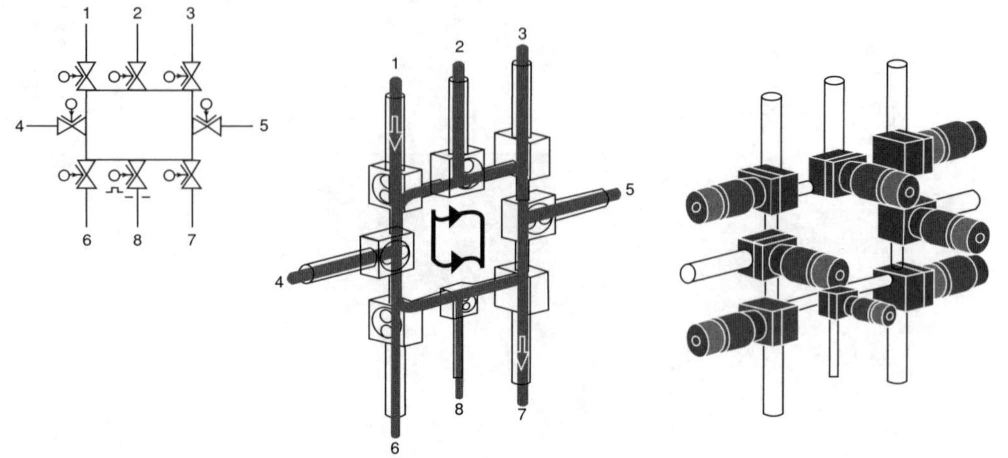

Figure 14.5 Example of a ring transfer panel. Special T-, Y-, X- and corner-type diaphragm valves allow a ring design without dead space. Any inlet can be connected to any outlet; for instance ports 1 to 5 can be inlets, port 6 can be connected to a drain, port 7 to the CIP return line and port 8 to a steam trap (courtesy of Bioengineering).

14.3.2.7 Pumps

Various pumps suitable for hygienic duties are available, the most common ones being diaphragm, centrifugal, rotary lobe, peristaltic and helical displacement pumps, all of the so-called sanitary type. Most pumps are equipped with mechanical seals between the impeller and the gearbox; this path may be steam traced for sterile applications or, alternatively, be replaced with a magnetic coupling. Ideally, these pumps should be self draining; centrifugal pumps, which lack this property, must be flushed with water. For sterile systems, fluid transfer via gravity or via pressure applied from the top of the source vessel is also commonly used.

14.3.3 Design of CIP Systems

14.3.3.1 Design of a CIP unit

A CIP system is typically made of one or several CIP units or skids, each being dedicated to a set of process equipment parts (e.g. tanks and associated piping) to be cleaned in place. The number of units in a plant should be based on the maximum rate of utilization of the process equipment and the estimated duration of one CIP cycle. Usually the units are independent although they are fed with concentrated cleaning and rinse solutions from common storage tanks.

A CIP unit consists of one or several tanks used to prepare and hold the various cleaning and rinse solutions. Each piece of process equipment to be cleaned is connected to the CIP unit via supply and return lines. Depending on the way the solutions are prepared and circulated, CIP units can be broken down into two main types.

In flow-through units, the solutions are discarded directly after a single passage through the equipment to be cleaned; these units are typically small and very simple. They are characterized by low capital investment but high operating costs, due to the large consumption of chemicals and energy for heating (Greene 2003). In practice, they are used when recirculation should be avoided for biocontainment reasons or when some size limitations or layout considerations of the facility make them the only practical choice.

In recirculating units, solutions are recirculated during one step or even reused, either in a subsequent step of the same CIP cycle or during the next one, thereby minimizing costs for water, chemicals and waste treatment. In fact, since a contact time between surfaces and the cleaning solution of up to 1 hour may be required, recirculation is essential for economical operations, particularly at large scale. Currently, recirculating units are thus the predominant type in the pharmaceutical and biotechnology industry. They are, however, designed so that some steps can also be run in a flow-through mode; for instance the solution used in the pre-rinse step (see Section 14.3.4), containing large amounts of cell debris, should flow only once through the equipment; in a second step, however, the cleaning solution can be recirculated several times to increase exposure time. The final rinse of a CIP cycle can also be recovered to prepare the pre-rinse solution of the next cycle.

Numerous versions of recirculating CIP units have been developed and are discussed in detail elsewhere (Adams & Agarwal 1990; Seiberling 1986, 1992; Stewart & Seiberling 1996). In the simplest version, one unit is made of one main tank to hold the cleaning solution, which is usually prepared directly in this tank by mixing water with the concentrated cleaning solution that is pumped from a main storage tank or from drums. The CIP tank should be designed in such a way that any air returned from the equipment being cleaned can be disengaged. It should also be equipped with some spray-balls (see Section 14.3.3.2), so that it can be self-cleaning. The same tank can be used in sequence to hold an acidic solution, if used during the rinse step (see Section 14.3.4). Multiple-tank CIP units have also been developed, to store solutions for different steps separately, so that each can be reused in a subsequent step of the same or of the next CIP cycle.

The other main components of CIP units are a circulating pump and a heat exchanger (or a tank with a heating jacket) to heat up the cleaning solution. The return of the cleaning and rinse solutions to the unit can be accomplished by gravity, a pump, an eductor or some combination thereof. These different options are discussed by Adams and Agarwal (1990), Greene (2003), and Stewart and Seiberling (1996). CIP units are typically highly automated, with a local or distributed controller for the sequence of all the operations, the flow rates and temperature as well as the monitoring of conductivity or pH during rinse steps. An example of a CIP system with one multiple-tank recirculating unit for the cleaning of a group of process tanks is shown in Figure 14.6.

14.3.3.2 Design of the CIP circuit

In traditional CIP systems, solutions were circulated via numerous supply and return lines, separate from process lines; however, as previously mentioned (Section 14.3.2.6), it is desirable, for economic and practical reasons, that a major portion of the tubing, fitting and valves be shared between the CIP, SIP and process circuits. CIP circuits have thus evolved towards a higher integration level with that of the process, as reviewed in Seiberling (1992); manual and ring transfer panels are particularly useful for this purpose (see Section 14.3.2.6). An example of shared CIP and SIP piping is illustrated in Figure 14.7.

A vessel is best cleaned using one or several 'spray balls' or 'spray heads', generally mounted on its upper internal part. The cleaning solution can thus reach and flow on internal surfaces without the need to fill up the vessel, which would take a very long time and be costly. Spray balls are typically sized for a flow rate of 3–5 m^3/h each, and a pressure drop of 1–2 bar (Greene 2003). Above 2.5 bar, aerosols can form, which should be absolutely avoided. Spray balls can be either static or dynamic; in this latter type, the head rotates during cleaning, thanks to a mechanical or hydraulic drive. Dynamic spray balls are typically more efficient, covering larger surface areas, but they are also more expensive and more prone to failure. Static spray balls have a smoother and

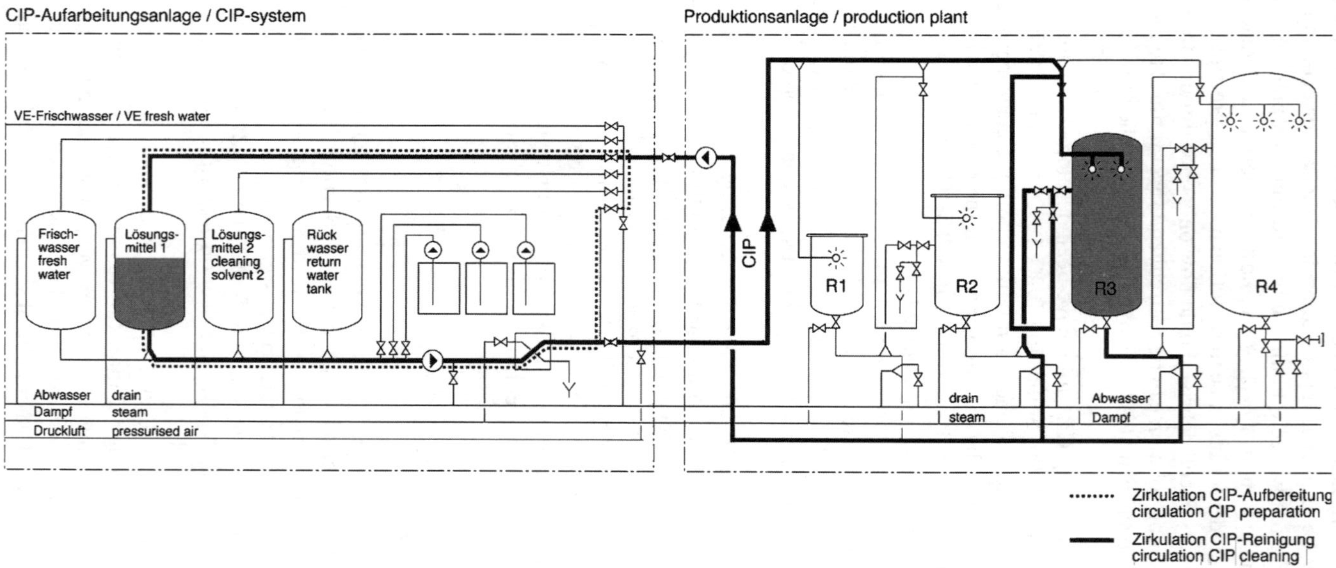

Figure 14.6 Example of a CIP system with one multiple-tank recirculating unit for the cleaning of a group of process tanks (e.g. bioreactors). The CIP unit consists of four tanks, for fresh water, two cleaning solutions and recycled water, as well as of three small containers for concentrated cleaning agents. The CIP circuit is designed in such a way that equipment can be cleaned in separate groups. This is illustrated here with the cleaning of tank R3 and part of the transfer line from tank R2; the thick lines show the circulation path of the CIP solutions (courtesy of Bioengineering).

smaller surface area and thus tend to be easier to sterilize. Design, selection criteria and operating conditions of various types are discussed in detail elsewhere (Adams & Agarwal 1990; Greene 2003; Haga *et al.* 1997). In large tanks (>1000 litres), additional spray balls or valves may be required in the central and lower parts of the tank to clean the stirrer, baffles, spargers and other side-mounted nozzles. Alternatively, the gas sparger itself may be used to spray the cleaning solution. The distribution efficiency of spray balls can be enhanced by operating the stirrer during cleaning operations. In some cases, nonetheless, it may be necessary to fill the vessel partially in order to immerse the impeller in the cleaning solution. Examples of spray devices for tank cleaning are shown in Figure 14.8.

The tank outlet valve should be carefully sized, as an undersized line might lead to a 'bathtubbing' effect, causing residues to accumulate on the sides of the tank; with an oversized line, the minimum velocity required for efficient cleaning might not be reached. For a proper control of the outlet flow rate, the use of a return pump is thus recommended (Greene 2003). To avoid excessive air incorporation in the return line, a removable flat-plate vortex breaker, 1 inch above the tank bottom, can be used (Seiberling 1992).

Multi-element filter housings, which were traditionally washed manually after dismantling, can now also be cleaned in place (Graf & Bernsley 2002). For this purpose, the solutions are circulated simultaneously through a spray-ball installed at the top of the housing and through the inlet and outlet piping while air pressure prevents the housing from filling (Figure 14.9).

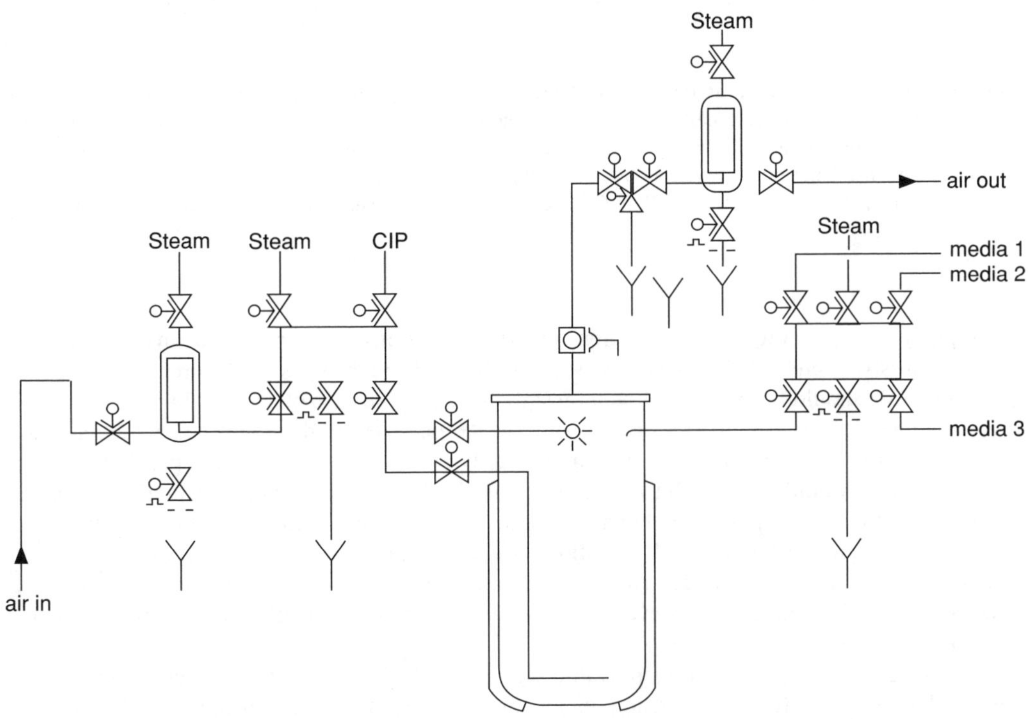

Figure 14.7 Example of a valve and piping configuration for the combined CIP and SIP of a bioreactor. The main steam inlets for the bioreactor are the spray ball and sparger; the various inlets (air, media) and outlet (air) can be sterilized separately. The figure also illustrates the high integration of process, SIP and CIP piping thanks to ring transfer panels (see Figure 14.5) (courtesy of Bioengineering).

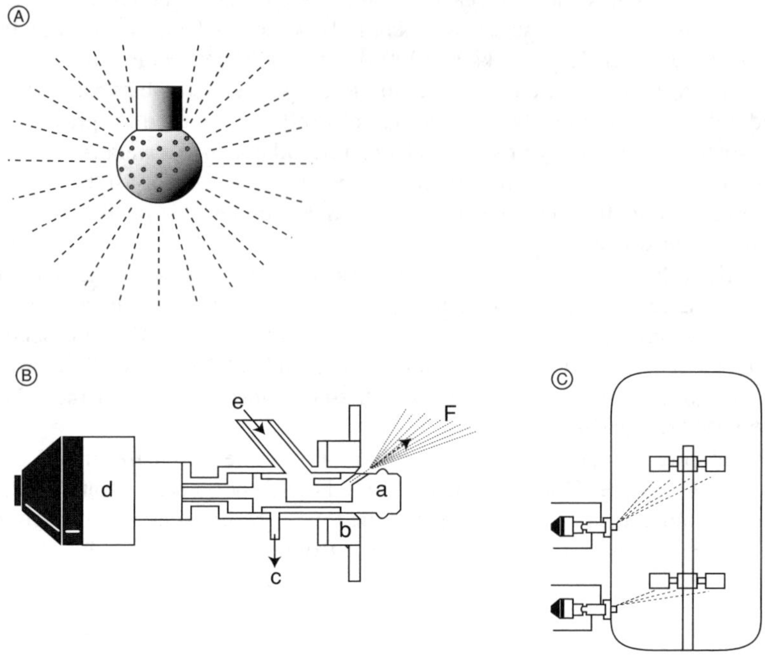

Figure 14.8 Examples of spray devices for the CIP of a tank. Ⓐ Static spray ball with action radius of 360 °C; one or several of these balls (depending on the size of the tank) are typically installed in the upper part of the tank, as shown schematically in Figure 14.6. Ⓑ Static spray valve ('e' shows the inlet of the cleaning and rinse solutions, as well as of steam, during SIP (as discussed in Section 14.4.2.2); 'c' shows the outlet of condensed water during SIP). Ⓒ Location of spray valves to clean a stirred tank (courtesy of Bioengineering).

In piping, the general recommendation for the velocity of the cleaning and rinse solutions is about 1.5 m/s (Greene 2003). When the solution flows through pipes of different diameters in series this rule should be applied to the largest diameter, in the case of several diameter changes, it is recommended the circuit be split to accommodate the different flow rates needed to reach the desired velocity. Lines should be cleaned individually or in series rather than in parallel, since in this latter case it would be difficult to attain the desired velocity in each path, and it might be possible for one path to back up into another and impede cleaning. Haga *et al.* (1997) have shown that if dead legs have a length/diameter (*L*/*D*) ratio above 2–3 (which is however not recommended, see Section 14.2.4), increasing the velocity up to 2–3 m/s may be beneficial in reducing the cleaning time. Above a *L*/*D* ratio of 6, however, the dead leg becomes practically uncleanable, no matter how fast the solution is circulated.

For mobile equipment, such as portable tanks, stand-alone CIP stations have been developed. They work on the same principles as above, except that the equipment to be cleaned has to be moved manually to the station and connected to the supply and return lines. Our recommendation is to use such stations even for bioreactors as small as 25–30 litres, since the weight (typically above 50 kg) makes their handling and transport into a washer very tedious.

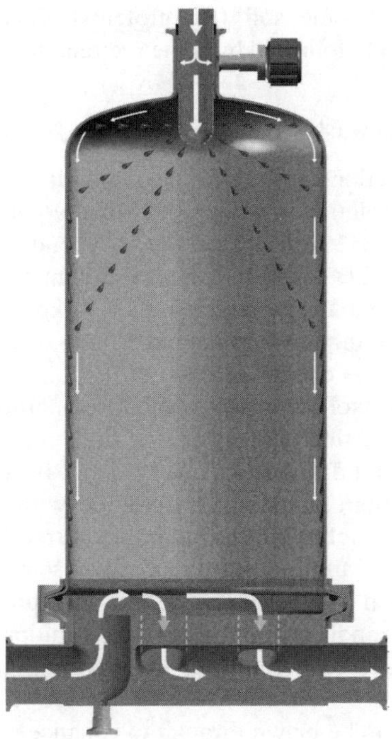

Figure 14.9 CIP of a filter housing. The housing is cleaned by the simultaneous circulation of solutions through a spray ball at the top and through the inlet and outlet process piping (reproduced with permission from Graf and Bernsley 2002).

14.3.4 Main Steps of a CIP Cycle

The main factors influencing the effectiveness of a CIP cycle are: (i) the characteristics (cleanability) of the surface (see Section 14.3.2); and (ii) the exposure time, temperature, composition and degree of turbulence of the cleaning and rinse solutions. The typical steps of a CIP cycle, together with recommended conditions, are summarized below (Greene, 2003; Stewart & Seiberling 1996; Thompson 1994). During each step below, each moving component (valve, agitator, pump, etc.) should be operated continuously or repeatedly, ideally under the same conditions as during production operations, to ensure that each solution successively contacts all surfaces to be cleaned; for instance, valves should be opened and closed five or six times. Cleaning should begin immediately after the completion of the process operation, because dried-on 'soil' may be significantly more difficult to remove.

14.3.4.1 Preparation of the Process Equipment

Parts that are washed manually or discarded (e.g. sensors, filter cartridges) are removed from the equipment to be cleaned. Jumpers are installed on transfer panels as needed, to select the proper CIP circuit.

14.3.4.2 Pre-rinse

This step is intended to remove the residual loose (cell) debris of the process as well as to prevent formation of precipitates. The solution can be fresh deionized water or the solution recovered from the final rinse of the previous CIP cycle. The temperature should be below 45 °C in order to avoid denaturation and precipitation of some 'soil' (e.g. proteins). This step should be performed in a 'flow-through' mode first, possibly followed by some recirculation.

14.3.4.3 Recirculated alkaline wash

This step uses a hot alkaline solution to solubilize the remaining proteins and other contaminants. A typical composition of this solution is sodium hydroxide, potassium hydroxide or a mixture thereof at a concentration of 0.5–3 % in deionized water; depending on the water quality and 'soil', species such as chelating agents (EDTA), surfactants or sodium hypochlorite (30–100 ppm) should be added. Temperature should be in the range of 60–85 °C. Exposure time for efficient cleaning is very variable (10 min–1 h) depending on the 'soil' and process circuit; recirculation of the alkaline solution is thus recommended for economic reasons.

Ideally, cleaning conditions (solution composition, temperature and time) should be optimized on the basis of the production process and equipment, with particular attention to the components of the cell culture fluid. Interestingly, Bird and Bartlett (1995) found that the most effective cleaning of whey protein at 50 °C was achieved with a NaOH concentration around 0.5 %, whereas both lower and higher concentrations required longer cleaning times. Similar findings were obtained for the cleaning of starch and glucose residues, but in each case, with a different optimum concentration of NaOH. For practical purposes, it is, however, desirable to select a single set of cleaning conditions for all pieces of equipment. For a detailed discussion on the chemistry of cleaning detergents for use in the pharmaceutical industry see Rohsner and Serve (1995).

At the end of this step, air can be blown through to enhance removal of the cleaning solution and thus make further rinsing easier.

14.3.4.4 Hot rinses

The equipment is then rinsed in several steps, to remove the alkaline solution. This can be achieved with fresh hot deionized water (60–85 °C), the final rinse of the previous CIP cycle or, preferably, first with a dilute acidic solution (typically phosphoric or citric acid) at room temperature followed by water. An acidic solution can remove the alkaline solution more efficiently than water, by neutralization; it also repassivates stainless steel surfaces (Adey & Pollan 1994), removing any mineral deposits left by the previous chemical solutions and rinses. The first rinse is usually recirculated. In any case (with or without acid), a second rinse, in this case in a flow-through mode, should then be applied, with fresh hot deionized water or with the final rinse of the previous CIP cycle. For the cleaning of non-vacuum-rated vessels, a cold rinse should not follow a hot wash, due to the danger of tank-collapse.

14.3.4.5 Final rinse

A final rinse with hot WFI removes the traces of the previous wash. It is typically monitored with conductivity, pH or resistivity to ensure complete removal of chemical solutions. The equipment is then ready to be sterilized; if not used for a long time, drying-in-place by hot air is recommended to remove any stagnant water.

14.3.5 Validation of CIP

All cleaning procedures need to be validated to demonstrate their efficacy in maintaining a plant in a hygienic condition and preventing cross-contamination. The validation of CIP implies both the qualification of the cleaning equipment and the validation of the cleaning cycle. Prior to starting the CIP validation, however, the qualification of the associated utilities (WFI, compressed air, etc.) should be completed. The discussion below focuses on the steps that are specific to CIP.

The first step of CIP qualification should be an installation qualification (IQ), to verify that the correct equipment was purchased, received and installed, and that the proper 'as built' drawings and engineering documentation is available. The specific items to verify for a CIP system are reviewed in Baseman (1992).

An operational qualification (OQ) should then be performed, to confirm and test the functional operation and adequacy of the CIP system and procedures. The CIP automation system should be checked for functionality, program description and logic flow; inputs and outputs as well as alarms should also be verified. Additionally, CIP test runs should be performed on each piece of equipment to be cleaned (Baseman 1992). The goal is to verify that the CIP system operates as defined in the design specifications, and to prove physical removal of materials from surfaces during cleaning cycles. For this purpose, a 'placebo contaminant' can be used for simulation; it should be easily detected and be as close as possible to the actual 'soil'. Fluorescent compounds such as riboflavin or fluorescein are often used; a milk solution or serum with fluorescent properties can also be applied to simulate protein 'soil', although the use of such compounds will require special care to prove complete removal before starting GMP production. Particular attention should be paid to verifying the ability of each spray ball to reach all the interior surfaces of the vessels effectively. During OQ, adjustments to the equipment or to the cycle parameters are often required. All modifications and new settings should be carefully recorded.

The validation of the cleaning cycle performance, often considered as a performance qualification (PQ) (Baseman 1992), should then verify that the CIP system is capable of removing the process 'soil' from all surfaces, both efficiently and reproducibly, under worst-case conditions. Cleaning tests should therefore be performed after the real process (media preparation, cell cultures, etc.) has been run in each piece of equipment. Tests should be carried out in triplicate and the equipment should meet the specifications each time.

Due to the wide variety of processes and products, regulatory health authorities do not set unique methods and specifications for determining whether a cleaning process is validated. Cleanliness should be assessed at the end of the CIP cycle by measuring any potential contaminant critical to the process and the final product, both in the final rinse solution and on equipment surfaces. Some useful guidance on selecting sampling techniques and setting limits is provided by the Parenteral Drug Association (PDA 1998b). For fill and finish pharmaceutical activities, the contaminants are usually easy to identify (e.g. active ingredient, excipients or detergents) and typical limits are a concentration below 10 ppm, a biological activity level below 1/1000 of the normal therapeutic dose, or no visible residue (FDA 1993). In cell culture processes, however, critical contaminants may be more difficult to identify and it may not be relevant to focus on the active substance. In practice, the rinse solution is tested for any critical media components, particulates, protein residues or simply total organic carbon. A case study for cleaning validation of bacterial fermentation equipment is discussed by McArthur and Vasilevsky (1995); similar principles can be applied to cell culture equipment. Specifications for the final rinse solution are usually set to be the same as the reference water (e.g. WFI). Some differences in non-critical properties can, however, be accepted; for instance, small traces of sodium or potassium ions from the cleaning solutions may have no significant effect on the quality of the next batch of culture medium (provided of course that the presence of any cytotoxic species such as detergents has been ruled out). A higher final conductivity value than in WFI could thus be acceptable. However, during validation, one cannot rely upon the analysis of the rinse solutions to verify thorough cleaning; equipment should be opened and interior surfaces visually inspected; swab tests should be performed on surfaces to detect the presence of any dirt, cell debris, product or other proteins, and bioburden, as well as residues of the cleaning solution (Lombardo *et al.* 1995). Cleaning validation should also be designed so that a correlation between cleanliness of the equipment and a parameter that is easy to monitor routinely (e.g. conductivity or pH of the final rinse solution) can be established. Tedious sampling and dismantling operations after a CIP cycle can then be avoided during production.

When establishing the tests and specification of a PQ plan, it is important to take into account the way the equipment will be used. Clearly, in the case of multi-product process equipment, the prevention of cross-contamination should become one of the main objectives of cleaning validation (McArthur & Vasilevsky 1995). A similar issue arises when the same CIP skid is used to clean several tanks at various steps of the process (PDA 1998b). Special cleaning precautions may also be required by biosafety regulations (see Section 14.5.5).

14.4 STERILIZATION OF UPSTREAM PROCESSING EQUIPMENT

14.4.1 Introduction

The most common sterilization technique for large-scale equipment is with clean steam under pressure; the goal is to kill both live microorganisms and their spores, the latter having a much greater resistance to heat. The commonly accepted goal of steam sterilization is to achieve a probability of microorganism survival below 10^{-6}. For this purpose, a typical combination of time and temperature for sterilization of large-scale equipment is 15 minutes and 121 °C (Oakley 1994); in practice, the hold time or temperature is extended to ensure a good safety margin. Different 'overkill' strategies, depending on the type of equipment and desired margin, are discussed in Pflug and Evans (2000). One should keep in mind the negative impact of such a treatment on heat-sensitive parts such as filters, membranes, gaskets and probes. Small-scale equipment is usually treated in an autoclave; large-scale equipment together with hard piping are typically sterilized in place (SIP), by direct injection of clean steam without any dismantling, offering similar benefits to CIP (see Section 14.3.1).

The primary challenge with an SIP system is that the process equipment must be designed for this purpose from the beginning; this is reviewed together with the design requirements for cleanable equipment in Section 14.3.2. The discussion below focuses on the design, operation and validation of SIP systems.

14.4.2 Design of SIP Systems

14.4.2.1 Steam generation

Steam is commonly supplied from a centralized generator, using purified water or even WFI as feed water and plant steam as the heating medium. It should then be distributed via 316L stainless steel piping (Roberts *et al.* 1995). Unlike in CIP systems, no recirculation (for instance recovery of condensate for clean steam generation) should be performed.

For maximum efficiency, steam should be saturated at the point of use, with non-condensable gases kept to a minimum (typically less than 100 ppm); superheated steam, i.e. heated above its saturation temperature, should be avoided. The various guidelines on the quality of steam for sterilization are discussed in Agalloco (2000).

Steam is usually injected into the equipment to be sterilized at a differential pressure of 1.5-2 bar, corresponding to a temperature of 127–134 °C; this gives a good safety margin and ensures that the equipment rapidly reaches the sterilizing temperature at all points.

14.4.2.2 Design of the steam circuit

During sterilization, a large number of valves have to be opened and closed in complex specific sequences. To ensure that this is performed reproducibly, automation should (as for CIP), be an essential part of a modern SIP system.

For the sterilization of vessels, the steam inlet should be located at the top of the bioreactor, whereas condensates should be eliminated from the bottom through a drain valve and diverted to a stream trap, or flow orifice, to regulate the flow (Oakley 1994). To ensure proper sterilization of all process inlets and outlets, such as sparger, inoculation dip tubes, side inlets, spray balls and secondary outlets, it is recommended that steam be supplied through all of them during the SIP cycle; this requires some additional valve configurations. Mechanical seals of the agitator must also be sterilized by direct injection of steam into the seal chamber. With respect to sterility maintenance, magnetically coupled stirrers are actually preferable.

SIP of hydrophobic filters must be performed so that they do not become wet, otherwise steam cannot cross the membrane, resulting in improper sterilization. Such filters are thus often the most difficult part of the system to sterilize (Agalloco 1990). The filter housing and the associated pipework should be installed so as to guarantee a supply of dry steam and a proper draining of the condensate via separate steam traps, on both sides of the filter membrane (Oakley, 1994). Care should be taken to minimize hydraulic stress on the membrane; this is best achieved by using two steam inlets, upstream and downstream of the filter, with a small and carefully controlled pressure difference (Rowe *et al.* 1996). Gas filters are typically sterilized during the same cycle as the corresponding tank. However for long-lasting cultures, during which the filter may have to be replaced, separate sterilization steps of the tank and of the filter housing(s) should be made possible; this requires a more complex valve configuration.

SIP of hydrophilic filters is less problematic since wetting does not block the passage of steam through the membrane. Each housing point should nonetheless have a drain valve to remove condensate efficiently. At the beginning of the sterilization cycle, the vent valve of the housing should be opened for a few minutes to purge air thoroughly. Regarding the sequence of sterilization, it is generally best to sterilize the vessel first and then the transfer line and filter together; in this way, at the end of the filter sterilization, the system steam pressure can be released into an already sterile vessel. Figure 14.7 shows an example of valve and piping configuration for the sterilization in place of a complete vessel.

In piping, the condensate produced must be removed promptly at *all* low points (Agalloco 1990). In practice, this is done either via steam traps, which are able to remove condensate from steam lines automatically, or via diaphragm valves that are controlled by the plant automation system to open for short periods at regular intervals. This second system tends to be preferred, being more reliable and better able to stabilize temperature at the sterile boundary (Wilde 1998). Steam should circulate in only one direction, in order to avoid the accumulation of a cool condensate at a point of conflicting streams. As previously discussed (Section 14.3.2.6), transfer lines should be designed to be sterilizable independently of adjacent vessels. Figure 14.10 shows one example of a piping configuration that provides this flexibility.

The only two parameters to be monitored during a SIP cycle are temperature and pressure. In vessels and pipes, temperature should be measured at various critical points, including near steam traps or near large heat sinks, in order to measure the lowest values during a SIP cycle. Typical probes for these measurements are discussed in Chapter 13. Temperature sensors (Pt-100) can be installed either inside the equipment or mounted on the external surface of pipes, in 'collars' (Wilde 1998).

For mobile tanks, an alternative to building a complete steam distribution network is to use stand-alone SIP stations, by analogy with CIP stations (see Section 14.3.3.2). In this case, the tank

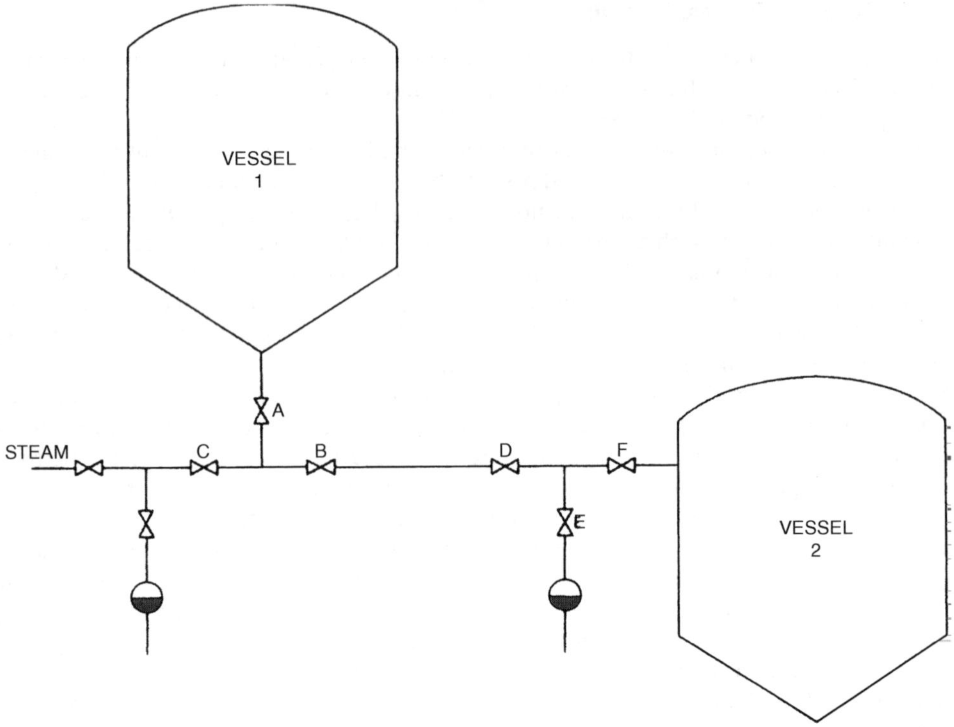

Figure 14.10 Example of a valve and piping configuration for the SIP of a transfer line between two vessels. This configuration enables the sterilization of the transfer line independently of the two vessels; vessel 2 is normally sterilized first, with valves E and F open and valve D closed. For the sterilization of the transfer line, the steam inlet valve is opened as well as valves C, B and D, while valves A and F are kept closed. At the end of the sterilization cycle, valves E and C are closed and valve F is opened to release the steam pressure and to avoid the risk of vacuum formation; vessel 2 and the transfer line are now ready to use (reproduced with permission from Oakley 1994).

is moved to the station and connected to it only for the duration of the sterilization cycle. As with CIP, our recommendation is to use SIP stations even for bioreactors as small as 25–30 litres.

14.4.3 Main Steps of a SIP Cycle

The main steps for the steam sterilization of empty equipment are summarized below (Wood 1999). Following the completion of a CIP cycle, a few manual operations are first required before a SIP cycle can be initiated:

(i) installation of hydrophilic and hydrophobic filter cartridges in their housings;

(ii) connection of jumpers on transfer panels, according to the desired process route to be sterilized;

(iii) installation of peripheral equipment, such as probes and connections for sampling.

The typical steps for an automatically controlled SIP cycle are as follows:

(i) A pressure-hold test is conducted with air to detect any leak; the bioreactor should therefore be equipped with a sanitary pressure gauge capable of being steam sterilized. A 24-hour test is actually required to be able to detect a minor leak or pinhole; in practice, this cannot be performed before every SIP cycle, so it is usually done only after major maintenance work. Routinely, a 30-min test will be able to detect any gross leak due, for instance, to valves left open or to incorrect assembly of probes or filters (Oakley 1994).

(ii) Equipment outlet valves and drain valves are opened to permit any remaining condensate to discharge through hydrostatic traps.

(iii) Steam is injected into the system in sequential steps, designed to discharge air (via filter bleed valves) and condensate, and to bring the whole equipment to the desired temperature.

(iv) When the temperature set–point (e.g. 121 °C) at all locations and the corresponding differential pressure (1.1 bar) have been reached, the time counter for the sterilization phase is started. Both temperature and pressure should be monitored to verify that steam is saturated and that no air is present in the equipment to be sterilized. The condensate is continuously eliminated via steam traps.

(v) At the end of the sterilization time, drain valves are shut first, followed by the steam inlet valves. When the pressure inside the vessel has dropped to a differential value of about 1 bar with all valves closed, high flow rates of sterile air at the same pressure should be introduced through the air inlet filter. This helps to cool down the equipment and to ensure that no vacuum is created, which would lead to the infiltration of non-sterile air. Positive pressure should then be maintained until the equipment is used.

14.4.4 Validation of SIP

Complete validation of a SIP system consists first of an IQ and OQ of the system, where proper operation of the various components (steam generator, distribution pipework and automation system) should be demonstrated (Baseman 1992; Wood 1999). The efficacy and reproducibility of the SIP process should then be verified by performing temperature distribution or thermal mapping studies under worst-case conditions. For this purpose, temperature probes, in addition to those of the SIP control system, and bioindicators, e.g. *B. stearothermophilus* spore strips, (Zimmermann & Bablok 1985) are placed at critical spots on the internal surface of the equipment to be steam-sterilized. Such spots are, for instance, the bottom of the vessel, filter housings, nozzles in the head of the vessel, low points in transfer lines and the farthest points from the steam source. A successful SIP cycle is one that achieves the specified combination of temperature and time together with the sterilization of all bioindicators.

In reality, thermal mapping studies are limited to surfaces that are easily accessible, and thus cannot ensure that all the equipment, particularly transfer lines and valves, are properly sterilized. Consequently, validation of SIP operations should be completed by media fill tests, where the equipment is filled and incubated with sterile medium (see Section 14.2.2.3); all internal surfaces claimed to be sterile can thus be tested.

14.5 GENERAL CONCEPTS FOR AN UPSTREAM PROCESSING PLANT

14.5.1 Introduction

This section presents the principles for designing an upstream processing plant, with focus on pilot- and large-scale cultivation equipment for production under GMP conditions. Applications for inoculum expansion and production bioreactors are discussed, as well as requirements for multiproduct equipment, room cleanliness and biocontainment. Design of individual components (tanks, pumps, tubing) are beyond the scope of this section and can be found elsewhere. Also, for a complete design of a whole facility (HVAC, room layouts, etc.), the reader should refer to Chapter 2.

14.5.2 Inoculum Expansion and Production Steps

Very limited information is available in the public domain regarding the general design and sizing of an upstream processing line for pilot- or large-scale production. Chu and Robinson (2001) present an overview of the technologies (bioreactor type and mode, medium type) for the various commercial products recently approved (recombinant proteins, vaccines, diagnostics and tissue culture), but few details on the scale of the cultivation vessels are actually available. In the following, typical approaches based upon our experience are discussed. Large-scale production of recombinant proteins has been selected as the example here, since it is currently the field of animal cell-derived products that is growing fastest (in terms of new molecules launched on the market), and that requires the largest amounts of drug substance (up to several hundred kilograms per year for monoclonal antibodies). Similar design principles can be applied for other smaller-volume products (vaccines, gene-therapy products), although with a simpler inoculum expansion train.

For commercial recombinant proteins, the production step is commonly run, at least for the processes developed during the last decade, in stirred-tank bioreactors, with a working volume of 10 000–15 000 litres in the fed-batch mode and 300–2000 litres in the perfusion mode. In the latter case, size is limited by the difficulties in scaling-up the cell-retention device.

The inoculum required for the production stage is prepared in a stepwise mode, using a series of cultivation systems of increasing size. The maximum allowable expansion factor between two steps of the inoculum expansion train is strongly process- and cell-line dependent; the goal is to keep cells at high viability, in the exponential growth phase. This typically occurs in the concentration range of 10^5–10^6 cells/ml, although much higher densities, up to 10^7 cells/ml, can be reached in optimized fed-batch processes. In practice, a conservative expansion factor of approximately 5 or 6 between two cultivation vessels is normally applied; each culture, if it is run in a batch mode, will last 3–4 days, including a possible lag phase after each transfer. At laboratory scale, cells can be harvested by centrifugation and resuspended in fresh medium to ensure optimal growth performance at each step. Above the litre scale, this becomes very difficult to perform while maintaining sterile conditions. For this reason, cells are normally transferred together with the spent medium into the next-step bioreactor. At laboratory scale, the inoculum is normally transferred by pipetting in a Class II (micro)biological safety cabinet. Transfers to and from small bioreactors (2–20 litres) are typically performed via flexible tubing that is connected by sterile welding; at larger scale, all transfers are performed via hard pipes, using transfer panels (see Section 14.3.2.6).

In the example of a *fed-batch* process with a production step of 10 000 litres, typically 1–2 ml of cell suspension at about 10^6 cell/ml from cryovials are expanded first in T and/or spinner flasks and then in a train of bioreactors with working volumes of, for instance, 20 litres, 80 litres, 400 litres and 2000 litres; the whole content of this last step can then serve as the inoculum for the production stage. A more efficient approach for inoculum expansion in the laboratory is to start with large cryobags (50–100 ml), containing frozen cells at high densities ($20–40 \times 10^6$ cells/ml); the first bioreactor can then be inoculated in one step and in a much shorter time (Heidemann *et al.* 2002). Bioreactors can also be run in a fed-batch mode, during inoculum expansion, with the addition of a large amount of fresh medium over several days, leading to a significant volume increase during the culture; this allows a reduction in the number of bioreactor steps required for inoculum expansion. In this case, the vessel must be designed such that it can operate over a broad range of liquid volumes (Heidemann *et al.* 2002). Bioreactors of 50 litres and larger are commonly made of stainless steel; below this, both glass and stainless steel are used although the latter is recommended above 20 litres for its better mechanical properties. Recently, disposable plastic bags have become available as an alternative to spinner flasks and tanks, with volumes of 100 ml to 500 litres (Wave Bioreactor®, Wave Biotech, Bridgewater, NJ, USA; Singh 1999) (see also Chapter 10); they can also be run in a perfusion mode. These bags have the typical advantages of disposable systems (minimal investment, no cleaning and sterilization, including validation); additionally, since stirring is achieved via a rocking movement, the range of working volume is much broader than in a stirred vessel. The culture volume can thus be increased by a factor of 5–10 in the same bag, via gradual feeding of fresh medium, thereby reducing the number of inoculum expansion steps. The drawbacks are relatively high running costs and the fact that pH and pO_2 cannot be controlled as accurately as in a stirred-tank bioreactor.

In a fed-batch process, the production step is typically extended to 10–15 days, in order to maximize accumulation of the recombinant product. For optimum utilization of the equipment, more than one production bioreactor per inoculum train should be installed. A ratio of 2 is typically applied; in this case, however, the production step is still the bottleneck if the culture lasts more than 8 days. Figure 14.11 illustrates the example discussed here, with four production bioreactors for a fed-batch process, two inoculum expansion trains and the associated media preparation tanks.

In *perfusion* processes, the same aforementioned principles can be applied to the inoculum expansion train. In this case, however, only one or two bioreactor steps are usually required to inoculate a production bioreactor, due to the lower working volume at this stage. In the inoculum train, cells can be grown in batch, fed-batch or perfusion mode. For instance for the ReFacto® process, a seed bioreactor is inoculated from the content of spinner flasks and is first run in fed-batch mode, with continuous addition of fresh medium (Eriksson *et al.* 2001). Once the desired working volume is reached, the perfusion mode is switched on and when a sufficient number of cells is available, the content of the seed bioreactor can be transferred to the production bioreactor. The production bioreactor is also operated first in a fed-batch mode until the desired working volume (500 litres) and cell population density are achieved; the perfusion mode is then switched on. Similar examples are given in Boedeker (2001) and Harrison (1998) where the production processes for the various forms of recombinant Factor VIII and IX are reviewed. For immobilized cells on microcarriers, inoculum expansion is difficult to achieve; Dürrschmidt *et al.* (1999) have reported strategies where they could expand the working volume in a fluidized-bed bioreactor by a factor of 6 via sequential addition of microcarriers and recolonization.

Since a perfusion culture usually lasts several weeks or months, one expansion train is sufficient to inoculate several production bioreactors, provided they are started sequentially. However it is common practice to install an excess of expansion trains (e.g. one for each three or four perfusion bioreactors); this ensures that a production bioreactor can be rapidly re-inoculated at any time, as needed. The additional capital costs are offset by a more efficient use of the production capacity.

In a pilot plant, similar principles can be applied; since the production scale is typically lower (1000–5000 litres for a fed-batch process), the inoculum expansion train should only contain two or three bioreactor steps. By definition, higher flexibility is required compared with manufacturing facilities for commercial products; consequently, the use of flexible tubing and disposable equipment (storage bags for solutions, disposable bioreactors) is recommended. An interesting concept of a flexible and modular pilot plant is discussed in Bardone *et al.* (1994).

14.5.3 Multi-product Cultivation Equipment

To contain costs and to respond quickly to the ever-changing needs of the market, it is desirable to build multi-product facilities, where two or more products can be produced with the same, or shared manufacturing equipment. The main concern of regulatory authorities is the risk of cross-contamination between the different products. The important issues to address with multi-product manufacturing facilities are therefore the engineering design, the procedural, temporal or spatial

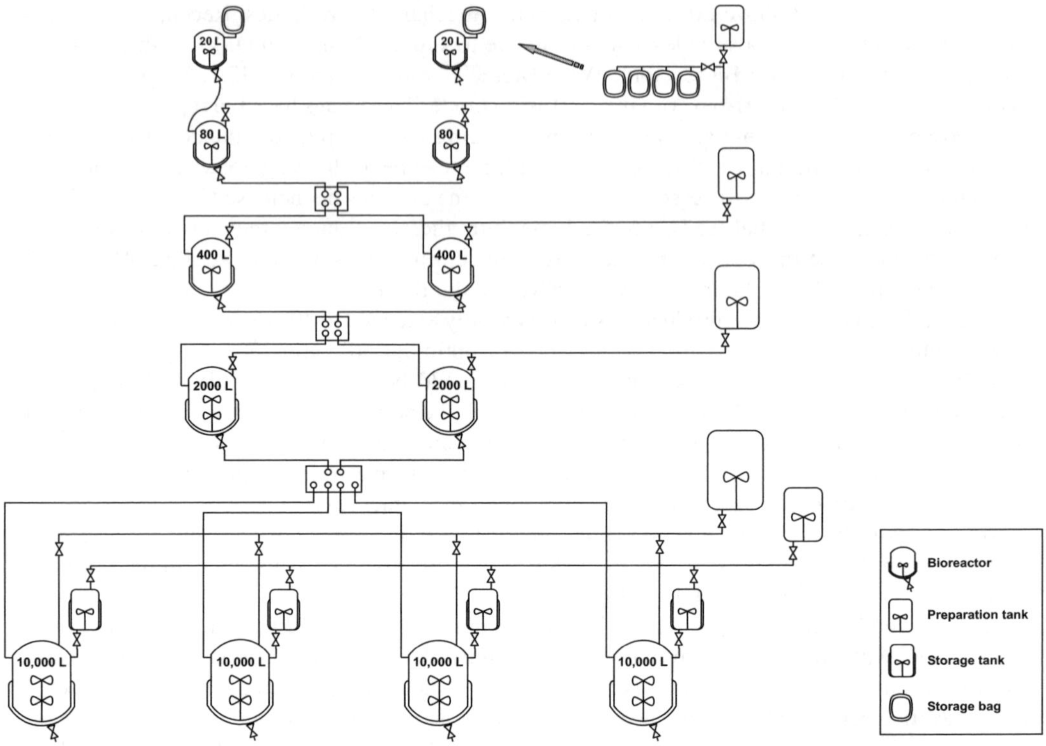

Figure 14.11 Schematic flowsheet of an upstream processing plant for a fed-batch process. The figure shows the main vessels for cultivation and media preparation. The production stage consists of four bioreactors, for which two inoculum expansion trains are required. Thawing of the working cell bank and laboratory cultivation are not represented. Manual transfer panels are used to transfer inoculum from one bioreactor to any bioreactor of the next step. At each scale, a single common tank is used for the preparation of the culture medium (see Section 14.2.1); in this example, the transfer route for medium is selected via valves (only the main ones are shown) instead of transfer panels; the medium is directly transferred into the bioreactors after preparation (sterilizing filters not shown), without storage, except at the 20 litre scale, where disposable bags are used. At the production stage (10 000 litres), the feed is prepared in a single common tank but is then transferred into hold tanks dedicated to each bioreactor, to allow maximum flexibility of the feeding regime.

separation of production activities, the use of validated closed systems, and the implementation of a validated cleaning/changeover program (Odum 1995).

One approach is to manufacture products in *campaigns*, i.e. sequentially. Only one product is then present in the facility at a time. Generally, campaigning involves the manufacturing of several batches of one product over an extended period of time (weeks or months), before the changeover to the next product takes place. In this case, a very comprehensive changeover procedure must be established and validated, particularly regarding cleaning of all common equipment. Cleaning studies are typically performed in triplicate and removal of previous product, process chemicals and cell material must be demonstrated by appropriate analytical procedures. Assays more specific than TOC measurements are recommended (Shahidi *et al.* 1995). For biocontainment areas, decontamination to remove viable organisms must be demonstrated. Good cleanability of the plant is thus essential. The use of disposable equipment, for instance plastic storage bags, can greatly simplify these cleaning requirements. For an accelerated changeover, one solution in upstream processing is to start growing cells for the new product in the laboratory while cells of the current product are still growing in the larger vessels. Since cell cultivation and handling in the laboratory are considered to take place in an open system, two segregated laboratories are required for this purpose. The relatively modest capital investment is, however, largely offset by a more efficient utilization of the plant capacity, thanks to faster changeover procedures.

Another approach is the simultaneous production of different products in segregated areas or systems within the same facility; this is often referred to as *concurrent manufacturing*. Segregation between the different processes may be achieved by the use of dedicated rooms (including HVAC), and/or by the use of validated closed systems. Typically, two or more suites with their own air handling module and personnel access will be required for the seed laboratories. For bioreactors, which are considered as closed systems, common rooms can, in principle, be used. It is, however, safe practice to vent all the off-gases directly to the outside of the facility in order to avoid risk of cross-contamination due to recirculation via the HVAC system. Physical separation of bioreactors (for instance in two groups in the same room) and operating procedures are also recommended to minimize the risk of mix-ups due to human error. Some equipment may however be shared, for instance media preparation vessels. Since operations are typically very short, one common area may be sufficient to serve several cell culture suites. In this case, media preparation should be performed in campaigns, and adequate cleaning validation is required for each 'changeover'.

In both cases (campaigning or concurrent manufacturing), cultivation equipment should be designed so that it can be readily used for different processes. However, the level of flexibility should be carefully evaluated because, in general, as the level of flexibility increases, so do the associated costs (DePalma 2003). As a general rule, bioreactors should be designed so that stirring (impeller type and number) and gassing methods (sparger type, flow rates) can be readily adapted. The location of peripheral equipment inside the vessel, such as CIP spray balls, probes and inoculum tubes, should be designed so that the bioreactor can be operated over a relatively broad range of working volumes. For the storage of culture media and feed solutions, the use of disposable bags is recommended at all scales, where possible, for both perfusion and fed-batch processes in order, to maximize flexibility. Further discussion on the design and general organization of multi-product plants can be found in Odum (1995) and Shahidi *et al.* (1995).

14.5.4 Room Classification

All cell culture handling operations that take place in an open system, for instance the transfer of cells between spinner flasks, should be performed in a Class II (micro)biological safety cabinet (which has an environment equivalent to US Class 100 or ISO 5). 'Open' cultivation systems, such as T and spinner flasks in an incubator, should be located in a Class 10 000 (ISO 7) laboratory, where only one cell type at a time should be cultivated and handled.

Bioreactors can be located in a lower-grade clean room, for instance Class 100 000 (ISO 8), since they are closed systems, although in principle even a non-classified area is acceptable, provided that adequate aseptic techniques are applied. The same comment can be made on media hold tanks. Media preparation tanks are normally located in a Class 100 000 or even a non-classified area, with the argument that all the solutions pass through a sterile-filter before entering bioreactors. Currently, due to the high investment and operational costs associated with clean rooms, the tendency in industry is to limit use of the classified areas to critical zones and to maximize 'grey space' (DePalma 2003). For instance, with large-scale bioreactors, the main part could be located in unclassified space, with a small clean room area in the form of a barrier alcove around the sample assembly (Marks 2003); an added advantage of such arrangements is that much of the maintenance can be performed without special gowning or environmental contamination concerns. For further discussion of room classification, see Chapter 12.

14.5.5 Containment Requirements

14.5.5.1 Principles of biosafety regulations

There are numerous regulations and guidelines on biosafety and containment requirements for the cultivation of mammalian cells, with some differences between the US, the EU and other countries. Their analysis is further complicated by the fact that they are regularly updated, reflecting the continuously expanding knowledge and experience with recombinant DNA technologies.

In the US, some seminal guidelines were published by the National Institutes of Health (NIH, 1976) to provide a framework for conducting genetic engineering research in a manner that protects employees from infection and prevents adverse impact on the environment. These have since been amended (relaxed), in particular regarding large-scale applications. The current key documents on biosafety in the US that are applicable, among others, to mammalian cell culture for research and production are *Guidelines for Research Involving Recombinant DNA Molecules* (NIH, 2002) and *Biosafety in Microbiological and Biomedical Laboratories* (US Department of Health and Human Services, 1999). In brief, biological agents are classified into four *risk groups* according to their relative pathogenicity for healthy adult humans. A comprehensive risk assessment should be performed, taking into account the risk group of the agent and the way it is manipulated (including the amount and dose) in order to determine the appropriate safety practices and containment conditions. The guidelines differentiate small-scale from large-scale activities (defined as research or production involving more than 10 litres of culture), the latter requiring additional precautions. At small scale, four *biosafety levels* (BL1–BL4) are defined, with increasing requirements (BL4 being the most stringent level). At large scale, there are four biosafety levels: Good Large-Scale Practice (GLSP), BL1 large scale (BL1-LS), BL2 large scale (BL2-LS) and BL3 large scale (BL3-LS). The GLSP level, i.e. the least stringent level, is applicable to large-scale cultivation of non-pathogenic organisms that have built-in limitations to their survival in the environment and for which there is an extended history of safe large-scale use. The BL1-LS level applies, at large scale, to viable organisms that would require a BL1 level in the laboratory and do not qualify for the GLSP level; the BL2-LS and BL3-LS levels correspond, at large-scale, to the BL2, and BL3 levels in the laboratory. No provisions are made for large-scale activities using viable recombinant organisms that would require BL4 containment. These would need to be established by NIH on an individual basis.

In the European Union, the current key biosafety documents applicable to mammalian cell cultures are the European Council Directive 98/81/EC on the contained use of genetically modified microorganisms (which amends the Directive 90/219/EEC) and the European Council Directive 2000/54/EC on the protection of workers from risks related to exposure to biological agents at work (which replaces the Directive 90/679/EEC); additionally, the European Norm EN 1620:1996 provides biosafety guidelines on large-scale biotechnology activities. Like in the US, genetically modified microorganisms, including mammalian cells, are classified into four risk groups according to their level of pathogenicity to humans. A risk assessment, taking into account the classification of the organism together with the operating conditions, leads to the determination of the required containment level. There are four levels (1-4, 4 being the most stringent), at both small- and large-scale, which essentially mirror, with small differences, the US biosafety levels. A good comparative analysis of the US and EC biosafety regulations can be found in van Houten and Fleming (1993), although the latest amendments are not included. In brief, these regulations require biocontainment measures of two types:

(i) *Operating procedures and practices* These include good microbiology practices as well as specific handling techniques, for which the personnel must be regularly trained.

(ii) *Physical containment barriers* These can be further divided into two classes: (a) primary barriers, which aim to prevent or minimize direct exposure of personnel to biological agents and are provided by safety equipment (e.g. biological safety cabinets), personal protection devices and closed containers; (b) secondary barriers, which aim to minimize or prevent the escape of organisms into the environment, i.e. outside the laboratory or the facility. These barriers are integral parts of the design of the facility and include special air handling systems, contained work areas, and systems for waste inactivation.

A complete discussion on the requirements of each US and EU class and the implication for the design of laboratories, facilities and production equipment can be found elsewhere (Meslar & Geoghegan 1991; Miller & Bergmann 1993, Pearson 2003, Sinclair & Ashley 1995). Only the key elements applicable to mammalian cell culture are discussed briefly below.

14.5.5.2 Physical containment barriers for mammalian cell culture

In the US, mammalian cells used for the production of recombinant proteins, such as CHO cells, normally require a BL1 level in the laboratory and a GLSP level at large-scale. In the EU, containment level 1 normally applies. Human and other primate cells, on the other hand, should be handled in a BL2/BL2-LS facility.

For both the US BL1 level and the EU containment level 1 at small scale, all primary and most secondary containment barriers are optional. Standard microbiological practice should be applied; inactivation of biological wastes is however required for the US BL1 level whereas it is optional for the EU level 1 (although where genetically modified (micro)organisms are used, as is the case for the production of recombinant proteins, national regulations may nevertheless demand inactivation). At large-scale, the US GLSP level and the EU level 1 do not require any primary or secondary containment barriers; the inactivation of biological wastes is in principle optional both in the US and in the EU but again should comply with any local regulations.

The tendency in the biopharmaceutical industry has been, however, to take a conservative approach with biosafety and to design facilities according to a BL1-LS or even a BL2-LS level. A BL1-LS level, would require, among others, inactivation of all biological wastes via validated methods, handling of cells in closed systems until inactivation, and treatment of exhaust gases by HEPA-type filters. The pros and cons of such a design approach have been discussed by Miller and Bergmann (1993). In brief, building a facility up front to comply with a containment level above requirements brings flexibility and makes it easier, later on, to accommodate different cell lines and processes. On the other hand, capital and operating costs can be higher. Additionally, overdesigning may raise questions about the confidence that a company has about the biosafety of the process, and may support perceptions in the public that biological risks are high.

In many cases, procedures and equipment required in cell culture to maintain sterility overlap with containment principles, so that the requirements of the BL1/BL1-LS levels are overachieved. Examples are the use of bioreactors equipped with sterile connections, sampling ports, leakage-free rotating seals and vent filters, as well as of Class II (micro)biological safety cabinets for handling cells in the laboratory. One exception is the positive pressure difference usually applied to clean rooms with respect to the environment: for biocontainment, a negative pressure difference should be applied instead, to prevent the spreading of organisms in the case of a failure of primary containment; this is however recommended only for the BL2/BL2-LS levels (and required for higher levels).

For the cultivation of mammalian cells, it is common practice to inactivate all biological wastes coming from the bioreactors and from the first recovery steps, including the CIP solutions, no matter whether this is required by the biosafety or some local regulations or it is simply optional. The paragraph below discusses some technical solutions for this operation. A retention basin, either in the basement of the facility or directly in the bioreactor room, also represents a common safety measure, even with organisms of the lowest classes.

14.5.5.3 Inactivation of biological wastes

At laboratory scale, manual autoclaving of biological solid and liquid waste is the simplest and most common method; alternatively, chemical sterilization using, for instance, iodine- or chlorine-containing agents can be performed. This second method is recommended for small-scale cultivation vessels, since heat treatment before washing may cause accumulation of cell debris as solid residues on surfaces. In a facility where a central liquid waste decontamination system is built for large-scale production (see below), a practical solution is to connect all the effluent from the laboratory sinks and automatic washers to this system, provided that dismantling prior to inactivation is allowed from a biosafety point of view; then only disposable solid waste will require autoclaving.

At large-scale, solid waste is commonly autoclaved whereas liquid waste is processed using a centralized and automatic liquid-waste decontamination system (LWDS). The LWDS should be located on the basement level so that the liquid waste from all process steps/CIP units can be drained by gravity to the system. The most common LWDS inactivation method is a heat treatment, either in batch or continuous mode (Carlson 2001; Miller & Bergmann 1993).

In a batch LWDS, the waste is collected in a so-called 'kill tank' and heated with plant steam, either via the tank jacket or via direct sparging, the latter being of course more efficient but leading to an increase in waste volume. It has been shown recently that exposing mammalian cells to 80 °C for 1 minute was sufficient to inactivate them all (Gregoriades *et al.* 2003). The waste is then cooled down to about 60 °C, either in the 'kill tank' or via an external heat exchanger. If neutralization is not performed in the LWDS, the waste must be directed to a neutralization system prior to discharge. A system with two 'kill tanks' in parallel, operating in a staggered way, is recommended to ensure uninterrupted service in a large-scale facility. To further enhance inactivation capacity, a sump tank for collecting liquid waste upstream of the two 'kill tanks' could be installed. With such a configuration, the waste, which is usually generated in a discontinuous and irregular way from process operations, accumulates in the sump tank. As soon as enough waste to fill one 'kill tank' is reached, it is rapidly transferred to one of the tanks and heating can follow

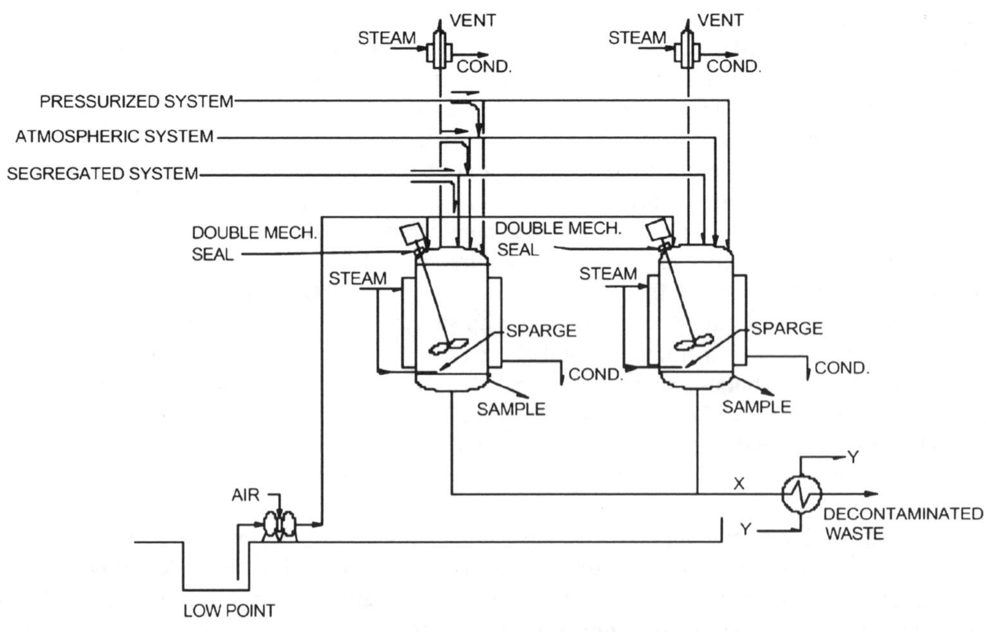

Figure 14.12 Example of a dual-tank batch LWDS without sump tank. The two tanks operate in parallel, in a staggered way. 'Pressurized, atmospheric and segregated systems' indicate various sources of biological liquid wastes. In this example, steam is injected both into the tank jacket and directly via a sparger; 'X' indicates the outlet for inactivated waste and 'Y' cooling water (reproduced with permission from Carlson 2001).

immediately. The idle time of the 'kill tanks' can thus be minimized. Figure 14.12 shows an example of a dual-tank batch LWDS without sump tank.

In a continuous LWDS, biological waste accumulates in a large collection tank connected to a continuous sterilizer; an additional sump tank upstream of the collection tank is sometimes installed, although in principle it is not required. Once a given level is reached in the tank, the liquid waste is pumped through a heat exchanger or mixed with steam, circulated through an insulated hold tube and then sent back to the tank during the heat-up phase. When the temperature set-point is reached over the whole hold tube, the inactivation step begins and the liquid is circulated in a flow-through mode. The aforementioned heat exchanger can be used at the same time for cooling down the inactivated wastes, thereby leading to significant energy savings. Figure 14.13 shows an example of a continuous LWDS.

The pros and cons of batch and continuous LWDS are discussed by Carlson (2001) and Miller and Bergmann (1993) with several configuration examples. In brief, batch systems tend to be more capital intensive (due to the requirement for larger equipment) and to consume more facility space and energy than do continuous systems; on the other hand, they are simpler to operate, require less maintenance and offer more flexibility if operating conditions, e.g. amounts of liquid waste, need to be changed. For these reasons, batch systems are the most common choice for pilot plants and multi-product facilities.

Alternatively, LWDS based upon chemical decontamination could also be used. The main problems, particularly at large-scale, are potential compatibility issues between the chemical agent and some materials, as well as the difficulty in ensuring adequate contact with all contaminated surfaces (Carlson 2001). For these reasons, this method is rarely used in the pharmaceutical and biotechnology industries.

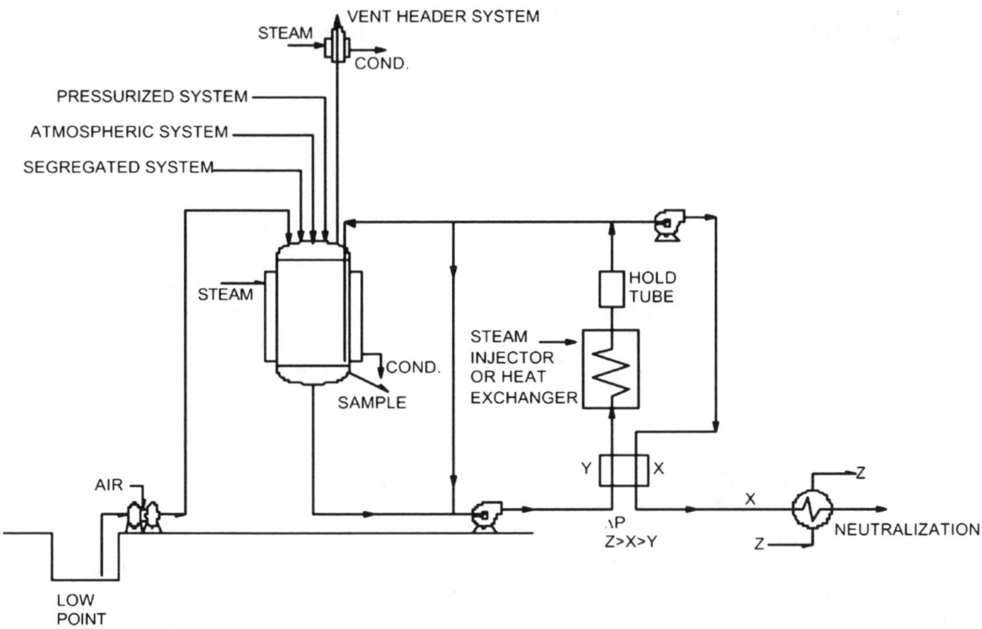

Figure 14.13 Example of a single-tank continuous LWDS. 'Pressurized, atmospheric and segregated systems' indicate various sources of biological liquid waste. Y indicates the stream of liquid waste, X the outlet of inactivated waste and Z cooling water. The liquid waste is recirculated until the deactivation temperature is reached throughout the whole hold tube. The waste is then eliminated, cooled down and sent to a neutralization tank. If needed, the tank itself and associated lines can also be decontaminated by heat (reproduced with permission from Carlson 2001).

Sterile-filtration of exhaust gases from bioreactors is considered to be an adequate containment barrier. If required, systematic integrity testing might be performed after each SIP cycle, whereas for the purpose of aseptic operations only, it is normal practice to rely upon a validated number of sterilization cycles from the vendor. In a more hazardous operation, the exhaust gas can be incinerated as an added precaution; exposure to 370 °C for a few seconds kills all known microorganisms and viruses (Dream 1993).

ACKNOWLEDGEMENT

The author thanks Zoltan Suemeghy for reviewing the manuscript and providing very valuable comments.

REFERENCES

Adams DG, Agarwal D (1990) *Pharm. Eng.*; 10(6): 9–15.
Adey H, Pollan MS (1994) In *Bioprocess Engineering: Systems, Equipment and Facilities.* Eds Lydersen BK, D'Elia NA, Nelson KL. John Wiley & Sons, New York; 473–497.
Agalloco J (1990) *J. Parenteral Sci. Technol.*; 44: 253–256.
Agalloco J (2000) *PDA J. Pharm. Sci. Technol.*; 54: 59–63.
Bardone E, Lani B, Cassani G (1994) *Pharm. Eng.*; 14(4): 34–38.
Baseman HJ (1992) *Pharm. Eng.*; 12: 37–46.
Bird MR, Bartlett M (1995) *Trans. IChemE.*; 73(C): 63–70.
Boedeker BGD (2001) *Seminars in Thrombosis and Hemostasis*; 27: 385–394.
Carlson CJ (2001) *Pharm. Eng.*; 21(3): 70–82.
Chu L, Robinson DK (2001) *Curr. Opinion Biotechnol.*; 12: 180–187.
DePalma A (2003) *Genetic Eng. News*; 23(3): 36–41.
Dillon CP, Rahoi DW, Tuthill AH (1992) *BioPharm.*; 5(May): 32–35.
Docksey S, Cappia JM, Rabine D (1999) *Eur. J. Parent. Sci.*; 4(3): 95–101.
Dream RF (1993) *Pharm. Eng.*; 13(6): 56–66.
Dürrschmid M, Landauer K, Simic G, Blüml G, Doblhoff-Dier O (2003) *Biotechnol. Bioeng.*; 83: 681–686.
Eriksson RK, Fenge C, Lindner-Olsson E *et al.* (2001) *Seminars Hematol.*; 38(Suppl. 4): 24–31.
European Council Directive 98/81/EC (26.10.1998) *Official J. Eur. Commun.;* L 330/13.
European Council Directive 2000/54/EC (18.10.2000) *Official J. Eur. Commun.;* L 262/21.
European Norm 1620 (1996) *Biotechnology – Large-scale Process and Production – Plant Building According to the Degree of Hazard.* AFNOR (Eds). Paris.
European Pharmacopoeia (2002) *2.6.1 Sterility.* Council of Europe, 67075 Strasbourg; fourth edition.
FDA (1987) *Guideline on Sterile Drug Products Produced by Aseptic Processing.* Center for Drugs Evaluation and Research, Rockville, MD.
FDA (1993) *Guide to Inspections – Validation of Cleaning Processes*, Rockville, MD.
FDA (2004) *Guidance for Industry – Sterile Drug Products Produced by Aseptic Processing – Current Good Manufacturing Practice.* Center for Drug Evaluation and Research, Center for Biologics Evaluation and Research and Office of Regulatory Affairs, Rockville, MD.
Garnick RL (1996) In *Viral Safety and Evaluation of Viral Clearance from Biopharmaceutical Products.* Eds Brown F and Lubiniecki AS. *Dev. Biol. Stand.* Karger, Basel; Vol. 88, 49–56.
Garnick RL (1998) In *Safety of Biological Products Prepared from Mammalian Cells.* Eds Brown F, Griffiths E, Horaud F, Petricciani JC. *Dev. Biol. Stand.* Karger, Basel; Vol 93, 21–29.
Gonzalez M (2001) *Pharm. Eng.*; 21(5): 48–63.
Graf EG, Bernsley J (2002) *Genetic Eng. News*; 22(18): 52–53.

Greene D (2003) *Pharm. Eng.*; 23: 120–130.
Gregoriades N, Luzardo M, Lucquet B, Ryll T (2003) *Biotechnol. Prog.*; 19: 14–20.
Haga R, Murakami S, Ostrove S, Weiss S (1997) *Pharm. Eng.*; 17(5): 8–21.
Harrison S, Adamson S, Bonam D *et al.* (1998) *Seminars Hematol.*; 35(2, Suppl. 2): 4–10.
Heidemann R, Mered M, Wand DG *et al.* (2002) *Cytotechnol*; 38: 99–108.
Huang W, Attenborough G, Cotter R (2000) *Pharm. Eng.*; May/June: 56–70.
Jornitz MW, Agalloco JP, Akers JE, Madsen RE Jr, Meltzer TH (2002a) *PDA J. Pharm. Sci. Technol.*; 56(1): 4–10.
Jornitz MW, Soelkner PG, Meltzer TH (2002b) *PDA J. Pharm. Sci. Technol.*; 56(4): 156–166.
Junker BH, Beshty B, Wilson J (1999) *Biotechnol. Bioeng.*; 62: 501–508.
Keating P, Levy R, Payne M, Proulx S, Rowe P, Pearl S (1992) *BioPharm.*; January/February: 36–41.
Kuriyel R, Zydney AL (2000) In *Methods in Biotechnology. Vol. 9: Downstream Processing of Proteins: Methods and Protocols.* Ed Desai MA. Humana Press Inc., Totowa; 185–194.
Leahy TJ, Gabler R (1984) *Biotechnol. Bioeng.*; 26: 836–843.
Liu S, Carroll M, Iverson R *et al.* (2000) *Biotechnol. Progr.*; 16: 425–434.
Lombardo S, Inampudi P, Scotton A, Ruezinsky G, Rupp R, Nigam S (1995) *Biotechnol. Bioeng.*; 48: 513–519.
Louie E, Williams B (2000) *Pharm. Eng.*; 20(4): 48–58.
Marks DM (1999) *Pharm. Eng.*; 19(2): 34–45.
Marks DM (2003) *Cytotechnol*; 42: 21–33.
Martin JM, Trotter AM, Schubert P, Katz H (1994) In *Bioprocess Engineering: Systems, Equipment and Facilities.* Eds Lydersen BK, D'Elia NA, Nelson KL. John Wiley & Sons, New York; 473–497.
McArthur PR, Vasilevsky M (1995) *Pharm. Eng.*; 15(6): 24–31.
Meeker JT, Hickey EW, Martin JM, Howard G Jr (1992) *BioPharm.*; March: 41–57.
Meslar HW, Geoghegan RF Jr (1991) *Pharm. Eng.*; 11(6): 27–33.
Miller SR, Bergmann D (1993) *J. Industrial Microbiol.*; 11: 223–234.
NIH (1976) Recombinant DNA research guidelines. *Fed. Regist.*; 41: 27902–27943.
NIH (2002) *Guidelines for Research Involving Recombinant DNA Molecules.* US Department of Health and Human Services, Bethesda, MD.
Oakley T (1994) In *Bioprocess Engineering: Systems, Equipment and Facilities.* Eds Lydersen BK, D'Elia NA, Nelson KL. John Wiley & Sons, New York; 473–497.
Odum J (1995) *Pharm. Eng.*; 15(5): 8–20.
Parenteral Drug Association (1998a) *PDA J. Pharm. Sci. Technol.*; 52(3): Supplement 1–31.
Parenteral Drug Association (1998b) *PDA J. Pharm. Sci. Technol.*; 52 (6): Supplement 1–23.
Pearson I (2003) *Pharm. Technol. Europe*; February, 1: 24–29.
Pfohl M, Stärk A (2003) *Pharm. Ind.*; 65: 1176–1183.
Pflug IJ, Evans KD (2000) *PDA J. Pharm. Sci. Technol.*; 54: 117–135.
Purnell S (2003) *Pharm. Technol. International*; 15(3): 47–50.
Raju GK, Cooney CL (1993) In *Biotechnology; Bioprocessing.* Eds Rehm HJ, Reed G. Vol. 3, VCH, Weinheim; 159–184.
Roberts TM, Kearns MJ, Latham TJ (1995). In *Bioreactor System Design.* Eds Asenjo JA, Merchuk JC. Marcel Dekker, New York; 589–613.
Rohsner D, Serve W (1995) *Pharm. Eng.*; 15(2): 20–28.
Rowe P, Tingley S, Walker S (1996) *Pharm. Eng.*; 16(1): 44–52.
Seiberling DA (1986) *Pharm. Eng.*; 6(6): 30–35.
Seiberling DA (1992) *Pharm. Eng.*; 12(2): 16–26.
Shahidi AJ, Torregrossa R, Zelmanovich Y (1995) *Pharm. Eng.*; 15(5): 72–83.
Sinclair A, Ashley MHJ (1995) In *Bioreactor System Design.* Eds Asenjo JA, Merchuk JC. Marcel Dekker, New York; 553–588.
Sinclair A, Monge M (2002) *Pharm. Eng.*; 22(3): 20–34.
Singh V (1999) *Cytotechnol.*; 30: 149–158.
Sofer G (2003) *BioPharm International*; January: 50–57.
Spanier HJ (2003) *Pharm. Technol. Europe*; February: 45–50.
Stewart JC, Seiberling DA (1996) *Chem. Eng.*; 103(1): 72–79.
Stinavage P (2003) *Am. Pharm. Rev.*; 6(1): 48–50.

Sundaram S, Eisenhuth J, Howard G Jr, Brandwein H (2001a) *PDA J. Pharm. Sci. Technol.*; 55(2): 65–86.

Sundaram S, Eisenhuth J, Howard G Jr, Brandwein H (2001b) *PDA J. Pharm. Sci. Technol.*; 55(2): 87–113.

Sundaram S, Eisenhuth J, Howard G Jr, Brandwein H (2001c) *PDA J. Pharm. Sci. Technol.*; 55(2): 114–126.

Sundaram S, Eisenhuth J, Steves M, Howard G Jr, Brandwein H (2001d) *PDA J. Pharm. Sci. Technol.*; 55: 373–392.

Thompson PW (1994) In *Bioprocess Engineering: Systems, Equipment and Facilities.* Eds Lydersen BK, D'Elia NA, Nelson KL. John Wiley & Sons, New York; 473–497.

US Department of Health and Human Services (1999), *Biosafety in microbiological and biomedical laboratories.* Public Health Service, Centers for Disease Control and Prevention, and National Institutes of Health, Bethesda, MD; fourth edition.

United States Pharmacopeia (2003) *71. Sterility Tests.* United States Pharmacopeial Convention, Inc., Rockville, MD; 26th revision.

van Houten J, Fleming DO (1993) *J. Industrial Microbiol.*; 11: 209–215.

Wilde F (1998) *Pharm. Technol. International*; 10(9): LV–LX.

Wood RT (1999) *PDA J. Pharm. Sci. Technol.*; 53: 231–234.

Zimmermann G, Bablok W (1985) *Pharm. Ind.*; 47: 1175–1181.

Relevant Web Sites

ASTM International (formerly American Society for Testing and Materials): Access (for a charge) to technical publications (standards, journals, books, etc.) on materials, piping, vessels, construction, etc.	www.astm.org
CDC (US Center for Disease Control and Prevention), Office of Health and Safety: Free access to guidelines and publications on biosafety, such as the manual *Biosafety in Microbiological and Biomedical Laboratories.*	www.cdc.gov/od/ohs/
EMEA (European Medicines Agency): Free access to guidance documents on human medicines.	www.emea.europa.eu
EU Regulations: Free access to various regulatory documents, for instance on biosafety.	http://europa.eu/scadplus/
FDA (US Food and Drug Administration): Free access to guidance documents of the Center for Biologics Evaluation and Research, and the Center for Drug Evaluation and Research, including documents of the International Conference on Harmonization (ICH).	www.fda.gov/cber/guidelines.htm; www.fda.gov/cder/guidance/index.htm
ISPE (International Society for Pharmaceutical Engineering): Access (for a charge) to technical and regulatory publications (technical guides, *Pharm. Eng.*, etc) in life sciences.	www.ispe.org

NIH (US National Institutes of Health), Office of Biotechnology Activities: Free access to *NIH Guidelines for Research Involving Recombinant DNA Molecules*, including the latest amendments. www4.od.nih.gov/oba/rdna.htm

PDA (Parenteral Drug Association): Access (for a charge) to technical and regulatory publications for the (bio)pharmaceutical industry (technical reports, *PDA J. Pharm. Sci. Technol.*, etc.) Free access to comments on regulatory draft documents. www.pda.org

15 System and Process Validation

N Chesterton

15.1 INTRODUCTION

Validation is an essential part of good manufacturing practice (GMP, see Chapter 34) for the production of pharmaceuticals. The *Oxford English Dictionary* defines validation as 'rendering or declaring legally valid', a useful definition in the context of producing pharmaceuticals in as much as it emphasizes the need for signed and approved documentary evidence that could be presented in a court of law. However, in more practical terms, validation is the process by which it is established and documented that facilities, utilities, equipment and processes both function and continue to function in accordance with their design/specification, and will consistently produce a product which meets its predetermined specification.

For the purposes of this chapter, the general term 'system' will be used to cover facilities, utilities and equipment. The definitions of other key terms can be found in the glossary at the end of the chapter. Note that this chapter will not cover cleaning validation, the qualification of starting materials (see Chapters 4, 14 and 32, as well as ICH 1997 and WHO 1998), nor validation exercises such as virus inactivation/removal (see Chapter 19) that must of necessity be performed in facilities and equipment (and usually at a scale) other than that used for manufacture.

15.2 THE COMPONENTS OF VALIDATION

In principle, there are either seven or eight components to a validation exercise (the eighth only applying when a complete process is being validated). The first of these is a planning document, the second two are design documents, and the remainder are qualification activities. They are:

- *The Validation Master Plan (VMP)* This is the over-arching document that defines the scope of the validation and its requirements, and specifies the relationship between the various documents and activities.
- *The User Requirement Specification (URS)* This document defines what the end-user requires of a system, and in most cases is the document against which proposed suppliers will submit tenders.
- *The Functional Specification (FS)* This is the response of the (proposed) supplier to the URS. It addresses the details of the way in which the supplier intends to fulfil the requirements set out in the URS.
- *Design Qualification (DQ)* This activity ensures that what has been asked for by the end-user is what the supplier has actually specified.

Medicines from Animal Cell Culture Edited by G. Stacey and J. Davis

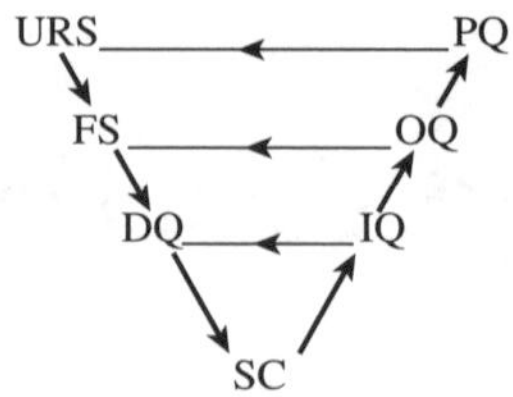

Figure 15.1 V-diagram of the interrelationship between validation activities for individual systems (SC = system construction; for other abbreviations, see text).

- *Installation Qualification (IQ)* This activity ensures that the system has been installed to the correct specifications.
- *Operational Qualification (OQ)* This activity ensures that, once the equipment is installed, it 'does what it says on the can', i.e. OQ tests the operation of the system against predetermined acceptance criteria.
- *Performance Qualification (PQ)* This activity ensures that, using relevant starting materials, the system can reliably and reproducibly perform as intended in its final operating environment/conditions.
- *Process Validation (PV)* This establishes that when all the separate systems (validated individually by the steps above) are brought together to turn starting materials into a finished product, the overall process will consistently produce a product meeting its predetermined specifications and quality attributes. The process must be clearly defined.

Each of these components will be dealt with individually below, but by considering the descriptions above it can be seen that, excluding PV, they relate to one another in a notional V-shape, with construction (of the system to be validated) at its point. This is shown in Figure 15.1. The bold diagonal arrows show the chronological order in which each component is generated or carried out, and the horizontal arrows show how these relate to one another. Thus the earliest stage, the URS, is checked out at the last formal qualification stage, PQ. The detailed FS for the system is checked during the OQ and the design and construction of the system is checked during the IQ. These relationships hold good in general, but in some projects the interrelationships may be rather more complex.

From the above, it can be seen that Qualification is part of Validation, but that the individual qualification steps alone do not constitute validation.

15.3 VALIDATION PLANNING

The very first step in validation planning is to define the boundary of the system to be validated. With a simple piece of equipment this boundary may be very obvious, but as the item gets more complex, so do the decisions defining the boundary. A system is defined in the ISPE baseline guide (ISPE 2001a) as 'An organization of engineering components that has a defined operational function (e.g. piping, instrumentation, equipment, facilities, computer hardware, computer software, etc.)' System boundary is defined as 'a limit drawn around a system logically to define what is and is not included in the system.' So, for example, if validating a system for cleaning vessels in-place, one must decide whether to include the water supply pipework and the cleaning solution pipework within the system boundary.

Once the system boundary has been defined, an impact assessment can be carried out. This evaluates the impact of a system on product quality, and identifies the critical components within the

Table 15.1 Questions to help assess the impact of a system.

(1) Does the system have direct contact with the product?
(2) Does the system provide an excipient, or produce an ingredient or solvent (e.g. water for injection)?
(3) Is the system used in cleaning or sterilizing?
(4) Does the system preserve product status?
(5) Does the system produce data that is used to accept or reject product?
(6) Is the system a process control system that may affect product quality, but without a mechanism in place whereby control system performance can be independently verified?

system. To help in determining whether systems are 'direct impact', 'indirect impact' or 'no impact' (these last two terms are interchangeable), the ISPE baseline guide (ISPE 2001a) lists six questions (Table 15.1). If any of these questions are answered in the affirmative, then the system is deemed a direct impact system. However, as the ISPE guide states, these six questions are there for help and should not replace the exercise of informed judgment by appropriately qualified personnel.

After the systems have been assessed with regard to their impact, the next stage is to assess the criticality of the components within any 'direct impact' systems.

The ISPE baseline guide (ISPE 2001a) identifies a critical component as 'a component within a system where the operation, contact with product, data, control alarm, or failure will have a direct impact on the quality of the product'. To help in this categorization, the guide lists seven questions to help distinguish critical components from non-critical components. These are shown in Table 15.2. If any of these questions is answered in the affirmative, then the component is deemed to be critical.

Once the system boundaries have been defined and the system and its associated components have been categorized, resources can be focused on those key parts of the system that require validation. This avoids the wasteful approach of simply validating everything, as was common practice in the past

15.3.1 Risk Assessment

Unlike impact assessments, the performance of which is simply best practice, the requirement to perform a risk assessment as part of the validation exercise is a regulatory requirement in Europe (European Commission 2001a). For the USA the situation is not quite so clear cut. The Food and Drug Administration (FDA) has been mandating risk management for medical devices since 1996 through the Quality System Regulation (QSR) 21 CFR Part 820 (FDA 2005a). Risk assessment is also a key element of the FDA's 21st century drug cGMP initiative (FDA 2002). Thus risk assessment is clearly encouraged by the FDA, but may not be mandatory.

Table 15.2 Questions to help distinguish critical from non-critical components.

(1) Is the component used to demonstrate compliance with a registered process?
(2) Does the normal operation or control of the component have an effect on product quality?
(3) Will failure or alarm of the component have a direct effect on product quality?
(4) Is information from this component recorded as part of the batch record, lot release data, or other GMP-related documentation?
(5) Does the component have direct contact with product or product components?
(6) Does the component control critical process elements that may affect product quality, and is it without independent verification of the control system performance?
(7) Is the component used to create or preserve a critical status of a system?

Risk assessment is composed of two parts: understanding the risk via risk analysis, and then evaluating the risk (and accepting it if minimal), so-called risk evaluation. These are followed by risk control, where any necessary controls are put in place, or strategies are developed to help minimize the risk. Risk management is a combination of these three separate parts.

There are many different methods of risk assessment, including hazard and operability study (HAZOP) (IEC, 2001) and hazard analysis and critical control point (HACCP) (ASQ, 2001) see also Chapter 31. However, failure modes and effects analysis (FMEA) is finding widespread use in the pharmaceutical industry, and its use can provide a structure for the validation activity, helping to identify risks associated with the critical components. Further information on the different types of risk assessment and their implementation can be found in the dedicated reference listing at the end of this chapter, but probably the best overview has been provided by the International Conference on Harmonization in their document *Quality Risk Management* Q9 (ICH 2005).

Through the use of the above processes the validation requirements of a system can be identified. These are normally described in the validation master plan.

15.3.2 The Validation Master Plan (VMP)

In the opinion of the author, this document is frequently made far too long. *The Rules Governing Medicinal Products in the European Union* Volume 4, under Annex 15, *Planning for Validation,* state that 'The VMP should be a summary document, which is brief, concise and clear.' (European Commission 2001a) Too often, VMPs contain large amounts of information and a great deal of repetition, little of which facilitates the intended function of the document, namely to lay out the requirements and methods for performance of the validation. Keep the VMP concise – put the detail in the other planning documents.

As a general rule, the VMP should contain the following types of information:

- *Organizational structure of the validation activities (validation teams)* For validation to be successful a team approach is normally the best practice. The team planning and executing the validation should include, as a minimum, representatives from the following departments: engineering, the end user department, quality assurance (QA), and quality control (QC), and research and development (R&D) as necessary. This is not an exhaustive list, but the VMP must clearly define team members and their expected roles and responsibilities throughout the validation exercise. When selecting team members it is vitally important that each team member is aware of the time commitment required to complete the project, and any training requirements. Too many validation projects fall apart due to the unavailabity of key team members, or the use of inappropriately trained individuals.
- *Summary of the facilities, systems, equipment and processes to be validated.* This can be in the form of diagrams, but must provide sufficient background information for the reader of the document to understand exactly the boundary of the validation activities.
- *Signature Registration.* The US FDA recommends (FDA 2005b) that all persons involved in testing and reviewing have their full names and signatures documented, to allow identification of the individuals who have performed any part of the qualification. The VMP should indicate how this will be achieved in individual protocols.
- *Documentation format.* This describes the format to be used for protocols and reports, and should provide the reader with a clear understanding of the interrelationships between the various parts of the qualification documentation.
- *Planning and scheduling.* It is the author's opinion that this should be the main focus of the VMP. It must be remembered that validation activities normally involve many different members

of different departments. Without a clearly agreed timetable incorporating agreed schedules that are realistic and achievable, then the validation activity is almost always doomed to failure. The importance of good planning, and the involvement of a multidisciplinary project validation team with the appropriate level of decision-making power to ensure that the planned schedule is both agreed and adequately resourced, cannot be over-emphasized.

Additionally, a VMP for a whole facility would include flow diagrams of the various components employed in the process, including such things as the movement of dirty vessels and their segregation from clean vessels to ensure there is no crossover. A flow diagram (or diagrams) of the process itself would also be essential. This/these would show important features such as how, for example, once product had been virally inactivated, the risk of recontamination would be minimized by segregation from untreated product and its related equipment.

The VMP should also clearly indicate how the validation exercise will integrate the other components of validation (see above). The VMP is the highest-level validation document, and requires authorization by senior members of relevant disciplines. It must be remembered that the VMP should be a living document and should be reviewed and updated as appropriate through the entirety of the validation project.

15.4 DESIGN DOCUMENTS

The Rules Governing Medicinal Products in the European Union Volume 4 covers Design Qualification in four lines of text (European Commission, 2001a) and the American Code of Federal Regulations (CFR) makes no specific reference to design documents (FDA 2005b). Nevertheless, the design documents create the bedrock on which the success of the validation will be built. Normally, poor design documentation is reflected in poor qualification documentation, and ultimately in a facility or equipment that is either different from that envisaged by the end user or, worse, is not up to the required job. Subsequent modification or replacement can be very, very expensive, so time and effort expended in generating good design documentation is an extremely good investment.

15.4.1 User Requirement Specification

Ultimately, validation is a service, normally performed for an end-user or users who actually require the functioning system – usually to a timescale that does not take into account the actual time required to create, execute, review and report the validation requirements!

One way to get buy-in from the end-users, and also ensure that they get the desired equipment/facility, is to involve them in the creation of the user requirement specification. This document should be prepared once the VMP has been completed, and is essential for the performance of the validation. There are a number of sources of excellent templates that can be used in the creation of a URS. These include the ISPE Baseline guides, especially Volume 5 (ISPE 2001a), and the good automated manufacturing practice (GAMP) guidelines. Although devised for computer system validation, the DQ template can be modified to fit other systems (ISPE 2001b).

It is always important to ensure that the URS contains the specifications with which the system is expected to comply. For example, if a facility requires an HVAC (heating, ventilation and air conditioning) system then the EU, FDA or other appropriate standard of air required in each room must be specified. It is always safer, and makes the document easier to comply with for the supplier, if all required standards are listed. For example, the URS may include the statement that the facility requires a EU Grade A, ISO 5 Isolator, the ISO specification (ISO 1999) being the engineering design standard (for a description of air grades see Chapter 12). This information provides the proposed supplier with details of the requirements in terms of filters, air flow and environment required.

It is vital that the URS makes it clear which regulatory standards must be complied with. So if a facility is expected to be involved in the manufacturing or testing of products that are intended for the American market, then the requirements listed in the URS must be in compliance with the FDA expectations as laid out in CFR 210 and 211 (FDA 2005b).

For clean room facilities, ease of cleaning must be considered when putting together the URS. It is essential that surfaces be crevice-free and manufactured from materials that are non-particle shedding and resistant to detergents. Some UPVC vinyl finishing, for example, is prone to staining when cleaned repeatedly with certain commercially available detergents. This can lead to the facility being cited for non-conformance during regulatory inspections, the argument being that the staining could be the result of product residue and not detergents.

The URS is the cornerstone of any successful validation. If the URS provides the required level of detail, and this is followed through with documents that provide a matrix to ensure that all user-required points have been covered, then the validation exercise has a good chance of success.

15.4.2 Functional Specification

Normally, the URS is sent to a number of potential suppliers who respond with either a functional specification (FS) or a functional design specification (FDS) (the latter normally incorporating the hardware and software design specifications that are required for computerised equipment). The FS should provide sufficient detail for the end-users to see exactly how the supplier intends to meet their requirements. It is normal for the FS to provide a greater level of technical detail than was provided in the URS, but it is vitally important that these details are carefully reviewed and commented on. For large-scale projects, the delivery of the FS by the supplier may be a payment point and constitute the end of the design phase for the supplier. This must be borne in mind by the validation group, and the supplier's FS should only be accepted once the end-users and validation team are happy that it has answered all the requirements of the URS and is sufficiently detailed to satisfy them that the supplier has understood the requirements and responded appropriately. It must however be remembered that in terms of sign-off, the authorship is that of the supplier, not the purchaser.

If the supplier does not show a good understanding of the requirements of the system through producing a detailed FS, then validation stands little chance of success, and it may be pragmatic to identify an alternative supplier.

15.5 QUALIFICATION ACTIVITIES

15.5.1 Design Qualification

In simple terms, the design qualification stage ensures that what was asked for by the end-user is what the supplier has actually specified. This stage can be completed in a number of different ways. Some validation sources (e.g. GAMP) (ISPE 2001b) define DQ as 'a series of stages taking place throughout the process of construction, during which progress against the design documents is reviewed'. In this case, DQ is only complete once IQ is ready to commence. A second approach treats DQ as a discrete validation stage. In this approach the DQ is performed as a review of all key design documents. Normally, a matrix is generated of the URS requirements against the functional specification to ensure that the FS addresses all the URS requirements. The DQ report highlights any areas where the URS requirements have not been met, and a written justification would be required for any discrepancy. Normally, execution should be completed by the validation team engineering representative.

15.5.2 Installation Qualification

Installation qualification (IQ) is defined as the performance and documentation of tests to ensure that equipment used in the manufacturing process is correctly installed in accordance with established specifications. In simple terms, IQ is the process of ensuring that the system, carefully designed with the help of the techniques listed above, has been constructed and installed to the correct specifications. The IQ can follow on from the supplier's checks of the system, and if performed at the supplier's premises is termed factory acceptance testing (FAT). Testing at the customer's site is termed site acceptance testing (SAT). These two types of testing are normally considered to be engineering functions and do not fall under the formal GMP qualification umbrella. In recent years there has been a move to use SAT, and sometimes even FAT, to replace or supplement the purchaser's own IQ/OQ. This has the benefit of reducing lead times and cutting down the cost of validation. In addition, an argument can be made that the supplier is best placed to write qualification protocols as they have a greater understanding of the system being supplied. This approach can work well, but there are a number of precautions that need to be taken.

- It is important to check that the supplier's qualification package conforms to GMP standards, as in some cases the tests covered can be very formulaic, i.e. they are generic and not specific to the actual system purchased by the customer. This can lead to issues where any customization, or changes to the basic system made for the customer, are not covered in the standard supplier test documentation.
- Care must be taken to ensure that the supplier's documentation fits with the purchaser's documentation and validation policy, so that things such as test sheets comply with the required way of representing data and of documenting the completion of testing. It should be noted that some larger suppliers are not prepared to make changes to standard FAT or SAT documents due to the costs of customizing them for each customer.
- It must be remembered that testing performed during FAT is being undertaken on equipment that has only just been completed, and may not even have been commissioned. Consequently, the results from the FAT may raise issues that would need to be formally documented and resolved to the satisfaction of the customer's QA department. All changes would require some form of customer approval, and this can be quite difficult if the supplier's premises are some distance from the customer's site. It must be remembered that if the FAT is to be used by the customer as evidence of formal qualification for GMP purposes, then the requirement for full documentation and customer input is greatly increased.
- Ultimately FAT is being undertaken in surroundings different from those intended for routine operation, and other than ensuring that the system actually functions, the test results may be of limited use. Consequently, most pharmaceutical companies class FAT as an engineering activity rather than a formal validation activity. That is not to say, however, that a regulatory auditor will not ask to see the results of any FAT testing. Thus it is always advisable for the purchaser to have a representative present during FAT.

SAT can be performed before or after the customer has performed the formal installation qualification. It can be performed as part of commissioning, or more formally as a separate activity. Normally SAT provides the assurance that the system has been installed in compliance with the URS and that the system operates as specified in the functional specification. It is useful, therefore, to ensure that the SAT protocol has a cross-reference table linking to the FS, to act as a check that all key specifications required in the FS have been fully tested. SAT is normally the point where the supplier hands over the equipment to the customer. (Some suppliers will help with the purchaser's IQ/OQ execution, but normally at a cost!) It is therefore vital to ensure that the system is functioning to the customer's satisfaction, as the balance of payment is usually made after completion of the SAT and getting the supplier back to fix issues after this point can be costly and time-consuming.

It is always important for the customer to remember that ultimately they are the party responsible for performing any validation that will be reviewed by a regulatory body, and that any savings made by just using supplier documentation may be lost if the outcome of a regulatory audit is poor.

15.5.2.1 IQ documentation

A good way to create an IQ protocol for equipment (although with some modifications it can also be used for utilities or facilities) is from a template made up of a number of suggested tests that can be removed or added to, depending on the complexity of the system being qualified. Table 15.3 lists examples of the tests that can be included in the IQ. The use of a template has the advantage that it brings a similar format to all qualification protocols and ensures that key tests are performed. The disadvantage is that using a template has a tendency to narrow focus onto only those tests listed, rather than to think about the actual requirements for the specific piece of equipment undergoing qualification.

Table 15.3 gives the key elements with some text to explain the testing requirements. Like any qualification, the key to obtaining meaningful results is to ensure that test acceptance criteria have a valid scientific rationale, and where possible are referenced back to the URS/FS.

Once the IQ protocol has been executed then it is normal practice for the data to be reviewed and a summary report completed, providing a concise review of the test results and describing any changes or deviations required as a result of testing. The report is normally circulated to the validation team members referenced in the VMP. It is recommended that QA sign off all qualification summary reports.

15.5.3 Operational Qualification

On completion of the IQ, the operational qualification (OQ) can commence. OQ tests the operation of the equipment against predetermined acceptance criteria.

The PIC/S guide offers the following requirements for OQ (PIC/S 2004b): 'Studies on the critical variables (parameters) of the operation of the equipment or systems will define the critical characteristics for operation of the system or sub-system'. Put simply, the OQ protocol should provide evidence to demonstrate that the system is capable of producing desired results in a reproducible manner. The testing can include checking the operation of the system at its operational extremes to ensure that it is robust. OQ can be viewed as a dry run of the system prior to the final end use of the system. PIC/S (PIC/S 2004b) makes the following observation: 'Where applicable, simulated product may be used to conduct the Operational Qualification'.

As shown in Figure 15.1, the OQ covers testing of the functional specification of the system. The main areas included in OQ are listed below in Sections 15.5.3.1 and 15.5.3.2.

15.5.3.1 Before testing commences

15.5.3.1a Required materials

A list must be made of all the instrumentation and consumables that will be required during the OQ of the particular system under test. This is used to ensure that the necessary items are available

Table 15.3 Tests and key elements to be included in installation qualification.

Pre-requisite Documents	This test ensures that the following key documents are present and approved prior to starting IQ. • URS/FS/ FDS • purchase order • supplier audit (by purchaser) • FAT documents • manufacturer-supplied training document/procedure • SAT documents • health and safety risk assessment
Equipment Manuals	List all applicable equipment manuals and verify they are filed and available for reference as needed.
Drawings and Schematics	Ensure that the latest revision of all drawings required through change control, or that contain GMP attributes, are present. Check through the 'as-built' drawings, and ascertain their accuracy and completeness by comparing them to the actual system.
Standard Operating Procedures (SOPs)/ Maintenance Programme	Ensure that all applicable equipment SOPs, maintenance procedures and equipment logbooks are present and contain the necessary content. It is not expected that SOPs will be QA-approved at this stage, but SOPs are expected to be in a draft form by the end of IQ.
Parts and Consumables List	A parts and consumables list based on the manufacturer's recommendations is necessary for ordering spare parts during routine maintenance.
Materials in Product Contact List	Ensure that the specified materials have been used for the equipment and sub-system components that come in contact with the product. Verify the actual material against that specified, and document the method of verification in each case. Verification can be confirmed from original documentation (e.g. manuals or certificates) supplied with the equipment or from documented testing carried out with calibrated instruments.
Control System Hardware/ Software List	For the key aspects of the software/hardware critical components, list the required version number, model types and serial numbers.
Environmental Requirements	Document the conditions and the specified target, tolerance and quality levels required. Ensure that these are met using calibrated instruments.
Equipment-generated Particles	Verify that the equipment does not generate any particles that may come in contact with the product or adversely affect the specified air grade of the environment.
Utilities	Document the specified utility requirements for the equipment, including target values, specified ranges and quality levels. Verify that acceptable utility supplies are available, using calibrated instruments.
Operating Specifications	Summarize the manufacturer-specified equipment and actual operating parameters/conditions of the equipment, i.e. operating ranges for instruments attached to the system, include applicable controller settings. Verify that the equipment/instrument is properly set to run within these approved parameters.
Calibration	List instruments associated with the equipment and document that each is appropriately labelled as critical or non-critical. Assure that critical instruments are entered into the calibration programme and have an acceptable range/accuracy.
Change Control Log/ Deviation Log	It is important to record any deviations that need to be raised as a result of the testing. These deviations are normally reviewed by QA prior to closure (see Glossary for terminology). Normally, deviations would be closed before moving on to the next stage of the qualification.

prior to the commencement of OQ, and that the test instrumentation has been calibrated to the required precision (and this has been formally documented in a calibration certificate).

15.5.3.1b Training records

It is essential to verify that all users/operatives have been fully trained in the execution of the relevant protocols and test equipment. It is also important to ensure that operators of systems have received training before or during the OQ, particularly if material intended for patient use is to be manufactured during the PQ. (It is a GMP requirement that all operators are fully trained when manufacturing any material for patient use (European Commission 2001b; FDA 2005c).)

15.5.3.2 Testing

Not all of the tests below may be necessary – it will depend on the nature of the system under qualification.

System Set-up Verification: This test verifies that the system start-up procedure has been followed, is complete, and the system is ready for operation.

Power Failure Recovery: The power failure recovery test demonstrates that the system can be restored to a required operational state after a power interruption and that any battery back-up or standby generators function as specified.

Temperature Mapping: It is important if dealing with a system where the operating temperature is important (e.g. a cold room, or an autoclave) that mapping exercises be performed. In this test, the area to be temperature mapped would be described, with dimensions, operating range, method of temperature control, etc. Relevant conditions for testing must also be defined. For an autoclave, for example, one must stipulate the number of runs to be performed, the time of each run, the type, number, and of thermocouple probes to be used, and the load. It is normal to test the system both empty and with the most challenging load that is likely to be used. Additional challenges may be required, such as an open-door study (for a cold room or incubator) or mapping after loss of power, to determine the length of time before a failing temperature is reached, and the recovery behaviour once power is restored.

As a guideline, the documentation of temperature mapping can have three parts. The first is a list of all the thermo-probes' identification numbers and placements, including pre- and post-calibration information, and run information (e.g. cycle parameters, set points, start and stop times, recording intervals). The second is a diagram of the area to be temperature mapped (for example, a chamber or heated surface), showing the placement of each thermo-probe along with its identification number. Placement must be in key areas (e.g. in parallel with the control point, and in likely hot and cold spots) and throughout the area dimensionally and spatially. The third is a summary of the data. Temperature graphs over the duration of each run are normally generated and attached to the raw data.

Alarms, Safety Devices and Interlocks: This test verifies that all alarms, safety devices and interlocks respond as designed, at all possible levels, categories or conditions. It ensures that all safety measures operate correctly prior to checking the operation of the equipment and its sub-systems.

Operator Panel and Screen Navigation Verification: This test verifies (with reference to the relevant specifications) that all operator-interactive devices/interfaces and indicator lights, e.g. computer driven menus, touch-screens, manual switches, push buttons, warning lights, etc., operate correctly. It also includes verification of the navigation through screen menus.

Password Testing and Security Testing: Regulatory expectation regarding validation and testing of computerized systems is covered in some detail in Annex 11 of *The Rules Governing Medicinal Products in the European Union* Volume 4 (European Commission 2001c), and for the US in CFR 21 Part 11 (FDA 1997). Both expect that computerized systems should be secure and that the data and any changes to it are fully traceable. Tests should be performed to check the various security system functions and the security measures utilized in controlling access to the equipment/system. For those systems that have both a user ID and password combination, testing should be undertaken to ensure the security of the system is adequately controlled. Testing can include checks that access is granted only to authorized individuals, that password issuances are periodically checked and revised, that access is only granted using a correct password and user ID combination, and that the user names of all accounts are unique.

Report Generation: Report generation testing checks that a report can be created as specified in the FS and, if the system uses electronic audit trails (see Glossary for definition), that the audit log records any changes made to electronic records, and can be printed out. The testing should also check that the reports, and data behind the reports, cannot be altered after the data has been stored.

Report/Data Archiving: Following on from security testing is the requirement that any data generated by the system, and which must be kept, is stored in a secure and retrievable manner. The data archiving system should be described, and then tests run to verify that data retained in the system on a specified day can be retrieved at a later date and shown to be the same.

Operational Sequence and Station Functional Inspections: By stepping through the operation of the system, this test should verify that the sequence and functions are correct in accordance with the FS.

On-line Inspection Devices: The operation and control of on-line inspection devices (such as those used on a packing line), and what each measures, should be described and tested, along with the acceptable range for each measurement. Devices should be challenged with a known set of unacceptable items dispersed amongst a large number of acceptable ones. Define how the testing will be done, including the method and sample size. Test-sample size must be based on a statistical rationale

On completion of OQ it is normal practice, as with IQ, for the data to be reviewed and a summary report completed providing a concise review of the test results, and describing any changes required or deviations that need to be raised as a result of testing. (For more information on the process of raising and resolving deviations, see 'Deviation' in the Glossary.) The summary report is normally circulated to the validation team members referenced in the VMP.

Operational Qualification, if performed correctly, will provide the end user with confidence that the system is robust and capable of operating in a reproducible manner.

15.5.4 Performance Qualification

In its simplest form PQ is the testing of the system in its intended normal operating environment and conditions.

The Rules Governing Medicinal Products in the European Union Volume 4 states that 'PQ should follow successful completion of IQ and OQ, and the key considerations should be testing using production materials, qualified substitutes or simulated products that have been developed from knowledge of the process and the facilities, systems or equipment. Testing is to include conditions or sets of conditions that encompass upper and lower operating limits.' (European Commission 2001d.)

From the first sentence above, it can be extrapolated that PQ should be performed on pre-qualified equipment using substances with well known properties. The knowledge of these substances can be built up during the previous qualification activities. To put it in simple terms, by the time PQ is under way there should be few surprises to uncover.

The second part of the statement is harder to interpret, as historically PQ has been seen to be about consistent performance. Normal practice has been that PQ involves (at least) three sequential runs to show that the system is performing in a consistent manner. To start including runs at upper and lower operating limits could not only lead to the requirement for additional runs, but can place additional demands on end-user staff who would be expected to run the equipment during PQ. To avoid this issue, it is worth remembering that during the OQ stage, tests at the upper and lower operating limits are performed. Providing the production parameters have been defined by the time OQ is performed, it should be possible to avoid lengthy delays in executing limit testing during PQ.

It must be remembered that if the material produced during PQ runs is intended for use in patients within Europe, then the requirement in *The Rules Governing Medicinal Products in the European Union* Volume 4 (European Commission 2001e) must be followed: 'If it is intended that validation batches be sold or supplied, the conditions under which they are produced should comply fully with the requirements of Good Manufacturing Practice, including the satisfactory outcome of the validation exercise and with the marketing authorisation'.

PQ can provide useful data on the operation of the system in its intended environment. Once the PQ is signed off, this is not the end of validation, just the completion of the qualification stages. It is a requirement in Europe (European Commission 2001f) that the validation status of a system must be reviewed if changes are made to it, and that any such changes and any consequent re-validation requirements be recorded under a system of change control. For the American market, revalidation is also required (FDA 1987). Normally the validation status of systems is also reviewed after a defined period of time governed by the validation policies of the company.

15.6 PROCESS VALIDATION

The facility, services, and other systems have been purchased, constructed and, by this stage, validated individually, in order to run a process to make a product. The final stage of validation, then, is process validation (PV), which is performed to ensure that the validated systems can be integrated and will work together to manufacture reproducibly product that complies with the specification and release criteria identified during process development.

PV can be completed by producing a minimum of three consecutive product batches in order to demonstrate the robustness of the process. PV batches should be manufactured at the same scale as that intended for commercial batches (European Commission 2001e) and the (validated) equipment used should be the same as that intended for use on subsequent commercial batches, and be run using the same control parameters (possibly with additional parameters measured for information). Where possible the PV batches should be manufactured by the same production staff, and using the same batch processing records and SOPs, as will be used for commercial batches. This final point is often overlooked. It is not unknown for R&D personnel, who have been heavily involved in the introduction and scale-up of the process, to be in charge of manufacturing the PV batches. However, this can lead to issues once commercial production starts, as the regular process personnel can in this situation be lacking in actual production experience in the new process, possibly leading to loss of commercial batches due to lack of familiarity with the new process.

In order to start performing the PV, it is necessary that the process itself is well defined, and has clear control points (see Glossary). This can be helped by performing a risk assessment to establish the critical control parameters to be monitored. Ideally, this risk assessment will have been performed during the preparation of the VMP, but may require modification prior to creation of the PV protocol. It is very useful to ensure that the R&D, production and engineering departments have an input into the creation of the risk assessment and agree on the critical parameters to be monitored. It is also important to ensure that the methods of measurement have all been validated to the required level of accuracy and specificity. When selecting these parameters it must be remembered that the PV should have limits set that are achievable and fall within a proven acceptable range (PAR), this range having been established during process development. Some companies perform PV at the normal operating range, where the parameters are tighter than the PAR. Whatever the range, it is important (in order to save a great deal of time and energy) to ensure that the process is capable of meeting the parameters to be used during the PV. This will reduce the need to raise deviations, and minimize investigations into failures.

Another key area to consider during planning is resources. For example, are enough of the correctly trained personnel available? Support personnel such as analytical laboratory staff, equipment preparation staff, engineers and QA are often overlooked in this respect. In a multi-product facility, liaison with production planning will be required to ensure that there are enough production slots available for the PV batches to be manufactured to a timetable that matches the plans for submitting the licence to the regulatory authority.

The final activity to perform before commencing PV is a review of facility readiness. This ensures that the production personnel, other staff, and the facility are in a state ready for production of the PV batches in a manner compatible with GMP.

Chapter 6 of the PIC/S guide (PIC/S 2004a) provides the following guidance on what is required in a PV protocol.

> 'Each experiment should be planned and documented fully in an authorized protocol. This document will have the following elements:
>
> (a) a description of the process;
> (b) a description of the experiment;
> (c) details of the equipment/facilities to be used (including measuring/recording equipment) together with its calibration status;
> (d) the variables to be monitored;
> (e) the samples to be taken – where, when, how and how many;
> (f) the product performance characteristics/attributes to be monitored, together with the test methods;
> (g) the acceptable limits;
> (h) time schedules;
> (i) personnel responsibilities;
> (j) details of methods for recording and evaluating results, including statistical analysis'.

Once again the best source of reference when writing the report on the PV is Chapter 6 of the PIC/S guide (PIC/S 2004a), which provides the following recommendations of items to be included in the PV report:

> (a) a description of the process – batch/packaging document, including details of critical steps;
> (b) a detailed summary of the results obtained from in-process and final testing, including data from failed tests. When raw data are not included reference should be made to the sources used and where it can be found;
> (c) any work done in addition to that specified in the protocol or any deviations from the protocol should be formally noted along with an explanation;
> (d) a review and comparison of the results with those expected;
> (e) formal acceptance/rejection of the work by the team/persons designated as being responsible for the validation, after completion of any corrective action or repeated work'.

For part (d), it should be remembered that details of any failed PV batches must be included in the report as these can provide vital information on the process. All failures must be investigated and the report must provide the reader with confidence that the issues have not only been understood but have been resolved in a satisfactory manner.

If PV is performed correctly, then the documentation created should provide a clear picture of the robustness and reproducibility of the process. PV is an important part of the validation lifecycle.

15.7 SUMMARY

For validation to be successful, it must be pre-planned, the system under validation must be well understood, and the boundaries of the system must be clearly defined. The design of the system must be documented not only by the user but also by the supplier. The design documents should be checked and differences agreed during the design phase. The IQ and OQ stages give assurance that the system has been installed and can operate in a manner that is both robust and reproducible. The operation of the system in its intended environment should be checked during PQ, and the ability of the interrelated systems to make product in a reproducible manner should be checked during process validation. Finally, to ensure that systems remain in a validated state, periodic revalidation is required. If all these stages are performed and (as importantly) documented, then the validation activities should promote a successful outcome to a facility inspection by a regulatory authority.

REFERENCES

ASQ (2001) *The Quality Auditor's HACCP Handbook*, American Society for Quality; Food, Drug, and Cosmetic Division, Milwaukee, USA.

European Commission (2001a) *The Rules Governing Medicinal Products in the European Union*. Volume 4 – Medicinal Products for Human and Veterinary Use: Good Manufacturing Practice. Annex 15, Qualification and Validation, Sept 2001. See http://ec.europa.eu/enterprise/pharmaceuticals/eudralex/vol-4/pdfs-en/v4an15.pdf

European Commission (2001b) *The Rules Governing Medicinal Products in the European Union*. Volume 4 – Medicinal Products for Human and Veterinary Use: Good Manufacturing Practice. Section 2.8–2.12. Sept 2001. See http://ec.europa.eu/enterprise/pharmaceuticals/eudralex/vol-4/pdfs-en/cap4en200408.pdf

European Commission (2001c) *The Rules Governing Medicinal Products in the European Union*. Volume 4 – Medicinal Products for Human and Veterinary Use: Good Manufacturing Practice. Annex 11. Sept 2001. See http://ec.europa.eu/enterprise/pharmaceuticals/eudralex/vol-4/pdfs-en/anx11en.pdf

European Commission (2001d) *The Rules Governing Medicinal Products in the European Union*. Volume 4 – Medicinal Products for Human and Veterinary Use: Good Manufacturing Practice. Annex 15, Section 16. Sept 2001. See http://ec.europa.eu/enterprise/pharmaceuticals/eudralex/vol-4/pdfs-en/v4an15.pdf

European Commission (2001e) *The Rules Governing Medicinal Products in the European Union*. Volume 4 – Medicinal Products for Human and Veterinary Use: Good Manufacturing Practice. Annex 15, Section 27. Sept 2001. See http://ec.europa.eu/enterprise/pharmaceuticals/eudralex/vol-4/pdfs-en/v4an15.pdf

European Commission (2001f) *The Rules Governing Medicinal Products in the European Union*. Volume 4 – Medicinal Products for Human and Veterinary Use: Good Manufacturing Practice. Chapter 5, Section 5.24. See http://ec.europa.eu/enterprise/pharmaceuticals/eudralex/vol-4/pdfs-en/cap5en.pdf

FDA (1987) *Guidelines on General Principles of Process Validation*, May 1987. Center For Drug and Biologics, and Center for Devices and Radiological Health FDA. (A revision of this is pending.) See web link http://www.fda.gov/cber/gdlns/validation0587.pdf

FDA (1997) Title 21, *Code of Federal Regulations*, Part 11.

Electronic Records; Electronic Signatures: Final Rule. Electronic Submissions; Establishment of Public Docket. Department of Health and Human Services. Date Effective: August 20, 1997. http://www.fda.gov/ora/compliance_ref/part11/frs/background/11cfr-fr.htm

FDA (2002) *FDA Pharmaceutical Current Good Manufacturing Practices* (CGMPs) for the 21st Century A risk based approach, March 2002 see website for up to date information on this initiative. http://www.fda.gov/cder/gmp/gmp2004/GMP_finalreport2004.htm

FDA (2005a) Title 21, *Code of Federal Regulations*, Part 820, Subchapter H, Subpart B – Quality System Requirements. Food and Drug Administration, Department of Health and Human Services. Revised April 1, 2005. See http://www.accessdata.fda.gov/scripts/cdrh/cfdocs/cfcfr/CFRSearch.cfm?CFRPart=820&showFR=1&subpartNode=21:8.0.1.1.12.2

FDA (2005b) Title 21, *Code of Federal Regulations*. Part 210 – Current Good Manufacturing Practice in Manufacturing, Processing, Packing, or Holding of Drugs; General Part 211 – Current Good Manufacturing Practice for Finished Pharmaceuticals. See http://www.fda.gov/cder/dmpq/cgmpregs.htm

FDA (2005c) Title 21 *Code of Federal Regulations*. Part 210 – Current Good Manufacturing Practice in Manufacturing, Processing, Packing, or Holding of Drugs; General Part 211.25 – Current Good Manufacturing Practice for Finished Pharmaceuticals. See http://www.fda.gov/cder/dmpq/cgmpregs.htm

IEC (2001) *Hazard and Operability Studies* (HAZOP Studies) – *Application Guide*. International Electrotechnical Commission, Geneva, Switzerland. See http://www.iec.ch

ICH (1997) *Derivation and Characterization of Cell Substrates Used for Production of Biotechnological/Biological Products Q5D*. International Conference on Harmonization of Technical Requirements for Registration of Pharmaceuticals for Human Use, Geneva, Switzerland. See http://www.ich.org/LOB/media/MEDIA429.pdf

ICH (2005) *Quality Risk Management* Q9. International Conference on Harmonization of Technical Requirements for Registration of Pharmaceuticals for Human Use, Geneva, Switzerland. See http://www.ich.org/LOB/media/MEDIA1957.pdf

ISO (1999) ISO 14644-1, *Cleanrooms and Associated Controlled Environments* – Part 1: *Classification of Air Cleanliness*. International Standards Organization, Geneva, Switzerland. Standard available from http://www.iso.org/iso/en/ISOOnline.frontpage

ISPE (2001a) *Pharmaceutical Engineering Guide for New and Renovated Facilities*. Volume 5, *Commissioning and Qualification*. 1st edition, March 2001, ISPE European Office 300, Brussels, Belgium. See http://www.ispe.org

ISPE (2001b) *The Good Automated Manufacturing Practice* (GAMP) *Guide for Validation of Automated Systems in Pharmaceutical Manufacture*, version 4. December 2001 ISPE Organisation. See http://www.ispe.org

PIC/S (2004a) *Recommendation PI006-2*: Validation master plan installation and operational qualification non-sterile process validation cleaning validation. Chapter 6. 1 July 2004, Pharmaceutical Inspection Convention Pharmaceutical Inspection Co-Operation Scheme. Available from http://www.picscheme.org

PIC/S (2004b) *Recommendation PI006-2*: Validation master plan installation and operational qualification non-sterile process validation cleaning validation. Chapter 5.2. 1 July 2004, Pharmaceutical Inspection Convention Pharmaceutical Inspection Co-Operation Scheme. Available from http://www.picscheme.org

WHO (1998) *Technical Report Series 878*, Annex 1, World Health Organization, Geneva, Switzerland. See http://www.who.int/biologicals/publications/trs/areas/vaccines/cells/WHO_TRS_878_A1Animalcells.pdf

Additional Risk Assessment References

Guidelines for Failure Modes and Effects Analysis (FMEA) for Medical Devices. Dyadem Press, Richmond Hill, ON, Canada, 2003.

Although primarily aimed at computer system validation, the risk assessment section of ISPE (2001b) – see above – provides a good starting point when contemplating validation and risk assessment.

Fault Tree Analysis (FTA) International Electrotechnical Commission, Geneva, Switzerland, 1990. See http://www.iec.ch

Useful Web Sites

European Medicines Agency http://www.emea.europa.eu

European Pharmaceutical Regulations http://pharmacos.eudra.org/F2/home.html

GAMP: Guide to Good Automated Manufacturing Practice Validation of Automated Systems http://www.ispe.org/gamp
International Conference on Harmonisation http://www.ich.org/
International Standards Organization http://www.iso.org/iso/en/ISOOnline.frontpage
International Society for Pharmaceutical Engineering http://www.ispe.org/
Institute of Validation Technology http://www.ivthome.com
Pharmaceutical Inspection Cooperation Scheme http://www.picscheme.org
US FDA Inspection Guides and Guidelines http://www.fda.gov/ora/inspect_ref/igs/iglist.html
US FDA Regulations http://www.fda.gov/

GLOSSARY

Active Pharmaceutical Ingredient (API): Any substance or mixture of substances to be used in the manufacture of a drug (medicinal) product and that, when used in the production of a drug, becomes an active ingredient of the drug product. Such substances are intended to furnish pharmacological activity or other direct effect in the diagnosis, cure, mitigation, treatment or prevention of disease, or to affect the structure and function of the body (Source: *The Rules Governing Medicinal Products in the European Union*. Volume 4 – *Medicinal Products for Human and Veterinary Use*: Good Manufacturing Practice Glossary).

Calibration: The set of operations that establish, under specified conditions, the relationship between values indicated by a measuring instrument or measuring system, or values represented by a material measure, and the corresponding known values of a reference standard Source: *The Rules Governing Medicinal Products in the European Union*. Volume 4 – *Medicinal Products for Human and Veterinary Use*: Good Manufacturing Practice Glossary)

Change control: A formal system by which qualified representatives of appropriate disciplines review proposed or actual changes that might affect the validated status of facilities, systems, equipment or processes. The intent is to determine the need for action that would ensure and document that the system is maintained in a validated state. (Source: *The Rules Governing Medicinal Products in the European Union*. Volume 4 – *Medicinal Products for Human and Veterinary Use*: Good Manufacturing Practice Glossary)

Control Points: This term describes a point in a process step, process condition, test requirement or other relevant parameter or item that must be controlled within predetermined criteria to ensure that the specifications are met (Source: *The Rules Governing Medicinal Products in the European Union*. Volume 4 – *Medicinal Products for Human and Veterinary Use*: Good Manufacturing Practice Glossary)

Deviation: A departure from an approved instruction or established standard. A deviation report is raised as a record of the deviation and the corrective actions to be taken to resolve the issue. Once these actions have been successfully completed and recorded, the deviation report can be closed. Closure is usually performed by the QA department. (Source: *The Rules Governing Medicinal Products in the European Union*. Volume 4 – *Medicinal Products for Human and Veterinary Use*: Good Manufacturing Practice Glossary)

Direct Impact System: A system that is expected to have a direct impact on product quality. These systems are designed and commissioned in line with good engineering practice and in addition, are subject to qualification. (Source: *ISPE Baseline Guide* Vol. 5)

Electronic Audit Trails: A record that electronically captures changes made to computerized records. The trail should remain part of the final computer record.

Factory Acceptance Testing (FAT): Inspection and static and/or dynamic testing of systems or major system components to support the qualification of an equipment system, conducted and documented at the supplier's site (Source: *ISPE Baseline Guide* Vol. 5)

Good Engineering Practice: Established engineering methods and standards that are applied throughout the project lifecycle to deliver appropriate, cost-effective solutions (Source: *ISPE Baseline Guide* Vol. 5)

Impact Assessment: The process of evaluating the impact of the operating, controlling, alarming and failure conditions of a system on the quality of a product (Source: *ISPE Baseline Guide* Vol. 5)

Indirect/Non-impact System: This is a system that is not expected to have a direct impact on product quality, but may support direct impact systems. These systems need only to be commissioned following good engineering practice. (Source: *ISPE Baseline Guide* Vol. 5)

Re-validation: A repeat process of validation to provide an assurance that changes in the process/equipment introduced in accordance with change control procedures do not adversely affect process characteristics and product quality (Source: *The Rules Governing Medicinal Products in the European Union.* Volume 4 – *Medicinal Products for Human and Veterinary Use*: Good Manufacturing Practice Glossary)

Risk Analysis: The method used to assess and characterize the crucial parameters in the functionality of equipment or a process (Source: *The Rules Governing Medicinal Products in the European Union.* Volume 4 – *Medicinal Products for Human and Veterinary Use*: Good Manufacturing Practice Glossary)

Site Acceptance Testing (SAT): Inspection and static and/or dynamic testing of systems or major system components to support the qualification of an equipment system, conducted and documented at the manufacturing site (Source – *ISPE Baseline Guide* Vol. 5)

System: An organization of engineering components having a defined operational function, e.g. piping, instrumentation, equipment, facilities, computer hardware, computer software, etc. (Source: *ISPE Baseline Guide* Vol. 5)

System boundary: A limit drawn around a system logically to define what is and is not included in the system (Source – *ISPE Baseline Guide* Vol. 5)

Validation Protocol: A written plan stating how validation will be conducted, and defining acceptance criteria. For example, the protocol for a manufacturing process identifies processing equipment, critical process parameters/operating ranges, product characteristics, sampling, test data to be collected, the number of validation runs and acceptable test results. (Source: *The Rules Governing Medicinal Products in the European Union.* Volume 4 – *Medicinal Products for Human and Veterinary Use*: Good Manufacturing Practice Glossary)

Processing and Preservation of Cells and Products

16 Cell Harvesting

P Hill and J Bender

16.1 INTRODUCTION

The first step in the purification of proteins and antibodies produced in animal cell cultures is the separation of the cells and debris from the product-containing culture fluid. Whether the bioreactor is operated in batch or continuous mode, it is desirable to achieve cell separation or retention with minimal cell lysis to prevent release of intracellular enzymes and additional impurities that can lower yield and complicate the downstream purification processes. For continuous bioreactors in which a portion of the culture is removed, it is critical that the cell retention method preserves cell viability. In addition to minimizing cell lysis, it is important that the harvest method achieve complete removal of cells and cellular debris to ensure a particle-free solution for subsequent filtration and purification steps. Since many biological solutions contain a wide size range of particulates, complete particulate removal is often only achieved through a combination of techniques.

The most commonly implemented harvest systems for animal cell cultures involve forms of filtration, centrifugation or a combination of both. Filtration methods separate cells and cellular debris based on size differences and include tangential-flow microfiltration, normal-flow depth filtration, and rotary-and vortex-flow filtration. Centrifugation methods utilize disc stack and tubular bowl centrifuges to achieve separation based on the density difference between the cellular solids and the cell culture fluid. In this chapter we will describe these harvest technologies, discuss operational challenges and benefits, present examples of industrial applications and review emerging harvest trends.

16.1.1 The Effect of Cell Culture

Processes utilizing a wide variety of animal cell lines have been approved by the Food and Drugs Administration (FDA) to produce therapeutics, vaccines, and diagnostics (Chu & Robinson 2001). Additionally, the production of recombinant proteins can be achieved utilizing several different modes of cell culture operation. Batch, fed-batch, repeated fed-batch, continuous and perfusion cell culture can all be implemented by the process development scientist, and all have advantages and disadvantages with respect to simplicity, volumetric productivity, contamination risk, product quality, and process control.

The choice of cell culture mode can also influence the challenge of particulate removal from the conditioned medium. When fed-batch bioreactors are harvested, the recovery system typically needs to be able to remove more than 10^{10} cells/l. While perfusion processes utilize retention devices to retain most of the viable cells inside the bioreactor, they are often operated to maximize the passage of dead cells and debris. Thus, perfusion processes produce huge volumes of conditioned medium (typically from 0.5 to 2 reactor volumes per day) with lower but significant quantities of dead and lysed cell material to be removed.

Medicines from Animal Cell Culture Edited by G. Stacey and J. Davis

More generally, bioreactor processes typically aim to maximize bioreactor productivity. To the degree that the bioreactor is operated in conditions not supportive of culture health, for example nutrient poor and waste product rich, the resulting harvest is likely to contain more fragile cells, more dead or fragmented cells, and more intracellular components. Thus, maximally productive bioreactors can result in low viability harvests and more challenging clarifications.

A less frequently encountered but perhaps greater challenge is the isolation of an intracellular product. While most cell-culture-derived proteins are targeted for secretion, occasionally the protein of interest can be located within the cell membrane. Lysis or permeabilization of cells releases significant amounts of intracellular materials, including contaminant proteins, lipids and DNA, which for obvious reasons further complicate the recovery of particle-free conditioned medium. Thus, the techniques used in cell culture can impact the solids load, particle size distribution, and even the fluid properties of the conditioned medium.

16.2 CELL RETENTION IN PERFUSION CULTURE

The first steps in the clarification of the conditioned media from these cultures will likely involve one or more of the unit operations described in Section 16.3 below. However, since perfusion cultures typically produce a product stream with reduced solids, a brief review of the larger-scale (>50 l/day) techniques available to retain cells in the bioreactor will be discussed. Additionally, an estimated cell population density (assuming a perfusion bioreactor culture at 10×10^6 cells/ml, 90% viable) of the resulting conditioned medium will be given based on published reports about device performance and the authors' experience. It should be noted, however, that the level of cellular debris can vary significantly from cell culture to cell culture as well as from run to run, and with the exception of membrane-mediated perfusion, the retention system is unlikely significantly to remove sub-micron debris. Thus, the given estimates of particulate load will only pertain to the amount of whole cells in the conditioned medium. Also, while not mentioned below, it is generally observed that the retention efficiency is somewhat lower for dead cells than for live cells.

16.2.1 Spinfilters

A spinfilter is a mesh cylinder that is closed at the bottom and rapidly rotated inside the bioreactor culture (Figure 16.1). Spinfilters can either be located inside the bioreactor or outside the bioreactor (requiring an external recirculation loop). Clarified conditioned medium seeps through the mesh and is withdrawn from inside the rotating cylinder through a diptube. In addition to the above-mentioned cell culture effects, the spinfilter rotational rate (which provides hydraulic 'sweeping'), mesh 'pore' size and weave type (which influence steric exclusion), and mesh material can all significantly influence the long-term performance of the spinfilter and therefore the particulate load in the effluent. Although cells appear to be excluded from the mesh principally by steric effects, hydraulic effects also play a role, and thus cells smaller than the mean pore size of the spinfilter can be significantly retained. For example, typical pore sizes reported are 10–25 μm for non-aggregating cell lines, but as large as 50 μm or more for clumped cell retention. For a given cell culture, as the mean pore size of the spinfilter increases, cell retention efficiency decreases. However, one advantage of larger pore sizes is the reduced risk of fouling.

Avoiding fouling of the spinfilters is one of the key operational challenges with these devices. Spinfilters can be fouled quickly via simple pore plugging if the medium withdrawal rate exceeds a critical level even for a brief period. Additionally, after prolonged use, biological materials and

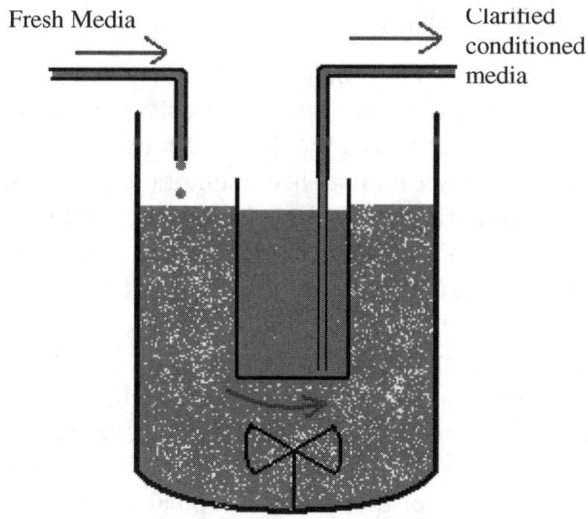

Figure 16.1 A simple drawing of a spinfilter mounted in a bioreactor (reproduced with permission from Woodside *et al.* (1998)).

even whole cells and clumps of cells can attach to the mesh, decreasing the achievable flux and reducing the medium dilution rate. This fouling is partially irreversible and leads to a significant decrease in the achievable flow rate and potentially to the termination of the run if the spinfilter is located inside the bioreactor. It has been reported that polymeric mesh spinfilters may be more resistant to fouling compared with stainless steel meshes, possibly due to their lower surface charge density; however, stainless steel meshes are still more commonly used.

Operationally, as the cell concentration, culture age, and desired perfusion rates increase, so does the risk of spinfilter clogging. Increases in spinfilter rotational rate can counteract these effects to some extent. The following relationship (Yabannavar *et al.* 1994; Deo *et al.* 1996) has been developed for the spinfilter effluent dilution rate (volume conditioned medium per day divided by the bioreactor volume) as it relates to cell concentration, rotational rate and bioreactor and spinfilter geometry

$$D \propto \frac{A_{\text{spinfilter}}}{V_{\text{bioreactor}}} \frac{\left[RPM_{\text{spinfiler}} \cdot R_{\text{spinfilter}}\right]^2}{\left[bioreactor\,total\,cell\,concentration\right]} \tag{1}$$

where: D = spinfilter effluent dilution rate (vol per tank vol per day), $A_{\text{spinfilter}}$ = mesh surface area, $V_{\text{bioreactor}}$ = bioreactor volume, $R_{\text{spinfilter}}$ = spinfilter cylinder radius.

Use of this relationship has been demonstrated to yield the same quality effluent in spinfilter cultures at different scales of operation. However, it should be noted that this relationship was developed to set operating conditions for equivalent cultures, and even under steady-state conditions, the spinfilter may foul within the perfusion vessel, leading to modification of the operating conditions. For example, it has been reported that the efficiency of cell retention rises with culture age, and thus the particulate level of the effluent improves (Iding *et al.* 2000).

Spinfilter operations have been reported at greater than 500 l/day scale, although the operational details and results are not well described (Deo *et al.* 1996; Scheirer 1988). Under the best operational conditions and with correctly optimized mesh pore sizes and areas, spinfilters are capable of achieving viable cell retention efficiencies of >95 %. For example, for a typical cell culture with 10×10^6 cells/ml and 90 % viability, a conditioned medium with approximately $1–6 \times 10^5$ cells/ml could be anticipated.

16.2.2 Tangential-Flow Filtration

While tangential-flow filtration (TFF) will be covered more extensively in the discussion of conditioned medium clarification, its use for perfusion cell culture is briefly described here. The use of tangential-flow microfiltration (pore sizes 0.2–10 μm) for perfusion is achieved by recirculating the culture through hollow-fibre or across flat-sheet membranes. This tangential fluid flow sweeps the membrane surface of accumulating particulates and increases flux. At small scale, the hollow-fibre units themselves are used as bioreactors. In this case, the cells are grown in the inter-lumen space of the unit and the medium is recirculated (see Chapter 10). At larger scale, the TFF devices retain the cells in a loop while conditioned medium is removed through the membrane. TFF perfusion systems are easily implemented and since the membranes are outside the bioreactor, they may be easily replaced as needed during the production run. However, as with external spinfilters, TFF perfusion systems require an external loop with a pump, seals, TFF device and other instrumentation that increases the relative complexity and the risk of contamination.

Because the membranes utilized for this purpose typically have pore sizes much smaller than the typical animal cell and most cellular debris, the biggest advantage of using TFF for perfusion culture is the quality of the conditioned medium. Generally, no further clarification steps may be required to remove cell debris. This advantage comes with a cost, though, and that is the eventual fouling of the membrane that is the most commonly reported problem with TFF-mediated perfusion cultures. Since there is little passage of dead cells or debris, these particulates and their larger components (DNA, membrane lipids) can accumulate in the bioreactor. This not only increases the burden on the TFF operation, but also possibly negatively impacts the quality of the cell culture. Additionally, over time, product passage through the membrane can be reduced due to concentration polarization at the membrane surface. Most small-scale (~10 l/day perfused and less) studies suggest that flux rates of 1–4 l per m^2 per hour (LMH) are supportable for at least several weeks, with larger pore-size membranes capable of supporting the higher flux rates. While there is little larger-scale (>50 l/day) data published, proper scaling should allow straightforward implementation. However, the higher feed flow rates required will likely subject the cells to increased shear and cell lysis within the membrane feed channels as well as within the pump and system piping. Thus, the ability to operate TFF systems successfully at larger scale may be limited by the ability to replace membranes, as well as the possible detrimental effects of pump-induced shear and system pressure drops on the culture viability.

A new technology may help to reduce the pump shear effects in scale-up. Figure 16.2 depicts a large diaphragm chamber (on the opposite side of a hollow fibre TFF unit) that can be used to 'pump' bioreactor culture (Furey 2002). Thus, this 'alternating tangential-flow' technology eliminates the pump shearing. Utilizing a 17 m^2 hollow fibre unit, this technology could theoretically allow TFF perfusion at the 1000 l/day scale. Although independent review of this technique has not been published, this technology also has the potential to provide particle-free conditioned media at production scale.

16.2.3 Acoustic Filters/Gravitational Settlers

Acoustic filters trap cells into settling channels within a standing acoustic wave. Differences in the compressibility and density between the cells and the medium force the cells toward the velocity antinodes of the resonance field (Pui *et al.* 1995). Figure 16.3 shows a resonance chamber in use where the cells are trapped in velocity antinode planes. Periodically the cells are returned to the bioreactor by removal of the resonance field and stoppages or reversals in the effluent flow.

Gravitational settlers are closely spaced inclined plates that promote the sedimentation of the animal cells. Since the flow pattern in the channels is laminar, a portion of the settled culture can return to the bioreactor in an underflow in the settled cell region. Over time, cells accumulate

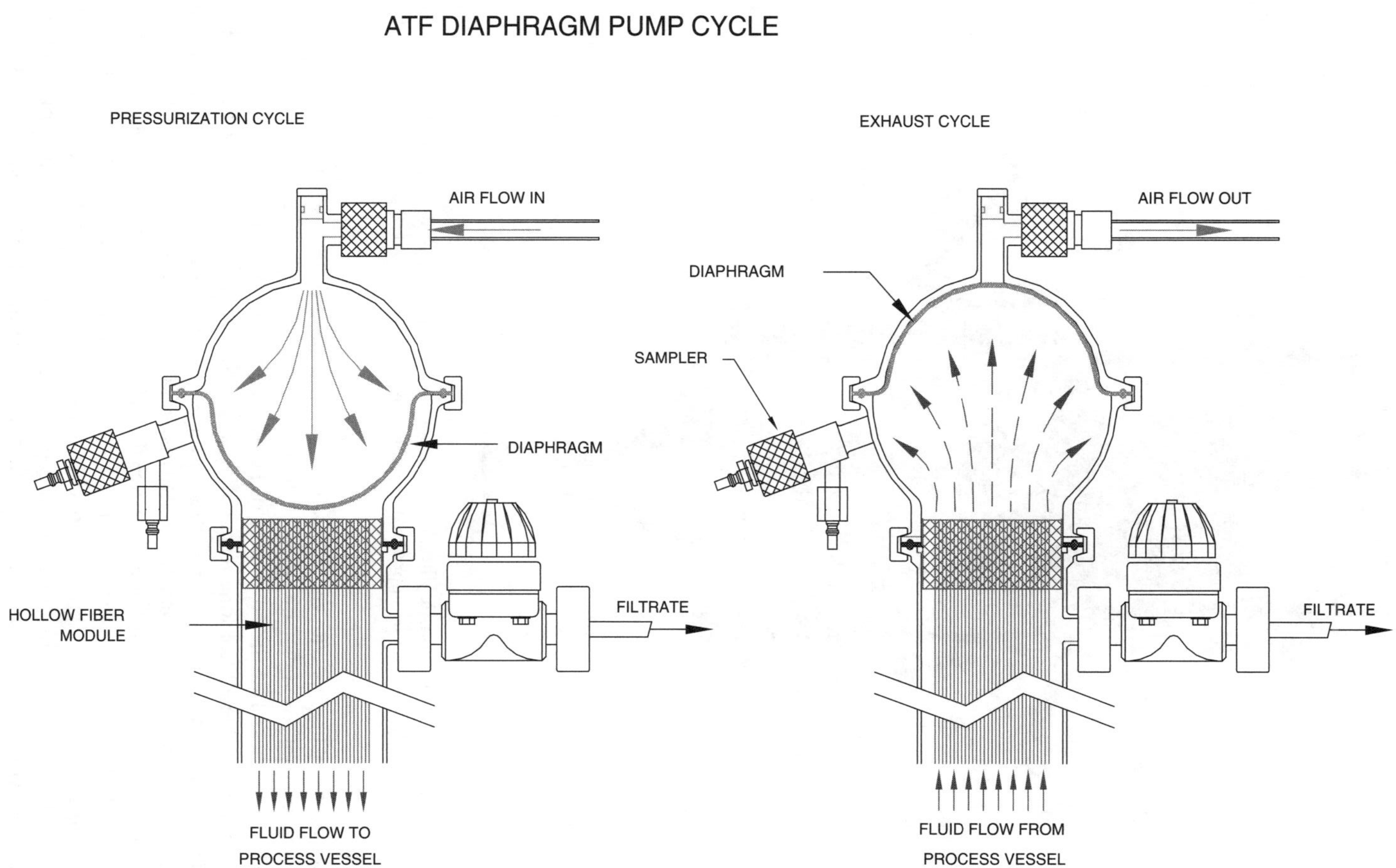

Figure 16.2 Schematic of the Alternating Tangential-Flow Device from Refine Technology (reproduced from Furey (2002) with permission of Genetic Engineering News Inc).

Figure 16.3 AppliSense 50L acoustic filter. (Reproduced with permission from Applisense product literature.) NB: Note the vertical lines of trapped cells. Liquid flow is vertically out of the resonance chamber.

on the tops of the plates; as with the acoustic filter, cells are then returned to the bioreactor via periodic flow reversals or via an air purge of the system.

These technologies offer the advantage of being able to operate indefinitely without fouling, which is a significant benefit for long-term cultures at high perfusion rates and/or high cell population densities. However, viable cell retention of these devices is highly dependent on the culture conditions and is generally less than 100 %. Perfusion rate is generally inversely related to retention efficiency for both units. Additionally, since the cells settle on the inclined plate and may not be returned to the bioreactor for several hours, it is possible that this suboptimal environment may negatively impact the culture viability or specific productivity. More frequent settler back flushes with air or conditioned medium can help mitigate this effect, but at the expense of a higher forward flow rate, which may reduce retention efficiency. Some investigators have successfully utilized chilling of the withdrawn culture in external settlers to slow cellular metabolism and reduce this environmental effect (Searles *et al.* 1994).

For acoustic filters, cavitation and temperature control (particularly at large scale) are concerns. Cavitation is more likely to be an issue at low cell population density, with serum-free cultures, and at high power settings. Larger-scale units such as the BioSep 200L are equipped with jackets to facilitate temperature control. The quality of the particulate removal for both devices will be influenced by the bioreactor conditions, the desired perfusion rate and sizing of the device, and the operational settings (for the acoustic filter). With gravitational settlers, careful control of temperature throughout the settler is critical for avoiding convective currents, which can lead to separation problems and reduced retention efficiency. The retention efficiency of both technologies will suffer if there is any gas bubble entrainment. For the acoustic filter, separation efficiency can be optimized to some extent by adjusting the power applied to the resonance chamber, the forward and stop/reverse run times, the recirculation rate in the loop, and the temperature of the chamber.

Acoustic filters capable of approximately 200 l/day are currently available from Applisense (Applikon). A 1000 l/day device, which is essentially a bundling of five 200 l/day units, is currently under development. Gravitational settlers have been used successfully at the 100–500 l/day scale with moderate retention efficiencies, but are typically custom-made and are difficult to scale further. For both acoustic filters and gravitational settlers, appropriately sized and operated devices may be expected to retain ~90 % of the viable cells. Considering a typical 90 % viable culture with 10×10^6 cells/ml, this would yield a conditioned medium with up to approximately 1×10^6 cells/ml.

16.2.4 Centrifugation

Similar to tangential-flow filtration, centrifugation is utilized for both perfusion and batch cultures. The focus of this section is those aspects of centrifugation that pertain to perfusion cultures. As described in more detail in the later section, centrifugation relies on the density differences between the cells and the medium to achieve cell retention. The key advantage of centrifugation technology for perfusion cultures is the rate at which the conditioned medium can be clarified and cells recycled. Efficient retention of viable cells should be feasible at scales in excess of 3000 l/day. Thus, centrifugation is really the only technology that can accommodate large perfusion rates (>500 l/day). If the centrifuge is oversized relative to the required perfusion rates, the centrifuge could be operated intermittently, allowing for sufficient downtime to perform maintenance between runs.

However centrifugation-mediated perfusion does have some challenges: centrifuges increase the operational complexity and risk of contamination. This can be a particular challenge to long-term perfusion cultures. Additionally, it is possible that the centrifugation process can negatively impact the cell culture. While properly designed centrifuges will minimize exposure to excessive shear and centrifugal forces, by their nature they subject the culture to periodic hyper-concentration. For example, the culture will spend some time in a very dense, probably oxygen-deprived, environment. While this periodic exposure may not affect the culture viability, a decrease in the specific productivity of the culture has been reported (Johnson *et al.* 1996). Detailed studies with representative cultures and operational conditions are recommended to ensure process robustness.

Kendro Laboratories manufactures two centrifuges designed for perfusion culture: the 300 l/day Lab II unit (Figure 16.4) and the larger 'Cell' unit for clarification up to 3000 l/day. Both centrifuges utilize the 'inverted J' or 'skip rope' technology in conjunction with disposable, aseptic plastic rotor inserts that help to simplify the technique and minimize the risk to aseptic processing. Centrifuge rotational speed, run time, cell concentrate discharge time, flow rates during centrifuge feeding and discharge can all be varied somewhat to tailor the operation. Viable cell retention efficiency, where reported, appears to be excellent (~99 %).

Various investigators have explored disc-stack and tubular bowl centrifuges modified specifically to enable perfusion culture (Tokashiki *et al.* 1990; Jager 1992: Bjorling *et al.* 1995; Takamatsu *et al.* 1996). Reports describing these changes demonstrate the feasibility of very large-scale

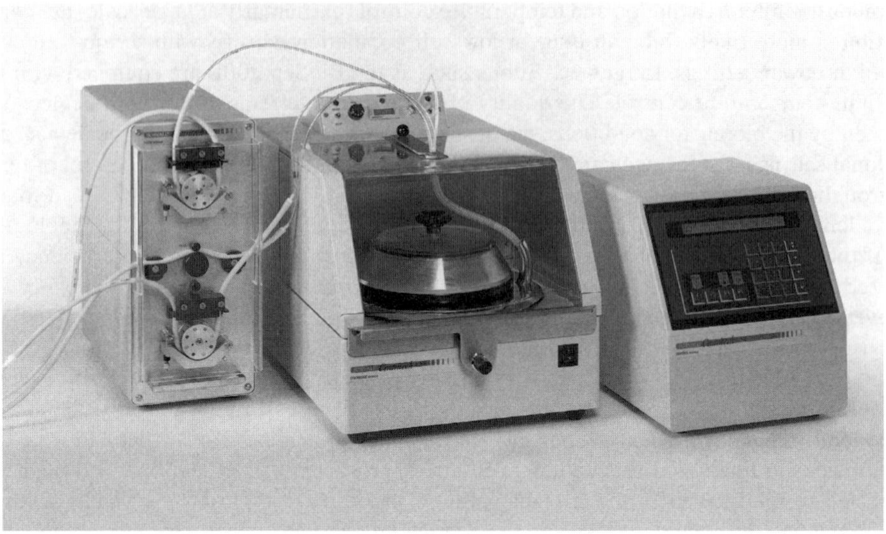

Figure 16.4 Centritech Lab II. Reproduced by permission of Kendro Laboratory Products

perfusion; however, long-term, high perfusion rate, demonstration may not be reported until this technology and the ability to process such large volumes, is required.

In general, retention efficiency and device fouling are not significant issues for centrifugation-mediated perfusion. However, complications surrounding the need for long-term aseptic processing, the level of maintenance required, and possible effects of culture productivity have yet to be addressed. Appropriately sized and operated centrifugal devices have the potential to retain roughly 99 % of the viable cells, unless lower efficiencies (and lower bioreactor viable cell concentrations) are desired to maximize dead cell passage. Thus, for a standard perfusion culture at 10×10^6 cells/ml, 90 % viable, a conditioned medium with up to 1×10^5 cells/ml would be expected.

16.2.5 Summary of Perfusion Devices

Several different technologies are available for use with perfusion cell cultures. While acoustic filters and gravitational settlers do not clog as readily, whole viable cells as well as smaller materials will pass into the conditioned medium. Depending on the mesh chosen and the operating/cell culture conditions, spin filters likely have the next cleanest outputs, typically retaining nearly all of the viable cells, though allowing some dead cells and debris to pass into the harvest. Centrifugal methods may retain all viable cells, allowing only the smaller debris and some dead cells to pass. Finally, tangential-flow membrane devices provide the lowest particulate conditioned medium, but could be costly to operate if frequent membrane replacement is required. Once a cell retention technology is chosen, the quality of the effluent stream will depend on appropriate scaling and operation of the device for the desired dilution rate, as well as the bioreactor process strategy. Optimal conditions could produce conditioned media with at least a 10- to 20-fold reduction in the total cell population density.

16.3 CONDITIONED MEDIUM CLARIFICATION TECHNOLOGIES

The principal methods for clarifying conditioned medium produced by animal cell cultures include depth filtration, tangential-flow microfiltration and centrifugation. Although the separation goals for each technology are similar, the mechanism by which separation is achieved differs. Depth filters use a combination of charge and size for particulate removal, whereas microfilters rely on differences in size and centrifuges on differences in density to achieve separation. Regardless of the harvest method, each technology strives for complete particulate removal to permit filtration prior to downstream purification, minimization of cell lysis that could introduce additional impurities and incur yield loss, and a reproducible process to ensure consistency from batch to batch. Often the scale of operation dictates the selection of the harvest method, and it is not uncommon to utilize one method at pilot scale and another method or combination of methods at production scale.

16.3.1 Depth Filtration

One of the more readily implemented cell separation techniques is normal-flow depth filtration. Depth filters employ the conventional filtration technique of separating cellular solids from the cell culture fluid by forcing the liquid through a porous medium in dead-end or normal-flow mode. The medium retains solids and the liquid flows through to a collection vessel. Especially well suited for smaller batch volumes (>2000 l), depth filtration can be an economical and easy-to-use primary cell removal method compared with tangential-flow microfiltration and centrifugation (Singhvi *et al.* 1996). Depth filtration is economical due to the lower capital costs as well as the low cost of the single-use cartridges. Ease of use is attributed to the normal-flow operation that can be performed either at constant flux with a single feed pump, or at constant pressure by pressurizing the bioreactor to drive the product into the filters. The process can be monitored simply with pressure or flow measurements, eliminating the need for a complicated control system and costly auxiliary instruments and equipment. Finally, the single-use nature of the cartridges eliminates the need for cleaning and reuse validation. Therefore depth filtration is an ideal harvest method for laboratory, pilot and smaller-scale production processes.

16.3.1.1 Depth filter media and cartridges

Depth filters are constructed from cellulosic fibres, filter aids and resins, which can impart positive charge on the media. Unlike membrane filters, depth filters are graded-density filters, meaning that the pore size gradually decreases across the thickness of the medium. The medium's thickness, positive charge and graded density combine to trap cells and adsorb negatively charged cellular debris and submicron particulates. This 'dirt-holding' capacity makes depth filtration an effective separation and clarification method for biological systems. Depth filter media are available in a range of 'grades' that reflect relative porosity as measured by clean water flow rate or permeability. Examples of typical media grades used for cell capture and debris removal are presented in Figure 16.5.

Media with higher permeabilities are used for primary cell removal and media with lower permeabilities are used for clarifying TFF filtrates and centrifuge supernatants. Although the media may be referred to by pore size, these designations should not be interpreted as absolute ratings, and care should be taken when evaluating the performance of 'equivalent' grades from multiple manufacturers, due to differences in charge and potential binding capacity.

The most commonly used form of the depth filter medium is made by layering media in flat sheets onto each side of a plastic support to form one cell. Multiple cells are then stacked together to form one cartridge and multiple cartridges of the same diameter are stacked together into stainless steel housings (Figure 16.6). The standard cartridge diameters are 8, 12 and 16 inches, each containing approximately 3.5, 18 and 50 ft^2 (0.3, 1.7 and 5.1 m^2) of effective filtration area. Since filter housings can be manufactured to accommodate an array of configuration options, depth-filtration harvest systems are relatively easy to customize for a particular application.

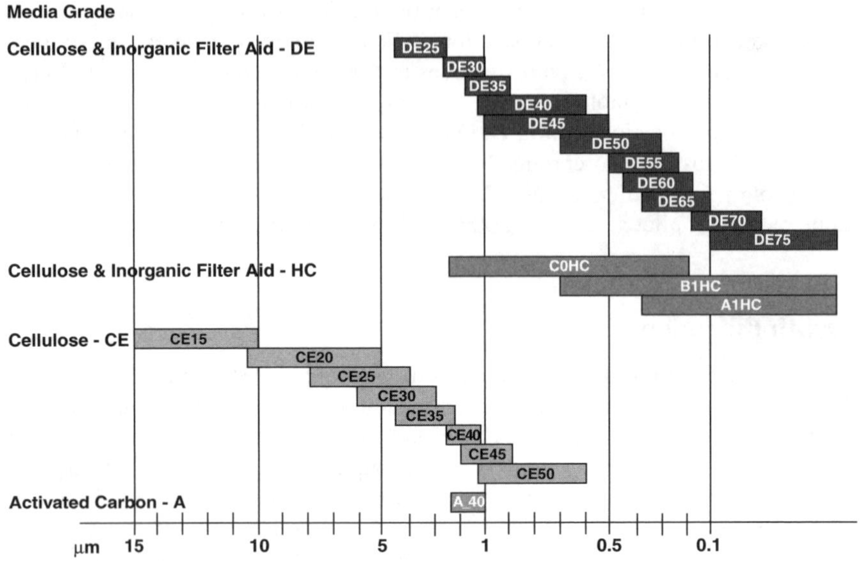

Figure 16.5 Depth filter retention ranges. (Reproduced with permission from Millipore Corporation.)

16.3.1.2 Depth filtration process development

One of the strategies for depth filtration harvest development is to determine the optimum configuration of media grades and filtration area to achieve the desired separation. Since some media grades are designed for trapping the large, intact cells and others are designed to clarify the smaller debris, often the optimum configuration is a multi-stage system with the more open filter upstream of the tighter-grade filter. Some manufacturers have developed cartridges containing multiple media grades to enable a dual-stage separation within one housing. These multiple-media cartridge options could lead to a reduction in the required filtration area and number of housings.

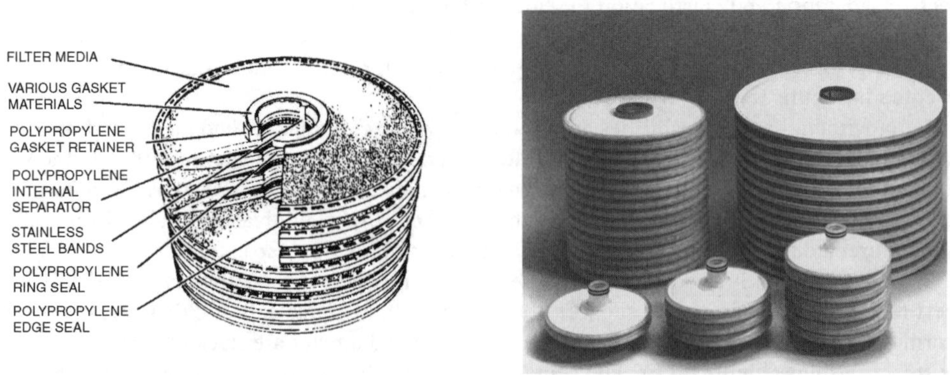

Figure 16.6 Cross-section of a Cuno Zeta Plus depth filter cartridge (left) and 8-inch, 12-inch and 16-inch depth filter cartridges (right) (reproduced with permission from Cuno Incorporated).

In addition to media selection, key process parameters for depth filtration harvesting are filtration flow rate and pressure drop as a function of throughput. In theory, depth filtration performance can be described by Darcy's relationship between flow rate and the pressure drop created by that flow through a porous solid (Geankoplis 1983):

$$\frac{\mathrm{d}V}{\mathrm{d}t} = \frac{kA\Delta P}{\mu L} \tag{2}$$

where $\mathrm{d}V/\mathrm{d}t$ is flow rate, k is bed permeability, μ is liquid viscosity, ΔP is pressure drop, L is bed thickness and A is filtration area.

One of the difficulties in applying Darcy's law to biological filtrations is the inability to calculate accurately the bed permeability or the resistance (k/L) for these complex biological fluids. While empirical relationships do exist for permeability (e.g. the Kozeny–Carman expression) and resistance, it is often more expedient to measure pressure drop experimentally as a function of flow rate for animal cell culture fluids (Carman 1937).

Small-scale experiments to determine depth filtration performance are relatively easy to conduct, and provide the most reliable data when the test feedstock closely resembles the actual feedstock in terms of solids volume, culture viability and culture density. Since flux and colloid formation are affected by temperature, small-scale studies may also need to be performed at the same temperature as the production-scale operation. The general experimental approach is to test different filter media at various flow rates to determine the optimum flux, filter area and media combination. Small-scale tests may be conducted using less than a few litres of culture fluid with 47-mm filter discs at either constant pressure (V_{max}) or constant flux (P_{max}) to evaluate clarity, throughput and product retention (Badmington *et al.* 1995; Ho & Zydney 2002). The commonly used V_{max} method uses the linear form of the pore-plugging model to predict filtration capacity:

$$\frac{t}{V} = \frac{1}{Q_i} + \frac{t}{V_{max}} \tag{3}$$

where t is time, V is volume, Q_i is initial filtrate flow rate and V_{max} is the maximum volume that can be filtered at the test pressure before the membrane is fouled.

The experiment is performed by filtering at a constant pressure and measuring the filtered volume as a function of time. If the plot of t/v versus t is linear, then the filtration follows the gradual pore-plugging model and V_{max} is calculated from the inverse of the slope. Finding the area needed at production scale is then linearly scaled from the V_{max} result. If a straight line is not obtained, then the solution does not follow the pore-plugging model and the constant flux method is recommended.

Constant flux experiments are conducted by controlling the flow rate and monitoring the rise in pressure across the filter as a function of filtrate volume. In addition to monitoring the decrease in flow rate for V_{max} and the increase in pressure for P_{max}, it may also be important to measure product yield throughout the filtration to ensure that the target molecule is not binding to the charged filter media. For example, filtration of a target protein produced in mammalian cell culture through two different depth filter media led to product yields of 95 % and 38 % under the same processing conditions (Bender 2001).

Typically for depth filtration media, V_{max} testing is a good screening tool for media selection, while the P_{max} approach provides more accurate results for predicting large-scale performance of a particular filtration configuration (Yavorsky & McGee 2002). Despite the reliability of small-scale

Figure 16.7 Industrial-scale depth filtration system for cell harvest and clarification. (Reproduced with permission from Cuno Incorporated).

data, the conservative process development approach is to select an appropriate 'safety factor' and oversize the production-scale filters to account for run-to-run variabilities, fluid dynamic differences and other system effects. Large-scale depth filtration harvest systems generally also include downstream bioburden reduction and sterilizing grade filters, which must also be sized appropriately in small-scale V_{max} or P_{max} experiments.

As shown in Figure 16.7, the largest industrial-scale depth filtration harvest systems consist of multiple housings containing multiple cartridges. The system shown is used to harvest cells from a 15000 l production bioreactor and includes housings for guard and sterile filters. Prior to the harvest, the depth filters are flushed with water or an appropriate buffer to remove loose particulates and extractables according to the manufacturer's recommendations. Once the harvest is completed, the filters are again flushed to recover the valuable product held up in the housings. By implementing a post-use flush and ensuring no product retention losses, depth filtration harvest yields exceeding 95 % are possible.

As process volumes increase, depth filtration may no longer be the most feasible harvest method due to the large filter housings, longer processing times, increased plant space requirements and laborious cartridge installation and disposal issues. Another limiting factor for depth filtration may be the increasing amount of cellular solids produced in the bioreactor that collect within the depth filter media and housings. Consider a 10 000-l bioreactor with a solids volume of 2.5 %: 250 l of cellular solids would need to be contained within the depth filter housings without blocking or fouling the effective filter surface area. These large solids volumes make depth filtration harvesting impractical relative to tangential-flow microfiltration where the cellular solids are returned to the bioreactor, or continuous centrifugation where the solids are discharged to a separate collection or to waste.

16.3.2 Tangential-Flow Microfiltration

One of the most broadly used harvest technologies for animal cell cultures is cross-flow or tangential-flow filtration (TFF). Unlike normal-flow filtration where the feed and filtrate flows are perpendicular to the filter media, in TFF operations the feed flows tangentially along the filter surface. The filtrate is forced through the membrane by applying a pressure drop across the surface while the cells and cellular debris are retained by the membrane (retentate) and flow back to the bioreactor. The shear caused by the feed flow across the membrane surface helps prevent formation of a concentration polarization layer which enables TFF operations to process volumes at higher fluxes than can NFF methods. Since the filtrate is continually removed, the solids are concentrated in the bioreactor until the desired end point. This concentration phase is typically followed by a diafiltration phase in which buffer enters the feed stream at the same rate as filtrate is removed to wash the cells of any residual product-containing fluid. During diafiltration, the volume of buffer used to exchange the entire volume remaining in the bioreactor (retentate) is called a diavolume. The theorectical product yield can be calculated as a function of the concentration factor and the number of diavolumes using (Cheryan 1986):

$$Y = 1 - CF^{(R-1)} \mathrm{e}^{DV(R-1)} \tag{4}$$

where Y is yield, CF is concentration factor, R is retention and DV is number of diavolumes. Retention is calculated using $R = 1 - C_{\mathrm{filt}}/C_{\mathrm{feed}}$ where C_{filt} and C_{feed} are the product concentrations in the filtrate and feed, respectively. Therefore products that completely pass through the filter have a retention equal to 0.

According to equation (4), a process with $CF = 10$, $R = 0$ and $DV = 1$ achieves a theoretical yield of 96 %. Increasing the DV to 3 improves the yield to 99.5 %.

Tangential-flow filtration is not only used as a cell harvest method, but also as a method for protein concentration and buffer exchange. Application of TFF for cell harvest is presented in this section and its application for protein concentration is presented in Chapter 17. Cell harvesting via TFF uses microfiltration membranes sized to retain intact cells and some portion of the cell debris. Since this method returns the cells to the bioreactor and removes the filtrate, TFF can be used to harvest both perfusion and batch bioreactors. Regardless of the mode of bioreactor operation, the principle of operation, membrane selection and key process development goals are similar.

16.3.2.1 Microfiltration membranes and devices

Microfiltration membranes are made from a variety of polymers, including cellulose nitrate and acetate, polyvinylidene fluoride (PVDF), polytetrafluoroethylene (PTFE), polyethylene,

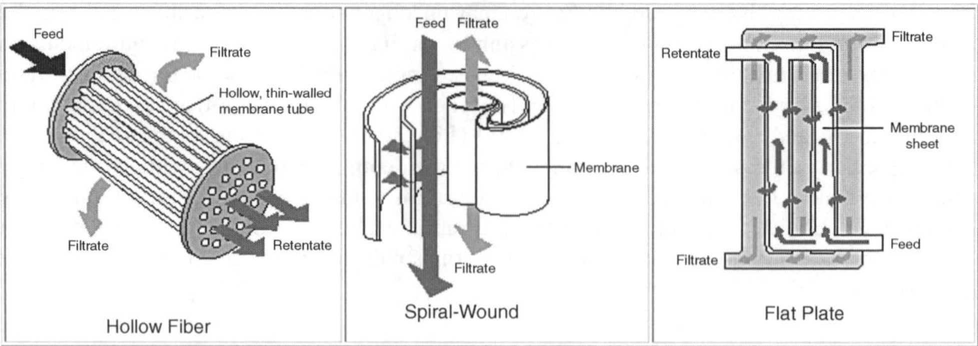

Figure 16.8 Fluid flow paths for hollow-fibre (left), spiral-wound (middle) and flat-plate (right) tangential-flow filtration devices. (Reproduced with permission from Millipore Corporation.)

polypropylene, polysulfone, polyethersulfone (PES), polycarbonate, polyvinylchoride (PVC) and polyamide (Zeman & Zydney, 1996). These polymers are cast into a thin layer and are bonded onto a support to form the membrane. Often the membrane surface is chemically modified to improve the performance. Unlike depth filters, microfiltration membranes have more defined pore structures with effective pore sizes ranging from 0.1 to 5 μm.

The membranes are then formed into hollow fibre, flat sheet or spiral wound modules from which tangential-flow devices of various filtration areas are constructed (Figure 16.8). These three geometries have all been implemented for harvesting biological solutions and the features of each will be described in this section. In addition, researchers are continually experimenting with new designs aimed at increasing the mass transfer rates by applying external fields, centrifugal forces or motion. These contributions to the traditional microfiltration approach will be presented later in this chapter.

16.3.2.1a Flat plate

One of the earliest devices manufactured for microfiltration was the flat plate or cassette device. Flat plate modules consist of layers of membrane, retentate (feed side) and filtrate (or permeate) channels. These layers are either held together with gaskets and fitted into a holder that is compressed to form a seal, or their edges are glued or heat-sealed using polyurethane or other hard polymer into a cassette. Cassettes can then be stacked together into large stainless steel holders for production-scale operations.

Cassettes are typically rectangular with the feedstock entering channels on one side, flowing across the membrane and exiting via retentate channels on the other side. The filtrate flows through the membrane and exits through filtrate channels. Figure 16.8 shows the flow paths through a flat plate TFF device. Most TFF device manufacturers offer cassettes with or without turbulence-promoting screens inserted into the feed and filtrate channels. These screens promote mixing by creating turbulent flow through the channel, improving the mass transfer rates without increasing the pumping requirements. Alternatively, unscreened or open-channel devices are noted for producing lower shear on the feedstock and are often the preferred choice for cell harvest operations.

Flat plate microfiltration systems can be used to harvest large (>10 000 l) production cultures as shown in Figure 16.9. This industrial-scale system contains 2000 ft^2 (186 m^2)of membrane area capable of harvesting 10 000 l of cell culture fluid in approximately 3 hours.

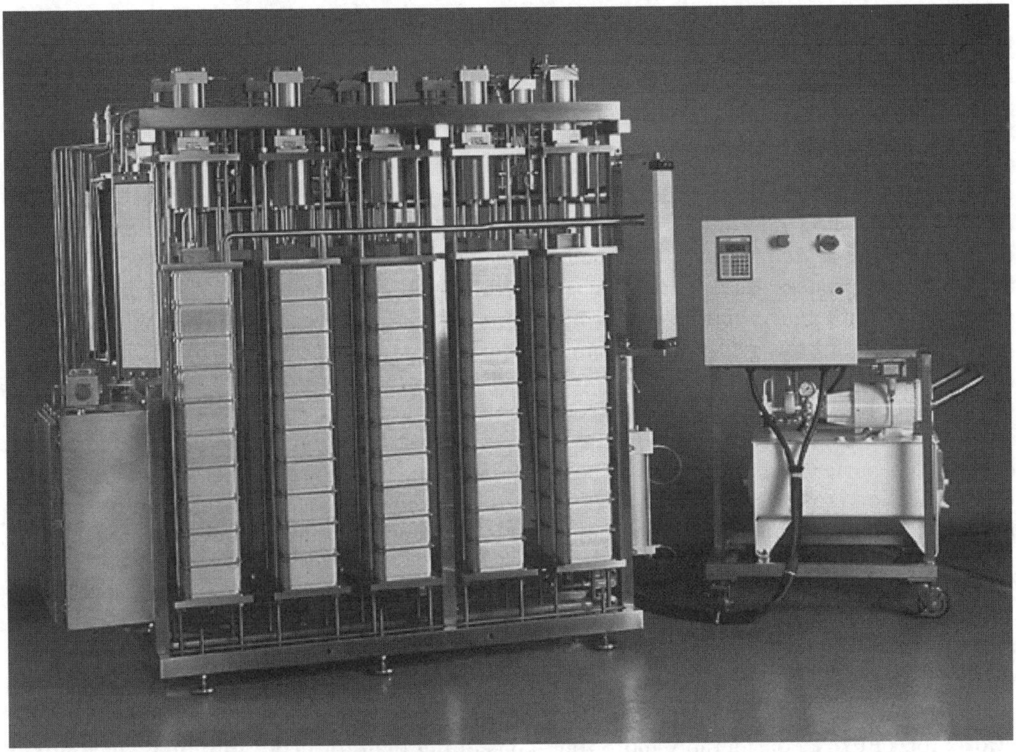

Figure 16.9 Industrial-scale microfiltration harvest system containing approximately 2000ft^2 of membrane area. (Reproduced with permission from Millipore Corporation.)

16.3.2.1b Hollow fibre

Another microfiltration device widely used for cell harvesting is the hollow-fibre module. Hollow-fibre devices group together many small diameter fibres that are packed together and bound on either end such that the inner diameters of the tubes are exposed. Usually, for microfiltration applications, the feed flows through the inner diameter of the tubes while the filtrate flows through the membrane into the 'shell' side. Gaskets on either end separate the feed and filtrate flow paths. Hollow-fibre devices are noted for their lower plant space requirements per-surface area, lower shear rates due to the open feed flow path and laminar flow through the inner fibre diameters, and ability to scale from bench to production harvest operations. Drawbacks to hollow-fibre systems include lower filtrate rates (or higher pumping requirements) relative to flat plates, fewer choices in membrane chemistries, and a greater tendency to foul when harvesting cells with higher debris loads.

16.3.2.1c Spiral-wound

Spiral-wound microfiltration devices are constructed by rolling a membrane layer and support screen around a hollow cylinder. The feed flows in at the top of the cylinder, along the membrane surface, and the retentate exits at the other end, similar to the design in a hollow-fibre module (Figure 16.8). The filtrate flows radially inward to the centre tube and is collected from one end of the module. Benefits of spiral-wound devices include high packing densities and relatively high filtrate fluxes, due to the incorporation of turbulence-promoting screens. Spiral-wound devices are not as easy to implement for production-scale operations due to their inherent inability to be scaled up linearly. Spiral-wound modules are also more susceptible to fouling and are more difficult to clean due to inaccessible regions within the module.

16.3.2.2 Microfiltration process development

Device selection does impact the process robustness and performance that can be achieved from tangential-flow microfiltration harvest systems. Key process development goals may only be met with one device format, given the feedstock characteristics and system constraints such as plant space, process time, operating costs, membrane lifetime and reuse. The primary design components for a tangential-flow microfiltration harvest system are system size (membrane area), feed rate, filtrate rate, filtrate control, transmembrane pressure (TMP), membrane pore size and permeability, concentration factor and number of diavolumes. Compared to depth filtration, TFF systems are more sophisticated and contain more equipment and instrumentation. A typical TFF system contains the membrane unit, feed and filtrate pumps, bioreactor, diafiltration buffer tank, downstream polishing and sterilizing-grade filter housings, and filtrate collection tank. Use of an automated control system at production scale permits process parameter control, data acquisition, electronic batch reporting and improved reproducibility from run to run.

TFF harvest process development is facilitated by the availability of small-scale devices whose performance is predictive of their large-scale counterparts. The benefits of linearly scaleable systems have been described by van Reis *et al.* (1997). As with depth filtration, small-scale testing with comparable feedstocks is an absolute requirement so that production-scale systems can be implemented with minimal risk. Unfortunately, process development efforts for TFF are considerably more time-consuming than for depth filtration, since flux is not the only parameter to evaluate and an efficient screening method such as V_{max} does not exist. Nevertheless, since TFF is primarily a size-based solid–liquid separation method, it is possible to develop standardized TFF harvest processes for multiple products expressed by similar cells.

Process development for TFF means determining the optimum feed and filtrate rates, membrane area, process time, amount of diafiltration, and downstream filtration requirements. Practical considerations such as collection tank volume and turn-around time typically dictate total process times and maximum diafiltration volumes. Once these constraints are in place, experiments are conducted with various membrane chemistries, pore sizes and device formats to assess separation performance. Selection of membrane chemistry may be driven as much by cleanability and compatibility with storage solutions as by permeability and cell retention. Selection of membrane pore size also depends on downstream filtration options. More open ($\geqslant$0.65 μm) membranes have higher fluxes but may require significant downstream depth filtration areas to protect the final sterilizing grade filters. Tighter ($\leqslant$0.2 μm) membranes may not require downstream filtration, but could foul more easily with higher density cultures.

During development, experiments with small-scale devices are conducted with representative feedstocks to determine the optimum feed and filtrate rates and the maximum TMP that can be tolerated by the system and the product. The filtrate rate is controlled using a pump or valve to ensure an adequate cross-flow rate and to minimize rapid membrane fouling during start-up. Ideally the goal is to maximize the filtrate flux and evaluate the filtrate clarity, product retention, TMP and cell lysis at various feed and filtrate rates. While higher feed rates lead to higher filtrate rates at the same TMP, the shear rate and number of pump passes also increase, possibly contributing to higher cell lysis and poorer product quality. Once the optimum filtrate rate and feed rate are determined for a given device, chemistry and pore size, the required membrane area for production scale can be determined.

Membrane area selection depends on the process volume and the desired process time. Larger systems reduce process times, but require higher capital and membrane costs, have higher volumetric hold-up and require more plant space. Too long a process time risks bioburden contamination due to longer non-sterile operation, and less efficient use of plant capacity. Once the TFF processing conditions are established, effluent from the TFF harvest system is used to size downstream depth and sterile filters. These filters are sized using the same P_{max} and V_{max} methods described previously. Since the filtrate is continually changing during the course of the microfiltration harvest process, it is important to test the downstream filters in-line with the TFF system by diverting some of the filtrate into small-scale housings. By extending the TFF and normal-flow testing beyond the expected volume at production scale, a safety margin can be added to the predicted filter area requirements.

Throughout the testing, consideration must be given both to the concentration of solids in the bioreactor and to cell lysis that occurs through multiple passes through the feed pump and system. For example, cell lysis measurements taken during multiple TFF harvests at production scale showed that the harvesting caused as much as 47 % additional cell lysis (George 1994). Another concern is if feed solids become too concentrated to pump through the feed channels at a reasonable pressure, the harvest operation could terminate early due to high pressures, resulting in low product yields. Although the maximum solids volumes can be experimentally determined for each culture, in general, solids volumes approaching 20–30 % become challenging for even the more open TFF membranes. As starting solids volumes increase, the researcher must evaluate the trade-off between concentration factor and the amount of diafiltration required to achieve a high (>95 %) yield. As the concentration factor decreases, the amount of diafiltration increases, the processing time increases and, importantly, the final filtrate volume increases.

16.3.2.3 Industrial-scale microfiltration systems

While microfiltration harvest systems are reliable and relatively easy to maintain, there are some key factors to consider when installing an industrial-scale system. First, the associated equipment, particularly the feed pump and piping, can be quite large and require significant volumes of cleaning chemicals. The capital costs of an industrial scale TFF system are far greater than those of a depth filtration harvest system. Second, the cost of the membranes encourages reuse, which must be well documented and validated to show that the first and last runs of the membranes perform identically. Finally, a separate set of membranes must be used for each product at each scale manufactured in a given facility. For example, harvests at 400 l, 5000 l and 10 000 l scale will require three complete TFF harvest systems, including pumps and other auxiliary equipment. Nevertheless, TFF continues to be used reliably for harvesting animal cell cultures and is a good option for processing large volumes of lower cell population density cultures. As the cell population densities increase, harvest methodologies that do not require concentration and recycling of the cell mass become more feasible than tangential-flow microfiltration.

16.3.4 Continuous Centrifugation

Continuous centrifugation has been used for biological separations for many years. Implementation of centrifugation for animal cell harvest applications is more recent, as normal-flow depth filtration and tangential-flow microfiltration technologies become less feasible as culture volumes and cell population densities increase. A key benefit of cell harvesting via centrifugation compared with tangential-flow microfiltration is the avoidance of concentration and recycling of the cell mass, leading to membrane fouling and cell lysis.

Among the many different types of centrifuge, the best suited for animal cell separations are the tubular bowl, imperforate bowl and solids-ejecting disc stack centrifuges.

16.3.4.1 Tubular-bowl centrifuges

Tubular-bowl centrifuges have been employed in the biotechnology and pharmaceutical industry for many years and are an ideal choice for high value, low solids, cell cultures. The key design feature of the tubular-bowl machine is the high length-to-diameter ratio, which enables long retention times despite the small bowl diameter. Tubular-bowl machines offer lower-shear separation and typically yield drier solids than disc-stack centrifuges. In a tubular-bowl centrifuge the feed enters from the bottom, the solids are retained along the bowl wall and the clarified liquid travels out the top. Paring discs can be used to pump the clarified liquid out of the machine. Once the solids space is filled, the centrifuge must be stopped and taken apart to remove the separated solids. Industrial-scale tubular-bowl machines are capable of achieving *g*-forces from 13 000–62 000 but are limited to smaller batch sizes due to the manual removal of solids (Letki 1998). The largest tubular-bowl machines have diameters up to 130 mm and operate at flow rates less than 100 l/min with solids capacities less than 10 l (Axelsson 1999). As feed solids increase, tubular-bowl machines become less attractive than solids-ejecting disc-stack centrifuges.

16.3.4.2 Imperforate-bowl centrifuges

Also called chamber-bowl centrifuges, the imperforate-bowl centrifuge typically has a cylindrical design in which the height and diameter are nearly equivalent. The classical design also requires manual removal of the solids; however a more recent design employs an automated scraping device to remove solids continually from the bowl during the harvest. As with the tubular-bowl machine, solids are collected at the bowl wall at high *g*-forces. Once the desired amount of solids has been captured, the bowl speed is reduced and a scraper arm is engaged to sweep the solids into a collection tank below the bowl. After the solids are discharged, the centrifuge returns to the operating speed and the separation cycle continues. This design is noted for its ability to produce extremely dry solids with minimal cell lysis (Betts *et al.* 2003).

Another adaptation of the imperforate-bowl machine has been tailored specifically for harvesting shear-sensitive animal cells. In this machine, shown in Figure 16.10, the feed tube is precisely centred on the axis of rotation so that the feed is introduced to the accelerator in a region of very low surface velocity, which minimizes shear forces as the cell culture enters the centrifuge bowl. Cells are then concentrated along a gentle path between the core and the bowl wall. The clarified supernatant is continuously discharged under atmospheric pressure, and the concentrated cells are periodically collected at 1 *g*. This machine is well suited for the collection of viable cells. Capable of operating at very high *g*-forces, this machine is available for bench to production (10 000 l) scale. Additional features include an automated control system, clean-in-place, sterilization and variable frequency drive for operation at lower *g*-forces.

16.3.4.3 Disc-stack centrifuges

Disc-stack centrifuges are among the more efficient separators used in the biotechnology industry, especially for separation of yeast and bacterial cells from the fermentation broth. Although, traditionally, disc-stack centrifuges have been found to cause a significant amount of cell lysis when processing sensitive cell cultures (Berthold & Kemken 1994), recent design changes have provided an opportunity to re-evaluate centrifugation of shear-sensitive cells (Tebbe *et al.* 1996). These design changes include modifications to the feed inlet zone to eliminate air–liquid interfaces, and incorporation of variable-frequency drives for operation at lower speeds. In addition,

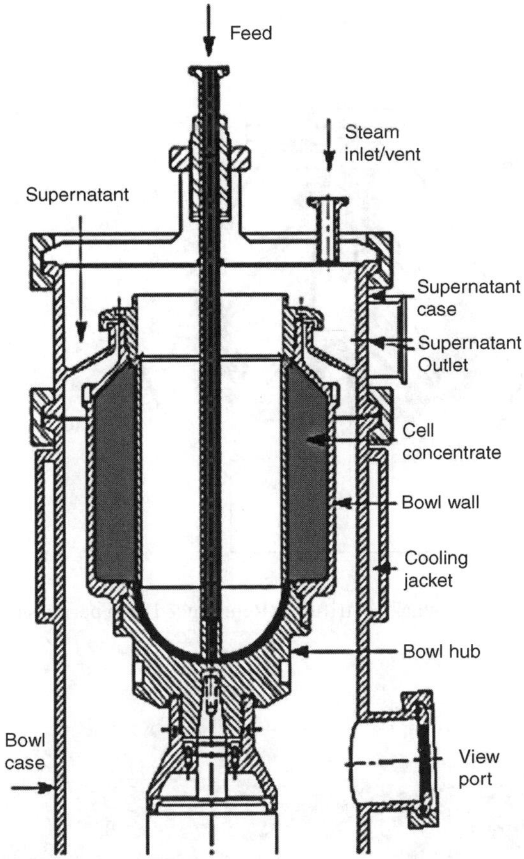

Figure 16.10 Imperforate-bowl centrifuge for low g-force recovery of viable cells. (Reproduced with permission from Kendro Laboratory Products.)

hermetic machine designs in which air is completely removed from the internal centrifuge volume are also available. However, hermetic machines tend to be more complex and contain an additional mechanical seal, which must be well maintained for proper operation.

Regardless of the choice to use a hermetic or non-hermetic machine, cell separation via centrifugation can be accomplished with minimal cell lysis with current machine designs. Figure 16.11 shows a schematic of a solids-ejecting disc-stack centrifuge. The separation occurs within the narrow spaces of the disc stack as the solids are captured at the periphery of the bowl and the clarified liquid travels upward to the centre of the bowl. The collected solids are discharged intermittently at the operating bowl speed under extremely high pressures, while the clarified liquid exits the centrifuge through a paring disc. The clarified liquid can then be collected or subsequently filtered using depth-and sterilizing-grade filters depending on the separation requirements.

Disc-stack centrifuges are readily available in sizes ranging from 1000 m^2 to greater than 200 000 m^2 of settling area. Harvest volumes from a few litres to thousands of litres can be processed with the variety of available machine options. Most production-scale centrifuge systems, such as that shown in Figure 16.12, feature automated clean-in-place and operation modes enabling robust processing and data collection. Although widely used in many industrial applications,

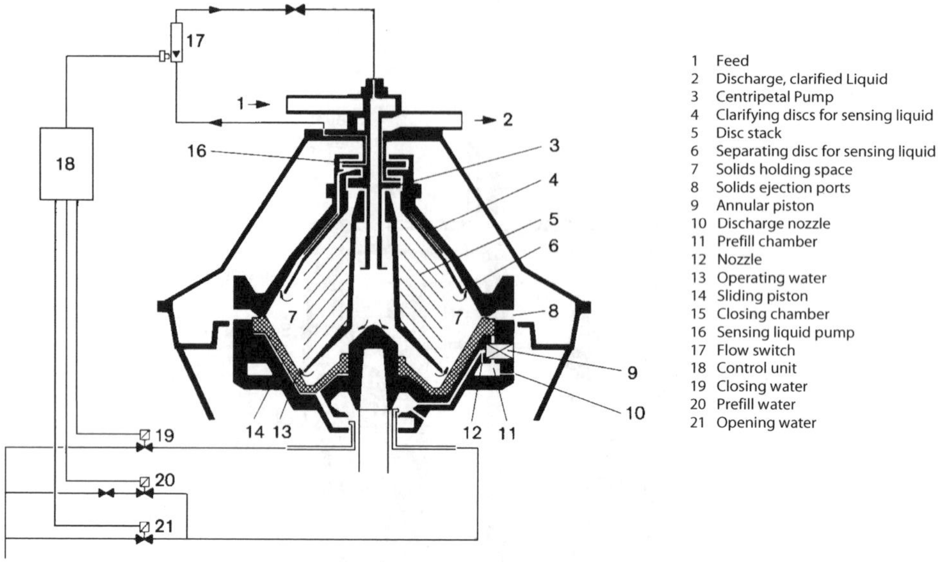

Figure 16.11 Schematic of a disc stack centrifuge. (Reproduced with permission from Westfalia Separator Inc.)

Figure 16.12 Production-scale disc-stack centrifuge system. (Reproduced with permission from Alfa Laval Incorporated.)

most centrifuges and accompanying process piping may be manufactured to conform to the specifications required for biopharmaceutical processing, especially highly polished surface finishes and capability for steam sanitization.

16.3.5 Centrifugation Theory and Principles

Centrifuges achieve separation of the cellular solids from the product-containing liquid by exploiting differences in density between the solids and the liquid. When the cell culture is allowed to stand with no mixing, eventually the solids will settle under the force of gravity in a process called sedimentation. The settling of particles of different densities under the influence of a centrifugal force is called centrifugation. Since centrifugal forces are thousands of times greater than the gravitational force, very small particles such as cells (10–20 μm) and cellular debris (<10 μm) settle much faster in a centrifuge. The rate at which particles settle, called the settling velocity, is a key factor for linking residence time within the centrifuge and the required settling area, or size, of the centrifuge. Stoke's law gives the settling velocity (u_c) of a spherical particle in low concentration solutions in a centrifugal field and is a good approximation for most biological systems:

$$u_c = \frac{d^2(\rho_s - \rho_l)\,\omega^2 r}{18\eta} \tag{5}$$

Where u_c is settling velocity, d is the particle diameter, ρ_s and ρ_l are the densities of the particle, and of the fluid, η is the fluid viscosity, r is the distance from the axis of rotation and ω is the angular velocity. The particle settling velocity (u) can be measured using laboratory-scale centrifuges and linked to the flow rate (Q) and size of the production centrifuge through the sigma factor relationship developed by Ambler (1959)

$$u_c = \frac{Q_{lab}}{\Sigma_{lab}} = \frac{Q_p}{\Sigma_p} \tag{6}$$

where the subscripts 'lab' and 'p' refer to the flow rates and sigma values for the laboratory-and production-scale centrifuges, respectively. Σ is defined using the generalized equation

$$\Sigma = \frac{u_c \omega^2 r}{sg} \tag{7}$$

where s is the effective settling distance, g is the gravitational constant and r is the radial distance from the axis of rotation. Sigma factors for the tubular-bowl, disc-stack and imperforate-bowl centrifuges were derived by Ambler (1959) and are shown in Table 16.1.

In theory, the sigma approach can be used to predict the separation performance of production-scale continuous centrifuges from laboratory spin tests in a bottle centrifuge. In practice, however, many researchers have found that the sigma relationship adapted with empirically derived correction factors more accurately predicts separation performance in large continuous machines from spin test data (Maybury *et al.* 2000; Mannweiler & Hoare 1992; Tomusiak 1992). In cases where similarly designed pilot-and production-scale centrifuges are evaluated, the sigma factor relationship may hold true without empirical adjustment and may provide a good tool for sizing a production-scale separator.

Table 16.1 Sigma relationships for tubular, multi-chamber (imperforate-bowl) and disc-bowl centrifuges.

Machine type	Sigma relationship	Definitions
Tubular bowl and single-chamber bowl	$\frac{\pi\,\omega^2\,L}{g}\left(\frac{r_2^2-r_1^2}{\ln\left(\frac{2r_2^2}{r_2^2+r_1^2}\right)}\right)$	L = inner length of bowl r_1 = inner radius of bowl r_2 = outer radius of bowl
Disc bowl	$\frac{2\pi\,\omega^2\,N}{3g}(r_2^3-r_1^3)\cot\alpha$	r_2 = maximum disc radius r_1 = minimum disc radius N = number of discs α = half cone angle of disc

16.3.6 Centrifuge Process Development

Although scaling centrifuge performance is not as robust as scaling tangential-flow filtration performance, the process development effort for centrifugation is more straightforward and less resource-intensive. One reason for this is that centrifuge performance can be assessed as soon as the first centrate material is processed because this is a single-pass operation in which the feed stream is constant. Another advantage that centrifugation has over filtration is the added flexibility of changing both feed flow rate and bowl speed to deal with feedstock variability. However, since centrifuges have more complex geometry and lack a characteristic length for linear scaling, it is critical that development runs be conducted on a machine similar in design to the industrial-scale centrifuge. Performance of a large-scale disc-stack centrifuge is best predicted by the results of a small-scale disc-stack centrifuge. Testing with very small volumes in a bottle centrifuge is not representative and should not be relied upon for predicting performance of a continuous machine. Not only is it important to minimize machine design differences, but also, as with the other harvest technologies, it is important to use representative feedstock during process development.

Key operational parameters to evaluate include feed flow rate, feed solids, centrifuge bowl speed, separation efficiency or centrate clarity, and solids concentration and handling. For solids-ejecting centrifuges, knowledge of the volumetric feed solids is needed to set the discharge frequency during harvest. Packed cell volume (PCV) is conveniently measured using a bottle-style centrifuge. If the PCV cannot be measured prior to harvest, it can be either estimated or determined by watching for cell breakthrough at the beginning of the harvest. Once the discharge frequency is established, the optimum feed flow rate and bowl speed may be determined experimentally.

In general, lower flow rates result in longer residence times within the bowl and better separation of cell debris. However, the lower the flow rate, the longer the processing time and the greater the opportunity for bioburden growth. The optimum flow rate is often based on the desired process time and the optimum flux through downstream depth and sterile filters. Bowl speed is the second key operational parameter to evaluate for centrifuge harvesting. The ability to vary centrifuge speed can be a great asset especially for harvesting shear-sensitive animal cells. In cases where the highest g-forces are not required to achieve the desired centrate clarity, operation at lower speeds can be advantageous. For a given feed inlet configuration, the required centrate pressures and the shear forces in the inlet zone decrease with decreasing bowl speed. Lower shear forces lead to lower cell lysis and less debris generation. On the other hand, operation at higher g-forces also has advantages. Since the centripetal pump capacity increases with increasing bowl speed, higher centrate flow rates are achievable at higher g-forces. In addition, high g-forces lead to higher settling velocities and potentially greater separation efficiency. Either way, the ability to vary centrifuge speed is an advantage when developing a centrifuge harvest process for animal cells.

Separation efficiency at various speed and flow conditions can be assessed using a variety of techniques including nephelometry, optical density, particle sizing or filterability (i.e. V_{max}). Cell lysis can be determined visually using a cell counting technique such as trypan blue exclusion or by measuring the concentration of a marker protein such as lactate dehydrogenase (LDH) in the feed supernatant and in the centrate. LDH is commonly used as a marker protein since the assay method is simple, rapid, extremely reliable and readily available (Decker & Lohman-Matthes 1998). Evaluating cell lysis as part of the harvest performance may be important, especially if the cell culture fluid contains proteases known to interact with the target molecule. In addition to monitoring cell lysis, measuring centrate clarity and feed flow rate are helpful for sizing downstream depth and sterile filters. These filters are sized using the same V_{max} or P_{max} procedures as described earlier with 47-mm filter discs or another small-scale normal-flow filtration device.

Although centrifugation is an extremely valuable technology for animal cell harvesting, especially at industrial scale, centrifuges are highly specialized and require careful attention and maintenance by skilled personnel. Development of a centrifuge cleaning protocol may often require a trial-and-error approach, and complete disassembly may be necessary in order to pass strict validation requirements. In addition, production-scale centrifuges often require a dedicated room to comply with noise regulations. Like tangential-flow microfiltration, the capital costs of a centrifuge harvest system can be significant considering the cost of the separator, pumps, downstream filtration equipment and utility requirements. However, the operating costs of a tangential-flow microfiltration system are typically higher than centrifugation due to the added cost of the MF membranes, and the additional validation studies covering membrane regeneration, membrane storage and membrane lifetime.

Choosing between tangential-flow microfiltration and centrifugation is often difficult since both technologies offer high throughput separation with high product yields. Generally as the solids volume and debris load increases, centrifugation becomes more attractive than tangential flow microfiltration. In many instances, both technologies can be used interchangeably with similar results. Additionally, given the trend toward higher cell population densities and debris loads, both centrifugation and tangential-flow microfiltration harvest systems benefit from using normal-flow depth filters downstream to capture sub-micron particles and reduce sterile filtration area requirements.

16.4 EMERGING HARVEST TECHNOLOGIES

In addition to the three traditional harvest techniques of depth filtration, tangential-flow microfiltration and continuous centrifugation, there are other methods under investigation that may provide higher mass transfer rates or purification coupled with separation. Many of these methods rely on applying some sort of energy to move the fluid away from the membrane surface. One promising method is high frequency reverse filtration in which a reverse pressure pulse is applied approximately every second using nitrogen or air pressure throughout the filtration process (Redkar & Davis 1995). Since the pulsations occur so often, the concentration polarization layer does not form as quickly as in conventional tangential-flow microfiltration processes. Researchers have found that permeate fluxes are significantly increased for wastewater processing (Gan 2001), and bacteria and yeast microfiltration (Levesley & Hoare 1999; Kuberkar *et al.* 1998). Although this approach has been successful in a number of microfiltration applications, its usage for animal cell harvesting has not been widely reported.

Another technology showing promise is the use of vibration to enhance mass transfer rates (Vigo *et al.* 1990). For example, the vibratory shear-enhanced processing (VSEP) design

(Culkin & Armando 1992) uses vibratory motion to prevent feed solids from depositing on the membrane. One of the drawbacks of this design is the limited ability to scale the equipment and predict the performance at larger scale. Finally, researchers have studied microfiltration modules incorporating the formation of either Taylor or Dean vortices in an effort to increase mass transfer rates and decrease membrane fouling (Winzeler & Belfort 1993; Chung *et al.* 1993). These modules have shown increases in filtration fluxes, but scale-up of these systems for industrial-scale processing remains a challenge.

In addition to exploring enhancements to tangential-flow microfiltration, other researchers have been exploring chromatographic methods in which the primary separation could be coupled with initial purification. Advantages to the chromatographic approach include potential to eliminate one processing step, leading to higher yields and shorter production cycles. Expanded-bed chromatography is one such approach in which the bed is fluidized or expanded during the load to allow cells and cell debris to pass through the column while the target molecule binds to the resin beads. Some of the key challenges with this technology are fouling and regeneration of the resin, elution of the target molecule without very large volume washes, and large-scale column design. While this method has been studied in a variety of ways by many researchers, implementation at industrial scale has not yet been achieved.

16.5 SUMMARY

Although cell harvesting methodologies are nearly as diverse as the cell cultures themselves, they each accomplish their goal of producing particulate-free solutions from either batch or perfusion reactors. Cell harvesting development efforts are not only focused on sizing filters, membranes and centrifuges based on desired processing times and volumes, but also require a good understanding of the nature and variability of the feedstock, the degree of separation required and the ability to scale performance predictably from a few to thousands of litres. Selection of the optimized cell harvesting method for a particular target molecule also involves consideration of the impact of the separation on cell viability, product quality and importantly, product yield. In addition to separation performance, equipment cleaning and validation studies must also be assessed when selecting a harvest method. Nevertheless, production-scale systems have been successfully implemented for depth filtration, tangential-flow microfiltration and continuous centrifugation for all types of animal cell culture processes. Often a mixture of methods is used across laboratory, pilot and production scales to accomplish the separation. Each technology offers its unique advantages and challenges, and newer methods are continually in development to meet the increasing demands of higher cell population densities, larger batch sizes, and regulatory compliance.

REFERENCES

Ambler C (1959) *J. Biochem. Microbial Tech. Engr.*; 1: 185–205.

Axelsson H (1999) In *Encyclopedia of Bioprocess Technology: Fermentation, Biocatalysis and Bioseparation*. Eds Flickinger M, Drew S.; 513–531.

Badmington F, Payne M, Wilkins R, Honig E (1995) *Pharmaceut. Technol.*; 19: 64–76.

Bender J (2001) Unpublished data collected at Genentech, Inc.

Berthold W, Kempken R (1994) *Cytotechnology*; 15: 229–242.

Betts M, Bracco C, Dellogio T (2003) CARR ViaFuge Reduces Downstream Filtration Requirements for Cell Culture Clarification. Kendro Laboratory Products.

Bjorling T, Dudel U, Fenge C (1995) In *Animal Cell Technology: Developments Toward the 21st Century*. Ed Beuvery EC.; 671–675.

Carman PC (1937) *Trans. Inst.Chem. Eng.*; 15: 150–167.

Cheryan M (1986) *Ultrafiltration Handbook*, Technomic, Lancaster PA.
Chu L, Robinson DK (2001) *Curr. Opin. Biotech.*; 12: 180–187.
Chung KY, Edelstein WA, Li X, Belfort G (1993) *AIChE J.*; 39: 1592–1602.
Culkin B, Armando AD (1992) *Filtr. Sep.*; 29: 425.
Decker T, Lohman-Matthes ML (1998) *J. Immunol. Meth.*; 15: 61–69.
Deo YM, Mahadevan MD, Fuchs R (1996) *Biotechnol. Prog.*; 12: 57–64.
Furey J (2002) *Genetic Engineering News*; 22: 62–63.
Gan Q (2001) *Chemical Eng. Processing*; 40: 413–419.
Geankoplis CJ (1983) *Transport Processes and Unit Operations.* Boston, Allyn and Bacon, Inc., Second edition.
George J (1994) Unpublished data collected at Genentech, Inc.
Ho CC, Zydney AL (2002) *Desalination*; 146: 75–81.
Iding K, Lutkemeyer D, Fraune E, Gerlach K, Lehmann J (2000) *Cytotechnology*; 34: 141–150.
Jager Y (1992) In *Animal Cell Technology: Basic and Applied Aspects.* Ed Murakami H.; 209–216.
Johnson M, Lanthier S, Massie B, Lefebvre G, Kamen AA (1996) *Biotech. Prog.*; 12: 855–864.
Kuberkar VT, Czekaj P, Davis RH (1998) *Biotechnol. Bioeng.*; 60: 77–87.
Letki A (1998) *Chem. Eng. Prog.* 94: 29–44.
Levesley JA, Hoare M (1999) *J. Memb. Sci.*; 158: 29–39.
Maybury JP, Hoare M, Dunnill P (2000) *Biotechnol. Bioeng.*; 67: 265–273.
Mannweiler K, Hoare M (1992) *Bioprocess Eng.*; 8: 19–25.
Pui PW, Trampler F, Sonderhoff SA, Groeschl M, Kilburn DG, Piret JM (1995) *Biotech. Prog.*; 11: 146–152.
Redkar SG, Davis RH (1995) *AIChE J.*; 41: 501.
Searles JA, Todd P, Kompala DS (1994) *Biotechnol. Prog.*; 10: 198–206.
Scheirer W (1988) *Anim. Cell Biotechnol.*; 3: 263–281.
Singhvi R, Schorr C, O'Hara C, Xie L, Wang D (1996) *Biopharm.*; 9: 35–41.
Takamatsu H, Hamamoto K, Ishimaru K, Yokoyama S, Tokashiki M (1996) *Appl. Microbiol. Biotechnol.*; 45: 454–457.
Tebbe H, Lutkemeyer D, Guderman F, Heidemann R, Lehman J (1996) *Cytotechnology*; 22: 119–127.
Tokashiki M, Arai T, Hamamoto K, Ishimaru K (1990) *Cytotechnology*; 3: 239–244.
Tomusiak M (1992) Abstr 105-BIOT, 203rd National Meeting of the American Chemical Society. Washington, DC.
van Reis R, Goodrich EM, Yson CL, Frautschy LN, Dzegeleski S, Lutz H (1997) *Biotechnol. Bioeng.*; 55: 737–746.
Vigo F, Uliana C, Ravina E (1990) *Sep. Sci. Tech.*; 25: 63.
Winzeler HB, Belfort G (1993) *J. Memb. Sci.*; 80: 35.
Woodside SM, Bowen BD, Piret JM (1998) *Cytotechnology*; 28: 163–175.
Yabannavar VM, Singh V, Connelly NV (1994) *Biotech. Bioeng.*; 43: 159–164.
Yavorsky DP, McGee S (2002) *Genet, Eng. News*; 22: 44–45.
Zeman LJ, Zydney AL (1996) *Microfiltration and Ultrafiltration: Principles and Applications.* Marcel Dekker, New York.

Additional Reading

Ambler C (1952) *Chem. Eng. Prog.*; 48: 150–158.
Axelsson H, Madsen B (2000) In *Ulmann's Encyclopedia of Industrial Chemistry.* Sixth edition, 1–28.
Banik GG, Heath CA (1995) *Biotechnol. Prog.*; 11: 584–588.
Banik GG, Heath CA (1996) *Appl. Biochem. Biotechnol.*; 61: 211–219.
Batt BC, Davis RH, Kompala DS (1990) *Biotechnol. Prog.*; 6: 458–464.
Bell DJ, Hoare M, Dunnill P (1983) *Adv. Biochem. Eng. Biotechnol.*; 26: 1–72.
Belfort G, Davis RH, Zydney AL (1994) *J. Memb. Sci.*; 96: 1.
Gehlert G, Luque S, Belfort G (1999) *Biotechnol. Progr.*; 14: 931–942.
Gorenflo VM, Smith L, Dedinsky B, Persson B, Piret JM (2002) *Biotech. Bioeng.*; 80: 438–444.
Güell C, Czekaj P, Davis RH (1999) *J. Memb. Sci.*; 155: 113–122.

Hansen HA, Damgaard B, Emborg C (1993) *Cytotechnology*; 21: 155–166.
Kempken R, Preissmann A, Berthold W (1994) *J. Indust. Microbiol.*; 14: 52–57.
Kempken R, Preissmann A, Berthold W (1995) *Biotechnol. Bioeng.*; 46: 132–138.
Lojkine MH, Field RW, Howell JA (1992) *Trans. Am. Inst. Chem. Eng.*; 70: 149.
Maiorella B, Dorin G, Carion A, Harano D (1991) *Biotechnol. Bioeng.*; 37: 121–126.
Mallubhotla H, Nunes E, Belfort G (1995) *Biotechnol. Bioeng.*; 48: 375–385.
Medronho RA, Castilho LR (2002) *Adv. Biochem. Eng/Biotech.*; 74: 129–169.
Morris BG (1966) *Br. Chem. Eng.*; *II* (5): 347–351.
Phylis PWS, Trampler F, Sonderhoff SA, Groeschl M, Kilburn DG, Piret JM (1995) *Biotechnol. Prog.*; 11: 146–152.
Rajiv VD, Rosen CG (1993) In *Biotechnology*. Eds Rehm *et al.*; 3: 469–487.
Redkar SG, Kuberkar V, Davis RH (1996) *J. Memb. Sci.*; 121: 229–236.
Rodgers VGJ, Sparks RE (1992) *J. Memb. Sci.*; 68: 149.
Shucosky AC (1993) *Filtr. Sep.* Jan/Feb.
Tarleton ES, Wakeman RJ (1992) *Filtr. Sep.*; 29: 425.
Titchener-Hooker NJ, Gritsis D, Mannweiler K, Olbrich R *et al.* (1991) *BioPharm.*; 4: 34–38.
Tutunjian RS (1984) *Dev. Ind. Microbiol.*; 25: 415.
Trowbridge M (1962) *Chem. Eng.*; August: A73–A87.
van Reis R, Leonard L, Hsu C, Builder SE (1991) *Biotechnol. Bioeng.*; 38: 413–422.
Voisard D, Meuwly F, Ruffieux P-A, Kadouri A (2003) *Biotech. Bioeng.*; 82: 751–764.
Waterson RM (1990) *Pharm. Eng.*; 10: 22–29.
Zydney AL, Ho CC (2003) *Biotech. Bioeng.*; 83: 537–543.
Zydney AL, Colton CK (1986) *Chem. Eng. Comm.*; 47: 1.

17 Protein Concentration

J Bender

17.1 INTRODUCTION

After harvesting, the final product of the batch or perfusion bioreactor is typically a dilute solution containing the target molecule, host cell proteins, fermentation media components and a variety of other impurities. The challenge of the downstream recovery process is to remove these impurities to virtually undetectable levels yet provide a high yield of the target molecule. These downstream purification processes or steps consist of a number of membrane and chromatographic separation methods. Chromatographic methods are used selectively to remove host proteins, impurities and viruses. Membrane separations are used for product concentration, buffer exchange, viral clearance, clarification and sterilization. Membranes may also have the potential to be used for purification as researchers continue to develop novel membrane chemistries and incorporate new processing strategies.

This chapter focuses primarily on the application of membrane processes for product concentration and describes the membrane devices and chemistry options, theory of operation, process development, scale-up considerations and membrane cleaning. In addition, a brief description of emerging technologies for concentration of biological molecules will be presented.

17.2 NORMAL-FLOW VERSUS TANGENTIAL-FLOW FILTRATION

Membrane separations are termed either normal-flow or tangential-flow, referring to the direction of feed flow relative to the membrane surface. Figure 17.1 illustrates the flow paths for normal (a) and tangential (b) filtration. In normal-flow operation, the feed and filtrate flow perpendicular to the surface where the retained molecules are deposited. In tangential-flow operation, the feed flow is parallel, or tangential, to the membrane surface and the filtrate flow is perpendicular to the membrane surface. For tangential-flow operation, the retained molecules flow parallel to the surface and exit the TFF device as the retentate. This sweeping flow path across the surface reduces particulate plugging and enables higher fluxes to be achieved than in normal-flow mode.

Normal-flow filters (NFF) are used for clarification of harvests by removal of sub-micron particulates prior to chromatography column loading, for virus removal, and for sterilization of buffers, product intermediates and final bulk product. Normal-flow filters primarily achieve separation based on size; however, charged membranes have also been used to assist in the removal of charged particulates.

Tangential-flow filtration (TFF) is employed in purification processes primarily for protein concentration, buffer exchange and viral clearance. TFF membranes used in the downstream recovery process are classified as ultrafiltration (UF) membranes with significantly smaller pore size rating than the TFF filters used for cell and cell debris removal. For example, the effective

Medicines from Animal Cell Culture Edited by G. Stacey and J. Davis

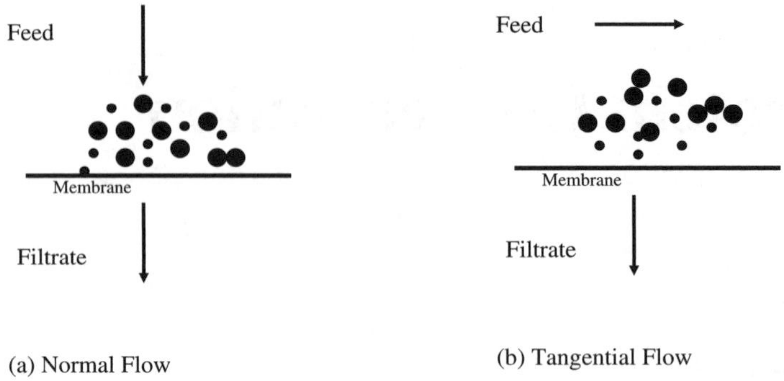

Figure 17.1 Normal-flow filtration (a) in which feed and filtrate flow paths are perpendicular to the membrane surface. Tangential-flow filtration (b) in which feed flows tangential and filtrate flows perpendicular to the membrane surface.

pore size range for microfiltration (MF) membranes is 0.05 to 10 μm, whereas UF membranes have pore sizes ranging from about 0.001 to 0.1 μm. In the classic UF operation, the selected membrane has a smaller pore size than the target molecule in order to retain and concentrate the product in the retentate while smaller molecules pass through the membrane into the permeate or filtrate. Like MF, the UF process is often combined with a process called diafiltration (DF), in which a new buffering species is introduced to exchange the product into the new solution without further dilution. UF/DF processes achieve separation due to the differences in filtration rates that occur among the solution components once pressure is applied to the retentate side of the membrane. Traditional UF processes are capable of achieving separation of solutes that differ by approximately tenfold in molecular weight. Whilst the effective pore size and relative size of the protein play a significant role in this separation, more recent studies have shown that membrane and protein charge interactions as well as manipulation of operational parameters can enable UF to be used for separation of components that differ by only twofold in molecular weight. These advances could enable UF to be used not only for protein concentration but also for protein purification (van Reis *et al.* 1997; 1999; Nakao 1988; Zydney & van Reis 2001).

17.3 ULTRAFILTRATION MEMBRANES: CHEMISTRY, PORE SIZE AND CHARACTERIZATION

Selecting the membrane chemistry is an important first decision when developing a UF/DF process. The choice of membrane chemistry depends not only on the required permeability and compatibility with the product, but also on the ability to clean the membrane effectively and repeatedly. Since most UF membranes are intended for repeated use, the compatibility with cleaning and storage solutions and conditions must be assessed. The membrane chemistry must also be compatible with the device format and able to withstand the processing pressures and temperatures at production scale.

UF membranes are cast from a variety of polymeric materials including cellulose acetate, regenerated cellulose, polyethersulfone, polypropylene, polysulfone and poly(vinylidene difluoride) (PVDF). Membrane chemistries differ in their permeabilities, surface charge, resistance to fouling and protein binding, optimum working pH and temperature ranges, compatibility with cleaning and storage agents and physical properties for withstanding pressures and flow rates at industrial scale. While the membrane manufacturer is the best source of information regarding chemical compatibility and mechanical stability of a given membrane and device, a few generalizations about membrane chemistry can be made. For example, PVDF membranes are noted for their broad chemical and thermal compatibility, but are susceptible to degradation at high pH. Cellulose membranes also feature broad chemical compatibility (except at high pH) and exhibit higher mechanical strength and lower protein binding, but are not recommended for high temperatures (McCray & Glater 1985). Polysulfone and polyethersulfone membranes exhibit greater resistance to high pH and high temperatures, but may be more susceptible to protein binding and fouling (Cheryan 1986). Regardless of which membrane chemistry is ultimately selected, studies to evaluate membrane fouling and regeneration, chemical and thermal compatibility and mechanical stability are critical in the development of a robust ultrafiltration process.

In addition to chemistry, important membrane characteristics to evaluate include permeability, charge density, solute and particulate rejection or retention, surface area, pore size distribution and porosity (Zeman & Zydney 1996). In particular, one parameter useful for selecting a UF membrane is nominal molecular weight cutoff (MWCO). MWCO is a measure of the membrane's ability to retain a protein of a given molecular weight. Although all UF membranes have a rated MWCO, the methods used to determine this value vary considerably from manufacturer to manufacturer, potentially leading to very different retention performance of the same target molecule. One method for obtaining the MWCO rating is to filter a marker protein, measure the protein concentration in the feed (C_{feed}) and filtrate (C_{filt}) and calculate the protein retention (R) using the equation given by Porter (1979):

$$R = \frac{C_{\text{feed}}}{C_{\text{filt}}} \tag{1}$$

However, using only one protein to determine MWCO is an oversimplification that may not provide reliable information about that membrane's retention of other proteins (McGregor 1986). Using combinations of proteins in one solution may also have limitations due to the effect of electrostatic interactions between the proteins and between the membrane and the proteins.

Instead of performing MWCO experiments with marker proteins, researchers have shown that using neutrally charged, polydisperse polymer solutions such as mixed dextrans and PEG solutions, give more reliable MWCO ratings (Tkacik & Michaels 1991) as well as providing information on pore size distribution (Meireles *et al.* 1991b; 1995). More recently, researchers have used charged molecules to evaluate the retention characteristics of charged and neutral membranes (Mulherkar & van Reis 2004). Regardless of the method used to determine the MWCO rating for a given ultrafilter, the true retention characteristics of that filter for a given biological product depend on protein size and charge, charge and interaction of the buffering species, and fluid dynamic conditions. Therefore a critical aspect of process development is to perform retention testing on the specific target molecule in the specific buffer components under similar fluid dynamic conditions to those to be used in the industrial-scale process.

Membrane permeability is another key characteristic to be evaluated when selecting a UF membrane. The hydraulic membrane permeability, L_p, of a membrane with uniform cylindrical pores is defined by

$$L_p = \frac{J}{\Delta P} = \frac{\varepsilon r^2}{8\mu\delta} \tag{2}$$

where J is filtrate flux, ΔP is transmembrane pressure, r is the radius of the pore, ε is porosity, μ is solvent viscosity, and δ is membrane thickness (Zeman & Zydney 1996). Membrane permeability is experimentally determined by measuring the filtrate flux at various transmembrane pressures. The clean membrane permeability, determined using water with a non-fouled membrane, is often measured before and after each protein ultrafiltration process to assess membrane cleaning and regeneration. It is important to note that the membrane permeability typically is lower after assembly into a device. The device permeability is measured in the same way that the membrane permeability is measured and is usually provided by the device manufacturer. Importantly, device manufacturers typically report a minimum and maximum permeability that could span a large range of values. Factoring in the range of device permeability and the differences in the permeability across devices of various scales improves the ability to predict large-scale performance from laboratory-scale data.

17.4 ULTRAFILTRATION DEVICES

Ultrafiltration modules are available in the same types of configurations used for microfiltration: flat-plate, hollow-fibre (including tubular) and spiral-wound. These devices are constructed so that they can be stacked or combined to allow for a range of filtration areas. The flow paths for these devices are described in the previous chapter; however, the advantages and disadvantages for protein ultrafiltration are briefly presented here.

Hollow-fibre devices are long cylindrical cases containing a large numbers of fibres (typically >10 000 fibres) bound to supports at either end. These devices have relatively high surface area-to-volume ratios and good mass transfer capabilities even at lower flow rates. Hollow-fibre modules have low hold-up volumes and can be flushed in reverse-flow mode to reduce the particulate fouling that tends to occur in their narrow channels. Disadvantages of this device format include limited operating pressures, high membrane replacement cost and tendency of the fibres to break. Detecting broken fibres is also often difficult.

Tubular modules are very similar in design to hollow-fibre devices except tubular devices have significantly larger tube diameters and smaller numbers of tubes in one bundle. Tube diameters ranging from 0.3 to 2.5 cm are common for tubular devices, compared with fibre diameters ranging from 200 to 2500 μm for hollow-fibre modules (van Reis & Zydney 1999). Tubular devices are available with single tubes or bundles of tubes within a cylindrical module. Advantages of these devices include low particulate plugging, good cleanability, and ease of membrane replacement. Disadvantages include large plant space requirements due to very low surface area to volume ratios, high feed flow rates and large hold-up volumes.

Spiral wound UF devices are typically used for large-volume processing due to their lower cost and high surface area-to-volume ratios. They also feature high mass transfer characteristics even at low feed flow rates. However, they are difficult to scale and can be difficult to clean due to their uneven flow paths and hard-to-reach spaces between the membrane and the housing. Since spiral wound modules are also prone to particulate fouling due to non-uniform flow distribution, they tend to require replacement more often than do hollow-fibre or flat-plate modules.

Flat-plate modules or cassettes are widely used for large-volume processing of high-value products due to their linear scalability, low hold-up volumes and high surface area-to-volume ratio. Flat-plate UF modules also offer either open or screened channels to improve mass transfer rates. The screened-channel devices are more prone to particulate fouling, but membrane regeneration is not as difficult as with spiral wound devices. Flat-plate cassettes are also relatively easy to customize into various size systems and membrane replacement is fairly simple.

Hollow-fibre and flat-plate devices are amongst the most widely used formats for industrial-scale protein concentration via ultrafiltration, due to their variety of membrane types, robust performance and lower plant space requirements for a given surface area. More detailed descriptions of each type of UF device are given in Zeman and Zydney (1996) and Dosmar and Brose (1998).

Regardless of the type of UF module chosen, the ability to predict industrial-scale performance from a small-scale device is necessary for developing a robust ultrafiltration process for high-value products such as therapeutic proteins. The creation of UF device formats, in which the fluid flow

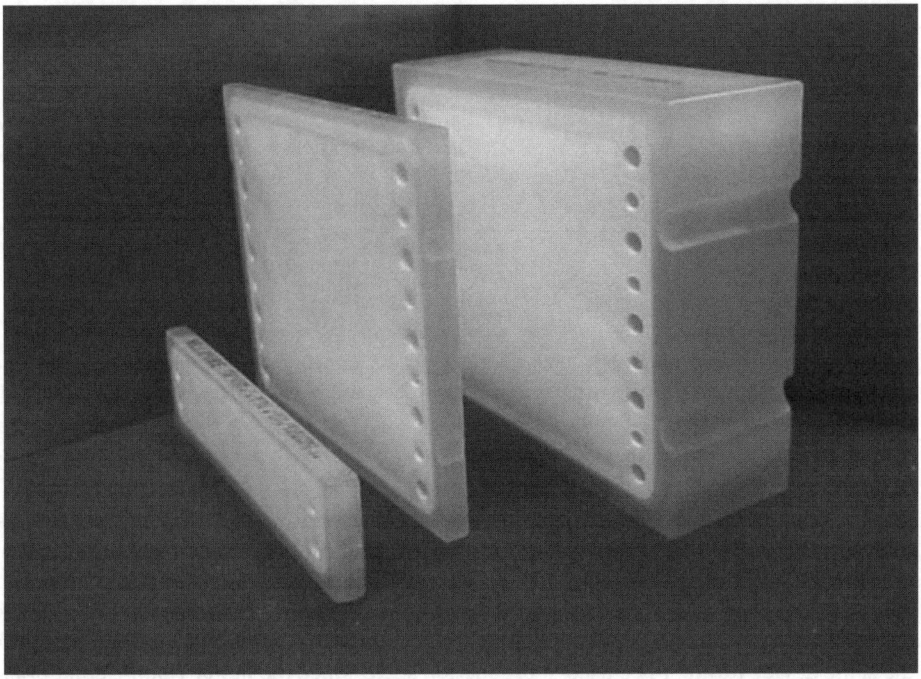

Figure 17.2 Flat-plate ultrafiltration devices containing 0.1, 0.5 and 2.5 m^2 of effective filtration area. These cassettes give linearly scalable performance and can be stacked together to create customized ultrafiltration systems. (Reproduced by permission from Millipore Corporation.)

paths and concentration and pressure profiles are constant as membrane area increases, has greatly improved the speed and accuracy of process development (van Reis *et al.* 1997). Figure 17.2 shows linearly scalable flat-plate membranes ranging in effective filtration area from 0.1 to 2.5 m^2. van Reis and coworkers (1997) demonstrated more than 400-fold scale-up in performance of ultrafiltration and diafiltration of monoclonal antibodies using flat-plate modules.

This linear scaling benefit also extends to characterization and validation studies that may be performed at small scale instead of industrial scale. The Food and Drug Administration (FDA) and other regulatory agencies are more likely to accept pilot-scale data in lieu of production-scale data when it can be proven that the studies are performed under representative conditions (O'Leary *et al.* 2001). Using small-scale systems to generate validation data not only reduces material and labour requirements, but also enables process scientists to explore a wider range of parameters without risking the quality of the final product.

17.5 ULTRAFILTRATION OPERATION AND PROCESS DEVELOPMENT

Since the concentration of proteins in biological systems involves fairly complex fluids with unknown fluid dynamic behaviour, process development depends almost entirely on empirical data collection. Although several models do exist to help the process scientist, it is typically too risky to rely exclusively on them when implementing a UF/DF process for high-value products. Therefore, process development involves conducting small-scale experiments with representative

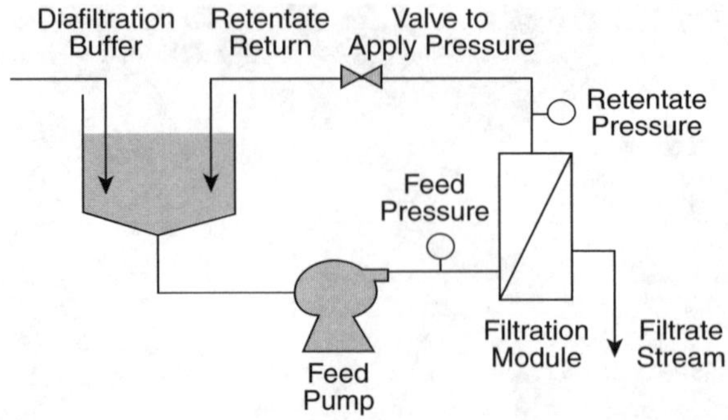

Figure 17.3 Schematic of a tangential-flow filtration process showing the feed/recycle tank, feed pump, UF device, diafiltration buffer, retentate control valve and feed and retentate pressure gauges. (Reproduced by permission of Millipore Corporation.)

feed stocks, membranes and devices and then scaling that performance to meet industrial-scale requirements. Generally for a protein concentration and diafiltration process, the only known elements are the starting and final product concentrations and buffer compositions. Membrane chemistry, device format, process control strategy and target process time are important first choices to make before exploring specific process parameters. It is important to conduct development studies with the same device format to be used at production scale, since the fluid dynamic conditions differ greatly among the available device formats. In addition, the critical process parameters for the concentration step must be identified and conserved across scales so that the development studies can reliably predict production-scale performance.

A typical TFF experimental system is shown in Figure 17.3. This simplified system contains a feed/recycle tank, feed pump, filtration module, diafiltration buffer inlet, filtrate outlet, retentate control valve, and feed and retentate pressure gauges. During a UF process, the feed flows tangentially along the membrane surface and pressure is applied to send a portion of the feed, called the filtrate or permeate, through the membrane pores. The remainder of the feed is called the retentate and returns to the recycle tank. For protein concentration, the product remains in the retentate where it is concentrated to the desired value through the removal of the filtrate. Unlike MF, UF applications rarely employ a filtrate pump and instead use the retentate pressure and feed rate to control the filtrate rate. Retentate pressure may be controlled either using a retentate valve as shown in Figure 17.3 or by using overlay pressure on the recycle tank. When the product is in the retentate, the diafiltration buffer may be added directly to the feed tank at the same rate as the filtrate is removed to exchange the buffering components. Processing parameters that are controlled or monitored during UF/DF operations include feed and filtrate flow rates; feed, retentate and filtrate pressures; transmembrane pressure (TMP); diafiltration volume and rate; and product concentration in the retentate and filtrate. TMP (ΔP) is defined as:

$$\Delta P = \frac{(P_{\text{feed}} + P_{\text{R}})}{2} - P_{\text{filt}} \tag{3}$$

where P_{feed} is the feed pressure, P_{R} is the retentate pressure, and P_{filt} is the filtrate pressure.

Key measured parameters include product yield, quality and filterability; process time; membrane regeneration (as measured by pre- and post-use flux); filtrate flux decline; and clearance of the initial buffer components. Development efforts typically focus on maximizing the filtrate flux

Figure 17.4 Automated Process Development System (APDS) manufactured by Millipore Corporation for small-scale ultrafiltration studies. (Reproduced by permission of Millipore Corporation.)

and minimizing the filtration area. Multiple lots of UF membranes/devices as well as multiple lots of product are tested to ensure reproducibility. Since process development studies require a significant investment of resources and material, the use of a small-scale automated system such as that shown in Figure 17.4 can greatly assist the process scientist. The unit shown incorporates the smallest UF devices and components into a programmable system able to control feed flow, temperature, recycle volume, retentate pressure and filtrate flow, and to simulate production-scale operations. This system, manufactured by Millipore Corporation, can be used to simulate even the more complicated control strategies used for industrial-scale ultrafiltration and diafiltration processing. Examples of industrial-scale ultrafiltration systems are shown in Figure 17.5. These systems show how the individual flat-plate cartridges can be stacked together to create a larger system area.

17.6 ULTRAFILTRATION PRINCIPLES AND THEORY

During UF, pressure (retentate pressure) is applied on the upstream side of the membrane that forces the filtrate to flow through the membrane. The flow rate per unit area, or flux, J, is

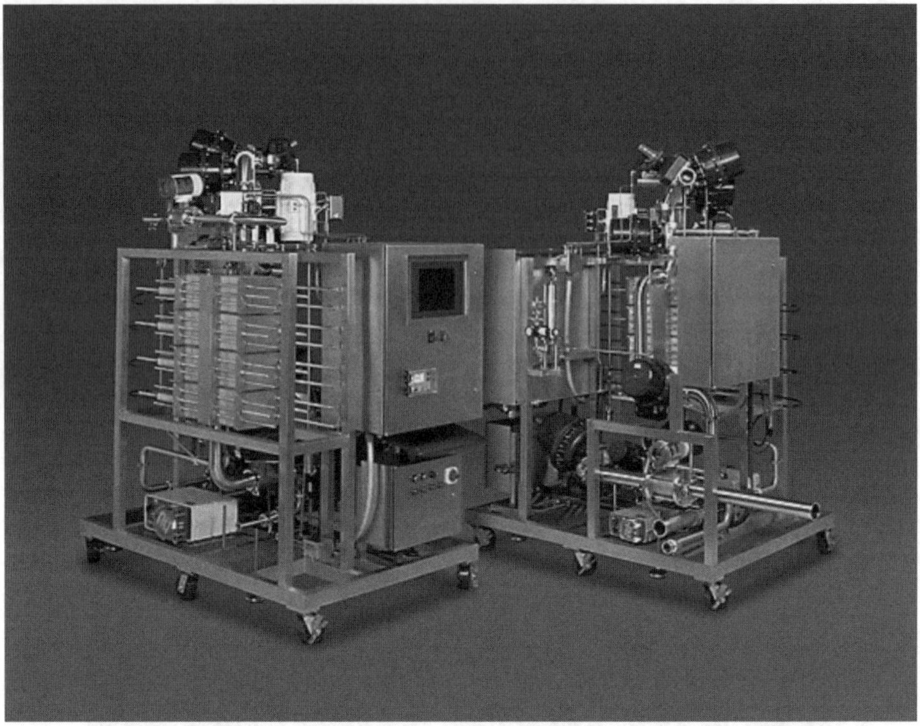

Figure 17.5 Industrial-scale ultrafiltration systems for concentration of proteins produced in cell culture fluid. (Reproduced by permission of Millipore Corporation.)

described by equation (2). Modifying equation (2) to include the effect of osmotic pressure ($\Delta\Pi$) yields:

$$J = L_p(\Delta P - \sigma\Delta\Pi) \tag{4}$$

where J is filtrate flux, L_p is membrane permeability, ΔP is transmembrane pressure and σ is the osmotic reflection coefficient (Belter *et al.* 1988). For solutes that completely pass through the membrane, $\sigma = 0$ and for solutes that are completely retained by the membrane, $\sigma = 1$. The osmotic pressure can be determined using the McMillan–Mayer (1945) theory for aqueous solutions:

$$\Delta\Pi = \alpha C + \beta C^2 + \gamma C^3 \tag{5}$$

where C is the protein concentration and α,β and γ are the first, second and third coefficients, respectively. Equation (5) can be simplified to include only the first two coefficients for most protein concentration applications. Additionally, this equation can be expanded to include the contributions of multiple components if needed; however, determining the individual coefficients for multicomponent solutions can be difficult.

Another key equation in the description of ultrafiltration is the rate of solute transport in the fluid near the membrane surface. Figure 17.6 shows that as filtrate flows through the membrane,

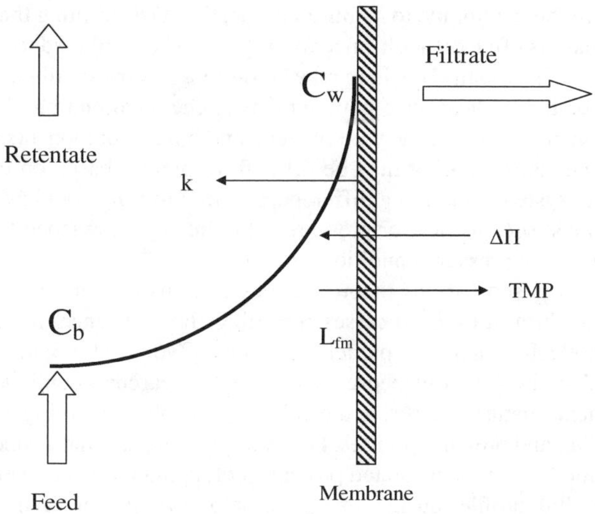

Figure 17.6 Concentration profiles in the bulk solution and at the membrane wall during ultrafiltration. (See text [pages 336–341] for explanation of terms.)

the retained molecules collect close to the membrane surface leading to an increased concentration of those molecules at the wall (C_w). Meanwhile, away from the membrane surface in the bulk solution, the concentration of these molecules (C_b) is lower due to the tangential flow of the retentate. When the protein concentration builds up within a boundary layer at the membrane surface, the resulting effect is called concentration polarization, and this leads to a reduction in the filtrate flux (J) which has been described by the stagnant-film model (Michaels 1968; Blatt *et al.* 1970):

$$J = k \ln\left(\frac{C_w - C_f}{C_b - C_f}\right) \tag{6}$$

where k is the solute mass transfer coefficient, C_w, C_b and C_f are the concentrations at the membrane wall (surface), in the bulk feed solution, and in the filtrate, respectively. Concentration polarization is described further in the following section.

17.7 CONCENTRATION POLARIZATION AND MEMBRANE FOULING

Concentration polarization and membrane fouling are inherent issues in any ultrafiltration process. They are often confused and difficult to distinguish between during processing, especially since both may occur rapidly and both result in flux decline. Basically, concentration polarization refers to the collection of retained solute particles near the surface of the membrane that can result in a lowering of the filtrate flux. On the other hand, membrane fouling is associated with the formation of a cake layer, which could be the result of protein adsorption to the membrane or formation of aggregates, precipitates or denatured proteins that deposit on the membrane surface. Membrane fouling has also been described as a change in the membrane performance as a result of specific interactions between the membrane and the feed components, and membrane regeneration is the process used to restore the membrane to its initial properties. Determining the cause of membrane fouling in a specific case is often difficult since fouling could be attributed to any one, or combination of, the feed components, many of which may be present in very small quantities. The rate and extent of fouling is related to the device flow dynamics, chemical and physical properties of the membrane, interaction of the membrane with the feed and buffer components and systemic effects such as pumping shear rates. Fouling may be difficult to predict based on development studies, especially if related to system and piping differences. Rapid fouling could be an indicator of non-ideal starting conditions, such as flow or pressure set points being reached too quickly or filtrate rates ramping up before the pressure conditions are met.

Researchers have proposed various theoretical descriptions of concentration polarization and membrane fouling in ultrafiltration processes including the resistance-in-series model, osmotic pressure model and the gel-polarization model. Zeman and Zydney (1996) and Belfort *et al.* (1994) provide more detailed reviews of these theoretical models. Researchers continue to expand upon and develop novel approaches better to describe the effects of membrane fouling and concentration polarization on filtrate flux and pressure profiles. For example, Saksena and Zydney (1997) proposed a bulk mass transport model that incorporated protein–protein interactions. Their approach more accurately predicted the flux profiles during ultrafiltration of mixtures of bovine serum albumin and IgG. Paris *et al.* (2002) modified the resistance-in-series model to include mean solute concentration and transmembrane pressure. They found that predicted filtrate fluxes were in good agreement with the model, except at low protein concentrations. Regardless of the theoretical approach used, it is still difficult to predict completely the behaviour of proteins during ultrafiltration.

Since validation of these theories relies on comparison with actual data, observation of flux decline and transmembrane pressure increase under various conditions continues to be critical in the assessment of ultrafiltration performance. Efforts to observe performance more rigorously have led researchers to explore visualization methods in which fouling and concentration can be directly observed. For example, scanning or transmission electron micrograph (SEM or TEM) analysis of fouled membranes has been used for many years to confirm the deposition of foulants and to evaluate physical changes in UF membrane structure. Additionally, researchers have explored methods to observe real-time occurrence of fouling and concentration polarization. For example, Yao *et al.* (1995) and Airey *et al.* (1998) employed nuclear magnetic resonance (NMR) to observe the formation of polarization layers and flow profiles in hollow-fibre and tubular modules. Oppenheim and coworkers (1994; 1996) and Khulbe *et al.* (2000) applied electron paramagnetic resonance spectroscopy to evaluate membrane fouling of polysulfone membranes during bovine serum albumin ultrafiltration. Li *et al.* (2002) used an ultrasonic method for studying fouling of a polysulfone UF membrane with paper mill effluent. This method allowed for non-invasive, real-time, detection of particle deposition during UF and particulate removal during cleaning. Chen *et al.* (2004) have recently reviewed these and other *in situ* methods for monitoring concentration polarization and membrane fouling during ultrafiltration processes.

17.8 ULTRAFILTRATION CONTROL STRATEGIES

Ultrafiltration processes offer some choices in control strategies, depending on the type of process control and the level of automation desired. One of the simplest strategies involving the least amount of automation is to operate at a constant retentate pressure or constant transmembrane pressure. Constant retentate pressure processes maintain selected feed flow rate and retentate pressures throughout the process while the filtrate rate and filtrate pressures are uncontrolled. Similarly, constant transmembrane pressure processes keep the feed and retentate pressures at a set value throughout the operation. Successful operation depends on the optimization of the feed rate or pressure, and the retentate pressure to avoid membrane fouling. A key drawback of this method is the risk that run-to-run variability in starting concentrations or volumes may lead to inconsistent processing times and significant flux decay.

Although more commonly used for microfiltration applications, another control option that may be used for some protein ultrafiltration processes is to control the filtrate flux by using a pump on the filtrate outlet or a control valve on the retentate outlet. For controlled filtrate flow operation, ideally the feed and retentate pressures remain constant and process development efforts focus on optimizing the feed and filtrate flow rates to achieve high product yields without fouling the membrane. This strategy is more often employed for MF operations since membrane fouling would occur quickly due to the high permeability unless the filtrate flux is reduced via a pump or valve (van Reis *et al.* 1991; Maiorella *et al.* 1991). In UF processes, controlling the filtrate flux is not as desirable because selecting too low a value for the flux leads to longer process times or greater membrane areas, and selecting too high a value leads to osmotic pressure limitations as protein concentration increases (van Reis *et al.* 1997).

A third control strategy, in which the solute concentration at the membrane wall is held constant throughout ultrafiltration and diafiltration, has also been proposed (Meireles *et al.* 1991a; van Reis *et al.* 1997). Van Reis and coworkers used the osmotic pressure model, the osmotic virial expansion and the stagnant film model to develop three control equations for maintaining a constant C_w throughout UF and DF:

$$\Delta P = \left(\frac{J}{L_{fm}}\right) + \alpha C_w + \beta C_w^2 \tag{7}$$

$$J = k \ln\left(\frac{C_w V}{C_0 V_0}\right) \tag{8}$$

$$\Delta P = \left(\frac{k}{L_{fm}}\right) \ln\left(\frac{C_w V}{C_0 V_0}\right) + \alpha C_w + \beta C_w^2 \tag{9}$$

where ΔP is transmembrane pressure, J is filtrate flux, L_{fm} is the fouled membrane resistance, α and β are the first and second coefficients, C_w is the wall concentration, C_0 is the initial concentration, V is volume, V_0 is initial volume, and k is the mass transfer coefficient.

van Reis and coworkers concluded that controlling the filtrate flux (8) rather than the transmembrane pressure to maintain a constant C_w was the preferred control strategy. Implementation of this control method requires careful optimization of the selected wall concentration. Too low a wall concentration leads to low filtrate fluxes, longer processing times and/or very large system sizes. Selection of a high wall concentration yields higher filtrate fluxes, but can also lead to solubility losses and filtrate losses. Solubility losses refer to lost product that has aggregated or denatured during the filtration process. Filtrate losses refer to product that has not been retained by the membrane during filtration. In addition to optimizing the wall concentration, since the mass transfer coefficient, k, may have different values depending on the specific buffering components, it may be beneficial to utilize two or more sets of parameters during the protein concentration and diafiltration processes. Implementation of this control strategy requires the use of an automated system both for development and production-scale operations.

17.9 MEMBRANE CLEANING AND SANITIZATION

Ultrafiltration process development does not end with the selection of the membrane, control strategy and processing parameters such as feed rate, wall concentration or retentate pressure. The development of membrane cleaning and sanitization procedures is also critical for the successful implementation of ultrafiltration at production scale. Membrane cleaning and regeneration studies are often done by trial-and-error. However, the use of statistical design may speed the development of an optimized process (Chen *et al.* 2003). Although, in some cases, membranes may be partially regenerated through backflushing or by raising the feed rates to provide more vigorous sweeping of the membrane surface during the processing, membrane cleaning is usually performed as a separate operation using different processing conditions.

Membrane cleaning may be accomplished using physical methods and/or chemical methods. Physical methods aim to remove particulate deposition by introducing some disturbance in the flow or by adding some abrasive materials to the feed stream. Al-Bastaki and Abbas (2001) have reviewed several of the physical techniques that have been implemented to reduce membrane fouling for ultrafiltration processes. Their review includes air sparging, backflushing, cyclic feed flow, and Dean vortices. Due to its effectiveness and simplicity, backflushing has been the most commonly implemented, either integrated into the filtration process or as a separate step in the cleaning and regeneration process (Crozes *et al.* 1997; Sondhi & Bhave 2001). In the typical backflushing procedure, the filtrate is flushed through the membrane in reverse direction at negative transmembrane pressures, disengaging the bound particulates to be carried away with the feed stream. The effectiveness of this procedure depends on the frequency and duration of the backflushing, the pressure, and type of rinse solution. Additional physical techniques include application of electrical fields (Bowen & Sabuni 1992), addition of abrasive components such as sponge balls to the feed (Al-Bakeri & El Hares 1993; Strohwald & Jacobs 1992; Burch 2001) and operation at high feed flow velocities.

Chemical cleaning of fouled membranes is usually performed as a separate operation following the protein concentration and diafiltration process. Once the protein is removed from the system, cleaning agents are introduced to restore the membrane to its initial state. Cleaning agents act in three basic ways to remove the foulant: chemical alteration (degradation, oxidation); solubilization; and displacement of the bound foulant. Often, cleaning strategies are developed to incorporate all three functions to ensure complete removal of all foulants. Regardless of the action of the cleaning agent, development of effective cleaning protocols is often done by trial and error and is highly dependent on the membrane chemistry and nature of the foulants.

Components found in cleaning protocols include acids, bases, chelating agents, detergents and enzymes. The most commonly used acids include phosphoric, citric, nitric, sulphuric and hydrochloric. These acids clean the membrane by reacting with salt deposits and metal oxides to create the chloride forms of the salts that are more soluble in the rinsing solutions. In addition to acids, bases are also found in many cleaning protocols. Bases commonly used for membrane cleaning include sodium hydroxide, phosphates (e.g. trisodium phosphate), and sodium hypochlorite (bleach). Sodium hydroxide is a particularly effective cleaning agent for removal of biological foulants, but not all membrane chemistries are compatible with repeated exposure to this caustic reagent. Removal of $NaOH$ from the system after cleaning is easily identified via pH measurement. Surfactants are also widely used for membrane cleaning by solubilizing the hydrophobic foulants. They may also displace or adsorb foulants remaining on the membrane surface. Although surfactants are readily available and highly effective for membrane cleaning, confirmation of their complete removal after cleaning may require additional testing. Finally, enzymes and enzymatic cleaners may be used for removal of proteins, lipids and carbohydrates from ultrafiltration membranes. The most commonly used enzymes are proteases, amylases and lipases. Proteases cleave proteins at specific amino-acid linkages to form smaller peptides. These smaller peptides are more easily removed from the membrane surface since they have a smaller number of binding sites. Amylases attack glucose sites in high molecular weight starches to create smaller molecular weight substances. Finally lipases break down large fatty acids into smaller fats that can be solubilized by basic solutions. Enzymes tend to be more expensive cleaning agents than acids, bases and surfactants, but can be highly effective for removal of specific foulants.

Many robust cleaning procedures will require flushing or recirculating multiple cleaning solutions through the membranes, so that a variety of foulants can be removed. Cleaning is often performed at higher temperatures to increase solubility, but care must be taken not to foul the membranes irreversibly via protein denaturation. Although a particular chemical agent may yield the desired results for membrane cleaning, studies to evaluate the effect of continued exposure on the membrane and system components might be needed to avoid degradation or corrosion. Selection of a cleaning agent that can also be used for storage of the membrane between processing runs minimizes the number of solutions to be prepared as well as reducing the turnaround time.

Cleaning performance may be assessed using a variety of methods. A common approach is to sample the final rinse water to ensure that both the product and the cleaning chemicals have been removed. For example, product removal can be confirmed by measuring the total organic carbon (TOC) level in the rinse water. Similarly, removal of cleaning agents can be confirmed by measuring pH, conductivity, TOC or concentration of a primary ingredient (such as phosphorus or potassium) in the rinse water. In addition, it is useful to measure the permeability (flux/TMP) of the membrane before and after cleaning to ensure that the values have not changed from run to run. Finally, membrane integrity is usually also measured before and after each use to demonstrate that there has been no change to the system or to the membrane that would result in reduced product yield.

In addition to cleaning, many production-scale ultrafiltration processes in pharmaceutical and biological plants require inclusion of sanitization or sterilization procedures. Sanitization solutions containing oxidizing agents can be flushed through the membrane as part of the pre-use procedures. Steam sterilization involves keeping the system at 121 °C for a minimum of 15 minutes. Less common for industrial-scale systems is sterilization using gamma-irradiation or gas sterilization with ethylene oxide. As with the cleaning procedures, the effect on membrane lifetime and performance after multiple sanitization and sterilization cycles must be determined.

17.10 EMERGING TECHNOLOGIES FOR PROTEIN CONCENTRATION

Although protein concentration is most commonly performed using ultrafiltration and diafiltration, there are other methods that can be used. Aqueous two-phase extraction is a technique in which two immiscible phases are used to induce partitioning of the target protein into only one of the phases. The phases are then separated via centrifugation or sedimentation. The most commonly reported two-phase systems for protein concentration and purification are polyethylene glycol (PEG) and dextran, or PEG and salt. PEG and dextran are commercially available in a wide range of molecular weights. Advantages of aqueous two-phase extraction include high product yield, ease of operation, and lower capital cost (depending on how the phases are separated). The main disadvantage of aqueous two-phase systems is the difficulty in removing excess PEG that can hinder downstream column chromatography. Additional information about aqueous two-phase extraction processes may be found in Walter (1985), Albertsson (1986) and Kula (1987).

Another promising method for protein concentration is crystallization. Although crystallization of proteins has been an important laboratory-scale procedure for the study of biological molecules, its application as an industrial-scale operation for concentration is rare. Researchers continue to explore scale-up of crystallization processes for purification and concentration. Crystallization of smaller proteins and peptides has been more readily achieved; however, crystallization of larger proteins such as monoclonal antibodies has not been as successful due to poor product recoveries. Defining conditions under which crystallization occurs has been made easier through the use of robotics and statistical design of experiments, but still takes considerable time and effort. Crystallization has the potential to reduce or replace chromatographic and ultrafiltration steps, depending on the level of purification and concentration required. Further information about crystallization may be found in Tavare (1995) and Ladisch (2001).

17.11 SUMMARY

Protein concentration via ultrafiltration and diafiltration has been widely implemented at industrial scale in the production of therapeutic proteins from animal cell cultures. Ultrafiltration process development is largely based on empirical studies. Device format, membrane chemistry, membrane retention or pore size, and permeability, are key parameters to evaluated when designing an ultrafiltration process. Concentration polarization and membrane fouling are inherent issues in any ultrafiltration process. Many researchers have used the osmotic pressure model, stagnant film theory or resistance-in-series models to describe protein fouling and concentration polarization during ultrafiltration. Several *in situ* techniques have been developed to observe directly particle deposition and formation of gel layers. These techniques have also been used to study membrane cleaning. A variety of control strategies may be used to optimize the process. Constant retentate pressure processes maintain a fixed feed flow rate and retentate pressure throughout the process, while the filtrate rate and filtrate pressures are uncontrolled. Successful operation depends on the optimization of the feed rate or pressure and the retentate pressure to avoid membrane fouling. For controlled filtrate flow operation, ideally the feed and retentate pressures remain constant and process development efforts focus on optimizing the feed and filtrate flow rates to achieve high product yields without fouling the membrane. Membrane cleaning may be accomplished using physical methods and/or chemical methods. Chemical cleaning uses acids, bases, detergents or enzymes to remove foulants bound to the membrane surface. Physical cleaning methods include adding abrasive materials to the flushing solutions, backflushing or operating at very high cross-flow rates. In addition to ultrafiltration, aqueous two-phase extraction and crystallization are finding increased use in production-scale concentration of therapeutic proteins produced in animal cell cultures.

Once the concentration process has been developed at laboratory scale, implementation of the process at industrial-scale is not complete without validation of the filter performance (process times, product yields, etc.), filter cleaning and sanitization, and membrane reuse or lifetime.

REFERENCES

Airey D, Yao S, Wu J, Chen V, Fane AG, Pope JM (1998) *J. Membr. Sci.*; 145: 145–158.

Al-Bakeri F, Hares EL (1993) *Desalination*; 92: 353–375.

Al-Bastaki N, Abbas A (2001) *Desalination*; 136: 255–262.

Albertsson PA (1986) *Partition of Cell Particles and Macromolecules*. John Wiley & Sons, New York; third edition.

Belfort G Davis RH, Zydney AL (1994) *J. Membr. Sci.*; 96:1.

Blatt WF, David A, Michaels AS, Nelson L (1970) In *Membrane Science and Technology*. Ed Flinn JE. Plenum Press, New York; 47–97.

Bowen WR, Sabuni HAM (1992) *Ind. Eng. Chem. Res.*; 31: 515–523.

Burch G (2001) In *Proceedings of the Fourth WISA-MTD Symposium on Membranes: Science and Engineering*, Stellenbosch, South Africa; 26–27.

Chen JP, Kim SL, Ting YP (2003) *J. Membr. Sci.*; 219: 27–45.
Chen JC, Li Q, Elimelech M (2004) *Adv. Coll. Interface Sci.*; 107: 83–108.
Cheryan M (1986) *Ultrafiltration Handbook*, Technomic, Lancaster, PA.
Crozes GF, Jacangelo JGC, Laîné JM (1997) *J. Membr. Sci.*; 124: 63–76.
Dosmar M, Brose D (1998) In *Filtration for the Biopharmaceutical Industry.* Eds Meltzer T, Jornitz M. Marcel Dekker, New York.
Kula MR (1987) In *Protein Purification: Micro to Macro.* Ed Burgess R. Alan R Liss, New York.
Ladisch M (2001) *Bioseparations Engineering: Principles, Practice and Economics*, John Wiley & Sons, New York.
Li J, Hallbauer DK, Sanderson RD (2002) *J. Membr. Sci.*; 215: 33–52.
Maiorella B, Dorin G, Carion A, Harano D (1991) *Biotechnol. Bioeng*; 38:121–126.
McCray SF, Glater J (1985) In *Reverse Osmosis and Ultrafiltration.* Eds Sourirajan S, Matsuura S. ACS Symposium Series, Washington, DC; 141.
McGregor WC (1986) In *Membrane Separations in Biotechnology.* Ed McGregor WC. Marcel Dekker, New York.
McMillan WG, Mayer JE (1945) *J. Chem. Phys.*; 13: 276–305.
Meireles M, Aimer P, Sanchez V (1991a) *Biotechnol. Bioeng.*; 38: 528–534.
Meireles M, Aimer P, Sanchez V (1991b) *J. Membr. Sci.*; 56: 13–28.
Michaels AS (1968) *Chem. Eng. Prof.*; 64: 31–43.
Mulherkar P, van Reis R (2004) *J. Membr. Sci.*; 236: 171–182.
Nakao S, Osada H, Kurata H, Tsuru T, Kimura S (1988) *Desalination*; 70: 191–205.
O'Leary RM, Etcheverry MT, Bezy P, Anicetti V, Burton LE (2001) *PDA J. Pharm. Sci. Technol.*; 55: 230–235.
Oppenheim SF, Buettner GR, Dordick J, Rodgers VGJ (1994) *J. Membr. Sci.*; 96: 289–297.
Oppenheim SF, Rich JO, Buettner GR, Rodgers VGJ (1996) *J. Coll. Interf. Sci.*; 183: 274–279.
Paris J Guichardon P, Charbit F (2002) *J. Membr. Sci.*; 207: 43–58.
Porter MC (1979) In *Handbook of Separation Techniques for Chemical Engineers.* Ed Schweitzer PA. McGraw Hill, New York.
Saksena S, Zydney AL (1997) *J. Membr. Sci.*; 125: 93–108.
Sondhi R, Bhave R (2001) *J. Membr. Sci.*; 186: 41–52.
Strohwald, NKH, Jacobs, EP (1992) *Wat. Sci. Technol.*; 25: 69–78.
Tavare NS (1995) *Industrial Crystallization, Process Simulation Analysis and Design*, Plenum Press, New York.
Tkacik G, Michaels S (1991) *Biotechnology*; 9: 941–946.
van Reis R, Leonard LC, Hsu C, Builder SE (1991) *Biotechnol. Bioeng.*; 38: 413–422.
van Reis R, Goodrich EM, Yson CL, Frautschy LN, Dzengeleski S, Lutz H (1997) *Biotechnol. Bioeng.*; 55: 737–746.
van Reis R, Gadam S, Frautschy LN *et al.* (1997) *Biotechnol. Bioeng.*; 56: 71–82.
van Reis R, Brake JM, Charkoudian J, Burns DB, Zydney AL (1999) *J. Membr. Sci.*; 159: 133–142.
van Reis R, Zydney AL (1999) In *Encyclopedia of Bioprocess Technology: Fermentation, Biocatalysis and Bioseparation.* Eds Flickinger M, Drew S. Wiley-Interscience, Chichester, UK; 2197–2214.
Walter H, Brooks D, Fisher D (1985) *Partitioning in Aqueous Two-phase Systems: Theory, Methods, Uses and Applications to Biotechnology.* Academic Press, Orlando, USA.
Yao S, Costello M, Fane AG, Pope JM (1995) *J. Membr. Sci.*; 99: 207–216.
Zeman LJ, Zydney AL (1996) *Microfiltration and Ultrafiltration: Principles and Applications.* Marcel Dekker, New York.
Zydney AL, van Reis R (2001) In *Membrane Separations in Biotechnology.* Ed Wang WK. Marcel Dekker Inc, New York; second edition; 277–298.

Additional Reading

Abel K (1997) *J. Membr. Sci.*; 133: 39–55.
Allie Z, Jacobs EP, Maartens A, Swart P (2003) *J. Membr. Sci.*; 218: 107–116.
Bargeman G, Houwing J, Recio I, Koops G-H, van der Horst C (2002) *Biotechnol. Bioeng.*; 80: 599–609.

Belfort G, Heath CA (1990) In *Membrane Processes in Separation and Purification*. Eds Crespo JG, Boddeker KW. Kluwer Academic, Boston; 1–35.
Belfort G (1988) *J. Membr. Sci.*; 35: 245–270.
Brose D, Dosmar M, Cates S, Hutchison F (1996) *J. Pharm. Sci. Technol.*; 50: 252–260.
Brou A, Jaffrin MY, Ding LH, Courtois J (2003) *Biotechnol. Bioeng.*; 82: 429–437.
Burns DB, Zydney AL (1999) *Biotechnol. Bioeng.*; 64: 27–37.
Cheang B, Zydney AL (2003) *Biotechnol. Bioeng.*; 83: 201–209.
Cho J, Amy G, Pelligrino J (2000) *J. Membr. Sci.*; 164: 89–110.
Delgado S, Diaz F, Vera L, Diaz R, Elmaleh S (2004) *J. Membr. Sci.*; 228: 55–63.
Derjani-Bayeh S, Rodgers VGJ (2002) *J. Membr. Sci.*; 209: 1–17.
Ebersold M, Zydney AL (2004) *Biotechnol. Bioeng.*; 85: 166–176.
Hallstrom B, Lopez-Leiva M (1978) *Desalination*. 24: 273–279.
Huisman I, Prádanos P, Hernández A (2000) *J. Membr. Sci.*; 179: 79–90.
Ghogomu JMC, Rouch JC, Clifton MJ, Aptel P (2001) *J. Membr. Sci.*; 181: 71–80.
Ghosh R, Cui ZF (2000) *Biotechnol Bioeng*; 68: 191–203.
Ghosh R (2002) *J. Membr. Sci.*; 195: 115–123.
Jones KL, O'Melia CR (2001) *J. Membr. Sci.*; 193: 163–173.
Jonsson G, Johansen PL (1991) *Filtr. Sep.*; 28: 21–23.
Kennedy M, Kim SM, Muteryo I, Broens L, Schippers J (1998) *Desalination*; 118: 175–188.
Khulbe KC, Matsuura T, Lamarche G, Lamarche A-M, Choi C, Noh SH (2000) *Polymer*; 42: 6479–6484
Kim KJ, Sun P, Chen V, Wiley DE, Fane AG (1993) *J. Membr. Sci.*; 80: 241–249.
Kim KJ, Fane AG (1995) *J. Membr. Sci.*; 99: 149–162.
Ko M, Pellegrino J (1992) *J. Membr. Sci.*; 74: 141–157.
Kuakuvi DN, Moulin P, Charbit F (2000) *J. Membr. Sci.*; 171: 59–65.
Levy P, Shehan J (1991) *Biopharm*; April: 24–33.
Laborie S, Cabassud C, Durand-Bourlier L, Laine JM (1997) *Filtr. Sep.*; 34: 886–891.
Marshall AD, Munro PA, Trägårdh G (1993) *Desalination*; 91: 65–108.
Matzinos P, Álvarez R (2002) *J. Membr. Sci.*; 208: 23–30.
McAlexander BL, Johnson DW (2003) *J. Membr. Sci.*; 227: 137–158.
McDonogh RM, Bauser H, Stroh H, Chmiel H (1990) *Desalination*; 79: 217–231.
McDonogh RM, Bauser H, Stroh H, Grauschoph U (1995) *J. Membr. Sci.*; 104: 51–63.
Meacle F, Aunins A, Thornton R, Lee A (1999) *J. Membr. Sci.*; 161: 171–184.
Miller KD, Weitzel S, Rodgers VGJ (1993) *J. Membr. Sci.*; 76: 77–83.
Möckel D, Staude E, Guiver MD (1999) *J. Membr. Sci.*; 158: 63–75.
Moulin P, Manno P, Rouch JC, Serra C, Clifton MJ, Aptel P (1999) *J. Membr. Sci.*; 156: 109–130.
Muñoz-Aguado MJ, Wiley DE, Fane AG (1996) *J. Membr. Sci.*; 117: 175–187.
Nabetani H, Nakajima M, Watanabe A, Nakao S, Kimura S (1990) *AIChE J.*; 36: 907–915.
Noordman TR, de Jonge A, Wesselingh JA *et al.* (2002) *J. Membr. Sci.*; 208: 157–169.
Raghavarao KSMS, Guinn MR, Todd P (1998) *Sep. Purif. Methods*; 27: 1–49.
Redkar S, Kuberkar V, Davis R (1996) *J. Membr. Sci.*; 121: 229–242.
Rodgers VGJ, Sparks RE (1992) *J. Membr. Sci.*; 68:149-168.
Rodgers VGJ, Sparks RE (1993) *J. Membr. Sci.*; 78: 163–180.
Romaro J, Zydney AL (2002) *Biotechnol. Bioeng.*; 77: 256–265.
Serra C, Clifton MJ, Moulin P, Rouch JC, Aptel P (1998) *J. Membr. Sci.*; 145:159–172.
Serra C, Wiesner M (2000) *J. Membr. Sci.*; 165:19–29.
Song L, Elimelech M (1995) *J. Chem. Soc. Faraday Trans.*; 91: 3389–3398.
Su TJ, Lu JR, Cui ZF, Thomas RK (2000) *J. Membr. Sci.*; 173: 167–178.
Vilker VL, Colton CK, Smith K, Green D (1984) *J. Membr. Sci.*; 20: 63–77.
Wilharm C, Rodgers VGJ (1996) *J. Membr. Sci.*; 121: 217–228.
Winzeler H, Belfort G (1993) *J. Membr. Sci.*; 80: 35–47.
Yousef MA, Datta R, Rodgers VGJ (1998) *J. Coll. Interf. Sci.*; 197: 108–118.
Yousef MA, Datta R, Rodgers VGJ (2002) *AIChE J.*; 48: 913–917.
Zumbusch P, Kulcke W, Brunner G (1998) *J. Membr. Sci.*; 142: 75–86.

18 Purification Methods

M Wilson

18.1 OVERVIEW OF THE AIMS OF PURIFICATION

The use of biological medicines in humans requires that they are purified to standards determined by regulatory authorities primarily to ensure patient safety, but also to ensure that product is consistent and effective. The standards relating to purification are used to determine the permissible levels of impurities in the administered product, and hence the specifications that are set for the product during development and manufacture (Seamon 1998). In beginning to devise a purification strategy for any biological medicine, it is vital to know the proposed dose, dose schedule, product concentration and route of administration, in order to interpret the regulatory guidelines and set limits for the expected impurities such as DNA, host cell protein and endotoxin.

Impurities, such as endotoxin or DNA, are usually limited to a certain specified amount per dose, so the product purity will be different depending on the amount of product in the dose. For example, a dose of erythropoietin may be only 10 μg, whereas the dose for albumin could be as high as 50 g. In either case the allowable level of endotoxin or DNA, per dose, would be the same. The purity per unit mass of product will be much higher where a greater amount of product is administered in a single dose.

It is important that assays for quality control testing of product impurities and process contaminants are available throughout purification development. A useful distinction can be made between impurities that exist in the process start material (fermentation harvest) such as host cell protein or DNA, and contaminants derived from processing materials that are introduced during the purification process such as leached protein A, or Tween used, perhaps, to prevent aggregation. Table 18.1 gives examples of quality control tests that may be employed at various stages in the downstream process of a product derived from animal cells.

Assays for product integrity and activity are also essential during process development. It is vital to know that the conditions and operations used to purify the product do not cause instability or inactivation. Temperature and pH may be critical to product breakdown or modification, and may cause the product to lose activity. Equally, high concentrations of product, either in solution or when loaded onto a chromatography column, could cause product aggregation. The presence of proteases from the cell culture may cause product degradation and loss of activity. It is important to have assays that can distinguish active, non-degraded product, and that these are used to determine the integrity of the product during the process. It will be important to define the conditions, for example temperature and pH ranges, within which product remains stable and for how long exposure to adverse conditions is acceptable.

Medicines from Animal Cell Culture Edited by G. Stacey and J. Davis

Table 18.1 Examples of Quality Control testing employed at various stages in the downstream process.

Process Stage	Description	Examples of QC Testing
Raw Materials	Column matrices, membranes, chemical reagents, buffers, excipients	Appearance, chemical identity, endotoxin content, package integrity, correct certification
Crude Harvest	Unpurified product from cell culture	Product concentration (e.g. by ELISA), total protein concentration (e.g. by Lowry or Bradford), SDS-PAGE, pH, conductivity, adventitious agents, mycoplasma, Reverse Transcriptase
Product Intermediates	Column eluates, eluate fractions, filtrates, concentrates	Product concentration (e.g. by ELISA), total protein concentration (e.g. by Lowry, Bradford or OD_{280}), SDS-PAGE, measurement of specific impurities for pooling of fractions, pH, conductivity
Drug Substance	Purified bulk product	Appearance, pH, HPLC (e.g. size exclusion or reverse phase), DNA, SDS-PAGE, Western blot, isoelectric focussing, N-terminal sequencing, endotoxin, bioburden, trace metals, potency
Drug Product	Formulated, filled product (e.g. in vials, ampoules or pre-filled syringes)	Sterility, pH, appearance, endotoxin, volume in container, strength (product concentration), osmolarity, potency

Purification processes are normally built from steps that can be categorized as follows (in sequential order of their use):

- clarification – the removal of cells and cell debris at the start of the process;
- capture – the removal of gross contamination and concentration of product, usually by chromatography;
- intermediate step – the conditioning of the product to prepare for the next process step, for example buffer exchange by diafiltration to remove salt prior to loading onto ion exchange, or to concentrate product prior to size exclusion chromatography;
- purification – removal of the bulk of the impurities, usually by chromatography;
- polishing – removal of trace impurities and contaminants, yielding drug substance (or bulk purified product);
- formulation and fill – putting the product into the right buffer and filling into the final containers for patient administration, yielding drug product.

These steps can be built into a coherent process, where each step has a purpose, or purposes. An example of such a process flow is given in Figure 18.1.

It is often difficult to categorize a purification step. For example, Protein A is often used as a capture step in antibody purification, but, as an affinity step, it also removes the bulk of the impurities. However the steps are categorized, each step should always have a defined purpose in the purification scheme, and should be justified by data supporting the need for, and effectiveness of, the step in the process.

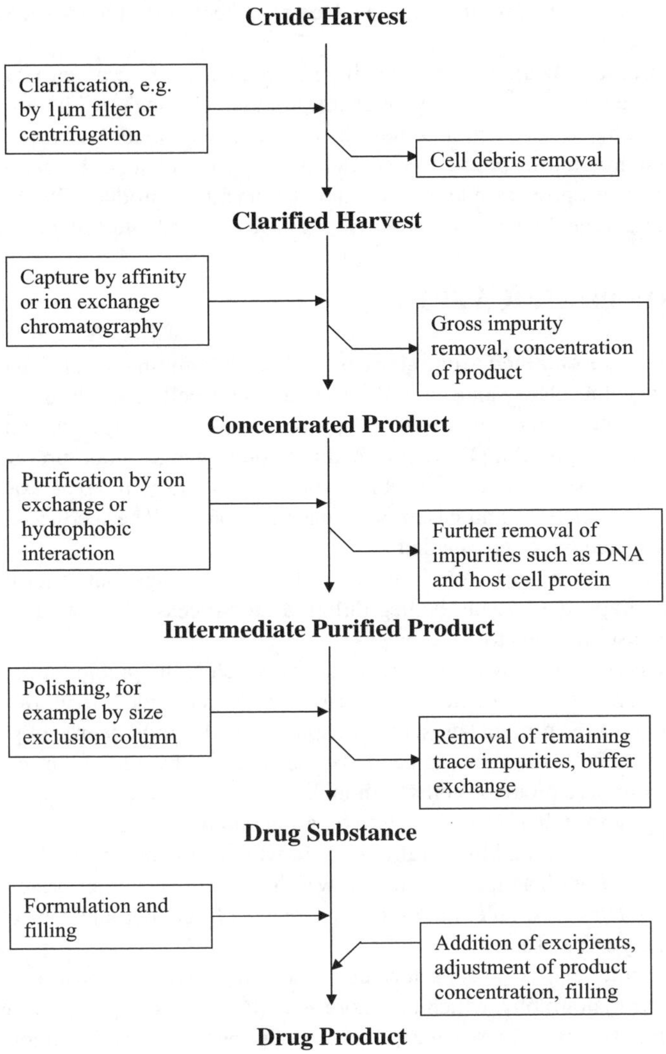

Figure 18.1 Example Process Flow Diagram.

Practical guides in the use of purification methods for biological products can be useful; some are intended for laboratory-scale work (Harris & Angal 1995a,b) and give guidance in the use of, for example, chromatographic techniques; others are aimed at large-scale bioprocessing (Subramanian 1998). Useful guides can be obtained from the suppliers of membranes and chromatography matrices, such as the series by GE Healthcare or *The Busy Researcher's Guide to Biomolecule Chromatography* by Applied Biosystems.

It is also necessary to know the scale at which product will need to be made for commercial supply, as this will influence the way the process has to be developed. The number of doses per year and the amount of product per dose should be known at the outset, and will be a key part of the business plan. A large volume product will generally need to be produced more cheaply, on a cost per gram basis, and the cost of goods will be vital to the commercial viability. A small volume product that can be sold at a high price will be less dependent on process economics, though the profitability of the product will be adversely affected if process economics are unduly neglected.

It is also important to bear in mind that improvements to the productivity in the upstream process will not necessarily help the downstream process. The scale of the downstream process will not be likely to change, even though the cell culture vessel may be a fraction of the size previously planned – the amount of product being purified will not change. Indeed, it is possible that changes in the upstream process may be beneficial in terms of productivity, but if the impurity profile is also changed the downstream process will have to be adapted accordingly.

18.2 TYPES OF PURIFICATION

Medicinal products are obtained from sources such as recombinant cells, blood from human donation, and transgenic plants and animals. Recombinant cells include bacteria, yeast, insect cells and mammalian cells, and the cell culture process produces a large volume of crude product also containing host cell protein, DNA, and medium components. Blood products must be separated from a large number of other blood-borne proteins. Transgenic expression is generally in the animal's milk, in which fat and casein are major impurities. The medicinal product must be extracted from the crude bulk and purified.

Chromatography plays a major role in most production schemes, but other methods are often employed in addition, particularly at the beginning of the process where a bulk impurity, such as serum albumin or casein, needs to be removed.

Human plasma is separated by Cohn fractionation, which is the precipitation of proteins using varying concentrations of cold ethanol. The various fractions made during this process contain different populations of the proteins present in the plasma and a good degree of purification can be achieved relatively cheaply. Fractionation can also be achieved by adding ammonium sulphate or polyethylene glycol to precipitate proteins, although this is less commonly used at industrial scale. The precipitate is generally harvested by centrifugation and dissolved.

Proteins can also be separated by ultrafiltration in which a membrane with a specific pore size is used to retain proteins with a higher molecular weight and allow smaller proteins into the filtrate (see Section 18.5.7). However, unless the size difference is large, perhaps tens of kilodaltons, the separation is unlikely to be efficient.

In most cases, crude separation is performed at the early stages of a process and the product is then purified by chromatography, which is a more selective means of purification.

Protein crystallization is an increasingly useful tool in purification. Different glycoforms, misfolded product or cleaved product, which are very difficult to separate by any other means, may be separable by crystallizing the required form of the protein and harvesting the crystals of product (Gulewicz *et al.* 1985; Alexander *et al.* 1995). This technique, however, may not be suitable for very large-scale manufacture.

It is likely that animal cell culture, particularly when serum-free or protein-free medium is used, will generate an unpurified bulk intermediate with a low level of contamination, and this in turn can make a purification scheme simple to develop. However, any animal cell culture inherently carries the risk of contamination from endogenous viruses and this adds an element of complexity to the production of medicines from animal cells (see Section 18.4.9).

18.3 PRINCIPLES OF CHROMATOGRAPHY

The fundamental principle behind the development of purification schemes for biological medicines should be to develop a robust process suitable for use at the scale required for manufacturing. Processes in manufacture should work the same each time, within predefined limits, and produce material meeting predetermined specifications. Yields need to be predictable and as high as possible, so that the scale can be set long before implementing transfer into manufacture. The key to ensuring robustness is to investigate limits around optimal values for parameters like pH, salt concentration, column loadings and flow rates. This should be done in the early stages, preferably during method screening.

Defining limits for variables involves testing different conditions, for example the salt concentration required to elute product from a column without eluting contaminants, and checking that the same result is achieved within a reasonable range either side of the optimum. It is useful to determine the effects of using parameters outside the range, for example overloading a column may cause product breakthrough and result in a low yield and suboptimal performance, whereas under-loading a column could result in heavy contamination of the product with impurities and a failed batch. Flow rates used may affect the elution volume and this may be important if the subsequent step relies on the loading volume, such as size exclusion. Variables that have a significant impact on product quality should be deemed to be critical variables (Seely *et al.* 1999). Critical variables are also those that are required to be very tightly controlled, such as the concentration of salt used for wash and elution of an ion exchange column, or variables that are difficult to control at scale, such as the temperature of the facility where, for example, hydrophobic interaction chromatography – which is temperature dependent – is performed.

Separation of product from host-cell protein impurities, DNA, and endotoxin can be achieved by a number of commonly used techniques. With the development of a purification scheme for biological medicines it is usual to end up with a process consisting of steps that function orthogonally, i.e. each with a different mechanism of separation to the others. The functions of these steps should be identified during screening, and the performance of each monitored in relation to its function as the process is put together.

Selectivity is the most important factor when choosing a chromatography matrix for use in a purification scheme. Selectivity is the ability of a matrix to interact with the product molecule in a different way to the impurities. A matrix that binds product and not the impurities has a high selectivity. A different matrix that has the same ligand chemistry but a different support, or a matrix from a different supplier, may have a different selectivity and might bind impurities differently. The selectivity is dependent on the properties of both product and impurities and is something that can only be determined empirically. The conditions yielding optimal selectivity on any given matrix also have to be determined experimentally.

A simple purification scheme may take clarified harvest material and use ion exchange as a capture step. Since salt is used to elute from the ion exchange matrix, hydrophobic interaction might be a useful choice as a second step because it requires salt to be present during loading. A polishing step might be size exclusion, which could simultaneously remove impurities from the product, and change the buffer into that required for formulation.

Any process step will be dependent on the performance of the preceding steps. Optimization of the performance of the process should be done as a complete unit by testing the effects of the optimization of any one step on the performance of the process as a whole. Inevitably, this will lead to compromises in the optimization of individual process steps (Ngiam 2003). A simple example may be where the optimization of the elution conditions of one step may affect the binding of the product onto the subsequent column. It is more likely that problems will occur where the optimization of one step alters the impurity profile and the challenge for a step further down the process; this can mean that new impurities or higher levels of impurities remain at the end of the process and it is important to know how the change arose. This can be resolved by running a full process with the changes to one step in place, along with full analytical support to test the purified drug substance. Changes to the process should, therefore, be introduced stepwise.

Such an approach should be taken to the purification development of any biological medicine, including proteins, viruses and gene therapy vectors, whatever cell type and cell culture method is used in the upstream process.

Chromatography matrices can be made of a variety of different ligand chemistries, with beads made of a number of different materials. Each manufacturer is likely to provide similar chemistries attached to their own type of bead. Most columns are run as packed beds, and the majority of resins are intended for use in this way. Beads are manufactured from soft materials such as agarose, or from materials that produce a more rigid bead such as methacrylate polymer, or poly(styrene divinyl benzene). Most of the leading manufacturers will be able to provide a drug master file (DMF, see also Section 18.4) for chromatography resins designed for use in biological medicines production; it is important to check that this is the case before using a resin intended for clinical production.

Bead sizes also vary, and in general a small bead size of, say, 20 μm, although giving higher resolution at small scale, is impractical for large-scale use, mainly because of the resultant high backpressures requiring expensive pumps that are impractical at industrial scales. Large beads of 150 μm or more can give poor dynamic binding capacity, due to the increased volume between functional groups on the large beads, and may require low flow rates to achieve high loading levels. Low flow rates can impact on the time a step will take, and hence adversely affect production efficiency. Low loading levels may make the step less economic, requiring more matrix for a given load. Larger beads may be of particular use in virus- or gene-therapy applications, as the large beads can allow the large virus particles to flow through the column without being filtered out by the matrix. Virus particles will not diffuse into the beads and will bind to the bead surface, and so binding capacities are likely to be low, though 10^9 particles per ml of packed bed can be achieved. Bead sizes of around 50 μm are most likely to suit production of most protein products; pressures of around 0.5 bar are normal, high flow rates can be achieved and protein loading levels of around 50 mg/ml packed bed are achievable, for example, with ion exchange resins.

Column packing efficiency is an important factor to take into account during process development for biological medicines. The normal measurements are number of theoretical plates (N) and asymmetry (As), the calculations of which are shown in Figure 18.2. The height equivalent to a theoretical plate (HETP) is often referred to where HETP = L/N, and L is the length of the column.

The number of theoretical plates and asymmetry are used to determine how much resolution a column of a certain bed height will give, and how much an elution peak will spread during elution, normally by peak tailing, respectively. It is important that a process in development does not rely on column packing efficiencies that cannot be replicated in manufacturing at large scale. For example, a size exclusion column with N of around 2000 and As of around 2 should be achievable at large scale. Although values of over 10 000 for N and under 1.5 for As may be possible at large scale, they are more readily achieved in the laboratory. A separation where the resolution is critical and is dependent on the packing may be unachievable upon scale-up, and may not be transferable, entailing expensive and time-consuming rework at a critical stage in the project.

A technique that has been increasingly applied to the capture of biological products is expanded bed adsorption (EBA) (Thömmes 1996). In such a system, the resin is not loaded as a packed bed. The upward flow of crude unclarified harvest material causes the beads to disperse up the length of a column about a metre high. The flow rate is balanced against the rate of the settling of the dense beads. The material is loaded onto the column and as the crude material passes through the column the beads remain suspended in the fluid. Once loaded, the matrix can be packed into a conventional bed, then washed and eluted as in conventional chromatography. The advantage of this technique is that unclarified material can be applied to the column, which may have advantages in some applications. For example, EBA can eliminate the need for clarification by centrifugation or large-scale filtration where unclarified material would otherwise block a conventional column.

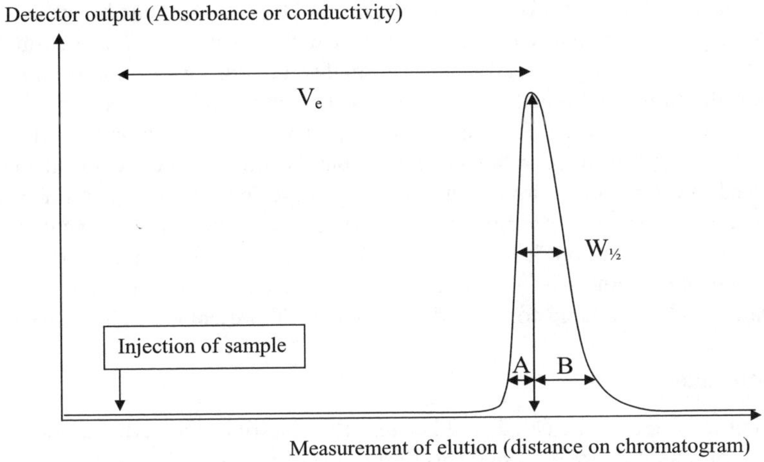

Number of theoretical plates:
The elution volume (V_e) can be measured on the chromatogram from the point of sample application to the apex of the peak.
The width at half the peak height is measured ($W_{½}$) on the chromatogram.

Number of theoretical plates (N) = 5.54 x (Ve / W½)

$$N = 5.54 \times (V_e / W_{½})$$

Asymmetry:
A vertical is drawn from the apex of the peak. At 10% of the peak height the distances from the vertical to the peak are measured; A at the front of the peak, B after the apex.
Asymmetry (As) = B / A

Figure 18.2 The calculation of Number of Theoretical Plates and Asymmetry of a column.

18.4 ISSUES TO BE CONSIDERED IN DESIGNING A PURIFICATION PROCESS

In developing a purification scheme for biological medicines (Berthold & Walters 1994), it is important (as mentioned previously) to ensure at the outset that the methods used are appropriate for use in the manufacture of clinical material at large scale. The choice of technique or selection of chromatography matrix, or other raw materials, will be influenced by a number of factors.

18.4.1 Raw Materials

The need to obtain raw materials, such as matrices and chemical reagents, that comply with cGMP (MCA 2002) is a major factor. The use of materials from any given supplier will ultimately require the supplier to be audited for cGMP compliance. A drug master file (DMF) is a collection of information lodged with regulatory authorities by the supplier to demonstrate the acceptability of process components, such as column matrices. The existence of a DMF is a useful way of gauging whether a supplier complies with regulations. Of particular importance is the certification of raw materials that should provide data on the lot supplied, including the date of manufacture, lot number, and expiry date. Information should be available from the supplier on the manufacturer, the origin of the raw material, and the origin of all materials used its manufacture. This applies particularly to any raw materials derived from animal sources, or which use animal products in their manufacture. It is a requirement of regulatory authorities that new medicines do not present a significant risk to patients from transmissable spongiform encephalopathies (TSEs). TSEs can be introduced into the process through any material that has been in contact with raw materials of mammalian origin. This is a potential problem for processes that use animal serum during cell storage or culture or, for example, human albumin as a product stabilizer. However, any raw material used in the downstream process should be also be assessed. Protein A and DNase are often used in biologicals manufacture and these can be obtained from recombinant sources, but it is important to be able demonstrate that all materials used in the recombinant production process are free of animal products. The amino acid cysteine is a good example; used in processes to control oxidation, cysteine can be obtained as a GMP-grade raw material, but it may take a lot of research to find that a GMP-grade cysteine is derived from chicken feathers or human hair with an associated (if extremely small) TSE risk. Cysteine is available as a synthetic compound, and this would present no such risk to the patient.

18.4.2 Economics

The cost of manufacture must be calculated because the viability of a medicinal product may depend on the cost of manufacture both in terms of the cost of raw materials and the cost of running the commercial manufacturing facility. Economic considerations must be taken into account in the selection of raw materials and suppliers. A major influence on production economics will be process productivity, i.e. the quantity of material made from each batch. An important factor in this is the recovery of product through the downstream process.

18.4.3 Scale

The scale of the process will affect the economics of the process, i.e. the overall cost and the cost per dose, and will also influence the choice of techniques used. It is necessary to determine the scale at which the product will be manufactured. This equation is dependent on the individual dose of product, the frequency of dosing (single or repeated dose), and the projected number of patients to be treated per year.

18.4.4 Batch Definition

Also relevant to the determination of the scale of the downstream process is the definition of a batch. The upstream harvest frequently defines the scale at which the downstream process operates. This is often the case with batch cultures, and is generally true for bacterial cultures where the upstream process takes 2–3 days. However, it may be more economical to pool a number of upstream harvests so that the downstream process train is used efficiently. This is particularly common where the upstream process is shorter than the downstream process, as can be the case with batch culture.

With animal- or insect-cell cultures, it may be economic to run large fed-batch or perfusion cultures that are much longer in duration than the downstream process. In such cases it may be beneficial to split culture harvests, or take-off harvests during perfusion, and purify these through smaller scale downstream processes.

However, such a production method that can take many weeks for the cell culture (for example, in the manufacture of a monoclonal antibody) could result in changes in the cells during the fermentation. This may cause changes to the glycosylation pattern of the product, and may also allow proteolytic cleavage of the product molecule.

Glycosylation patterns on proteins are dependent on the state of the cells that produce them. It is known that culture conditions can influence the glycosylation of proteins (Yuk & Wang 2002), e.g. recombinant tPA (Senger & Karim 2003), and it is important that the product behaves the same from batch to batch in the downstream process. The product from each batch should be characterized with respect to its glycosylation whenever the process is scaled-up or transferred to new equipment or facilities.

So, it is important to define a downstream process that consistently delivers product meeting the specification from harvests throughout the upstream process.

18.4.5 Reuse

The cost of matrices and membranes also has an impact on the economics and thus the strategy for reuse. It is clearly cheaper to reuse any components if possible, but matrices and membranes may be relatively inexpensive compared with the overall cost of manufacture or the value of the medicinal product. Cleaning validation is a relatively difficult and expensive task, and reuse may introduce risks such as transfer of old, degraded product or contaminants from one batch to another. Thus single use may prove more practical and economic. It is, however, most likely that some of the more expensive process components will be cleaned and reused. The reuse of a process component requires demonstration that its reuse does not affect the process over the useful life of the component, and this should be done as early in the development of the purification process as practicable.

18.4.6 Cleaning

Where reuse is necessary, cleaning methods must be employed that remove impurities from the processing of one batch and restore the function of the column or membrane for the processing of the next batch, and ensure limited carry-over between batches. Many cleaning methods involve the use of incubation in sodium hydroxide. However, silica or glass beads cannot be exposed to alkaline reagents and are sanitized with acid/ethanol. Protein A and Protein G also do not withstand harsh sanitization. Validation must be performed to demonstrate that the cleaning method will enable the material to be reused beyond the required number of cycles without significant deterioration of its performance, by testing the product of the first and last cycles and demonstrating equivalence in terms of purity, recovery, and potency.

Membranes and matrices may be reused a hundred or more times depending on their durability and chemical resistance, and the level to which they are contaminated during the process. Cleaning may be much more difficult for membranes and matrices used in the initial steps of a process, and which are consequently contaminated with cell debris, endotoxin, DNA and host-cell protein. In such cases, single use may be advisable. The need to reuse expensive ultrafiltration membranes, for example, coupled with the difficulty of cleaning them, may favour the use of desalting chromatography for large-scale buffer exchange.

The testing of cleaned materials is also complicated. Total organic carbon in rinse water is used to check that glass or stainless steel has been cleaned, but column matrices and membranes contain organic carbon, which interferes with the test. Impurities may need to be tested for by total protein or silver staining, DNA, and product ELISA assay. It is acceptable for there to be as much as 0.1 % carry over from one batch to another provided that the product is identical (Institute of Quality Assurance Monograph on Cleaning Validation, see the IQA web site). Between different products cleaning must be far more thorough, and it is usual for all product contact parts, such as column tubes, matrices and membranes, to be dedicated to a single product. Fixed vessels and pipework and other stainless steel hardware can be used for different products, but must be cleaned by validated methods, and shown to be clean, before the new product can be processed (see Chapter 31).

18.4.7 Equipment and Scale-up

The availability of hardware at the required scale, or the complexity of its deployment, may influence which techniques are used and how they are implemented.

Large-scale engineering, for example of flow distributors, can be a critical issue. The difficulties of large-scale operation can be balanced against the cost of running multiple cycles at a

Figure 18.3 Pilot-scale columns used in Phase II clinical manufacture. A 30-litre size exclusion column and a 2-litre ion exchange column used for production of clinical lots for a protein vaccine product at Xenova. In these Millipore Vantage A columns, the headspace above the top adjustable adapter can be pressurised to enable rapid and consistent column packing. Photo courtesy of Jim Mills, reproduced with permission of Xenova.

smaller scale. This approach is often used for size exclusion, but as the technique is slow, multiple cycles may take days to complete and this in turn will impact on the economics of the facility. This technique can be difficult to implement at commercial scale, and scale-up can cause a loss of the required resolution (see Section 18.5.5).

The selection of chromatography hardware will be influenced by the scale of operation. Columns of up to about 100l such as the BioProcess Glass columns from GE Healthcare, can be packed using traditional hardware. Other column ranges offer zero dead space around the adapter seals, such as the Vantage columns from Millipore (Figure 18.3), and these are potentially more easily cleaned or sanitized. Such columns are also available with air packing, which allows the column to be packed by applying air pressure in the headspace above the adjustable adapter.

Columns of 500l or more can be custom made from stainless steel. At this scale, it is very costly to have adjustable adaptors, but it is very difficult to pack columns of the correct bed volume using fixed headplates, and to do so without headspace voids.

Figure 18.4 A large-scale, 1.5m diameter Chromaflow column, when used with a semi-automated packing station, enables clean and efficient column packing at large scale. Photo courtesy of GE Healthcare. Bio-sciences AB

Chromaflow columns, from GE Healthcare Bio-Sciences AB, offer a semi-automatic method of packing that can be of benefit when packing large columns of up to 900l in a consistent manner (Figure 18.4). The columns are packed by pumping a matrix slurry into a fixed column tube. These columns do not require adjustable adaptors, and the column tube is made to the required length to accommodate the required bed volume. These columns can also be unpacked without dismantling the column hardware, which can be an advantage in minimizing production downtime for column packing.

A similar packing method is used in radial-flow columns, e.g. from Sepragen. In radial-flow columns, the flow is applied to the column around the circumference of the column tube and liquid flows through the matrix towards the centre of the column where outflow is collected along the axis at the centre of the column. This is different to axial-flow columns where flow is applied to the top cross section, flow is parallel to the column axis, and the outflow is collected at the bottom cross section.

Axial column scale-up is generally achieved by maintaining the linear flow rate and the column bed height, whilst increasing the cross-sectional area and volumetric flow rate. In an axial flow column, the linear flow rate is constant along the flow path.

Radial-flow columns are scaled up simply by increasing the length of the column tube and the volumetric flow rate, and this minimizes the footprint required by the column at manufacturing scale. However, the linear and volumetric flow rates across the radial flow path are not constant, and the flow rate increases as the flow nears the centre of the column.

It is necessary to ensure that scale-up of the process will be possible (Sofer & Hagel 1997). The commercial scale will influence how the process will work; for example any operations involving

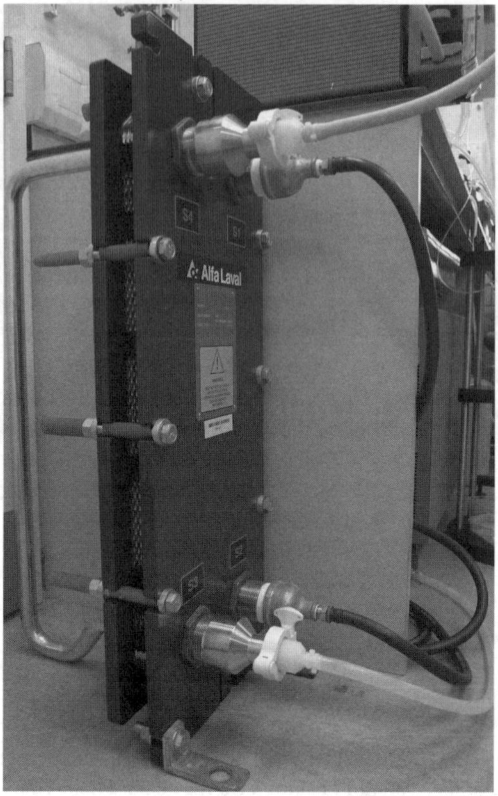

Figure 18.5 A 1/10-scale heat exchanger used in the development of a process to cool 50L of clarified harvest in the production of a live virus vaccine. Photo courtesy of Jim Mills, reproduced with permission of Xenova.

mixing, changes in temperature, or low temperatures, will be more difficult, or take longer, the larger the scale.

It is important to attempt to model the anticipated large-scale equipment capability during development. For example, in a particular process requiring the cooling of 500 l of clarified product for the manufacture of a live-virus vaccine, a heat exchanger had to be employed to reduce the temperature from 34 °C in the culture vessel to 8 °C, within one hour, for the subsequent downstream process step. A 1/10-scale process was used in development and, rather than rely on variable cooling in a refrigerator, the manufacturer identified for the large-scale heat exchanger supplied a 1/10-scale model to achieve the same cooling in the same time for 50 l of harvest. The 1/10-scale heat exchanger is shown in Figure 18.5. The implementation in the laboratory of this scaled-down model of the full-scale equipment ensured that scale-up would not change the process.

The stainless steel employed for process hardware, buffer tanks or pipework needs to conform to the relevant standards, for example the FDA regulation 21 CFR 211.65 states that 'equipment shall be constructed so that surfaces that contact components, in-process materials, or drug products shall not be reactive, additive, or absorptive'; further information can be found at http://www.fda.gov/cder/dmpq/cgmpregs.html. 316L stainless steel is accepted in the industry as the standard for any metal in contact with product or process fluids. The design of the steelwork needs to be sanitary and cleanable, for example with no dead-legs and with smooth flow paths (see Chapter 14).

It is an increasing expectation of regulatory agencies that manufacturers develop good working relationships with suppliers of equipment (and raw materials), and it is commonplace for suppliers to provide information to the manufacturer for equipment validation, for installation and operation qualification, and this will help minimize the work required of the manufacturer and optimize the validation process for cGMP purposes.

Disposables are increasingly used in manufacture to avoid the difficulties of cleaning and cleaning validation. Polypropylene is a common non-leaching plastic used for bioprocess containers. Disposable flexible bags or plastic liners for tanks in volumes up to about 1500 l, such as those from combrex, Sartorius, HyClone and Stedim, are now commonly used for large volumes of buffers or product handling. It is important to assess disposables for leaching of chemicals, such as plasticizers and anti-oxidants. For example, water for injection with a known total organic carbon (TOC) content can be placed in a flexible bag and sampled at intervals for TOC analysis. Leaching may occur, in the order of thousands of ppm, and the water may fail a <500 ppm TOC limit within a few days. It may be required by regulatory authorities to test leaching from bioprocess containers in contact with the process fluids, such as buffers and column eluates, under the conditions used in the process. It is possible that the plastic film itself does not leach under conditions that comply with the USP Class VI test, but plasticizers may be liberated when the disposable product is gamma-irradiated and a build up of TOC may be present when the bag is used.

As stated before, it is important to consider the large-scale implementation of the process during development, such as the types of column, UF membrane or disposables to be used, so that changes introduced later in the project are minimized. For example, the need to repeat toxicology studies can arise if new product-contact materials are used that have the potential to leach different amounts or types of components into the product, and this will inevitably cause costly delays to a project.

18.4.8 Secure Supply

Because the manufacture of a successful medicinal product will continue for many years, it is advisable to ensure a secure supply of all materials required. A secure supply means that a vital process component can be guaranteed to be available for manufacture into the future. Reliance on a single supplier carries a high risk, as even if that supplier is a successful business, the company may go bankrupt, or their production may fail, or they may discontinue supply of any given raw material that is not economic for them to make. Materials that can be obtained from a number of suppliers are preferable, so that supply can be assured if one source is not able to continue to provide them, but this can be a difficult issue where, for example, a process relies on a specific column matrix from a certain manufacturer.

18.4.9 Viral Clearance

An increasingly demanding area of compliance is viral clearance from mammalian- or insect-cell-derived products (Burstyn & Hageman 1996). Selection of orthogonal techniques is essential to the ability of a process to achieve sufficient viral clearance across the process. Nanofiltration (Burnouf & Radesovich 2003) can be used to remove potential virus contamination from the process stream. Such filters have pore sizes of 50 nm or 15–20 nm. Viruses can alternatively, or additionally, be inactivated by low pH hold (at around pH 3.5), heat or solvent/detergent. These methods are usually effective against enveloped viruses. In addition, virus removal can be achieved by partitioning during chromatography, separating virus from product (Uren 2000). Such chromatography steps can be included in viral clearance studies (see Chapter 19).

It is necessary to validate virus clearance/inactivation by demonstrating the capacity of multiple steps in the process to reduce the titre of spiked virus by a factor of up to 10^{12}. In the absence of a dedicated virology laboratory, it is usual to contract this work out to a specialist company.

When considering a mammalian cell-line such as CHO cells, certain viruses are particularly relevant. CHO cells are known to produce non-infectious retrovirus particles (Adamson 1998), so clearance studies for Phase 1 clinical submissions may best be performed using a retrovirus such as murine leukaemia virus (MLV) as a model virus. However, since MLV is easily cleared, the study would be more convincing and would better support a regulatory submission if clearance of the more robust minute virus of mouse (MVM) could also be demonstrated, though MVM is a parvovirus and represents an adventitious contaminant. It may complicate matters if MVM clearance cannot be demonstrated; other process steps will need to be developed subsequently and validated. For example, it would be useful to include a 15–20 nm virus filter because a 50 nm filter will not effectively remove the 24 μm MVM.

Virus clearance validation for Phase 3 will need to be done on the entire purification process. Clearance studies at this stage will additionally include two or three of the following viruses: parainfluenza, a reovirus or adenovirus, pseudorabiesvirus or herpes virus, polio, or SV40 virus. Such a variety of virus types in a clearance validation study would cover viruses of differing sizes, RNA and DNA viruses, and differing robustness to inactivation(see Chapter 19).

18.4.10 Process Validation

The process will need to be validated (Parenteral Drug Association 1992a,b TR14 and TR15) to demonstrate that the process will deliver a consistent dose of product, with the same potency and impurity profile. Process validation for purification of biological medicines relies largely on the robustness of the process, i.e. how well the process copes with variations in conditions. Examples include impurity levels in the feedstock, the pH of process buffers or temperature fluctuations in the facility. As stated before, the ability of a process to work consistently and reliably depends, from the outset, on screening a variety of methods successfully and determining conditions for operation that can readily be met at large scale and which preferably avoid critical conditions that need to be controlled precisely.

In practice, validation is a major undertaking to document that a process consistently generates product meeting its predetermined specification. A scaled-down model process must be established and shown to function in the same way as the large-scale process. This process model is then used to systematically test the ranges of all critical variables and parameters, such as pH and ionic strength in the wash and elution of an ion exchange chromatography, or volume of load onto a size exclusion column. The aim is to demonstrate that the product quality and the consistency of the process are not affected by any allowed variation of the process parameters. The large-scale process must also be run to demonstrate that, within the permitted parameter ranges, the process delivers product consistently meeting the specification.

To avoid changing the process at a critical stage in the project, it is useful to take into account the regulatory, economic, scale and validation factors from the outset, and to attempt to build them into the process during the initial small-scale method screening.

18.5 PROTEIN PURIFICATION METHODS

18.5.1 Method Screening

Process robustness in chromatography relies heavily on the selection of the matrix. A matrix-screening program can be efficiently achieved using automated ‘scouting’ systems such as the BioCad (Applied BioSystems) or the Akta (GE Healthcare, previously Amersham Biosciences). These machines can be connected to several columns. The machine can then run a variety of methods through each column in turn. Different methods at different pHs can be run, gradients are used to elute the product and fractions are collected. Many permutations can be tested on the

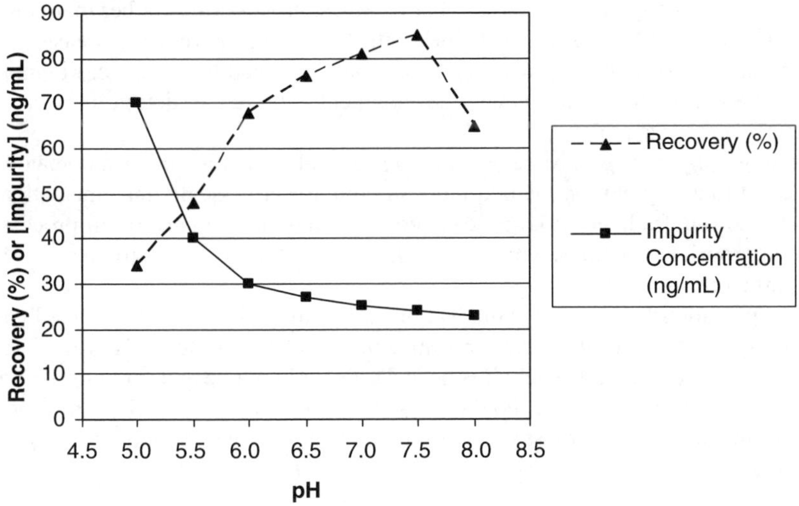

At pH 7.5 the recovery is at its highest and the impurities near their lowest. However, a range on the pH must be allowed for. A slight rise in pH above pH 7.5 could cause a significant drop in recovery. pH 7.5 should, in this example, be the high end of the range. From this data it would be useful to define a set-point for pH of 7.0, with a tolerance of ± 0.5. Thus, a range from pH 6.5 to 7.5 would be allowed within which neither recovery nor impurity levels would vary greatly. However, it would be useful to investigate the effect of pH between pH 7 and pH 8 at intervals of 0.2 to investigate this more thoroughly since pH can be controlled to a tolerance of ± 0.2. It may then be possible to define a set-point for pH of 7.3 with a range of ± 0.2, i.e. from pH 7.1 to pH 7.5, depending on how the recovery performs between pH 7.5 and 8.0. Optimisation for robustness is key to the success of a process in manufacture, but this should be done at the method screening stage so that the other steps in the process are developed coherently and re-work can be avoided.

Figure 18.6 The possible effect of pH on recovery and impurity removal, showing the selection of a set-point that allows for process control and enables a robust method.

different columns and this is achieved automatically. This highly efficient method of chromatography screening will generate a huge number of fractions for analysis, and this will very quickly generate a great deal of data, to facilitate the comparison of different matrices.

It is important during the screening phase to build in robustness. Do not necessarily select the conditions that give the highest yield or best purification. The problem with the best, or 'optimal', conditions is that they may only be achieved under a narrow range of conditions. These conditions may be relatively easily achieved in the laboratory where high resolution techniques on sophisticated equipment and perfectly packed columns are available, but at a large scale they may be difficult to achieve. It is preferable to rely on conditions that are easily controlled; find methods where broad ranges of pH or conductivity, for example, can be used. Figure 18.6 shows how the optimal conditions for recovery or impurity removal may not necessarily be optimal for robustness, but that the robust process need not be poor in performance.

During method screening, select the conditions where a consistent result can be achieved assuming that the parameter will vary within a specified range. A set point within this range may not give the maximum yield or best purification, but, as long as sufficient yield and purification are achieved, such a set point should be viewed as one optimized for robustness.

It is probably better to rely on a greater number of robust process steps rather than fewer steps that involve control of critical parameters. If the control of a given parameter is easy to achieve within the necessary limits then the process will be robust and readily validated. Manufacturing expertise should be consulted during process development activities to determine what level of control is practicable at large scale.

Most chromatography steps will be developed using a gradient elution on an automated system and then converted to a step elution method more suitable for large-scale implementation. However, it is possible to run gradients at large scale, collect fractions and perform in-process assays before pooling appropriate fractions. This may make a critical separation sufficiently robust for biologicals manufacture.

There is an expectation in science for innovation and in the field of biological medicines this may be particularly true as a result of the innovative nature of the products. However, whilst innovation can be useful in a process where it is needed, it should not be an aim in itself. Tried and tested methods, if they will do the job, should be used; not only are they likely to be quicker and easier to get working, and probably less costly to implement, they will be more readily accepted by regulatory authorities.

18.5.2 Ion Exchange

Ion exchange chromatography separates molecules on the basis of the strength of the interaction between charges on the molecule and the charges on the matrix (Wang 1990). The two types of ion exchange resin, positively charged anion exchangers and negatively charged cation exchangers, are available in strong and weak forms. (This is not a reference to the strength of binding of biomolecules at neutral pH, but is an indication of the pH range within which the resin is charged.)

Both strong and weak exchangers should be screened, as the selectivity will be very different. Quaternary amines (strong anion exchangers) can be particularly useful in DNA or endotoxin removal (Petsch & Anspach 2000), but tertiary amines (weak anion exchangers, such as DEAE) may well be better in separating host cell protein. Weak cation exchangers, such as COO^-, may also be useful tools as their selectivities are different from those of strong cation exchangers, such as SO_3^{2-}. As many of each different type of matrix should be screened, and as many manufacturers' matrices should be tested, as possible, provided that a DMF is, or will be, available for each matrix tested.

Ion exchange is the backbone of most purification processes used in biological medicines production. Even where an affinity step is used, e.g. for the purification of antibodies, ion exchange is also likely to be employed (Graf *et al.* 1994; Duffy *et al.* 1989; Corthier *et al.* 1984; Mao & Hearn 1996). Ion exchange is particularly useful for medicinal products because leached ligands are generally non-toxic and non-immunogenic, in contrast to, for example, dye ligands or Protein A.

The optimization of an ion exchange method with the selected matrix will involve investigation of the effects of pH and salt concentrations during loading, washing and elution. It is common to use an increasing salt gradient during method development to identify how the product and the impurities elute from the column, and then to change the method to a stepwise increase in salt concentration. The first step is normally a wash that removes some of the impurities bound to the column without causing the product to elute. The second step is then used to elute product, leaving further impurities bound to the matrix. The column can then be cleaned with, for example, a high concentration of salt and/or sodium hydroxide.

Column length, flow rate and amount of product loaded are also important parameters that will affect the performance of the purification. It is vital that each parameter is set during method screening (Dasarathy *et al.* 1996) within a sufficiently broad range that permits robust running of the column.

18.5.3 Affinity Chromatography

Affinity purification has the advantage of very high selectivity. It is extremely useful to employ a step in which only the product will bind to the matrix and a stringent wash, such as a high salt wash, can be applied to the column. In this way, a high degree of purification can be achieved in a single step and, although usually expensive, an affinity matrix can be very cost effective.

Affinity matrices are often used in monoclonal antibody purification (Gagnon 1996; Fahrner *et al.* 1999, 2001). Depending on the IgG subclass of the antibody, Protein A (Schmerr 1985) or Protein G (Bill *et al.* 1995) is used. These are normally linked to a bead such as Sepharose. Other antibody types, IgA, IgD, IgE, and IgM, can be purified on Protein B, Protein L and Protein P (Gagnon 1996), but these ligands may be difficult to source as large-scale process matrices. One disadvantage to such protein ligands is that they cannot be sanitized under stringent conditions, such as 0.5 molar sodium hydroxide (Hale *et al.* 1994), in the same way as most other matrix types, such as ion exchange matrices.

It is usual to elute from Protein A at low pH, which can be useful in that the low pH can then be adjusted to around pH 3.5 and maintained for a time, usually about 30 minutes, as a virus inactivation step. While antibodies are normally stable under such conditions, it is important to ensure that the product is not adversely affected by the low pH hold.

A technique called hydrophobic charge induction is an affinity purification tool for antibodies that enables loading and elution without harsh conditions (Boschetti 2002; Burton & Harding 1998). The matrices used in this technique are also sanitizable using reagents such as 1 molar sodium hydroxide.

There are other useful affinity matrices, most of which are specific to particular products, for example heparin can be used as a ligand for cell-surface binding proteins or viruses.

The major disadvantage of any affinity matrix is the potential for leaching of the ligand from the matrix into the product. Once a ligand like Protein A has become detached from the bead it is very likely to bind and stay bound to the product and is, therefore, likely to co-purify (Bloom *et al.* 1989; Fuglistaller 1989). It may be advantageous to avoid the use of Protein A, for example, and to purify antibodies by employing sanitizable matrices of synthetic origin, such as the hydrophobic charge induction chromatography matrix MEP-Hypercel supplied by BioSepra (Schwartz *et al.* 2001).

Another potentially useful class of affinity matrix has a mimetic dye as the ligand, such as the range available from ProMetic Biosciences. The many different dyes may mimic molecules like, for example, NADH and bind to active sites in the protein, while others may mimic proteins like Protein A (Kabir 2002). Dyes can be useful high affinity ligands in specific applications, such as the blue dyes in albumin (Allary *et al.* 1991) or alkaline phosphatase (Pozidis & Bouriotis 1995) purification. These matrices can be readily sanitized with sodium hydroxide. However, the dyes do leach from the matrix and their toxicity should be assessed when used for biological medicines production.

Reverse elution can be useful in affinity chromatography, and can be performed by applying the eluent in the opposite direction to that of the flow during loading and washing of the column. This technique can be used to minimize product contact with the ligand, to enable collection at a higher product concentration, and possibly also to improve step recovery.

18.5.4 Hydrophobic Interaction Chromatography

Hydrophobic interaction chromatography (HIC) is possibly one of the most useful methods available for the removal of host cell proteins (Queiroz *et al.* 2001), DNA and endotoxin (Wilson *et al.* 2001). Proteins that co-purify by charge or size are likely to be separable by HIC.

The principle of salting-out or ammonium sulphate precipitation to separate proteins crudely on the basis of their solubility in high salt solutions, has been used for many years. The HIC principle uses conditions where the protein is almost 'salted-out', where the salt concentration has caused the protein to be less soluble and more likely to bind to a hydrophobic surface, and for the charges on its surface to be negated by the high ionic strength. Loading can be achieved in different salts such as ammonium sulphate, sodium sulphate or high concentrations of sodium chloride. For example, a solution of 0.5 to 1.0 molar ammonium sulphate should suffice to cause most proteins to bind to a hydrophobic resin.

HIC resins consist of aliphatic or aromatic carbon structures linked to the chromatography bead. Butyl, octyl and phenyl ligands are common. HIC ligands are less hydrophobic than those used under the more severe denaturing conditions of reverse-phase chromatography. Solvents should not be needed to remove the product from a HIC matrix; normally a reduction in the salt concentration is all that is required. A decreasing salt gradient may be used, but it is most common to use a step decrease to remove impurities, and then a further step decrease to elute the product (similar to ion exchange). It may be that a particularly hydrophobic protein may require elution with 10 % ethylene glycol in order to be recovered fully, and ethylene glycol is often used to clean HIC columns.

HIC can potentially be denaturing, and may in some cases cause proteins to aggregate. However, this is a powerful technique that is readily implemented at scale and is often used along with ion exchange chromatography, and in preference to size exclusion. It can be extremely useful in allowing DNA to be removed in the column flow-through during loading. In addition, endotoxin can bind very tightly and high endotoxin clearance can be achieved (Wilson *et al.* 2001).

18.5.5 Gel Filtration and Size Exclusion

It is useful to use the term size exclusion to describe the type of gel filtration used to separate proteins by molecular weight. Size exclusion chromatography can be a useful purification tool in separating by size two proteins that may have similar charges and so co-purify on ion exchange. However, for this technique to be efficient, it is reliant on a concentrated feedstock, and a very large column may be required as the load volume is often only about 1–2 % of the column volume.

This technique is generally difficult to scale-up and the column bed height is an important factor in this. To separate well, a size exclusion matrix may need a bed height of 70 to 90 cm. Columns for production scale may need to be split into sections in order to give the matrix sufficient support. There are rigid size exclusion matrices, such as Toyopearl from Tosoh Bioscience, which may be scaled more easily than agarose-based matrices.

As columns are scaled-up, the bed height is kept constant and their diameter is increased. This can cause the flow distribution to be uneven over the cross section of the column, with flow at the edges lagging behind. The resolution between protein peaks is reduced and the required separation at large scale may not be achieved, see Figure 18.7.

Separation is determined by the size and shape of the proteins, and selecting the matrix by its nominal separation range for model globular proteins may be misleading. A variety of matrices should be screened.

Size exclusion is also normally a lengthy process step due to the low flow rates employed, and in terms of process economics in manufacture, it can be costly.

Gel filtration can also be used to separate protein from buffer solutes to effect a buffer exchange, also called desalting, and is a useful alternative to diafiltration. This method requires columns with relatively short bed-heights of approximately 20 cm. It is generally scaleable and the column load can be around 30 % of the column volume. The relatively high flow rates that can be used contribute to making this a useful method. However, when used in this mode it does not separate different proteins. Buffer exchange can also be achieved during size exclusion chromatography, as above. Gel filtration is a more rapid and more convenient method of buffer exchange when size separation is not required.

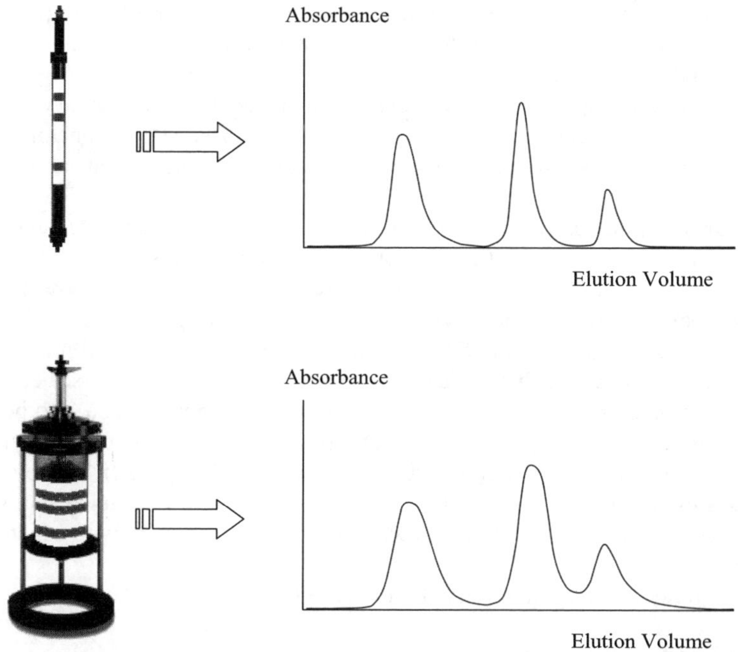

Figure 18.7 The scale-up of a size exclusion step can cause loss of resolution. The flow distribution on a small column allows the protein to be uniformly applied to the matrix. With a much wider column it is likely that the flow distribution will be imperfect, spreading the peak and reducing resolution. This can be critical in size exclusion separations where resolution is a key factor for purification. Loss of purity can result, or, if less of the product peak is collected where overlap occurs, recovery will be reduced.

18.5.6 IMAC

Recombinant proteins are often made with multiple histidine residues, such as his-tag motifs (Muller *et al.* 1998; Schmitt *et al.* 1993), attached in order to permit the use of immobilized metal affinity chromatography (IMAC) (Gaerc-Porekar & Menart 2001), also referred to as metal chelate chromatography. A variety of chelating matrices are available whose purpose is to bind divalent metal ions such as zinc, nickel, cobalt, or copper. It is the metal ions to which the his-tag will bind. Most often his-tagged proteins are purified on nickel or zinc. Once loaded with product, the column can be washed with a salt-containing buffer to remove impurities that may be bound ionically. The product can then be eluted with imidazole or histidine. Imidazole may cause product to denature or aggregate, particularly if present when the product is subjected to freeze/thawing. Non-animal derived histidine can be used as a more expensive alternative eluent.

This method of chromatography can also be useful with proteins that are not his-tagged, for example factor IX or protein C from human plasma. It is important with any IMAC process to determine selectivity of different matrices in combination with different metal ions.

18.5.7 Ultrafiltration

Ultrafiltration is a technique in which porous membranes are normally used to retain proteins and allow the passage of solutes and small molecular weight impurities into the filtrate. This method can be used with a continuous buffer feed to diafilter the retained proteins into a different buffer.

It is likely that a purification scheme will involve the use of ultrafiltration; to remove water and concentrate product, to perform buffer exchange by diafiltration or to remove low molecular weight material and small proteins.

The use of buffer exchange steps is time consuming and is often unnecessary. pH titration and dilution are often sufficient conditioning for loading onto a subsequent chromatography step. It is worth testing this as soon as a sequence of steps is developed in order to avoid unnecessary unit operations.

Buffer exchange is often used at the end of a process to provide product formulated in the right buffer. This is normally done by diafiltration or gel filtration (Kurnik *et al.* 1995). At this stage, when the buffer components used will be administered to the patient, it is most important to ensure that all components are acceptable with respect to regulatory guidelines.

Companies such as Millipore, Pall, Sartorius, and Schleicher and Schuell produce membranes and hardware for ultrafiltration. It is essential to use a system that can be scaled-up for the requirements of commercial manufacture. It is ideal to use a system that can also be scaled down to laboratory scale for use as a model for such activities as process validation and virus clearance validation. One such system is the Centrasette system from Pall (Figure 18.8), which can be used at very large scales (in the Centrastak) but is also supplied as a low volume Centramate system, using identical materials of construction and with an identical flow path, with a membrane size as small as 0.01 m^2.

For a more detailed discussion of ultrafiltration, see Chapter 17.

18.5.8 Membrane Adsorbers

Some suppliers, such as Sartorius, Pall and Millipore, now sell membranes that have been given surface properties like those of chromatography beads. These membrane adsorbers (Gottschalk

Figure 18.8 The Pall Centrasette system can be directly scaled from a 0.01m^2 scale-down model to industrial production scale with membrane areas of 80m^2 or more, and capable of processing around 25,000L. (*Above*) (a) Pilot scale ultrafiltration equipment used in the concentration and diafiltration of a live virus herpes vaccine. The Pall Centramate system shown houses 0.5m^2 of membrane and is driven by a 4-head diaphragm Quattro pump, also supplied by Pall. Photo courtesy of Jim Mills, rerpoduced with permission of Xenova.

Figure 18.8 (*Continued*) (b) Two duplex Pall Centrastak™ 300 ultrafiltration systems, for use with Pall Centrasette™ cassettes, designed for large-scale downstream processing of biopharmaceutical products. Photo courtesy Pall Corporation.

et al. 2004) offer the advantage of very high flow rates because the binding of product relies much less on diffusion than is the case with chromatography beads. In addition, the breakthrough curve at the point where the membrane saturates with product is very steep, which is useful from the point of view of efficient use of the medium – where a chromatography column may only be partially saturated before significant product begins to breakthrough, the membranes should be almost completely saturated before breakthrough is detected.

Scale-up of membrane adsorbers is becoming increasingly common in the manufacture of biological medicines. Membrane adsorbers have been shown to contribute significantly to viral clearance in processes for production of biological medicines from animal cells (Gottschalk *et al.* 2004).

18.6 PURIFICATION OF VIRUSES AND GENE THERAPY VECTORS

Virus products and gene therapy vectors can generally be purified using the same techniques as used for protein products (O'Neil & Balkovic 1993; Lydiatt & O'Sullivan 1998). Ion exchange, affinity and hydrophobic interaction methods can all be used efficiently to purify such products. Size exclusion may be achieved with a matrix such as GE Healthcare's Sephacryl 1000 SF (Hewish & Shukla 1983).

Ultrafiltration with membrane pore sizes of 10^5 to 10^6 daltons can be a very useful technique with this type of product as the large particles can easily be retained while diafiltration can be used to remove proteins.

The removal of DNA and RNA from such products can be achieved quickly and easily using a DNase/RNase enzyme such as Benzonase® (Hagen *et al.* 1996), an endonuclease isolated from *Serratia marcescens*, which is available as a GMP-grade reagent from Merck. The DNA fragments and the Benzonase® enzyme can then simply be removed by diafiltration.

There should be no need to rely on centrifugation through caesium chloride or sucrose; these techniques are difficult to scale-up and time-consuming to run. It is possible to develop simple and scaleable process methods for the purification of viruses and gene therapy vectors, such as herpesvirus, adenovirus, or vaccinia. However, it is likely that virus products, particularly enveloped viruses, will be labile and particularly susceptible to damage by high shear forces (generated by high flow rates), by changes in ionic strength during elution or dilution (O'Neil & Balkovic 1993), or by interaction with surfaces (including column matrix surfaces). Losses of up to 1000-fold can occur especially with enveloped viruses, like herpesvirus, if the process is not carefully controlled, but recoveries of around 75 % are possible for ultrafiltration and chromatography steps.

Large chromatography beads and slow flow rates may assist virus recovery. Ultrafiltration should be performed with minimal turbulent flow, and air bubbles or foaming should be avoided. Problems with inactivation of virus not only cause low recoveries, but also increase process variability.

Detergents can be particularly damaging to enveloped viruses and so rinsing and preparation of process glassware, for example, may be critical to the process. Plasticizers in disposables may also be problematic.

Processes may need to be kept cold, and pH conditions may need to be kept neutral throughout the process. However, during process development the limits within which the viral product remains viable should be determined so that unfavourable conditions can be avoided and acceptable conditions can be exploited.

Biologics production usually relies on the use of aseptic 0.22-μm filtration at the end of the process to ensure sterility of a product. However, sterile filtration may be problematic with large viruses. For example, the recovery after passage through a 0.22-μm filter of a herpes virus with a diameter around 0.18-μm may be as low as 1 %. The optimal filter type, ionic strength, and flow rate may give recoveries of around 50–70 %.

Virus titres are usually measured using biological assays that are prone to a high degree of assay variance of around 30–50 %. This makes the estimation of viral recoveries through the process very difficult. The recovery from any given step may only be known accurately after a large number of runs have been performed. This makes the incremental optimization of the process difficult, because an improvement may not appear to be significant until further large numbers of runs have been performed.

It is likely that the downstream processing of such viral products will require containment in a similar manner to the live production during the upstream process. Operations should be carried out either in closed systems, or in a microbiological safety cabinet (MSC), to protect both operator and environment. Air exhaust from the MSC will need to be HEPA filtered before it can leave the clean room. Cleaning of the room using validated methods will also be required before new products can be introduced.

18.7 CONCLUSION

However biological medicines are produced and from whatever production system, almost all must be purified. The development of a purification process for biological medicines is a multifacetted discipline. The technical aspects of purifying a protein from a complex mixture must result in a reliable process delivering product that meets an appropriate specification. The process must be scaleable and economic to operate at the scale required for the provision of commercial supply, and must be validatable and comply with regulatory guidelines.

REFERENCES

Adamson SR (1998) *Dev. Biol. Stand.*; 93: 89–96.
Alexander C, Bilgin N, Lindschau C *et al.* (1995) *J. Biol. Chem.*; 270: 14541–14547.

Allary M, Saint-Blancard J, Boschetti E, Girot P (1991) *Bioseparation*; 2: 167–175.
Berthold W, Walters J (1994) *Biologicals*; 22: 135–150.
Bill E, Lutz U, Karlsson B-M, Sparrman M, Allgaier H (1995) *J. Mol. Recognition*; 8: 90–94.
Bloom JW, Wong MF, Mitra G (1989) *J. Immunol. Methods*; 117: 83–89.
Boschetti E (2002) *Trends Biotechnol.*; 20: 333–337.
Burnouf T, Radosevich M (2003) *Haemophilia*; 9: 24–37.
Burstyn DG, Hageman TC (1996) *Dev. Biol. Stand.*; 88: 73–79.
Burton SC, Harding DRK (1998) *J. Chromatogr.* A; 814: 71–81.
Corthier G, Boschetti E, Charlet-Poulain J (1984) *J. Immunol. Methods*; 66: 75–79.
Dasarathy Y, Ramberg M, Andersson M (1996) *BioPharm.*; September: 42–45.
Duffy SA, Moellering BJ, Prior GM, Doyle KR, Prior CP (1989) *BioPharm.*; June: 34–47.
Fahrner RL, Whitney DH, Vanderlaan M, Blank GS (1999) *Biotechnol. Appl. Biochem.*; 30: 121–128.
Fahrner RL, Knudsen HL, Basey CD, Galan W *et al.* (2001) *Biotechnol. Genet. Eng. Rev.*; 18: 301–327.
Fuglistaller P (1989) *J. Immunol. Meth.*; 124: 171–177.
Gaerc-Porekar V, Menart V (2001) *J. Biochem. Biophys. Meth.*; 49: 335–360.
Gagnon PS (1996) *Purification Tools for Monoclonal Antibodies.* Validated Biosystems Inc., Tucson, AZ.
Gottschalk U, Fischer-Fruehholz S, Reir O (2004) *BioProc. Int.*; 2(5): 56–65.
Graf H, Rabaud JN, Egly JM (1994) *Bioseparation*; 4: 7–29.
Gulewicz K, Adamiak D, Sprinzl M (1985) *FEBS Lett.*; 189: 179–182.
Hagen A, Aboud RA, DePhillips PA, Oliver CN, Orella CJ, Sitrin RD (1996) *Biotechnol. Appl. Biochem.*; 23: 209–215.
Hale G, Drumm A, Harrison P, Phillips J (1994) *J. Immunol. Methods*; 171: 15–21.
Harris ELV, Angal S (1995a) *Protein Purification Applications, A Practical Approach.*; IRL Press, Oxford.
Harris ELV, Angal S (1995b) *Protein Purification Methods, A Practical Approach.* IRL Press, Oxford.
Hewish, DR, Shukla DD (1983) *J. Virol. Methods*; 7: 223–228.
Kabir S (2002) *Immunol. Invest.*; 31: 263–278.
Kurnik RT, Yu AW, Blank GS *et al.* (1995) *Biotechnol. Bioeng.*; 45: 149–157.
Lydiatt A, O'Sullivan DA (1998) *Curr. Opinion Biotechnol.*; 9: 177–185.
Mao QM, Hearn MTW (1996) *Biotechnol. Bioeng*; 52: 204–222.
MCA (2002) *Rules and Guidance for Pharmaceutical Manufacturers and Distributors.* HMSO, Norwich, UK.
Muller KM, Arndt KM, Bauer K, Pluckthun A (1998) *Anal. Biochem.*; 59: 54–61.
Ngiam SH, Bracewell DG, Zhou Y, Titchener-Hooker NJ (2003) *Biotechnol. Prog.*; 19: 1315–1322.
O'Neil PFO, Balkovic ES (1993) *Bio-Technology*; 11: 173–178.
Parenteral Drug Association (1992a) *Industrial Perspective on Validation of Tangential Flow Filtration in Biopharmaceutical Applications.* Technical Report No. 15, Supplement, Vol. 46, No. S1.
Parenteral Drug Association (1992b) *Industry Perspective on the Validation of Column-based Separation Processes for the Purification of Proteins.* Technical Report No. 14, Supplement, Vol. 46, No. S3.
Petsch D, Anspach FB (2000) *J. Biotechnol.*; 76: 97–119.
Pozidis C, Bouriotis V (1995) *Bioseparation*; 5(2): 89–93.
Queiroz JA, Tomaz CT, Cabral JMS (2001) *J. Biotechnol.*; 87: 143–159.
Schmerr MJF, Patterson JM, Van Der Maaten MJ, Miller JM (1985) *Molec. Immunol.*; 22: 613–616.
Schmitt J, Hess H, Stunnenberg HG (1993) *Molec. Biol. Reps*; 18: 223–230.
Schwartz W, Judd D, Wysocki M, Guerrier L, Birck-Wilson E, Boschetti E (2001) *J. Chromatogr. A*; 908: 251–263.
Seamon KB (1998) *Curr. Opinion Biotechnol.*; 9: 319–325.
Seely RJ, Hutchins HV, Luscher MP, Sniff KS, Hassler R (1999) *BioPharm.*; April: 33–36.
Senger RS, Karim MN (2003) *Biotechnol. Prog.* 19(4); 1199–1209.
Sofer G, Hagel L (1997) *Handbook of Process Chromatography: A Guide to Optimization, Scale-up, and Validation.* Academic Press, London.
Subramanian G (1998) In *Bioseparation and Bioprocessing, A Handbook.* Ed Subramanian G. Wiley-VCH, Weinheim; Vol. 1.
Thömmes J, Bader A, Halfar M, Karau A, Kula M-R (1996) *J. Chromatogr. A*; 752: 111–122.
Uren E (2000) *Downstream*; 31: 24–25. Amersham Biosciences, Uppsala.

Wang NW (1990) *Bioproc. Technol.*; 9: 359–400.
Wilson MJ, Haggart CL, Gallagher S, Walsh DJ (2001) *Biotechnol.*; 88: 67–75.
Yuk IHY, Wang DIC (2002) *Biotechnol. Appl. Biochem.*; 36: 133–140.

Useful Web Sites

Regulatory authorities

EMEA	www.emea.europa.eu
FDA	www.fda.gov
ICH	www.ifpma.org/
MHRA	www.mhra.gov.uk/

Suppliers

Applied Biosystems (Column matrices, chromatography equipment) www.appliedbiosystems.com
Asahi (Planova nanofilters) www.asahi-kasei.co.jp/planova/en/
Bio-Rad (Columns, matrices, chromatography equipment) www.bio-rad.com
BioReliance (Contract virus clearance validation) www.bioreliance.com
BioSepra (Columns, matrices) www.pall.com
Covance (Contract virus clearance validation) www.covance.com
GE Healthcare (Columns, matrices, chromatography equipment, UF membranes): www.gehealthcare.com/worldwide.html
Hyclone (Process liquids, bioprocess containers) www.hyclone.com
IQA (Institute if Quality Assurance) www.iqa.org
Mallinckrodt Baker (GMP grade chemicals and reagents) www.mallbakes.com
Merck (Column matrices, Benzonase®, GMP grade chemical reagents) www.emerck.com
Millipore (Vantage chromatography columns, matrices, filters, UF membranes and hardware) www.millipore.com
Pall (Filters, Centrasette UF membranes and hardware, nanofilters) www.pall.com/biopharm
ProMetic Biosciences (Mimetic dye matrices) www.prometic.com
Sartorius (Filters, UF membranes, disposable bioprocess containers) www.sartorius.com
Schleicher & Schuell (UF membranes and hardware) www.schleicher-schuell.com
Sepragen (Radial flow columns) www.sepragen.com
Stedim (Disposable bioprocess containers) www.stedim.com
Tosoh Biosciences (Column matrices) www.tosohbiosep.com
Validated Biosystems Inc www.validated.com
Whatman (Chromatography matrices, filtration) www.whatman.com

19 Virus Safety of Cell-derived Biological Products

PL Roberts

19.1 INTRODUCTION

Advances in gene cloning technology and the development of methods for the large-scale culture of cells *in vitro* have permitted the exploitation of cell-based systems for producing biological products for therapeutic use. One potential risk associated with the use of such material is the transmission of infectious agents to the recipients. While cellular microorganisms such as bacteria, fungi or mycoplasma can generally be controlled by standard sterile filtration procedures, this is not straightforward with smaller agents such as viruses. In addition, a group of novel infectious agents known as prions, believed to be the causal agents of transmissible spongiform encephalopathies (TSEs), are also important. The main concerns with cell-derived biological products have been endogenous viruses, i.e. those originating from the animal tissue at the time of cell preparation. There have also been cases of exogenous contamination arising from the use of media components of biological origin, such as calf-serum, media, or from the individuals handling the cell cultures. Such concerns have lead to strategies for both the banking of cells and the testing of these banks for viruses in order to ensure their consistent virus safety. Furthermore, steps must be included in the product purification process physically to remove or inactivate viruses further to ensure the safety of the product.

Cell-derived biological products generally have a good microbiological safety record. This is in contrast to biological products directly derived from humans, such as cellular blood products, i.e. red blood cells and platelets, and plasma products such as coagulation factors, albumin and immunoglobulin, which have transmitted hepatitis viruses and the human immunodeficiency virus (HIV) in the past. In the case of plasma-products, virus reduction methods have been developed and incorporated into protein purification processes for a number of years. These methods have, in many cases, been adapted for use with cell-culture derived biological products.

The testing of cell-lines for viruses and the validation of purification processes for virus removal are required by the regulatory agencies that licence therapeutic products e.g. European Medicines Agency and the US Food and Drug Administration (FDA). Various guidance documents have been published on these topics (Center for Biologics 1993, 1997; Committee for Proprietary Medicinal Products 1996, 1997, 2001; European Commission 1994; World Health Organization 1998, 2003; International Conference on Harmonisation 1999) – see also Chapter 34. In addition the viral safety of biological products has been a continuing topic of debate (Horaud & Brown 1991; Brown 1993; Brown & Lubiniecki 1996; Brown *et al.* 1998, Brown *et al.* 1999; Robertson 2004).

Medicines from Animal Cell Culture Edited by G. Stacey and J. Davis

19.2 VIRUS CHARACTERISTICS

Viruses are obligate intracellular pathogens, i.e. they only replicate inside living cells. While in most cases this leads to cell disruption, in some cases the viruses can be passed on from cell to cell at the time of cell division without any obvious cytopathic effect. Some viruses have the capacity to infect certain types of cell culture persistently, either by multiplying at only a low level and/or only infecting a proportion of the cells. In addition some can cause a latent infection by integrating into the cell genome, e.g. herpes viruses, and retroviruses such as HIV.

Various aspects of virus structure are relevant to virus inactivation/removal from biological products. These include such factors as the presence or absence of an envelope, the type of nucleic acid they contain and the size of the viral particle. The physicochemical resistance to specific conditions such as heat or pH is also relevant.

19.3 VIRUS DETECTION

The absence of infectious virus in cell cultures used for the preparation of therapeutic products is generally essential. Various approaches are used for virus detection/identification. Some of these detect viruses that are able to replicate, e.g. using cell culture, and others may detect the presence of virus that is not necessarily infectious, e.g. using the polymerase chain reaction (PCR) or electron microscopy (EM). In addition, detection methods may be of a general nature, i.e. able to detect a range of diverse virus types by growth in cell culture or haemagglutination/haemadsorption for example, or of a specific type, i.e. only able to detect one type of virus by techniques such as enzyme-linked immunosorbant assay (ELISA) or PCR.

The viruses that are of concern in a particular product depend on the species from which the cells were originally derived and the nature of the cell-growth additives that are used during culture, e.g. bovine serum or feeder cells. Virus contamination may inadvertently occur during handling or transport, or due to the deliberate introduction of virus, e.g. Epstein-Barr Virus for cell immortalization or Sendai virus for cell fusion. Those viruses that cause gross cytopathic effects on the cells may be easy to recognize. However, those viruses that are able to cause persistent or latent infections in cell culture, without any apparent cytopathic effect, are more difficult to detect. A list of the main viruses of concern in human cells is given in Table 19.1. In the case of rodent cells, screening for a range of viruses is routinely carried out (see Table 19.2). Some

Table 19.1 Potential viral contaminants of human cells.

Virus	Abbreviation	Type(-virus)
Human immunodeficiency virus type -1, -2	HIV-1,-2	Retro
Human T-cell lymphotropic virus type -1, -2	HTLV-1, -2	Retro
Squirrel monkey retrovirus	SMRV	Retro
Herpes simplex virus type -1, -2	HSV-1, -2	Herpes
Epstein–Barr virus	EBV	Herpes
Human herpes virus type -6, -7, -8	HHV -6, -7, -8	Herpes
Hepatitis B virus	HBV	Hepadna
Hepatitis C virus	HCV	Flavi
Human adenovirus	HAdV	Adeno
Human papillomavirus	HPV	Papilloma
Human parvovirus B19	B19	Parvo
Human polyomavirus JC, BK	JC, BK	Polyoma

Table 19.2 Virus contaminants of rodent cells.

Virus[a]	Abbreviation	Type	Host[b]
Ectromelia virus	Ectro	Pox	M
Epizootic diarrhoea of infant mice virus	EDIMV	Rota	M
Hantavirus	HTNV	Hanta	M, R
H-1 virus	H-1	Parvo	R
K virus (mouse pneumonitis virus)	K (MPV)	Polyoma	M
Kilham rat virus	KRV	Parvo	R
Lactate dehydrogenase virus	LDV	Toga	M
Lymphocytic choriomeningitis virus	LCM	Arena	M, H
Minute virus of mice	MVM	Parvo	M
Mouse adenovirus	MAV	Adeno	M
Mouse cytomegalovirus	MCMV	Herpes	M
Mouse hepatitis virus	MHV	Corona	M
Mouse poliovirus (Theiler's murine encephalomyelitis virus/GDVII)	TMEM/GD7	Picorna	M, R
Mouse parvovirus	MPV	Parvo	M
Mouse thymic virus	MTV	Herpes	M
Polyomavirus	Poly	Polyoma	M
Pneumonia virus of mice	PVM	Paramyxo	M, R,H
Rat coronavirus/sialodacryoadenitis	RCV/SDAV	Corona	R
Reovirus type 3	Reo-3	Reo	M, R, H
Sendai virus	Sendai	Paramyxo	M, R, H
Simian parainfluenza	SV-5	Paramyxo	H

[a]*Virus detected by antibody production test in mice (MAP), rat (RAP) or hamster (HAP). Alternatively a panel of PCR tests can be used to detect the virus directly.*
[b]*Host: mouse (M), rat (R) or hamster (H)*

of these are able to infect humans: hantavirus, lymphocytic choriomeningitis virus (LCMV), rat rotavirus, reovirus type 3 and Sendai virus. In addition, retroviruses are also of concern (see Section 19.3.3). Defective endogenous retroviruses are found in Chinese hamster ovary cells, a cell-line commonly used for producing biological products. Bovine viruses are an issue where foetal-calf serum has been used to culture the cells. This concern still remains even where the cells are later grown in serum-free medium. The main bovine viruses of concern are bovine viral diarrhoea (BVDV), infectious bovine rhinotracheitis virus (IBRV), parainfluenza, and bovine polyoma virus. This last virus is of particular concern because it can cause tumours in animals. If porcine trypsin has been used then porcine viruses such as porcine parvovirus may also be present.

19.3.1 Electron Microscopy

Transmission electron microscopy is a useful method for detecting viruses and virus-like particles within cells, and in the cell-culture medium (Liptrot & Gull 1996). In the case of cells, ultra-thin sections are prepared and stained. Cell-free virus is detected by negative straining after concentration by ultracentrifugation. The sensitivity of the method is not very high, with about 10^5 particles/ml required for virus detection. The concentration of virus can be estimated by the addition of a known concentration of latex beads to the preparation. Details of virus structure and the cytopathology of infected cells can allow some degree of virus identification to be made. Also the percentage of cells infected can be estimated.

19.3.2 *In Vivo* Detection

A range of rodent viruses can be detected by the inoculation of mice, rats or hamsters, due to the production of virus-specific antibody after infection. These tests are known as the mouse antibody production (MAP) test, rat antibody production (RAP) test and hamster antibody production (HAP) test. ELISA or other sensitive serological tests can be used to identify any specific antibodies that are produced.

Hens' eggs and animals are also used for virus detection. Suckling mice and guinea pigs are inoculated and followed for clinical signs of infection. In the case of embryonated hens' eggs, inoculations are made via the allantoic and amniotic cavities and the yolk sac, and virus subsequently detected by the agglutination of red blood cells.

19.3.3 *In Vitro* Detection

A range of indicator cells can be used for virus screening. Those most commonly used have a relatively wide spectrum of virus susceptibility and can include, for instance, HeLa (human cervical epithelial carcinoma), MRC-5 (human diploid fibroblast) and Vero (African green monkey kidney) cells for detecting viruses capable of replicating in humans/primates. Other cells may be used for detecting specific viruses, e.g. H9 (human T-cells) for HIV or freshly harvested T-cells from peripheral blood for HTLV-1. An indicator cell-line of a similar species/tissue to the cell-line under test should also be included. Bovine, porcine and murine cells can be used where appropriate. Co-cultivation methods are also used in order to increase the sensitivity of virus detection. These involve culturing the mixture of test and indicator cells for several passages.

In the case of murine retroviruses such as murine leukaemia virus (MLV), a number of specific cell-based methods can be used to detect infectious virus. Various types of MLV exist, i.e. those that can replicate in rodent cells (ecotropic), in cells other than rodent (xenotropic) or in both (amphotropic). In addition to amphotropic viruses, another group of MLVs with dual tropic properties also exists, i.e. the mink focus forming viruses. Ecotropic MLV can be detected by its ability to form syncitial plaques in layers of UV-treated XC cells (Rous sarcoma virus-induced rat tumour cells). The virus can first be amplified by culturing in SC-1 mouse embryo cells. Xenotropic MLV can be detected using S+L− cells, i.e. mink cells containing a defective murine sarcoma virus genome. Rescue of this virus by MLV results in the loss of contact inhibition by the cells and thus the formation of foci. All types of MLV are able to infect *M. duni* cells. Infection is recognized by a cytopathic effect (CPE) or detected by immunofluorescence. In all cases co-cultivation and/or the passage of cell extracts/supernatants can be used to increase the sensitivity of virus detection.

19.3.4 Reverse Transcriptase

A general method that can be used for the detection of retroviruses is assaying for the presence of the viral reverse transcriptase enzyme. This method involves concentrating the virus from the cell-culture medium by ultracentrifugation and then solubilizing the virus particles with detergent. Reverse transcriptase activity, i.e. RNA-dependent DNA synthesis, is then detected by the incorporation of radio-labelled thymidine triphosphate into acid-precipitable material using an RNA template in the presence of magnesium or manganese ions. A negative control using a DNA template is also included. More sensitive methods that are based on PCR, e.g. PERT (product-enhanced reverse transcriptase), can also be used. It should be noted that the detection of reverse transcriptase activity does not necessarily indicate the presence of infectious virus. Low levels of reverse transcriptase have been found associated with chick embryo cells, but this was not considered to be an issue for concern as no infectious agent was detected (Robertson 1997).

19.3.5 Polymerase Chain Reaction

The use of PCR for the amplification of viral nucleic acid is now the most commonly used method for the sensitive detection and identification of viruses. The nucleic acid sequence of the virus of interest needs to be known so that appropriate primer pairs can be identified and prepared. Confirmation of the identity of the virus detected can be performed using Southern blotting or by a second PCR using internal primers, i.e. a 'nested PCR'. One possible problem for some viruses is strain variation. This can be overcome by the use of redundant primers directed against conserved viral sequences, in order to ensure that a wide range of virus strains is detected. The detection of several different viruses can be carried out in one reaction by the use of several primers in a multiplex. PCR systems for the automated handling of samples and for real-time quantitative PCR are now available. However, it should be noted that the PCR method can be very sensitive and thus precautions must be taken to prevent cross-contamination, particularly by the products of previous amplification reactions. PCR can be used as an alternative to the MAP test and has been shown to be equally as sensitive (Bauer *et al.* 2004; Blank *et al.* 2004). Again, it should be noted that the detection of viral nucleic acid by PCR does not necessarily indicate the presence of infectious virus.

19.4 VIRUS REMOVAL/INACTIVATION METHODS

A complementary approach to screening and testing for viruses is to treat the product in order to remove physically or inactivate any viruses that might be present (Roberts 1994, 1996). This approach (hereafter termed 'virus reduction' for convenience) has the potential advantage that it can be effective against a wide spectrum of viruses. A virus-reduction step may be part of the standard purification process for the product that may fortuitously also remove viruses. However, it may not be possible to modify such a step to maximise virus reduction. Alternatively, a specific or dedicated step may be used which has no other purpose.

The ideal method would bring about a high level of virus reduction, be effective against all types of virus, and be robust, i.e. effective over a wide range of process conditions. In practice such a goal is not easily accomplished. For instance, most non-enveloped viruses are relatively resistant to inactivation and are not easy to remove by virus filtration due to their small size. In 'spiking' studies (see Section 19.4.1) the level of virus titre reduction that is considered substantial for a single virus reduction step is about 10^4-fold (commonly referred to as 4 logs). The titre of the virus stock and the toxicity of the samples to be assayed may mean that this is difficult to demonstrate in practice. The reduction values from several process steps can be combined to give an estimate for the total purification process, although this is only strictly applicable where different reduction mechanisms are involved. This can give figures in the order of, for example, around 10 logs for small resistant viruses and 20 logs or more for viruses of low to medium resistance. Quantitative risk assessment, taking into account the level of virus that is likely to be present in the cell supernatant harvest, is used to determine if the level of virus reduction is sufficient. The target is to ensure that the concentration of infectious virus in a therapeutic dose of product reaches a level several orders of magnitude below that which could result in an infection. Some further guidance on the residual levels that may be considered acceptable is given in the European Pharmacopoeia, which requires that containers of final sterile products contain less than 10^{-6} viable microorganisms per container. In addition, the application for which the product is to be used is also important. For instance, where the product is to be used for treating a life-threatening condition for which no current treatment exists, e.g. some cancers, a higher degree of risk may be acceptable. Thus the acceptability of any virus reduction procedure for a specific product or application must be judged on a case-by-case basis. Because of the limitation of any particular removal or inactivation method with regard to the range of viruses affected, there has been pressure to include two or more specific virus-reduction steps in a purification process. In addition, the methods used should act by different mechanisms. The inclusion of several virus reduction steps also has the advantage that, in the unlikely case of one step failing, there should still remain an adequate margin of virus safety. The use of specific virus reduction steps, rather than just the use of standard process steps, is advocated. This is because these are generally more readily controlled, can be specifically optimized for virus reduction and can be more readily validated. An example of a protein purification process, incorporating several virus reduction steps is shown in Figure 19.1. The virus reduction values obtained for various viruses in virus spiking studies is given in Table 19.3.

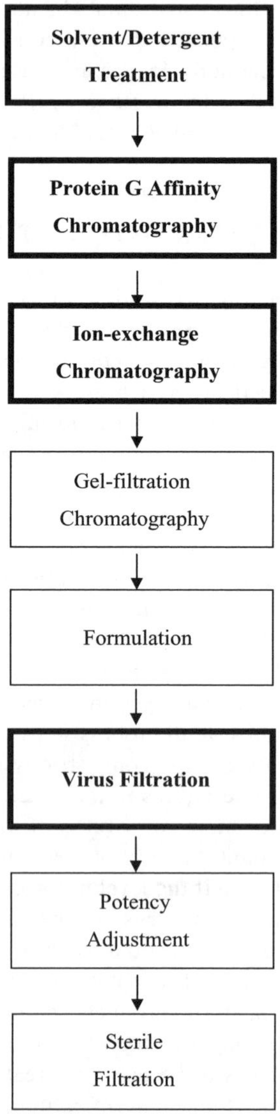

Figure 19.1 Monoclonal antibody purification process with virus reduction steps. Virus reduction steps are shown in bold boxes. These include dedicated virus reduction steps as well as purification steps shown, in validation studies (Table 19.3), also to make a significant contribution to virus reduction.

Table 19.3 Example of virus reduction during the purification of a monoclonal antibody.

	Virus reduction (log_{10})					
Step[a]	HSV-1[b]	Sindbis	SFV[b]	Vaccinia	MMV[b]	Polio-1
Solvent/detergent[c]	>5.7	>5.7	>6.7	nd	>3.6	0[d]
Protein-G affinity chromatography[e]	7.1	6.0	nd	nd	nd	3.1
Ion-exchange chromatography	5.5	nd	nd	nd	nd	2.4
Size-exclusion chromatography	2.7	nd	nd	nd	nd	1.2
Virus filtration (50 nm)[f]	>7.6	6.7	6.7	6.8	nd	1.0
Total reduction	>28.6	>18.4	>13.4	6.8	>3.6	7.7

[a]*Purification of human anti-D monoclonal antibody BRAD-3*
[b]*HSV-1, herpes simplex virus-1; SFV, Semliki Forest virus; MMV, minute virus of mice*
[c]*0.3 % tri-n-butyl phosphate/1 % Triton X-100 for 1 hr*
[d]*Non-enveloped virus, thus not susceptible to solvent/detergent treatment*
[e]*Step includes elution at acid pH which contributes to virus reduction by causing inactivation*
[f]*Ultipore DV-50 (Pall) or Viresolve-180 (Millipore) or equivalent, e.g. Planova 35 (Asahi)*
nd = Not Done

In addition to the inclusion of effective virus reduction steps, it is essential to ensure that virus safety is not subsequently compromised by recontaminating the product during the manufacturing process. The various precautions taken to prevent recontamination are part of good manufacturing practice (GMP) and include the use of closed processing systems and/or physical separation in different rooms or areas (see Chapters 12 and 34). Effective cleaning and sterilization procedures for equipment are also used (see Chapter 14). One approach commonly employed is to process product intermediate, after a major virus reduction step, in a segregated area commonly known as a 'virus-secure area'. One advantage of using a terminal virus inactivation step on the product in the final container is that there is no possibility of recontamination.

A major factor that must be considered with any virus removal/inactivation method is the possibility that the step may have adversely affected the product. Most effective virus inactivation methods are relatively severe, and a compromise between maximum virus inactivation and maximum product yield must be reached. The activity, structure, immunogenicity or thrombogenicity of the product can also be affected. For instance, the heat-treatment of coagulation factors has, in some cases, led to the production of neoantigens and thus the induction of inhibitors, i.e. antibodies directed to the protein, in recipients (Rosentaal et al. 1993). This has been a problem with some heat-treated factor VIII products. In the case of antibodies, the product can be altered in such a way that anti-complement activity can result. While virus filtration is a very gentle procedure, product losses on the filter itself can be high if the size of the protein is too near to that of the filter pores.

19.4.1 Evaluating Virus Reduction Steps

The ability of specific steps in the purification process to remove or activate viruses can be evaluated or validated by virus 'spiking' studies. This involves the deliberate addition of virus to a small-scale version of the purification process, and following the virus through the process. Such studies are a regulatory requirement for any purification process for a new medicinal product. Guidance on the approach to be taken has been published (Committee for Proprietary Medicinal Products 1996). Although these studies would ideally be done using the actual full-scale manufacturing process, this would not be acceptable because the manufacturing facility would become contaminated. Instead, laboratory studies, using a scaled-down process model where appropriate, are used. The laboratory model must be tested/validated, using various physical, chemical and

Table 19.4 Examples of viruses used in virus reduction studies.

Virus	Type (-virus)	Envelope	Size(nm)	Physico-Chemical Resistance	Specific Model For[a]
Murine leukaemia Human immunodeficiency	Retro	+	80–100	Low	
Herpes simplex-1 Pseudorabies Bovine herpes	Herpes	+	120–200	Medium	
Bovine viral diarrhoea	Flavi	+	60–70	Low	HCV
Sindbis	Toga	+	60–70	Low	HCV
Polio-1 Encephalomyocarditis Hepatitis A	Picorna	–	25–30	Medium	
Minute virus of mice Canine parvovirus Porcine parvovirus	Parvo	–	18–24	High	B19
Reo–3	Reo	–	60–80	Medium	Bluetongue
Parainfluenza 3	Paramyxo	+	100–200	Low	

[a]*All can act as specific, relevant or general models, depending on the situation. Viral attributes such as the possession of an envelope, small-size or high physico-chemical resistance can be useful properties when assessing a particular virus reduction method.*

biochemical criteria, to confirm that it adequately represents the full-scale process. Product intermediate is then spiked with virus and the process step performed. The amount of virus present at the start and end of the process is then quantified by physico-chemical methods, such as the detection of radionuclides, immunochemistry, nucleic acid detection, or by assaying infectivity.

Determining infectivity is the most commonly used virus assay method. This is because other methods do not necessarily just detect infectious virus but also inactivated or fragmented virus. So, for example, detection and quantification by PCR can be used for virus removal methods such as virus filtration, but in the case of an effective inactivation step, the use of this method might lead to the conclusion that virus inactivation had not occurred.

Various viruses can be used for spiking studies (Table 19.4) including 'relevant viruses', i.e. those that are of particular concern in the product. In some cases where *in vitro* studies are not possible, studies with animals have been carried out. For instance chimpanzees have been used for virus inactivation studies with hepatitis B and C (Prince *et al.* 1984, 1986). However this approach is not generally advocated for ethical reasons, the scarcity of suitable animals, and the limited amount of experimental data that can be obtained. In addition to relevant viruses, so called 'model viruses' are used. Viruses that can be cultivated to high titre in cell culture are preferred. Model viruses that are related to a virus of concern are considered 'specific model' viruses, e.g. BVDV or Sindbis is a specific model for HCV. In addition, other general model viruses are used in order to increase the range of virus types being tested, and to include viruses with particular properties, e.g. enveloped or non-enveloped, small or large size, or high resistance to physicochemical damage. Possible model viruses have been suggested by regulatory authorities (Committee for Proprietary Medicinal Products 1996) and their general use allows a comparison of different reduction technologies to be made. However the final selection of viruses with which to challenge a process will also depend on the specific product/process that needs evaluation. Even where it is possible to use a relevant virus for evaluating virus reduction, the properties of such an agent produced in the laboratory may not be the same as those of the virus naturally contaminating the product. The strain of the virus involved *in vivo* and *in vitro* may vary, and adapting a virus to growth in cell culture can lead to changes in its properties. Thus even relevant viruses should be considered models for the purpose of evaluating virus reduction.

When performing viral validation studies, excessive manipulation of the virus-spiked pre- and post-treatment samples such as freeze-thawing, centrifugation or chromatographic processing should be avoided where possible. This is in order to ensure that virus recovery is consistent for all the samples. Control studies must be carried out in order to confirm that the product is not toxic to the cells or assay system at the highest concentration assayed. It is often necessary to dilute the sample before assay, e.g. 1 in 10, to ensure the product has no inhibitory effects on the assay system. Where a chemical inactivation method is being evaluated, greater dilution, perhaps 1/100 or 1/1000, may be needed. Alternatively the chemical may have to be removed by centrifugation or chromatography before assay. The virus reduction value for the step is traditionally expressed as a log value, determined by subtracting the $\log_{10}$ of the total amount of virus found at the end of the process from that present initially. Where the virus reaches undetectable levels, the reduction value is expressed as a 'greater than' value. In this situation, further information on the effectiveness of the reduction process can be obtained by following the kinetics of virus inactivation. In such studies, inactivation is usually seen to be a multiphase process rather than a simple exponential decrease. This makes extrapolation beyond the limits of virus detection unreliable and also means that it is not possible to calculate D values accurately, i.e. the time required to inactivate 1 log of virus, as commonly used to describe the kinetics of bacterial inactivation by disinfectants or heat treatment. The significance of this lack of an exponential decrease is not clear although the presence of virus aggregates or mixed virus strains may be important. In order to increase the sensitivity of virus detection and to allow a higher inactivation value to be demonstrated, larger volumes of material can be assayed and virus preparations of a higher titre can be used. The virus can be concentrated for this purpose by ultracentrifugation or polyethylene glycol precipitation, but care should be taken to ensure that the method used does not inactivate the virus and does not cause the virus to aggregate. Enveloped viruses are fairly fragile and can be inactivated by ultracentrifugation. To ensure that the addition of the virus spike to the product does not significantly alter the properties of the starting material, the volume used must be limited to no more than about 1 in 10 (v/v). The absence of any effect on the protein purification process should be confirmed by comparing relevant process parameters with an 'unspiked' control. A greater dilution, e.g. >1 in 50 (v/v), should be used where necessary. For instance, in the case of virus filtration, filtration rate and filter capacity may be reduced by virus spiking. To prevent this, the volume of the virus spike used can be reduced, or virus of higher purity used.

An estimate of the total virus reduction capability of a manufacturing process can be made. While this is best based on an estimate for the total process in one study, this is only really practical for processes with relatively low virus reduction capabilities, due to limitations imposed by the maximum concentration of the spike virus that can be attained. However, adding up the individual reduction values for each stage of the process can give an estimate for the total process. Although this approach is only strictly applicable where the individual steps are orthogonal, i.e. based on different principles, this method has become widely accepted for summarizing virus reduction data. The data derived from virus reduction studies can be used in various ways. For instance the effectiveness of any particular virus reduction procedure can be compared with alternative methods. Also the effect of various parameters on the process, e.g. protein concentration, residual moisture content, stabilizer concentration, pH, ionic strength, temperature and chromatographic parameters should be investigated. From such studies, an idea of the robustness of the inactivation process can be obtained and suitable process parameters defined within which virus inactivation is effective.

19.4.2 Methods Used for Virus Reduction

The various methods used for virus inactivation or removal are considered below. Their effectiveness may vary and can only be fully determined for a particular product by virus spiking studies. In general, the methods are of limited effectiveness against small, resistant, non-enveloped viruses.

19.4.2.1 Downstream processing

Many of the purification steps used during the product purification process may provide some degree of virus reduction (see Table 19.3). For instance, ultrafiltration steps used to concentrate protein products may contribute to virus removal if the protein is substantially larger than the virus. If this is not the case then virus may in fact be concentrated with the protein. Filtration steps using a pore size of 0.2 μm, used for removing cellular microorganism and debris, will remove larger viruses to some extent. Filters of a smaller pore size, e.g. 0.1 μm, used for removing mycoplasma, are capable of removing larger viruses even more effectively as well as removing some smaller viruses. Precipitation methods used to concentrate the product from the bulk harvest can also separate virus into the waste material. For instance, this occurs during the cold-ethanol fractionation steps that are used during the preparation of albumin and immunoglobulin from plasma (Yei *et al.* 1992; Morgenthaler *et al.* 1993). The chemical reagents that are used may themselves inactivate viruses (Morgenthaler 1989). Storage of process intermediates at −40 °C, 4 °C or 25 °C can lead to some degree of virus inactivation that may be enhanced by the chemical composition of the solution involved. However all these procedures generally only provide a limited contribution to virus safety and are not considered reliable virus reduction steps.

Chromatographic methods are used in the preparation of high purity products and are likely to make a significant contribution to the virus safety of the product (Burnouf 1993). As described above, virus reduction has been evaluated using laboratory-scale chromatographic models that have been shown to mimic accurately the full-scale manufacturing system. Measuring the various physical, chemical and biochemical parameters of the chromatographic process at both scales can confirm this. Column packing can be tested using a salt solution to determine the HETP value (see Chapter 18). The elution profile can be monitored and the characteristics of the peaks used to confirm column performance during the actual purification run. Alternatively, the elution of a trace contaminant can be monitored and used as a marker for column performance. Affinity or ion-exchange methods have been reported to give significant virus reduction in some cases. Affinity methods using, for example, monoclonal antibody or metal ligands to capture the protein product, can be very affective (Hrinda *et al.* 1991; Lawrence 1993; Roberts *et al.* 1994). However, for a given chromatographic system used for the purification of a specific protein, the degree of virus reduction cannot be predicted. Reduction values ranging from 2–7 logs have been reported for different viruses in different chromatography systems. For instance, the reduction of 7 logs of Sindbis and 5 logs of polio was obtained while purifying factor IX on a copper chelate affinity column (Roberts *et al.* 1994). However, for an immunoglobulin purified on a CM-Sepharose ion-exchange resin, the reduction levels were lower, i.e. 4 logs of Sindbis and 2 logs of polio-1. This variation is not surprising in view of the diversity of surface characteristics associated with different viruses and column resins. Virus reduction during chromatography is largely due to virus removal, and procedures that enhance this, such as extensive washing or the use of buffers of varying ionic strength and pH, can be important. In addition, virus inactivation can occur, either during washing or at elution, due to the direct effect of the physical nature (e.g. pH) or chemical composition (e.g. isothiocyanate) of the solution involved. In manufacturing it is essential fully to control and monitor the chromatographic process to ensure reproducibility and performance to ensure both consistent protein purification and virus removal.

Table 19.5 Examples of virus filters.

Filter type	Manufacturer	Grade	Pore-Size[a]	Material	Mode of use[b]	Mechanism	Integrity testing
Viresolve	Millipore	70/180	30nm	PVDF[c]	T	Membrane	Gold particle
		NFP	20nm		DE		
Planova	Asahi	15N	15 nm	Regenerated cellulose	DE or T	Depth	Liquid/ liquid Intrusion
		35N	35 nm				
Ultipor VF	Pall	DV-20 DV-50	20nm 50nm	PVDF[d]	DE	Membrane	Forward-flow

[a]*Approximate pore-size only*
[b]*T, tangential or DE, dead-end*
[c]*Polyvinyl deoxyfluoride*
[d]*Triple layer membrane*

19.4.2.2 Virus filtration

Virus filters have now become widely used in the pharmaceutical industry with a range of cell culture-derived and other products (see Chapter 16). These filters, as the name implies, have a pore-size small enough for effective virus removal (Levy *et al.* 1998; Roberts 2000a; Aranha 2001; Carter 2002; Burnouf *et al.* 2003). Ultipore VF (Oshima *et al.* 1996; Roberts 1997), Planova (Hamamoto *et al.* 1989; Burnouf-Radosevich *et al.* 1994; Manabe 1996; Oshima *et al.* 1996; Yokoyama *et al.* 2004) and Viresolve (DiLeo *et al.* 1993; Hughes *et al.* 1996; Maerz *et al.* 1996; Levy *et al.* 1998) filters have been used (Table 19.5). These filters are available in a range of formats and have been shown to remove viruses in a size-dependent fashion. Some filter types come in a range of pore-sizes and the most appropriate type that is predicted to give the highest virus reduction without removing significant levels of product should be selected. This must then be confirmed in laboratory studies. With virus filters of the smallest pore sizes available, removal of large viruses (>80 nm), e.g. infectious bovine rhinotracheitis virus (IBRV), Epstein–Barr virus, and retroviruses, is most effective. Medium sized viruses (~50–60 nm), e.g. parainfluenza 3 and HCV, and small viruses (~20–30 nm), e.g. parvoviruses and picornaviruses, will also be removed although to a lesser extent. If virus filters with pores of medium size are used, then virus removal is restricted to viruses of medium to large size. In addition, the ionic conditions used may affect virus removal (Yokoyama 2004). However these are predictions only and virus validation studies must bc carried out for each specific combination of product, filter and operating conditions. Using multiple filter units in series can increase the efficiency of virus removal supstantially (Over 2000).

As with sterilizing filters that are designed to remove bacteria, it is essential that integrity testing is carried out on viral filters. Different test methods are available depending on the type of filter involved. The filter manufacturer may also use supplementary test methods for quality control and release of the product. In all cases the performance of the filter in the integrity tests has been correlated with the removal of a representative virus that has been selected to have a size close to that of the filter pore-size. The filter manufacturers have generally used bacteriophages for this purpose because they are non-pathogenic and can be grown to very high titres. The filter manufacturer provides an acceptable pass value and/or correlation curve of virus removal against integrity test value.

The gold-particle integrity test involves evaluating the removal of gold-particles, slightly larger than the pore-size of the filter, from a solution. The optical absorbance of pre- and post-filtration samples is measured on a spectrophotometer and their ratio calculated. This method is destructive, and can therefore only be performed after filter use and allows no possibility of retesting the filter. In the liquid/liquid intrusion test, the filter is first treated with a liquid, e.g. polyethylglycol 800, to lower the pressure required to force a second immiscible liquid, e.g. ammonium sulphate solution, through the filter. The second immiscible liquid is then introduced and the pressure increased to a specified level. The flow rate under these conditions is then determined. The forward-flow test is a variation on the familiar bubble-point method. It involves treating the filter with isopropanol in order to reduce the pressure required to force air through the filter. The pressure is then increased and the pressure needed to force air through the filter is measured.

Virus-removal filters have found application, not only during the purification of biological medicines but also in the elimination of viruses such as BVDV and other larger viruses such as IBRV and parainfluenza-3 from bovine serum used in cell culture. One manufacturer of bovine serum has used six filter units of 0.04 µm in series in an attempt to remove BVDV and other possible viral contaminants (Hyclone Laboratories 1987; Pall Process Filtration 1991).

19.4.2.3 Solvent/Detergent treatment

The most widely used method for virus inactivation is the solvent/detergent procedure. This involves treating products with a mixture of an organic solvent and a non-ionic detergent in order to destroy enveloped viruses. The New York Blood Center originally developed this procedure (Horowitz *et al.* 1985, 1993; Prince *et al.* 1984, 1986). Many coagulation factors are treated in this way, as well as other plasma-derived products and plasma itself, and the method is increasingly being used for biological products derived from cell culture. The method as first developed used ether as the organic solvent but 0.3 % tri-n-butyl phosphate (TNBP) was subsequently adopted for safety reasons. Also, the detergent sodium cholate was used because it occurs naturally in the human body and was thus known to be of low toxicity. Polysorbate 80 (Tween® 80) or Triton® X-100 at 1 % are the detergents most commonly used at present. It should be noted that polysorbate 80 may be manufactured from bovine sources, but the use of vegetable-derived polysorbate 80 is equally effective (Roberts & Sims 1999) and is to be preferred in view of concerns over prions (see Section 19.4.6). Solvent/detergent treatment is best carried out early in the standard purification process, so that subsequent purification steps, particularly those involving chromatography, may be sufficient to remove the chemicals to acceptably low levels. Initial treatment with soybean oil is sometimes used for the partial removal of TNBP, and specific chromatographic resins are available to remove the detergent if needed. The solvent/detergent procedure is gentle, generally having no apparent effect on structure, activity or immunogenicity of proteins. When polysorbate 80 is used, the inactivation of >4 logs of a typical enveloped virus can occur in about the first 30 minutes of the typical 6-hour incubation period used for this solvent/detergent system. With Triton X-100 or Polysorbate 20, this level of inactivation occurs even more rapidly, e.g. within the first 2 minutes of a 30 minute incubation period. An example of the virus inactivation kinetics during a solvent/detergent treatment is given in Figure 19.2. Where virus of a very high titre has been used, the inactivation of >9–11 logs has been demonstrated. The procedure has also been shown to be effective for the inactivation of hepatitis B and C in chimpanzee studies, although for general laboratory validation studies other model enveloped viruses are used. While solvent/detergent treatment is effective for most enveloped viruses, it has been reported that an enveloped virus of a more unusal type, i.e. the pox virus vaccinia, is relatively resistant (Roberts 2000b). The method is robust and relatively insensitive to small changes in parameters such as temperature, solvent/detergent concentration, pH, ionic strength and protein concentration (Roberts & Dunkerley, 2003). Thus relatively wide limits can be set for this inactivation step in a manufacturing environment.

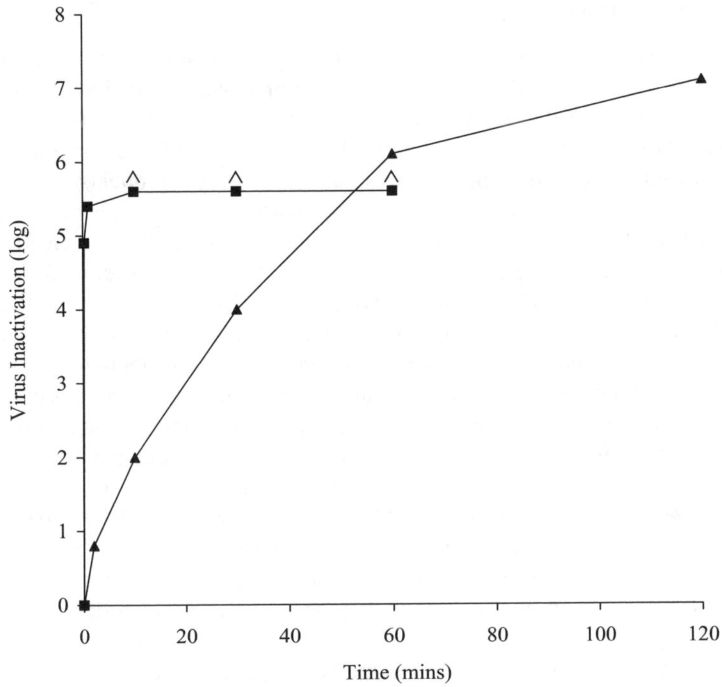

Figure 19.2 Inactivation of Sindbis, a representative enveloped virus, by two commonly used versions of the solvent/detergent procedure. The conditions were 1 % Triton X-100 (■) or 1 % polysorbate 80 (▲), in both cases with 0.3 % tri-n-butyl phosphate. In a manufacturing situation, incubation is for 30 mins or 6 h in the case of Triton X-100 or polysorbate 80 respectively. Where a virus reached levels below the sensitivity of the assay, this is indicated (^).

19.4.2.4 Pasteurization

Heat-treatment in solution at 60 °C for 10 h was the first specific virus inactivation method incorporated into the manufacturing process for a biological product (Edsall 1984). It was introduced in the 1940s to reduce the risks of transmitting hepatitis viruses via albumin preparations, and was later confirmed to be effective for inactivating a wide range of viruses (Nowak 1993). This procedure is carried out on the final bottled product by using conventional ovens or ovens in which the bottles are continuously sprayed with water to give better temperature control. The fatty acid caprylate (octanoate) and/or acetyl tryptophan is used as a stabilizer for albumin and does not subsequently need to be removed. Other products may not need any stabilizer and could also be treated in the final container. In other cases it may be necessary to use a stabilizer, such as a sugar, amino acid or salt, at a high concentration such that it will have to be removed after pasteurization. Under pasteurization conditions, the inactivation of various typical model viruses occurs rapidly, e.g. >4 logs in 10–30 min. However inactivation is slower for the heat-resistant hepatitis A virus, taking 5 h to reach similar levels. In the case of the highly resistant parvoviruses, inactivation may be partially or completely ineffective even after 10 h treatment. Pasteurization has also been applied to other plasma products such as coagulation factors, immunoglobulin and whole plasma. An alternative approach to the use of standard pasteurization conditions is the use of the high-temperature short-time method (Charm *et al.* 1992). This involves treatment at 65–85 °C for a very short period of time, i.e. 0.01 s, using a microwave heat-treatment system.

19.4.2.5 Dry-heat treatment

Dry-heat treatment was originally developed for treating coagulation factor concentrates in order to prevent the spread of HIV. Although products were originally treated at about 60 °C for 30 h, cases of HIV transmission still occurred. Even when higher temperatures were used, such as 60–68 °C for 24–72 h, although the transmission of HIV was prevented, the transmission of hepatitis viruses still occurred. Still higher temperatures can be used, and products that are freeze dried and treated at 80 °C for 72 h have a good safety record with no transmission of HBV, HCV or HIV (Winkelman *et al.* 1989; Hart *et al.* 1994; Roberts *et al.* 2000b, 2006). Inactivation studies have shown that representative enveloped viruses such as HIV, and non-enveloped viruses of low to medium resistance, undergo 3–5 logs of inactivation in 24 h. However others, such as vaccinia and parvoviruses, are very resistant, yielding only 3–4 logs inactivation after 72 h. Some loss in virus infectivity occurs during the freeze-drying process, which will contribute to the overall safety of the product, but should not be considered a reliable virus inactivation step in its own right. Several factors may influence virus inactivation by this method, e.g. residual moisture content, and protein and stabilizer concentrations. While a high residual moisture content can lead to an increase in virus inactivation, product denaturation may reach unacceptable levels. Higher temperatures, e.g. 100 °C (Roberts 1995; Dichtelmuller *et al.* 1996; Santagostino *et al.* 1997) have been evaluated with the aim of improving the inactivation of highly resistant, non-enveloped, viruses such as parvoviruses. However, the possible production of neo-antigens in the product by such extreme conditions, and thus the induction of an immune response against the product in recipients, needs to be considered.

In addition to terminal heat treatment, vapour heating has also been used for virus inactivation (Dorner *et al.* 1996; Barrett *et al.* 1997;). This is an in-process method involving heating the freeze dried product under conditions of controlled moisture that are somewhat higher than those normally associated with freeze-dried products. Temperatures of 60 °C for 10 h, in some cases followed by 80 °C for 1–3 h, have been used.

19.4.2.6 Extreme pH

Treatment at low pH, usually about pH 4 or below, has been used for virus inactivation in various plasma- and cell-derived biological products (Reid *et al.* 1988; Kempf *et al.* 1991; Hamalainen *et al.* 1992; Louie *et al.* 1994; Bos *et al.* 1998). In the case of intravenous immunoglobulin prepared from human plasma, treatment at pH 4 is sometimes carried out in the presence of trypsin in order to remove any anticomplement activity associated with aggregated and/or denatured immunoglobulin. The low pH, rather than the presence of trypsin, is largely responsible for virus inactivation. Some non-enveloped viruses, such as hepatitis A and parvoviruses, are relatively resistant to acid conditions and may not be inactivated by this method. Lowering the pH and/or increasing the temperature may enhance virus inactivation. At 37 °C incubation periods of 1–18 h have been used, whereas at 4 °C periods of about 7 days are needed. Under appropriate conditions many immunoglobulin products can be kept at low pH and 4 °C almost indefinitely.

Low pH has also been used for virus inactivation in cell-derived biological products such as monoclonal antibodies (Baker *et al.* 1994; Brorson *et al.* 2003). The conditions generally used are pH 3.5–4.0 for 30 min–2 h. Such conditions have been shown to inactivate such viruses as murine leukaemia virus and pseudorabies virus. Buffers of low pH are often used to elute the product from affinity chromatography gels such as protein-A or -G Sepharose used for purifying immunoglobulin. Holding the low pH eluate under set conditions of temperature and time can further standardize and optimize this process for virus inactivation. During protein purification by some chromatographic methods, low pH buffers may be used to wash off impurities prior to eluting the bound product. This process will also contribute to virus inactivation.

19.4.2.7 β-Propiolactone

Treatment with β-propiolactone is a well-known chemical technique that has been extensively used for inactivating live virus in human and veterinary viral vaccines (see Section 19.4.3). However it has found only limited application with other biological products, although the method has been used for viral inactivation in some immunoglobulin products (Dichtelmuller *et al.* 1993). It has been shown that virus inactivation can be influenced by the exact conditions used, and may be of limited effectiveness for some viruses including HIV (Norley *et al.* 1993; Scheidler *et al.* 1998).

19.4.2.8 Photochemical inactivation

This approach to virus inactivation involves the addition of a photosensitizing chemical followed by irradiation with UV or visible light (Prodouz *et al.* 1994; Council of Europe 2001; Wainwright 2002). Inactivation may be mediated directly via free radials, or indirectly by singlet oxygen produced after irradiation. Macromolecules including nucleic acid, lipid and protein, can act as targets to a greater or lesser degree, depending on the particular photosensitizer involved. There is considerable interest in this approach at present, particularly with a view to increasing the safety of cellular blood components, i.e. red blood cells and platelets (see Section 19.4.4), and the approach may also find application with protein products. Some chemicals have the advantage that they preferentially target nucleic acid and this should increase their specificity for viruses rather than proteins.

In Germany, methylene blue has been used as a photosensitizing agent for virus inactivation in individual plasma donations. Because of the carcinogenic potential of this compound, it is essential to remove the dye from the product completely after treatment.

19.4.2.9 Ultraviolet light

Irradiation with UV light has been widely investigated as a method for virus inactivation (Kallenbach 1989; Hart *et al.* 1993; Caillet-Fauquet 2004; MacLeod 2004; Wang *et al.* 2004). Because of the poor penetrating power of UV light, it is necessary to use a system that ensures all the liquid is effectively exposed to the light. This may involve the use of a thin film of liquid or a static mixer. Light of 200–290 nm, i.e. UV-C, appears to be the most effective wavelength and is able to inactivate both enveloped and non-enveloped viruses. If damage or alteration to proteins by free radicals and reactive oxygen species is a problem, this can be prevented by the inclusion of a quencher such as the flavonoid rutin (Chin *et al.* 1995, 1997).

An alternative approach to the use of conventional UV irradiation systems for virus inactivation is the use of high intensity broad spectrum while light (Cover *et al.* 2001; Roberts *et al.* 2003b). The inactivating component is again the UV component of the light. This is delivered in relatively short bursts of around 0.3 msec. This procedure has been used for treatment of water and has shown potential, at the laboratory scale, for virus inactivation in protein products.

19.4.2.10 Gamma irradiation

This method has been used extensively for the sterilization of equipment but has only recently found some application to biological products. Irradiation from a cobalt-60 gamma irradiation source is required, and thus access to a commercial irradiation facility is needed. Early studies showed that losses of protein activity occurred at the high doses required for effective virus inactivation (Kitchen *et al.* 1989; Hiemstra *et al.* 1991). However it has been reported that if the dose rate and total exposure dose are controlled, effective virus inactivation can be achieved with only a limited effect on the product (Reid 1998). This approach can be further optimized and forms the basis of the Clearant Process™ for pathogen inactivation. The inclusion of protectants, such as the antioxidant ascorbate, removes the free radicals and reactive oxygen that can damage the product (Grieb *et al.* 2002; Amareld *et al.* 2003; Miekka *et al.* 2003; Tran *et al.* 2004). The selection of a protectant such as ascorbate for this purpose has the advantage that it is not toxic and does not need to be removed. Freezing, or lowering the moisture content of the product by freeze drying, prior to irradiation can further reduce the level of protein damage. This method has been used with cell-derived monoclonal antibodies (Grieb *et al.* 2002). Viral inactivation in serum using this method is discussed in Chapter 4.

19.4.2.11 Iodine

Iodine has long been used as a general disinfectant, and its potential for virus inactivation in biological products has also been investigated. Free iodine can be used, but linkage to starch or polyvinylpyrrolidone particles allows for its simple removal by centrifugation or filtration after treatment (Highsmith *et al.* 1993, 1994). The protein can be treated by passage through a filter (Feifel *et al.* 2001; Skladanek 2001) or a chromatographic gel matrix (Miekka *et al.* 1998) to which iodine has been bound. Further research has lead to the identification of iodoacetic acid as a possible candidate for virus inactivation in solution (Bergvall *et al.* 2001).

19.4.2.12 Pressure

Treatment at high hydrostatic pressure has been shown to inactivate viruses (Nikagami *et al.* 1996; Bradley *et al.* 2000). A further development of this approach is the critical fluid inactivation technique developed by the Aphios Corporation. This involves the use of a suitable gas that, when compressed by pressure, exhibits superfluid properties and is able to penetrate virus particles. Upon reducing the pressure, virus disruption and inactivation occur. The process can be repeated several times in order to increase virus inactivation. It is most effective against non-enveloped viruses.

19.4.2.13 Caprylate

The unsaturated fatty acid caprylate (octanoate) has been used to inactivate enveloped viruses and has been incorporated into the manufacturing process for several plasma products (Lundblad & Seng 1991; Dichtelmuller *et al.* 2002; Korneyeva *et al.* 2002; Johnston *et al.* 2003). Treatment conditions of about 20 mmol caprylate under acidic conditions, i.e. pH <6, for 1–10 h have been used. The non-ionized form of this fatty acid is the active agent for virus inactivation and so conditions of low pH are required.

19.4.3 Viral Vaccines

Modern viral vaccines are prepared in cell culture and, in the case of killed vaccines, an appropriate inactivation method is used. However in contrast to the low viral safety risks associated with cell-derived protein products, the risk of virus transmission to the recipient is in this case much greater. Current human and animal viral vaccines commonly use formaldehyde, β-propiolactone or ethylenimine as the inactivating agent. In the case of split virus vaccines, detergent or an organic solvent is included that solubilizes the envelope and inactivates the virus. Over the years there have been rare cases of infection associated with inactivated viral vaccines, and so the methods used for virus inactivation continue to be an issue (Brown 1992).

19.4.4 Cellular Products

Attempts to develop methods for the inactivation of viruses in cellular products have largely been confined to red blood cells and platelets. In addition to extracellular virus, the inactivation of intracellular virus is also required. The methods currently being investigated mostly involve the use of photosensitizers of one sort or another, e.g. porphyrins, phenothiazines, cyanines, psoralens or riboflavin. Irradiation with visible or UV light is then used to complete the virus inactivation process by the mechanisms described in Section 19.4.2.8.

Based on virus inactivation capabilities and the lack of any significant effect on cell function, some compounds have been investigated more extensively and are being tested in clinical trials with a view to their commercial use. These are Psoralen S-59 from Cerus (Corash 2000; Dupuis *et al.* 2004), riboflavin from Gambro (Goodrich 2000; Ruane *et al.* 2004), FRALE (Frangible Anchor Linked Effector) S-303 from Cerus (Corash 2000) and Inactine PEN110 from Vitex (Lazo *et al.* 2002). All of these compounds are able to specifically target nucleic acids and to bind irreversibly either directly (PEN110), after irradiation (riboflavin, S-59), or after a pH shift (S-303). They are able to inactivate cell-free virus as well as cell-associated forms of the virus. However, as neither red blood cells nor platelets possess a nucleus, DNA or the capacity to divide, any effects on these aspects (critical to the viability of nucleated cells) will not be apparent. In fact, where the effect on other cell-types, such as lymphocytes, has been investigated, they have been found to inhibit cell proliferation. These compounds may also have application with protein products.

19.4.5 Sterilization of Equipment and Media

In addition to viruses originating from inside the cells themselves, i.e. endogenous viruses, viral contamination may also arise from an exogenous source. In order to prevent this from occurring, all the equipment and media used for growing the cells and purifying the product must be sterilized as far as possible before use (Roberts 2002). This is relatively straightforward for standard items of equipment where procedures such as autoclaving or dry-heat treatment can be used. Most consumable items can be purchased having been sterilized by gamma-irradiation. Sanitization procedures that involve the use of sodium hydroxide to clean and decontaminate equipment between uses are also routinely employed. This may be accomplished with dedicated clean-in-place systems using conditions such as 0.5 molar NaOH at 60 °C for 10 min or 25 °C for 1 h.

In the case of basal media, 0.2 μm filtration to remove cellular microorganisms is usually considered sufficient. However, as a further safeguard, raw materials should be treated, as far as possible, to inactivate viruses. Methods such as autoclaving, pasteurization or virus filtration can be used. Contamination of large-scale cell-culture facilities by MVM has been reported on a number of occasions and this is believed to be due to contamination of raw materials (Garnick 1996; Potts 2003). This illustrates the importance of instigating an effective rodent control program throughout the reagent supply chain.

Supplements of animal origin, particularly human or bovine, e.g. calf serum or other reagents used in cell culture such as trypsin, should preferably not be used in the production of biological medicines because of the risks of transmitting viruses and prions (see Section 19.4.6). In the case of calf serum, procedures such as virus filtration, UV treatment or gamma irradiation have been used for virus inactivation (see Chapter 4). However as all these methods are only partially effective, the use of serum-free or protein-free media would seem a better option (see Chapter 3).

19.4.6 Prions

The infectious agents that cause transmissible spongiform encephalopathes (TSEs) (Prusiner 1991) – Creutzfeldt–Jakob disease (CJD) and the variant form of CJD (vCJD) in man, as well as bovine spongiform encephalopathy (BSE), the agent that is believed to have originally given rise to vCJD (Bruce *et al.* 1997; Hill *et al.* 1997) – have become of increasing concern in biological products. CJD has been transmitted by transplanted tissue such as brain or cornea, or by the administration of human growth hormone obtained from cadavers. However there is no evidence to date for transmission by products derived from cell-culture.

Prions differ considerably from conventional viruses in that a nucleic acid component has not been detected. Instead, infectivity is associated with the presence of the prion protein (PrP^{Sc}), an abnormal conformational form of a normal cellular protein (PrP^{C}) of unknown function (Weissmann 2004). This abnormal form is relatively resistant to digestion by proteinase K, is tightly membrane bound and aggregated, and forms characteristic amyloid plaques in the brain.

Suitable detection and screening methods for prions that are simple, sensitive, and specific enough for rapid detection and routine use, have yet to be developed (Minor 2004). Currently the most sensitive detection methods rely on infectivity titration using mice or hamsters and commonly take about a year to complete. Although the immunochemical detection of PrP^{Sc} can be used as a TSE marker, this is less sensitive than infectivity. More sensitive and rapid cell culture-based methods for detecting infectivity are under development. Consequently, at present, health screening of (animal or human) donors and careful sourcing of biological materials is the main approach recommended for controlling the transmission of this agent. Therefore, in the case of cell culture, it is recommended that alternatives to bovine-derived serum be used. Where their use is unavoidable, material from countries that are free of BSE and vCJD should be used and be obtained from a reputable supplier. In the case of human material, health screening and donor history should be used to confirm suitability. In the UK, the decision to exclude individuals that have received a transfusion in the UK since 1980 from donating blood has been made in order to reduce the risk of transmitting vCJD. Material from such individuals should not be used for preparing cell cultures. Some reagents such as detergents, e.g. polysorbate 80, may also be manufactured from cattle, but vegetable-derived alternatives are available.

Validation of prion reduction steps can be carried out in an essentially similar way to that described previously for viruses. The prion can be assayed by determining PrP^{Sc} levels in initial studies and by infectivity studies in animals for those steps found to be particularly effective. Guidelines have been published to aid in the design of such studies (Committee for Proprietary Medicinal Products for Human Use 2004). While many protein purification steps, e.g. filtration, centrifugation, precipitation, and chromatography steps, may contribute to prion removal, there are unlikely to be any that are effective enough to inactivate this highly stable agent. Virus filters will remove the agent, particularly when those of the smallest pore size available are used (Tateishi *et al.* 2001; Van Holten *et al.* 2002). However when the prion material is reduced to its smallest size using detergent, it is able to pass through such filters to some extent.

19.4.6.1 Prion decontamination of equipment

Prions are extremely resistant to inactivation by convention methods (Taylor 1991). The sterilization methods recommended for prion inactivation include treatment with 2 M sodium hydroxide for 1 h or 20 000 ppm hypochlorite for 1 h. The recommended autoclave cycles are 132 °C for 1 h, or a porous load cycle of 134–138 °C for 18 min. Incineration is also effective. Other chemical treatments such as potassium permanganate, urea or formic acid are at least partially effective. Even those methods that have been recommended may only be partially effective against the more resistant strains of prion. Also, the most effective methods do not provide a level of inactivation comparable to those obtained with conventional microorganisms using standard sterilization methods. One particularly severe approach that has been proposed involves treatment with sodium hydroxide followed by, or in combination with, autoclaving. Such a treatment would destroy any normal biological product. A vigorous and repeated washing procedure, using detergent or sodium hydroxide, should be carried out prior to the use of a specific disinfection method. This is in order to reduce the infectious load present on the item and also to remove organic material which may protect prion infectivity from inactivation. Under some conditions e.g. high pH (Kasermann & Kempf 2003), or high pressure (Garcia *et al.* 2004), prions are converted to a form that is susceptible to protease digestion. Also, protease-based disinfectants that digest prions under conditions of high temperature and high pH have been identified (McLeod *et al.* 2004). Thus it is likely that more effective conditions for prion inactivation/decontamination can be developed for routine use.

19.5 SUMMARY

The screening of cells using a range of methods of varying specificities and sensitivities is one approach used to ensure the virus safety of cell-derived biological products. Increasingly, molecular methods such as PCR are being used for virus detection. However, to increase further the degree of safety, and because of the risk of exogenous contamination, virus inactivation or removal steps are also included in the protein purification process. Specific steps such as virus filtration, solvent/detergent or low pH treatment are used. These supplement the virus reduction obtained during the purification process by critical steps such as chromatography. More extreme measures may be needed in the case of resistant viruses and prions, although they may lead to unacceptable effects on the product. The virus testing and reduction approach currently favoured for cell-derived biological products provides a high level of assurance regarding the freedom of products from risks posed by viruses.

ACKNOWLEDGEMENT

I should like to thank colleagues in the R&D Department at BPL for useful discussions on product manufacture and safety, and Gillian Tinn for diligent help in preparing the manuscript.

REFERENCES

Amareld RW Jr, Wersocki M, Drohan M *et al.* (2003) *Biologicals*; 31: 261–264.
Aranha H (2001) *BioPharm.*; Feb: 32–42.
Barrett PN, Meyer H, Wachtel I, Eibl J, Dorner F (1997) *Transfusion*; 37: 215–216.
Baker RM, Brady A-M, Combridge BS *et al.* (1994) In *Separations for Biotechnology 3.* Ed Pyle DL Royal Society of Chemistry, Cambridge; 53–59.
Bauer BA, Besch-Williford B, Riley LK (2004) *Biologicals*; 32: 177–182.
Bergvall P, Ahmad T, Baumann H, Wadensten H, Towner M (2001) *Plasma Products Biotechnology Meeting* 2001, Downstream PPB 2001. Amersham Biosciences; 55–57.
Blank WA, Henderson KS, White LA (2004) *Lab. Anim.*; 33: 26–31.
Bos OJ, Sunye DG, Nieuweboer CE, van Engelenburg FA, Schuitemaker H, Over J (1998) *Biologicals*; 4: 267–276.
Bradley DW, Hess RA, Tao F *et al.* (2000) *Transfusion*; 40: 193–200.
Brorson K, Krejci S, Lee K, Hamilton E, Stein K, Xu Y (2003) *Biotechnol. Bioeng.*; 82: 321–329.
Brown F (1992) *Dev. Biol. Stand*; 81: 103-107.
Brown F (Ed) (1993) *Virological Safety Aspects of Plasma Derivatives. Dev. Biol. Stand. 81.*
Brown F, Lubiniecki A (Eds) (1996) *Viral Safety and Evaluation of Viral Clearance from Biopharmaceutical Products. Dev. Biol. Stand. 88.*
Brown F, Gartwright T, Horaud F, Spieser J-M (Eds) (1999) Animal Sera, Animal Sera Derivatives and Substitutes used in the *Manufacture of Pharmaceuticals: Viral Safety and Regulatory Aspects.* Dev. Biol. (Basel) 99.
Brown F, Griffiths E, Horand F, Petriccian JC (Eds) (1998) Safety of Biological Products Prepared from Mammalian cell Cultures. *Dev. Biol.* (*Basel*) 93.
Bruce ME, Will RG, Ironside JW *et al.* (1997) *Nature*; 389: 498–501.
Burnouf T (1993) *Dev. Biol. Stand.*; 81: 197–199.
Burnouf T, Radosevich M (2003) *Haemophilia*; 9: 24–37.
Burnouf-Radosevich M, Appourchaux P, Huart JJ, Burnouf T (1994) *Vox Sang.* 67: 132–138.

Caillet-Fauquet P, Di Giambattistra M, Draps ML, Sandras F *et al.* (2004) *J. Virol Methods*; 15: 131–139.
Carter J, Lutz H (2002) *Euro. J. Parenter. Sci.*; 7: 72–78.
Center for Biologics Evaluation and Research (1993) *Points to Consider in the Characterisation of Cell Lines used to Produce Biologicals.*
Center for Biologics Evaluation and Research (1997) *Points to Consider in the Manufacture and Testing of Monoclonal Antibody Products for Human Use.*
Charm SE, Landua S, Williams B, Horowitz B, Prince AM, Pascual D (1992) *Vox Sang.*; 62: 12–20.
Chin S, Williams B, Gottlieb P *et al.* (1995) *Blood*; 86: 4431–4436.
Chin S, Jin R, Wang XL *et al.* (1997) *Photochem. Photobiol.*; 65: 432–435.
Committee for Proprietary Medicinal Products, European Agency for the Evaluation of Medicinal Products (1996) *Notes for Guidance on Virus Validation Studies: The Design, Contribution and Interpretation of Studies Validating the Inactivation and Removal of Viruses.* CPMP/BWP/268/95.
Committee for Proprietary Medicinal Products and Committee for Veterinary Medicinal Products, European Agency for the Evaluation of Medicinal Products (2001) *Note for Guidance on Minimising the Risk of Transmitting Animal Spongiform Encephalopathy Agents via Human and Veterinary Medicinal Products.* EMEA/410/01.
Committee for Proprietary Medicinal Products for Human Use, European Agency for the Evaluation of Medicinal Products (2004) *Guideline on the Investigation of Manufacturing Processes for Plasma-derived Medicinal Products with Regard to vCJD Risk.* CPMP/BWP/CPMP/5136/03.
Corash L (2000) *Vox Sang.*; 78(suppl 2): 205–210.
Council of Europe Expert Committee in Blood Transfusion Study Group on Pathogen Inactivation of Labile Blood Components, Pathogen Inactivation of Labile Blood Products, (2001) *Transfus. Med.*; 11: 149–175.
Cover WH, Holloway JM, Xue H, Busby TF (2001) *Plasma Products Biotechnology Meeting.* Downstream PPB 2001. Amersham Biosciences; 42–44.
Dichtelmuller H, Rudnick D, Breuer B, Granshirt K (1993) *Biologicals*; 21: 259–268.
Dichtelmuller H, Rudnick D, Breuer B *et al*, (1996) *Biologicals*; 24: 125–130.
Dichtelmuller H, Rudnick D, Kloft M (2002) *Biologicals*; 30: 135–142.
DiLeo AJ, Vacante DA, Deane EF (1993) *Biologicals*; 21: 287–293.
Dorner F, Barrett N (1996) *Hamostaseologie*; 16: 282–285.
Dupuis K, Sampson-Johanne A, Bernard K (2004) *Vox Sang*; 67(suppl.3): 68
Edsall JT (1984) *Vox Sang*; 46: 338–340.
European Commission (1994) *Note for Guidance on the Production and Quality Control of Monoclonal Antibodies*, III/5271/94.
Feifel K, Zeiler M (2001) *Pharma + Food Int.*; 21–25.
Garcia AF, Heindl P, Voigt H *et al.* (2004) *J. Gen. Virol.*; 85: 261–264.
Garnick RL (1996) *Dev. Biol. Stand.*; 88: 49–56.
Goodrich RP (2000) *Vox Sang*; 78 (suppl 2): 211–215.
Grieb T, Forng R-Y, Brown R *et al.* (2002) *Biologicals*; 30: 207–216.
Hamalainen E, Suomela H, Ukkonen P (1992) *Vox Sang*; 63: 6–11.
Hamamoto Y, Harada S, Kobayashi S *et al.* (1989) *Vox Sang*; 56: 230–236.
Hart H, Reid K, Hart W (1993) *Vox Sang*; 64: 82–88.
Hart HF, Hart WG, Crossley J *et al.* (1994) *Vox Sang*; 67: 345–350.
Hiemstra H, Tersmette M, Vox AHV, Over J, van Berkel MP, DeBree H (1991) *Transfusion*; 31: 32–39.
Highsmith FA, Caple M, Walthall B, Shanbrom E, Drohan WN (1993) *J. Infect. Dis.*; 167, 1027–1033.
Highsmith FA, Xue H, Caple M, Walthall B, Drohan WN, Shanbrom E (1994) *Transfusion*; 34: 322–326.
Hill AF, Desbruslais M, Joiner S *et al.* (1997) *Nature*; 389: 448–450.
Horaud F, Brown F (Eds) (1991) *Virological Aspects of the Safety of Biological Products.* Dev Biol Stand 75.
Horowitz B, Weibe ME, Lippin A, Stryker MH (1985) *Transfusion*; 25: 516–522.
Horowitz B, Prince AM, Horowitz MS, Watklevicz C (1993) *Dev. Biol. Stand.*; 81: 147–161.
Hrinda ME, Huang C, Tarr GC, Weeks R (1991) *Semin. Hematol.*; 28: 6–14.
Hughes B, Bradburne A, Sheppard D, Young D (1996) *Dev. Biol. Stand.*; 88: 91–98.
HyClone Laboratories (1987) *Art to Science in Tissue Culture*; 5: 1–5.
International Conference on Harmonisation (1999). Q5A (R.1) *Viral Safety Evaluation of Biotechnology Products Drived from Cell Lines of Human or Animal Origin.*

Johnston A, Uren E, Johnstone D, Wu J (2003) *Biologicals*; 31: 213–221.

Kallenbach NR, Cornelium PA, Negus D, Montgomerie D, Englander S (1989) *Curr. Stud. Haematol. Blood Transfus.*; 56: 70–82.

Kasermann F, Kempf C (2003) *J. Gen. Virol.*; 84: 3173–3176.

Kempf C, Jentsch P, Poirier B (1991) *Transfusion*; 31: 423–427.

Kitchen AD, Mann GF, Harrison JF, Zuckerman AJ (1989) *Vox Sang*; 56: 223–229.

Korneyeva M, Hotta J, Lebing W, Rosenthal RS, Franks L, Petteway Jr SR (2002) *Biologicals*; 30: 153–156.

Lawrence JE (1993) *Dev. Biol. Stand.*; 81: 191–197.

Lazo A, Tassello J, Jayarama V *et al.* (2002) *Vox Sang*; 83: 313–323.

Levy RV, Phillips MW, Lutz H (1998) In *Filtration in the Biopharmaceutical Industry.* Eds Meltzer TH, Jornitz MW. Marcel Dekker, New York; 619–646

Liptrot C, Gull K (1996) In *Cell and Tissue Culture: Laboratory Procedures.* Eds Doyle A, Griffiths JB, Newell DG. John Wiley & Sons, Chichester; 7B:2 1–14.

Louie RE, Galloway CJ, Dumas ML, Wong MF, Mitra G (1994) *Biologicals*; 22: 13–19.

Lundblad JL, Seng RL (1991) *Vox. Sang.*; 60: 75–81.

MacLeod AJ (2004) *Dev. Biol.*; 118: 139–147.

Maerz H, Hahn SO, Maassen A *et al.* (1996) *Nat. Biotechnol.*; 14: 651–652.

Manabe S (1996) *Dev. Biol. Stand.*; 88: 81–90.

McLeod AH, Murdoch H, Dickinson J *et al.* (2004) *Biochem. Biophys. Res. Commun.*; 4: 1165–1170.

Miekka SI, Busby TF, Redi B, Pollock R, Ralston A, Drohan WD (1998) *Haemophilia*; 4: 402–408.

Miekka SI, Forng R-O, Rohwer RG *et al.* (2003) *Vox. Sang.*; 84: 36–44.

Minor PD (2004) *Vox Sang.*; 86: 164–170.

Morgenthaler JJ (1989) *Curr. Stud. Hematol. Blood Transfus.*; 56: 109–121.

Morgenthaler JJ, Omar A (1993) *Dev. Biol. Stand.*; 81: 185–190.

Nikagami T, Ohno H, Shigehisa T *et al.* (1996) *Transfusion*; 36: 575–576.

Norley SG, Lower J, Kurth R (1993) *Biologicals*; 21: 251–258.

Nowak T, Niedrig M, Bernhardt D, Hilfenhaus J (1993) *Dev. Biol. Stand.*; 81: 169–176.

Oshima KH, Evans-Strickfaden TT, Hightsmith AK, Ades EW (1996) *Biologicals*; 24: 137–145.

Over J (2000) *Downstream*; 31:26, Amersham Biosciences, Little Chalfont,UK.

Pall Process Filtration Ltd (1991) *Pall Scientific and Technical Reports* STR1358. Pall Process Filtration Ltd, Portsmouth, UK.

Potts BT (2003) *PDA-EMEA European Virus Safety Forum.* Langen, Germany, 29 Sept–1 Oct 2003. PDA, Bethesda, USA.

Prince AM, Horowitz B, Brotman B, Huima T, Richardson L van der Ende MC (1984) *Vox Sang*; 46: 36–43.

Prince AM, Horowitz B, Brotman B (1986) *Lancet*; 1: 706–710.

Prodouz KN, Fratanton JC (1994) In *Scientific Basis of Transfusion Medicine.* Eds Anderson KC, Ness PM. Saunders, Philadelphia; 852–871.

Prusiner SB (1991) *Dev. Bio. Stand.*; 75: 55–74.

Reid BD (1998) *Biologicals*; 26: 125–130.

Reid KG, Cuthbertson B, Jones ADL, McIntosh RV (1988) *Vox Sang.*; 55: 75–80.

Roberts PL (1994) In *Separations for Biotechnology 3.* Ed Pyle DL Royal Society of Chemistry, Cambridge; 420–426.

Roberts PL (1995) *Vox Sang.*; 68: 61–62.

Roberts PL, Walker CP, Feldman PA (1994) *Vox Sang.*; 67 (supp l): 132–138.

Roberts P (1996) *Rev. Med. Virol.*; 6: 25–38.

Roberts P (1997) *J. Virol. Methods*; 65: 27–31.

Roberts P (2000a) *Eur. J. Parenter. Sci.*; 5: 3–9.

Roberts P (2000b) *Biologicals*; 28: 29–32.

Roberts P, Sims S (1999) *Biologicals*; 27: 263–264.

Roberts PL (2002) In *Basic Cell Culture: A Practical Approach.* Ed Davis JM Oxford University Press, Oxford; second edition: 29–67.

Roberts PL, Dunkerley C (2003) *Vox Sang.*; 84: 170–175.

Roberts PL, Dunkerley C, McAuley A, Winkelman L (2007) Vox Sang.; 92:56–63.

Roberts PL, Hart H (2000) *Biologicals*; 28: 185–188.
Roberts P, Hope A (2003) *J. Virol. Methods*; 110: 61–65.
Robertson JS (ED) (2004) PDA/EMEA European Virus Safety Forum Dev. Biol (Basel) 118.
Robertson JS, Nicolson C, Riley AM *et al.* (1997) *Biologicals*; 25: 403–414.
Rosendaal FR, Nieuwenhuis HK, van den Berg et al. (1993) Blood; 81:2180–2186.
Ruane PH, Edrich R, Campp D, Keil SD, Leonard RL, Goodrich RP (2004) *Transfusion*; 44: 877–885.
Santagostino E, Mannucci PM, Gringeri AA *et al.* (1997) *Transfusion*; 37: 517–522.
Scheidler A, Rokos K, Reuter T, Ebermann R, Pauli G (1998) *Biologicals*; 26: 135–144.
Skladanek M (2001) *PDA/FDA Viral Clearance Forum* (Bethesda, MD, 1–3 October 2001) PDA, Bethesda, USA.
Tateishi J, Kitamoto T, Mohri S *et al.* (2001) *Biologicals*; 29: 17–25.
Taylor DM (1991) *Dev. Biol. Stand.*; 75: 97–102.
Tran H, Marlowe K, McKenney K *et al.* (2004) *Biologicals*; 32: 94–104.
Van Holten RW, Autenrieth S, Boose JA, Hsieh WT, Dolan S (2002) *Transfusion*; 8: 999–1004.
Wainwright M (2002) *Curr. Med. Chem.*; 9: 127–143.
Wang J, Mauser A, Chao SF, Remington K *et al.* (2004) *Vox Sang.*; 86: 230–238.
Weissmann C (2004) *Nat. Rev. Microbiol.*; 2: 861–871.
Winkelman L, Feldman PA, Evans DR (1989) *Curr. Stud. Hematol. Blood Transfus.*; 56: 55–69.
World Health Organization (1998), WHO Requirements for the Use of Animal Cells as In Vitro Substrates for the Production of Biologicals (Requirements for Biological Substances No. 50); *Biologicals* 26: 175–193.
World Health Organization (2003) *WHO Guidelines on Transmissible Spongiform Encephalopathies in Relation to Biological and Pharmaceutical Products.* WHO/BCT/QSD/03.01.
Yei S, Yu MW, Tankersley DL (1992) *Transfusion*; 32: 824–828.
Yokoyama T, Murai K, Murozuka A, Tanifuji M, Fuji N, Tomono T (2004) *Vox Sang.*; 86: 225–229.

Useful Web Sites

Pall Corporation	http://www.pall.com
Millipore Corporation	http://www.millipore.com
Asahi Kasei Planova® Filters	http://www.asahi-kasei.co.jp/planova/en/product/filters.html
European Medicines Agency	http://www.emea.europa.eu
CDC Centres for Disease Control and Prevention	http://www.cdc.gov/page.do
PDA, Parenteral Drug Association	http://www.pda.org
FDA, US Food and Drug Administration, Center for Biologics Evaluation and Research	http://www.fda.gov
All the virology on the www	http://www.virology.net/garryfavweb.html
The Universal Virus Database of the International Committee on Taxonomy of Viruses	http://www.ncbi.nlm.nih.gov/ICTVdb/index.htm
Vitex	http://www.vitechnologies.com
New York Blood Centre	http://www.nybloodcenter.org
Aphios Corporation	http://www.aphios.com
Therm Systems	http://www.utherm.com/about_us.htm
Clearant	http://www.clearant.com/
Gambro BCT	http://www.gambrobct.com/Default.asp?bhcp = 1
Cerus	http://www.cerus.com/

20 Formulation and Freeze Drying for Lyophilized Biological Medicines

P Matejtschuk and P Phillips

20.1 INTRODUCTION

In the development of new biotechnology cell-derived therapeutics much attention is paid to the means of biosynthesis, together with the choice of cell line, its species, the history of the cell line (including freedom from endogenous viruses), expression capabilities and yield. In addition, the ability to provide appropriate glycosylation and other post-translational modifications for the proteins produced is vital (see Chapter 24). The means of purification must be robust, reproducible, scalable, capable of validation, and retain the required biological activity and potency.

However, in terms of the appearance, storage and shipment of the biotherapeutic, the later stages of formulation and product presentation are equally important for a successful outcome, i.e. to yield a stable potent product with an appropriate formulation for use. The aim of this chapter is to provide a brief introduction to the final formulation of therapeutic products and the principles behind terminal lyophilization of the product in its final delivery format.

From the start of the formulation development, the desired route of administration must be defined and the formulation and final presentation tailored to deliver an effective product fit for the desired application. For many, but not all, products this will be a freeze-dried formulation as this offers advantages of improved stability, ease of shipment and convenience of storage. Key criteria for an acceptable product are given in Table 20.1. However, the additional substantial processing costs introduced by the requirement for pharmaceutical-grade lyophilization equipment and by the extended manufacturing process time should not be under-estimated.

20.2 FORMULATION

In developing new biological medicines the required mode of delivery (Davis 2002) must be considered at the outset of the project. From this criterion the required formulation will be defined. Typically, most biologicals have been delivered by injection, either intramuscularly or intravenously. This requires them to be presented as a sterile preparation, free of endotoxin, bacterial or viral contamination. In addition, agents likely to cause inflammation or irritation at the point of injection should be avoided, with osmolarity and pH chosen to be compatible with the body tissue or fluid into which they will be injected, and excipients used that are compatible with normal biological function. For example, it would be inappropriate for buffer salts known to inhibit the biological activity of the product to remain from earlier purification steps. Also, residual chelators

Medicines from Animal Cell Culture Edited by G. Stacey and J. Davis

Table 20.1 Key product criteria for a lyophilized biological medicine.

Criterion	Desirable properties
Appearance	Acceptable to customer, consistent throughout batch, stable throughout shelf life
Reconstitution	Rapid, complete, consistent throughout batch, no particulates or gel remaining. Requiring no excessive agitation or manipulation
Residual moisture content	Consistent with long-term product stability in the lyophilized state, consistent throughout batch
Biological activity	Acceptable level of recovery, consistent throughout batch, consistent throughout storage under specified shelf-life conditions
Freeze-dried cake qualities	Robust, sufficient to permit handling of container during labeling and shippage without collapse, homogeneous throughout batch

or heavy metal ions should be avoided. The delivery volume is very important in setting limits for such contaminants. For instance, high volume parenterals such as human albumin (which may be administered in litre quantities) are strictly controlled in terms of the permissible levels of aluminium ions, endotoxin and vasoactive contaminants (Matejtschuk et al. 2000).

20.2.1 Mode of Delivery

The choice of the mode of delivery will depend upon the action of the drug, its toxicity, and other patient-related factors. Some general guidelines are given below.

20.2.1.1 Intramuscular injection

Although less stringent than products intended for intravenous administration, the formulation must still be compatible with direct introduction into the body, i.e. the choice and concentration of excipients that might prove irritating or harmful should be controlled. Isotonic and neutral pH preparations should be used to prevent damage to the tissue at the site of introduction. This said, some of the criteria might be less demanding than for intravenous application, for instance aggregate levels are of lesser importance. However, uptake from the site of intramuscular injection can be far less efficient for large molecules (such as antibodies) and therefore the effective dose in the circulation can be less than would be initially envisaged from the quantity administered. In addition, since release from the site of injection may be protracted, intramuscularly administered materials reach the maximum circulatory concentration later than the intravenous analogue. For example, intramuscular immunoglobulin attains a maximum circulatory concentration after 3–7 days (Scheiermann & Kuwert 1983). Finally, there is a practical limit to the injection volume that can be administered by this route as excessive volume can make such injections quite painful.

20.2.1.2 Intravenous injection

Intravenous injection provides the best option for introducing biologicals rapidly and with maximum uptake. However, it is also more demanding in terms of both the molecular properties of the biological (aggregates/fragments, etc.) and also the permissible excipients and impurities. Although there is potential for a greater injection volume, this must be balanced against the impact of even minor levels of interfering materials such as aggregated or denatured material, vasoactive contaminants and adverse reactions caused by excipients/residuals (Levy *et al.* 1986).

20.2.1.3 Inhalation delivery

Inhalation delivery is not a common delivery method as yet for biological medicines, but is widely used for small molecules and has obviously been considered for biologicals with direct application to the lungs (e.g. alpha-1-protease inhibitor for emphysema). Some of the studies on this route for large molecules have not been encouraging (Edwards & Dunbar 2002) but Fineberg *et al.* (2005) have reported promising trials in human diabetics with inhaled insulin, and Alpar *et al.* (2005) with protein antigen and DNA delivery via the intranasal/intrapulmonary route in animals. The First licenced inhalable insulin, Exubera, was approved in 2006. Intranasal delivery of vaccines has also been reported (Mills *et al.* 2003). Hussain *et al.* (2004) have reviewed recent studies aimed at enhancing delivery of proteins via the pulmonary route.

20.2.1.4 Topical application

Another relatively new means of applying biopharmaceuticals is via the skin. Although the skin presents an effective barrier against uptake of most biologicals, subcutaneous injection is used for some biologicals (for example immunoglobulins and insulin). Promising novel delivery methods include transdermal patches and also needle-free injection through the skin using high pressure compressed air delivery of finely dispersed aerosols (see the web site of the Association of Needle Free Injection Manufacturers www.anfim.com). Transdermal delivery is being actively investigated for delivery of both proteins (Benedek *et al.* 2005) and DNA vaccines (Raviprakesh *et al.* 2003). Petersen and Jani (2001) reported equivalent delivery of recombinant erythropoeitin by needle-less and low volume traditional syringe during haemodialysis. Problems were envisaged due to the size of macromolecules compared with the small molecule drugs delivered by this technology (Russo *et al.* 1999) but trials of the delivery method seem promising, especially in the vaccine area (Brave *et al.* 2005; Meseda *et al.* 2005), and systems to deliver both insulin and human growth hormone by needle free injection pens are available.

20.2.1.5 Oral delivery

Most biologicals cannot be delivered by the oral route as they cannot withstand the harsh conditions of the stomach and gut, with low pH in the former and a proteolytic cocktail in the latter. However, there are exceptions, for instance the oral polio vaccine where the live viral vaccine is introduced via its normal route of infection (Burke *et al.* 1999). Insulin in combination with a novel delivery agent has been shown to be absorbed across the gastrointestinal tract when administered orally (Kidron *et al.* 2004). A potential alternative route, popular because it avoids hepatic blood flow and hence removal by the liver, is sublingual or rectal delivery, although these have not been used extensively for biopharmaceuticals.

20.2.2 Formulation Process

20.2.2.1 Achievement of the desired excipient composition

The formulation process can assist in removing unwanted components persisting from the culture environment of the biological or introduced as part of the purification process. Examples might include traces of metal ions such as copper from immobilized metal ion affinity chromatography (IMAC), high salt concentrations (from ion-exchange chromatographic purification), lyotropes (from hydrophobic interaction stages) and traces of ligands (from affinity separations). The desired formulation can sometimes be achieved as part of the final stage of purification (e.g. a chromatographic stage) or can be brought about by use of a buffer exchange method (ultrafiltration/diafiltration).

20.2.2.2 Achievement of stability during storage and shippage

The product must remain fully potent and stable during the stated shelf life and the formulation should be tailored to assure this. The formulation should contain appropriate stabilizers to prevent or minimize molecular degradation. Darkened glass containers may be required to minimize factors such as photolysis. There should be no detectable contamination leaching from the container or its components during the storage period. A review of common leachates from container systems has been written recently (Jenke 2002). Such leaching may well be exacerbated by inappropriate formulation, for example components such as chelators may induce aluminium loss from glass containers, and inappropriate pH may increase leakage of minor trace components of the container, such as rubber closure components.

20.2.2.3 Provision of a medium compatible with the injection environment

This means that the product should have appropriate osmolarity and pH and should only contain stabilizers compatible with the mode of injection. A comprehensive treatise on pharmaceutical excipients has recently been updated (Rowe *et al.* 2005).

20.2.2.4 Facilitation of rapid and full reconstitution and uptake from site of introduction

The product should be 'ready for injection' or, if lyophilized, must reconstitute completely and without aggregation in as short a time as possible after reconstitution, to prevent delay in treatment time.

20.2.3 Liquid versus Freeze-dried Formulation

The choice of liquid or lyophilized presentation will depend upon the stability of the biotherapeutic in terms of its molecular properties and therapeutic efficacy achievable with optimum formulation. Many of the degradation pathways that occur in biologicals are water catalysed. Some biologicals are inherently unstable and so will only be presentable if lyophilized, as removal of available water will reduce the degradation rates. The desired shelf life for the product should be determined for its likely market niche, and the formulation and the presentation format chosen should be tailored to meet this specification. The particular problems associated with the lyophilization of DNA preparations such as those used in gene therapy has been reviewed (Allison & Anchordoquy 2001).

For some biologicals a liquid presentation with room temperature storage may be possible, delivering a shelf life of several years (e.g. human albumin solution), whereas for others lyophilization will be necessary and even then the shelf life may have to be relatively short. Oxidation by head-space (atmospheric) oxygen may be a problem for product stability. Use of an inert environment (nitrogen) needs to be evaluated against using an evacuated container; this choice may also have implications for reconstitution time if the product is lyophilized, as a partial vacuum can aid water uptake and dissolution.

20.2.4 Degradation Mechanisms in Biologicals

A number of degradation pathways have been identified in biologicals (Lai & Topp 1999). Although the rate of degradation may be reduced, it should not be assumed that these pathways are suspended totally in the lyophilized state, since with any freeze-dried product some residual moisture will be present. The level of residual moisture present can influence the long-term stability of the biological material (Breen *et al.* 2001). Oxidation of amino acid residues (Fransson *et al.* 1996) and other changes can occur; chemical modifications involving reactions between excipient sugars and free amino groups can also be a problem (Li *et al.* 1996).

Aggregation or degradation of biological molecules can also occur during downstream processing and on storage. The stability of biological medicines, and methods for its determination, is discussed in Chapter 27.

20.3 LYOPHILIZATION OF BIOLOGICALS

The remainder of this chapter will focus on the lyophilization process.

20.3.1 Freeze Drying: Principles

Lyophilization is the sublimation of ice from a frozen material to leave behind a stable dessicated product. For practical purposes most of the lyophilization process is performed at sub-ambient temperatures under vacuum, to ensure that the material remains completely frozen and to enhance the sublimation rate while protecting the biological material from thermal damage. The lyophilization process can be seen as comprised of three basic steps (Figure 20.1):

- freezing, under conditions that will result in the conversion of the maximum amount of the water present in the sample to ice crystals. The biological material together with any excipients and some water is freeze concentrated to form an extremely viscous liquor, a glass, containing an amount of non-frozen (amorphous) water. This is discussed further below.
- primary drying under vacuum, at a temperature below the collapse or glass transition temperature (see Section 20.4 below for explanation). The frozen component of the water will be removed by sublimation during this stage. The amount of water remaining in the amorphous glass will vary from one formulation to another and may be relatively high.
- secondary drying, at an elevated temperature, still under vacuum, often a deeper vacuum than for the primary drying. This step removes much of the remaining amorphous water from the glass.

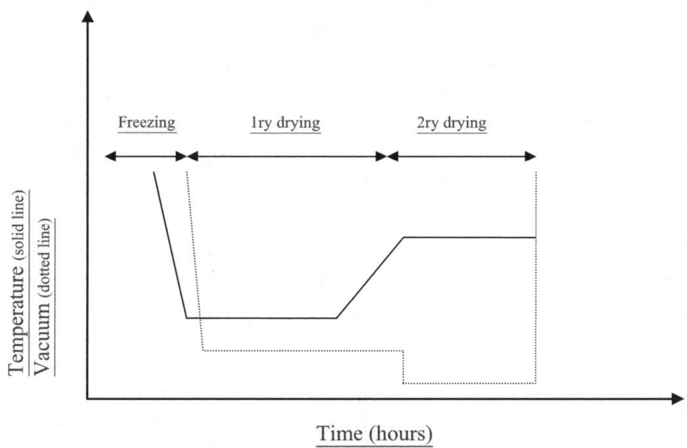

Figure 20.1 Schematic of a freeze drying profile.

20.3.1.1 Freezing

The successful freezing of biological materials prior to commencing sublimation is of crucial importance in obtaining a stable and visually acceptable freeze-dried product. Since reviews of this topic are available (Patapoff & Overcashier 2002) this stage will be mentioned only briefly here. As a mixture of biological macromolecules, inorganic salts and water are reduced in temperature below the freezing point (or more accurately, the melting point as the freezing point may be depressed by supercooling) of the mixture, water will be removed from the liquid phase as it crystallizes to ice. The temperature at which ice nucleation occurs is usually below the freezing point, the amount below being termed the degree of supercooling, and will depend upon a number of factors including the presence of nucleating agents, the composition of the solute material and influence of particulate matter. The degree of supercooling has been shown to be important in influencing the length of primary drying that is required (Searles *et al.* 2001a). This is thought to be due to the size and structure of the ice crystals formed. A small degree of supercooling results in slow crystallization with large crystals forming through crystal growth. Larger crystals allow large pores to remain as ice sublimes, and so there is less resistance to further sublimation from within the product cake and drying proceeds more quickly. Conversely, small pores may restrict the lyophilization rate and can result in poor product appearance. However, some materials may be insufficiently robust to withstand long freezing steps and so may require snap freezing.

Most proteins and carbohydrates do not crystallize on cooling. They are increasingly concentrated by the removal of water as ice. Some materials will crystallize when maximally concentrated and the temperature at which crystallization is maximized is termed a eutectic point. An isotonic saline solution will reach an effective concentration of 3M NaCl when frozen to its eutectic point and similar concentration of other components will also occur (Franks 1990). Other materials, including most biological materials, will continue to become concentrated as the temperature falls until they reach a maximally concentrated state and form a glass, stable at that temperature. The temperature at which this glass formation occurs is termed the glass transition temperature (T_g). This glass will still contain some water, which remains uncrystallized and is amorphous, being a constituent of the glass (see Figure 20.2)

Primary freeze drying has to be undertaken at a product temperature (or rather the freeze drying interface temperature within the product) below T_g, otherwise the glass will be unstable and release fluid water that may be sufficient in quantity to 'collapse' the freeze-drying product. It is important to note that there may well be heterogeneity across the shelves and between shelves of the freeze drier in terms of temperature during freezing and sublimation. The shelf temperature in itself is not the critical factor; it is the temperature within the frozen product, which will be affected by the efficiency of heat transfer between the shelf and the product (i.e. the quality of the vial finish). Temperature mapping within the freeze dryer and examination of the distribution of complete freezing should be part of the process validation. However, temperature probes in containers cause the product in that container to behave in an atypical manner as the probe can be a route for additional heat and can induce nucleation. Typically, the duration of the freezing step should be continued for some time after freezing occurs in the containers with temperature probes.

Since crystallization occurs from the bottom of a container, slow freezing also gives rise to the exclusion of non-freezing material to the top of the liquid. In the extreme case, a meniscus skin can form a barrier to sublimation which compromises freeze-drying.

20.3.1.2 Annealing

During the freezing stage the product temperature can be raised, held for a period, and subsequently lowered. This procedure is known as annealing. Some authors have suggested the use of 'annealing' stages to ensure greater uniformity in achieving a fully frozen state across the

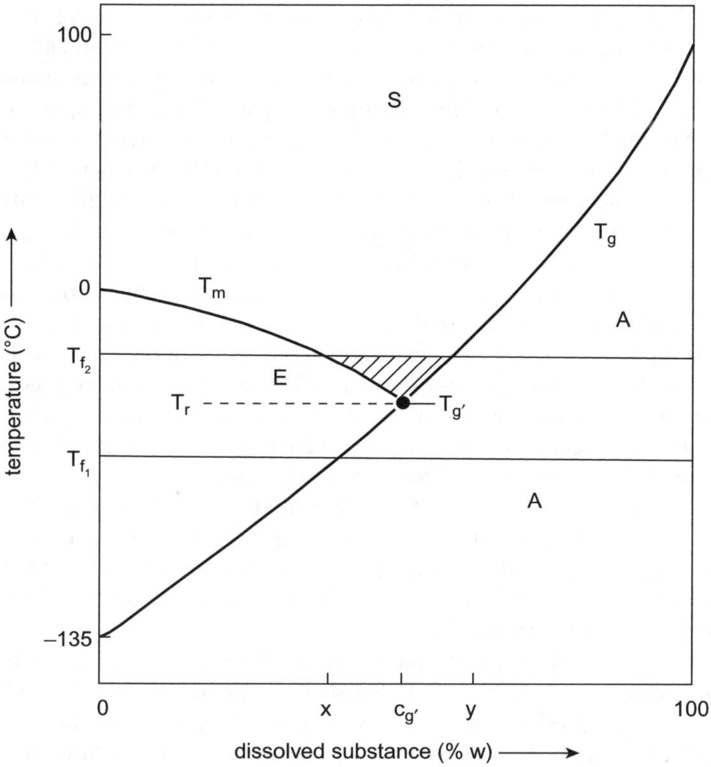

Figure 20.2 Idealized phase diagram for a hypothetical small carbohydrate to show the relationship of phase to temperature and concentration. If a temperature between Tf1 and Tf2 is applied, the product can, above the Tg, recrystallize, start melting, or remain in the amorphous phase, depending upon the concentration of the dissolved substance. Below Tg, or at concentrations smaller than Cg, crystallization is possible. Key: A: amorphous solid, E: ice, S solution/liquid. Reproduced from Oetjen (1999) with kind permission from Wiley VCH

batch (Searles *et al.* 2001b). Annealing has been recommended by several practitioners and has the advantage that it can induce larger ice crystal formation through crystal growth – Ostwald ripening (Searles 2004) – which can speed the rate of subsequent sublimation. It can allow cycles to be shortened as the better pore structure allows faster primary drying and so this stage can be completed more quickly. By crystallizing out troublesome formulants the overall product T_g can be raised to make lyophilization a practical possibility. However, if these same formulants also provide lyoprotection to the biological (see Section 20.3.2.2), then the risk of increased lyophilization-induced damage must be weighed against these benefits.

20.3.1.3 Primary drying

Once the sample to be freeze dried is fully frozen, a vacuum may be applied and the primary sublimation begun. As the sublimation rate is related to the vapour pressure of water above the product compared with that above the condenser where the ice re-forms, it is essential to have cooled the condenser initially to a temperature below that of the product, typically a temperature of −70 °C. In practice, most production-scale freeze-driers cannot attain a shelf temperature below −55 °C. This limits the freeze drying to formulations with a T_g' greater than about −50 °C, although the use of liquid nitrogen to cool condensers may give some additional flexibility.

The choice of vacuum conditions is a compromise, as a deep vacuum will induce fast sublimation rates but this will be tempered eventually by the impact of a high vacuum in inhibiting heat transfer within the chamber and so reducing the heating capability. Indeed, it is preferable to have a weaker vacuum but to maintain efficient heat transfer within the chamber. This can be achieved by means of a vacuum bleed needle valve which allows air (or nitrogen if preferred) into the chamber to maintain the internal pressure at the programmed value (Rowe 2005).

For a given container, the heat transfer into the product depends on the shelf temperature and chamber vacuum. At 'low' pressures, the principal heat transfer is by conductance through the container's points of contact with the shelf, together with radiation from the metalwork within the chamber. At 'higher' pressures, heat transfer by convection of the gaseous molecules within the chamber is significant. The materials slowest to freeze dry are typically those in containers located in the middle of the shelf; a door with a large viewing port can also induce heterogeneity in speed of drying. Modification in the heat flow to the product can be affected more quickly by chamber pressure changes than by shelf temperature changes.

The temperature required to maintain the product in an immobile state (frozen glass) is determined by the glass transition point of the formulation, as explained in the section above. Product temperature should be maintained below the glass transition temperature, but for economic reasons should not be very much below this temperature, as the cost of maintaining production-scale freeze dryers at these temperatures is high.

The heat required for sublimation is taken from the frozen product. This is replaced by heat transfer from the shelves and shelf fluid. If the shelf temperature is maintained at the product freezing temperature, the product temperature will typically be less than that of the shelf due to the loss of heat of sublimation. To compensate, more heat can be put into the product by raising the temperature of the shelves. Heat will be transferred to the product by conduction from the shelves, by radiant heat from shelves and sides of the drier, and carried both to and from the product by convection via what atmosphere (of water vapour and gases) remains within the chamber.

As the sublimation proceeds so more heat can be put into the product, and for a larger bulk of freeze drying material the temperature of the shelf can be maintained above the glass transition as a gradient of heat will occur in the product cake itself. The rate at which the shelf temperature is raised (referred to as ramping), which can be used safely without causing collapse of the product cake, will need to be determined empirically for each depth of fill and container type. It may well vary as the size of the batch and possibly the dimensions of the freeze drier chamber are changed, and also the capacity and load on the condenser.

Once the crystalline water ice has sublimed, the product temperature may be allowed to rise above the original glass transition temperature. The end of primary drying can be determined in a number of ways. In the pressure rise (or barometric) test, the isolator valve between chamber and condenser is closed for a short period of time during primary drying, and the rise in chamber pressure measured as an indication of the rate of sublimation of water vapour. As primary drying reaches its end, the rate and amount of sublimation occurring will fall and this will be indicated by a negligible change in the chamber barometric pressure on closing the valve. This test is non-invasive but it does perturb the freeze drying process if primary drying is not yet complete, and as such is seen as suitable for validation tests but may be less appropriate in routine manufacture.

Another test for the completion of primary drying is to follow the temperature profile of containers fitted with thermocouples or resistance thermometers. As stated above, data from these containers must be treated with caution; they are atypical, freeze drying proceeding more quickly than in a container with no probe in a similar position. The temperature of the product is initially below that of the shelf due to cooling from the sublimation. However, as the primary drying nears completion the temperature rises to the shelf temperature or even exceeds it. An additional safety period should be implemented to allow the slower-drying containers of the batch to catch up.

Other more involved monitoring methodologies have been suggested including the monitoring of the vapour phase (via the vacuum pump exhaust) by mass spectrometry (Connelly & Welch 1993) and the comparison of pressure measurements as indicated by different barometric monitoring devices (Milton *et al.* 1997).

20.3.1.4 Secondary drying

A product at the end of primary drying may still contain as much as 10 % residual moisture (Franks 1990) as amorphous water. If brought directly to room temperature and pressure this product would be unsuitable for long term storage. Collapse would be likely or hydrolytic degradation might proceed at an unacceptable rate. Such collapse may not be seen immediately as the release of water from the glass, although thermodynamically favoured, is kinetically slow. For this reason moisture content is further reduced by the use of secondary drying under vacuum at near ambient or even elevated temperatures, and hard vacuum used to remove the remaining water down to very low residual levels. Residual moisture in the lyophilized product can be a source of product instability, and dependent upon the material in question, and the intended shelf life and storage conditions, a maximum residual moisture level should be set and the secondary drying should be optimized to achieve this level or lower. For therapeutic products the acceptable levels of moisture may be 1–3 % by weight. For some applications even lower levels may be required, as for instance with biological standards, which need a long shelf life, where a level 1 % or lower is preferred.

As a guideline it has been suggested that a secondary drying period of one-third the length of primary drying be used (Murgatroyd 2001) with a deeper vacuum than is used in the primary drying, typically 1–30 μbar. Some have claimed that pressure rise tests can be used as indicators of the end of secondary drying as in primary drying, although the pressure difference will be far smaller.

20.3.2 Formulation Choices in Freeze Drying

20.3.2.1 Cryoprotectants

Damage during lyophilization can be of two fundamental types, damage during the freezing process and damage either during the dehydration process or on reconstitution. Freezing introduces a concentration of the biological and the excipients surrounding it as the available water is preferentially compartmentalized into ice crystals. Eventually a glassy state is achieved where the biological is maximally concentrated. This may result in marked changes in the local concentration of excipients (for instance NaCl) and alteration in the microenvironment (e.g. pH) of the medium. Selective crystallization of mixed phosphate components can cause the pH of a sodium phosphate-buffered solution to shift from neutral pH to pH 3 during freezing (Anchordoquy & Carpenter 1996). Such changes may induce denaturation of the biological or loss of functional activity. Biologicals may also be inherently less stable at higher concentrations, and there may be an increase in aggregation. Cryoprotectants are molecules that reduce the deterioration in biologicals during this freeze-concentration phase. Examples include amino acids, detergent molecules such as polysorbate, and saccharides such as sucrose or sorbitol (Chang *et al.* 1996a, b; Nena & Avis 1992).

20.3.2.2. Lyoprotectants

Denaturation and/or inactivation may not occur during repeated cycles of freeze-thawing but may occur once a product is freeze-dried. This damage is thought to be due to the loss of 'stabilizing' water from around the essentially hydrated biological molecules, with consequent partial or significant complete loss of the tertiary structure and hence aggregation or loss of functional activity. Many cryoprotectants are unable to prevent such lyophilization-induced damage. Molecules that do protect against these changes are termed lyoprotectants. Saccharides such as sucrose or trehalose are particularly good lyoprotectants, and they may also offer significant cryoprotection (Wang 2000, Carpenter *et al.* 1997).

20.3.2.3 General comments on formulant choice

An ideal formulation should have a relatively high T_g, so that a higher shelf temperature can be used in the freeze drier, hence reducing running costs. The concentration of the biological material present should not be excessive (i.e. in practice not above 10 % w/v), in order to avoid 'skin' formation as concentrated proteins are lyophilized. The use of pH buffers comprised of components that behave differently on freezing should be avoided, as should the use of reducing sugars or those that readily form reducing sugars on hydrolysis. The level of inorganic salts should be minimized as these lower T_g and expose the non-crystallizing materials to high osmolalities, which may be damaging. Suitable excipients should be included that help to preserve biological activity and others that may assist reconstitution. Finally, there should be sufficient dry weight remaining after lyophilization to ensure that a substantial freeze-dried cake is formed. Salts may be included in the reconstituting medium rather than in the freeze dried product if this means that the formulation for lyophilization is more advantagous in terms of the freeze drying conditions that can be applied.

20.3.2.4 Examples of formulations used for biological medicines

Table 20.2 illustrates common formulations used in biological medicines. This data was obtained from publicly accessible sources and is meant to be illustrative only, being by no means exhaustive.

Table 20.2 Typical formulations in commercial biotechnology products. Data compiled from Internet sources, order of components does not indicate their respective proportions in the formulations concerned.

Product type	Format	Formulation components
Monoclonal antibodies	Lyophilized	• Histidine, glycine, mannitol
		• Mono- and dibasic sodium phosphate, NaCl, sucrose, Polysorbate 80
		• Histidine, trehalose, Polysorbate 20
Interferons	Lyophilized	• Human albumin, NaCl, glucose
		• Human albumin glycine, mono- and dibasic sodium phosphate
	Liquid	• Ammonium acetate, NaCl, benzyl alcohol, Polysorbate 80
Erythropoietin	Lyophilized	• NaCl, urea, mono- and dibasic sodium phosphate, calcium chloride, glycine, leucine, isoleucine, threonine, phenylalanine, benzylalkonium chloride, benzyl alcohol
	Liquid	• Albumin, citric acid, sodium citrate, NaCl
Insulin	Liquid	• Isotonic phosphate(PBS?), glycerol, cresol
Somatropin	Lyophilized	• Glycine, mannitol, *m*-cresol, mono- and dibasic sodium phosphate, mannitol, glycine, sodium hydrogen carbonate, benzyl alcohol
Vaccines	Liquid suspension	• Aluminium hydroxide, aluminium phosphate, 2-phenoxy-ethanol, formaldehyde, Polysorbate 20, Polysorbate 80, NaCl
		• Aluminium hydroxide, mannitol
		• Potassium chloride, potassium dihydrogen phosphate, NaCl, calcium chloride, magnesium chloride, thiomersal
Recombinant clotting factors	Lyophilized	• Glycine, NaCl, calcium chloride, human albumin
		• NaCl, sucrose, histidine, calcium chloride, Polysorbate 80
		• Histidine, sucrose, glycine, Polysorbate 80.

There are many common features between the formulations included in Table 20.2. There is a need for the inclusion of osmolarity regulators (typically sodium chloride, glycine or sugars) and for buffers to control pH. Amino acids, proteins such as albumin, and saccharides are also often included, reinforcing their applicability as stabilizers for the lyophilization process. Detergents and buffers are included in some but not all formulations.

20.4 FREEZE DRYING CYCLE DESIGN

The production-scale freeze drying of biopharmaceuticals is a significant component of the overall production costs. Inefficient cycles can tie up very costly large batch freeze-driers for additional days per batch, at substantial power and maintenance costs and reduction in the number of batches that can be processed each year. Suboptimal cycles may also be less robust, resulting in a high percentage of rejects per batch and even occasional failed batches. Inadequately designed cycles may also generate products of inferior visual appearance that may have impact on product sales and customer satisfaction. Hence it is important to optimize the freeze drying cycle during initial development and to verify each scale-up in process with sufficient validation runs so that quality can be assured for the routine production process. In order to design economic and effective freeze drying cycles, the thermal properties and freeze drying characteristics of the chosen formulation should be assessed as part of the cycle development. Using this information the intended product should be freeze-dried at pilot scale and the resultant material tested for residual moisture content, activity, appearance, and reconstitution time.

20.4.1 Analysis of Thermal Properties

The thermal properties of a formulation to be used for freeze-drying may be determined by a number of analytical methods. Several common methodologies are reviewed here but this list is not exhaustive.

20.4.1.1 Differential scanning calorimetry

Differential scanning calorimetry (DSC) is a powerful method for the detection of eutectic points and some glass transition temperatures. Although the quantity of sample that can be tested is small (typically less than 100 μl), and hence the thermal event signal from an aqueous solution will be much weaker than for a dry lyophilized powder, improvements on the basic technology have been suggested to allow greater sensitivity and resolution. These include modulated temperature DSC, where the weak transitions can be resolved from the stronger relaxation events which may otherwise obscure them (Kett 2001, Kett *et al.* 2004), and rapid heating rates as used in Hyper™ DSC (Pijpers *et al.* 2002) and T zero™ technology (Danley 2003).

Samples are typically frozen rapidly to below the temperature at which lyophilization is proposed (e.g. −50 °C) and then warmed at a controlled rate back to ambient conditions, using either a single oven, containing reference and sample pans, or in matched thermal ovens. The reference pan is empty and the heat required to match the temperature of the sample and reference pans is used to derive the heat flow profile of the sample. The eutectic point, glass transition point and melting point of the sample may be identified, if present.

DSC on frozen liquid samples is perhaps most informative where a clear transition or eutectic crystallization is discernible. Where no such event occurs the technique may be unrevealing (see Figure 20.3). However, for the determination of glass transition temperatures for the lyophilized product, DSC is unrivalled in its applicability. Such analyses have been shown to be of value in predicting the shelf stability of lyophilized formulation variants, hence indicating superior

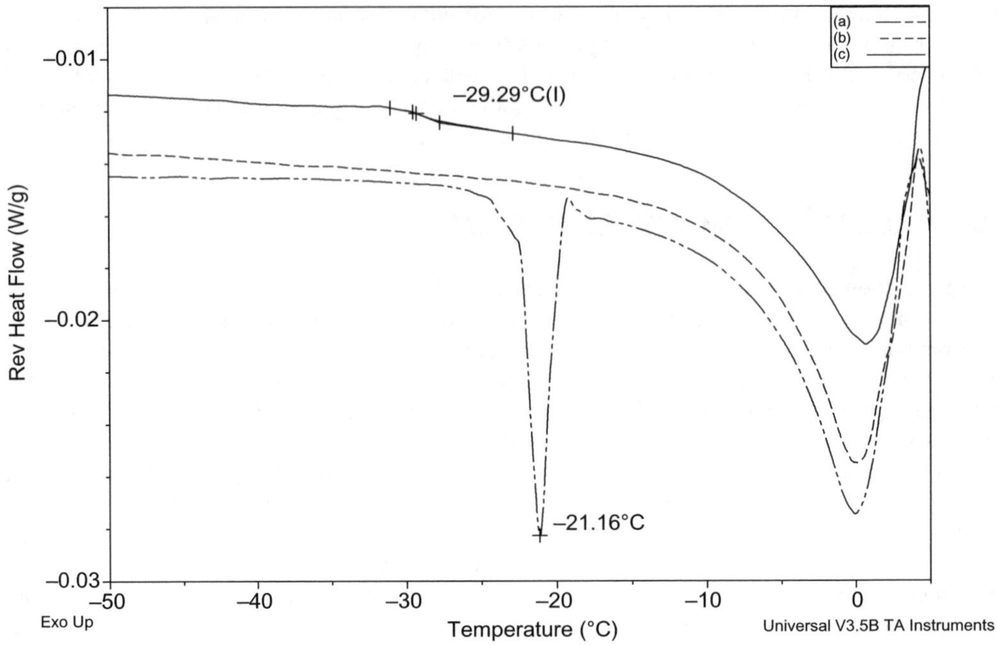

Figure 20.3 DSC profile showing materials with (a) no obvious thermal event; (b) a sodium chloride eutectic (−21 °C) with possibly additional T_g at shoulder of the peak; (c) glass transition (−29 °C). Samples (80 µl) run in steel pans against empty reference pan at 1.5 °C/min ramping temperature with heat-only modulation on TA 2920 DSC (TA Instruments, Crawley, UK).

formulations for maximizing shelf life and providing convenient storage/shippage conditions (Duddu *et al.* 1997).

20.4.1.2 Electrical resistance

The electrical resistance properties of an aqueous sample change markedly on warming from the deeply frozen state (Rey 1961). The large change in electrical resistance that occurs on mobilization of water within the frozen matrix, although well below the melting temperature, can be illustrative of a suitable product temperature below which primary drying must proceed if collapse is to be avoided (Figure 20.4). In our experience this technique has wider applicability than DSC but critical temperatures will still be influenced heavily by the ionic components of the formulation.

20.4.1.3 Differential thermal analysis

Differential thermal analysis (DTA) compares the thermal properties of a sample to that of a reference material frozen and warmed in parallel with it. These techniques have long been used in the analysis of materials for freeze drying (Jennings 1999; Chang *et al.* 1996 a, b) and are sometimes performed with home-made systems, although commercial equipment is also available. The heat energy differences between sample and reference material are compared during warming from frozen, and again marked changes can be indicative of the maximum successful product primary drying temperature. For an example of a DTA profile, see Figure 20.4. Although simpler than DSC, electrical resistance, and DTA methods can provide useful data from a wide variety of samples.

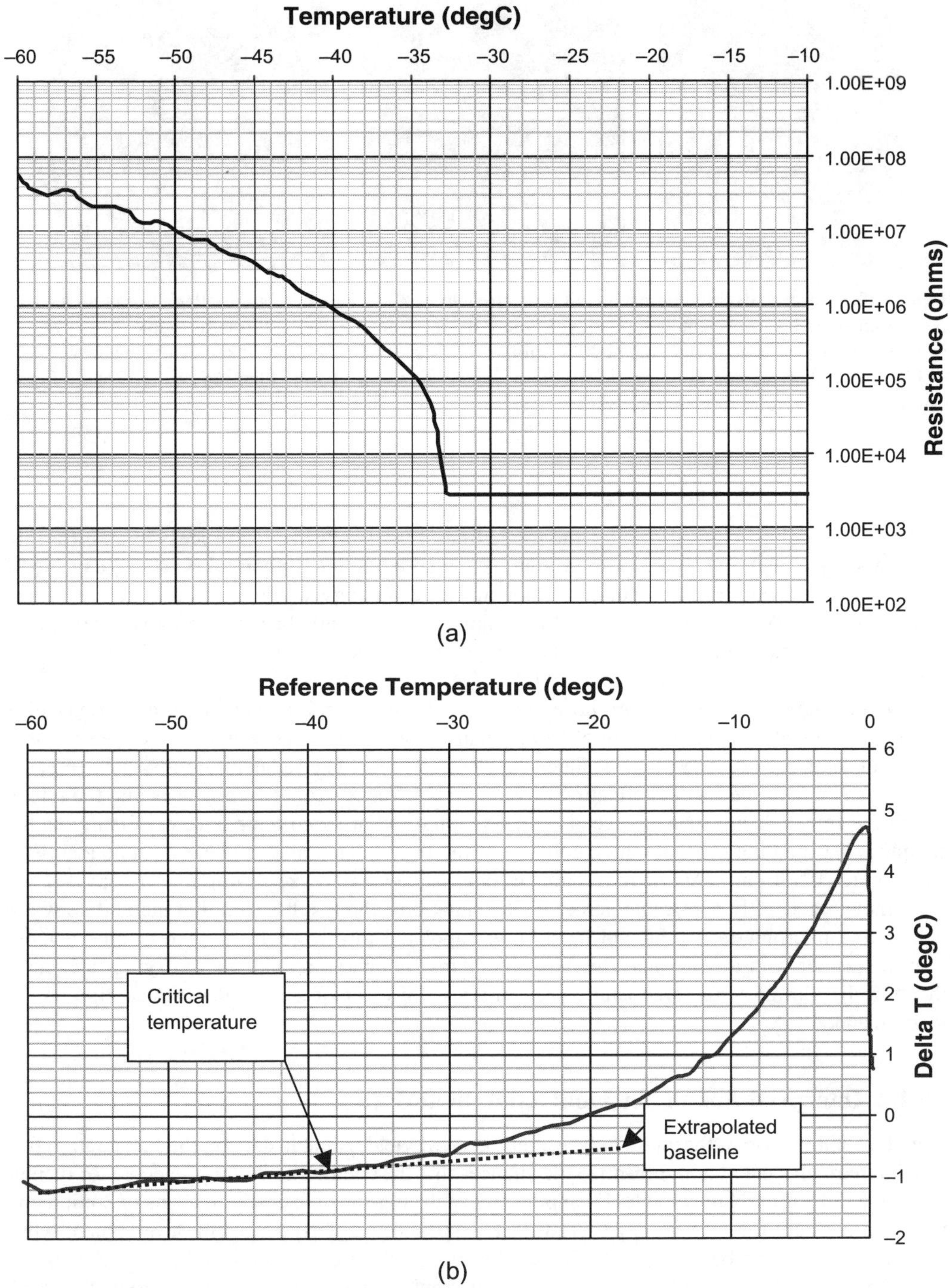

Figure 20.4 Thermal analysis of a monoclonal antibody-containing tissue culture supernatant. (a) Conductivity showing a transition at −32 °C. (b) DTA profile showing a change in the gradient of delta T at −38 °C. Runs performed on Lyotherm (Biopharma Technology Limited, Winchester, UK).

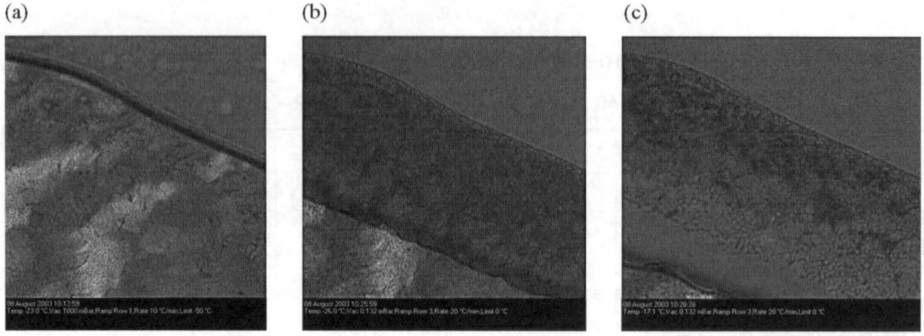

Figure 20.5 Freeze drying microscopy of 0.2 % trehalose, 0.5 % human albumin, 0.9 % sodium chloride. (a) Frozen at −23 °C, no vacuum applied; (b) Freeze drying front advancing through sample, below $T_{collapse}$ (−26 °C); 132 μbar vacuum, ramping at 20 °C/min. (c) Collapse of freeze dried cake occurring above $T_{collapse}$ (−17 °C); 132 μbar vacuum ramping at 20 °C/min. Images courtesy of Julie Fleming and Dr Roland Fleck, Cell Biology & Imaging Division, NIBSC.

20.4.1.4 Freeze drying microscopy

Freeze drying microscopy allows for the modelling of the lyophilization process in an ultra-thin layer of sample enclosed in a modified microscope slide mounted upon a cryostage, capable of evacuation to partial or full vacuum. Although the technology was described many years ago (MacKenzie 1964), commercially available equipment has recently become available. Freeze drying microscopy can add valuable insights into suitable freeze drying parameters for a given formulation (Zhai *et al.* 2003; Fleck *et al.* 2003).

The technique is different from those previously discussed in that it can be used to determine the temperature at which a thin film of frozen matrix undergoes collapse and so provides a visual record of the freeze drying process, albeit on a micro-scale. The temperature at which freezing occurs can be identified. On application of vacuum the freeze-drying commences and the rate at which the freeze drying front proceeds can be monitored at given temperatures. Finally as the sample is warmed the point at which collapse of the freeze dried matrix occurs can be identified (see Figure 20.5 for an example of the images obtainable from freeze-drying microscopy).

Although generally applicable across all formulations, the results from this methodology are obtained at micrometre depths and must be extrapolated to a lyophilized cake of one or more centimetres depth in the product; the data on collapse temperature and freeze drying rates must therefore be viewed with some caution and margins of error of several degrees centigrade are recommended.

20.4.2 Interpretation of Thermal Analytical Data

Use of a combination of thermal analysis methods will yield valuable data on the formulation that is to be freeze dried and will minimize the chances that the conditions initially selected for lyophilization will be grossly unsuitable. Even pilot freeze drying runs are costly in terms of equipment, time and possibly the biological material itself (especially at early stages in development when scale-up may not yet be optimized). Hence the time and effort invested by the use of multiple analytical methods may well be rewarded by more rapid progress of a candidate product through freeze drying development to final process conditions.

The thermal profile of a given formulation may contain a single event or a series of discrete events. Some formulants may exert an influence modifying the glass transition values observed

Table 20.3 Thermal properties of some common lyophilization excipients.

Excipient	T_g (°C)	$T_{collapse}$ (°C)
Lactose	−28[b]	−31[a]
Trehalose		−29/−34[a]
Glucose	−43[b]	−40[b]
Sucrose	−32[b]	−32[b]
Sorbitol	−44[b]	−45[b]
Mannitol	−27/−33[a]	
Glycine	−37[a]	
Histidine	−32[a]	
Sodium dihydrogen phosphate	−45[a]	
Potassium dihydrogen phosphate		−55[a]
Albumin	−11/−13(bovine)[a]	−9.5 (human)[a]
Dextran (10 kDa)	−13.5[b]	−9[a,b]
Sodium citrate	−41[a]	
Sodium acetate	−64[a]	

[a]*Wang, 2000.*
[b]*Levine & Slade, 1988.*

for other components. The biological component, dependent upon its concentration, may be a minor contributor to the overall thermal properties of the formulation. Therefore, experience with a small number of well-characterized and acceptable formulations may prove useful across a wide range of biological agents. Whereas, as previously discussed, some formulants may be required to prevent loss of activity or minimize denaturation during lyophilization, others may be chosen to ensure that the composite formulation is capable of being processed under economically viable conditions. From this viewpoint the thermal properties of some common formulants are presented in Table 20.3. The inclusion of a polymer such as human albumin or dextran will result in a rise in the glass transition (or collapse temperature) of the formulation. The inclusion of formulants with low T_g values will have the effect of lowering the overall glass transition temperature; this may therefore require more demanding lyophilization parameters or the development of alternative strategies such as annealing (see Section 20.3.1.2).

20.4.3 Application of Thermal Analysis Properties to Freeze Drying Cycle Design

Although thermal analysis data is of great value in choosing the freezing and primary drying temperatures, each method described provides only approximations to the real conditions that a material will experience when undergoing lyophilization. As such it is better to obtain a consensus temperature from data derived by several different analytical techniques, as this will give greater assurance than that derived from any single technique in isolation. It is common practice to allow a safety margin of several degrees centigrade to accommodate the differences between analytical and production systems.

Even with the effective safe freezing and primary drying temperatures identified, the length of each freeze drying step must be determined empirically and indeed may change as the process is scaled up. For this reason, development runs should be performed to optimize the freeze-drying cycle. The lengths of time required for all of the samples to be fully frozen and for primary and secondary drying to be completed can only be determined from trial freeze drying runs. A well planned programme early in development to determine the thermal properties of the formulation to be processed will reduce the number of test lyophilization runs required and the overall development time.

20.5 RESIDUAL MOISTURE

20.5.1 Impact of Residual Moisture Levels on Stability

As mentioned previously, many of the degradative processes that affect biologicals are water catalysed. Thus the level of residual moisture remaining in a biological product is crucially important in terms of its long-term storage stability. High levels of residual moisture can cause aggregation of the biological and even loss of functional activity (Pikal *et al.* 1992; Lueckel *et al.* 1998). One might assume, therefore, that the drier the lyophilized product the greater its stability will be. Lyophilization is an expensive process and so the benefit of lower moisture levels must be balanced against the impact and cost implications of longer cycle times. Some biologicals have been shown to maintain good functional stability even where the moisture levels have been higher than might be thought ideal (Yoshioka *et al.* 1997). There is also evidence that some biological materials can become inactivated by excessively low residual moisture levels (Chang *et al.* 1996a, b). The danger of over-drying products has been recorded for lactate dehydrogenase and β-galactosidase (Jiang & Nail 1998). Denaturation on drying may be prevented in some systems by addition of suitable stabilizers. Damage due to overdrying lyophilized insulin has been noted (Bristow *et al.* 1988), and also loss of biological activity in viruses (Grieff & Rightseal 1968) and bacteria (Cammack & Adams 1985) have been reported where moisture content has been too low. Therefore, the optimal moisture content should be determined on a product-by-product basis, supported by accelerated degradation and real-time stability studies.

Even when material is prepared so as to yield low moisture levels at the time of lyophilization, these can rise during storage. Even in perfectly sealed vials water can be desorbed from the closures, eventually reaching equilibrium with the product (Pikal & Shah 1992; Ford & Dawson 1994, Matejtschuk *et al.* 2005). Effective pretreatment of closures to reduce the absorbed water content should minimise such effects (May *et al.* 1992).

20.5.2 Methods for Determination of Residual Moisture

A number of methods exist for the determination of residual moisture in lyophilized products and these have been reviewed elsewhere (Savage *et al.* 1998). Thermogravimetric methods are popular, especially in the USA, and when supplemented with mass spectrometry to allow determination of the onset of thermal degradation (with consequent release of carbon dioxide) this is a powerful if expensive means of demonstrating final water content. Karl Fischer analysis, and in particular coulometric Karl Fischer analysis that requires smaller quantities of material, is also widely practised. This method relies upon the determination of iodide production from a chemical reaction of iodine with sulphur dioxide and water (from the sample). The iodide can be reconverted to iodine electrochemically and so the quantity of water initially present can be determined from the electrical current used. Some formulants can interfere with the Karl Fischer reactions, and so lyophilized samples containing these formulants may give rise to anomalous water contents when analysed by this method. Both of the above methods are destructive and so a small proportion of the batch must be sacrificed for water measurement.

More recently, several reports of infrared spectrophotometric methods for non-destructive moisture determination have appeared (Savage *et al.* 1998; Lin & Hsu 2002). Although the infrared method must be calibrated against a series of samples of known moisture content (determined by Karl Fischer or TGA), once the relationship between infrared absorbance value and residual moisture content has been determined it can be applied to test the entirety of a batch non-destructively. Therefore it can be very useful to assess within-batch heterogeneity or indeed to compare batch-to-batch variation. This is a great advantage over laborious and destructive Karl Fischer testing. However, the calibration is only valid for a given container size and shape, a given formulation and a given product concentration. Variation in such parameters will require a new calibration to be performed.

20.6 FUNCTIONAL ACTIVITY

The functional activity of any biological medicine must be maintained throughout its production, purification and presentation in its final state. Degraded molecules may cause unwanted immunogenicity and may interfere by competition with biologically active product. As methods for determination of functional activity are product-specific they will not be discussed further here: reviews in this area have been published (for example Robinson 1999; Thorpe *et al.* 1999).

20.7 EQUIPMENT AND PROCESSING ISSUES

20.7.1 Processing Equipment

Lyophilization of biological medicines is performed as a terminal processing stage, and so the product must remain free of microbial contamination and endotoxin during the process. This can only be achieved by strict adherence to good manufacturing practice (GMP). (For more details on GMP see Chapter 34; MCA 2002; WHO 2003; Federal Register entries at www.fda.gov.) Containers should be chosen so as to allow optimal freeze-drying rates (i.e. to maximize the interfacial area between the product and the gas phase). Ideally, the depth of product cake should be kept to 1 cm or less so as to avoid significant temperature gradients through the cake. In reality, this ideal guideline to fill depth is moderated due to other concerns, such as dose size, stability of the biological at given concentrations, operational considerations and optimum reconstitution.

Glassware should preferably be washed using equipment that will allow automation of acid and water rinsing. Water-for-injection (WFI) grade water (see Chapter 2) should be used for all wash steps. The containers should then be dried, sterilized and used immediately or stored in a secure manner so as to prevent recontamination from the air.

Closures should similarly be washed, treated (siliconized if appropriate) and bagged so as to prevent recontamination. Moisture levels should be minimized by optimized baking of the closures and storing them appropriately to reduce reabsorption of moisture thereafter.

20.7.2 Dispensing

The dispensing process should be performed with suitable automated filling equipment for all but the smallest of trial fills so as to remove the risk and variability of operator-introduced errors. Dispensing equipment should be cleaned by validated clean-in-place (CIP) protocols (see Chapter 14) to prevent any contamination or cross-over between fills. Wherever possible, fluid paths should be single use to prevent possibility of cross-contamination.

The ability to deliver an accurate and reproducible fill volume is essential. The coefficient of variation typically achievable with automated equipment during the course of fills of at least several thousand containers can be of the order of 0.25 %, but this will be influenced by the viscosity of the material to be filled. It is necessary to avoid splashing on the sides of the containers. Such splashes may introduce variability in terms of the dose of product recovered on reconstitution and the appearance of the product subsequent to lyophilization. Frothing during the filling process should also be minimized, as it will result in a distorted lyophilized cake surface.

Material intended for medicinal application should be terminally sterile filtered and the filling apparatus should be operated in a suitable environment to prevent any contamination after dispensing. However, not only should the material be sterile, but the level of endotoxin detectable in the product should also be minimal as it could cause endotoxic shock in patients. Endotoxin contamination can arise by shedding from bacteria even though they are subsequently removed by sterile filtration. Therefore, the storage of material to be freeze dried prior to dispensing and during dispensing and freezing should be tightly controlled in order to prevent bacterial contamination.

Human operators are a major potential source of contamination and the performance of freeze drying activities should be carefully controlled so as to minimize the risk of contamination from the operators. Suitable protective clothing and gowning procedures, combined with strictly controlled operating practices and well-defined cleaning and monitoring procedures are a minimum. Some manufacturers choose to operate within an isolator barrier to protect the product further. A detailed discussion of such equipment is outside of the scope of this review – interested readers are referred to relevant literature (Midcalf *et al.* 2004).

20.7.3 Freeze driers

Freeze driers come in all sizes from units that fit on a bench top and have limited (if any) shelf temperature control, to units for which the process can be carefully monitored and controlled and that can have in excess of $50\,m^2$ of shelving. In addition to scale and complexity of operation, if the freeze drier is to be used to process material for therapeutic application, issues of Good Manufacturing Practice (GMP) will also apply (Cameron & Murgatroyd 1997). The intended use of a freeze drier must be known before the unit is designed, as retrofitting for a therapeutic application will probably not be practical. The cost of a GMP production-scale freeze drier is a sizeable component for any project budget and so it is vital that the system is planned to fit its intended use. A system for large-scale production of therapeutics with its associated filling equipment and controlled dispensing environment will be a multimillion pound project.

A schematic of a production-scale freeze drier is given in Figure 20.6. The same basic components are common to all large freeze driers. There is a sample chamber with temperature-controlled

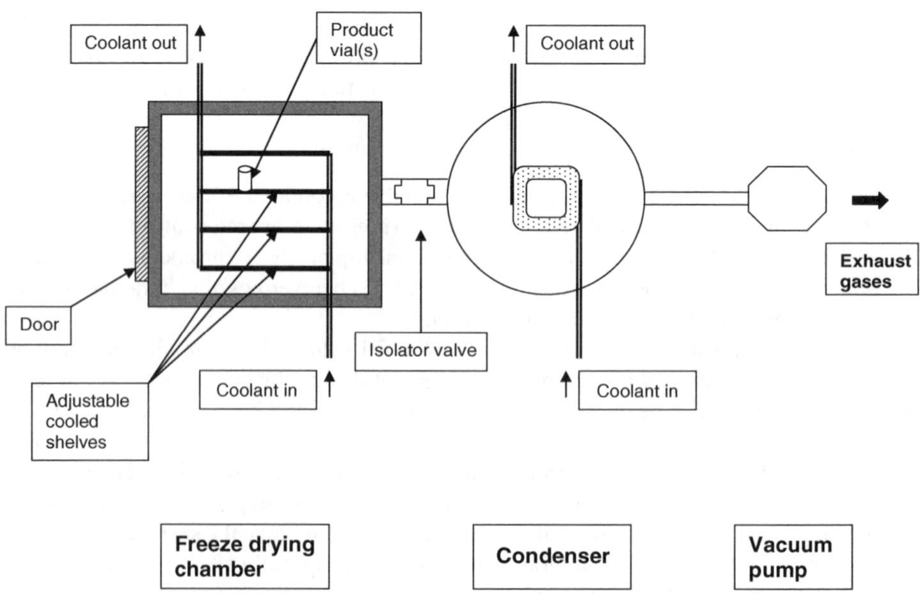

Figure 20.6 Schematic of a freeze dryer.

shelves, along with a mechanism for stoppering of the containers *in situ*. This chamber is connected via a closable valve to the condenser chamber (although in some models the condenser can be integral with the main chamber). The water vapour sublimed from the product collects on the cooled condenser coils. Connected to the coolant pathway are compressors that provide cooling and warming to the shelves and condenser coils, and vacuum pumps that deliver the required vacuum. Internal pressure is monitored by pressure gauges and shelf temperature by thermocouples or resistance thermometers. The chamber can be flushed with inert gas and often there are means for the condenser coils to be rapidly thawed. On modern systems all of these items are controlled by a programmable Logic controller (PLC) and this interfaces to PC-based software, so that all of the operations are computer controlled and the parameter data collected and stored.

For therapeutic applications there will be basic requirements, irrespective of scale and whether filling is to be an automated or manual process. All parts in atmospheric contact with the product should be of 316-grade stainless steel. At least steam-in-place (SIP) and possibly chemical clean-in-place (CIP) systems should be built into the system to allow decontamination and sterilization between runs. The filling environment must be assured by use of downwards flow laminar air for loading and unloading operations. Suitable safety measures to prevent contamination from the operators (protective clothing/equipment) and equipment used (dedicated, sanitizable, pyrogen free) must be used. Vacuum Leak rate testing and filter integrity testing must be performed and meet specification. Operations should be well planned and controlled by standard operating procedures. Validation of the processes and equipment will be required. Validation should be of the freeze drier itself, its operating parameters, consistency of operation and limits of variability. Also, the cycle being used for any given product should be validated (including the acceptable range of variation permissible within individual containers of product in a single batch) and the limits set in terms of the variation in operating parameters that can occur at any given cycle step. In addition, the limit of variation in critical parameters must be determined for at least three and preferably more batches so that consistency of the freeze-dried product can be demonstrated. The temperature variation within a freeze drier will need to be determined by temperature mapping using calibrated thermocouples, across the shelves and between shelves. The coldest points within a freeze drier will need to be identified and used during validation of the steam sterilization cycles. The validation required for a pharmaceutical freeze drier has been outlined in a number of publications (Parenteral Society 1997; Bindschaedler 2004). Other useful information on GMP lyophilization applications is available on the FDA website (FDA 1998).

20.7.4 Sealing and Labeling

Stoppering of vials should be performed within the freeze drier prior to product removal. This will maintain the environment within the vial and also prevent any bacterial contamination upon removal. Many products are sealed under partial or total vacuum as this improves the reconstitution rate of the product by drawing in the reconstituting fluid. Reconstitution is often by connection to a bottle of reconstituting medium by means of a two-ended needle assembly.

20.7.4.1 Closures

Halobutyl rubber is the most widely adopted closure material for lyophilized biologicals. It has superior properties to butyl rubber in terms of its moisture retention. The nature of elastomeric closures has been widely discussed elsewhere (Corveleyen *et al.* 1997; Hora & Wolfe 2004) and will not be dwelt on here.

20.7.4.2 Two-part injection devices

These two part containers consist of the lyophilized agent in one part of the unit separated by an impervious closure from the reconstitution fluid, in such a way that when desired the closure is breached and the product instantly resuspends, ready for injection. This format minimizes the risk of exposure to contamination from the environment and provides greater convenience for the end-user (Maa & Prestrelski 2000).

20.7.4.3 Details of labelling

This should include an unambiguous description of the product, the batch identity, its mode of administration and the dosage. The address details of the manufacturer and the expiry date of the batch should also appear as well as the storage conditions for the product. For full details refer to GMP guidelines already referenced above.

20.7.5 Scale-up Issues

As has already been mentioned, lyophilization processes are not necessarily identical in freeze driers of different design, dimensions, load size and performance capabilities. Hence in any scale-up, the process should be validated in terms of demonstrating the consistency of the batches (Table 20.4). Unexpected failures can be minimized if the thermal properties of the material to be dried are well understood. It must be noted that changes in the dimensions of the container or closure, their composition and preparation must also be validated and cannot be assumed to have no impact on the freeze drying. Any changes in formulation, concentration or fill size will require a return to cycle validation to confirm their impact on the process.

20.7.6 Spray Drying

Spray drying has been widely used in the food industry and is now gaining popularity in some pharmaceutical applications (Jovanovic *et al.* 2004) The liquid material is snap frozen and extruded through a series of fine constrictions such that a sprayed powder of freeze-dried material is produced. It has been applied successfully even to large biomolecules such as monoclonal antibodies (Maa *et al.* 1998). Although no freeze-dried cake is produced, the finely divided freeze-dried powder has superior properties in terms of reconstitution and may be preferable where novel delivery applications (such as inhalation therapy) are required (Maa *et al.* 2004).

As alternatives to freeze drying, supercritical fluid concentration (Moshashaee *et al.* 2000) and foam drying (Burapatana *et al.* 2004) may hold promise if the biological being processed can tolerate these treatments.

Table 20.4 Issues in freeze drying scale-up.

Scale of production	Focus
Laboratory scale	Selection of formulation parameters, optimization of freeze drying parameters
Pilot scale	Scale-up of fill volume/container dimensions, consistency of freeze-drying across batch
Pre-production	Final formulation and container dimensions/fill volume Consistency across batch and between batches

20.8 CONCLUSIONS

In this chapter we have aimed to give an introduction to the issues surrounding the formulation of lyophilized biologicals. The choice of liquid or lyophilized format and the selection and inclusion of stabilizing excipients are parameters that must be considered for each product early in its development. Similarly, the early use of thermal analytical methods (which are very powerful for idenitifying suitable freeze-drying parameters) is recommended in order to minimize failures during freeze-drying trials, shorten development times and optimise freeze-drying conditions. However, there is still a need to perform small-scale trials and to assess these rigorously, considering parameters such as biological efficacy, stability and moisture content. Degradation pathways under stressing environmental conditions (temperature and humidity) should be determined in order to arrive at a suitable formulation and presentation that will meet the requirements of the pharmaceutical marketplace. Once established, this formulation/presentation will need to be successfully scaled up and a sufficient number of validation runs performed to ensure that the same quality can be consistently delivered in production batches. The development of the production process and scale-up must precede, or at least run in parallel with, the preclinical development. Only then can problems due to poor formulation development or inadequate process control, which may otherwise only become apparent later in routine production-scale manufacture, be avoided.

ACKNOWLEDGEMENT

We thank Julie Fleming and Dr Roland Fleck for the freeze-drying microscopic images.

REFERENCES

Allison SD, Anchordoquy TJ (2001) In *Methods in Molecular Medicine*. Ed Findeis MA. Humana Press Inc., Totowa, NJ; Vol. 65: 225–252.

Alpar HO, Somavarapu S, Atuah KN, Bramwell V (2005) *Adv. Drug Delivery Rev.*; 57: 411–30.

Anchordoquy TJ, Carpenter JF (1996) *Arch. Biochem. Biophys.*; 332: 231–236.

Benedek K, Walker E, Doshier LA, Stout R (2005) *J. Chromtogr. A*; 1079: 397–407.

Bindschaedler E (2004) In *Freeze Drying/Lyophilization of Pharmaceutical and Biological Products*. Eds Rey L, May JC. Marcel Dekker, New York; second edition: 535–574.

Brave A, Ljungberg K, Boberg A *et al.* (2005) *Mol. Ther.*; 12: 1197–1205.

Breen ED, Curley JG, Overcashier DE, Hsu CC, Shire SJ (2001) *Pharm. Res.*; 18: 1345–1353.

Bristow AF, Dunn D, Tarelli E (1988) *J. Biol. Stand.*; 16: 55–61.

Burapatana V, Prokop A, Tanner RD (2004) *Appl. Biochem. Biotechnol.*; 114: 619–626.

Burke CJ, Hfu T-A, Volkin DB (1999) *Crit. Rev. Ther. Drug Car. Syst.*; 16(1): 1–80.

Cameron P, Murgatroyd K (Eds)(1997) *Good Pharmaceutical Freeze Drying Practice*. Interpharm/CRC Press, Boca Raton, FL, USA.

Cammack KA, Adams GDJ (1985) In *Animal Cell Biotechnology* Eds Spier RE, Griffiths BJ. Academic Press, London; Vol. 2: 251–288.

Carpenter JF, Pikal MJ, Chang BS, Randolph TW (1997) *Pharm. Res.*; 14: 969–975.

Chang BS, Kendrick BS, Carpenter JF (1996a) *J. Pharm. Sci.*; 85: 1325–1330.

Chang BS, Reeder G, Carpenter JF (1996b) *Pharm. Res.*; 13: 243–249.

Connelly JP, Welch JV (1993) *J. Parenter. Sci. Technol.*; 47: 70–75.

Corveleyn S, de Smedt S, Remon JP (1997) *Int. J. Pharm.*; 159: 57–65.

Danley RL (2003) *Thermochim. Acta*; 402: 91–98.

Davis SS (2002) *Am. Pharm. Rev.*; 5: 29–36.

Duddu SP, Zhang G, Dal Monte PR (1997) *Pharm. Res.*; 14: 596–600.

Edwards DA, Dunbar C (2002) *Annu. Rev. Biomed. Eng.*; 4: 93–107.

Eriksson JHC, Hinrich WLJ, de Jong GJ, Somsen GW, Frijlink HW (2003) *Pharm. Res.*; 20: 1437–1443.

FDA (1998) *Guide to Inspections of Lyophilization of Parenterals*, available at http://www.fda.gov/ora/inspect_ref/igs/lyophi.html

Fineberg SE, Kawabata T, Finco-Kent D, Liu C, Krasner A (2005) *J. Clin. Endocrinol. Metab.*; 90: 3287–3294.

Fleck RA, Green E, Matejtschuk P (2003) *GIT Imag. Microsc.*; 5(1): 11–13.

Ford AW, Dawson PJ (1994) *Biologicals*; 22: 191–187.

Franks F (1990) *CryoLetters*; 11: 93–110.

Fransson J, Florin-Robertsson E, Axelsson K, Nylhen C (1996) *Pharm. Res.*; 13: 1252–1257.

Grieff D, Rightseal WA (1968) *Appl. Microbiology*; 16: 835–840.

Hora MS, Wolfe SN (2004) In *Freeze Drying/lLyophilization of Pharmaceutical and Biological Products*. Eds Rey L, May JC. Marcel Dekker, New York; second edition: 309–324.

Hussain A, Arnold JJ, Khan MA, Ahsan F (2004) *J. Contr. Rel.*; 94: 15–24.

Jenke D (2002) *PDA J. Pharm. Sci. Tech.*; 56: 332–371.

Jennings TA (1999) *Lyophilization: Introduction and Basic Principles.* Interpharm Press. Denver, Colorado, USA; 165–252.

Jiang S, Nail SL (1998) *Eur. J. Pharm. Biopharm.*; 45: 245–257.

Jovanovic N, Bouchard A, Hofland GW, Witkamp GJ, Crommelin DJA, Jiskoot W (2004) *Pharm. Res.*; 21: 1955–1969.

Kett V (2001) *Eur. J Parenteral Sci.*; 6: 95–99.

Kett V, McMahon D, Ward K (2004) *Cryoletters*; 25: 389–404.

Kidron M, Dinh S, Menachem Y *et al.* (2004) *Diabetic Medicine*; 21: 354–357.

Lai MC, Topp EM (1999) *J. Pharm. Sci.*; 88: 489–500.

Lueckel B, Helk B, Bodmer D, Leuenberger H (1998) *Pharm. Dev. Technol.*; 3: 337–346.

Levine H, Slade L (1988) *Wat. Sci. Rev.*; 3: 79–185.

Levy J, Roizen MF, Morris JM (1986) *Spine*; 11: 282–291.

Li S, Patapoff TW, Overcashier D, Hsu C, Nguyen TH, Borchardt RT (1996) *J. Pharm. Sci.*; 85: 873–877.

Lin TP, Hsu CC (2002) *PDA J. Pharm. Sci. Technol.*; 56(4): 196–205.

Maa Y, Prestrelski SJ (2000) *Curr. Biopharm. Biotechnol.*; 1: 283–302.

Maa YF, Nguyen PA, Andya JD *et al.* (1998) *Pharm. Res.*; 15: 768–775.

Maa Y-F, Ameri M, Shu C, Payne LG, Chen D (2004) *J. Pharm. Sci.*; 93: 1912–1923.

MacKenzie AP (1964) *Biodynamics*; 9: 223–231.

Matejtschuk P, Gascoigne E, Dash CH (2000) *Br. J. Anaes.;* 85: 887–895.

Matejtschuk P, Rafiq S, Johnnes S, Gaines Das R (2005) *Biologicals*; 33: 63–70.

May JC, Wheeler RM, Elz N, DelGrons A (1992) *Dev. Biol. Stand.*; 74: 153–164.

Medicines Control Agency, MCA (2002) *Rules and Guidance for Pharmaceutical Manufacturers and Distributors.* HMSO, London.

Meseda CA, Garcia AD, Kumar A, Mayer AE *et al.* (2005) *Virology*; 339: 164–175.

Midcalf B, Phillips MW, Neiger JS, Coles TJ (Eds) (2004) *Pharmaceutical Isolators: A Guide to Their Application and Design.* Pharmaceutical Press, London.

Mills KHG, Cosgrove C, McNeela EA, Sexton A *et al.* (2003) *Infect. Immun.*; 71: 726–732.

Milton N, Pikal MJ, Roy ML, Nail SL (1997) *J. Parent. Sci. Tech.*; 51: 7–16.

Moshashaee S, Bisrat M, Forbes RT, Nyqvist H, York P (2000) *Eur. J Pharm. Sci.*; 11: 239–245.

Murgatroyd K (2001) *Eur. J Parenteral Sci.*; 6:(1) 21–26.

Murgatroyd K (1997) In *Good Pharamaceutical Freeze Drying Practice.* Eds Cameron P, Murgatroyd K. Interpharm/CRC Press, Boca Raton, FL, USA; 1–58.

Nena S, Avis KE (1992) *J. Parent. Sci. Technol.*; 47: 76–83.

Oetjen GW (1999) *Freeze Drying.* VCH Wiley Press, Weinheim.

Parenteral Society (1997) *Specification and Validation of Freeze Dryers*, Technical Monograph No 9, The Parenteral Society, Swindon, UK.

Patapoff TA, Overcashier DE (2002) *Biopharm.*; 3: 16–21.

Petersen J, Jani A (2001) *ASAIO J.* 47: 608–610.

Pijpers TFJ, Mathot VBF, Goderis B, Scherrenberg RL, van der Vegte EW (2002) *Macromolecules*; 35: 3601–3613.

Pikal MJ, Dellerman K, Roy ML (1992) *Dev. Biol. Stand.*; 74: 21–37.
Pikal MJ, Shah S (1992) *Dev. Biol. Stand.*; 74: 165–177. Discussion, 74: 147–179.
Raviprakash K, Ewing D, Simmons M *et al.* (2003) *Virology*; 315: 345–352.
Rey LR (1961) *Biodynamica*; 8: 241–260.
Robinson CJ (1999) *Dev. Biol. Stand.*; 97: 21–27.
Rowe RC, Sheskey PJ, Weller PJ (Eds) (2005) *Handbook of Pharmaceutical Excipients* Pharmaceutical Press/American Pharmaceutical Association, Grayslake, USA; fourth edition.
Rowe T (2005) *Cryoletters*; 26: 297–304.
Russo PL, Harrington GA, Spellman DW (1999) *Am. J. Inf. Control*; 27: 431–434.
Savage M, Torres J, Franks L, Masecar B, Hotta JA (1998) *Biologicals*; 26: 119–224.
Scheiermann N, Kuwert EK (1983) *Dev. Biol. Stand.*; 54: 347–355.
Searles JA (2004) In *Freeze Drying/Lyophilization of Pharmaceutical and Biological Products.* Eds Rey L, May JC. Marcel Dekker, New York; second edition: 109–145.
Searles JA, Carpenter JF, Randolph TW (2001a) *J. Pharm. Sci.*; 90: 860–871.
Searles JA, Carpenter JF, Randolph TW (2001b) *J. Pharm. Sci.*; 90: 872–877.
Thorpe R, Wadhwa M, Page C, Mire-Sluis A (1999) *Dev. Biol. Stand.*; 97: 61–71.
United States Food and Drug Administration (1987) *Guideline on Sterile Drug Products Produced by Aseptic Processing.*
United States Food and Drug Administration. *Guide to Inspections of Lyophilization of Parenterals* at http://www.fda.gov/ora/inspect_ref/igs/lyophi.html
Wang W (2000) *Int. J. Pharm.*; 203: 1–60.
WHO (2003) Good Manufacturing Practices for Pharmaceutical Products: Main Principles, *WHO Tech. Report Ser.*; 908: 37–89.
Willemer H (1999) In *Freeze Drying/Lyophilization of Pharmaceutical and Biological Products.* Eds Rey L, May JC Marcel Dekker, New York; 79–122.
Yoshioka S, Aso Y, Kojima S (1997) *J. Pharm. Sci.*; 86: 470–474.
Zhai S, Taylor R, Sanches R, Slater NKH (2003) *Chem. Eng. Sci.*; 58: 2313–2323.

21 Cell Preservation

R Fleck and B Fuller

21.1 INTRODUCTION

Establishment of international tissue banking initiatives, progress in tissue engineering, and stem cell and gene therapies place the medical community at the advent of an explosion in the range and number of biological medicines derived from cell cultures. International regulatory authorities will need to consider their control and regulation, including maintenance, preservation and supply of cell cultures and tissue-based products. Currently, cryopreservation is the only accepted method capable of delivering long-term stable preservation of viable nucleated mammalian or plant cells. This chapter is thus pertinent to these developments and will review what is known about cryopreservation, the benefits and limitations of the current methods, and new developments, some of which may be suitable for routine application. We will also discuss some of the new ideas on alternative approaches to cell storage, and describe the progress made so far.

21.2 REGULATORY PERSPECTIVES

The overall process of cell preservation and storage of cells for transplantation involves numerous stages requiring careful quality assurance (Figure 21.1). Cryopreserved stocks for the provision of human cells, tissues, and cellular tissue-based products have recently (2001) been considered by the United States Food and Drug Administration (US FDA) during discussions preceding the adoption of 21 CFR Part 1271 – *Current Good Tissue Practice for Manufacturers of Human Cellular and Tissue-based Products* (HCT/P); *Inspection and Enforcement.* In response to the draft rules a number of concerns were raised, notably by the Foundation for the Accreditation of Hematopoietic Cell Therapy (FAHCT). In essence the FAHCT disputed US FDA statements regarding morbidity and costs incurred following use of contaminated haematopoietic transplant products. Although, the prerequisite for safe sterile product is clear, the claims of contamination were considered to be exaggerated, with contamination of the haematopoietic transplant rarely occurring within the transplant collection itself, and generally being restricted to relatively non-pathogenic skin flora, which can be readily treated with appropriate antibiotics during the cell infusion. The US FDA also questioned the potential toxicity of dimethyl sulfoxide (DMSO, also commonly abbreviated to Me_2SO), a common cryoprotectant. The perceived DMSO toxicity can, however, be readily managed by limiting the amount of DMSO infused and, in any case, in most 'optimum' cryopreservation protocols removal of the cryoprotectant is common and often essential (see below).

In the final draft of the bill, sterilizing, preserving, and storage agents were exempt from the proposed regulatory controls, if the addition of the agent was not considered to raise new clinical safety concerns. The FDA and the European Directorate for the Quality of Medicines (EDQM)

Medicines from Animal Cell Culture Edited by G. Stacey and J. Davis

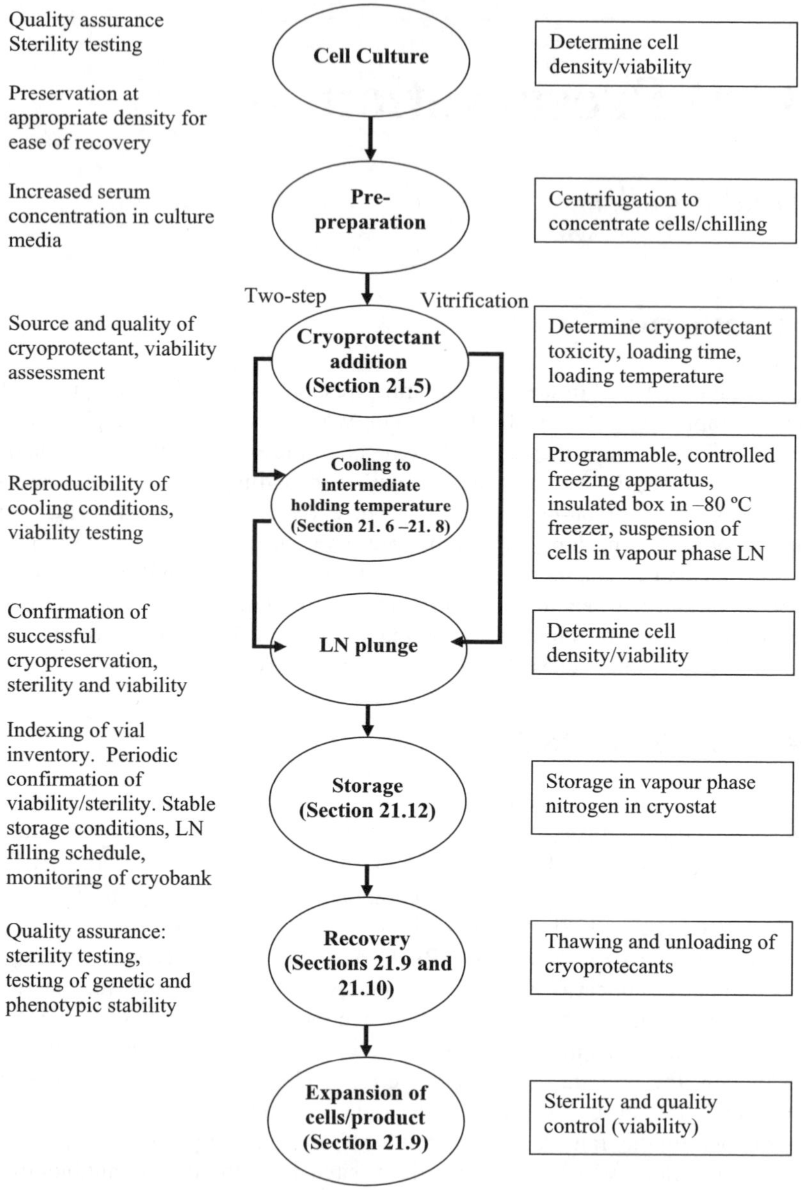

Figure 21.1 'Generalized' cryopreservation processes, a crib sheet for developing cryopreservation regimes for human cellular and tissue-based products.

do, however, hope to encourage the development of industry standards for safe use of sterilization, preserving, and storage agents critical for the prevention of contamination of HCT/P. At present, cryoprotectants (e.g. DMSO) are considered to be examples of substances that would generally be acceptable. Furthermore, in the draft guidance for reviewers of *Human Somatic Cell Therapy Investigational New Drug Applications*, the US FDA suggests that processes critical to product safety be described, e.g. cryopreservation, storage, and recovery of the master cell bank (MCB), 'including information pertaining to cell density, number of vials frozen, storage temperature, and cell bank location'. In addition, procedures to ensure and test genetic and phenotypic stability of cells from the MCB after multiple passages, as well as the viability of cells after cryopreservation, should be documented, together with a stability protocol to ensure that the product is stable during the period of cryopreservation. With regard to the final use of the product, the US FDA suggests, 'if the final product is frozen prior to use, the sponsor should perform testing on the product prior to cryopreservation with results available prior to patient administration. However, if the product undergoes manipulation (e.g. washing, culturing) after thawing, particularly if procedures are performed in an open system, the sponsor might need to repeat sterility testing'. The issues of sterility are clearly important, particularly in the case of 'Medicines from Cell Culture'. Further to considerations of HCT/P, consultations by the Human Fertilisation and Embryology Authority (HFEA) have also emphasized the need for a review of how fertility clinics in the UK ensure safe, effective cryopreservation and storage of potentially infective material (e.g. gametes and embryos) (Tomlinson & Sakkas 2000).

More recently, the British Standards Institute, Publicly Available Specification (PAS), PAS 83:2006 came into effect (BSI 2006). This document draws together in one place "guidance on codes of Practice, standardized methods and regulations for cell-based therapeutics – from basic research to clinical application". Although covering a diverse range of processes and codes of pratice, preservation of cell lines is discussed. Of particular interest is Directive 2004/23/EC of the European Parliament and of the Council of 31st. March 2004 on setting standards of quality and safety for the donation, procurment, testing, processing, preservation, storage and distribution of human tissues and cells. The directive does not diescribe cryopreseravation specifically. However, the directive is applied to tissues and cells including haematopoietic peripheral blood, umbilical-cord (blood) and bone-marrow stem cells, reproductive cells (eggs, sperm), foetal tissues and cells, and adult and embryonic stem cells, and in specifically indentifying perservation and storage, the directive can be considered to apply to cryopreservation of human cell cultures in general. The directive is largely concerned with "accreditation designation, authorisation or licensing of tissue and cells preparation processes". Article 21 of the directive deals specifically with "tissue and cells storage conditions", but it does not specifically mention cryopreservation, cryoprotectants or processes. Never the less, the improtance of purity, sterility, absence of contaminating agents, storage vessel integrity and the processes by which practices and methdologies are controlled are critically improtant if cryopreservation is to adopted as the preservation method of choice across the biopharmaceutical and regenerative medicine sectors. It must therefore be assumed that increased regulatory processes targeted at cryopreservation will evolve in the near future to meet the demands of emerging medicines from animal cell culture.

21.3 HISTORICAL PERSPECTIVES ON CRYOPRESERVATION

For eukaryotic cells, life processes are largely dependent upon water. The processes of energy production, synthesis and repair in cells are carried out in aqueous environments separated into ordered domains by hydrophobic barriers presented by cell membranes. These structures and processes must be protected by the storage technique if viability, in the frozen system, is to be maintained.

At normal body temperatures, a positive energy of activation exists for biochemical reactions, so the simplest way to slow life processes is to lower the free energy of the reactants by cooling. Cooling to refrigeration temperatures (down to 0 °C without freezing) is effective in maintaining viability only for a matter of hours or days, depending on the complexity of the cells. The biochemical reactions do not stop at these temperatures, and metabolic changes (for example, production of lactate from glucose) can be easily detected in cooled cell cultures. There is a wealth of information on the effects of hypothermia on mammalian systems, which is beyond the scope of the current discussion (Fuller 1999, Taylor 2006), but we have to proceed to far lower temperatures to be able to retain cell viability. The concept of 'frozen in time' or 'suspended animation' is very apt, since we must aim for storage at temperatures where molecular interactions are inhibited over the storage periods required (years). This can only be achieved at deep subzero temperatures; the over-arching problem is the central role of water in cell processes, and the physical laws that explain the phase transition of water to ice.

21.4 THE PROBLEMS OF ICE FORMATION FOR CELLS

The thermodynamics of dilute solutions can be used to describe the changes occurring in suspensions of cells during freezing. The chemical potential of an ideal, dilute solution is related to temperature, pressure and concentration of dissolved solutes. On cooling, ice invariably nucleates outside cells in the surrounding medium. (The formation of ice inside cells is a delayed event in the process of cooling, as will be described below, and may never happen in some circumstances.) The temperature at which this occurs will be dictated by the content of solutes, and the process of nucleation itself. The physical state of a particular medium at a range of temperatures may be expressed in a 'phase diagram' (Figure 21.2). For cell cultures in basal culture medium, the equilibrium freezing (or melting) point of the saline mixture is about $-0.55\,^{\circ}C$ (this is also the temperature at which the enthalpy or heat release of melting can be detected), but at this temperature nucleation events are infrequent (Mazur 2004). In practical terms, solutions often will not freeze until cooled several degrees below this equilibrium melting point – the solution is said to be 'super-cooled', or more correctly, 'under-cooled' (Franks 1982). This is because ice formation requires a nucleation event that creates a solid–liquid interface. The reduction in temperature increases the statistical likelihood of ice nucleation such that the solution will inevitably form ice, as the super-cooled state is metastable and the cooled water will not remain in the liquid form indefinitely (unless other physical parameters in the system are altered). Nucleation events are most likely to be catalysed by small particles in the solution that assist the ordering of the water molecules, an event called heterogeneous nucleation. In specially prepared samples subjected to filtration and maintained in small droplets, it is possible to cool pure water to much lower temperatures close to $-40\,^{\circ}C$. At this temperature homogeneous nucleation occurs, catalysed by molecular interactions between the water molecules themselves (Franks 1982). However, this is not a practical way of cooling cell suspensions.

The changes in physical orientation of water molecules in liquid water and in ice are beyond the current discussion (see Taylor 1987, and Franks 2003, for descriptions) but in simple terms, water molecules in ice become hydrogen bonded to a group of four other water molecules in a tetrahedral orientation, in such a way that the solvent interactions with other solutes in the original solution are restricted – ice crystallizes as pure water, excluding the ions and other solutes, and concentrating them in the residual partially frozen solution. This increases the osmolality of the residual solution in the extracellular compartment, lowering the chemical potential of the external solution relative to that of the intracellular water. Water then moves down this chemical potential gradient by osmosis, causing osmotic stress to the cells (due to the high ionic strength of the extracellular medium) and a reduction in cell volume. As cooling proceeds further, ice growth continues, increasing the solute content of the residual solution such that concentrations in excess of 5 molar can be reached as the temperature descends below $-25\,^{\circ}C$ (Lovelock 1953, Mazur 2004, Fowler & Toner 2005).

Mazur (1963) frist formulated four simultaneous equations to describe this process, and a full discussion appears in Mazur (1990). The dominant factors that dictate how a particular cell type will respond to subzero cooling include those describing the osmotic characteristics of the cell (the water permeability of the cell membrane and the surface area/volume ratio of the cell), the change

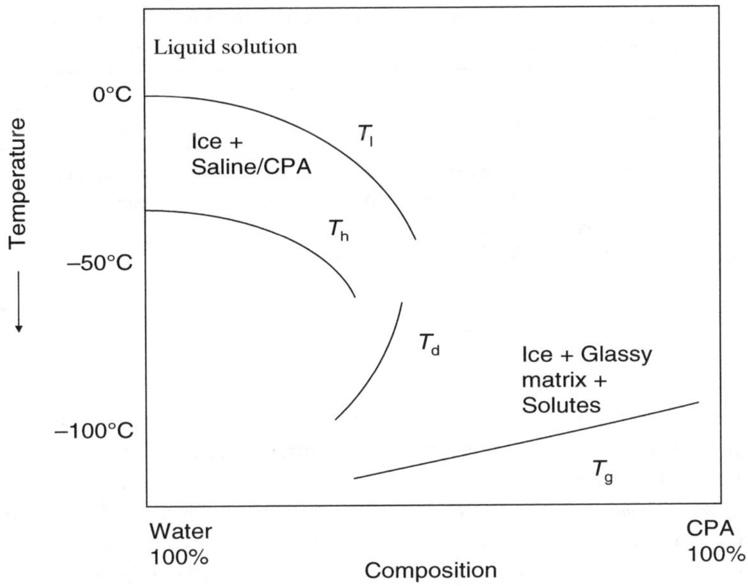

Figure 21.2 A hypothetical phase diagram for an aqueous solution of cryoprotectant (CPA) during cooling to low temperatures. T_l – liquidus (or equilibrium melting) curve; T_h – homogeneous nucleation curve; T_g – glass transition curve; T_d – devitrification curve on warming. For explanations, see Section 21.5.

of water permeability with cooling, the time and temperature of the cooling event, the change in vapour pressure and the change in solute contents of intracellular and extracellular solutions as the fraction of ice increases. By measuring or computing these factors and inserting their values into the equations, Mazur has shown that the predicted outcomes during cell cooling can, to a reasonable degree, match the observed changes (Mazur 1990). If cooling rates are low, and the water permeability of the cell membrane remains balanced with the changing vapour pressure of the solution, the cells can maintain an effective osmotic balance with the concentrated ice matrix. If the cooling is faster, such that super-cooled water remains inside the cell, then there is a strong likelihood that this will nucleate intracellular ice, which is commonly associated with lethal cell damage. These are the 'horns of the dilemma'; cool slowly and the cells are exposed for prolonged periods to a hypertonic medium more than 30 times the osmolality of media needed for normal survival (such osmotic stress can cause irreversible breakdown in membrane structures and destabilize proteins); cool rapidly and intracellular ice can be formed. This has come to be known as the '2-factor' hypothesis for cell injury during cryopreservation (Leibo *et al.* 1970) and will be described below.

On further cooling, physical changes in the frozen solution can be detected even below −80 °C (see Figure 21.2). For simple saline solutions, the increase in local sodium chloride concentration as ice forms leads to a point of solidification of the residual salt/liquid/ice mix at the eutectic temperature (in this case, –21.6 °C the temperature at which the eutectic mixture melts). However, in biological samples, such as cell suspensions, containing complex mixtures of phospholipids in membranes, proteins, and other high molecular weight solutes, it is unlikely that such a clear eutectic exists. The evidence from calorimetric and thermal analyses (Boutron *et al.* 1986; MacFarlane 1987), and from observations using a cryomicroscope (Rall *et al.* 1980) suggests that the residual, highly concentrated, viscous mixture undergoes a transition to a glassy matrix at temperatures between −100 to −120 °C (Figure 21.2). It is not until this state has finally been achieved that true, long-term preservation of viability can be maintained. This is thought to be one reason why many

Table 21.1 A list of common cryoprotectants. Other solutes have been used in occasional circumstances, but those on the list have been used in more than one application.

Low molecular weight –OH solutes	Sugars	Polymers	Other compounds
Glycerol	Sucrose	Hydroxyethyl starch (HES)	Dimethyl sulphoxide (DMSO, Me_2SO)
Ethylene glycol	Raffinose	Polyvinyl pyrrolidone (PVP)	Acetamide
Propylene glycol	Trehalose	Dextrans	
Methanol	Sorbitol	Bovine serum albumin	
Ethanol			

cells can be stored only for an intermediate time (few weeks or months) at −80 °C; there is still slow but continuous molecular degradation at that temperature, which is incompatible with cell viability. From these descriptions it will be obvious that control of the conditions of cooling are an important part of successful cryopreservation. However, before discussing this in more detail, it should be pointed out that mammalian cells cannot survive either fast or slow cooling without the addition of protective or compatible solutes, also known as cryoprotective agents (CPA), except in a very few rare and specialised cases.

21.5 THE ROLE OF CRYOPROTECTANTS IN CELL FREEZING

The history of successful cryopreservation is relatively short. It is less than 60 years since the seminal paper on recovery of viable cells (in this case chicken semen) was published (Polge *et al.* 1949), and the story of that success is also the story of the identification of cryoprotectants. It was only by the use of relatively high concentrations of glycerol that this first success was achieved. Since that time, numerous investigations have focused on agents that can be used as CPA. Working on cryopreservation of red cells, Lovelock (1954) identified a central mechanism of protection. He hypothesized that agents such as glycerol act as colligative effectors (dependant on the ratio of the number of particles of solute and solvent in the solution and not the identity of the solute) at the concentrations used in cryopreservation; they could be added in high concentrations (in excess of 10 % w/v). At these levels they would significantly reduce the quantity of ice formed at a given sub-zero temperature (i.e. act by a solute depression of freezing point). Lovelock identified a range of neutral solutes, including DMSO, which possess such activity, and which (equally importantly) could be tolerated by cells without lethal toxic effects during short pre-freezing exposure. Another important point was that the CPA needed to be present in both intra- and extracellular solutions, and thus the cell membrane permeability for a given CPA is of great significance. Agents such as glycerol or various sugars (including sucrose) also increase the viscosity of the solution at low temperatures, and this may effectively inhibit ice crystal growth on a kinetic basis, whilst enhancing the likelihood of achieving a vitreous transformation. This effect may also be achieved by using high concentrations of polymeric agents such as polyethylene glycol, and is termed 'non-ideal solution' since the ice content at a given low temperature is still lower than predicted, but the effect cannot be assigned to a true molar depression of the freezing point since these solutes are of high molecular weight (Fuller 2004). Thus, for mammalian cells, CPA can be divided into high and low molecular weight categories (Table 21.1), but it is essentially true for standard cryopreservation procedures that a permeating low molecular weight CPA is essential, even if it is combined with other high molecular weight agents. No routine cryopreservation protocols for nucleated mammalian cells have been formulated using only extracellular CPA.

Figure 21.2 is a hypothetical phase diagram for the changes occurring during freezing with a CPA. The liquidus curve (T_l) follows the increasing concentration of the residual solution as ice forms in the system under equilibrium conditions. The homogeneous nucleation temperature (T_h) is lowered by increasing solute concentration in the mixture, but ice formation will occur by this route at temperatures close to −50 °C. At high concentrations of CPA, the glass transition temperature (T_g) increases, but devitrification can occur during warming (T_d). It is theoretically possible to cool and produce a stable glass without any ice formation, but this requires CPA concentrations in excess of 80% w/v, which are very toxic to cells.' In practical cryopreservation (using initial CPA concentrations of 10–20 % w/v), the final mixture is thought largely to comprise ice with a small residual glassy matrix of CPA solutes and hydrates.

It can be beneficial to use additive combinations of CPA to maximize survival. For example, when cryopreserving some mammalian cells types, such as pre-implantation embryos, a cell-penetrating CPA such as DMSO may be used in combination with a sugar such as sucrose (Stein *et al.* 1993). This sugar may act in several ways beyond the simple colligative effect. As a non-penetrating solute, the sugar will act as an osmotic agent, reducing the water content in the intracellular compartment even before ice formation begins. As a solute with a high viscosity at low temperatures, the sugar will increase overall viscosity of the mixture during freezing and facilitate the transition to a glassy state (through its high T_g', maximally freeze-concentrated T_g of the frozen system). A similar additive effect can be gained using polymers such as hydroxyethyl starch (HES), in combination with traditional CPA such as DMSO in cryopreservation of bone marrow stem cells (Stiff *et al.* 1983). In some cases, addition of polymeric agents may permit a reduction in the required concentration of cell-penetrating CPA, which is of benefit if toxic effects of the low molecular weight agent might be causing problems in the system (Stiff *et al.* 1983; Stiff 1995).

Another effect of CPA (indirectly related to amelioration of dehydration) has been the ability to stabilize structures by preferential exclusion from the hydration layers of important cellular components, such as proteins or phospholipid bilayers. The theory was originally described by Timasheff and colleagues (Timasheff *et al.* 1976, Arakawa & Timasheff 1985) and applied to the effects of CPA by Crowe and colleagues (Crowe *et al.* 1990). Because exclusion is entropically unfavourable, it drives the stabilization of the macromolecules, despite extreme dehydration during slow freezing which would act to cause molecular destabilization. Another effect, related again to extreme dehydration, has been linked to the hypothesis that some sugars, such as trehalose, can act as 'water replacement' molecules, protecting phospholipid bilayers from undergoing destabilization during slow dehydrative freezing (Anchordoguy *et al.* 1987). Sugars such as trehalose have been identified in the natural world in highly selected organisms (such as polar insect species) which can survive freezing (Block 1990), and presumably their effects will be similar, at least in part, during laboratory cryopreservation. However, a note of caution (and a probable reason why trehalose cannot be used as sole CPA for routine cell banking) is sounded by the observation that in nature trehalose must be present in the intracellular compartment, not only in the surrounding environment, for survival – the insects, observed by Block, synthesize the sugar as part of a cold hardening process. Since mammalian cells are only poorly permeable to such sugars, simply adding them to cells in suspension cannot provide full protection against freezing damage. However, novel approaches to this problem are being considered and will be discussed later.

Yet another benefit from addition of CPA may arise from their ability to act as free radical scavengers; for example DMSO is a potent radical scavenging agent, and free radical 'signatures' have been detected in plant cell cultures following cryopreservation (Fleck *et al.* 2000, Benson & Bremner 2004). It is likely that such free radical formation results from disrupted metabolism in cells recovering from cryopreservation, rather than from a primary effect of ice formation *per se*, but this added effect of CPA may play a consistent small part in successful recovery in the immediate post-thawing period. It will be seen that CPA have a diverse range of effects, which can be used in additive fashion for specific cell populations. However, although many of these documented effects are at least partly understood, there is at present no specific predictive test that can be used prospectively to identify an optimal CPA regime, without entering into preliminary testing on cell recovery before full-scale application. It will also be noted that many of the protective effects have been documented during slow dehydrative freezing. The role of CPA in protection during rapid cooling, when intracellular ice may be formed, will be discussed later.

Whatever protective effects are sought by use of chosen CPA, osmotic and chemical toxicity are generic problems with the use of these agents at the necessary high concentrations. The addition of the permeating CPA (usually at concentrations of between 10–15 % w/v) imposes an osmotic stress on the cells, which although not as extreme as that experienced during the cooling itself, can exceed the tolerable limits for the cell. For osmotic excursion (dehydration) of a mammalian cell through the addition of glycerol (for this example a murine oocyte), there was a large initial shrinkage as water flowed out of the cell, to be replaced more gradually during the period when the glycerol started to permeate into the cell. This biphasic effect results from the fact that, in general, cells are much more permeable to water than to neutral solutes such as those used for CPA (Paynter *et al.* 1999, Fuller & Paynter 2004). Having protected the cell during cryopreservation, the opposite procedure (removal of the CPA and return of the cells to normal culture medium) will be the desired goal (this will be discussed later). In this instance, a mirror-image effect will occur; water will now flow rapidly by osmosis into the CPA-loaded cell, causing acute cell swelling, which is only gradually reversed as the CPA diffuses out of the cell. It has been established by empirical observation that these osmotic transients are an additional damaging factor in the overall cryopreservation regime. Cell volume excursions beyond about plus or minus 30 % of normal isotonic volume are poorly tolerated (Armitage & Mazur 1984; Wusteman & Pegg 2001). Thus an integral part of designing a successful protocol is to be aware of these changes. The problem can be reduced by diluting CPA out in small sequential steps rather than by direct immersion in the full

Figure 21.3 Examples of cryotechnologies, equipment for cryopreservation and cryopreservation protocol development. Top left: cryomicroscope; top right: cryostat with vapour phase cooling device; bottom left: programmable freezer; bottom right: cryopreserved cells stored on cryorods in a vapour phase LN Dewar. (Photographs by kind permission of N.I.B.S.C.).

final volume. The limits of the steps can be measured for a given cell type by undertaking osmotic measurements of water and solute permeability, and prospectively calculating the optimal range of changes (Paynter *et al.* 2001; Wusteman & Pegg 2001). A 'rule of thumb' that can be used for most cell types is to avoid addition or removal of CPA in steps of greater than 0.5 M, allowing time (again as a rough guide – 5 min at temperatures of 20 °C) between steps. Another way to reduce osmotic damage during CPA removal is to include an osmotic buffer in the initial washing steps. For the commonly used concentrations of CPA (around 10 % w/v), addition of a non-permeating sugar, such as sucrose at between 0.1 to 0.3 M, can avoid excessive cell swelling (Paynter *et al.* 2001). It is certainly true that the effects of CPA exposure and removal should be checked against the specific functional recovery of the cell line of interest before any freezing steps are introduced. It is also true that the avoidance of these large transient osmotic stresses may be even more critical after the cooling and rewarming procedures, which will already have subjected cell components such as the plasma membrane to damage (see below).

21.6 SLOW COOLING AND ITS EFFECT ON CELL RECOVERY

Most routine cryopreservation procedures employed in culture collections (e.g., American Type Culture Collection, ATCC; Deutsche Sammlung von Mikroorganismen und Zellkulturen, DSMZ, and European Collection of Cell Cultures, ECACC) have been based on use of cooling rates in the region of 1 °C per min. Such rates are relatively easy to achieve on a consistent basis, using either programmable freezing machines (Figure 21.3), or passive cooling in insulated boxes in a −80 °C freezer, or by suspending vials in a cryogen such as the vapour phase above liquid nitrogen, but in fact their success results from more than practical convenience alone. For most nucleated mammalian cells, the cell membrane permeability to water and the changes of this parameter during cooling dictate that by cooling at these rates, the cell water content remains effectively in equilibrium with the increasing osmotic potential of the ice–water matrix. Therefore, the cells become dehydrated and there is little remaining intracellular water to participate in intracellular ice formation. (This is not true for some specialized or large cells such as mammalian oocytes, which require even slower rates of cooling.) For example, McGann *et al.* (1987) made measurements of cell membrane transport parameters on cooling for bone marrow progenitor cells and, using coupled equations (similar to those described by Mazur 1990), found that rates of between −1 °C to −3 °C/min permitted this type of equilibrium freezing during cooling. Rates faster than −10 °C/min were predicted to have a high incidence of intracellular freezing and, in practice, cryopreservation with these faster cooling rates was associated with poor cell survival. Thus slow cooling over the temperature range where cell dehydration can proceed (down to about −50 °C, although this end temperature is not known exactly) allows successful cryopreservation when these samples are eventually transferred to liquid nitrogen. Many effective slow cooling regimes employ cooling at 1 °C/min to a 'safe' intermediate temperature (typically −70 °C), from which the ampoules can be directly transferred through the final glass transition temperature (at around −100 °C) to the storage temperature (usually in either the vapour or liquid phase of nitrogen). Thus, even if cell survival is achieved by careful selection of CPA and cooling conditions, the cells will experience the effects of dehydrative freezing. It will be the reversibility or irreversibility of these events, the 'damage load', which will dictate survival or death on an individual cell basis.

The early studies on injury from slow freezing focused on the possibility that the high external salt concentration would drive a low temperature 'salt loading' of the cells, resulting in osmotic imbalance, excessive swelling and lysis on thawing (Lovelock 1953). The protection offered by CPA such as glycerol was afforded by the colligative effect reducing the salt concentration and thus avoiding the damage. It has subsequently been argued that under the conditions of slow equilibrium cooling, there would be little difference between internal and external solute concentrations, so salt loading would be unlikely (Pegg & Daiper 1988). Merryman (1970) suggested that damage could result from dehydration to osmotic potentials that would effectively cause maximal shrinkage beyond a minimum cell volume, but again this has not been supported by subsequent biophysical modelling. Another postulated effect of dehydrative freezing has been linked to the large volume of external ice formed during slow cooling. This has been shown to leave the cells trapped in narrow channels between large ice crystals, causing dense packing of cells and close apposition to the ice; this has been called rheological damage (Mazur & Cole 1989). Again, by reducing the amount of ice during slow cooling, CPA reduce the narrowing of inter-ice channels. There is certainly some effect of 'cell packing' during slow cooling (which translates at a practical level to selection of optimized cell concentrations during routine cryopreservation), since it has been shown in studies on erythrocytes (red blood corpuscles) (Nei 1981, Pegg & Daiper 1983) and nucleated cells such as hepatocytes (De Loecker *et al.* 1998) that increasing the packed cell volume of the suspension to be cryopreserved reduces the success of the procedure.

Other studies on slow-cooling injury have focused on the plasma membrane as a major site of injury (given that freezing damage on thawed cells can often be detected very soon after warming by microscopy or staining techniques that pick up an increase in membrane permeability). In work on isolated plant protoplasts, Steponkus and colleagues (1992) have shown that dehydrative freezing can result in 'blebbing' and 'budding off' of the plasma membrane, caused by endocytotic vesiculation of the plasma membrane when the cells become excessively shrunken due to osmotic contraction. (NB: this is also a morphological feature of advanced cell death by apoptosis). In more extreme dehydration, changes in the molecular architecture of the phospholipid bilayer can result in loss of the lamellar structure of the membrane, resulting in appearance of non-bilayer 'tubes' or micelles, which destroy the essential selective permeability of the membrane (Wolfe *et al.* 1983, Webb & Steponkus 1990, Steponkus *et al.* 1992, Pearce 2004). It has also been shown that when cells are compressed by extreme dehydrative freezing into stacks of 'contorted bilayers', there can be fusion events between outer and inner bilayers and between bilayers of adjacent cells, resulting in irreparable damage on thawing (Anchordoguy *et al.* 1987; Crowe *et al.* 1990; Uemara & Steponkus 1995). It is more than likely that several of these effects will impinge on cells during slow freezing, so care must be taken at all stages of the cryopreservation regime (including post-thaw dilution steps, handling and centrifugation) to maximize recovery of these partially stressed cells. (The conditions of warming are also important and will be discussed separately below.) A comprehensive discussion of the events in slow cooling injury during cryopreservation has recently been provided by Mazur (2004) and Muldrew *et al.* (2004).

21.7 RAPID COOLING AND ITS EFFECT ON CELL RECOVERY

From the above, it will have become obvious that for cell survival during routine cryopreservation, judicious selection of CPA and slow cooling will be the norm. Thus, it is only necessary to discuss rapid cooling briefly to give an insight into what is known of the damage mechanisms. Rapid cooling is employed in the technique commonly referred to as 'vitrification' (see below), but in that situation many aspects of the protocol are different, and should not be confused with the changes induced by cooling cells exposed to traditional CPA (at the usual concentrations of about 10% w/v) by, for example, plunging an ampoule of cells directly into liquid nitrogen. As mentioned above, the major problem with rapid cooling and departure from the equilibrium conditions of cell osmotic response to external high solute in the ice matrix, is the residual unfrozen cell water, which will be in the supercooled state. The likelihood of intracellular ice formation increases with the degree of supercooling, so it is not possible to pass through the nucleation 'danger zone' between about −20 and −50 °C without ice forming inside the cells. The presence of ice in cells under conditions of rapid cooling has been demonstrated by freeze-fracture electron microscopy (Farrant *et al.* 1977) and by cryomicroscopy (Rall *et al.* 1983, Toner *et al.* 1990, Mazur *et al.* 2005) where intracellular 'black flashing' has been ascribed to refraction of light by ice crystals. Since the weight of evidence suggests that cell membranes inhibit the passage of ice crystals from the external ice matrix, the debate in this area has centred on how ice can nucleate inside the cell. In the 1960s, Mazur proposed that ice could nucleate through existing pores in the plasma membrane (Mazur 1965). For many years this was discounted, based on calculations of the necessary sizes of pores and the surface of curvature of growing ice crystals, but the more recent discovery of protein 'water pores' or aquaporins (Venkmann *et al.* 1996) in some mammalian cells has revitalized this concept. Another theory has focused on the ultrastructural and molecular changes in severely dehydrated membranes (see above) that have been proposed to expose 'nucleation sites'. This has been called the membrane-catalysed nucleation theory (Karlsson *et al.* 1994). Yet another group (Muldrew & McGann 1994, Muldrew *et al.* 2004) presented evidence that at high cooling rates (and rapid growth of the extracellular ice matrix) the osmotically driven flux of water out of the cell is so rapid that its movement can impose frictional forces ($>1 \times 10^{-4}$ atm) capable of rupturing the membrane (allowing ice crystals then to grow in). At a practical level, all these theories have been linked with cell death under the studied conditions, so avoidance of these in routine cryopreservation is strongly recommended. There is some evidence that cells may tolerate a very small fraction of intracellular ice without death (Rall *et al.* 1980), but it is such a small amount that it cannot be accurately predicted across a cell population, so effective cell recoveries would be likely to vary considerably. However, an important development will be in linking the traditional cryobiological parameters of cell death with current and rapidly developing knowledge of cell biology.

21.8 VITRIFICATION

A large part of the previous discussion centred on the achievement of a 'glassy' state at low temperatures for long-term maintenance of viability. It has been known for a long time from a theoretical basis that it should be possible to vitrify aqueous solutions (Luyet & Gehenio 1940), but only at extremely high rates of cooling (tens of thousands of °C/min), which would be impractical for routine cell banking, and may also be unrealistic with liquid nitrogen due to the insulation provided by nitrogen gas 'boil off'. The concept was revitalised by Fahy *et al.* (1984) and Rall and Fahy (1985), who made extensive studies of the use of high concentrations of CPA to permit vitreous transformation at cooling rates that could be used in everyday applications. It has been shown that for many CPA such as DMSO, glycerol and other polyols, concentrations in the region of 40–60 %w/v will allow low temperature vitrification, but still require cooling rates of several hundred °C/min. These high solute mixes increase viscosity (favouring glass formation) and have a strong colligative action (reducing the statistical likelihood of ice nucleation during rapid cooling). This has been successfully applied to the cryopreservation of small numbers of valuable cells such as animal or human embryos in reproductive medicine (Bernard & Fuller 1996, Fuller & Paynter 2004), where very small volumes of medium can be cooled, for example, in thin plastic straws (Chen *et al.* 2000). The technique has not yet been applied (to our knowledge) for storage of cell lines in banking or biotechnology but is now being used for embryonic stem cells (Reubinoff 2001, Fujioka *et al.* 2004). Also, in the past few years, new techniques for rapid cooling of cells in very small droplets have been developed (Lieberman *et al.* 2002) and vitrification is becoming more widely used in reproductive cryobiology. One major issue is the toxicity of the CPA to particular cell types at these high initial concentrations (around 40 % w/v), which undoubtedly have osmotic and chemical effects on cells. Additionally, insufficiently rapid warming cycles may permit growth of ice crystals, which in turn may reduce post-thaw viabilities (see below).

21.9 THE IMPORTANCE OF WARMING RATES

The importance of the control of warming rates stems from the necessity to bring the sample of cryopreserved cells back through the biophysical events as the glassy matrix 'devitrifies' (around −100 to −80 °C), water molecules begin to mobilize within the ice–solute matrix (above −60 °C), and ice recrystallization becomes significant (particularly inside the cells) on the time scale of the thawing process (above −40 °C). These temperature ranges are only approximate, but physical events dictated by the phase diagram of the particular ice/solute/CPA mixture in a given cell sample are unavoidable. For samples that have been cooled 'rapidly' (i.e. under non-equilibrium conditions), and where there is a high probability that ice nucleation centres have formed inside the cell (even though these may be down at the level of molecular clusters too small to detect by microscopy), then there is a high statistical likelihood that ice crystals will grow during recovery from very low temperatures. This can be described as 'ice formation during warming' (Rall *et al.* 1980). The obvious approach is to attempt to bring the cell sample up through the 'danger zone' for ice crystal growth as fast as possible, and it has been known for some time that rapid thawing rates improve cell survival after rapid cooling (Taylor *et al.* 1987, Fowler & Toner 2005). There is strong supporting evidence for the harmful effects of ice crystal growth during warming in cells (such as murine embryos) that are large enough to permit visualization by cryomicroscopy (Rall *et al.* 1984, Mazur *et al.* 2005) (Figure 21.3). However, factors such as the geometry and volume of the cell sample, and the heat transfer properties of the ampoule or tube used for cryopreservation, combine to restrict the absolute rate of rewarming achievable. The most common method of warming is to plunge the cell samples into a water bath at 37 °C, with shaking to avoid local thermal gradients in the water bath, but for samples of about 1 ml volume, this restricts the maximum warming rates to a few hundred degrees celsius per minute. It has been calculated for some mixtures of solutes that to avoid safely both the formation of ice nuclei and significant ice crystal growth in the temperature zone above −60 °C, samples should be warmed at rates in excess of 1200 °C/min (Pegg 1988). Thus it will be obvious that the warming rates that can be achieved in routine cell banking are right on 'the edge' of potential failure, reinforcing the view that care and vigilance must be taken in all steps of the protocol.

The need for rapid warming is even more acute where samples have been stored using vitrification protocols. We have discussed the fact that practical cell vitrification methods are limited to what is effectively 'quasi-vitrification'; establishment of a low temperature glass within which is a potentially high number of ice nucleation centres. In this situation, thawing must be even more rapid, and this is one reason why sample volumes are kept to a minimum (a few μl), whilst often the samples have been cooled, not in traditional cryo-tubes, but on receptacles such as electron microscope grids (Martino *et al.* 1996) or tiny plastic Loop, which have a very high heat transfer capacity in comparison with glass or plastic ampoules. There have also been attempts to use agents such as 'antifreeze' or 'thermal hysteresis proteins' (proteins that occur in freeze-tolerant species of insects or plants (Zachariassen & Zachariassen 2000) and have an ability to block the addition of water molecules to the surface of growing ice crystals on a kinetic basis) as additives to vitrification media. These have achieved some improvement of recovery (O'Neil *et al.* 1998), but the problems have so far not been fully resolved.

The optimal selection of warming rates for 'slowly' cooled cells are more difficult to predict. The reasons for this are not fully understood, but may in part relate to biophysical characteristics of a given cell type, including the membrane water permeability (and the rate of increase of this at sub-zero temperatures during warming) and the cell surface-to-volume ratio. For most cells (including those commonly encountered in routine cell banking) preserved by slow dehydrative 'equilibrium' cooling, cell recoveries are improved by rapid warming (Mazur 1963; McGann & Farrant 1976, Muldrew *et al.* 2004). This improvement has been linked to factors such as avoidance of time for excessive intracellular ice growth (see above) or an additional period of dehydration damage (such as membrane reorganization or fusion – see above) during slow warming, but unequivocal evidence for either theory is at present lacking. For some selected cell types (such as mammalian embryos or oocytes), there is evidence that an intermediate slow warming rate (in the region of 8 °C/min) is beneficial (Whittingham *et al.* 1972). The reason for this may lie in the rate of movement of water or solutes across the plasma membrane in the situation where the cell has been excessively shrunken during cooling, with a resultant high osmotic potential for the remaining intracellular contents. Rapid warming (liberating free water molecules into the diminishing ice matrix around the cells) could encourage a dramatic influx of water into the cell, causing membrane damage itself (see the 'membrane rupture' theory of rapid cooling damage, Muldrew *et al.* 2004), or lysis by 'overhydration', particularly if there has been a solute loading effect during the slow cooling. Warming more slowly may permit rehydration to proceed at rates that avoid these damaging effects. However, this slow warming benefit appears to be a characteristic for a limited number of large cells, as far as we are currently aware.

21.10 CPA UNLOADING AND RECOVERY

The processes of preservation by freezing/low temperature and the thawing of material have been described in detail above. Following thawing, the cryoprotectant must, however, be unloaded and in the case of a cell line, the cell culture expanded in number. Commonly, cryoprotectant can be immediately diluted out of the cell suspension by adding fresh tissue culture medium (e.g 20 ml into 1 ml of sample) to dilute the cryoprotective compound. Alternatively, the cryopreservative can be removed by centrifugation of the thawed cell suspension and resuspension in fresh medium. However, these methods risk osmotic injury (Wolf *et al.* 1983). This can be reduced by the adoption of stepwise unloading, where fresh medium is added to the cryovial to dilute the cryoprotectant by about 50 %. The partially unloaded cells can then be further diluted through resuspension in fresh medium.

In the cases highlighted above, dilution of the cryoprotectant, either in a single step or by stepwise dilution, may result in culture media contaminated with low levels of the cryoprotectant. This may be worthy of consideration where further manipulation of the cells is required, e.g. differentiation to a more mature cell phenotype. The period of time required for expansion of cells also needs to be considered. It has been noted for cell lines preserved for the purpose of later differentiation, that differentiation should only be attempted once the cells have readjusted to the growth conditions and active cell growth is obtained. This process can take on average between 2 to 3 weeks.

21.11 NOVEL APPROACHES TO CELL PRESERVATION

The requirement to store cells at ultra-low temperatures below the glass transition of the solute mixture has some drawbacks, including cost and availability of cryogens, and the need for continued surveillance to ensure adequate cryogen in the cell bank containers at all times (this usually necessitates temperature alarm devices and routine 'topping up' with liquid nitrogen on a weekly basis). It also means that cell samples for despatch have to be kept at the same low temperatures (for example in a cryogen-cooled 'dry shipper'). If nucleated cells could be either lyophilized or 'dried at ambient temperatures', many of these problems would be avoided. Having said this, the problems in these approaches remain formidable. However, based on the molecular studies of Crowe and his colleagues (Crowe *et al.* 1983, 1984) into the mechanisms of the protective effects of disaccharides, especially (but not exclusively) trehalose, in organisms such as the brine shrimp (*Artemia* sp.) which can exist in a dry state at ambient temperatures for many years, there has been a recent revival of research in this area. What has become apparent from the natural systems is that trehalose is required on both sides of the plasma membrane of cells to exert its full protection, so one major issue has been how to introduce the sugar into the cells of interest for banking (this type of disaccharide would not normally cross the plasma membrane in sufficient concentrations). One approach has been to try to engineer a 'switchable pore' by introducing α-haemolysin (from *Staphylococcus aureus*), which can be controlled to 'open' by changing the ion composition of the medium, allowing trehalose to enter by diffusion (Russo *et al.* 1997). Controlled 'permeability breaches' using 'thermal poration' (Wolkers *et al.* 2001) or 'electro-poration', (Djuzenova *et al.* 1996) have been suggested as methods for loading extracellular protectants into cells, whilst micro-injection has been studied (Eroglu *et al.* 2002) in situations where a small number of cells (such as individual oocytes) might be stored. Having used these approaches, freeze drying of mammalian blood platelets has recently been reported, with successful preservation of some physiological functions (Wolkers *et al.* 2001). Assisted 'ambient drying' by exposure to conditions such as a flow of desiccated air has also been reported to permit survival of normal mammalian cells, at least for a period of a few days (Puhlev *et al.* 2001). Again, the intention in using such approaches is to achieve a 'glassy state', to 'freeze in time' the metabolism and ultrastructure of the cells. However this technology remains firmly in the future for routine cell banking. A great deal of further research is required into how sugars such as trehalose interact with nucleated cells in the desiccated state, whether glass transition of the cell constituents can truly be achieved, and the stability of such glasses to physical and chemical (especially oxidative) degradation at elevated temperatures.

21.12 SAFETY ISSUES IN CELL BANKING

The ready availability of cryogens such as liquid nitrogen has greatly increased the application of cryo-banking for cell resources. However, the use of these agents introduces considerations of safety and cross-infection, which need to be foremost in the minds of those involved in cryopreservation activities. The obvious dangers when handling cryopreserved specimens relate to skin 'burns' resulting from touching extremely cold materials, or potential explosions of vials or containers caused by rapid expansion of the liquid nitrogen into the gas phase on warming. These problems can be readily overcome and accidents avoided, as long as correct use of safety gloves and eye protection are routinely employed. The problems themselves can also be avoided or risk of accident mitigated through the use of appropriate techniques and equipment. Where large volumes of liquid nitrogen are used in single or multiple containers, consideration must be given to potential oxygen deprivation in the local atmosphere if the liquid cryogen spills and vaporizes. Good ventilation and use of an oxygen depletion monitor in the storage room should minimize any risk to staff. However, provision of personal oxygen depletion monitors can provide additional warning of reduced oxygen levels. Another issue, perhaps not so obvious, is the need to ensure that samples stored at low temperatures cannot be infected with agents, especially viral agents, inadvertently released into the storage tanks from other (infected) samples. There has been a report of a viral transmission between samples of bone marrow stored in the same liquid nitrogen storage tank (Tedder *et al.* 1995), which may have arisen from failure of the filling ports on the storage bags. Viruses in general may survive immersion in liquid nitrogen (Gould 1999). There are also concerns that some types of plastic container, such as freezing straws used in reproductive medicine, may be susceptible to small leakages of potentially infectious samples during freezing (Letur-Konirsch *et al.* 2003), due to deficiencies in some steps in the handling procedures, particularly the sealing procedure of the straws. Thus, as a general warning, staff handling cryopreserved samples should treat them with the same precautions as fresh samples, and remain vigilant about procedures for filling and rewarming the samples. It has also been suggested that storage in the vapour (rather than the liquid) phase may reduce chances of cross infection (Figure 21.3). However, such use of vapour phase storage requires very good and continuous temperature monitoring to ensure that there are no risks of inadvertent warming and consequent destruction of the cell bank. Achievement of long-term cryostorage is well established for cryopreserved cells but has yet to be validated for vitrified cell preparations. Given the need to carry out time-consuming safety testing on cell therapy products, cryobiology and the ability to hold cell products in suspended animation provides the window to perform such testing and improve the economics of cell therapy.

REFERENCES

Anchordoguy T, Rudolph A, Carpenter J, Crowe JH (1987) *Cryobiology*; 24: 324–331.

Arakawa T, Timasheff SN (1985) *Biophys. J.*; 47: 411–414.

Armitage J, Mazur P (1984) *Amer. J. Physiol.*; 247: C373–C381.

Benson EE, Bremner D (2004). In *Life in the Frozen State*. Eds Fuller BJ, Lane N, Benson EE. CRC Pres, Boca Raton. Pp 205–242.

Bernard A, Fuller B (1996) *Hum. Reprod. Update*; 2: 193–207.

Block W (1990) *Phil. Trans. Roy. Soc. Ser. B*; 326: 613–631.

Boutron P, Mehl P, Kaufmann A, Angibaud P (1986) *Cryobiology*; 23: 453–469.

British Standards Institute(2006) *Publicly Available Specification* 83; Guidance on codes of practice, standardized Methods and Regulation for Cell-based Therapeutics; www.bsi-global.com.

Chen S-U, Lien Y-R, Chen H-F, Chao K-H, Ho H-N, Yang Y-S (2000) *Hum. Reprod.*; 15: 2598–2603.

Crowe JH, Jackson S, Crowe L (1983) *Mol. Physiol.*; 3: 99–105.

Crowe JH, Crowe L, Chapman D (1984) *Science*; 223: 701–703.

Crowe JH, Carpenter J, Crowe L, Anchordoguy T (1990) *Cryobiology*; 27: 219–231.

Crowe J, Crowe L, Wolker W, Tsvetkova N, Oliver A, Torok Z, et al. (2006). In *Advances in Biopreservation*. Eds Baust JG, Baust JM. CRC Press, Boca Raton. Pp 383–412.

De Loecker W, Koptelov V, Grischenko V, De Loecker P (1998) *Cryobiology*; 37: 103–109.

Djuzenova CS, Zimmermann U, Frank H, Sukhorukov VL, Richter E, Fuhr G (1996) *Biochim. Biophys. Acta*; 1284: 143–152.

Eroglu A, Toner M, Toth TL (2002) *Fertil. Steril.*; 77: 152–158.

Fahy G, MacFarlane D, Angell C, Meryman H (1984) *Cryobiology*; 21: 407–426.

Farrant J, Walter C, Lee H, Morris J, Clarke A (1977) *J. Microsc.*; 123: 17–34.

Fleck RA, Benson E, Bremner D, Day JG (2000) *Free Radic. Res.*; 32: 157–70.

Fowler A, Toner M (2005). *Ann N Y Acad Sci*; **1066**: 119–135.

Franks F (1982) In *Water: A Comprehensive Treatise*. Ed Franks F. Plenum Press, New York; Vol. 7: 215–338.

Franks F (2003). *Philos Transact A Math Phys Eng Sci*; **361**(1804): 557–574.

Fujioka T, Yasuchika K, Nakamura Y, Nakatsuji N, Suemori H (2004). *Int J Dev Biol*; **48**: 1149–1154.

Fuller B (1999) *Transplant. Revs.*; 13: 55–66.

Fuller BJ (2004). *CryoLett*; **25**: 375–388.

Fuller BJ, Paynter S (2004); *Reprod. Biomed. OnLine*; **9**: 680–691.

Gould EA (1999) *Molec. Biotechnol.*; 13: 57–66.

Karlsson JOM, Cravahlo EG, Toner M (1994) *J. Appl. Physiol.*; 75: 4442–4455.

Leibo SP, Farrant J, Mazur P, Hanna M, Smith L (1970) *Cryobiology*; 6: 315–332.

Letur-Konirsch H, Collin G, Sifer C, *et al.* (2003) *Hum. Reprod.*; 18: 140–144.

Liebermann J, Nawroth F, Isachenko V, Isachenko E, Rahimi G, 'I'ucker MJ (2002); *Biol. Reprod.* **67**: 1671–1680.

Lovelock J (1953) *Biochim. Biophys. Acta*; 10: 414–426.

Lovelock J (1954) *Biochem. J.*; 56: 265–270.

Luyet B, Gehenio PM (1940) *Life and Death at Low Temperatures*. Biodynamica Press, Normandy.

MacFarlane D (1987) *Cryobiology*; 24: 181–195.

Martino A. Songsasen N, Leibo SP (1996) *Biol. Reprod.*; 54: 1059–1069.

Mazur P (1963) *J. Gen. Physiol.*; 47: 347–369.

Mazur P (1965) *Ann. NY Acad. Sci.*; 125: 658–676.

Mazur P (1990) *Cell Biophys.*; 17: 53–92.
Mazur P (2004) In *Life in the Frozen State*. Eds Fuller BJ, Lane N, Benson EE. CRC Press, Boca Raton, FL, USA. Pp 3–66.
Mazur P, Cole KW (1989) *Cryobiology*; 26: 1–29.
Mazur P, Pinn IL, Seki S, Kleinhans FW, Edashige K (2005) *Cryobiology*; **51**: 235–239.
McGann LE, Farrant J (1976) *Cryobiology*; 13: 269–273.
McGann LE, Janowska-Weiczorek A, Turner A, Hogg L, Muldrew K, Turc J (1987) *Cryobiology*; 24: 112–119.
Meryman HT (1970) In *The Frozen Cell, CIBA Foundation Symposium, London*. Ed O'Connor GEW. Churchill Press; 565–569.
Muldrew K. Acker J. Elliot G, McGann LE (2004) In *Life in the Frozen State*. Eds Fuller BJ, Lane N, Benson EE. CRC Press, Boca Raton, FL, USA. Pp 67–108.
Muldrew K, McGann LE (1994) *Biophys. J.*; 66: 532–541.
Nei T (1981) *Cryobiology*; 18: 229–237.
O'Neil L, Paynter S, Fuller B, Shaw R, De Vries A (1998) *Cryobiology*; 37: 59–66.
Paynter S, O'Neil L, Fuller B, Shaw R (2001) *Ferti. Steril.*; 75: 532–538.
Paynter S, McGrath J, Fuller B, Shaw R (1999) *Cryobiology*; 39: 205–214.
Pearce R (2004). In *Life in the Frozen State*. Eds Fuller BJ, Lane N, Benson EE. CRC Press, Boca Raton. Pp 171–204.
Pegg DE (1988) In *Low Temperature Biotechnology: Emerging Applications and Engineering Contributions*. Eds McGrath JJ, Diller K. American Society of Mechanical Engineers; 3–21.
Pegg DE, Daiper M (1983) *CryoLetters*; 4: 129–136.
Pegg DE, Daiper M (1988) *Biophysical J.*; 54: 471–488.
Polge C, Smith A, Parkes A (1949) *Nature (Lond.)*; 164: 666–667.
Puhlev I, Guo N, Brown DR, Levine F (2001) *Cryobiology*; 42: 207–217.
Rall W, Fahy G (1985) *Nature (Lond.)*; 220: 1315–1317.
Rall W, Reid D, Farrant J (1980) *Nature (Lond.)*; 286: 511–514.
Rall W, Mazur P, McGrath J.J (1983) *Biophysical J.*; 41: 1–12.
Rall W, Reid D, Polge C (1984) *Cryobiology*; 21: 106–121.
Reubinoff BE, Pera MF, Vajta G, Trounson AO (2001) *Human Reprod.*; 16(10): 2187–2194.
Russo MJ, Bayley H, Toner M (1997) *Nature Biotech.*; 15: 278–282.
Stein A, Fisch B, Tadir Y, Ovadia J, Kraicer PF (1993) *Cryobiology*; 30: 128–134.
Steponkus PL, Langis R and Fujikawa S (1992) In *Advances in Low Temperature Biology*. Ed Steponkus PL. JAI Press, London; 1–61.
Stiff PJ (1995) In *Marrow and Stem Cell Processing for Transplantation*. Eds Lasky LC, Warkentin PI. Bethesda: American Association of Blood Banks; 69–82.
Stiff PJ, Murgo JA, Zaroulis CG, DeRisi MF, Clarkson BD (1983) *Cryobiology*; 20: 17–24.
Taylor M.J (1987) In *The Effects of Low Temperatures on Biological Systems*. Edward Arnold, London and Baltimore; 3–71.
Taylor M, Bank H, Benton M (1987) *Cryobiology*; 24: 91–102.
Taylor MJ (2006). In Advances in Biopreservation. Eds Baust JG, Baust JM. CRC Press, Boca Raton. Pp 15–62.
Tedder R, Zuckerman M, Goldstone A, *et al.* (1995) *Lancet*; 346: 137–140.
Timasheff SN, Lee J, Pitz E, Tweedy N (1976) *J. Colloid Interface Sci.*; 55: 658–663.
Tomlinson M, Sakkas D (2000) *Hum. Reprod.*; 15: 2460–2463.
Toner M, Cravahlo E, Karel M (1990) *J. Appl. Phys.*; 67: 1582–1593.
Uemara M, Steponkus PL (1995) *Plant Physiol.*; 109: 15–30.
Verkmann A, Hoek A, Ma T, Frigeri A (1996) *Amer. J. Physiol.*; 270: C12–C30.
Webb M, Steponkus PL (1990) *Cryobiology*; 27: 666–667.
Whittingham D, Leibo S, Mazur P (1972) *Science*; 178: 411–414.
Wolfe J, Dowgert M, Steponkus PL (1983) *Plant Pysiol.*; 71: 276–285.
Wolkers WF, Walker NJ, Tablin F, Crowe JH (2001) *Cryobiology*; 42: 79–87.
Wusteman M, Pegg DE (2001) *Tissue Engineer.*; 7: 507–518.
Zachariassen K, Zachariassen E (2000) *Cryobiology*; 41: 257–279.

Properties of Cell Products

22 Product Characterization from Gene to Therapeutic Product

K Baker, S Flatman and J Birch

22.1 INTRODUCTION

The market for recombinant therapeutic proteins has expanded rapidly over the last decade, and the regulatory agency demands for control and characterization of these products has evolved to meet international safety requirements in response to increasing experience with the products. The production of recombinant biologicals using animal cell expression systems for medicinal products requires tight control and testing throughout the product's lifecycle, from the gene to the final therapeutic product, to guarantee product efficacy, purity and safety. This encompasses characterization requirements at all stages of production, and includes raw materials, choice of cell line, cell-line creation and history, cell-line stability, genetic characterization to ensure correct sequence (e.g. absence of mutations), product glycosylation, and other post-translational modification profiles. The product should be characterized during cell-line construction and selection, process development/optimization (preferably in a chemically defined and animal-component-free or protein-free medium), process scale-up, and during fermentation and purification prior to final product supply (Figure 22.1). Appropriate specifications must be set and agreed prior to approved release for human use. As a general rule, the level of product characterization and stability testing required increases with increasing proximity to licensing of the product. For example, the level of validated assay characterization and stability testing required for an investigative new drug is less than for a licensed product, which would be subject to the strictest control measures using fully validated characterization and lot-release test methods.

22.2 TARGET SPECIFICATIONS

It is important to recognize that, in its broadest sense, the term 'specification' includes public, regulatory and clinical acceptability, and some of these issues are dealt with in Chapters 34 and 35. However, for the purposes of this discussion 'specification' refers to the physicochemical and biochemical characteristics of each recombinant therapeutic protein. Specifications are critical quality standards applied by the manufacturer and must be justified (i.e. validated) prior to the approval of a therapeutic drug for in-market supply by the regulatory authorities. A specification is defined as the list of tests (analytical procedures) with appropriate acceptance criteria described (numerical limits, ranges or other criteria). The specifications establish the set of criteria to which a drug substance (bulk material), drug product (finished product in desired dosage form) or other materials at other stages of its manufacture must conform before it can be accepted for its

Medicines from Animal Cell Culture Edited by G. Stacey and J. Davis

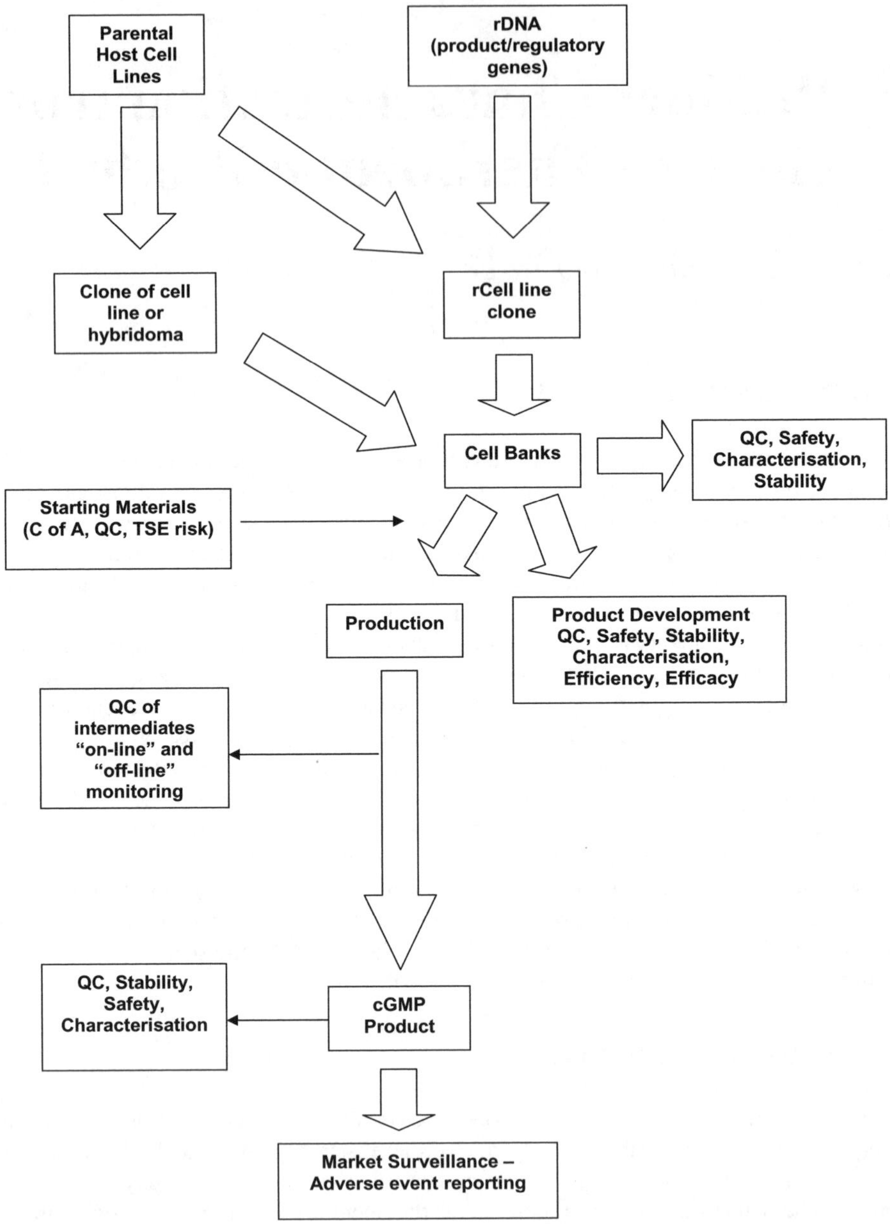

Figure 22.1 Characterisation of biological products from gene to market.

intended use (ICH Steering Committee 1999). Specifications, together with product characterization throughout development ensure that adequate product quality and consistency are maintained. However, unlike the thorough product characterization undertaken during product development, specifications are chosen to confirm that the quality characteristics of the drug substance or drug product, which may affect product efficacy and safety, are met during good manufacturing practice (GMP) production which is described in more detail in Chapter 34.

22.3 SETTING SPECIFICATIONS

During the ongoing product development process, the specifications for the release of material for clinical use generally become tighter. For example, whereas an investigative new drug (IND) destined for initial clinical trials may have purity specification criteria of >95 % evaluated from a minimal set of experiments, the specification for in-market supply material might be >98 %, based on a defined consistent production process and robust fully validated test methods. The specifications should be established by the manufacturing process (based on demonstrating consistency between batches). For example, given the above purity specifications, it is imperative that the manufacturing process is capable of consistently delivering product of >98 % purity for this specification to be valid. These specifications should also be linked to clinical trial data, i.e. in-market supply material should be representative in quality of that used in the clinical trials. Analytical data (e.g. potency, purity, impurities) including data for the stability of the drug substance or drug product should also be taken into account. Characterization of new products falls broadly into four separate areas:

- determination of physicochemical properties;
- biological activity;
- immunochemical properties;
- purity and impurities (including contaminants).

This ongoing product characterization not only confirms the identity of the product, but also ensures controlled levels of process-derived impurities (e.g. host-cell proteins, host-cell DNA, antifoam agents, proteins if medium is not protein-free, antibiotics or other selection agents, cleansing agents and purification substrates). It also ensures the absence of contaminants such as viruses, bacteria, fungi, mycoplasma or transmissible spongiform encephalopathy agents. However, it should be noted that characterization is only part of the process of assuring product safety, which is much more than simply performing investigations for adventitious agents. Each of the specification parameters described above is captured prior to the release of each lot of product and together they form a portion of the post-market surveillance specifications for the product (FDA Final Ruling 2001).

22.4 PHYSICOCHEMICAL PROPERTIES

Physicochemical characterization encompasses the chemical composition, physical properties and primary structure of the desired product. It may also, where biologically relevant, include characterization of product heterogeneity and analysis of post-translational modifications. Recombinant proteins produced using animal cell expression systems are complex molecules that undergo a variety of post-translational modifications that can result in structural and functional heterogeneity (James & Baker 1999) (see also Chapter 23). Changes to the post-translational modifications may result in the addition or masking of an immunogenic site, changes to product stability or immunogenicity through product aggregation, or changes in solubility. The *in vivo* pharmacokinetics of a product may also be sensitive to very small changes in product structure or glycosylation, which may affect the circulatory half-life, efficacy and required dosage of product. This does not, however, preclude a drug product demonstrating an inherent degree of structural heterogeneity due to the biosynthetic processes used by recombinant organisms. Nonetheless, the pattern of heterogeneity should be defined by the manufacturer and lot-to-lot consistency demonstrated prior to in-market supply. This may negate the need to evaluate each isoform separately for activity, efficacy and safety. It is, however, imperative that the quality of each batch of the final drug product be assessed to confirm that it is comparable with the original (intended) product in terms of composition, primary and secondary/tertiary structure, function, bioactivity and long-term stability. If significant differences are found in the biochemical analysis, this could warrant the collection of additional *in vivo* data or the failure of batches.

Drug substance characterization is undertaken during various stages of process development utilizing a number of studies including: reference characterization, product stability, product equivalence testing and consistency studies, in addition to primary amino acid sequence determination and identification of degradation products [e.g. monitoring glycosylation, protein clipping (C- and N-terminal), deamidation, oxidation, sulphation and disulphide bonding]. A number of techniques are readily available for achieving detailed product reference characterization for initial clinical supply, and these are outlined in greater detail below (e.g. in depth characterization of initial GMP material). Product stability testing is undertaken at each stage of process development to ensure continuing stable product quality and to provide sufficient stability data for the duration of any ongoing and planned clinical trials (see Chapter 27). Product equivalence studies are performed when there is a defined change in process parameters (e.g. change in manufacturing scale from that used for Phase I/II material to the scale used for Phase III/in-market supply). Finally, product/process consistency is demonstrated on validation lots (for regulatory licence approval) this product often being used for Phase III clinical trials.

Characterisation of the drug substance during process development and for clinical supply can be achieved using a variety of techniques, each of which provides data revealing different aspects of the product. The analytical methods used for testing clinical material all require validation prior to in-market supply. Many of these methods are listed in Table 22.1 and are described in greater detail in Chapter 23.

Table 22.1 Analytical methods used to characterize product.

Primary Sequence

Amino Acid Composition Amino acid hydrolysis followed by analysis and comparison with the theoretical amino acid sequence

Terminal Amino Acid Sequence To identify the homogeneity of the amino- and carboxy- terminal amino acids. The relative amounts of any variant species should be determined

Peptide Mapping Most common technique for identification of product structure, using selective fragmentation and separation by HPLC

Disulphide Bond Assignment By peptide mapping under reducing and non-reducing conditions followed by LC-MS or MALDI-MS

Glycosylation Analysis Monosaccharides, size and charge profiling, antennary profiling using a variety of techniques including HPLC and mass spectrometry – see Chapter 24

Physicochemical Properties

Molecular Weight or Size Determined using size exclusion chromatography, SDS PAGE (under both reducing and non-reducing conditions), mass spectrometry or analytical ultracentrifugation

Charge Isoform Pattern Determined using isoelectric focusing (IEF), ion exchange HPLC or capillary IEF

Extinction Coefficient Determination (Or Molar Absorptivity) Determined via a variety of techniques including amino acid composition and nitrogen determination

Electrophoretic Patterns Provides data on homogeneity, purity and identity obtained via SDS PAGE, IEF, Western blotting, capillary electrophoresis or other suitable procedures

Liquid Chromatography Patterns Provide information on the identity, homogeneity and purity using size exclusion chromatography (SEC), reversed-phase liquid chromatography (rp HPLC), affinity chromatography, ion-exchange chromatography or other suitable procedures

Spectroscopic Profiles Used for secondary structure determination using techniques including circular dichroism and nuclear magnetic resonance (NMR)

22.5 BIOLOGICAL ACTIVITY

To characterize a product fully, it needs to be assessed not only in terms of its structural and physicochemical properties but also in terms of its biological properties, i.e. its ability to achieve a specific physiological effect. Assays for such properties may be biochemical (e.g. enzymatic reactions, ligand-receptor binding assays), cell culture-based 'bio-assays' measuring responses at the cellular level, or animal-based assays that measure the animal's biological response to the product. The 'potency' of a product is the quantitative measure of its specific biological activity (often compared to a reference material).

22.6 IMMUNOCHEMICAL PROPERTIES

Antibody-based products form the majority of recombinant therapeutics currently in late-stage clinical trials or in in-market supply (Walsh 2003). The immunological properties of these products should be well characterized, either using biological antigen binding assays or other immunochemical techniques. For recombinant therapeutics, the choice of cell line used to manufacture the product may become important as post-translational modifications may vary between species. For example, the industrial standard murine cell line NS0 produces terminal carbohydrate epitopes (galactose-α-1,3-galactose) that are immunogenic in humans (approximately 1 % of circulating antibodies in humans recognize this epitope). Therefore, the choice of cell line for highly glycosylated recombinant proteins may be important (e.g. non-murine cell lines such as Chinese hamster ovary cells may provide a better alternative for some recombinant therapeutics). The immunogenicity of impurities in cell derived vaccines is covered in detail in Chapter 25.

22.7 PURITY AND IMPURITIES

The purity of a drug substance and a drug product is assessed using a combination of analytical methods (e.g. electrophoretic and/or chromatographic separations, and/or biological assay). When a drug substance or drug product comprises several desirable isoforms (e.g. due to natural product heterogeneity), they are considered to be product-related and not an impurity.

Impurities can be product- or manufacturing process-derived and may be identified or unknown. Product-related impurities may include product aggregates (that can be detected by size exclusion chromatography, SEC, or light scattering methods), fragments, or other truncated product components such as subunits (e.g. free heavy and light chains of immunoglobulins, analysed using HPLC or SDS PAGE and/or peptide mapping). Undesirable heteroforms of the drug substance or drug product may be analysed using the appropriate characterization tools (e.g. HPLC, capillary electrophoresis or mass spectrometry).

Process-derived impurities comprise those inherent to the manufacturing process, including host cell DNA (analysed by hybridization, immunoassay or PCR), endogenous host cell viruses (analysed by electron microscopy, PCR or other methods - see Chapter 19) or proteins (analysed by immunoassay and/or immunodetection by Western blotting), and culture medium components (method of analysis dependent on component). Substances may also be introduced during downstream processing, e.g. purification substrates, enzymes, heavy metals, solvents, etc., which can be determined by an appropriate method. Acceptance criteria should be specified based on historical data obtained throughout the process development lifecycle of the product, but should also include defined limits for impurities or contaminants to ensure the safety of the product at the intended dosage.

22.8 STABILITY ANALYSIS

Following the release of a batch of recombinant therapeutic product for clinical trials or in-market supply, the stability of the protein is continually monitored for an appropriate time (commonly up to 3 years) to support the assigned shelf life of the product. In general, protein therapeutics are very complex molecules and most contain many potential sites for degradation, which may in turn affect the efficacy of the product. The rate of degradation at each site is dependent on both the structure of the protein and the environment in which it is stored, including formulation and storage temperature (Wright 1997). This subject is covered in detail in Chapter 27.

22.9 THE ROLE OF CHARACTERIZATION DURING PROCESS DEVELOPMENT

There is a growing trend to use product characterization data to define the choice of final manufacturing process. In the past, the primary goal of process development was to improve volumetric productivity. More recently, awareness of the potential impact of process changes on product structure and function has led to increased emphasis on detailed characterization of product throughout process development. In some cases this may limit the process changes or lead to optimization studies directed specifically at product quality.

Thus, characterization of product is required at the earliest stages of process development from cell line creation onwards. Selection of cloned transfectant cell lines should ideally be based not only on product titre, but also on product quality data and potentially (product-dependent) pharmacokinetic (PK) data. Recent advances in high-throughput purification and mass spectrometry (e.g. MALDI-MS) permit rapid characterization of product quality during cell line construction and selection when the numbers of cell clones to be screened are very high.

In addition, the development of assays for product and process characterization early in production process development provides useful tools to drive process development and optimization of cell culture and purification processes. It is also important to establish at an early stage the criteria and assays which will be used to support the specifications for the final product. The examples in the following section illustrate the need for early and ongoing characterization during product development.

22.10 EFFECT OF PROCESS CHANGES ON PRODUCT CHARACTERISTICS: CASE STUDIES

22.10.1 Cell Culture Process Change Characterized by IEF: OKT3

One of the first examples of the importance of product characterization during process development and following a process change, was for the therapeutic antibody product OKT3. OKT3 is a well-defined monoclonal antibody produced by Ortho Biotech and was approved for marketing in 1986. It was produced using mouse ascites-derived cultures (A-OKT3). However, as fermentation of hybridoma cells permits cellular growth under more defined conditions, and in general offers more flexibility to meet market demand, this method was evaluated for OKT3 production and demonstrated a change in the OKT3 product characterization profile. Upon switching production of OKT3 from the original ascites production method to air-lift fermentation (TC-OKT3), 5 to 10% of the antibody molecules retained the heavy chain C-terminal lysine residue, leading to an increase in product isoelectric point (pI) that was identified by isoelectric focusing. Nonetheless, there was negligible effect on product efficacy (Rao *et al.* 1991) and the product remained acceptable. This example, however, demonstrated the importance of ongoing characterization during process development.

22.10.2 Amino Acid Mutation-induced Changes to Glycosylation and Pharmacokinetic Behaviour Characterized by IEF and Potency Assays: ARANESP™

As a result of improvements in characterization technology, numerous therapeutics on the market now list post-translational modification specifications as criteria for in-market supply, including ARANESP™(darbepoetin alfa; Amgen, Inc.) for the treatment of anaemia associated with chronic renal failure and chemotherapy. The glycan structures in glycoproteins frequently have sialic acid as the terminal residues. The presence of sialic acid can have a major influence on the *in vivo* half-life of a therapeutic protein and its therapeutic efficacy. In the absence of sialic acid, a therapeutic protein may be more rapidly cleared via hepatic asialogalactose receptors. However glycoprotein populations are frequently heterogeneous with respect to the number of sialic acid residues, a characteristic that may be influenced by culture conditions.

ARANESP™'s activity is greater than the natural human erythropoietin (EPO) protein from which it is derived and is dependent on five amino acid substitutions (Ala30Asn, His32Thr, Pro87Val, Trp88Asn and Pro90Thr). These substitutions have been exploited to yield two additional N-linked, sialic acid-containing, carbohydrate chains compared with the existing recombinant human EPO marketed by Amgen (EPOGEN™; epoetin alfa). The potential total number of sialic acid residues on ARANESP™ is 22 compared with EPOGEN™'s potential 14. This resulted in a markedly different IEF profile (due to the relative change in glycoprotein charge) but demonstrated a consistent number of predominant isoforms. This provided a good tool for monitoring ARANESP™ both 'on-line' during production and during treatment (as it can be detected in patient urine against the background of endogenous EPO, Catlin *et al.* 2002).

As a result of changes in the glycosylation profile, the hyperglycosylated rhEPO demonstrated that both the ARANESP™ half-life and potency were dependent on terminal sialylation (Egrie & Brown 2001). ARANESP™, which contained five *N*-linked sialic acid-bearing glycosylation sites and one O-linked glycosylation site demonstrated an approximately threefold longer serum half-life than the traditional rhEPO (containing only three *N*-linked and one *O*-linked sialic acid-bearing glycosylation sites). In addition, ARANESP demonstrated increased potency *in vivo* (based on receptor-binding potency assays) and could be administered less frequently to obtain the same biological response (Egrie & Brown 2001).

22.10.3 Choice of Cell Line: Glycosylation and Activity Differences Characterized by HPLC/MS and Potency Assay: Xigris™

The post-translational modifications of recombinant human activated protein C (Xigris™) produced by Eli Lilly and Company, are critical to its anti-coagulant activity. In addition, as Xigris™ is a treatment for patients with severe sepsis, it is imperative that the recombinant protein does not contain immunogenic glycan epitopes that may invoke an immune response *in vivo*. There are several post-translational modifications present on Xigris™ that require characterization prior to batch release. These include amino acid modifications, including nine gamma-carboxyglutamic acid (GLA) residues and one beta-hydroxyaspartic acid residue (which are required for optimal anticoagulant activity) and four *N*-linked glycosylation sites, 12 di-sulfide bonds and various proteolytic processing events including N-terminal clipping of both heavy and light chain components (Yan *et al.* 1990).

Recombinant human protein C (rHPC) expressed in human kidney 293 (HK-293) cells demonstrates higher anti-coagulant activity than endogenous plasma-derived human protein C while maintaining a comparable circulatory half-life. The rHPC has highly fucosylated glycans and a novel *N*-acetylgalactosamine-containing *N*-linked glycan (PC-238), unique to the rHPC product. In addition, the sialic acid content of this unique PC-238 glycan was lower than that observed in native HPC (Yan *et al.* 1990, 1993), a factor that potentially regulates its circulatory half-life. This unique PC-238 glycan was not present on rHPC produced by two alternative cell lines (Syrian baby hamster kidney, BHK-Ad, and Syrian hamster AV12-664), confirming how the choice of cell line can influence the final product characteristics as well (Yan *et al.* 1990).

The post-translational processing complexity of rHPC necessitates both structural characterization of the Xigris™ protein structure, and characterization of its glycosylation profiles and activity/potency. This involves utilizing a number of techniques outlined in the following chapters, including HPLC-based assays, mass spectrometry, peptide mapping, glycosylation analysis and potency assays. These analytical characterization tools provide the detailed information required to guarantee the efficacy and safety of Xigris™ for use *in vivo*.

22.11 CONCLUSIONS

The examples above highlight the requirement for ongoing characterization of protein products during both process development and in-market supply. Although there is emphasis on the manufacturing environment to yield homogeneous therapeutic products, some product heterogeneity is acceptable as long as the heterogeneity remains consistent between lots, and characterization data support the efficacy (potency) and safety of the product. Nonetheless, the case studies exemplify the need for supporting pharmacokinetic and characterization data for the release of therapeutics dependent on post-translational modifications for their activity.

Therefore, for consistency, sufficient characterization data are required to set the relevant specifications for recombinant therapeutics based on both clinical experience and manufacturing history. Critical post-translational modifications, however, require testing on a lot-to-lot basis and as part of any ongoing stability trials.

REFERENCES

Catlin DH, Breidbach A, Elliott S, Glaspy J (2002) *Clin. Chem.*; 48: 2057–2079.
Egrie JC, Browne JK (2001) *Br. J. Cancer*; 84(Suppl): 3–10.
Food and Drug Administration (FDA) Final Rule on 21 CFR Parts 314 and 601: 'Postmarketing Studies for Approved Human Drug and Licensed Biological Products'; Department of Health and Human Services, FDA; 27 February 2001) *Federal Register*, Vol. 65, No 210, October 30, 2000/*Rules and Regulations*.
ICH Steering Committee (10 March 1999) *ICH Harmonized Tripartite Guideline Q6B: Specifications: Test Procedures and Acceptance Criteria for Biotechnological/Biological Products*.
James DC, Baker KN (1999) In *Encyclopedia of Bioprocess Technology*. Eds Flickinger MC, Drew SW John Wiley & Sons, Inc. New York; 1336–1349.
Lucore CL, Fry ET, Nachowiak DA, Sobel BE (1998) *Circulation*; 77: 906–914.
Rao P, Williams A, Baldwin-Ferro A *et al.* (1991) *BioPharm.*; Nov/Dec: 38–43.
Walsh G (2003) *Nature Biotechnol.*; 21: 865–870.
Wright JF (1997) *Eur. J. Par. Sci.*; 2(4): 103–108.
Yan SC, Razzano P, Chao YB *et al.* (1990) *Bio/Technol.*; 8: 655–661.
Yan SC, Chao YB, van Halbeek H (1993) *Glycobiology*; 3: 597–608.

23 Protein Analysis

K Baker and S Flatman

23.1 INTRODUCTION

Recombinant therapeutics cover a wide range of relatively complex proteins that require detailed product characterization prior to use for clinical trials or in-market supply. The four main aims of protein characterization are to ensure that the recombinant therapeutic destined for clinical use demonstrates adequate purity, efficacy, safety and strength.

A range of analytical methods is required to satisfy these demands at various stages of production, including process development, during manufacture, quality control testing and stability testing. The regulatory authorities produce a number of guidelines for quality control of recombinant therapeutics (Code of Federal Regulations, ICH guidelines).

There are two main subsets of analytical protein techniques. The first are standard techniques routinely used for protein characterisation (Table 23.1) and the second subset of techniques is a group of specialist techniques for applications such as stability analysis and secondary structure determination (Table 23.2). Although the list of techniques in Table 23.1 and Table 23.2 are not exhaustive, they represent the predominant range of techniques employed routinely for therapeutic protein analysis and characterisation.

23.2 THE STANDARD TOOLBOX

Each of the techniques listed in Tables 23.1 and 23.2 are discussed in more detail in the following sections. The principles of each of the test methods are described and where appropriate, an example of their application to recombinant protein analysis provided. To demonstrate the use of these analytical methods, a chimeric human IgG_4 monoclonal antibody cB72.3 recognizing the TAG72 antigen is used as an example of an animal-cell-derived recombinant therapeutic protein, and data for other recombinant proteins are described where appropriate.

23.2.1 Sodium Dodecyl Sulphate Polyacrylamide Gel Electrophoresis

The sodium dodecyl sulphate polyacrylamide gel electrophoresis (SDS-PAGE) method originally described by Laemmli (Laemmli 1970), is primarily used as a test for recombinant protein purity (Table 23.1). Test samples are analysed under reducing or non-reducing conditions. SDS-PAGE provides information on product homogeneity and permits identification of proteolytic clipping, product dissociation, cross-linking and/or the presence of impurities in the product formulation.

Applying the method to analysis of the cB72.3 antibody, we would expect to observe, under non-reducing conditions, a single band at approximately 150 kDa. In practice, however,

Medicines from Animal Cell Culture Edited by G. Stacey and J. Davis

Table 23.1 Standard toolbox for product characterization.

	Analysis Applications				
Analytical technique	Purity	Identity	Post-translational modification	Conformation	Aggregation
Sodium dodecyl sulphate polyacrylamide gel electrophoresis (SDS PAGE)	✓	✗	✓	✗	✗
Western blotting	✓	✓	✗	✗	✗
Isoelectric focusing (IEF)	✗	✓	✓	✗	✗
Capillary electrophoresis	✓	✗	✓	✗	✗
Capillary isoelectric focusing	✗	✓	✓	✗	✗
Two-dimensional SDS PAGE	✓	✓	✓	✗	✗
Native PAGE	✗	✗	✗	✓	✓
Protein lab-on-a-chip	✓	✗	✗	✗	✗
Enzyme-linked immunosorbent assay (ELISA) or other immunological assay	✗	✓	✗	✗	✗
Size exclusion chromatography (SEC)	✓	✗	✗	✗	✓
Reversed-phase chromatography	✓	✓	✓	✗	✗
Affinity chromatography	✗	✓	✗	✗	✗
Ion exchange chromatography (cation or anion)	✗	✓	✓	✗	✗
Peptide mapping	✗	✓	✓	✗	✗

post-translational modifications such as glycosylation can affect the apparent molecular weight disproportionately and produce multiple banding patterns. This is particularly obvious in non-reducing gels (lanes 4 and 5, Figure 23.1), where the protein appears to migrate with a higher than expected molecular weight. The second predominant species is a half antibody at approximately 80 kDa, comprising a single heavy and light chain (Angal *et al.* 1993), a dissociation product often observed during the antibody production process.

Analysis of cB72.3 antibody in the presence of a reducing agent gives a banding profile consisting of two components: the heavy chain with a molecular weight of approximately 50 kDa and the light chain with a molecular weight of approximately 25 kDa (Figure 23.1, lanes 1 and 2). Proteolytic degradation products appear at low molecular weights compared with the respective intact antibody bands, whilst intermolecular cross-linked degradation products appear as high molecular weight bands.

Protein bands are routinely visualized using two methods. The first is Coomassie blue (as shown in Figure 23.1), which provides reliable quantitation of purity of most proteins due to equivalent binding of dye over a relatively wide mass load range (e.g. 10 ng to 10 μg). The limit of detection of coomassie blue is approximately 10 ng per stained band.

The second method of staining (silver stain) is used to increase the sensitivity of detection to ≤1 ng of protein band. However, as the staining intensity of different proteins can vary markedly, the method is not appropriate for accurate quantitation of overall product purity. Nonetheless, individual proteins demonstrate a proportional increase in staining with increased mass range.

Table 23.2 Specialist toolbox for product (and aggregate) characterization.

Specialist analytical technique	Analysis applications: Identity	Post-translational modification	Conformation	Aggregation	Structure determination
Liquid chromatography – mass spectrometry (LC – MS)	✓	✓	×	×	✓
Matrix-assisted laser desorption ionization MS (MALDI MS)	✓	✓	×	×	×
Electrospray ionization MS (MALDI MS)	✓	✓	×	×	✓
Carbohydrate analysis (various methods)	✓	✓	×	×	×
Circular dichroism	×	×	✓	×	✓
Analytical ultracentrifugation	×	×	✓	✓	×
Differential scanning calorimetry	×	×	✓	✓	×
Infared spectroscopy	×	×	✓	✓	✓
Light scattering	×	×	✓	✓	×
Tomography	×	×	✓	✓	✓
Potency/activity assays	✓	×	×	×	×

Therefore, the levels of a particular process impurity can be determined by applying the relevant standards and controls to each SDS-PAGE gel. More recently, fluorescent probes and stains have become available that offer quantitation over 4 to 5 orders of magnitude (e.g. 0.1 ng to 10 μg).

Quantitation is usually achieved using densitometry with either a monochromatic light source or a red laser of an optimal wavelength. The absorbance units of each protein band component detected are totalled and expressed as a relative percentage of the total value.

The main advantages to SDS-PAGE include its simplicity, low cost and mainstream acceptance for purity determination by the regulatory authorities. The primary disadvantages are relatively high variability (1 % RSD) compared with other techniques (e.g. HPLC, 0.1 % RSD) and lack of specificity for detection of components with similar molecular weights.

23.3 WESTERN BLOTTING

Used in conjunction with SDS-PAGE, Western blotting provides a simple method for identification of both product and impurity species using specific antibody reagents. Following SDS-PAGE, the proteins are electrophoretically transferred ('blotted') to an inert membrane that can then be probed with product-specific probes (e.g. sheep anti-product probe). Product bands are then visualized following incubation with a secondary anti-species reagent-linked probe (e.g. horseradish peroxidase labelled anti-sheep IgG) and detection using reagent substrate (e.g. tetramethyl benzidine).

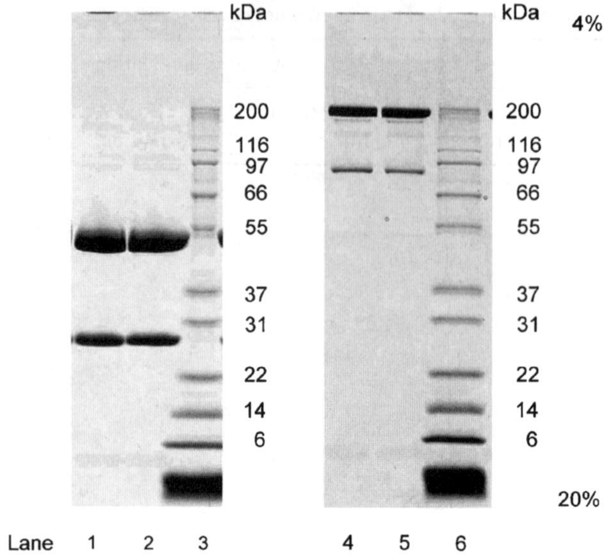

Figure 23.1 SDS-PAGE analysis of monoclonal antibody samples under both reducing and non-reducing conditions.
Key–Reducing SDS PAGE: lane 1, cB72.3 IgG_4, sample A; lane 2, cB72.3 IgG_4, sample B; lane 3, molecular weight markers.
Non-reducing SDS PAGE: lane 4, cB72.3 IgG_4, sample A; lane 5, cB72.3 IgG_4, sample B; lane 6, molecular weight markers. SDS-PAGE was performed under standard Laemmli conditions over a 4 % to 20 % (w/v) polyacrylamide gradient.

Specific antibody probes against numerous recombinant therapeutics can be obtained commercially and may recognize either the whole protein or subunit components of the recombinant protein. The specificity of each probe should be confirmed prior to routine use and cross-reactivity minimized during assay development.

Product-stained blots are then compared against identical SDS-PAGE gels that have been silver stained to permit identification of product and impurity species. Protein bands identified by Western blotting can be quantified by densitometry using an appropriate standard.

An example of the application of this technique is the identification of impurity species. An example of a product-specific Western blot for cB72.3 is presented in Figure 23.2. Comparison of the Western blot to the silver stained gel leads to the identification of a number of product-associated protein species, confirmed to be either derived from the heavy chain or light chain of the antibody.

23.3.1 Host Cell Protein (Western Blotting Method)

Host cell proteins (HCPs) are the most important group of non-product-related components found as impurities of bulk purified cell-derived therapeutic products. The HCPs comprise a wide range of molecular weight species that may co-elute with product components, making them difficult to remove during product purification processes. Because many of these components may be antigenic to humans, it is vital during product analysis to assay for these impurities prior to release of any therapeutic product for clinical use, in order to demonstrate the safety of the intended product.

Polyclonal antisera specific for the null host cell line (i.e. for a recombinant product, the host cell line transfected with a vector that does not contain the product DNA) grown under identical conditions to the product, are generated in at least two different species and then pooled to maximize

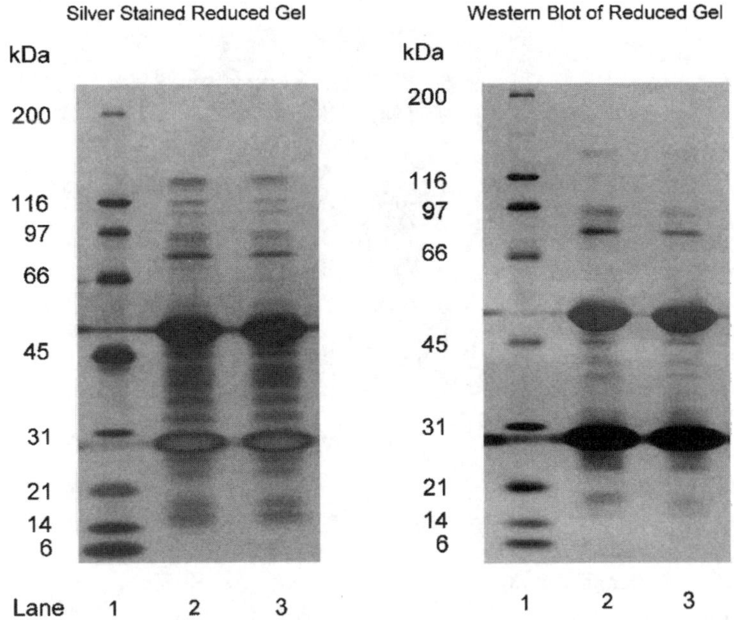

Figure 23.2 Antibody fragment assay for the cB72.3 antibody.
Key–Lane 1, molecular weight markers (reducing); lane 2, cB72.3 IgG_4, sample A; lane 3, cB72.3 IgG_4, sample B. SDS-PAGE was performed under standard Laemmli conditions over a 4 % to 20 % (w/v) polyacrylamide gradient. Samples of cB72.3 were analysed under reducing (lanes 2 to 3) conditions. One gel was silver stained and one gel was probed with anti-heavy chain and anti-light chain antibodies.

the response range against the multitude of host cell protein components. Total cell lysates from the product-containing and null cell lines are checked by two-dimensional SDS-PAGE (2D PAGE; see Section 23.3.4) for >90 % homology (in absence of product species) prior to raising antiserum.

Although the total host cell protein content of a bulk therapeutic product can be determined by an enzyme-linked immunosorbent assay (ELISA; see Section 23.3.7), identification of individual host cell protein impurities can be undertaken using Western blot analysis of either normal SDS-PAGE gels or 2D PAGE gels, depending on the level of identification and characterization required.

23.3.2 Isoelectric Focusing

Isoelectric focusing (IEF) provides high resolution of differentially charged isoforms of proteins with net charge (pI) differences equivalent to 0.01 pH units under optimized conditions.

There are two main methodologies used for IEF using either immobilines or carrier ampholytes to form a pH gradient on either agarose or polyacrylamide gel matrices. The pH gradient over which the protein isoforms resolve can be optimized following the addition of extra ampholytes in the desired pH range. Exogenous protein stabilization reagents (e.g. urea, detergents) are often incorporated to prevent precipitation or to prevent interaction of the desired protein with the gel matrix. Suitable precast gels are available from commercial sources and offer the benefit of increased stability of pH gradients, which in turn leads to improved resolution of protein isoforms. Gradient profiles can then be modified to provide increased flexibility to fit resolution requirements.

Antibodies usually demonstrate a degree of microheterogeneity and resolve with a number of bands. This microheterogeneity can be caused by a variety of chemical changes (e.g. deamidation,

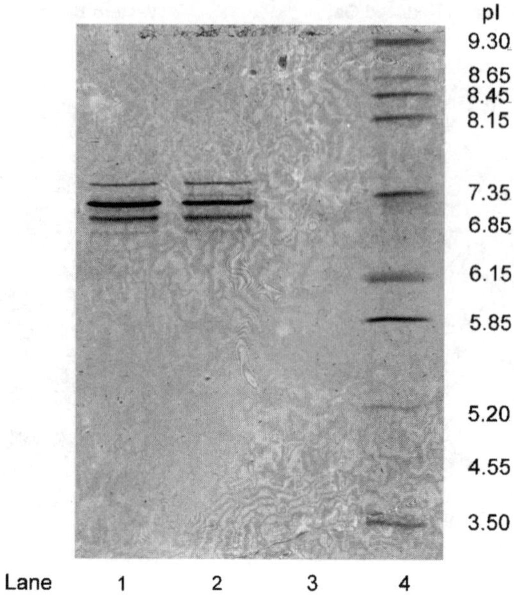

Figure 23.3 IEF profile for cB72.3 antibody samples produced using different fermentation processes. *Key*–Lane 1, cB72.3 IgG_4, sample A, process A; lane 2, cB72.3 IgG_4, sample B, process B; lane 3, blank; lane 4, pI markers.

oxidation, changes to glycosylation, C-terminal clipping etc.) which result in a change in net charge of the protein. The heterogeneity observed can also depend on both the cellular expression system used and the manufacturing process. Nonetheless, the banding patterns observed for a well controlled specific antibody process remain consistent (although the intensity of different bands may alter) and the IEF technique is widely used as an identification tool.

Figure 23.3 demonstrates the IEF profile for the cB72.3 antibody cultured using different manufacturing processes. Although the banding pattern maintains a high degree of homology, the intensity of each band differs depending on the process used to produce it. There is a feint band with an approximate pI of pH 7.35, which is readily visible in cB72.3 antibody derived from process B. The same band is present in the cB72.3 antibody derived from process A, but at a much lower intensity.

23.3.3 Capillary Electrophoresis

Capillary electrophoresis (CE) is a rapidly advancing technique that offers multiple separation methods including electrophoresis, isoelectric focusing, isotachophoresis (for non-proteinaceous ionic separations) and micellular electrokinetic chromatography (for the separation of both neutral and charged protein species simultaneously).

Separation of proteins is undertaken in a fused silica capillary, usually filled with linear (non-cross-linked) polyacrylamide, 20 to 30 cm long and with an internal diameter of 10 to 100 μm. Both ends of the capillary are immersed in electrode reservoirs in which the reagent buffer is placed (e.g. 20 mM to 30 mM sodium phosphate buffer pH 2.6 for electrophoresis of peptides).

Identification of individual species is undertaken by comparing the mobility or molecular weight of the sample relative to reference standards or molecular weight markers. The resultant profiles resemble HPLC chromatograms and can be integrated using the supplied instrumentation software.

For capillary isoelectric focusing, the resolution conferred by CE is similar to IEF but with the advantage of shorter analysis times and direct quantitation of isoforms (see Figure 23.4). In

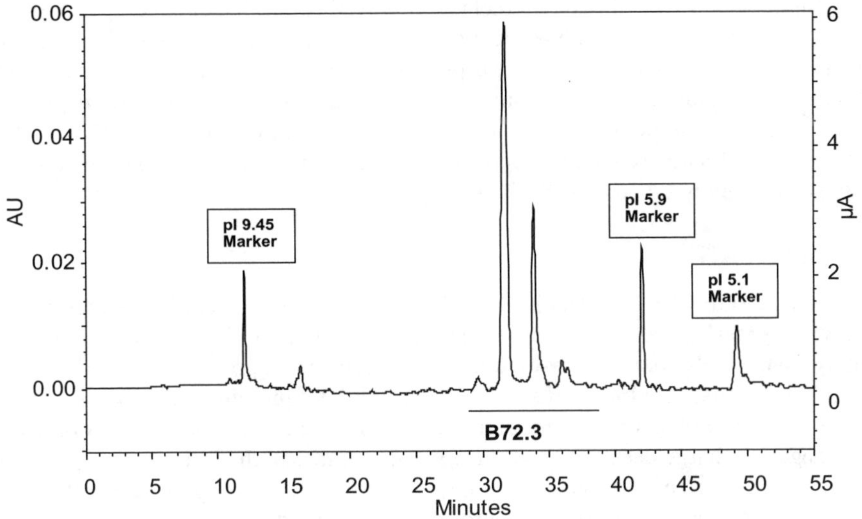

Figure 23.4 Capillary electrophoresis profile for cB72.3 antibody sample (4 mg/ml).

addition, and in contrast to many HPLC-based analyses, proteins are not damaged during CE separation.

The advantages of using CE over traditional gel-based analyses include a reduced amount of chemicals and protein sample required (usually ≤4 nl injected), rapid separation (usually 10 to 20 minutes), availability of a number of commercial capillary reagent kits for protein analyses, which typically include the capillary, complete protocols, buffers and standards, and the ability to couple the CE instrumentation with a variety of detection systems, including UV/Vis, fluorescence, electrochemical detection and conductivity. In addition, a variety of different capillaries are available for different analyses and the instruments can be linked with other analytical instruments including HPLC and MS. The systems are usually semi- to fully automated and most CE systems routinely come fitted with autosampler units and/or fraction collection units (for preparative separations). The main disadvantage is the relatively high initial capital outlay compared with conventional electrophoretic equipment.

23.3.4 Two-dimensional SDS PAGE

Routinely employed for proteomic analyses, two-dimensional SDS-PAGE offers the ability to resolve proteins not only on the basis of size, as in traditional one-dimensional SDS-PAGE, but on the basis of charge as well. Therefore, when multiple protein species co-migrate with the same apparent molecular weight, they can often be resolved using 2-D SDS-PAGE for identification and quantitation.

In the first dimension, proteins are focused along an immobilized pH gradient (the usual method employed now) on a IPG 'strip' into very narrow pH intervals on the basis of charge, followed by separation of the proteins in the second dimension on the basis of size, similar to routine SDS-PAGE.

Protein species are detected using radioactive labeling, traditional protein stains (e.g. coomassie blue, silver staining, fluorescence staining) or by immunodetection following blotting of the 2-D gel. Following normal detection, the individual species (which appear as single or multiple spots) can be excised, undergo in-gel tryptic digestion and the spot identification confirmed by mass spectrometry combined with database searching for sequence tag homology.

The advantages of this methodology include improved resolution of species with similar molecular weights, coupled with the advanced bioinformatics tools available to elucidate and confirm identification of both known and unknown protein species. The main disadvantage is that the method relies on denaturation of the sample for analysis and, therefore, the structural characterisation of the protein or proteins in general relies on other methodologies. In addition, it is still a relatively slow process and requires a degree of skill to master the image analysis tools and only one sample can be analysed per 2-D SDS-PAGE gel.

23.3.5 Native PAGE

Non-denaturing PAGE, also called native PAGE, separates proteins according to their mass:charge ratio. Native polyacrylamide gels contain a combination of polymerized bis-acrylamide and acrylamide only. Native gels are typically used to permit recovery of proteins (or other biomolecules) from the gel in their native, biologically active, form. As most proteins carry a net negative charge at slightly basic pH, proteins can be separated under electric current on the basis of charge. The higher the negative charge density (more charges per unit mass), the faster a protein will migrate. The migration of the proteins through the gel is also dependent on the pore size of the acrylamide matrix (usually a gradient is used). Large proteins are retarded by the gel matrix compared with small proteins and therefore migrate more slowly through the gel matrix. Thus, native PAGE separates proteins based upon both their charge and mass.

The advantages of the method include the ease of preparation of the gels and ability to recover biologically active proteins that have maintained their quaternary structure. The disadvantages include only being able to analyse sub-mg quantities and thus native PAGE gels are not routinely utilized for the preparative separation of proteins.

23.3.6 Protein Lab-on-a-chip

The protein lab-on-a-chip (chip) is a miniaturized micro-fluidic replacement for traditional SDS-PAGE analysis. It employs separation of proteins across a linear non-crosslinked polymer matrix (similar to a 4 to 20 % gradient gel). During chip preparation, a dye concentrate is mixed with the gel, followed by filling of the chip channels with the gel–dye mix. Detection is based on the interaction of a fluorescent dye with the SDS-protein complexes, coupled with laser-induced fluorescence detection (Figure 23.5 and Figure 23.6).

Advantages of the chip method include low cost disposable chips (compared to SDS-PAGE reagents), fast analyses (1 to 2 minutes per sample), analysis of up to ten samples per chip, and a choice of digital data output including traditional electropherograms (similar to those obtained following densitometry analysis as shown above) or gel-like format (see Figures 23.5 and 23.6). High throughput systems are currently in development. Disadvantages include the requirement for sample denaturation and dilution prior to analysis. In addition, some sample matrices can interfere with detection and may require additional sample preparation. There is currently no miniaturized 'on-a-chip' equivalent to IEF.

23.3.7 Enzyme-Linked Immunosorbent Assay

Enzyme-linked immunosorbent assay (ELISA) assays are routinely employed for a number of applications including identification and quantitation of specific recombinant proteins or impurities using, for example, anti-protein or anti-impurity specific antibody reagents. In addition, ELISAs are often employed for determination of protein potency. This is covered in more detail in Section 23.3.17.

Most ELISAs employ a similar format in which one biospecific protein (or impurity) reagent is bound to a solid phase and a second biospecific reagent (usually an antibody) is bound to an

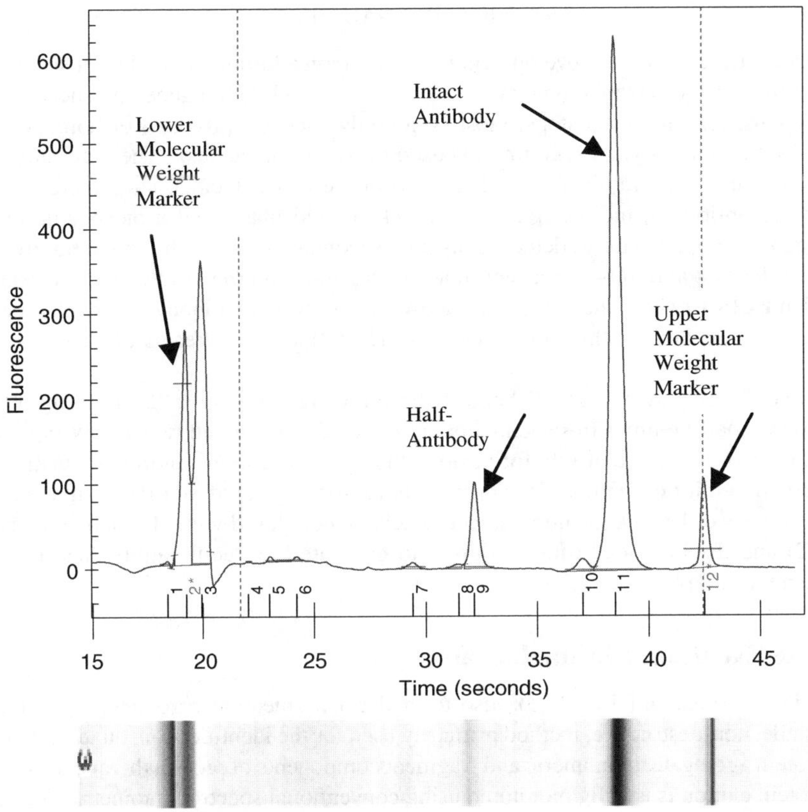

Figure 23.5 Non-reduced electropherogram and digitally generated 'gel image' demonstrating separation of intact cB72.3 antibody using the 2100 Bioanalyser (lab-on-a-chip).

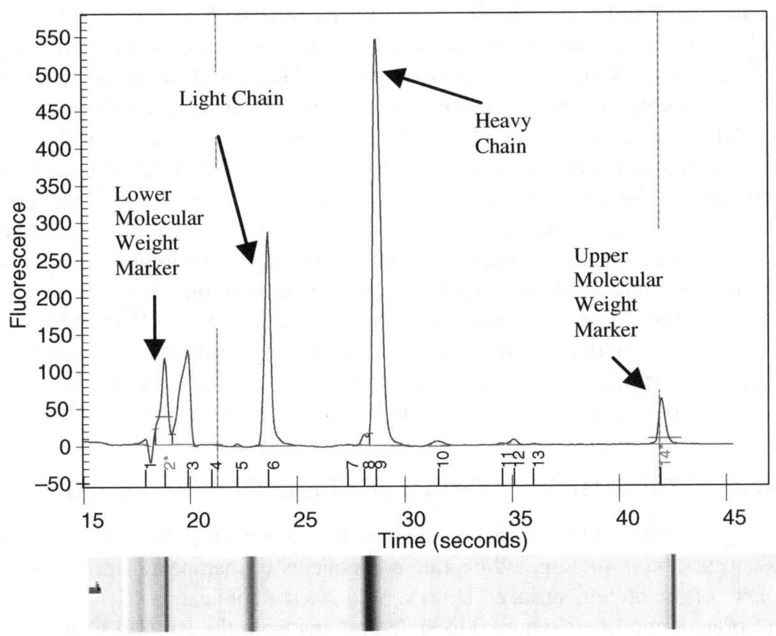

Figure 23.6 Reduced electropherogram and digitally generated 'gel image' demonstrating separation of cB72.3 antibody heavy and light chains using the 2100 Bioanalyser (lab-on-a-chip).

enzyme that catalyses a colour development reaction, upon addition of, a relevant substrate, which can be quantified by spectrophotometry. ELISAs can be both heterogeneous (successive reagent additions, incubation and wash steps; most commonly used for proteins) or homogeneous (reagents added simultaneously; most commonly used for small molecules). The four main subgroups of ELISAs are the direct (antigen bound to solid phase and detected using a single biospecific enzyme-linked antibody), indirect [antigen bound to solid phase and a biospecific antibody or reagent added to antigen prior to detection using a secondary enzyme-linked antibody or reagent (recognizing the first antibody or reagent, not the antigen)] and then direct and indirect forms of the sandwich ELISA [where the antigen is sandwiched between antibody or reagent coated to the solid phase and a second (or third in the case of indirect) antibody/reagent bound to enzyme for detection].

ELISAs confer the advantages of being biospecific, very sensitive (LOD in ng/ml), simple, requiring very small volumes of reagents, being relatively cheap and reasonably rapid and there being a number of commercial kits for various therapeutic proteins and/or impurities available off the shelf. In addition, method development is reasonably rapid and the format is adaptable. The disadvantages of ELISA include reduced precision compared with alternative methodologies (e.g. HPLC) and the long lead times required to generate biospecific antibodies if commercial options are not available.

23.3.8 Size Exclusion Chromatography

Size exclusion chromatography (SEC), also termed gel permeation chromatography (GPC), is a highly versatile, non-destructive, method primarily used for the identification, quantitation and separation of protein aggregate, monomeric and fragment components of protein therapeutics on the basis of size. Protein elution is usually monitored using conventional spectrophotometric detection at selected wavelengths (e.g. 215 nm or 280 nm). The sensitivity of the method is dependent on the choice of column and the detection system employed (e.g. diode array detection) but the technique should in general provide a limit of detection of $\leq$0.1 % for any impurity (e.g. aggregate; Figure 23.7).

SEC is usually employed in parallel with other purity tests such as SDS-PAGE. It can be automated for analysis of a large number of samples, and offers good reproducibility and precision of quantitation. SEC examines the native molecule whereas SDS-PAGE relies on denaturation of the protein. The tests are complementary. Therefore, different impurities are often detected (see also reversed-phase HPLC in Section 23.3.9). In addition, the pore size of the column matrix can be chosen to optimize the separation of particular protein components of different molecular weights.

SEC is also used for the determination of apparent molecular weight. However, for non-globular proteins, migration through the SEC matrix is not linear with respect to the routine molecular weight standards available, and thus may give grossly inaccurate apparent molecular weights.

In addition, hydrophobic and basic proteins often interact with the column matrix, resulting in delayed elution, peak broadening and reduced resolution between peaks. Often, this can be resolved through manipulation of the mobile phase to include a small concentration of detergent or organic solvent, or by altering the salt concentration or pH of the mobile phase, or by adding certain components (e.g. basic amino acids such as arginine and lysine) to minimize undesirable interactions.

23.3.9 Reversed-Phase High Performance Liquid Chromatography

Reversed-phase HPLC (RP HPLC) is regarded as a highly versatile technique for the isolation, quantitation and identification of protein and peptide components of therapeutics primarily on the basis of hydrophobicity of the protein surface. However, due to the application of a solvent gradient and the addition of organic modifiers required to elute proteins from the RP HPLC matrices, it may or may not denature (and in some cases, may precipitate) the protein of interest, and/or lead to a loss

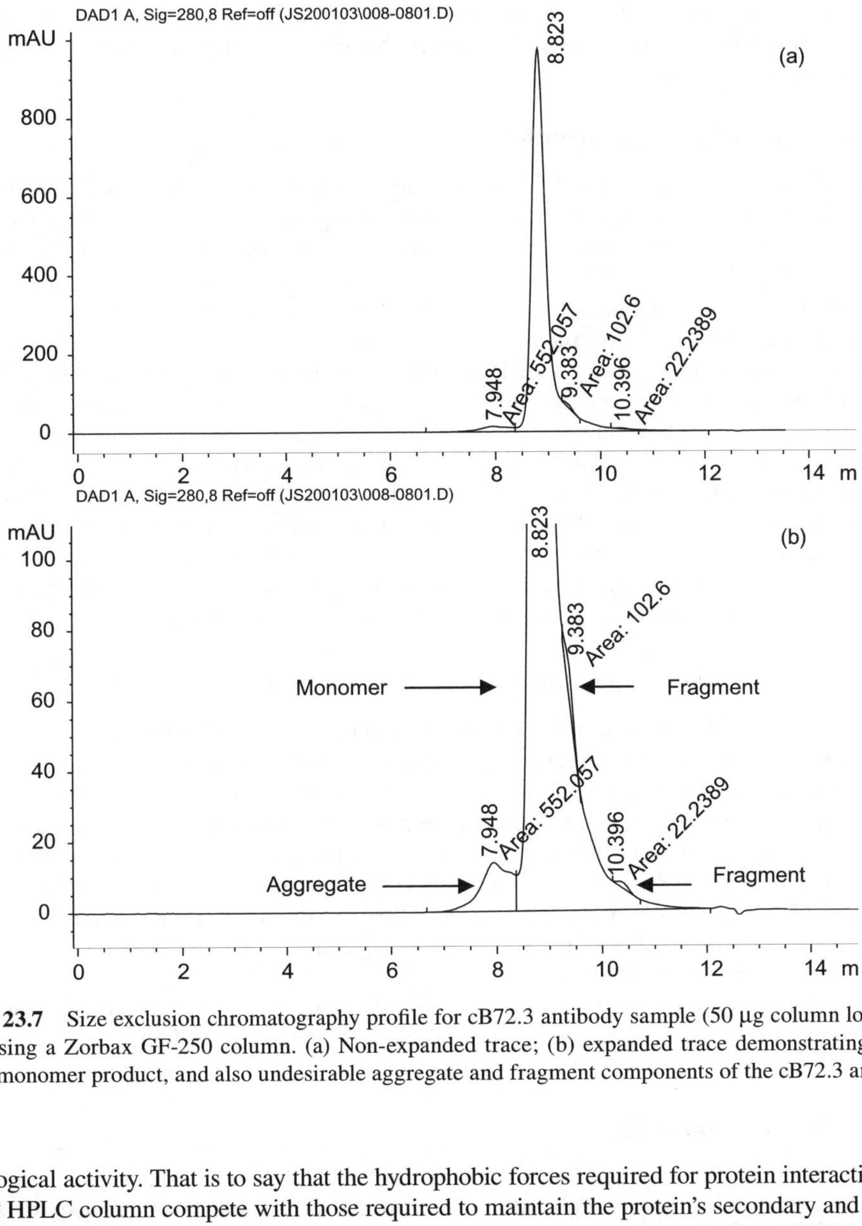

Figure 23.7 Size exclusion chromatography profile for cB72.3 antibody sample (50 μg column load) analysed using a Zorbax GF-250 column. (a) Non-expanded trace; (b) expanded trace demonstrating the required monomer product, and also undesirable aggregate and fragment components of the cB72.3 antibody.

in biological activity. That is to say that the hydrophobic forces required for protein interaction with the RP HPLC column compete with those required to maintain the protein's secondary and tertiary structure. Nonetheless, it is widely used as a complementary tool alongside SEC and SDS-PAGE for process monitoring, purity and stability determinations. Each protein, in general, demonstrates a reproducible RP HPLC 'footprint'. Optimization of RP HPLC chromatographic separation is dependent on the choice of column matrix, the mobile phase composition and the temperature of separation.

Like SEC, RP HPLC can be automated for analysis of a large number of samples at once and offers good reproducibility and precision of quantitation. Protein elution is usually monitored using conventional spectrophotometric (UV) detection at selected wavelengths (e.g. 215 nm or 280 nm). As for SEC, the sensitivity of the method is dependent on the choice of column and the detection system employed (e.g. diode array detection, fluorescence). The assay in general should provide a limit of detection of ≤0.1 % for impurities. Alternatively, intrinsic fluorescence [excitation 280 nm, emission at 320 nm (tyrosine) or 340 nm (tryptophan)] may be employed where an increase in sensitivity is required to differentiate between protein peaks and peaks originating from non-protein impurities.

23.3.10 Affinity Chromatography

Primarily used for identification and in-process purification of proteins, affinity chromatography offers a high degree of purification in a single step, with good associated recoveries even in the presence of a large number of impurities. For antibodies, the use of immobilized protein A or protein G is very popular and available on a commercial scale. Affinity matrices are most amenable to identification and quantitation of proteins, as the resultant protein is generally >95 % pure (for proteins separated using high specificity affinity ligands).

Affinity chromatography using an HPLC system can be automated for analysis of a large number of samples in sequence, and offers good reproducibility and precision of quantitation (typically <1 % CV). Protein elution is usually monitored using conventional spectrophotometric (UV) detection at selected wavelengths (e.g. 215 nm or 280 nm). The sensitivity of the method is dependent on the affinity of the column matrix for the protein of choice and the detection system employed (e.g. variable wavelength versus diode array detection). The major disadvantage of affinity columns is their cost and the potential leaching of ligand from the column matrix. However, very sensitive commercially available assays have been developed for ligands such as protein A and protein G, and thus potential problems of this nature can be monitored.

23.3.11 Ion Exchange Chromatography (Cation or Anion)

Another method for the identification of proteins is ion exchange chromatography. Proteins are separated based on the binding of negatively charged molecules to a positively charged matrix, or *vice versa*. The method is particularly useful for the identification of proteins with an isoelectric point (pI) which is significantly different from that of most of the impurities with which it may otherwise co-elute using other identification methodologies. It is also a very useful tool for monitoring of different isoforms of proteins that may not traditionally be amenable to IEF analysis due to their pI. Although resolution of different isoforms is often inferior to IEF, it is more amenable to quantitation than IEF and like all HPLC-based assays, can be automated for analysis of a large number of samples in sequence.

Protein elution is usually monitored using conventional spectrophotometric (UV) detection at selected wavelengths (e.g. 215 nm or 280 nm). The sensitivity of the method is dependent on the elution conditions, the choice of column matrix, and the detection system employed.

23.3.12 Peptide Mapping

Peptide mapping is a key method for monitoring product composition, homogeneity, and essentially as an identification tool through separation of proteolytic peptides of the product (most often tryptic) by reversed phase HPLC to generate a unique 'fingerprint' specific to its primary amino acid sequence.

Proteins are analysed in a native or denatured (reduced and alkylated) form followed by digestion using proteolytic enzymes (e.g. trypsin, chymotrypsin) into specific peptide fragments. These peptides are then routinely separated based on the peptides' hydrophobicity by RP HPLC (Section 23.3.9). The efficiency of digestion is assessed by SDS-PAGE. Example chromatograms demonstrating comparable peptide map profiles for a model antibody produced under two different fermentation conditions are shown in Figure 23.8. These show almost identical chromatograms indicating that in this case the differences in culture conditions did not appear to affect the product although other analytical techniques may reveal differences.

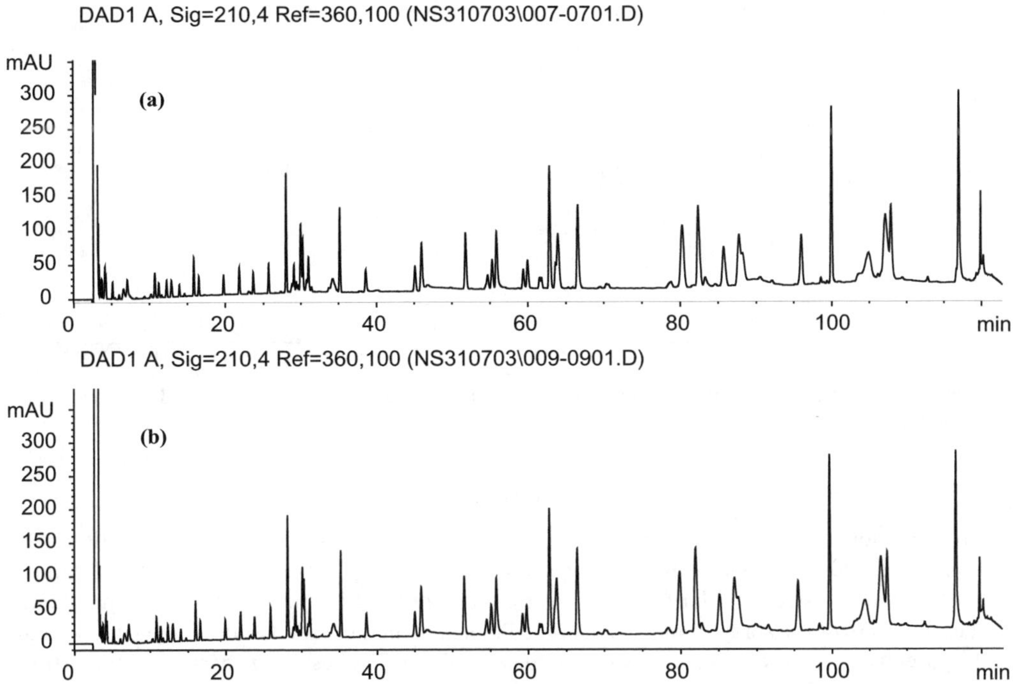

Figure 23.8 RP HPLC tryptic peptide map profiles (absorbance 210 nm) for the cB72.3 antibody demonstrating comparable peptide map profiles for two different fermentation processes (a) and (b)

Although not a simple task for very large proteins, it is a sensitive tool for assessing protein heterogeneity including variations in glycosylation and amino acid substitutions (which lead to a change in peptide hydrophobicity etc.; Harris *et al.* 1993). Peptide mapping is routinely used to identify post-translational modifications, including deamidation and oxidation, and is an important method to consider for stability study analyses.

23.3.13 Liquid Chromatography–Mass Spectrometry

In addition to routine peptide mapping of proteins, further characterization may be required at various stages of a product's life cycle to confirm product composition unequivocally. Mass spectrometry has been used directly, interfaced with liquid chromatography (LC-MS) systems, to support conventional N-terminal sequencing (which is limited to the first 20 to 30 residues) for identification purposes to determine the primary structure of proteolytic fragments. LC–MS permits accurate mass determination of peptide fragments with high precision (± 0.01 %) permitting exact confirmation of primary structure (also see Section 23.3.14.1).

23.3.14 Mass Spectrometry

Mass spectrometry (MS) is a powerful tool for characterization of the primary and secondary structure of proteins. The two main techniques used routinely for the analysis of proteins include electrospray ionization mass spectrometry (ESI-MS; Section 23.3.14.1) and matrix-assisted laser desorption ionization time-of-flight mass spectrometry (MALDI-TOF-MS; Section 23.3.14.2). These have different advantages and disadvantages but both offer unsurpassed sensitivity, precision of measurement ($\leq \pm 0.01$ %) and adaptability to multiple assay formats.

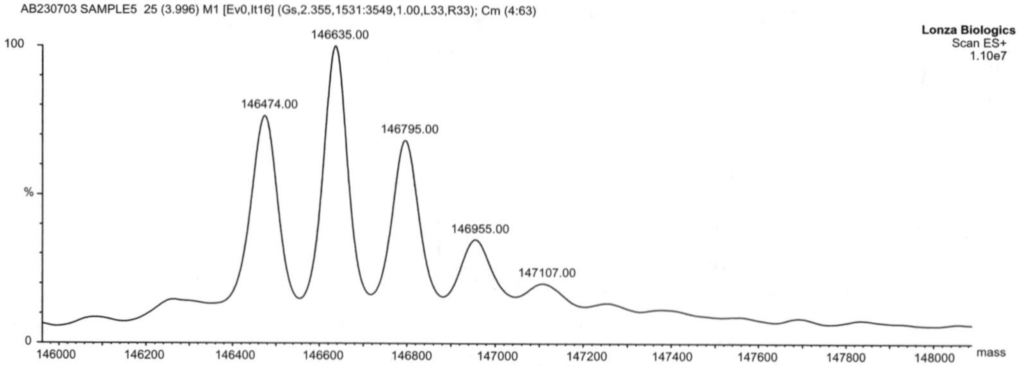

Figure 23.9 Electrospray ionization mass spectrometry profile of the cB72.3 antibody produced in a GS-CHOK1SV cell line.

23.3.14.1 Electrospray ionization mass spectrometry

Electrospray ionization mass spectrometry (ESI-MS) is an atmospheric-pressure ionization (API) technique whereby a sample is dissolved in a mobile phase (e.g. water/acetonitrile) and pumped through a stainless steel capillary at atmospheric pressure and high potential. This high potential creates an electrostatic spray of multiple charged ions that are propelled under increasing vacuum in the gaseous phase, towards a high vacuum mass analyser. The resulting spectrum demonstrates multiple charge peaks for each component of an identifiable mass. The spectrum can then be transformed ('deconvoluted') to demonstrate the relative mass of a 'zero-charged' mass species.

It is considered a 'soft' ionization technique, using the minimum energy required for ionization, thus permitting analysis of proteins without extensive fragmentation. This permits a degree of secondary conformation analysis of proteins unlike MALDI-MS. In addition, the technique is suitable for interfacing with standard chromatographic techniques such as RP HPLC and CE. The main disadvantages include the requirements for complex data interpretation (although modern software has alleviated many of these issues), and its reduced sensitivity and general mass range compared with MALDI-MS. However, with the advancement in column technology and reduced flow rates, the sensitivity of ESI-MS can be improved. It is often more suitable than MALDI-MS for the analysis of very high molecular weight proteins, where the tendency to produce multiply charged ions permits the mass-to-charge ratios of these large proteins to fall within the range of mass spectrometers.

Figure 23.9 demonstrates an ESI-MS profile for the cB72.3 antibody produced by a CHO cell line. A number of different mass isoforms are readily visible.

23.3.14.2 Matrix-assisted laser desorption ionization time-of-flight mass spectrometry

Unlike ESI, which uses a liquid matrix for ionization, MALDI-TOF employs a crystalline matrix. In addition, samples are ionized by laser rather than voltage. The direct result is a dramatic increase in both sensitivity and mass range of target proteins and peptides. The protein (or peptide) sample is dissolved in a matrix (e.g. 2,5-dihydroxybenzoic acid and sinapinic acid) and permitted to crystallize onto a stainless steel template. The template surface is then bombarded with a pulsed laser beam (typically generated at 337 nm using a nitrogen laser), and molecules are ionized either in positive-ion mode or in negative-ion mode, depending on the matrix and analyte used. The time taken for the ion to travel a set distance to the detector is measured with great accuracy and is directly related to the mass-to-charge ratio of the ion. Figure 23.10 demonstrates a typical MALDI-TOF MS profile for the recombinant cB72.3 antibody produced at Lonza Biologics.

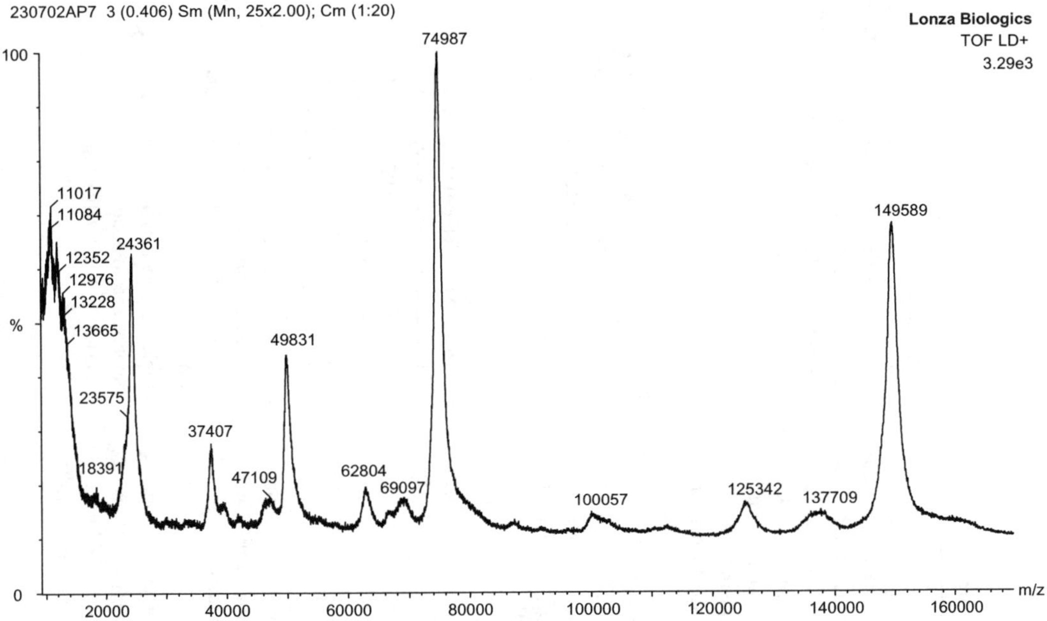

Figure 23.10 MALDI-TOF mass spectrometry profile of intact cB72.3 antibody produced in a GS-NS0 cell line.

The advantages of MALDI-TOF-MS include a very wide mass range for the analysis of proteins (e.g. proteins with molecular weights >300 kDa have been ionized successfully), and high sensitivity. Fragments and multiply charged ions are generally of low abundance in this ionization mode. The main disadvantage of MALDI-TOF-MS is the incompatibility of a crystalline matrix to other analytical interfaces.

23.3.15 Carbohydrate Analysis

Recombinant proteins may contain a variable amount of carbohydrate per recombinant protein molecule ranging from approximately 1 % to 40 % (w/w), and this can be critical for therapeutic protein efficacy *in vivo*. Carbohydrate content is also very dependent on the cell line chosen for expression and is often very sensitive to very small changes in process variables (James & Baker 1999).

It is increasingly common to monitor and measure the relative proportions of oligosaccharides bound to recombinant therapeutics, both in terms of their monosaccharide content, sialic acid content and site occupancy. For early-phase evaluations and routine monitoring purposes, glyco-fingerprinting techniques such as MALDI-MS analysis may offer sufficient information. The choice of analytical strategy employed will be dependent on the level of structural information required, the sample composition and the amount of protein available for analysis (Anderson *et al.* 1996; Harvey 2001; Davies & Hounsell 1998). A number of techniques are routinely employed to help analyse and quantitate the bound oligosaccharides of therapeutic proteins and they are summarized in Table 23.3 and are elaborated on in greater detail in Chapter 24. Examples of carbohydrate profiles generated for the cB72.3 antibody using MALDI-MS, normal phase HPLC, cation exchange HPLC and gel electrophoresis are demonstrated in Figures 23.11 to 23.16 respectively.

Figures 23.11 and 23.12 demonstrate MALDI-TOF MS analysis of released *N*-linked glycans from the cB72.3 antibody generated in different cell lines (GS-NS0; Figure 23.11 and GS-CHOK1SV; Figure 23.12). Table 23.4 outlines the predicted glycan structure for each mass species. An example ESI-MS spectrum of oligosaccharides released from bovine fetuin is shown in Figure 23.13.

Table 23.3 Summary of common techniques used for carbohydrate analysis of recombinant therapeutics.

Technique	Assay format characteristics	Advantages	Disadvantages
MALDI-MS	Relatively simple technique whereby samples are co-crystallized on a metal surface with an excess of low molecular weight high UV-absorbing matrix.	Simple. Amenable to analysis of either whole glycoproteins or released glycans. Easy tool for glycan sequencing with exoglycosidases. Useful as early-phase screening tool. Can be interfaced with tryptic peptide analysis (PMAP). Sensitivity can be increased by derivatization with fluorescent labels prior to analysis. Rapid analysis.	High capital outlay. Not suitable for analysis of very low levels of glycosylation. Quantitation less reliable than HPLC.
ESI-MS	Samples are ionized in volatile solvents resulting in multiply charged ions which can be 'deconvoluted' to yield a single zero-charged peak.	Easily coupled on-line with both HPLC and CE applications (e.g. PMAP or CE). Amenable to analysis of either whole glycoprotein or released glycans. Easy tool for glycan sequencing with exoglycosidases. Low picomole to femtomole sensitivity.	High capital outlay. Data analysis more complex than MALDI-MS techniques. Derivatization of oligosaccharides required for glycoanalysis.
RP HPLC	Method of choice for separation of glycopeptides (i.e. with PMAP) for identification of gross changes in glycosylation.	Simple. Usually no additional capital outlay required. Routine technology available to most laboratories.	Samples require derivatization (eg. fluorescent) to obtain acceptable sensitivity. Large glycoproteins often demonstrate over-complex 'glycopeptide maps' to delineate any useful data. Relatively extensive analysis time.
Normal phase HPLC	Derivatized (usually) samples are separated using a normal phase column on the basis of (oligosaccharide) size.	Simple. Usually no additional capital outlay required. Routine technology available to most laboratories. Commercial kits available.	Samples require derivatization (e.g. fluorescent) to obtain acceptable sensitivity. Relatively extensive analysis time.
Cation exchange HPLC	Derivatized (usually) samples are separated using an inert anion-exchange column on the basis of (oligosaccharide) charge.	Simple. Usually no additional capital outlay required. Routine technology available to most laboratories. Commercial kits available. Excellent tool for charge profiling of glycoproteins.	Samples require derivatization (e.g. fluorescent) to obtain acceptable sensitivity. Relatively extensive analysis time.

High pH anion exchange chromatography with pulsed amperometric detection	The most sensitive HPLC-based method for detection and resolution of non-derivatized charged (sialylated) oligosaccharides.	Sensitive. Does not require labelling. Commercial assay kits available. Amenable to measurement of both acidic and neutral monosaccharides and oligosaccharides.	Moderate capital outlay. Requires specific LC system. Reduced resolution of neutral and sulphated oligosaccharides. Aggressive eluents (eg. 1 M NaOAc and 0.1 M NaOH). Relatively extensive analysis time.
CE	Variety of CE applications have been used to separate intact glycoproteins, glycopeptides, oligosaccharides and monosaccharides.	Can be combined with more in-depth tools such as ESI-MS for in-depth characterization. Versatile tool for 'problematic' glycoproteins that are not amenable to other separation techniques.	Moderate capital outlay. Some acidic CE buffers unsuitable for the separation of sialylated species. Resolution may be inferior to alternative methods.
Gel electrophoresis (e.g. FACE)	Fluorescently labelled glycans are separated on the basis of size through a high density polyacrylamide matrix.	Simple. Sensitive. Low capital outlay. Commercial kits available. In combination with exoglycosidase digestion, method offers limited structural information suitable for screening early-phase material.	Samples require fluorescent derivatization. Quantitation less precise compared with HPLC.

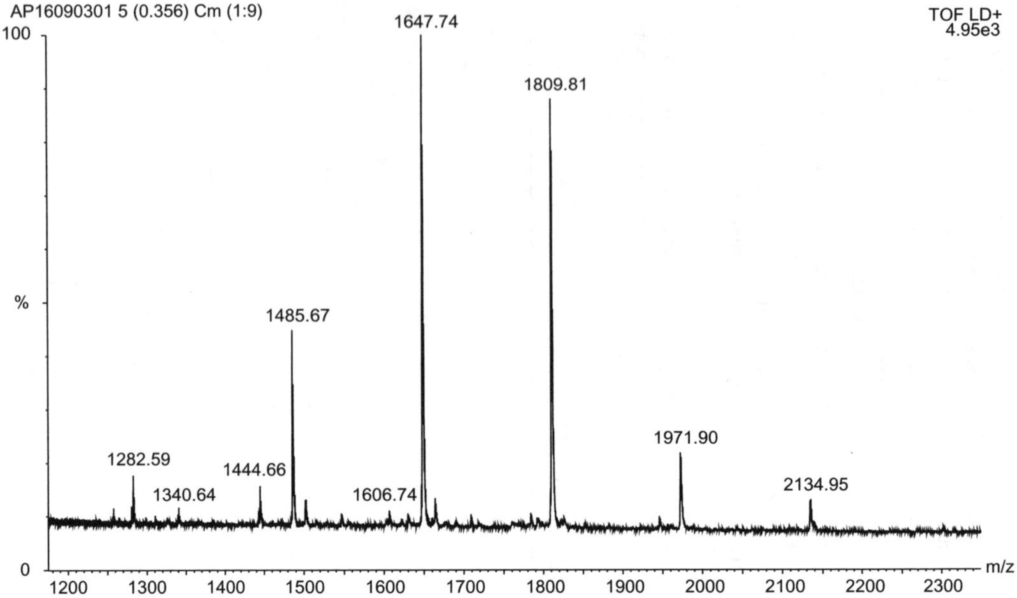

Figure 23.11 MALDI-TOF mass spectrometry of the cB72.3 antibody derived oligosaccharides produced in a GS-NS0 cell line.

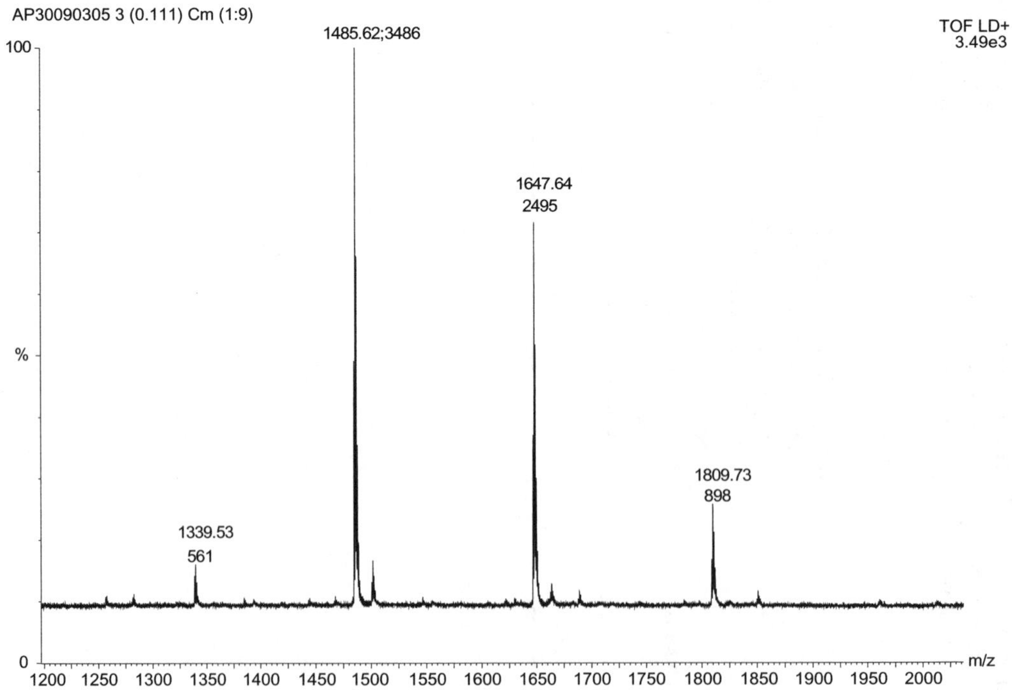

Figure 23.12 MALDI-TOF MS oligosaccharide profile of the cB72.3 antibody produced in a GS-CHOK1SV cell line.

Table 23.4 Oligosaccharide species identified routinely for the cB72.3 antibody.

Structure	Mass (Da)
G2F	1810
G1F	1648
G0F	1486
G1F-GN	1445
G0	1341
G0F-GN	1283
Man-5	1258

Key. G: Galactose; F: Fucose; GN: Glucosamine; Man-5: oligomannose-5. e.g. G2F refers to a glycan with core fucosylation and 2 terminal galactose residues.

An example of a normal phase separation of desialylated cB72.3 antibody derived glycans produced in different cell lines is demonstrated in Figure 23.14. Figure 23.15 demonstrates an example of cation exchange HPLC separation of 2-AA labelled charged (sialylated) species of the cB72.3 antibody produced in both GS-NS0 and GS-CHO cell lines. An example of gel electrophoresis separation of fluorescently labelled glycans is shown in Figure 23.16.

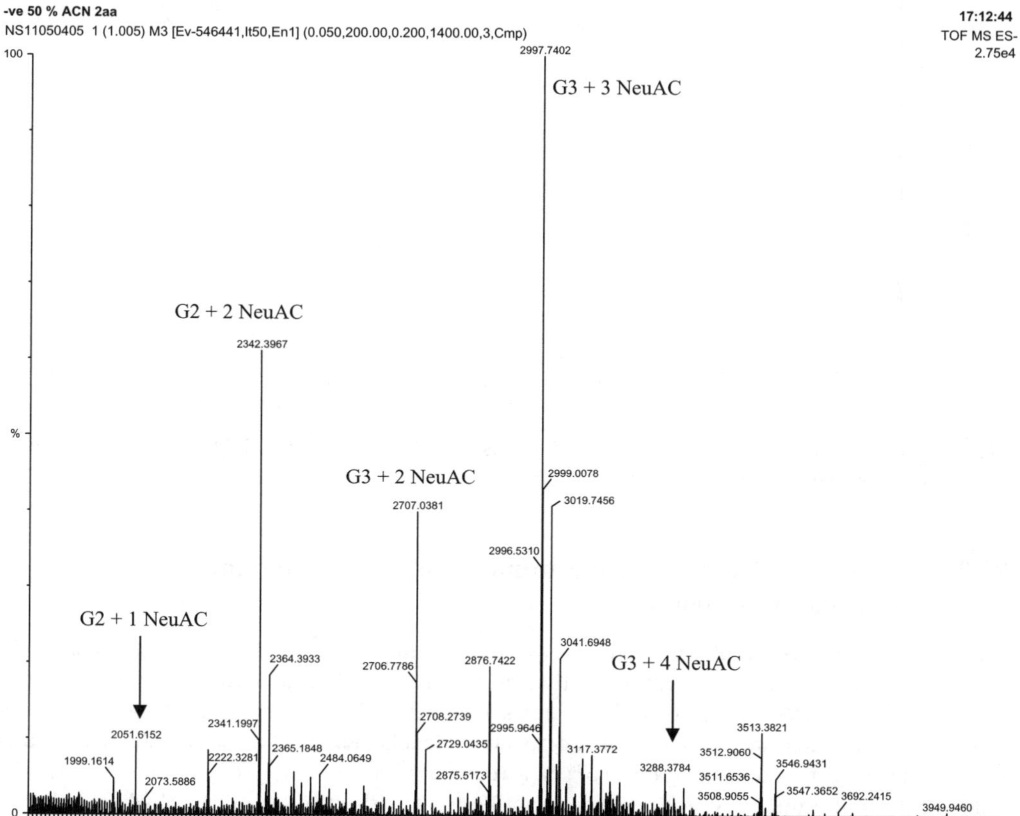

Figure 23.13 Electrospray ionization mass spectrometry of 2-AA labeled oligosaccharides from bovine fetuin. 'G' refers to the number of galactose residues, and 'NeuAc' refers to the number of sialic acid (*N*-acetylneuraminic acid) residues.

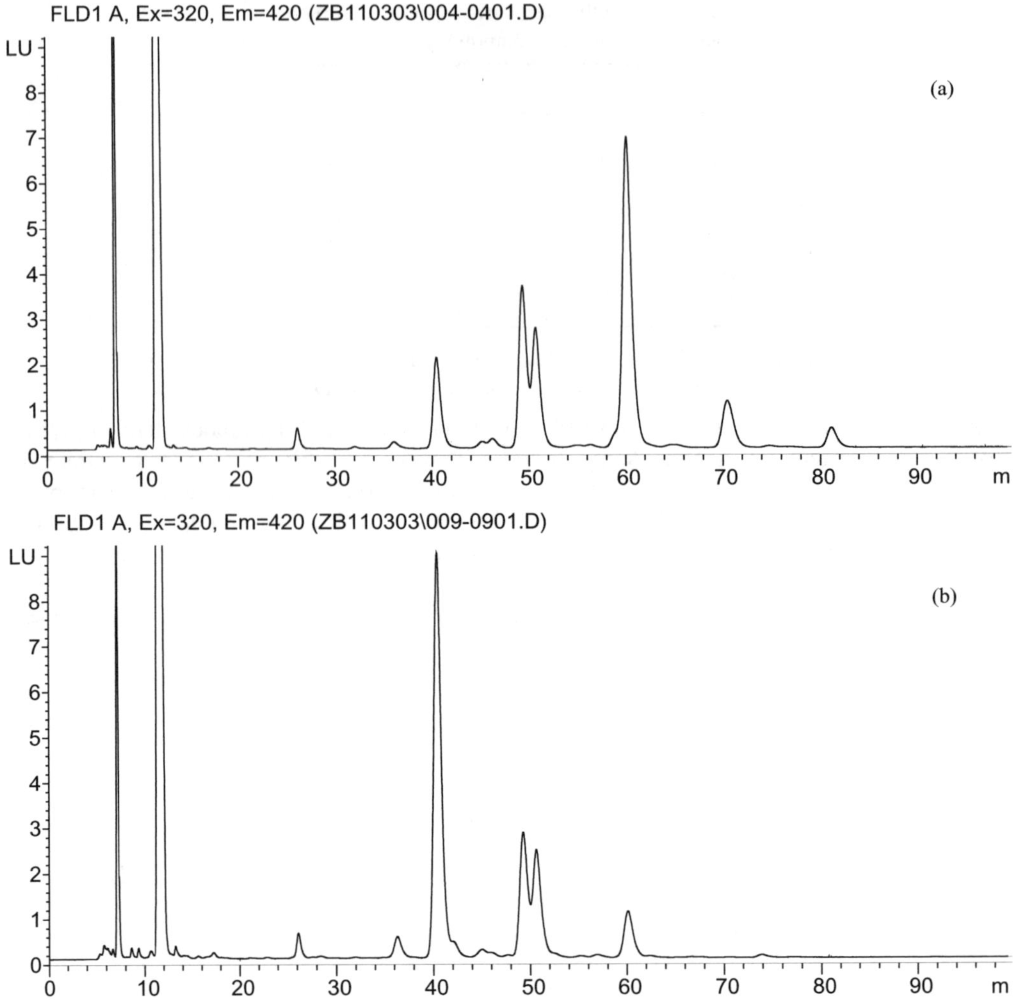

Figure 23.14 NP HPLC neutral (2-AA labelled) oligosaccharide profile of the cB72.3 antibody from an (a) GS-NS0 cell line and a (b) GS-CHO cell line.

23.3.16 Structural Assays for Determination of Conformation, Size and Aggregation

The correct conformation of a protein is key to maintaining its particular function. A change in this specific spatial orientation can not only lead to a loss of specific activity or potency but to other unwelcome affects for protein therapeutics. There have been several reports of patients developing an immune response to protein therapeutics (Schellekens, 2003). A possible cause of this has been a change in size (aggregation) or shape (denaturation/unfolding) of the protein molecule, inducing the host immune system to generate antibodies to either protein or carbohydrate epitopes of the protein drug that previously had not been presented to the immune system. This has caused severe effects to patients including fatalities (Casadevall *et al.*, 2002).

It is, therefore, necessary to employ analytical techniques to determine the size, shape and specific structural elements (secondary structure, folding) of protein therapeutics during development

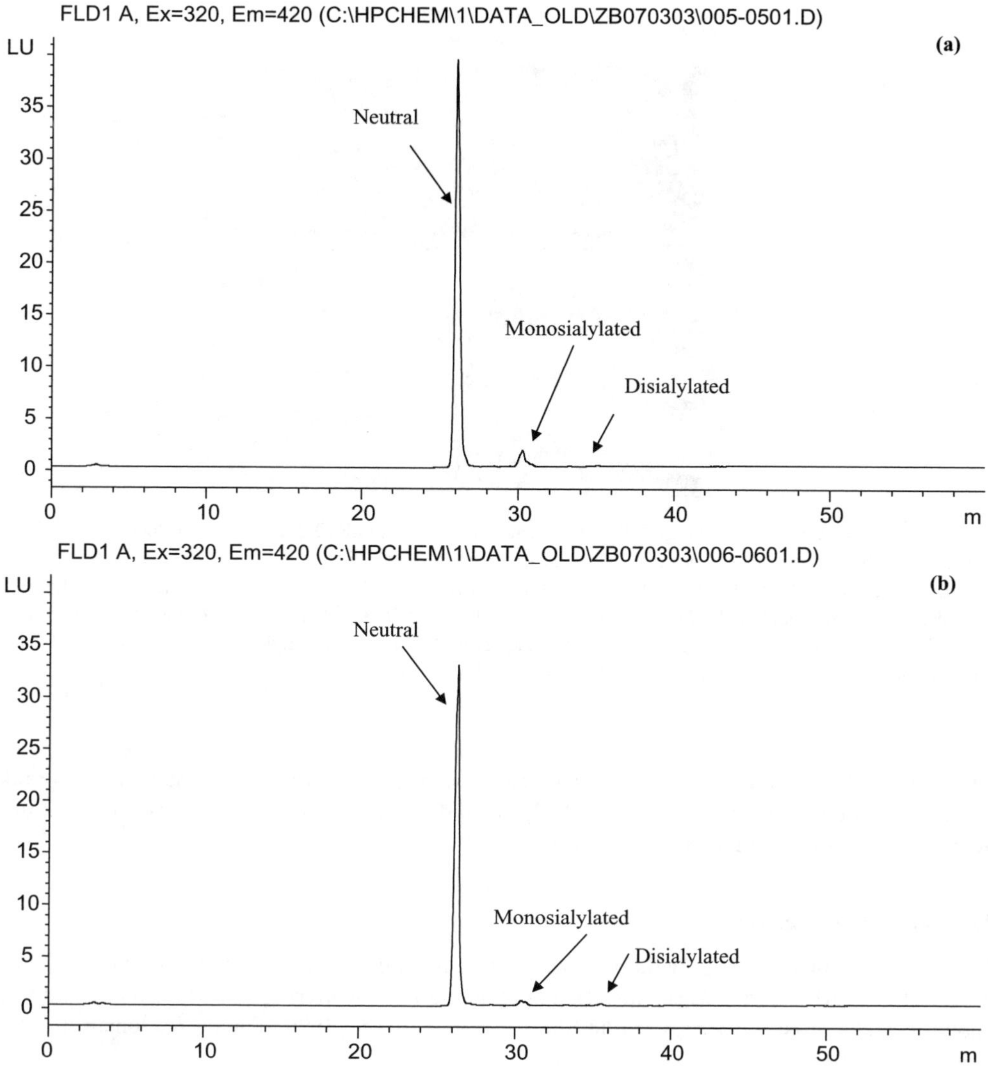

Figure 23.15 Comparison by cation exchange HPLC of 2-AA labelled oligosaccharide charge profiles for the cB72.3 antibody from an (a) GS-NS0 cell line and a (b) GS-CHO cell line.

programmes as well as following any significant changes to the production process or formulation. There are a large number of techniques available for this purpose. These include well-established traditional methods and also new emerging techniques. A summary (by no means exhaustive) of the more widely used analytical methods for structural determination is presented in Table 23.5. These techniques are most often employed to confirm structural integrity, stability, and process and product consistency.

23.3.16.1 Size exclusion chromatography

An example of size exclusion chromatography (SEC) has been provided in Figure 23.7. Its application to structural determination includes molecular weight (MW) determination, based on elution

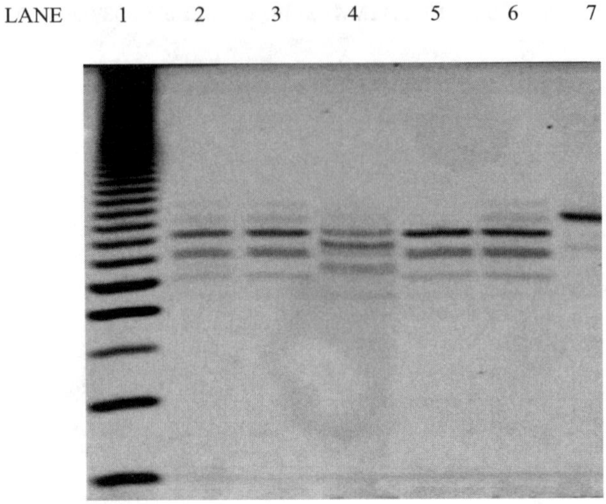

Figure 23.16 FACE sequencing gel of oligosaccharides from the cB72.3 antibody produced in a GS NS0 cell line and digested using various exoglycosidases.
Key–Lane 1, calibration standard; lane 2, intact (sialylated) cB72.3 oligosaccharides; lane 3, desialylated cB72.3 oligosaccharides; lane 4, desialylated cB72.3 oligosaccharides (fucosidase treated); lane 5, desialylated cB72.3 oligosaccharides (galactosidase treated); lane 6, desialylated cB72.3 oligosaccharides (mannosidase treated); lane 7, desialylated fetuin profiling control.

relative to protein markers with defined molecular weights. However, the technique does not take account of protein conformation (Table 23.3) or binding of the protein of interest to the column matrix, and additional analyses are required.

23.3.16.2 Light scattering techniques

A number of light scattering methods are available, including low angle and multiple angle laser light scatter (LALLS/MALLS) analyses and dynamic light scattering (DLS), also referred to as photon correlation spectroscopy (Folta-Stogniew & Williams 1999; Tianbo & Benjamin 2002; Wen *et al.* 1996).

LALLS/MALLS is an independent determination of hydrodynamic radius and MW. The technique is dependent on accurate protein concentration and requires either a UV or retractive index (RI) detector. Essentially, the angle of light scatter is dependent on the protein size, and is routinely coupled with more traditional analyses such as SEC. An example of LALLS/MALLS compared to traditional UV detection for the cB72.3 antibody is demonstrated in Figure 23.17.

DLS measures time-dependent light scattering intensity fluctuations due to the Brownian motion of particles in solution. It is a rapid tool for measuring the hydrodynamic size (i.e. Stokes's radius) directly (i.e. it does not require chromatographic separation of components), making it a technology available to most laboratories. Although the resolution between, for example, monomeric and dimeric aggregates may not be definitive, the technique is ideal for size measurement across a wide range of hydrodynamic diameters and has therefore been used for the characterization of heterogeneous protein conjugates (e.g. 'pegylated' proteins). The intensity of response is proportional to mass (an amount of a large aggregate will give a larger response than the same amount of monomer), An example is shown in Figure 23.18.

Table 23.5 Analytical Techniques Commonly Employed to Confirm Structural Integrity, Stability and Consistency

Analytical method	Aggregation	Conformation	Comments
Size exclusion (SE) HPLC	✓✓✓✓✓	✓	Simple method with excellent precision and sensitivity for aggregates. Less sensitive for the detection of conformational differences
Light scattering	✓✓✓	✓✓✓	Range of simple techniques available. Good sensitivity and can identify changes to shape/size of the molecule.
Analytical ultracentrifugation (AUC)	✓✓✓	✓✓✓	Good sensitivity and can identify changes to shape/size of the molecule. Reference technique for size determination of aggregated and monomeric product. Quite specialized.
X-Ray crystallography	X	✓✓✓✓✓	Powerful technique for establishing conformation, but dependent on the ability of protein to form crystals. Very specialized.
Nuclear magnetic resonance (NMR)	X	✓✓✓✓✓	Powerful technique for establishing conformation, but range limited to protein up to approximately 25 kDa. Quite specialized.
Circular dichroism (CD)	✓	✓✓✓	Good for the determination of secondary structure, particularly helices. Quite specialized.
Infrared (IR)	✓✓	✓✓✓✓	Good for the determination of secondary structure, particularly sheet structures. Current instrumentation more sensitive than CD and developed for routine operation. Specific area of the spectrum characteristic of aggregation.
Viscometry	X	✓✓✓	Simple instruments can be used with SEC to measure density of protein and consequently differences in conformation.
Ultrasonics	✓✓✓	✓✓✓	An emerging technique that measures the elasticity of proteins (affected by aggregation and conformation).
Electron tomography	✓✓✓✓	✓✓✓✓	Similar to X-ray crystallography but more widely applicable. Provides measure of aggregation and conformation.
Fluorescence	X	✓✓✓✓	Simple and rapid method for determination of changes in conformation.
Differential scanning calorimetry (DSC)	X	✓✓✓✓	Simple and rapid method for determining changes in conformation. Extensively applied to stability and formulation development. See Section 20.4.1.1.

Key: ✓ *Increasing number of ticks indicates an increasing value of analysis.*
X *Not recommended for analysis of the respective characteristic.*

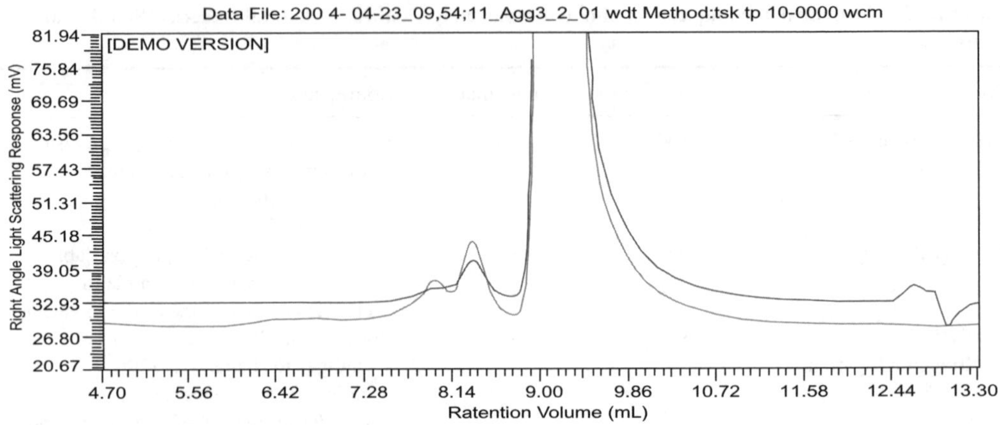

Figure 23.17 The application of LALLS/MALLS (lower trace) compared to traditional UV detection (upper trace) for testing cB72.3 antibody samples.

23.3.16.3 Viscometry

There are a number of simple viscometers on the market that can be coupled with traditional techniques such as SEC to measure the density of a protein, and consequently detect changes in secondary conformation. Put simply, the macromolecular size and shape of a molecule determines its relative resistance to flow (Harding 1997). An example trace demonstrating replicate viscometry signals for the cB72.3 antibody is shown in Figure 23.19. In combination with traditional techniques such as SEC, viscometry presents a more complete picture of molecular structure and conformation.

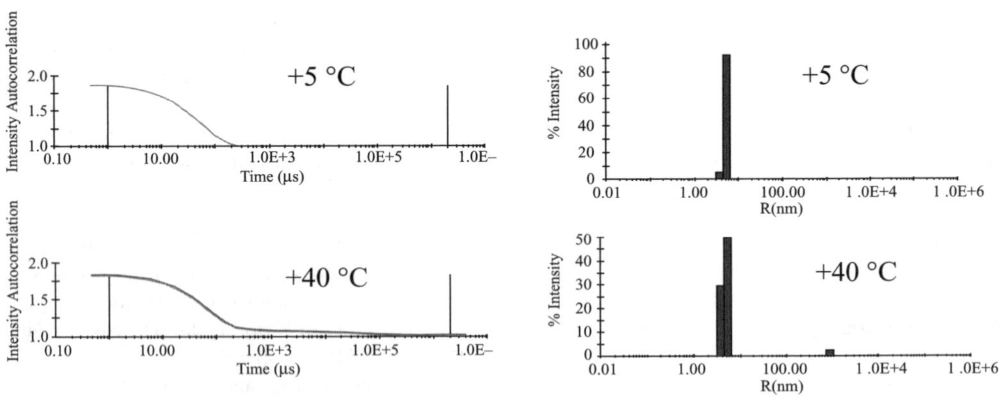

Sample	Rh (nm)	MW (kDa)	Comments
+5 °C	5.6	188	No apparent large aggregates
+40 °C	5.5	184	Trace amount large aggregates
	881.9	2623230	

Figure 23.18 Example of DLS for the detection of trace amounts of cB72.3 aggregates that were below the limit of detection by traditional methodologies.

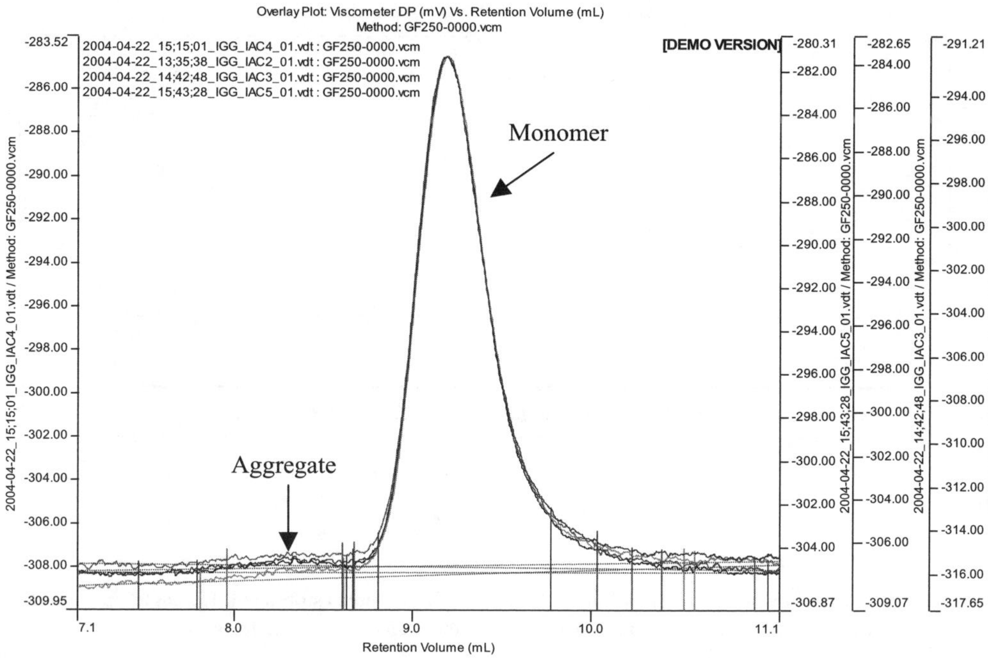

Figure 23.19 Example of cB72.3 antibody replicate viscometry traces. The observed signal is very similar to a traditional UV (SEC) trace.

23.3.16.4 Analytical ultracentrifugation

Analytical ultracentrifugation (AUC) is a well-established technique that is not widely used, as it is both expensive and relatively complicated to implement. The sedimentation velocity of a protein is measured using an analytical ultracentrifuge and the determined sedimentation coefficient (in Svedbergs) is used for the calculation of protein monomer, dimer and fragment profiles (Huang *et al.* 2000). It offers good sensitivity to changes in shape and size. AUC can be used as a reference technique to verify the reliability of a SEC method. An example is shown in Figure 23.20.

23.3.16.5 X-Ray Crystallography

X-Ray crystallography is the most powerful technique used to determine the precise macromolecular three-dimensional atomic structure of large molecules such as proteins. This highly specialized technique is dependent on the ability of the protein to form crystals (membrane proteins, for example, crystallize poorly). Crystals are used because the diffraction pattern from one single molecule could be insignificant, but the many individual, identical molecules in a crystal amplify the pattern. Although crystallization sometimes distorts portions of a structure due to contacts between neighbouring molecules in the crystal, the protein crystals as used for diffraction studies are highly hydrated ('wet and gelatinous') so the structures determined from crystals by X-ray diffraction are usually not much different from the structures of soluble proteins in aqueous solution, making it suitable for testing many therapeutic proteins.

One disadvantage is that X-ray crystal diffraction cannot usually resolve the positions of hydrogen atoms or reliably distinguish nitrogen from oxygen from carbon. This means that the chemical identity of the terminal side-chain atoms is uncertain for aspartate (Asp), glutamine (Gln) and threonine (Thr) and is usually inferred from the protein environment of the side chain (i.e. the

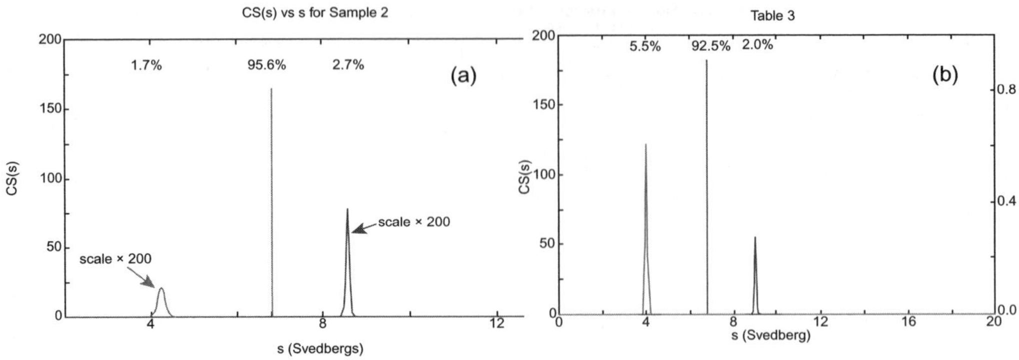

Sedimentation coefficient (Svedbergs)	Calculated MW (kDa)
4	50.5
6.8	165.6
9	330.3

Figure 23.20 Analytical ultracentifugation of cB72.3 stability samples stored at different temperatures (a) 5 ± 3 °C, and (b) 40 ± 3 °C, demonstrating sensitivity for detection of aggregate formation.

side chain orientation that forms the most hydrogen bonds or makes the best electrostatic interactions is selected and built by the crystallographer as the most plausible choice). Sometimes there is also uncertainty about whether an atom that is not part of the protein is a bound water oxygen or a metal ion. Many crystallographers rely on access to a small number of extremely powerful radiation synchrotrons, available in a small number of locations throughout the world, to gain the high degree of electron acceleration necessary for sensitive structural elucidation. Nonetheless, it remains a powerful, if specialized, tool for structural determination (Rhodes 2000).

23.3.16.6 Nuclear magnetic resonance

NMR determines the structure of proteins in solution, but is limited to molecules not much greater than 25 to 30 kDa in size, although increased magnetic field strengths have the potential perhaps to double this limit. In principle, all molecules tumble and vibrate with thermal motion. NMR detects the chemical shifts of atomic nuclei with 'non-zero' spin. The detected shifts depend on the identities and distances of nearby atoms relative to the nuclei.

Although NMR is the method of choice for small proteins that are not readily crystallized, it is not suitable for large therapeutics such as antibodies. The results of NMR analysis are an ensemble of alternative models, in contrast to the unique model obtained by crystallography. However, it offers the advantage of yielding the positions of some hydrogen atoms. Proteins greater than approximately 15 kDa in size require labelling with ^{13}C and ^{15}N prior to analysis. In addition, the protein must be soluble at high concentration (0.2–1 mM, 6–30 mg/ml) and stable for days without aggregation under experimental conditions. Once again, it is quite a specialized technique and the capital costs are quite high.

Alternative NMR techniques, such as solid-state NMR, are useful for difficult-to-crystallize proteins such as membrane proteins. Solid-state NMR can yield time-averaged structures of proteins in the fluid membrane environment, the milieu that is critically important for the function of membrane proteins. However, the technique lacks the sensitivity of solution NMR. In addition, new advances in NMR to increase sensitivity and reduce signal-to-noise problems have been developed that are demonstrating some success for the structural characterization of large protein–protein interaction sites, making it ideal for the determination of structure–activity relationships ('SAR-by-NMR') (Fiaux *et al.* 2002; Shuker *et al.* 1996).

23.3.16.7 Circular dichroism

Circular dichroism (CD) is a light-based spectroscopic technique using polarized light. As therapeutic proteins possess both suitable chromaphores (e.g. amide or aromatic groups) and are chiral (i.e. have distinct left- and right-handed forms of the same molecular structure), they are suitable for CD analysis. There are two regions in the ultraviolet (UV) spectrum that can be investigated in CD. The far-UV spectrum provides useful information on the secondary structure of a protein and the near-UV spectrum can provide disulfide bonding and other tertiary and quaternary structural information on a protein.

23.3.16.7a Far-UV Circular Dichroism (190 nm to 240 nm)

The far-UV spectrum provides useful information on the secondary structure of a protein and provides an estimation of α-helical and β-pleated sheet content of a protein. Therefore, each structural motif will give a characteristic profile, with α-helices dominating the spectra if present. For therapeutic proteins such as IgGs, there is a very low helical content and so the β-pleated sheet structures are clearly definable. This permits the determination of the percentage of each structure relative to the protein's known structure. The far-UV response represents an average solution structure. It remains quite a specialised technique. An example is shown in Figure 23.21(a).

23.3.16.7b Near-UV Circular Dichroism (240 nm to 350 nm)

The near-UV spectra can provide disulfide bonding and other tertiary and quaternary structural information on a protein. The response observed is due to a combination of aromatic side chains and disulfide bonds. An example trace is shown in Figure 23.21(b).

Other than some differences detected at 195 nm in Figure 23.21(a), which were determined to be non-significant, there were no discernable structural differences detected between the samples by CD analysis.

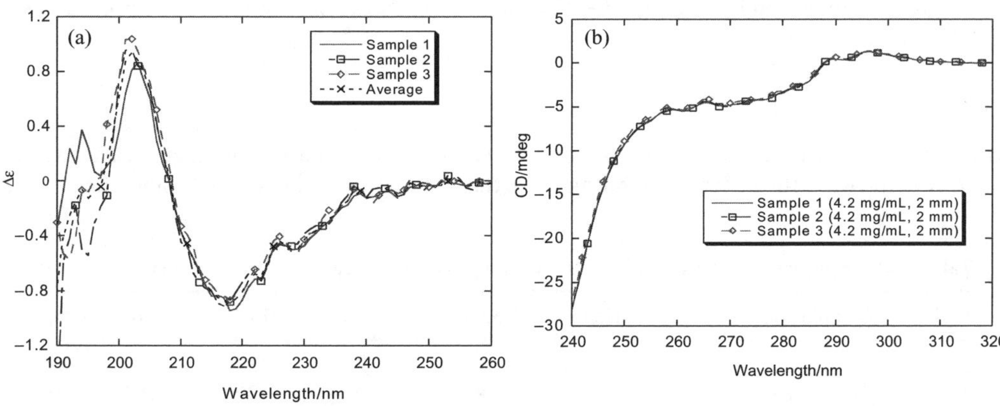

Figure 23.21 Far-UV spectrum (a) and near-UV spectrum (b) for the cB72.3 antibody stored at different temperatures. Sample 1 was stored at 5 ± 3 °C; sample 2 was stored at 40 ± 3 °C and sample 3 was stored at −65 °C or below.

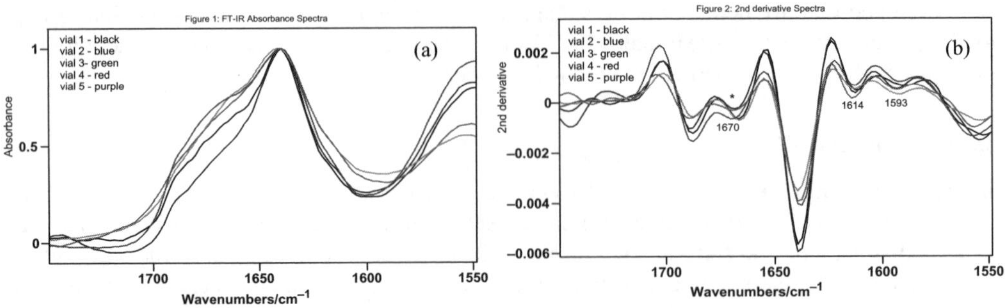

Figure 23.22 (a) FTIR absorbance spectrum and (b) second derivative spectrum for the cB72.3 antibody stored at different temperatures.

23.3.16.8 Fourier transform infrared spectroscopy

Fourier transform infrared spectroscopy (FTIR) is a useful tool for secondary structure determination and for the estimation of α-helical and β-pleated sheet content of a protein. Light at the infrared end of the spectrum interacts with the chemical bonds within a protein. Specifically for FTIR analysis, the amide I band caused by stretching of the carbonyl bond adjacent to the peptide bond is weakly coupled to the bending of the N–H bond. The frequency at which the carbonyl bond stretches is dependent upon the environment it is in. Therefore, different types of secondary structure give rise to different frequencies, which can be measured at the infrared end of the spectrum.

Each structural motif will give a characteristic profile. In addition, analysis of the second derivative spectrum will highlight any particular structural features, permitting the percentage of each structure to be determined relative to known protein structures. The response generated represents an average solution structure. The technique is particularly appropriate for proteins with a high β-pleated sheet content. Beta sheet structures give signals at ~1638 cm^{-1} and ~1690 cm^{-1}. Signals detected at ~1620 cm^{-1} are indicative of product aggregation. See Figure 23.22 for example traces of an absorbance spectrum (Figure 23.22a) and a second derivative spectrum (Figure 23.22b) for the cB72.3 antibody, which had been stored at different temperatures or had undergone induced freeze/thaw cycles. The structure types observed are summarised in Table 23.6.

From these data, changes in both alpha-helical content (indicated by changes at ~1652 cm^{-1}) and aggregate (indicated by changes at ~1620 cm^{-1} or by an increase in β-pleated sheet at ~1700 cm^{-1}) were detectable.

23.3.16.9 Ultrasonics

This technique is based upon passing sound waves through the sample and measuring the velocity of the wave through the sample and also the attenuation of the sound through the sample.

Table 23.6 Summary of the secondary structure types observed in cB72.3 test samples.

Vial	Sample condition	α-Helix	β-Sheet	Bend	Turn	Coil	Sum
1	40 ± 3 °C	5	37	17	13	26	98
2	Freeze/thawed	2	38	19	13	23	95
3	5 ± 3 °C	15	35	10	13	23	97
4	−65 °C or below	2	36	20	12	22	92
5	5 ± 3 °C	10	38	12	13	24	97

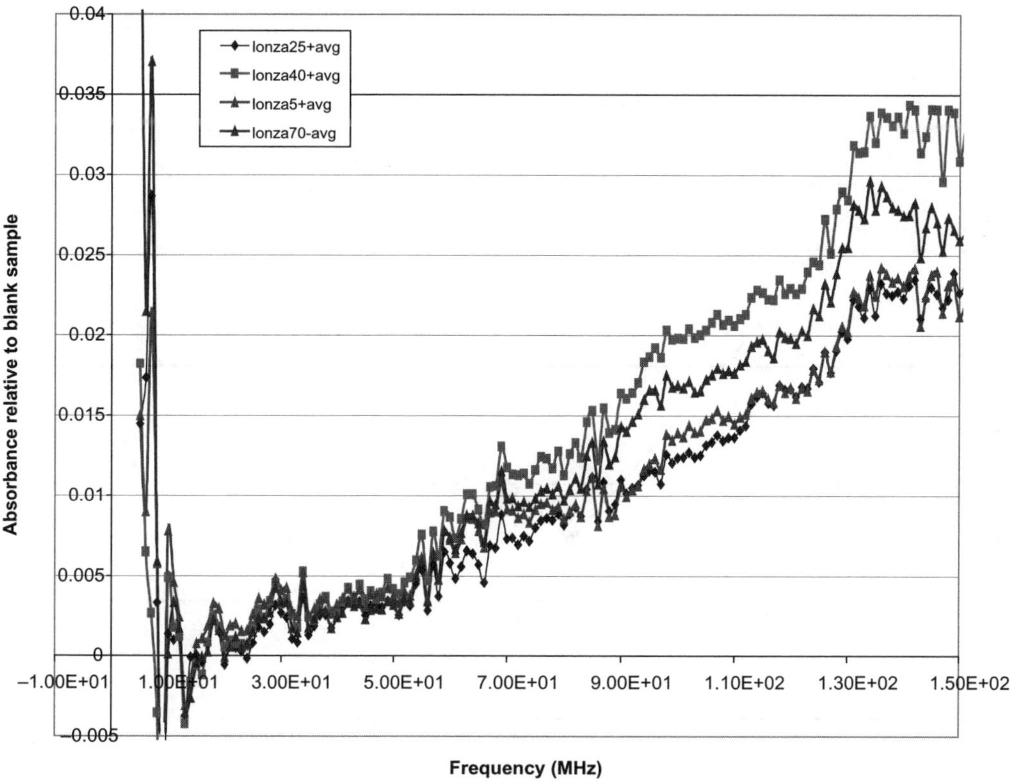

Figure 23.23 Normalized ultrasonic signals for the cB72.3 antibody samples stored at different temperatures. (40 ± 3 °C; −65 °C or below; 5 ± 3 °C; and 25 ± 3 °C).

Potentially, this emerging technique can yield information about the physical properties and the viscoelasticity of a protein sample as well as particle size. Attentuation is thought to have more potential for biomolecular structural ‘fingerprinting’, permitting conformational changes to a protein structure to be detected. As the viscoelasticity of a protein may change with physical state (e.g. unfolding), these changes affect the ultrasound signal (see Figure 23.23 and 23.24). However, the technique is still in an early stage of development for biomolecule research, and as such it is difficult to attribute discrete changes in structure to experimental observations.

23.3.16.10 Fluorescence

There are two main methods of measuring fluorescence that are useful tools for determination of changes to protein conformation. Intrinsic fluorescence measures the natural fluorescence properties of a protein via naturally fluorescing amino acids, primarily tryptophan, but also to a minor degree tyrosine and phenylalanine. The emission spectrum of tryptophan is environmentally dependent, and as solution polarity decreases the fluorescent intensity increases so, for example, following exposure of core protein tryptophan molecules following denaturation, there would be an increase in solution polarity, and thus a decrease in fluorescence intensity. This is a useful tool for monitoring protein conformational changes over time (e.g. stability studies).

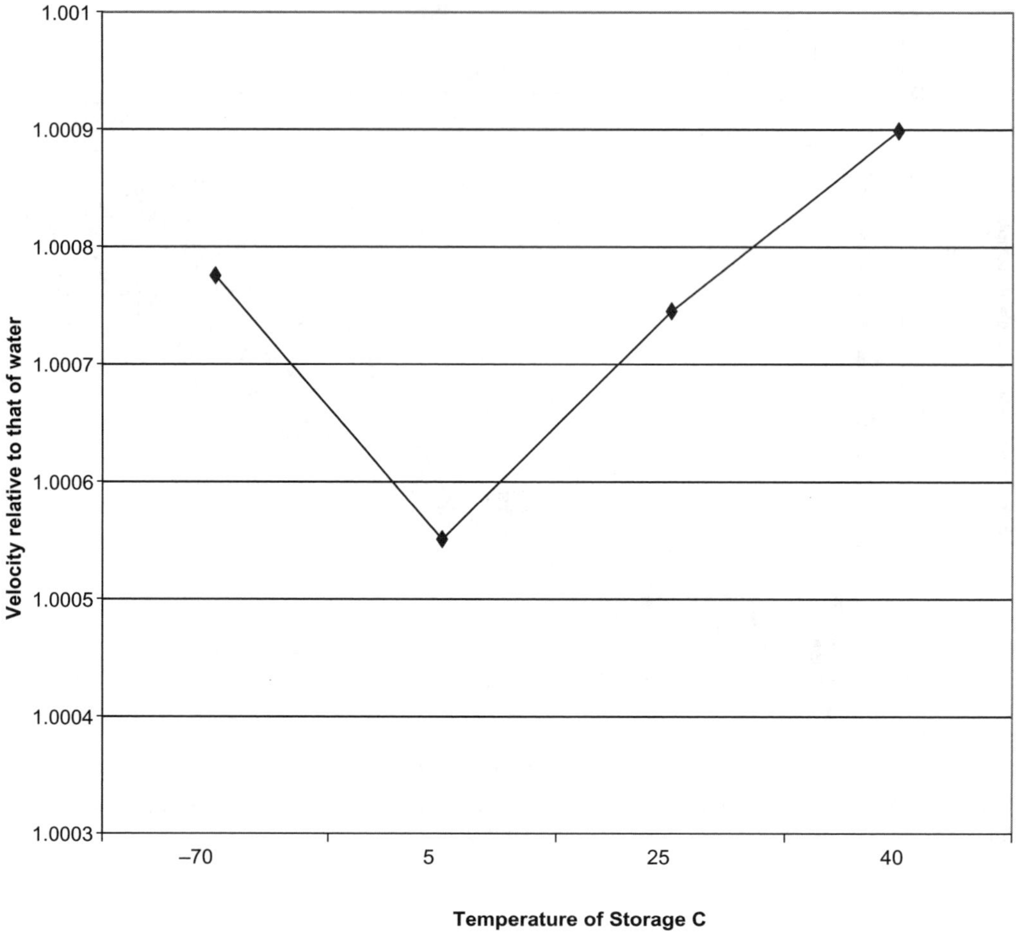

Figure 23.24 Normalized ultrasonic velocity plot for cB72.3 antibody samples stored at different temperatures (−65 °C or below; 5 ± 3 °C; 40 ± 3 °C and 25 ± 3 °C).

Intrinsic fluorescence is generally more robust, although not as sensitive as measurement of extrinsic fluorescence, utilizing probes that bind to exposed hydrophobic areas. The fluorescent signal is proportional to the amount of probe bound (and hence the exposed hydrophobic areas). It is a relatively sensitive technique although the binding of the fluorophore can often lead to differences in folding kinetics of the protein of interest. However, it remains a useful tool for real-time analysis of protein conformation.

An example of an extrinsic fluorescence profile for the cB72.3 antibody using the fluorescent label 1,8-anilinonaphthalene sulphonic acid (1,8-ANS) is shown in Figure 23.25.

23.3.16.11 Differential scanning calorimetry

Differential scanning calorimetry (DSC) is a heat-based denaturation technique that measures the amount of heat required to denature a protein. This information can then be used to

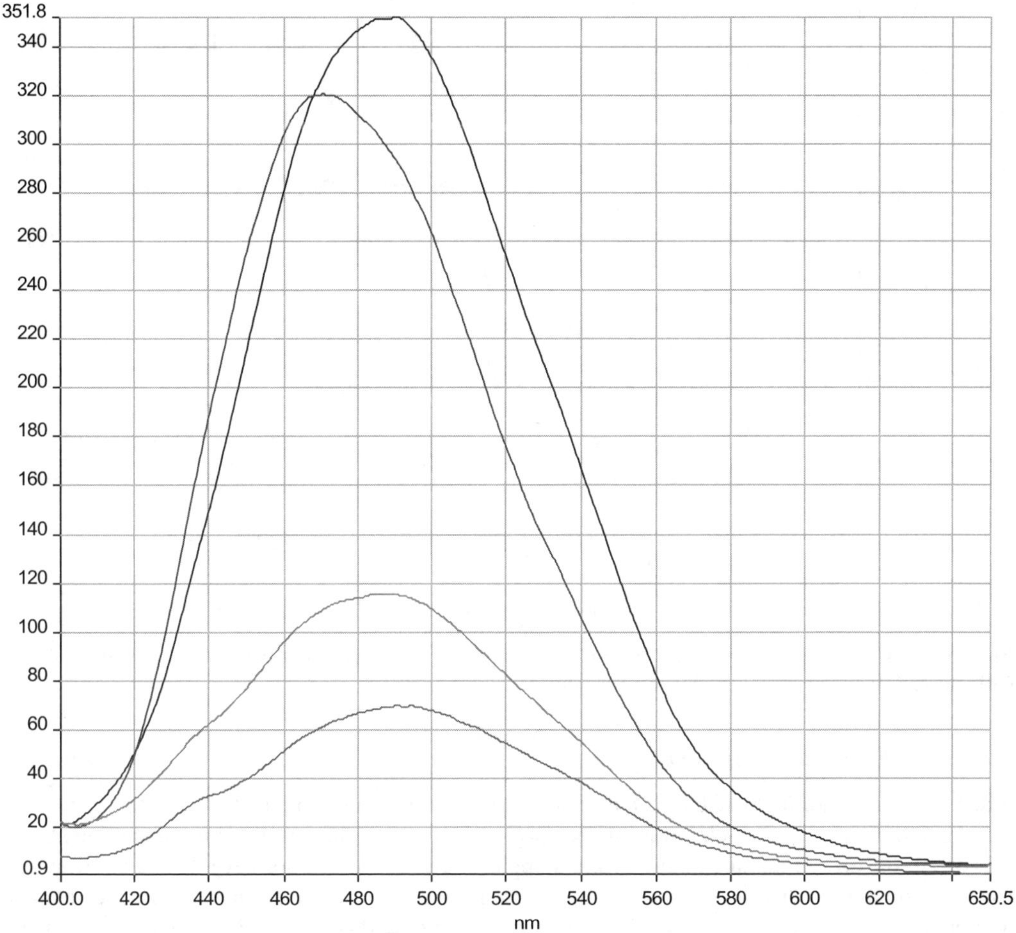

Figure 23.25 Extrinsic fluorescence profile of 1,8-ANS-incubated cB72.3 antibody samples stored at different temperatures: 5 ± 3 °C; (highest trace) −65 °C or below; (2nd lowest trace); 40 ± 3 °C (2nd highest trace) compared with the ANS signal (lowest trace).

determine the relative stability of a protein of interest (Vermeer *et al.* 2001; Hartmann *et al.* 2004). It is typically applied to formulation development. A higher transition midpoint (T_m) indicates greater stability of the folded protein conformation tested. In addition, the T_m profiles often correlate with aggregation profiles. Therefore, a formulation with a high T_m may aggregate more slowly than a low T_m formulation. An example of DSC applied to the cB72.3 antibody is shown in Figure 23.26.

In Figure 23.26, the three cB72.3 antibody samples formulated at different pHs demonstrate an increase in T_m (71.2 °C to 71.5 °C) following a shift from pH 6.8 to pH 8.0, indicative of a slight increase in conformational stability. However, as the change in T_m was negligible and there are known chemical modifications that may also occur at pH 8.0, it would be unlikely that the cB72.3 antibody would be stored at this pH. However, the TMs of samples stored at pH 6.8 and pH 3.0 are very different. In this example, a reduction in pH denatures the protein to the extent that no T_m could be measured, confirming the unsuitability of a pH 3.0 formulation for the cB72.3 antibody.

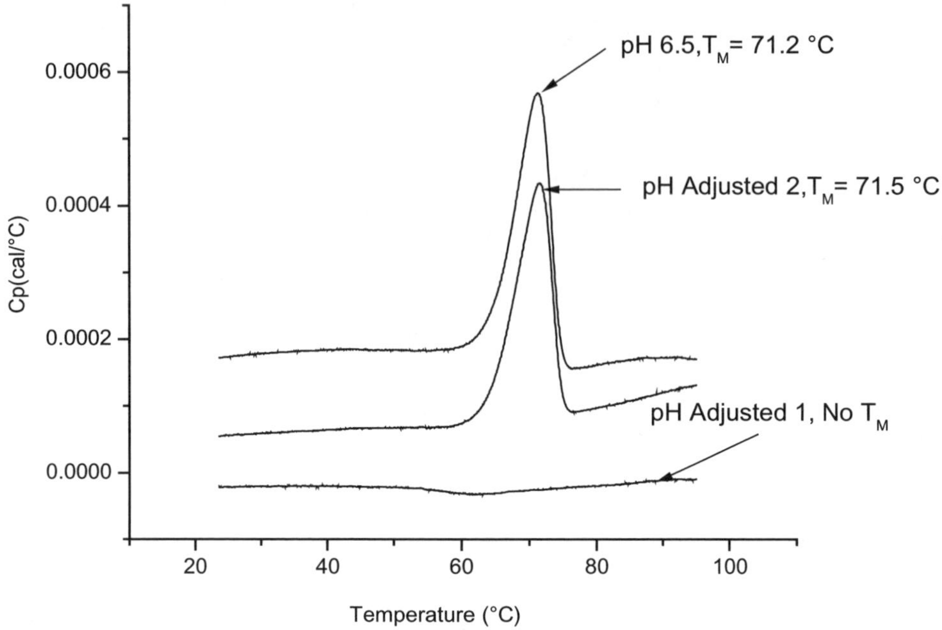

Figure 23.26 DSC analysis of cB72.3 antibody samples stored at different pHs [pH 6.5 control; pH 8.0 (pH adjusted 2) and pH 3.0 (pH adjusted 1) samples].

23.3.16.12 Electron tomography

Sidec electron tomography (SET) is a unique method for imaging of molecular events *in situ* or *in vitro*, combining cryo-EM, electron tomography and advanced data processing algorithms. Samples are flash frozen to create 'solid liquid' samples on cryo-grids to preserve their ultrastructure. Electron tomography uses a tilt-series of electron micrographs collected into 3-D images using a low dose cryo transmission electron microscope (Savage 2004). This 3-D reconstruction using SET technology (essentially measuring constrained maximum entropy) permits visualization and structure analysis with a resolution of ~20 Å (i.e. SET permits discernment of molecular structures). In combination with X-ray crystallography structural information, it can provide a powerful three-dimensional map of a protein's tertiary and quaternary structure. However, SET is more widely applicable than X-ray crystallography, and can be applied to membrane proteins and the direct analysis of proteins both *in vitro* and *in situ*. SET is becoming more popular for the analysis of receptor complex or ion channel formation, binding site analysis and subunit assembly.

Disadvantages include no averaging of solution structures, as each image is a direct snapshot of that moment in time. Nonetheless, it provides a measure of protein aggregation and conformation and within a single analysis can provide structural information of protein monomer and potential aggregate profiles as well as overall protein conformation.

Figure 23.27 demonstrates a SET solution snapshot for the cB72.3 antibody while Figure 23.28 shows a difference in 3-D conformation (aggregation) visualized by SET technology.

23.3.16.13 A brief summary of assays for testing structural integrity, stability and conformation

A range of test methods are required effectively to determine the structural integrity, conformation and stability of therapeutic proteins. These must include analytical methods for monitoring

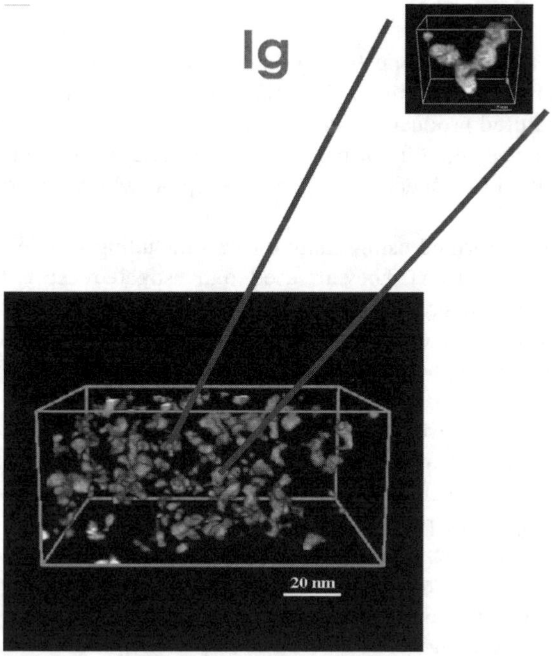

Figure 23.27 SET snapshot of a cB72.3 antibody solution demonstrating the characteristic 'Y' arm conformation of an antibody.

chemical and physical changes. The methods for analysing physical changes are improving to an extent that their sensitivity and precision are suitable for use as stability-indicating assays. The application of these test methods will increase in importance as any regulatory concerns over adverse immunogenicity in patients increase.

The overall aim of instrument manufacturers is to make these types of analyses more widely available for use by simplifying the instrumentation and improving the sensitivity. Therefore it is expected that analysis for structural integrity, composition and stability will become a greater part of the characterization of protein therapeutics in future years.

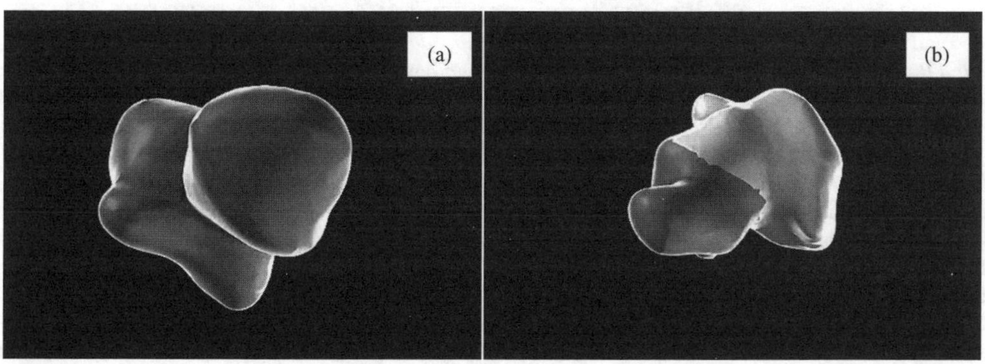

Figure 23.28 Examples of cB72.3 antibody 3-D structure: (a) cB72.3 antibody dimer (b) cB72.3 antibody demonstrating higher aggregate.

23.3.17 Potency

Potency assays are essential for demonstrating the efficacy of a therapeutic product *in vitro* or *in vivo* prior to its application in the clinic. They form a critical tool to ensure the continued consistency of the manufactured product of interest. Potency assays must be specific and relevant to the intended therapeutic use and are normally defined as assays used to measure the ability of a specific therapeutic to bind to a specific antigen or receptor (which may or may not then elicit a functional response *in vivo*).

These assays can be performed using many formats including simple ELISA assays (Section 23.3.7) or radioimmunoassay (RIA). They are used to investigate receptor binding capability, and are typically termed activity assays. More complex cell-based *in vitro* assays are mainly used to measure the functional response of an antigen/receptor binding event. Pharmacokinetic studies using animal models are employed to measure the *in vivo* response following the administration of the therapeutic material.

For early-phase products in development, routine monitoring (and screening) of therapeutic potency is often performed using simple activity-based assays, but as the product reaches the clinic, the requirement for a more relevant model increases, and cell-based assays become more critical for demonstrating efficacy of the product *in vitro*.

Not suprisingly, as the complexity of the model format increases (ELISA→ cell based *in vitro* assay→ animal model) and the relevance of the assay for its intended clinical application increases, there is often an inherently lower assay precision due to the natural heterogeneity of cellular systems over the more homogeneous antigen preparations used in ELISA-type formats. In addition, the amount of the therapeutic product required for analysis increases from the nanogram to microgram (and possibly even milligram) scale. Concomitantly, so do the other reagents, and where applicable, animal costs.

REFERENCES

Anderson JS, Svensson B, Roepstorff P (1996) *Nature Biotechnol.*; 14: 449–456.

Angal S, King DJ, Bodmer MW *et al.* (1993) *Mol. Immunol.*; 30(1): 105–108.

Casaderall N, Nataf J, Viron B *et al.* (2002) *New. E. J. Med.*; 346: 469–475.

Creighton TE (1992) *Proteins: Structures and Molecular Properties.* Second edition, WH Freeman & Co. New York.

Davies MJ, Hounsell EF (1998) *Meth. Molec. Biol.*; 76: 79–100.

Fiaux J, Bertelsen EB, Horwich AL, Wüthrich K (2002) *Nature*; 418: 207–211.

Folta-Stogniew E, Williams KR (1999) *J. Biomolec. Techn.*; 10: 51–63.

Food and Drug Administration (FDA) *Final Rule on 21 CFR*, Chapter I, Subchapter F.

Food and Drug Administration (FDA) (1985) *Points to Consider in the Production and Testing of New Drugs and Biologicals Produced by Recombinant DNA Technology.*

Food and Drug Administration (FDA) (1984) *Points to Consider in the Manufacture of Injectable Monoclonal Antibody Products Intended for Human Use In Vivo.* Office of Biologics Research and Review, Center for Drug and Biologics, FDA. Federal Register, Vol. 49, 1138.

Harding SE (1997) *Prog. Biophys. Mol. Biol.*; 68: 207–262.

Harris RJ, Murnane AA, Utter SL *et al.* (1993) *BioTechnology*; 11:1293–1297.

Hartmann WK, Saptharishi N, Xiao Yi Yang, Mitra G, Soman G (2004) *Anal. Biochem.*; 325: 227–239.

Harvey DJ (2001) *Proteomics*; 1: 311–328.

Huang TH, Yang D, Plaskos NP *et al.* (2000) *J. Mol. Biol.*; 297(1): 73–87.

ICH Steering Committee (1999) *ICH Harmonized Tripartite Guidelines.* March: Q6B and Q5A to 5E.

James DC, Baker KN (1999) In *Encyclopedia of Bioprocess Technology.* Eds Flickinger MC, Drew SW. John Wiley & Sons, Inc., New York; 1336–1349.

Laemmli LK (1970) *Nature*; 227: 680–685.

Rhodes G (2000) In *Crystallography Made Crystal Clear.* Academic Press, New York; second edition: 34.
Savage C (2004) *Bio Tech International;* 16 (October): 8–11.
Schellekens H (2003). *Nephrol. Dial. Transplant*; 18: 1257–1259.
Shuker SB, Hajduk PJ, Meadows RP, Fesik SW (1996) *Science*; 274: 1531–1534.
Tianbo L, Benjamin C (2002) In *Encyclopaedia of Surface and Colloid Science,* Marcer Dekker, New York; 3023–3043.
Vermeer AW, Giacomelli CE, Norde W (2001) *Biochim. Biophys. Acta*; 1476: 139–148.
Wen JT Arakawa, Philo JS (1996) *Anal. Biochem.*; 240: 155–166.

24 Glycosylation of Medicinal Products

E Tarelli

24.1 INTRODUCTION

Polypeptide therapeutics such as erythropoietin, tissue plasminogen activator, immunoglobulins, etc., are usually glycosylated in their native state. Recombinant forms, despite possessing identical amino-acid sequences are, however, often glycosylated differently to the native material because of the variation in the expressions of glyco-processing enzymes that are species and cell-type dependent. Cell culture conditions such as pH, nutrients and their concentrations, as well as the physical nature of cell suspension/adherence may also influence glycosylation (Gawlitzek *et al.* 1995). The precise functions of polypeptide glycosylation are not fully understood, however the pharmacological effects include modulation of biological activity and circulatory lifetime, distribution of the product *in vivo* and immunogenicity (Rademacher *et al.* 1988). Consequently glycosylation of recombinant medicinal products requires careful assessment to ensure that the material produced is suitable for patient use, and such evaluations are now becoming licensing requirements. The constituent monosaccharides usually comprise the neutral sugars D-mannose (Man), D-galactose (Gal), L-fucose (Fuc), the acylamino sugars *N*-acetyl-D-glucosamine (GlcNAc), *N*-acetyl-D-galactosamine (GalNAc), and the acidic residues *N*-acetylneuraminic acid (Neu5Ac) and, from non-human cell-lines, *N*-glycolylneuraminic acid (Neu5Gc). In some instances sulphate esters are present and may replace Neu5Ac. These monosaccharides are combined in oligosaccharide chains (glycans) and the glycans are then covalently bound to the polypeptide in one of two main ways, either through *N*-glycosylation via an asparagine side-chain amide function (–GlcNAc–NH–CO–CH_2–) or through O-glycosylation usually via serine or threonine hydroxyls (–GalNAc–O–CH_2– or –GalNAc–O–CH(CH_3)–). There are also other types of attachment such as *C*-glycosylation, exemplified by mannose linked through a carbon–carbon bond to tryptophan in ribonuclease (Hofsteenge *et al.* 1994), complement (Hofsteenge *et al.* 1999) and recombinant IL-12 (Doucey *et al.* 1999). In addition, reducing sugars such as glucose, lactose, etc., can bind covalently to polypeptides following condensation between the aldehydo function of the sugar and an amino group of the polypeptide. This is often referred to as non-enzymatic glycosylation or glycation, and is the first stage of the Maillard reaction (Maillard 1912). *C*-Glycosylation is rarely encountered and so this chapter will confine itself to discussing *N*- and *O*-glycosylation, with a brief reference to glycation.

24.2 *N*-GLYCOSYLATION

The biosynthesis of *N*-glycans takes place in two stages within the endoplasmic reticulum and Golgi apparatus. The first stage is the transfer of a lipid-linked $Glc_3Man_9GlcNAc_2$ oligosaccharide

Medicines from Animal Cell Culture Edited by G. Stacey and J. Davis

(Glc = D-glucose) to an asparagine acceptor within the nascent polypeptide. The transferred oligosaccharide then undergoes processing by glycosidases, followed by the stepwise addition of monosaccharides, (usually from nucleotide sugar donors) by the action of glycosyltranferases (Kornfeld & Kornfeld 1985; Natsuka & Lowe 1994; Trombetta 2003). More than 16 glycosyltransferases have been characterized and many of these have been cloned (Spiro 2002). Only asparagines present in the sequons Asn-X-Ser or Asn-X-Thr (X = any amino acid except proline) and occasionally Asn-Ala-Cys (Vance *et al.* 1997, 1999) undergo *N*-glycosylation. Such sequons do not necessarily glycosylate but, when they do so, there may be several different glycan structures at the same site generating a family of molecules, referred to as glycoforms and the phenomenon itself as microheterogeneity. All *N*-glycans possess the common pentasaccharide core structure Manα1-6(Manα1-3)Manβ1-4GlcNAcβ1-4GlcNAcβ1- (shown in bold in Figure 24.1). Additional saccharides are then attached to the core in three main ways (Figure 24.1) to give the following types of *N*-glycan:

(i) *high mannose*, which contain only further Man residues;

(ii) *complex*, which usually possess two, three or four (and in some instances five) antennae each of which can comprise (in sequence from Man) GlcNAc, Gal, Neu5Ac (occasionally GalNAc may replace Gal);

(iii) *hybrid*, having additional Man residues on the Manα1-6 arm of the core, and one or two antennae, as in (ii), on the Manα1-3 arm.

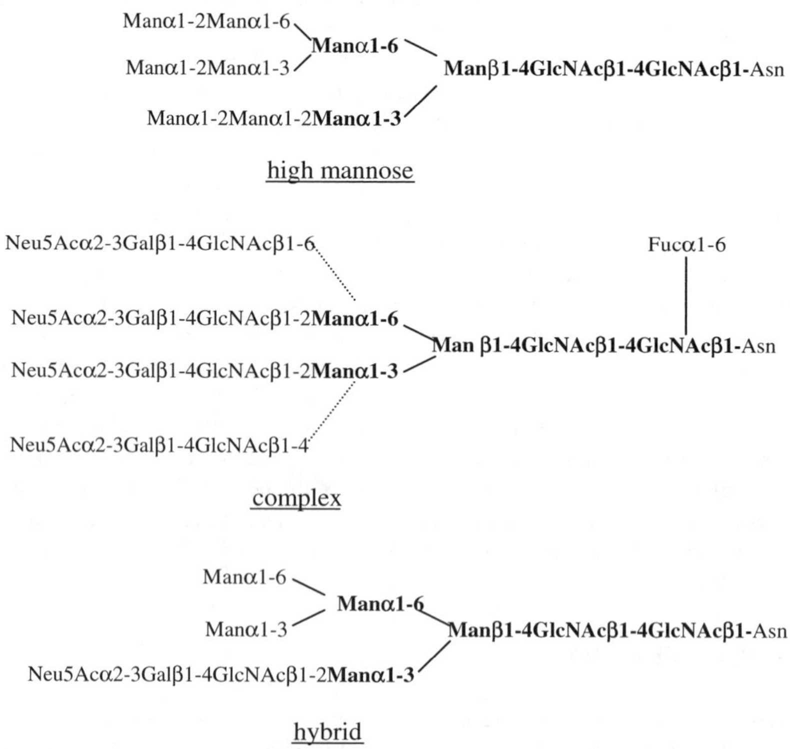

Figure 24.1 Representative structures of high mannose, complex (with proximal fucose) and hybrid *N*-linked glycans. The core pentasaccharide is shown in bold. For the complex type structure, dotted lines indicate the tri- and tetra-antennae (a fifth antenna, if present, may be linked β1–4 to the Manα1–6 residue).

Except for type (i) above, the core may be extended with *polylactosamine* (repeating Galβ1-4GlcNAcβ1-3 units) structures, or possess *bisecting GlcNAc*, that is GlcNAcβ1–4 linked to the β-Man residue. The Asn-linked GlcNAc may also be substituted as shown in Figure 24.1 with a Fucα1–6 (Fucα1–3 in non-mammalian glycoproteins) proximal residue. Fucose may also be substituted on an outer arm GlcNAc and in plants xylose may be substituted β1–2 in the β-Man residue. Despite the common core structure, it is obvious that huge structural diversity for *N*-linked glycans is possible and indeed this is found to be the case.

24.3 *O*-GLYCOSYLATION

O-Glycans are biosynthesized by the sequential addition of monosaccharides, from nucleotide sugar donors, to Ser or Thr residues by the action of glycosyltransfersases. In contrast to *N*-linked glycans there does not appear to be a specific sequon requirement for *O*-glycosylation, however proline residues are often abundant in peptide sequences that contain *O*-glycans. At least eight different core structures have been identified for *O*-linked glycans (van den Steen *et al.* 1998). However in recombinant glyoproteins, *O*-linked oligosaccharides usually possess the type 1 core structure, that is, Galβ1–3GalNAcα1– (Grabenhorst *et al.* 1999). This core structure can also possess Neu5Ac linked α2,6-GalNAc and/or α2,3-Gal. Occasionally structures such as *O*-Fuc or *O*-Glc are encountered as are other hydroxy-amino acids that are *O*-glycosylated.

24.4 GLYCOSYLATION AND CELL TYPE

The glycosylation characteristics of some hamster and murine cell lines frequently used to produce human glycoprotein therapeutics are summarized in Table 24.1. CHO and BHK-21 cell lines produce essentially the same *N*- and *O*-glycosylation. Murine cells generally exhibit a higher glycoprotein microheterogeneity than do the hamster cells (Grabenhorst *et al.* 1999).

To be effective, glycoprotein therapeutics require adequate *in vivo* circulatory lifetimes and an important determinant of this property can be the presence/absence, in *N*-glycans, of chain-terminal Neu5Ac (Morell *et al.* 1971). The circulatory lifetime can be dramatically reduced sometimes from several hours (when Neu5Ac is present) to just a few minutes (when Neu5Ac absent), a result of exposure of chain terminal (unsubstituted) Gal (or GlcNAc when Gal is also absent) residues that are recognized by hepatic asialoglycoprotein (and other) receptors (Ashwell & Harford 1982). Such asialoglycoproteins are taken up by endocytosis for degradation in the lysosomes. Other lectin-like proteins involved in pre-immune responses and clearance mechanisms may also recognize chain terminal Fuc and Man glycans. Some asialoglycoproteins (notably IgG) are, however, not affected in this manner because their glycans lie buried within the polypeptide and as a result are unavailable to the receptors.

In human glycoproteins, Neu5Ac is attached to Gal in α2,6- or as a mixture of α2,6- and α2,3-linkages. Some cell lines used to express recombinant glycoproteins are however only capable of producing Neu5Acα2,3-linkages (Table 24.1) and this can result in undersialylation and consequently affect circulatory lifetime and other properties. In addition, in non-human cell lines, Neu5Gc can replace Neu5Ac (see Table 24.1, in which they are obbreviated as NeuGly and NeuAc respectively). It has also been noted that Neu5Gc can be incorporated into human stem cells when cultured using animal-derived nutrients (Martin *et al.* 2005), demonstrating the importance of growth conditions upon protein glycosylation. Human antibodies directed against Neu5Gc have been reported (Hokke *et al.* 1990). Other antigenic determinants may also be expressed in non-human cell lines such as Galα1–3Gal (the Galili antigen) which is recognized by around 1 % of circulating human IgG (Galili *et al.* 1984). Other undesirable or antigenic saccharide structures may also be expressed, such as proximal Fucα1,3 or mannose

Table 24.1 Structural features of *N*-linked oligosaccharides from recombinant glycoproteins expressed in mammalian host cells. Data are based on structural analysis of the recombinant human glycoproteins IFN-β, Epo, AT III, IL-6, tissue-plasminogen activator and β-TP as well as recombinant humanized antibodies, soluble receptor proteins and *N*-glycosylation mutants of human IL-2. (Reproduced from Grabenhorst *et al.* 1999 with kind permission of Kluwer Academic Publishers and the authors).

	host cell line			
carbohydrate structure	CHO	BHK-21	C127	Ltk$^-$
proximal fucose	+	+	+	+
Fuc(α1-2)Gal-R*	+	+	?	?
α2,6-NeuAc	−	−	+	+
α2,3-NeuAc	+	+	+	+
NeuAc(α2-8)NeuAcα2-3R	+	+	−	+
NeuGly*	+	+/−	+	+
tri/tetra-antennarity	+	+	+	+
Gal(β1-4)GlcNAc repeats	+	+	+	+
Gal(β1-3)GlcNAc-R	−	+	−	−
sulfated glycans	+	+	+	+
Gal(α1-3)Gal	+	−	+	+
branched repeats	?	−	+	−
mannose 6-phosphate*	+	+	?	?
bisecting GlcNAc	−	−	+	+
GalNAc(β1-4)GlcNAc	−	+**		

**detectable only in trace amounts*
***detected in large amounts in the BHK-21A variant cell line.*

6-phosphate, which can trigger lysosomal phagocytosis or sialyl Lewis X (Neu5Acα2,3Galβ1,4 (Fucα1,3)GlcNAcβ-R) which is associated with leukocyte-mediated inflammation processes. The presence of these and other blood-group structures therefore requires consideration.

Insect cell lines are also used to produce recombinant glycoproteins. Insect cells, in general, are unable to biosynthesize lactosamine complex-type *N*-glycans and produce only truncated oligomannosidic structures that may contain additional proximal Fucα1,3; they are also unable to biosynthesize sialylated core 1 *O*-glycans (Grabenhorst *et al.* 1999). Insect cells can however be cultured in serum-free media, and this is claimed to be beneficial on grounds of both cost and biosafety (Altmann *et al.* 1999). Glycoproteins can also be produced in transgenic plants (Ma *et al.* 1998; Ko *et al.* 2003), but may contain glycan motifs that are not immediately compatible with use as pharmaceutical glycoproteins (Lerouge *et al.* 2000). To overcome such glycosylation shortcomings, methods have been developed for *in vitro* glycan remodelling and for redesigning (humanizing) the cellular glycosylation machinery. For example, additional *N*- and *O*-glycans have been incorporated into recombinant glycoproteins by the inclusion into the expressed polypeptide of additional peptide domains that are recognized by glycosyltranferases. Strategies have also been described for retaining polypeptides longer in the endoplasmic reticulum in order to increase the degree of glycosylation, e.g. by the addition to the polypeptide of the *C*-terminal sequence Lys–Asp–Glu–Leu (Tekoah *et al.* 2004). Other strategies employ the inhibition of Golgi glycosyltransferases or the incorporation of additional glycosyltransferases (e.g. α2–6 neuraminyltranferase into hamster cells) (Grabenhorst *et al.* 1995; Lerouge *et al.* 2000; Seo *et al.* 2001; Hollister *et al.* 2002). As well as affecting biological properties, glycosylation can also influence properties

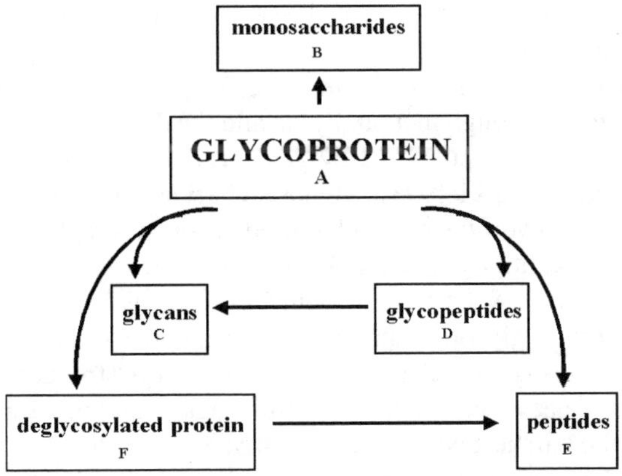

Figure 24.2 General strategies for characterizing glycoprotein glycosylation indicating specific degradations that are often employed.

such as solubility, chemical stability and cellular secretion. These factors may well have a bearing on the downstream processing of recombinant products and consequently need to be considered in the overall production process.

24.5 GLYCOSYLATION ANALYSIS

Figure 24.2 summarizes the general strategies employed in characterising glycoprotein glycosylation. For full experimental details readers are also referred to the many excellent literature reports describing glyco-analysis of specific glycoproteins.

Full characterization is not a trivial matter and requires the combined application of a number of sophisticated physico-chemical and biochemical techniques. Full glycosylation analysis needs to address the monosaccharide composition and ring size, the identification of glycan types, their sequences, and positional attachment of the monosaccharide residues within the glycan, their anomericity and absolute configuration, and also the sites of glycan attachment within the polypeptide chain. For routine analysis, such as process monitoring or quality control of batches, full characterization is neither feasible nor necessary and particular methods are often chosen for these applications.

Some structural information can be generated from analysis of the intact glycoprotein itself (Figure 24.2, A). That it contains carbohydrate, together with an indication of the components and their amounts can be directly obtained from simple laboratory colorimetric tests and/or by using sensitive kits available from commercial sources (Manzi & Varki 1993). Analysis of the glycoprotein by SDS-PAGE or mass spectrometry in combination with glycosidase treatment can indicate glycan types present and their amounts (Mechref & Novotny 1988), and direct ESI mass spectrometric analysis of the intact glycoprotein can identify glycoform composition (Wan *et al.* 2001; Greer & Morris 2003). Isoelectric focusing provides information on microheterogeneity with respect to charged glycoform composition (Catlin *et al.* 2002) as does capillary zone electrophoresis (Watson & Yao 1993, Zhou *et al.* 2004, Neususs *et al.* 2005). Sensitive lectin-binding assays may be used to identify the presence of specific sugar motifs (Kim *et al.* 1992) and such assays have been used, for example, to evaluate recombinant erythropoietin produced in different cell lines (Storring *et al.* 1998). Structural information obtained from an intact glycoprotein, although useful, is not however particularly detailed and in order to probe glycosylation in greater depth, the glycoprotein is required to be cleaved selectively with the derived fragments then undergoing more detailed analysis.

Monosaccharide release and analysis (Figure 24.2, B) provides compositional data. Cleavage to monomers can be achieved through methanolysis, and the derived methyl glycosides (as volatile derivatives) can be identified and quantified by gas chromatography combined with mass spectrometry (GC-MS) (Manzi & Varki 1993). Monosaccharides may also be liberated by aqueous acid (e.g.trifluoroacetic acid) hydrolysis and identified chromatographically or by electrophoresis (see below). The reaction conditions required for hydrolysis are dependent upon the nature of the sugar linkage (Biermann 1988). Neu5Ac (and Neu5Gc) residues are relatively acid-labile and can be selectively removed from a glycoprotein by hydrolysis under mild conditions (such as 0.05 M acid at 80 °C for 1 h). Neutral sugars such as Gal and Man require stronger conditions (e.g. 2 M acid at 100 °C for 2 h) for their release with *O*-linked aminosugars, and the Asn amide-linked GlcNAc in particular, requiring the harshest conditions (e.g. 4 M acid at 120 °C for >2 h) for complete liberation. Acid hydrolysis causes some monosaccharide degradation and therefore suitable controls are necessary if quantitation is required.

The absolute configuration (D or L) of the released monosaccharides can be determined by GC analysis of (−) 2-butyl glycosides (Gerwig *et al.* 1979). Once available, monosaccharide composition is a good indicator of the type(s) and relative amounts of glycans present and this information is valuable for the rational planning of further investigations. Thus GlcNAc, Man and Fuc would suggest the presence of *N*-glycans, whereas GalNAc indicates *O*-glycosylation (GalNAc is however occasionally also a component of *N*-glycans). *N*- and/or *O*-glycosylation can be confirmed from the susceptibility of their linkages to endoglycosidases and/or chemical cleavage (discussed later).

The linkage positions of individual monosaccharides within the glycan can be determined from methylation analysis, a widely used and well-established method (Kobata & Takasaki 1993). The procedure first methylates all exposed glycan hydroxyls and then, after sequential acid hydrolysis and reduction of the now free aldehydo functions (i.e. R—CHO → R—CH_2OH), acetylates all of the newly exposed hydroxyls of the so-formed monomers. The exact position of each methyl and acetyl group (the latter corresponding to a linkage position) can then be readily determined from GC-MS analysis of these derived methylated alditol acetates.

Further information on the glycan structures can be obtained following deglycosylation of the glycoprotein (or its derived glycopeptides) (Figure 24.2, C). Deglycosylation may be carried out enzymatically using endoglycosidases. For *N*-glycans, peptide-*N*-glycosidase F (PNGaseF) which cleaves between GlcNAc–Asn is commonly used since it liberates all *N*-glycan types except those containing proximal Fucα1–3, for which PNGase A can be used (Tarentino & Plummer, 1994). A number of other endoglycosidases that cleave between the core GlcNAc–GlcNAc residues may also be used to liberate *N*-glycans lacking the terminal GlcNAc, which remains attached to the peptide and thus still defines the glycosylation site. PNGase F also leaves a signature at the glycosylation site, in this case conversion of the deglycosylated Asn to Asp which labels with ^{18}O when the PNGase F digest is performed in $H_2{}^{18}O$ (Xiong & Regnier 2002). These 'footprints' are extremely useful for identifying (e.g. by mass spectrometry) the precise location of the glycosylated amino acids within the polypeptide chain. In contrast to *N*-glycans, the enzymatic cleavage of *O*-glycans is of limited practical use. Only a single peptide-*O*-glycosidase specificity is available and this only liberates the peptide-linked disaccharide Galβ1–3GalNAcα1–Ser/Thr (Ishii-Karakasa *et al.* 1992) and therefore chemical cleavage is mainly employed to prepare intact *O*-glycans.

O-Glycans can be liberated chemically by alkaline β-elimination initiated by abstraction of the acidic α-hydrogen of serine or threonine residues. Under alkaline conditions however, *O*-glycans are susceptible to degradation ('peeling') and to prevent this, the reaction can be performed with added borohydride, which rapidly reduces the liberated glycans to alditols which are alkali stable (Carlson 1966). The reducing *O*-glycans themselves can however be isolated from alkaline β-elimination when the glycoprotein is immobilized (e.g. on reversed-phase resin) and the alkali neutralized immediately following the release of the glycans (Oh-eda et al. 1996; Karlsson & Packer 2002). Reducing oligosaccharides can also be prepared from hydrazinolysis (Takasaki *et al.* 1982). Alkali, and especially hydrazine are also used to liberate *N*-glycans, reagent specificity depending upon the reaction conditions employed (Patel *et al.* 1993). Hydrazinolysis is probably the method of choice for the chemical preparation of *N*-glycans. Cleavage of *O*-glycans by ammonia has also been described. The glycosylated Ser and Thr residues are converted to their dehydro analogues, which can be identified within the peptide sequence by MS analysis (Rademaker *et al.* 1998); the glycans are released as glycosylamines, which are hydrolysed to reducing oligosaccharides (Huang *et al.* 2001). It is noteworthy that under these reaction conditions, ammonia also selectively cleaves polypeptides between Asn–Pro and furthermore it would appear that some *O*-glycans are not released (Tarelli & Corran 2003 and unpublished results). A protocol for total deglycosylation has been described and involves the sequential treatment of the glycoprotein with PNGase F, isolating the *N*-glycans and then subjecting the residue to reductive alkaline β-elimination and then isolating the *O*-glycans as their alditols (Damm *et al.* 1987). It should be noted that enzymatic methods may result in incomplete deglycosylation and that chemical methods may produce side products.

Once the oligosaccharides are obtained, detailed analysis can be carried out (the following methods can also be used for monosaccharide analyses–see above). In the case of reducing oligosaccharides, high detection sensitivity can be achieved by introducing a fluorescent label (Hase 1996). These labels, e.g. 2-aminobenzamide (Bigge *et al.* 1995), contain a free amino group that can undergo reductive amination with the aldehydo function of the saccharide. The labelled oligosaccharides may then be separated on the basis of their charge and size and quantified (since there is one fluorescent group per glycan). This can be performed using commercially available fluorophore-assisted carbohydrate electrophoresis (FACE) kits (Starr *et al.* 1996), or by using, in combination, normal and reversed phase HPLC (Townsend *et al.* 1996; Guile *et al.* 1996). Underivatized (free) oligosaccharides can be analysed by using high pH anion-exchange chromatography with pulsed amperometric detection (HPAEC-PAD) (Townsend & Hardy 1991). Apart from the advantage that no sample derivatization is required, HPAEC possesses high resolution (Figure 24.3) and is also able to analyse alditols (which since they do not possess an aldehyde group cannot undergo reductive amination). A disadvantage is that quantitation requires the use of suitable oligosaccharide standards that may not be readily available.

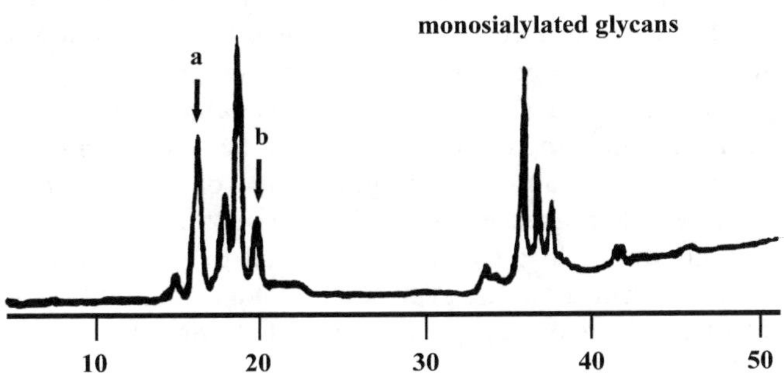

Figure 24.3 HPAEC-PAD analysis of the neutral (14–21 minutes) and monosialylated (32–38 minutes) glycans released by PNGase F treatment of a recombinant monoclonal antibody. a = $GalGlcNAc_2Man_3GlcN\text{-}Ac_2$ with proximal Fuc. b = $Gal_2GlcNAc_2Man_3GlcNAc_2$ with proximal Fuc and bisecting GlcNAc.

The sequence of monosaccharides in an oligosaccharide can be determined using exoglycosidases, either singly or in combination (Edge *et al.* 1992), followed by analysis using any of the separatory methods described here. Sequence information can also be obtained from mass spectrometric analysis, especially using soft-ionization techniques such as fast atom bombardment (FAB), matrix assisted laser desorption (MALDI) and electrospray (ESI). These methods of ionization tend to produce molecular ions that then fragment in a structurally dependent manner (Dell 1989; Dell *et al.* 1994; Settineri & Burlingame 1995; Dell & Morris 2001; Harvey 2003, 2005; Harvey *et al.* 2004). Both underivatized and derivatized (e.g. fluorescently labeled or permethylated) oligosaccharides are amenable to such analysis, derivatization resulting in greater sensitivity. Alditols prepared from glycoproteins may be separated and characterized by LC-ESMS using a graphitized carbon column (Kawasaki *et al.* 2003; Karlsson *et al.* 2004). MS techniques are now finding wide application for probing the fine structure of oligosaccharides and glycopeptides, e.g. for differentiating α2,3- and α2,6-linked Neu5Ac residues (Wheeler & Harvey 2000). Oligosaccharide analyses using HPLC and MS have been described as being suitable for routine stability and process monitoring of recombinant glycoproteins (Field *et al.* 1996; Yuen *et al.* 2002).

Other useful analytical techniques include high resolution NMR spectroscopy, which is capable of providing detailed structural information on, among other things, the constituent sugars of a glycan, their anomeric configurations and their linkage positions. Interpretation of ^1H spectra of oligosaccharides may be carried out using structural reporter group strategy (Vliegenthart *et al.* 1981), which compares experimental results with data from libraries of reference compounds in order to assign signals within a spectrum and thereby identify specific structural features. These libraries are accessible via the Internet (Gerwig & Vliegenthart 2000). Although NMR is non-destructive, high nmol to μmol amounts of material are required and this may prohibit its routine use. It is nevertheless probably one of the most powerful analytical tools available and provides structural information that would be difficult to obtain directly by other means.

Glycopeptides, together with peptides, (Figure 24.2, D), are the result of digestion of glycoproteins with proteases such as trypsin, chymotrypsin or V-8. Pronase may be used when there is a requirement for a very short peptide sequence (e.g. for NMR analysis). Fractionation of digests on reversed phase media enables the separation and isolation of the proteolytic fragments and lectin affinity chromatography can be used as a means of isolating glycopeptide families (Merkle & Cummings 1987). Glycopeptides are amenable to analysis by the procedures described earlier and may be used instead of, or as well as, glycans for structural elucidation studies. Glycopeptides are particularly useful for determining glycan composition and glycosylation sites within a polypeptide (Dell & Morris 2001; Pouria *et al.* 2004; Tarelli *et al.* 2004) including recombinant proteins. The recombinant glycoprotein tissue plasminogen activator (TPA), for example, after trypsin digestion produced, in liquid chromatography-electrospray mass spectrometry (LC-ESMS) analysis, a triply protonated ion $[M + 3H]^{3+}$ with a mass/charge (m/z) value of 1064.0 (therefore a mass of 3189) that did not correspond to any expected peptide but could be attributed to a glycopeptide. Fragmentation of this ion initially resulted in fission of the glycosidic (weakest) bonds and, from analysis of these fragment ions, the glycan component could be identified as a monosialylated biantennary structure with proximal fucose. Subsequent fragmentation of the so-formed GlcNAc-peptide ion then confirmed the peptide sequence as ^{441}CTSQHLLNR^{449}. A monosialylated biantennary with proximal fucose glycan could therefore be assigned to Asn 448 in this preparation of TPA. This analysis was performed on 2 μg of the digest (Wu *et al.* 2002). Selective detection of glycopeptides in proteolytically digested glycoproteins by LC-ESMS has been described (Sullivan *et al.* 2004).

MALDI-TOF MS in combination with protease and endoglycosidase digestions has also been used to assign *N*-glycosylation structures and peptide site occupancy (Mills *et al.* 2000), as has Edman sequencing (Gooley *et al.* 1994). The molecular weight of a deglycosylated glycoprotein (Figure 24.2, F), e.g. by SDS-PAGE or MALDI-TOF MS analysis, can be useful in assessing the

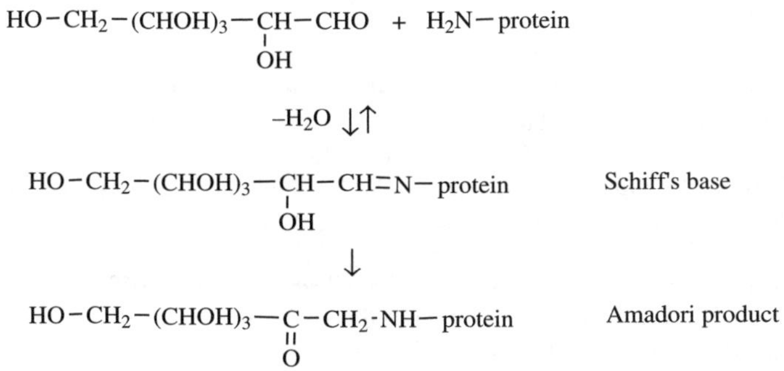

Figure 24.4 Protein glycation resulting from the condensation of a hexose with an amino group of a protein. The Schiff's base formed initially undergoes an irreversible rearrangement to an Amadori product.

degree of glycosylation. Subsequent protease digestion of the deglycosylated protein and peptide analysis (Figure 24.2, E) can then be used to obtain additional structural information including amino acid sequence data and confirmation of the sites of glycosylation.

24.6 GLYCATION

Glycation results when an aldehydo function of a reducing sugar condenses with an amino group of a polypeptide with the elimination of water (Maillard 1912; Reynolds 1965). Initially a Schiff's base is formed but this then undergoes irreversible rearrangement to the so-called Amadori product (Figure 24.4). The Amadori product may undergo further complex reactions yielding a wide variety of products, the precise composition of which depends upon pH, temperature, types and concentrations of salts, etc. Such reactions have, for many years, been studied in relation to food, but it is only relatively recently that the importance of glycation of proteins in physiological environments has been appreciated, e.g. the formation *in vivo* of glycated haemoglobin, albumin, etc. (Bunn 1981). Certain chemical structures, such as the N-terminus of vasopressin hormones are particularly reactive towards glycation (Tarelli *et al.* 1994). The possible modification of recombinant proteins through glycation should therefore be considered if they are exposed to reducing sugars (for example if produced alongside milk, which contains substantial amounts of the reducing disaccharide, lactose). In addition, excipients used in subsequent formulations should be selected so that possible glycation is avoided. This is one reason why the non-reducing disaccharide $\alpha\alpha'$ trehalose was introduced instead of lactose as an excipient for International Biological Standards (Tarelli & Wood 1981).

24.7 FUTURE PROSPECTS

Further developments in modifying the cellular machinery of host cell lines should enable the manufacture of recombinant glycoprotein therapeutics with more finely tuned glycosylation. This should improve their safety, efficacy and stability and be a means of modulating their biological activity, circulatory lifetimes and their *in vivo* tissue targeting characteristics. These developments will rely heavily on the ability routinely and accurately to perform detailed structural characterization of the glycoforms so produced, and it is likely that improved methods of physico-chemical analysis, especially mass spectrometric analysis, will play a pivotal role in allowing rapid progress to be made. These exciting new developments should then provide a wider and improved repertoire of specifically glyco-designed therapeutic products.

REFERENCES

Altmann F, Staudacher E, Wilson IB, Marz L (1999) *Glycoconj. J.*; 16: 109–123.

Ashwell G, Harford J (1982) *Ann. Rev. Biochem.*; 51: 531–554.

Biermann CJ (1988) *Advan. Carbohydr. Chem. Biochem.*; 46: 251–272.

Bigge JC, Patel TP, Bruce JA, Goulding PN, Charles SM, Parekh RB (1995) *Anal. Biochem.*; 230: 229–238.

Bunn HF (1981) *Prog. Clin. Biol. Res.*; 51: 223–239.

Carlson DM (1966) *J. Biol. Chem.*; 241: 2984–2986.

Catlin DH, Breidbach A, Elliott S, Glaspy J (2002) *Clin. Chem.*; 48: 2057–2059.

Damm JBL, Kamerling JP, Van Dedem GWK, Vliegenthart JFG (1987) *Glycoconjugate. J.*; 4: 129–144.

Dell A (1989) *Adv. Carbohydr. Chem. Biochem.*; 45: 19–72.

Dell A, Morris HR (2001) *Science*; 291: 2351–2356.

Dell A, Reason AJ, Khoo KH, Panico M, McDowell RA, Morris HR (1994) *Meth. Enzymol.*; 230: 108–132.

Doucey M-A, Hess D, Blommers MJ, Hofsteenge J (1999) *Glycobiology*; 9: 435–441.

Edge CJ, Rademacher TW, Wormald MR *et al.* (1992) *Proc. Natl. Acad. Sci. USA*; 89: 6338–6342.

Field M, Papac D, Jones A (1996) *Anal. Biochem.*; 239: 92–98.

Galili U, Rachmilewitz EA, Peleg A, Flechner I (1984) *J.Exp. Med.*; 160: 1519–1531.

Gawlitzek M, Valley U, Nimtz M, Wagner R, Conradt HS (1995) *J. Biotechnol.*; 42: 117–131.

Gerwig GJ, Kamerling JP, Vliegenthart JFG (1979) *Carbohydr. Res.*; 77: 10–17.

Gerwig GJ, Vliegenthart JFG (2000) In *Proteomics in Functional Genomics*. Eds Jolles P, Jornvall H Birkhauser Verlag, Basel, Switzerland.

Gooley AA, Pisano A, Packer NH *et al.* (1994) *Glycoconjugate J.*; 11: 180–186.

Grabenhorst E, Hoffmann A, Nimtz M, Zettlmeissl G, Conradt HS (1995) *Eur. J. Biochem.*; 232: 718–725.

Grabenhorst E, Schlenke P, Pohl S, Nimtz M, Conradt HS (1999) *Glycoconjugate. J.*; 16: 81–97.

Greer FM, Morris HR (2003) *Methods Mol. Biol.*; 211: 195–210.

Guile GR, Rudd PM, Wing DR, Prime SB, Dwek RA (1996) *Anal. Biochem.*; 240: 210–226.

Harvey DJ (2003) *Int. J. Mass. Spectrom.*; 226: 1–35.

Harvey DJ (2005) *Expert Rev. Proteomics*; 2: 87–101.

Harvey DJ, Martin RL, Jackson KA, Sutton CW (2004) *Rapid Commun. Mass Spectrom.*; 18: 2997–3007.

Hase S (1996) *J. Chromatogr. A*; 720: 173–182.

Hofsteenge J, Muller DR, de Beer T, Loffler A, Richter WJ, Vliegenthart JFG (1994) *Biochemistry*; 33: 13524–13530.

Hofsteenge J, Blommers M, Hess D, Furmanek A, Miroshnichenko O (1999) *J. Biol. Chem.*; 274: 32786–32794.

Hokke CH, Bergwerff AA, van Dedem GW, van Oostrum J, Kamerling JP, Vliegenthart JFG (1990) *FEBS Lett.*; 275: 9–14.

Hollister J, Grabenhorst E, Nimtz M, Conradt H, Jarvis DL (2002) *Biochemistry*; 41: 15093–15104.

Huang Y, Mechref Y, Novotny MV (2001) *Anal. Chem.*; 73: 6063–6069.

Ishii-Karakasa I, Iwase H, Hotta K, Tanaka Y, Omura S (1992) *Biochem. J.*; 288: 475–482.

Karlsson NG, Packer NH *Anal. Biochem.*; (2002) 305:173–185.

Karlsson NG, Wilson NL, Wirth HJ, Dawes P, Joshi H, Packer NH (2004) *Rapid Commun. Mass Spectrom.*; 18: 2282–2292.

Kawasaki N, Itoh S, Ohta M, Hayakawa T (2003) *Anal. Biochem.*; 316: 15–22.

Kim B, Behbahani I, Meyerhoff ME (1992) *Anal. Biochem.*; 202: 166–171.

Ko K, Tekoah Y, Rudd PM *et al.* (2003) *Proc. Natl. Acad. Sci. USA*; 100: 8013–8018.

Kobata A, Takasaki S (1993) In *Glycobiology: A Practical Approach*. Eds Fukuda M, Kobata A. Oxford University Press Inc. New York; 179–181.

Kornfeld R, Kornfeld S (1985) *Ann. Rev. Biochem.*; 54: 631–664.

Lerouge P, Bardor M, Pagny S, Gomord V, Faye L (2000) *Curr. Pharm. Biotechnol.*; 1: 347–354.

Ma JK, Hikmat BY, Wycoff K *et al.* (1998) *Nat. Med.*; 4: 601–606.

Maillard LC (1912) *C.R. Acad. Sci. (Paris)*; 154: 66–68.

Manzi AE, Varki A (1993) In *Glycobiology: A Practical Approach*. Eds Fukuda M, Kobata A. Oxford University Press Inc. New York; 27–76.

Martin MJ, Muotri A, Gage F, Varki A (2005) *Nat. Med.*; 11: 228–232.
Mechref Y, Novotny MV (1988) *Anal. Chem.*; 70: 455–463.
Merkle RK, Cummings RD (1987) *Meth. Enzymol.*; 138: 232–259.
Mills K, Johnson AW, Diettrich O, Clayton PT, Winchester BG (2000) *Tetrahed. Asymm.*; 11: 75–93.
Morell AG, Gregoriadis G, Scheinberg IH, Hickman J, Ashwell G (1971) *J. Biol. Chem.*; 246: 1461–1467.
Natsuka S, Lowe JB (1994) *Curr. Opin. Struct. Biol.*; 4: 683–691.
Neususs C, Demelbauer U, Pelzing M (2005) *Electrophoresis*; 26: 1442–1450.
Oh-eda M, Tominaga E, Nabuchi Y *et al.* (1996) *Anal. Biochem.*; 236: 369–371.
Patel T, Bruce J, Merry A *et al.* (1993) *Biochemistry*; 32: 679–693.
Pouria S, Corran PH, Smith AC *et al.* (2004) *Anal. Biochem.*; 330: 257–263.
Rademacher TW, Parekh RB, Dwek RA (1988) *Ann. Rev. Biochem.*; 57: 785–838.
Rademaker CJ, Pergantis SA, Blok-Tip L, Langridge JI, Kleen A, Thomas-Oates JE (1998) *Anal. Biochem.*; 257: 149–160.
Reynolds T M (1965) *Adv. Food Res.*; 14: 167–283.
Seo NS, Hollister JR, Jarvis DL (2001) *Protein Expr. Purif.*; 232: 234–241.
Settineri CA, Burlingame AL (1995) In *Carbohydrate Analysis: High Performance Liquid Chromatography and Capillary Electrophoresis.* Ed El Rassi Z. Elsevier Science, Amsterdam; 447–514.
Spiro RG (2002) *Glycobiology*; 12: 43R–56R.
Starr CM, Masada RI, Hague C, Skop E, Klock JC (1996) *J. Chromatogr. A*; 720: 295–321.
Storring PL, Tiplady RJ, Gaines Das RE *et al.* (1998) *Br. J. Haematol.*; 100: 79–89.
Sullivan B, Addona TA, Carr SA (2004) *Anal. Chem.*; 76: 3112–3118.
Takasaki S, Mizuochi T, Kobata A (1982) *Meth. Enzymol.*; 83: 263–268.
Tarelli E, Corran PH (2003) *J. Peptide Res.*; 62: 245–251.
Tarelli E, Wood JM (1981) *J. Biol. Stand.*; 9: 121–129.
Tarelli E, Corran PH, Bingham BR, Mollison H, Wait R (1994) *J. Pharm. Biomed. Anal.*; 12: 1355–1361.
Tarelli E, Smith AC, Hendry BM, Challacombe SJ, Pouria S (2004) *Carbohydr. Res.*; 339: 2329–2335.
Tarentino AL, Plummer TH (1994) *Meth. Enzymol.*; 230: 44–77.
Tekoah Y, Ko K, Koprowski H *et al.* (2004) *Arch. Biochem. Biophys.*; 426: 266–278.
Townsend RR, Hardy MR (1991) *Glycobiology*; 1: 139–147.
Townsend RR, Lipniunas PH, Bigge C, Ventom A, Parekh R (1996) *Anal. Biochem.*; 239: 200–207.
Trombetta ES (2003) *Glycobiology*; 13 77R–91R.
Vance BA, Wu W, Ribaudo RK, Segal DM, Kearse KP (1997) *J. Biol. Chem.*; 272: 23117–23122.
Vance BA, Bennett MJ, Ward Y, Gress RG, Kearse KP (1999) *Arch. Biochem. Biophys.*; 368: 214–220.
van den Steen P, Rudd PM, Dwek RA, Opdenakker G (1998) *Crit. Rev. Biochem. Mol. Biol.*; 33: 151–208.
Vliegenthart JFG, Van Halbeek H, Dorland L (1981) *Pure Appl. Chem.*; 53: 45–77.
Wan HZ, Kaneshiro S, Frenz J, Cacia J (2001) *J. Chromatogr. A*; 913: 437–446.
Watson E, Yao F (1993) *Anal. Biochem.*; 210: 389–393.
Wheeler SF, Harvey DJ (2000) *Anal. Chem.*; 72: 5027–5039.
Wu S-L, Bondarenko P, Shaler T, Shieh P, Hancock WS (2002) *ThermoFinnigan Application Report*; No 300.
Xiong L, Regnier FE (2002) *J. Chromatogr. B Analyt. Technol. Biomed. Life Sci.*; 782: 405–418.
Yuen C-T, Gee CK, Jones C (2002) *Biomed. Chromatogr.*; 16: 247–254.
Zhou G-H, Luo G-A, Sun G-Q, Cao Y-C, Zhang X-D, Zhang X (2004) *J. Pharm. Biomed. Anal.*; 35: 425–432.

Useful Web Sites

http://www.functionalglycomics.org/static/consortium/
http://us.expasy.org/
http://www.abrf.org
http://dionex.com
http://www.kratos.co.uk
http://www.spectroscopynow.com

http://www.thermo.com
http://boc.chem.uu.nl/sugabase/databases.html
http://www.prozyme.com
http://www.glycosuite.com
http://www.dextralabs.co.uk
http://www.ludger.com
http://www.glycosciences.org.uk

25 Immunogenicity of Impurities in Cell-Derived Vaccines

M Duchene, J Descamps and I Pierard

25.1 INTRODUCTION

In the fight against infectious diseases, vaccines are being developed to treat in a prophylactic or therapeutic way. In both cases, the induction of an adequate immune response by the protective antigen is fundamental. However, the presence of cellular by-products or impurities could, theoretically, jeopardize the safety and efficacy of the vaccine.

In this chapter, the different vaccines will be reviewed with respect to their compositions and immunological characteristics.

25.2 OVERVIEW OF EXISTING VACCINES

On a general classical basis, vaccines can be grouped into four classes:

(i) *Inactivated vaccines* composed of whole killed microorganisms such as whole cell pertussis, hepatitis A vaccine, inactivated polio vaccine (IPV), influenza vaccine, rabies vaccine.

(ii) *Live attenuated vaccines* containing live attenuated microorganisms such as polio (Sabin), measles, mumps, rubella, varicella, and salmonella.

(iii) *Subunit vaccines* containing proteins or toxins extracted from microorganisms such as diphtheria toxoid, tetanus toxoid, acellular pertussis antigens such as PT (pertussis toxin), FHA (filamentous haemagglutinin) and pertactin (69KD protein);
polysaccharide (PS) vaccines such as meningococcal vaccines;
conjugate vaccines where polysaccharides are coupled to carrier proteins such as haemophilus influenzae type B PS coupled to tetanus toxoid, and pneumococcal PS-protein conjugates.

(iv) *Recombinant vaccines* such as hepatitis B surface antigen produced in yeast cells, or Lyme OSP (outer surface protein) produced in *Escherichia coli*.

Other vaccines such as DNA vaccines, vector vaccines, or anti-idiotype vaccines are still in the development phase or entering clinical trials.

In addition to the active principle, all vaccines except for the live attenuated ones contain an adjuvant in order to stimulate the immune response. However, as this stimulation is not specific for the active principle, all immune responses towards possible contaminants could also be enhanced.

Medicines from Animal Cell Culture Edited by G. Stacey and J. Davis

25.3 POSSIBLE CONTAMINANTS

According to their purity level, vaccines can also be classified as:

- *Purified vaccines* such as hepatitis surface Ag vaccine or acellular pertussis. This class of vaccine is usually well characterized.
- *Semi-purified vaccines* such as inactivated polio vaccines and inactivated hepatitis A vaccine. Here a basic purification is done such as column chromatography and ultrafiltration.
- *Non-purified vaccines* such as the live attenuated vaccines (e.g. oral polio vaccine, measles, mumps, rubella). As these vaccines contain live attenuated strains, a simple cell clarification step is done when harvesting.

The possible contaminants present in the final vaccine depend mostly on the production process and the substrate or cells used for production. Generally speaking, contaminants can be of protein, lipid, polysaccharide or DNA origin, the level depending on the degree of purification. For instance, in the case of well-characterized vaccines such as HBs Ag, the protein purity level is above 95 % and may be as high as 99 %.

25.4 IMMUNE RESPONSES

The efficacy of the vaccine will depend on the immune response induced by the active principle. The time response, type of immunity induced, level and class of antibodies induced, and the affinity and avidity of the induced antibodies, are among the most important parameters to be followed during vaccine development and clinical trials. It is still vital that product batches are consistent in the composition of the active principle and any contaminants so that any development data is representative of the final product. Accordingly current Good Manufacturing Practice is required to ensure the necessery consistency (see Chapter 34) and that the efficacy seen during the vaccine trials can be reproduced by the final product.

25.4.1 Immune Response to Active Principle of Vaccine

The efficacy of a vaccine is evaluated by the reduction of clinical disease. This reduction is correlated with relevant immunological parameters, such as the appearance of the desired antibodies in sufficient quantities or the appearance of cell-mediated immunity depending on the class of antigens. A general overview of the different vaccines, including achievements and future promises, has recently been presented by André (2002, 2003) and Hilleman (2002). A review of DNA vaccines has been published by Liu (2003), and a general review on new vaccine technologies has been published by Wood (2002).

25.4.2 Immune Response to Contaminants or Impurities

During the production and purification process, the possibility exists that contaminant proteins or DNA from the cells used or impurities induced in the process will co-purify with the active principle. These impurities could originate from the cells, raw materials, stabilizers or excipients used. Since the adjuvant present in the vaccine will stimulate the immunogenic properties of all molecules present, the possibility exists that antibodies will be formed against these contaminant proteins or impurities. If these antibodies are directed against biological entities of the host, important safety issues could arise. Since, in general, polysaccharides and lipids are poor immunogens, the discussion below will focus on protein and DNA contaminants.

Numerous studies have been made of unwanted immunological responses elicited by vaccines. We will take the cases of hepatitis B surface (HBs) antigen produced in recombinant yeast (*Saccharomyces cerevisiae*) cells, hepatitis A vaccine produced on MRC-5 cells, and inactivated polio vaccine (IPV) produced on Vero cells as examples.

For HBs, the level of protein impurities is in the range of less than 1 µg per dose of 20 µg Ag. For residual yeast DNA, the level is less than 10 pg per vaccine dose. The appearance of antibodies against yeast cell proteins was demonstrated during the clinical development of this vaccine. In order to evaluate the human response to yeast-derived impurities, the sera of subjects vaccinated with the recombinant vaccine were evaluated by immunoblotting. Sera were obtained from subjects prior to and after three doses of vaccine. All pre-vaccination sera displayed a complex pattern of yeast-reactive antibodies unique to each subject. The pattern observed for each individual was not modified after the three doses. There was no significant change in the relative intensities of the antibody species detected, nor was any new antibody species found. These data thus indicate that the vaccine did not induce an immune response to yeast-derived impurities or raise the level of pre-existing antibody species. Also, the level of anti-yeast IgE antibodies was evaluated finding that, although anti-yeast IgG is detectable, no significant levels of anti-yeast IgE are present. These data are in agreement with guinea pig sensitization studies carried out according to a procedure described by Cox (1976): animals injected with five human doses of HBs vaccine were not sensitized to yeast components (1987).

For hepatitis A vaccine, the level of MRC-5 derived impurities was followed: per human dose, a maximum of 5 µg of protein could be detected, most of it of MRC-5 origin. The viral proteins constituted less than 100 ng per dose. The level of bovine serum albumin was below the detection level and no bovine immunoglobulin could be detected in the final purified antigen. As the product thus contained a 50:1 ratio of MRC-5 proteins to viral proteins, the possibility exists that vaccinees could develop anti-MRC 5 protein antibodies that could cross-react with self-antigens. To investigate this possibility, 30 paired (pre- and post-vaccination) serum samples were tested by Western blotting. No additional band could be observed in post-vaccination samples, and there was no increase in intensity of the existing bands. No differences between different lots were observed.

In another study, a total of 145 paired serum samples were tested by an inhibition enzyme-linked immunosorbent assay (ELISA). This was chosen rather than a direct binding assay in order to avoid non-specific binding. The results indicated that no difference in the OD response between pre- and post-vaccination samples could be detected. The fact that none of the samples tested could inhibit the binding of rabbit anti-MRC 5 antibodies to MRC 5 proteins immobilized on microtitre-plates suggests that no detectable levels of anti-MRC 5 antibodies were induced.

By using two different methods, ELISA and Western blotting, the presence of anti-MRC 5 antibodies could not be detected in vaccinees who had received three doses of vaccine.

Also the level of contaminating DNA of MRC-5 origin was measured in hepatitis A vaccines: using a DNA–DNA slot-blot assay, the DNA content was less than 125 pg per dose.

For IPV produced in Vero cells, the total level of cell proteins is less than 10 µg per human dose as specified by WHO Requirements. The level of DNA is less than 10 pg per dose. The present WHO Guidelines require that the vaccine does not contain more than 10 ng of Vero cell DNA per dose.

To estimate the capability of the purification process to remove host cell proteins to an acceptable level, Western blot analysis was performed on three batches of IPV monovalent bulk (one of each serotype). Each batch was tested at the different purification steps. Vero cell contaminants were present in the first step but no Vero cell contaminants could be detected by Western blot in the purified bulk. This experiment demonstrates that the purification process removes or denatures to a high degree the Vero cell proteins that are present in the clarified harvests. DNA clearance by the process was also evaluated. The total DNA content as measured by the Threshold system (Molecular Devices Corp.) for the three serotypes can be reduced by a factor exceeding 10^7. The DNA content in the purified bulks is consistently below or close to the detection limit (2 pg).

The same level of clearance is achieved for residual bovine serum proteins. The BSA content in purified bulks is consistently below or close to the detection limit of the ELISA test (2 ng/ml). These data demonstrate that the production process is capable of clearing to a significant degree the contaminant proteins and DNA.

In the case of non-purified vaccines such as measles, mumps, rubella or varicella, the level of contaminants from cellular origin varies with the viral production yield and the titre needed to vaccinate: values of protein content can range from 0.1 μg per vaccine dose to above 10 μg per vaccine dose. For the level of DNA content in these non-purified vaccines, the range varies between 0.1 to 3 μg per dose, again depending on the specific vaccine and the related viral yield and titre needed per vaccine dose. These data show that, depending on the specific vaccines, the level of contaminants in these non-purified vaccines is not very different from the levels found in the other vaccines: this is mainly due to the very low level of antigen content present as compared with the purified vaccines.

Furthermore, for these live attenuated vaccines, the vaccination schedule is generally different from that of killed or inactivated vaccines: currently, only one dose of attenuated vaccine is needed as compared with three to four doses for killed vaccines. Also, in some countries there is a trend to recommend a second dose of live vaccines but with an interval of many years between the two doses. This implies that, in spite of the non-purified status, the probability of induction of a hypersensitivity reaction is significantly lower than in the case of repeated standard inoculations. Nevertheless, some vaccination reactions to egg proteins have been reported in individuals with a history of egg allergy, but the vaccine itself did not appear to induce this allergy.

25.5 DISCUSSION

At present, vaccines are used to protect healthy people by inoculating one, two or three doses in order to induce an appropriate protective immune response. Also, vaccines are under development to treat people already having a disease or carrier state: here multiple injections can be needed in order to overcome the disease.

The presence of unwanted contaminant proteins or impurities in vaccines should thus be evaluated from the following perspectives: will there be an immune response induced against these contaminants after one, two or three doses or more in the case of therapeutic vaccines and if so, will this cause any safety issues? Similarly, is the level of contaminating DNA of cellular origin sufficient to induce an immunogenic response?

In the case of therapeutic proteins produced in cell culture systems or by biotechnological methods, some examples have been described (Lupker 1998).

Adverse effects of treatment with recombinant human erythropoietin (EPO) have been published: anaphylactic reactions have been observed in very rare cases. This has been described as being linked to the development of antibodies against EPO but not to host cell–derived impurities (Garcia *et al.* 1993). No antibodies have been observed to be induced by recombinant human granulocyte colony stimulating factor, follicle stimulating hormone, or to residual host-cell proteins (Garcia *et al.* 1993).

The interferons are the best characterized group of therapeutic proteins with respect to their immunogenicity: interferon-alpha induces neutralizing antibodies in over 50 % of recipients whereas interferon-beta showed a seroconversion rate of over 45 %. In all cases, multiple doses at concentrations much higher than the levels of contaminant proteins found in vaccines were needed to induce an immune response.

A good overview of immunogenicity of therapeutic biological products has been published recently by Brown and Mire-Sluis (2005).

From the data found with HBs vaccine produced in yeast, hepatitis A vaccine produced in MRC 5 cells or IPV produced in Vero cells, one can conclude that in these vaccines no immune memory is raised against protein impurities of cellular origin.

For the level of residual Vero cell DNA, the World Health Organization (WHO) currently accepts a limit of residual DNA from continuous cell lines of 10 ng per dose for products administered parenterally. All values found for the different IPV vaccines on the market that are derived from continuous cell lines were in the pg per dose range. This is far below the levels that should be needed for DNA vaccination (Liu 2003). For instance, experiments in mice have shown that doses of plasmid DNA needed to vaccinate against influenza virus ranged from 1 to 200 μg depending on the number of doses injected.

The levels of contaminating DNA found in cell culture-derived products are in agreement with published data. In the production of IPV, values ranging from 1.8 pg per dose to 15 pg per dose are common. More than 100 million doses of this vaccine have been administered with only very rare adverse events (Montagnon & Vincent-Falquet 1998). One can thus conclude that the level of contaminating DNA in vaccines is not a serious safety issue: the levels obtained after clearance by a consistent production process are well below the levels needed for biological activity. Additionally, the new trend in different guidelines to take into account the size of DNA as well as the content will further increase the safety margin of the product.

Some vaccines such as measles, mumps, influenza or yellow fever, are produced on embryonated egg substrate. It is known that vaccines containing small quantities of egg protein can cause hypersensitivity reactions in some people with egg allergy. Anaphylaxis after administrating measles-containing vaccines is a rare event but has been reported in persons with pre-existing hypersensitivity to eggs as well as in persons with no history of egg allergy. In some cases, allergy to residual contaminants from cell cultures, e.g. neomycin, was suspected, or reactions to excipients, e.g. gelatine, but in most cases no source of allergen was identified (Canadian Immunization Guides 1996). A recent article by Bohlke *et al.* calculates the risk of anaphylaxis after vaccination of children and adolescents at 0.65 cases per million doses administered (Bohlke *et al.* 2003).

In general, reactions to vaccines are rare events. In the case of a highly purified hepatitis B vaccine (Engerix B), after 500 million doses administered the incidence of the most common reaction (fever) was 0.4 per 100 000 doses, and for allergic reactions (urticaria), the incidence rate was 0.2 per 100 000 doses (Keating & Noble 2003). Over 1 billion doses have now been injected without any change in the safety profile. In the case of a semi-purified vaccine such as that for hepatitis A (Havrix), the incidence rates after 69 million doses administered over 10 years were similar: 0.6 per 100 000 doses for fever, and 0.23 per 100 000 doses for allergic reactions (André *et al.* 2002). Thus the degree of purity of the vaccine *per se* does not seem to influence the reaction rate, as similar reaction rates are observed for highly purified vaccines as for semi-purified vaccines.

In the case of combined non-purified vaccines such as measles–mumps–rubella, local adverse events such as injection-site pain, redness and swelling, and systemic adverse events such as fever, rash and parotid gland swelling, have been described at a low level. The incidences do not increase after the second dose and are similar between different vaccine producers (Wellington & Goa 2003). When these vaccines are combined with varicella, similar rates of pain, redness or swelling at the injection sites are observed (Nolan *et al.* 2002). However, there are no indications that the reactions are related to the level of impurity present.

25.6 CONCLUSION

The overall quality of a vaccine is determined by full characterization of the product in laboratory testing and clinical trials, and by the guarantee of consistent production through the application of cGMP. The inclusion of very sensitive immunological, biochemical and electrophoretic methods to detect protein, polysaccharide, lipid or DNA impurities, and the rigorous process clearance validation studies to demonstrate removal or clearance of contaminant proteins or host cell DNA are a further basis for confidence in the quality and safety of all vaccines used.

A review of more than 20 years of vaccination can conclude that the present levels of purity of the existing vaccines and the low content of contaminants derived from cell culture systems do not pose any safety issues.

REFERENCES

André FE (2002) *Dev. Biol. (Basel)*; 110: 25–29.
André FE (2003) *Vaccine*; 21: 593–595.
André F, Van Damme P, Safary A, Banatvala J (2002) *Exp. Rev. Vaccines*; 1(1): 9–23.
Bohlke K, Davis RL, Marcy SM *et al.* (2003) *Pediatrics*; 112: 815–820.
Brown F, Mire-Sluis AR (2005) *Immunogenicity of Therapeutic Biological Products. Development in Biologicals 112,* Basel, Karger.
Canadian Immunization Guide (1996). *Supplementary Statement MMR Vaccine and Anaphylactic Hypersensitivity to Egg or Egg-related Antigens.* Canada Communicable Disease Report 22-14, F1-2.
Cox CB (1976) *J. Biol. Stand.*; 4: 81–89.
Garcia JE, Senent C, Pascual C *et al.* (1993) *Nephron*; 65: 636–637.
Hilleman MR (2002) *Intervirology*; 45(4–6); 199–211.
Keating GM, Noble S (2003) *Drugs*; 63: 1021–1051.
Liu MA (2003) *J. Intern. Med.*; 253: 402–410.
Lupker JH (1998) *Dev. Biol. Stand.*; 93: 61–64.
Montagnon BJ, Vincent-Falquet JC (1998) *Dev. Biol. Stand.*; 93: 119–123.
Nolan T, McIntyre P, Roberton D, Descamps D (2002) *Vaccine*; 21: 281–289.
Petre J, Van Wijnendaele F, De Neys B *et al.* (1987) *Postgrad. Med. J.*; 63(suppl. 2): 73–81.
Wellington K, Goa KL (2003) *Drugs*; 63: 2107–2126.
Wood DJ (2002) *Dev. Biol. Stand.*; 111: 285–290.

26 Potency and Safety Assessment of Vaccines and Antitoxins: Use of Cell-based Assays

D Sesardic

26.1 INTRODUCTION

Traditionally animals are used for quality control of biological products, and this is particularly true for vaccines containing chemically attenuated toxins, such as diphtheria, tetanus and pertussis, and immunosera against specific toxins (European Pharmacopoeia 2005a). Most of the animals are used for confirmation of absence of toxin, freedom from reversion to toxin, specific toxicity in final product and for potency tests (Weisser & Hechler 1997). *In vitro,* human and animal cell lines can be used to demonstrate particular aspects of product safety and efficacy, but regulatory acceptance of this technology is limited, to date.

26.2 SPECIFIC SAFETY TESTS

Better understanding of the mode of action of some of the bacterial toxins has led to the introduction of cell-based assays. The best example is provided by diphtheria toxin. Up to 1997 the European Pharmacopoeia monograph for diphtheria vaccine (European Pharmacopoeia 1997) together with the WHO requirement for diphtheria, pertussis, tetanus and combined vaccines (WHO 1977, 1990) stipulated the use of guinea pigs to confirm absence and freedom from reversion of diphtheria toxin. Basic studies in the 1970s and 1980s (Miyamura *et al.* 1974; Mekada *et al.* 1988) provided evidence that Vero cells have specific diphtheria toxin receptors and that toxin is selectively cytotoxic to cells by ADP-ribosyl-transferase activity. The A domain of the toxin catalyses the ADP-ribosylation of elongation factor EF-2, a critical factor in the protein synthetic machinery. The reasons for using suitable cell lines to confirm safety are obvious, as all functional entities of the toxin are detected, i.e. binding and internalization as well as intracellular events, and in the case of diphtheria the presence of the toxin leads to cell death. First attempts to adopt the tissue culture assay for toxicity of diphtheria toxin were reported as early as 1986 (Abreco & Stainer 1986). However, it took almost another 20 years of development (Efstratiou *et al.* 1998; Jooren *et al.* 2001; Hoy & Sesardic 1994) and validation studies (Brown *et al.* 2002; Hendricksen & Gardoff 1994; Jooren *et al.* 2001) to adopt the approach and amend regulatory requirements for toxicity testing of chemically inactivated diphtheria toxin prior to use in human vaccine. The European Pharmacopoeia monograph for diphtheria vaccine (adsorbed) effective from January 2005 (01/2005:0443) (European Pharmacopoeia 2005c) now includes only the Vero cell assay to

Medicines from Animal Cell Culture Edited by G. Stacey and J. Davis

confirm absence of toxin and irreversibility of toxoid for bulk purified diphtheria toxoid. The cell assay sensitivity is monitored by including a reference toxin in the test (Sesardic *et al.* 2003). This is the first example where an *in vitro* model has entirely replaced a traditional *in vivo* model for specific toxicity in the European Pharmacopoeia.

The replacement of animals for toxicity testing of other toxins such as tetanus and pertussis, is still at an early stage of development and these methods have yet to be approved by the regulatory authorities. Clustering of Chinese hamster ovary cells by pertussis toxin (Isawa & Fujiwara 1990; Gillenius *et al.* 1985) has been considered as a Suitable method of confirming absence of toxin and irreversibility of toxoid, and as a possible replacement for the *in vivo* histamine sensitization activiey which is included in current WHO guidelines (WHO 1999) and in relevant Pharmacopoeia monographs (2005a). However, recently published evidence suggests that such an approach may not be suitable for aldehyde-inactivated pertussis toxin and is unlikely to be taken further for validation (Kataoka *et al.* 2002). It is recognized that any future cell-based assay will also require the use of a suitable reference toxin (Xing *et al.* 2002; Dobbelaer *et al.* 2001).

Future generations of genetically attenuated toxins, for example CRM_{197} of diphtheria toxin and pertussis toxins, are likely to replace current chemically attenuated products. Quality control to confirm safety of such proteins will require different strategies and will be entirely dependent on an *in vitro* approach. For example, the lack of toxicity and activity, and production of cAMP by Y1 adrenal cells in response to genetically modified adjuvants such as cholera toxin and *E. coli* heat-labile toxin, will be key elements in supporting attenuation of those toxins (Yamamoto *et al.* 2003).

The demonstration of replication deficiency of live attenuated virus vaccines and vectors can be performed *in vitro* with human and animal cell lines. An example is the demonstration of the inability of a novel live attenuated *Salmonella enterica* serovar Typhi vaccine to replicate in human macrophages, a reservoir for virulent *S. typhi* in humans, compared with the wild type strain or the licensed *S. typhi* Ty21a (Vivotif™). This would provide a suitable cellular basis for safety confirmation of the strain (Tamara & Kurata 2000).

26.3 GENERAL SAFETY

It is more difficult, if not impossible, to replace general safety tests by a single cell line assay and most preclinical safety testing is performed on whole animals (CPMP 1995; WHO 1969; Brennan & Dougan 2005). The pyrogen test in rabbits has been replaced for some products by a validated bacterial endotoxin test, but this has proved to be ineffective for many products, particularly for antigenic components derived from bacteria used in the production of vaccines (Gummar & Vitanen 2005). A sensitive and reliable cell line assay for detection of endotoxin contamination in biological products has been described (Yamamoto *et al.* 2002, 2003), using human monocyte cell lines such as THP-1, 28SC or Mono Mac 6. However use of a single human cell line approach for safety testing is considered limited. Mouse macrophage cell line RAW264.7 may in time replace the rabbit model for pyrogenic substances. Human whole blood and isolated primary human mononuclear cells have been used in developing models for detection of bacterial endotoxins derived from Gram-negative bacteria, Gram positive toxins, fungi or viruses. Such products and bacterial LPS will stimulate human monocytes to produce the endoenous pyrogenic cytokine IL-6 (Gaines Das et al., 2004). While this method can detect both endotoxin and non-endotoxin contaminants it is unsuitable for preparations that have the intrinsic property of stimulating of inhibiting the release of IL-6.

26.4 POTENCY OF VACCINES

The majority of animal models for assessing the potency of bacterial and viral vaccines are dependent on the induction of specific antibody responses in animals and subsequent detection of antibodies in suitable *in vitro* or *in vivo* models. As only a proportion of antibodies induced by antigen in a vaccine will have specific ability to neutralize toxin or virulence factors, the most suitable models are those with the ability to detect functional antibodies, and therefore rely on seroneutralization. The best example of a potency assay relying on cells in culture is that of poliomyelitis vaccines in which antibodies induced in rats, guinea pigs or chicks are used to protect Vero or Hep2 cells from a defined challenge dose of each poliovirus (European Pharmacopoeia 2005b). Potency testing of traditional tetanus, diphtheria and whole cell pertussis vaccines, on the other hand, involves immunization of animals and subsequent direct challenge of animals with native toxin or a virulent strain of bacteria (European Pharmacopoeia 2005a, c). Such challenge models, because of the severity of the procedure to live animals, have been considered a high priority for replacement from an animal welfare perspective (Weisser & Hechler 1997; Brown *et al.* 2002; Hendricksen & Gardoff 1994). As part of the initiative to replace the European Pharmacopoeia direct challenge assay for diphtheria vaccines, validation studies were initiated by ECVAM and EDQM and supported by the European Commission (Sesardic *et al.* 2001, 2004; Winsnes & Sesardic 2002; Winsnes *et al.* 2004). Successful conclusion of these studies has led to revision of the European Pharmacopoeia general chapter for potency testing of diphtheria vaccine (Pharmeuropa Bio 2005) which was adopted by the European Pharmacopoeia Commission in March 2006 and will become effective from March 2007. The method includes seroneutralization on Vero cells by antibodies induced in animals following vaccination.

In vitro cell-based studies are also increasingly considered of use in demonstrating the pharmacological activity and efficacy of new vaccines and adjuvants, particularly in situations where there are no relevant or suitable animal models (Brennan & Dougan 2005). For example, the ability of modified LPS-based adjuvants such as MPL or CpG-containing adjuvants to effect activation of human or animal lymphocytes and dendritic cells can be tested by measurement of proliferation and cytokine production *in vitro,* and may provide a basis for future potency assays for vaccines (De Becker *et al.* 2000; Bohle *et al.* 1999). Similarly, cellular uptake, replication competence and nuclear localization of gene-therapy-based vaccines, and the transformation efficacy, and expression of DNA vaccines, as well as the effect of immunomodulatory cytokines, can all be conveniently investigated on cells *in vitro* (Guan *et al.* 2001; Diogo *et al.* 2001).

26.5 POTENCY OF ANTITOXINS

Immunosera and specific antitoxins (which can be polyclonal or monoclonal) form another set of biological products. Where the therapeutic product is an antitoxin, neutralizing potency is generally measured by passive protection *in vivo*, where the toxin's effect on animals is neutralized by the antibody (Weisser & Hechler 1997). Several therapeutic antitoxin products are widely available, such as tetanus antitoxin, human anti-tetanus IgG and botulinum antitoxin trivalent, but many new antitoxin products against anthrox, botulinum and ricin toxins are considered as front-line treatment for potential bio-terror agenst, and their popularity as well as production is on the increase.

Potency of antitoxin proparations relies on toxin neutralisation principles and cell culture models can be used provided that cells are responsive with sufficient sensitivity to the toxin. Successful examples include neutralisation of diphtheria toxin induced cytotoxicity on Vero cells (Miyamura et al., 1974), ricin induced cytotoxicity of Hela or Vero cells (Argent et al., 1994) and anthrax lethal toxin neutralisation using mouse macrophage-like cell lines RAW264.7 or J774A1 (Hering et al., 2004.)

Although clostridial neurotoxins potently and specifically inhibit the release of neurotransmitters in certain established cell lines of neuronal lineage such as PC-12 and SH5Y, providing an opportunity for *in vitro* cell-based assays for antitoxin, to date the established cell lines have provided insufficient sensitivity and robustness for this application (Shone & Melling 1992; Sesardic *et al.* 2004a,b). At present the only promising cells remain primary rat spinal cord cells, which have specificity as well as sensitivity for this application (Hall *et al.* 2004) but their use in routine testing and as part of regulatory requirements for potency of botulinum antitoxin remain to be validated.

In vitro cell-based tests for detection of other clostridial toxins, toxoids and antitoxins mainly include veterinary products, in particular toxins produced by *Clostridium septicum* and *Clostridium perfringens.* The MDCK cell line, an established dog kidney cell line, has been successfully used for titration of toxins and *in vitro* toxin neutralization of the corresponding antitoxin products (Knight *et al.* 1990a,b). However, to date these methods have not been validated sufficiently to replace established *in vivo* pharmacopoeia tests.

REFERENCES

Abreo CB, Stainer DW (1986) *Dev. Biol. Stand.*; 64: 33–37.

Argent R.H, Roberts 1.M Wales r Robertus J.D and Lord J.M. (1994) Journal of Biological Chemistry, 269; 26705–26710.

Bohle B, Jahn-Schmid B, Maurer D, Kraft D, Ebner C (1999) *Eur. J. Immunol.*; 29: 2344–2353.

Brennan F, Dougan G (2005) *Vaccine*; 23: 3210–3222.

Brown F, Hendriksen CFM, Cussler C, Sesardic D (Eds) (2002) *Advancing Science and Elimination of the Use of Laboratory Animals for Development and Control of Vaccines and Hormones. Developments in Biological Standardization* Vol. 111. Karger Press, Basel. Switzerland.

CPMP (1995) Notes for guidance on preclinical pharmaceutical and toxicological testing of vaccines (CPM/SWP/465/95).

De Becker G, Moulin V, Pajak B *et al.* (2000) *Int. Immunol.*; 12: 807–815.

Diogo MM, Ribeiro SC, Queiroz JA *et al.* (2001) *J. Gen. Med.*; 3: 577–584.

Dobbelaer R, Daas A, Esposito-Farese, M-E (2001) *Pharmaeuropa BIO*; 2001-1: 15–25.

Efstratiou A, Engler KH, Dawes CS, Sesardic D (1998) *J. Clin. Microbiol.*; 36: 3173–3177.

European Pharmacopoeia (1997) *Diphtheria vaccine (adsorbed) monograph 0443.* Council of Europe, Strasbourg; third edition.

European Pharmacopoeia (2005a) Diphtheria Vaccine (adsorbed) Monograph 0443; Diphtheria and Tetanus vaccine (adsorbed) Monograph 0444; Diptheria, Tetanus and pertussis vaccine (adsorbed) Monograph 1931. Counsil of Europe, Strasbourg; fifth edition.

European Pharmacopoeia (2005b) *In vivo Assay of Poliomyelitis Vaccine (Inactivated), Monograph 2.7.20.* Council of Europe, *Strasbourg; fifth edition.*

European Pharmacopoeia (2005c) *Assay of Diphtheria Vaccine Adsorbed.* General chapter 2.7.6, *Assay of Tetanus Vaccine Adsorbed.* General chapter 2.7.8. Council of Europe, Strasbourg; fifth edition.

Gaines Das RE, Brugger P, Patel M, Mistry Y and Poole S. (2004). Journal of Immunological Methods 288; 165–177.

Gillenius P, Jaatmaa E, Askelof P, Granstrom M, Tiru M (1985) *J. Biol. Stand.*; 13: 61–66.

Guan Y, Whitney JB, Detorio M, Wainberg MA (2001) *J. Virol.*; 75: 4056–67.

Gummar C, Vitanen E (2005) *Vaccine*; 23: 3709–3715.

Hall YH, Chaddock JA, Moulsdale HJ *et al.* (2004) *J. Immunol. Methods*; 288 (1-2): 55–60.

Hendricksen CFM, Gardoff B (1994) *ALTA*; 22: 420–434.

Hering D., Thompson W., Hewetson J., Little S., Norris S. and Pace-Templeton J. (2004). Biologicals, 32: 17–27.

Hoy CS, Sesardic D (1994) *Toxicol. in vitro*; 8: 693–695.

Isawa S, Fujiwara H (1990) *Biologicals*; 18: 127–130.

Jooren J, Peyre M, van der Gun J, Hendriksen C, Sesardic D (2001) Validation of Vero cell assay for specific toxicity of diphtheria toxoid component in vaccines. IABS meeting '*Advancing Science and Elimination of the Use of Laboratory Animals for Development and Control of Vaccines and Hormones*'. Utrecht 2001, Abstract p. 69.

Kataoka M, Toyoizumi H, Yamamoto A, Ochiai M, Horiuchi Y (2002) *Biologicals*; 30: 297–302.
Knight PA, Tilleray JH, Queminet J (1990a) *Biologicals*; 18: 181–189.
Knight PA, Queminet J, Blanchard JH, Tilleray JH (1990b) *Biologicals*; 18: 263–270.
Mekada E, Okada Y, Uchida T (1988) *J. Cell Biol.*; 107: 511–519.
Miyamura K, Nishio S, Ito A, Muata R, Kono R (1974) *J. Biol. Stand.*; 2: 189–201.
Pharmeurapa Bio (2005); 17.3 July 2005: General chapter 2.7.6. Assay of diphtheria vaccine adsorbed; Proposed revised draft.
Rainey G.J.A. and Young J.A.T. (2004). Nature Reviews Microbiology 2: 721–726.
Sesardic D, Winsnes R, Rigsby P, Gaines Das R (2001) *Biologicals*; 29: 107–122.
Sesardic D, Prior C, Daas A, Buchheit K-H (2003) *Pharmeuropa BIO*; 2003-1: 7–21.
Sesardic D, Jones RGA, Leung T, Alsop T, Tirney R (2004a) *Movement Disorders*; 19: 85–91.
Sesardic D, Winsnes R, Rigsby P, Behr-Gross M-E (2004) *Pharmeuropa Bio*; 2003–2: 69–75.
Shone CC, Melling J (1992) *Eur. J. Biochem.*; 207: 1009–1016.
Tamara SI, Kurata TA (2000) *Jpn. J. Infect. Dis.*; 53: 98–106.
Weisser K, Hechler U (1997) *Animal Welfare Aspects in the Quality Control of Immunobiologicals: A Critical Evaluation of the Animal Tests in Pharmacopoeial Monographs. FRAME, Nottingham, UK.*
WHO (1969) *Guidance on Nonclinical Evaluation of Vaccines* (WHO/BS/03).
WHO (1977) *Manual for the Production of and Control of Vaccines – Diphtheria Toxoid.* WHO BLG/UNDP/77.1.
WHO (1990) *Tech. Rep. Ser.*;800: 87–189.
Winsnes R, Sesardic D (2002) In Advancing Science and Elimination of the Use of Laboratory Animals for Development and Control of Vaccines and Hormones. Eds Brown F, Hendriksen C, Sesardic D, Cussler K. *Developments in Biological Standardisation*, 111: 143–151.
Winsnes R, Sesardic D, Dass A, Behr-Gross M-E (2004) *Pharmeuropa Bio*; 2003-2: 35–44.
Yamamoto A, Ochiai M, Kataoka M, Toyoizumi H, Horiuchi Y (2002) *Biologicals*; 30: 85–92.
Yamamoto A, Sakai T, Kamachi K, Kataoka M, Toyoizumi H, Horiuchi Y (2003) *Jpn J. Infect. Dis.* 56: 93–100.
Xing D, Gaines-Das R, Newland P, Corbel M (2002) *Vaccine*; 20: 3535–3542.

27 Product Stability and Accelerated Degradation Studies

P Matejtschuk and P Phillips

27.1 INTRODUCTION

Biological materials are subject to an array of degradative processes that may impact on product quality, reduce potency, or indeed give rise to fragmented materials that would inhibit the intended biological activity. Preservation in a lyophilized format can significantly reduce these degradative processes. However, there can still be deterioration in the product, especially when it is stored for long periods of time. Clearly, the determination of stability is an important issue. The optimum storage temperature and the most suitable final container (such as ampoule, syringe or vial) must be determined in order to maximize the useful shelf life of a biological medicine. Long shelf life under low-cost storage conditions is an important element in the delivery of economically viable products. Accordingly, the goal is storage under ambient conditions with no significant loss of potency throughout a shelf life of several years.

27.1.2. The Degradative Processes

Degradation of biological materials occurs in both the liquid and the lyophilized states, and although processes may be slower in the dry state, it cannot be taken for granted that lyophilized materials necessarily have excellent storage stability, and studies of the deterioration processes in the dry state have been made (Lai & Topp 1999; Li *et al.* 1996).

Stability studies, especially accelerated studies, can provide useful insight into degradation pathways for the molecule under study. High-resolution techniques such as peptide mapping (Kroon *et al.* 1992) and matrix-assisted laser desorption ionization time of flight (MALDI-TOF) mass spectrometry can be used to identify these degradation events. These methods must be carefully validated (see ICH guidelines Q2(R1)) in order to allow detailed comparisons for large biomolecules, which are usually enzymatically degraded and analysed as the constituent peptides (Matejtschuk *et al.* 1998).

The degradative changes that may occur in proteins have been studied extensively (Lai & Topp 1999; Cleland *et al.* 1993). Some examples of these are given below. However other types of biomolecule, e.g. DNA, can also be damaged by downstream processing and lyophilization (Poxon & Hughes 2000). Some degradation may occur during purification but other deterioration may only be revealed on long-term storage. Many degradative processes are water catalysed and therefore occur more rapidly in the liquid state and more slowly when the biomolecule is stabilized in the dry state, although the optimum level of moisture for long term stability may vary from one

Medicines from Animal Cell Culture Edited by G. Stacey and J. Davis

preparation to another. The impact of moisture content on lyophilized product stability has been discussed (Hageman 1992). Hsu *et al.* (1992) reported that there is an optimal moisture content for recombinant tissue plasminogen activator below which the protein shows reduced potency and stability, and an intermediate optimum moisture content was also demonstrated for varicella vaccine (Bennett *et al.* 1992) and bacteria (Nei *et al.* 1966). Uneven distribution of moisture in a non-homogeneous lyophilized cake may result in biphasic losses in functional activity (Franks 1990).

27.1.2.1 Oxidation of methionine/cysteine/tryptophan

Methionine, cysteine and tryptophan residues can all undergo oxidation reactions in the liquid state. This may result in modification in the activity of a biological medicine. Methionine can be oxidized to the sulphone and sulphoxide, inducing a change in charge and potentially modifying a useful electron donor centre in the sequence (Nguyen 1994).

Oxidation of methionine in lyophilized somatotropin has been described in the presence of oxygen or intense light (Becker *et al.* 1988). Trace metals can cause oxidation (Farber & Levine 1986). In the lyophilized state the percentage amorphous content, moisture content and molecular mobility may affect oxidation. Human growth hormone can be freeze dried without substantial degradation but can undergo degradation on storage via oxidation of methionine and deamidation of asparagines (Pikal *et al.* 1992).

Cysteine can be oxidized to cystine with the formation of a disulfide bond. Mixed disulphide interchange may result in disulphide bonds between residues not normally paired, so resulting in aggregation (Liu *et al.* 1991); such disulphide exchange may be facilitated by lyophilization-induced structural perturbation (Costantino 1998c).

In the liquid state tryptophan residues can also be oxidized, disrupting its heterocyclic structure (Wang & Hanson 1988). If the affected residues are in the active site of the protein or induce some destabilization of the tertiary structure of the molecule, then these changes may result in deterioration of the biological activity or a propensity for increased aggregation or fragmentation. Formulation pH can greatly affect oxidation, cysteine oxidation proceeding faster, but oxidation of methionine less quickly, above neutral pH. Histidine is less prone to oxidation below pH 6 as the nitrogen is not protonated in this amino acid (Hageman 1992).

27.1.2.2 Deamidation

Deamidation, the conversion of asparagine to aspartate and glutamine to glutamate, can occur in both the liquid and freeze dried states (Manning *et al.* 1989). Such deterioration is known to occur during processing and so it is unsurprising that it can also occur during storage. Indeed, it is quite common in cell-culture-derived therapeutics and may not cause significant loss of activity, unless the residue modified is located in a functionally important sequence or contributes to an active site. In order to combat these modifications, pH should be carefully controlled and water content kept to a minimum. The occurrence of deamidated residues in recombinant therapeutic proteins has been reported (Breen *et al.* 2001; Becker *et al.* 1988; Harris *et al.* 2001) and will also occur in materials derived from non-recombinant sources. Where N-terminal glutamine occurs it can undergo spontaneous rearrangement to give N-terminal pyroglutamate, although this may not necessarily adversely affect the functional activity of the therapeutic (Roberts *et al.* 1995).

27.1.2.3 Hydrolysis at susceptible peptides

Peptides with Asn–Gly and Asp–Pro sequences are particularly prone to hydrolysis reactions at these bonds, the cleavage resulting in a breakage in the peptide bond, best seen by fragmentation analysis (Clarke *et al.* 1992). Such cleavage reactions can be pH sensitive, aspartyl residues being prone to hydrolysis under acidic conditions (Joshi & Kirsch 2004).

27.1.2.4 Isomerization of aspartate

As an alternative to the more drastic bond lysis, there is also the possibility of aspartate isomerization to give isoaspartate or β-aspartate (Zhang & Czupryn 2003). This may induce conformational change and disruption in the tertiary structure.

27.1.2.5 Photodecomposition

This affects tryptophan residues in particular, and may cause discolouration in protein preparations, in both liquid and dried states (Cleland *et al.* 1993).

27.1.2.6 Aggregation

Aggregation and/or fragmentation can occur, either due to molecular modification, exposure of normally buried hydrophobic pockets, or to physical denaturation (such as is induced at air–liquid interfaces). Aggregation of biological molecules can be highly deleterious. The aggregates may lack biological activity, may become immunogenic or induce biochemical cascades (such as complement fixation) and so their level must be carefully monitored and controlled in biological products, especially parenterals. Once aggregation begins, it may accelerate with time so that initially insignificant levels may rise to more substantial levels on storage. This 'seeding' effect may be seen with some aggregation in the liquid state, which may return on storage even though the initially present aggregates were removed by filtration. Changes in molecular structure may induce instability that can be seen in terms of the energy required to induce denaturation of macromolecules by techniques such as isothermal calorimetry. Conversely, proteolytic contamination (such as by cathepsins from the cell culture) has been known to cause problems with fragmentation of biological molecules, which can occur after synthesis but prior to purification, and/or (if inadequately removed by the purification process) during product storage.

Denaturation may occur initially as a reversible process but then progress to irreversible aggregation and concomitant loss of functional activity. From studies by FTIR, a loss of α-helical structures and rise in β-sheet often occurs on lyophilization, but these changes are usually reversible on reconstitution. More recently, some authors have found a correlation between structural perturbations detected by Raman spectroscopy and aggregation on longer-term storage (Sane *et al.* 2004).

High performance size-exclusion chromatography is a powerful method for the detection and quantification of aggregates and fragments in biological macromolecules. Other methods may include SDS polyacrylamide gel electrophoresis (PAGE) and SDS capillary electrophoresis (Matejtschuk *et al.* 1998). By use of reducing and non-reducing conditions in parallel tests, fragmentation occurring within disulphide bonded polypeptides may be identified, whereas these may be undetected by analyses based on non-reducing conditions.

27.1.2.7 Maillard reactions

Maillard reactions may occur between primary amino groups and free aldehyde functions, even in the dried state (Andya *et al.* 1999). This may be particularly of concern where protein products are formulated with reducing sugars and are held at above ambient temperatures. Such reactions have been the topic of much study in the food industry, where such complexes may influence flavour and colouration of heated foodstuffs. In biopharmaceuticals, discoloration may also occur and, more importantly, Maillard modification may result in loss of activity or potential immunogenicity. To minimize such reactions, reducing ('active aldehyde') sugars such as glucose should be avoided in formulations of biologicals. Those dissacharides that might be subsequently broken down to yield reducing sugars should also be suspect (O'Brien 1996).

27.2 STABILITY IN THE LYOPHILIZED STATE

Damage to lyophilized materials may occur at a number of different stages. Denaturation may result from the freezing itself (this type of damage can be identified by repeated freeze–thaw studies) or due to the dehydration that occurs during the freeze-drying stage. Deterioration may only become apparent when the lyophilized material is reconstituted, such as incomplete reconstitution or aggregate formation on reconstitution. Finally, some instability may not be apparent initially but may occur on prolonged storage in the lyophilized state.

As previously noted, lyophilization in general will improve the storage stability of biological molecules. However, the choice of excipient will impact on the physical appearance, the stabilization against lyophilization-induced damage and the longer-term storage stability. Lyophilized materials may contain both amorphous materials, held within a glassy state with little molecular mobility, and also crystalline materials, dependent upon the formulation used. Amorphous materials present in the lyophilized formulation may undergo increased mobilization when the temperature rises above the glass transition temperature (T_g). The T_g value will be reduced by the presence of substantial levels of residual moisture. Adams and Ramsay (1996) reported storage instability above the T_g and also the lowering of effective T_g by transfer of moisture from the closure to the product during storage, with subsequent deleterious implications for storage stability.

Chang *et al.* (1996) made a comparison of T_g and accelerated stability for a range of lyophilized formulations of recombinant human interleukin-1 receptor antagonist (rhIL-1ra). Even when stored below the T_g of their lyophilized materials, some formulations resulted in protein inactivation. It was concluded that it is important to prevent damage occurring during lyophilization as this may have an adverse impact on storage stability.

Bell *et al.* (1995) studied lyzozyme and recombinant somatotropin using differential scanning calorimetry (DSC) and showed the effect of excipient choice on resultant instability after lyophilization. The stability appeared to correlate with the resultant T_g values and residual moisture content is also an important factor influencing instability.

Modulated DSC has also been applied to follow the long-term physical stability of freeze-dried formulations. On exposure to elevated temperature or humidity, contraction of the lyophilized cake may occur when the storage temperature exceeds the T_g of the system. This information can then be used to develop a more robust system by the rational selection of excipients that increase or decrease the T_g (Fitzpatrick & Saklatvala 2003).

Izutsu *et al.* (1994) studied β-galactosidase lyophilized from inositol phosphate buffer formulations. In some formulations, collapse of the freeze-dried cake occurred on storage. Inositol was protective when in the amorphous state but lost it stabilizing effect on crystallization.

Although it is visually unacceptable, collapse (also known as glassy melt) of the freeze-dried cake may not necessarily indicate poorer stability. Wang *et al.* (2004) reported on the lyophilization of recombinant Factor VIII and alpha-amylase as model systems. Although, when using aggressive freeze drying cycles, collapse was seen for the Factor VIII preparation, the stability of the material was unaffected on storage at elevated temperatures and was better than that of the formulations not showing collapse.

Costantino *et al.* (1998a) showed that the stability of a recombinant monoclonal antibody in the spray-dried state was influenced by the concentration of mannitol as excipient. Although the monoclonal was stable when formulated with 10–20 % mannitol, at a concentration of 30 % mannitol stability declined and aggregation of antibody occurred at both 5 °C and 30 °C, due to crystallization of the mannitol excipient. The amorphous mannitol conferred a protective effect that the crystalline state did not. This instability could be addressed by addition of sodium phosphate buffer that inhibited mannitol crystallization. The same author (Costantino *et al.* 1998b) demonstrated the usefulness of Fourier transform infrared (FTIR) spectroscopy to demonstrate changes in secondary structure in lyophilized biologicals. In general there is a loss of α-helix and a rise in β-sheet on lyophilization, although these changes could be reversed on dissolution where no loss of biological activity was noted. Non-crystallizing excipients help to prevent protein aggregation. Stabilization of encapsulated proteins for use in slow-release dosage forms has been recently reviewed (Perez *et al.* 2002).

27.3 STABILITY TESTING

Degradation at the temperature of storage may be too slow for real-time studies to be practical in determining whether a particular material or formulation is satisfactory for use. In such instances, accelerated degradation studies are useful in obtaining a relatively quick prediction of stability. These accelerated studies can be useful in a research and development environment to determine, for example, which of several otherwise equally satisfactory options yields the best stability and should be selected for further development.

Stability studies are also important for determining safe shipment conditions as, in transportation, products may encounter widely differing environmental conditions that may prove deleterious and result in the need for 'cold chain' procedures, which assure limits to temperature and humidity exposure. These topics are outside of the scope of this chapter.

Ultimately, accelerated degradation studies are no substitute for real-time studies where material is stored at the recommended temperature of storage and loss of activity monitored. In addition, they are no substitute for other studies on stability including:

- the effect of time/temperature on the final product, to mimic the conditions expected during shipment by the intended method;
- real shipment studies;
- studies on the stability of the reconstituted material, including mimics of the typical method of use.

A number of regulatory guidelines are available on the topic of stability testing. Most recently, an agreement between the European, the US and the Japanese regulatory bodies (The International Conference on Harmonization or ICH) has issued guidelines with the approval of all three zones (see ICH website www.ich.org). In these guidelines the minimum requirement in terms of stability testing for submission to these authorities is outlined. At least three batches, produced at the intended manufacturing process scale, in a representative fill volume and presentation, should be subjected to real-time stability testing. Testing intervals should be specified in a protocol by the manufacturer but should include at least 3-monthly testing for the first year and 6-monthly thereafter. Accelerated degradation testing can be included. Conditions that subject the product to chemical stress, such as humidity and temperature, are especially useful for establishing the likely breakdown pathways and products of the active material and any excipients.

If bulk materials (either cell culture supernatant or purified but unformulated bulk intermediates) are to be kept for any period during processing, then the stability of these materials should also be demonstrated. The sterility of the material on prolonged storage should be demonstrated as well as the impact, if any, of long term contact with the container/closure chosen and any changes in physical appearance, colour, opacity and pH of the material (if lyophilized, once reconstituted). For lyophilized material, any changes in the moisture content after prolonged storage should also be assessed.

Once sampled, material should be tested with an appropriate range of stability-indicating analytical methods. The biological molecule, as a concentrate and also in its intended formulation, should be exposed to extreme chemical stress (in terms of temperature and relative humidity for instance) and should be assessed by a range of assays and the degradation products identified, so as to define the pathways by which deterioration is likely to occur. The assays selected as stability-indicating should be capable of detecting and quantifying the degradation products or degraded biological. These assays will then be used to test samples generated from the formal stability studies.

Biological potency should be followed using an assay calibrated against a national or international reference material. Simple specificity and immunogenicity may be tested by immunoassay. Structural integrity may be demonstrated by such methods as size-exclusion HPLC and gel or capillary electrophoresis under native and reducing conditions. Finally, subtle chemical modifications may require assays such as isoelectric focusing, reverse phase and hydrophobic interaction chromatography, and possibly peptide mapping, which are sensitive to particular types of molecular change. These assays must be validated to show that they can reproducibly detect and quantitate degradation in the tested material. Here again, material degraded during stress studies is particularly helpful. ICH guidelines for such analytical method validation are available (ICH guideline Q2(R1): Validation of Analytical Procedures: Text and Methodology).

27.4 DESIGN OF STABILITY STUDIES

27.4.1 Reference Temperature

Often, the activity of a biological material cannot be determined in absolute units; the activity has to be determined relative to some reference material. For key therapeutic biologicals, there are World Health Organization (WHO) International Biological Standards of defined activity, which are the primary unit of biological activity in the same way as the kilogram is the primary reference for mass. The activity of a biological preparation is determined in a test in which the appropriate international standard (or derived secondary standard) is tested in parallel.

For degradation studies, the loss of activity of a biological is expressed relative to the identical material stored at some low temperature where degradation is (or is assumed to be) insignificant – this is the reference material stored at the reference temperature.

In an accelerated degradation test, material is stored at a range of elevated temperatures, and/or a range of relative humidity conditions, and at the reference temperature/humidity where deterioration is deemed to be insignificant. The activity of these samples is determined at known time intervals, relative to the activity of the reference material stored for the same duration.

The reference temperature should be less than the intended temperature of storage. The study yields no direct information on the stability of the material stored at the reference temperature. However, some indirect information can be obtained from or from a maximum likelihood analysis of degradation data (Kirkwood 1984), of from a trend analysis of the particular assay response (for example, peak height, ring diameter, time to form clot) of the reference material; a change with time could suggest instability at the reference temperature.

27.4.2 Initial Testing Protocol

Initially, samples from higher temperature storage (for example, at 56 °C) are removed periodically until significant degradation (20 % to 30 %) is observed, relative to the reference material. Attention is then focused on the next highest storage temperature (for example, 45 °C and 37 °C). Samples are removed periodically until significant degradation (15 % to 25 %) is observed. The study proceeds for a suitable period until degradation (5 % to 10 %) is observed at the lower storage temperatures (for example, 20 °C and 4 °C).

Discolouration of the lyophilized material may occur at higher temperatures, especially when sugars are present in the formulation. The reaction of proteins with reducing sugars are well documented and again may be seen to occur *in vivo* as an indicator of molecular aging

Table 27.1 Typical accelerated degradation program for a biological undergoing stability testing.

	Temperature of storage site					
Samples removed from storage site at time	Reference temperature	4 °C	20 °C	37 °C	45 °C	56 °C
1 month	✓			✓	✓	✓
2 months	✓			✓	✓	✓
3 months	✓		✓	✓	✓	(✓)
6 months	✓		✓	✓	(✓)	
12 months	✓	✓	✓	✓		
Subsequent times	✓	✓	✓	✓		

(Meli *et al.* 2003). Where an important amino function is thus modified, then again inactivation or impairment may occur in the biological so modified. Modifications may also result in unwanted immunogenicity, or exposure of crypto-antigens (i.e. antigenic components not normally exposed in the native state). The reactions proceed more quickly at elevated temperatures and so may be first identified in accelerated degradation studies (see later in chapter). Samples from higher temperature conditions may be difficult to reconstitute fully due to aggregation.

The testing protocol given in Table 27.1 may be a helpful model when little is known about the stability of the material under study. The frequency of testing will depend on the rate of degradation observed at the higher storage temperatures. It is important not to use samples from the critical lower temperature storage early in the study. Such samples would probably show no significant degradation and would waste irreplaceable material. It is the material stored at the lower temperatures that yields the most valuable information in terms of the eventual shelf life.

27.4.3 Formal Stability Study

For a formal stability study in support of a licensing application, the selected assay methods should be applied to finished product, manufactured according to the intended production process and scale, prepared in the intended formulation, container and closure. At least three batches of finished product should be tested in the stability study. The published guidelines should be referred to (www.ich.org) when preparing a biological with a range of different dosages and presentations.

It must be emphasized that while accelerated testing is a useful indicator of likely deterioration pathways, and may be useful when the real time storage data is initially limited, there is no substitute for stability data at the intended storage temperature when looking to establish a definitive product shelf life.

The intended market location and transport considerations should be taken into account when building the stability study. Guidelines describe the environmental requirements for different intended distribution zones. Depending upon these and the intended storage temperature and method, the temperature range and relative humidity conditions for which studies should be undertaken are described in the ICH guidelines (www.ich.org).

27.4.4 Mathematical Models of Degradation

A review of statistical methods as used in accelerated degradation testing is given in Nelson (2004).

27.4.4.1 Mathematical relationships

Chemical reaction rates can be approximated by use of differential equations that address the demands of the rate law. The order of the rate law is defined by the number of terms in the rate equations. Rate equations of zero order or third order are rare and most chemical reactions follow first or second order kinetics. For the degradation process to fit well to the Arrhenius equation it should follow first order kinetics. It is usual to model the degradation of materials using the Arrhenius equation:

$$\ln(K_{\{t\}}) = A + B/T$$

or Eyring equation:

$$\ln(K_{\{t\}}) = A + B/T + \ln(T)$$

where: $K_{\{t\}}$ is the degradation rate at the absolute temperature T (Kelvin), relative to the material stored at the reference temperature; A and B are constants; and ln is the logarithm to base e.

The Eyring equation is said to have a slightly stronger theoretical basis. Both models assume a single mechanism of degradation of first-order kinetics. By determining the relationship between the relative degradation rate and temperature using samples stored at a range of higher temperatures of storage, the relative degradation rate at lower, conventional, storage temperature can be predicted (Kirkwood 1977; Tydeman & Kirkwood 1984; Kirkwood & Tydeman 1984).

Statistical prediction of drug stability based on non-linear parameter estimation has been suggested by King *et al.* (1984) who advocated this method as yielding smaller and more symmetrical 95 % confidence intervals than the classical Arrhenius approach.

The kinetics of the degradation of the peptide gonadorelin in solution were described by Hoitink *et al.* (1996), who found that the overall rate obeyed Arrhenius-type kinetics; and neither the concentration of gonadorelin nor the buffer components influenced the decomposition rate. Jang *et al.* (1997) found that the degradation of a somatostatin analogue cyclic peptide in aqueous solution followed first-order kinetics. Various buffer species showed differing effects on the degradation of the octapeptide, degradation being faster in citrate- or phosphate-containing buffers than in acetate or glutamate.

Sun and Davidson (1998) compared the efficacy of trehalose and sucrose for the stabilization of glucose-6-phosphate dehydrogenase using Williams Landel Ferry (WLF) kinetics (Williams *et al.* 1955). The WLF and Arrhenius models were compared and the Arrhenius model was found to overestimate the true degradation rate that is no longer first order. WLF gave better fit for amorphous glasses when above their T_g. They justified the improved performance with trehalose/glucose as opposed to sucrose/glucose as being due to the properties of its glass, in terms of its low free volume, restricted molecular mobility and the ability to resist phase separation and crystallization during storage. Sun and Davidson (2001) also described the stabilizing effect of dextrans on lyophilized gamma globulins, showing a relationship between molecular weight of the dextran and the T_g of the lyophilized material. Activity loss in the lyophilized material followed biphasic first order kinetics.

Magari (2002) studied the effect of lot-to-lot variability levels on the prediction of stability using two statistical models for estimating degradation in real time and accelerated stability tests. Several data sets were simulated, and the 95 % confidence intervals for the degradation rate depended on the amount of lot-to-lot variability. When lot-to-lot degradation rate variability was relatively large (CV $\geq$ 8 %) he recommended that each lot be assayed individually.

27.4.4.2 Confidence limits of prediction

The random error in potency estimates is assumed to be log-normally distributed.

The upper 95 % confidence limit is derived from the equation:

$$K_{\{t\}}'' = K_{\{t\}} + (C.\mathrm{se}K_{\{t\}})$$

where: $K_{\{t\}}''$ is the upper 95 % confidence limit of the degradation rate $K_{\{t\}}$; C is a constant; and se$K_{\{t\}}$ is the standard error of $K_{\{t\}}$.

The value of C depends on the total statistical uncertainty of the estimate of the standard error (the reciprocal of the variance of $\log_{10}$ estimates of relative potency, from which the mean relative potency is obtained), that is:

- For a total statistical weight (see below) of 30 000 or more, and at least 25 % degradation at temperatures of 37 °C or greater, a value for C of 4 (with three elevated temperatures) or of 3 (with four elevated temperatures) or of 2 (with five elevated temperatures). The value of C used is smaller as data from more temperatures are available, as the uncertainty becomes less.
- For a total statistical weight of less than 30 000, a value for C of 5 should be used.

Statistical weight is the reciprocal of the individual $\log_{10}$ estimates of relative potency from which the mean relative potency was determined. The precision of estimates of degradation rate from accelerated degradation studies is improved by:

- increasing the range of temperatures at which the samples are stored, even if the total statistical weight is kept constant;
- increasing the total statistical weight by having replicate samples tested; and
- increasing the duration of the study.

27.4.4.3 Data acquisition and analysis

The observed potency of the preparation under study should be measured during the course of a period of time stored at a range of temperatures. Multiple assays at each point can be helpful for the reasons given in 27.4.4.4 below.

As the decay is not time-dependent, potency data (or the loss in potency) from each temperature collected across the course of the study can be plotted in one graph of temperature against activity. This data can then be fitted to the Arrhenius equation and the expected degradation rate at a given interpolated temperature derived. From a stability study over a period of at least 6 months, the goodness of fit of the observed data to that predicted from the equation can be derived. On longer term incubation at the intended storage temperature the actual observed degradation rate should be compared to that predicted from the Arrhenius estimation.

27.4.4.4 Variation between and within assays

A separate analysis of the measured response from the reference samples (for example, if a radial diffusion assay, the ring diameter; if an ELISA assay, the absorbance; or if an HPLC assay, the peak height) will indicate the between-assay variation. The analysis of at least duplicate samples of the reference preparation will indicate the within-assay variation.

27.4.4.5 Stability and expiry date

The stability data when accumulated should give an indication of a suitable expiry date that can be assigned. A detailed discussion of expiry dates is not within the scope of the chapter and would be considered on a case-by-case basis, but one useful measure of expiry may be the period of storage by which mean potency falls to below the 95 % confidence limit of the initially determined potency.

27.4.5 Common Problems Seen with Accelerated Degradation Studies

27.4.5.1 An initial high rate of degradation followed by a second phase of slower degradation

This may be seen with freeze-dried preparations and is likely to be caused by an irreversible consumption of the residual oxygen or moisture present as a contaminant of the atmosphere within the container or within the freeze-dried material itself.

27.4.5.2 Discontinuity in the relationship between degradation rate and temperature

Both Arrhenius and Eyring models assume that the rate of molecular diffusion (i.e. viscosity) in the stored samples does not significantly alter over the range of temperatures studied. A significant change in viscosity can occur around the glass transition temperature of freeze-dried preparations containing a large proportion of substances that do not crystallize on freezing, but form a glass with a defined proportion of amorphous water. Such substances include plasmas or formulations containing high levels of proteins and carbohydrates as bulking materials or cryoprotectants. At these temperatures the WLF model will fit the data better.

The crystalline water (ice) is removed during the sublimation stage of freeze drying; a proportion of water remains, amorphous, as a constituent of the glass. The large majority of the amorphous water is removed during the 'secondary drying' stage. This is when the temperature of the freeze-dried material is raised to room temperature or higher, under vacuum. Thermodynamically, the glass, stable at the lower temperature used to sublime the ice, becomes unstable and would progress, depending on the amount of amorphous water released, to a mobile liquid via a viscous toffee/syrup stage. However, kinetically, this process is slow and the water released from the glass is removed by desorption before collapse of the freeze-dried material occurs.

Assuming equilibrium conditions, at the end of secondary drying, the freeze-dried material is essentially a protein/carbohydrate glass stable at the secondary drying temperature. Subsequently, if the temperature of the freeze dried material exceeds this temperature, for example by storage at some elevated degradation/storage temperature, there can be progressive collapse of the previously stable glass by further amorphous water released from the glass. Kinetically, this process can be slow. In addition, the amount of amorphous water released from the glass can be insufficient to produced noticeable collapse of the product.

Over the temperature range where this progressive collapse of the glassy structure occurs, there will be a marked change in the rate of molecular diffusion when the Arrhenius/Eyring equations are not valid. This may be revealed by a discontinuity in the relationship between ln(relative potency) or degradation rate, and the reciprocal of absolute temperature.

27.4.5.3 Additional degradation processes at higher temperatures

At higher storage temperatures, additional degradation processes can become significant, which are irrelevant at the lower temperatures at which the material will normally be stored. These additional degradation processes can include Maillard reactions and often result in a coloured or even charred appearance of the material stored at the elevated temperatures, and/or a resistance to reconstitution.

In addition, as stated above, if, on storage at an elevated temperature, the temperature of the freeze-dried material exceeds the temperature of secondary drying, there can be a progressive collapse of the previously stable freeze-dried structure by the water released from the glass. As water is released, the glass progressively collapses to a more mobile phase. This mobile phase will probably be less stable than the freeze-dried glassy material found at lower temperatures.

If statistically significant, these additional processes can prevent the experimental data from being fitted to the degradation equation, since there is no longer a linear relationship between ln(degradation rate) and the reciprocal of absolute temperature.

Even when not statistically significant, such additional degradation processes can lead to an overestimation of the rate of degradation at conventional storage temperatures when based on the degradation data obtained only at the higher temperatures. This is particularly the case if the degradation study was not continued long enough at lower storage temperatures for significant degradation to occur, in which case undue emphasis may be placed on the data from the higher temperatures (Franks 1994).

The glass transition of the lyophilized material can be determined by DSC and this can provide very useful information for indicating at which temperature the lyophilized material will remain stable and above which there is increased likelihood of molecular mobility of the water and resulting deterioration (Duddu *et al.* 1997). The changes that occur above the T_g can be described by the WLF equation (Williams *et al.* 1955).

However, T_g alone is not the only criterion for stability and a formulation resulting in a higher T_g may not necessarily give the most stable preparation (Wang 2000). Some workers have used NMR to determine increasing mobilization in glassy states and have defined a critical temperature T_{mc} based upon the relaxation events occurring in the sample as being more predictive of stability than T_g (Yoshioka *et al.* 1998). Loss of activity occurring in stability studies where the temperature approaches the T_g of the material may not be well fitted to the Arrhenius equation and in such situations approximation to the WLF model may be more appropriate (Sun & Davidson 1998).

The degradation kinetics at higher temperatures may not be first order as additional degradation mechanisms may have come into play. When these extreme temperature data are omitted, the fit to first order kinetics becomes better.

27.4.5.4 Effects of temperature on the integrity of the container

If vials are used as the final product container, high or low temperatures may affect the physical properties of the stopper and thus the integrity of the seal it makes with the neck of the container (Ford & Dawson 1994). This can result in ingress of the external atmosphere, which may affect stability. In addition, the increased temperature can desorb moisture from the inner surfaces of the stopper to the vial's headspace for absorption by the product (Pikal *et al.* 1992). These are further examples of degradation processes irrelevant at the lower, conventional, temperatures of storage.

27.5 EXAMPLES OF ACCELERATED DEGRADATION STUDIES

27.5.1 Example Where a Clear Differential Accelerated Degradation Rate Occurs

This example deals with degradation studies on two blood coagulation Factor VIII preparations. Primary data (Table 27.2) were kindly provided by A Hubbard, NIBSC, and the degradation data is given in Table 27.3. For a detailed explanation of the statistical methods involved refer to Nelson (2004) or other statistical textbooks.

Primary data

Table 27.2 Primary data.

Material	Factor VIII concentrate preparations	
Date Study commenced	Material A 12 October 1999	Material B 24 June 1999
Reference temperature	−20°C	
Assay method	European Pharmacopoeia Chromogenic coagulation assay	

Table 27.3 Accelerated degradation data on two blood coagulation Factor VIII preparations.

Date of removal from storage	Storage temperature (°C)	Relative potency[a]	Assay weight[b]
Material A			
07 February 2000	4	0.9300	7693.74
07 February 2000	20	0.8025	3480.52
07 February 2000	37	0.4600	5727.70
04 July 2000	4	0.9100	858.41
04 July 2000	20	0.7275	923.21
04 July 2000	37	0.1800	195.01
04 July 2000	45	0.0475	137.82
Material B			
04 February 2000	20	0.9650	7905.49
04 February 2000	37	0.8800	3617.27
04 February 2000	45	0.7300	4233.09
23 January 2001	4	0.9732	15489.78
23 January 2001	20	0.9455	25401.52
23 January 2001	37	0.7755	9964.62
23 January 2001	45	0.5438	896.92
15 January 2002	4	1.0025	7845.68
15 January 2002	20	0.9550	13296.65
15 January 2002	37	0.6900	6125.46
13 March 2003	4	0.9800	36186.92
13 March 2003	20	0.9450	4217.53

[a]*Mean potency relative to the potency of samples stored at the reference temperature.*
[b]*Reciprocal of the variance of $\log_{10}$ estimates of relative potency, from which the mean relative potency is obtained.*

Fitted parameters

Data was fitted using the Arrhenius equation (see Section 27.4.4.1)
Maximum likelihood estimates of the constants A and B are given in Table 27.4.

Goodness of fit of predicted to observed data

Following fitting of the data to the Arrhenius equation the predicted degradation rate was calculated and compared, as the study continued, to the observed degradation rate at the given temperatures (Table 27.5).

Table 27.4 Estimation of Arrhenius equation constants for the two blood coagulation Factor VIII preparations.

Item	Material A	Material B
A	34.72	35.15
Asymptotic error of A	2.88	3.21
B	−11332.7	−11467.68
Asymptotic error of B	897.94	999.78
Asymptotic covariance of A and B	−2589.12	−3206.94

Predicted degradation rates

The predicted degradation rates are given in Table 27.6.

The clear conclusion from this data is that preparation B is the more stable.

The degradation process with time for each preparation is displayed in Figure 27.1. Clearly, these two preparations differ in their predicted stability at the 4 °C temperature of storage. One preparation has a faster degradation rate than the other. In this case the two Factor VIII preparations are not the same and this example is illustrative of how accelerated degradation studies can be applied to compare different candidate materials.

Where a difference in degradation rate is detected between alternative formulations of the same material, possible causes for such enhanced degradation may include unsuitable excipients (absence of stabilizers, presence of inhibitors), presence of contaminants such as proteases, or sub-optimal freeze drying conditions.

The importance of stabilizers in preserving biological activity, and preventing loss of active material on the container surface was demonstrated by Page *et al.* (2000). In their study, eight

Table 27.5 Predicted and observed degradation data for two preparations of blood coagulation Factor VIII.

Storage temperature (°C)	Time elapsed (years)	Activity remaining (%)[a]		Chi-square contribution
		Observed	Predicted	
Material A				
4	0.3231	93.00	95.74	1.225
20	0.3231	80.25	82.36	0.441
37	0.3231	46.00	46.55	0.150
4	0.7283	91.00	90.66	0.002
20	0.7283	72.75	64.56	2.484
37	0.7283	18.00	17.84	0.003
45	0.7283	4.75	4.46	0.105
Chi-square test statistic = 4.41 for 5 degrees of freedom. At the 5 % level, the predicted activities remaining are not significantly different from those observed.				
Material B				
20	0.6160	96.50	98.86	0.873
37	0.6160	88.00	90.67	0.610
45	0.6160	73.00	78.00	3.510
4	1.5852	97.32	99.70	1.704
20	1.5852	94.55	97.10	3.398
37	1.5852	77.55	77.72	0.009
45	1.5852	54.38	57.77	0.153
4	2.5627	100.25	99.51	0.080
20	2.5627	95.50	95.36	0.006
37	2.5627	69.00	66.54	1.525
4	3.7181	98.0	99.26	1.171
20	3.7181	94.5	93.16	0.163
Chi-square test statistic = 13.218 for 10 degrees of freedom (degrees of freedom = $n-2$ where n is the number of observations). At the 5 % level, the predicted remaining activities are not significantly different from those observed.				

[a] *Relative to the potency of samples stored at the reference temperature.*

Table 27.6 Predicted degradation rates for two blood coagulation Factor VIII preparations.

Storage temperature(°C)	Degradation rate $K_{\{t\}}$	Standard error of $K_{\{t\}}$	Potency loss(% per year[a])	95 % confidence limit[b] of potency loss (% per year)
Material A				
4	0.14698	0.02956	13	± 12
20	0.61314	0.06400	46	± 15
37	2.37935	0.07451	91	± 3
Material B				
4	0.00194	0.00078	0	± 1
20	0.01859	0.00390	2	± 4
37	0.15901	0.00802	15	± 18

[a] *Relative to the potency of samples stored at the reference temperature.*
[b] *Based on 5 times the standard error of* $K_{\{t\}}$.

formulations of the cytokine, Interleukin-11, were studied and an optimized formulation developed for long-term storage. After selection of the best conditions for preservation of biological activity, two potential formulations were compared in an accelerated degradation study at tempratures from −70 °C to 56 °C over a 6-month storage period. The choice of isotonic saline or water as the formulation base resulted in a marked difference in the predicted long-term stability based on accelerated degradation studies. The predicted annual reduction in potency at −20 °C and +4 °C are given in Table 27.7. In this example, the saline-containing formulation appeared to have superior stability. Although general principles can be followed for selecting excipients, in terms of avoiding reducing sugar, avoiding materials that crystallize unpredictably, avoiding high salt concentrations, etc., the impact of excipients must be determined for each product on a case-by-case basis.

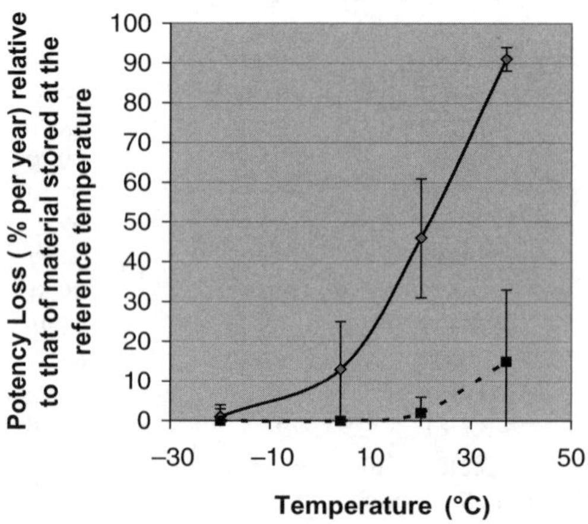

Figure 27.1 Potency loss (per year) with temperature of storage (±95 % confidence limits) of two blood coagulation Factor VIII preparations. Material A (◆), shows less stability than material B (■).

Table 27.7 Predicted annual percentage reduction in potency for a recombinant cytokine recombinant human interleukin-11 (data adapted with permission from Page *et al.*(2000).

Formulation	Predicted reduction in potency at −20 °C (%)	Predicted reduction in potency at +4 °C (%)
IL-11: 1 µg/ml in phosphate, water, 0.5 % human albumin,0.1 % trehalose, 0.02 % Tween 20	1.456	7.109
IL-11: 1 µg/ml in phosphate, 0.9 % sodium chloride, 0.5 % human albumin,0.1 % trehalose, 0.02 % Tween 20	0.631	4.362

27.5.2 Predicting Stability when Accelerated Degradation Study Shows No Degradation

Some biological materials exhibit little significant degradation over a 3-month period at 37 °C. Assuming a doubling of reaction rate with every 10 °C, this would suggest the material would be equally stable at 4 °C (i.e. some three 10 °C reductions) for 24 months. Continuation of the degradation study and, as always, real time studies, would nevertheless still be carried out. Accelerated data can be useful for predicting the long-term storage of biological molecules at lower temperatures as is illustrated by the examples below. On some occasions, no degradation is apparent at elevated temperatures over short periods of time. In these situations little can be inferred as to the stability at lower temperatures, except that it can be assumed to be as great or greater than the periods of study at the higher temperatures that resulted in no deterioration.

Saldanha *et al.* (1999) studied the deterioration of hepatitis C viral RNA in plasma across a range of temperatures. Here the marked stability indicated by short periods at 45 °C supported a far longer shelf life, in excess of 200 days, at lower operational temperatures (4 °C and −20 °C) (Table 27.8). However, prolonged periods at 45 °C resulted in the RNA becoming inextractable due to heat-induced changes in the sample material. Therefore if insufficient deterioration is seen at higher temperatures to be able to predict the stability at the designated storage temperature then shelf life assignment will need to be based upon real-time studies at lower temperatures as they become available.

27.5.3 Example Where Accelerated Degradation Studies Provided a Good Estimation of the Deterioration Seen in Real-Time Studies

In a similar study of PCR reactivity of a HIV-1 RNA reference material, accelerated degradation testing indicated deterioration to be less than 0.2 % per annum at −20 °C and less than 0.6 % per

Table 27.8 Reactivity of hepatitis C RNA in PCR assays on prolonged storage at elevated temperatures (relative to −20 °C control sample). (Data adapted from Saldanha *et al.* 1999 with permission of Blackwell Press).

Storage temperature	45 °C	37 °C	20 °C	4 °C
Stability (RNA titre by PCR)	No loss after 21 days	Some loss after 56 days	No appreciable loss after 200 days	No appreciable loss after 200 days

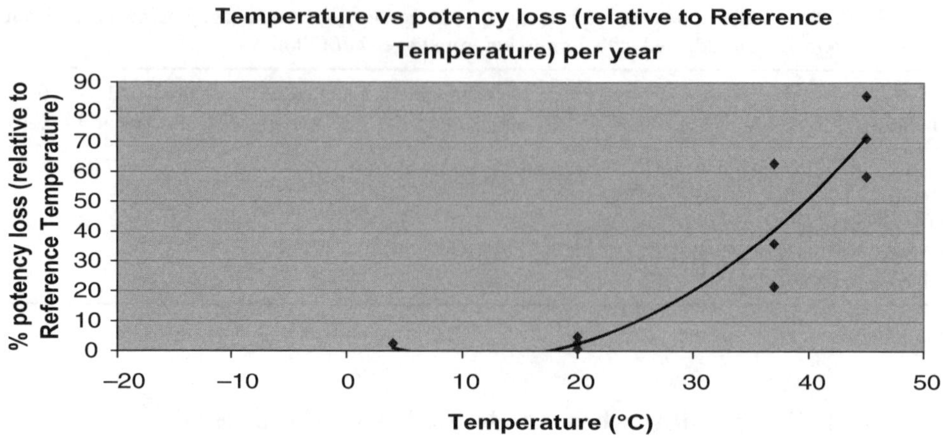

Figure 27.2 Accelerated degradation stability study of a Factor VIII preparation.

annum at 4 °C (Holmes *et al.* 2001). These deterioration rates have been confirmed over longer periods of storage at these temperatures (C Davis, NIBSC, unpublished results).

In the studies on the Factor VIII preparations given in Section 27.5.1 above, the stability at −20 °C was predicted using the Arrhenius equation from accelerated degradation studies (Figure 27.2). The predicted deterioration rate was then shown to match that observed over a 2-year real-time period (Figure 27.3).

27.5.4 Example Where Accelerated Degradation Studies Over-estimate Deterioration in Lower Temperature Real-time Studies

In some cases the accelerated degradation data can overestimate the deterioration rate at lower temperatures as illustrated for a study of gamma-glutamyl transferase reference material.

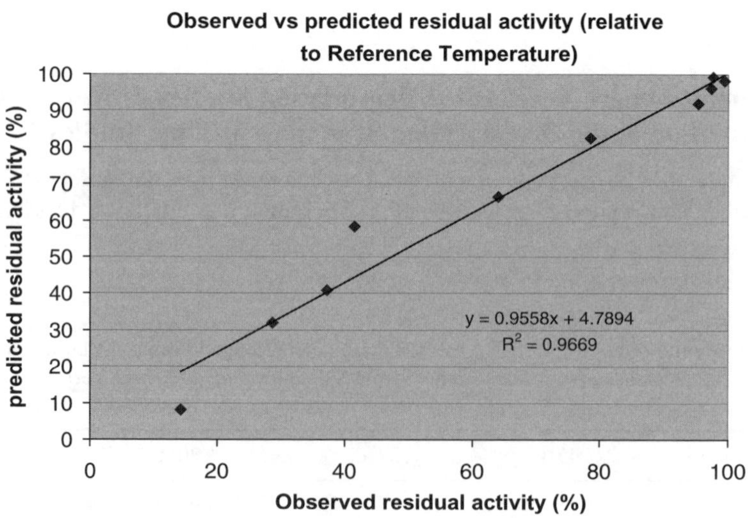

Figure 27.3 Predicted stability against observed stability at −20 °C over two years for the Factor VIII preparation referred to in Figure 27.2.

Table 27.9 Predicted and observed degradation rates of gamma-glutamyl transferase at 4° C (adapted from Moss *et al.* 1994 with permission of John Libbey Eurotext)

Storage time (months)	Predicted loss per year (%)	Observed loss per year (%)
1	6.4	—
9.6	0.84	0.38
45.6	0.76	0.76
81.6	0.59	0.51

Moss *et al.* (1994) demonstrated that the predicted rate of degradation at 4 °C (compared with −20 °C where degradation was considered negligible) was overestimated based upon data from early on in the stability studies and reduces to match the observed deterioration rate as the study progresses (see Table 27.9). Degradation pathways active at the higher temperatures used in accelerated studies may not occur in samples stored at lower temperatures. Observed deterioration may be due to a number of different factors, some of which may be more temperature-dependent than others.

27.5.5 Example of Arrhenius Analysis of the Stability of a Liquid Product

The Arrhenius equation modelling has so far in this chapter been applied to lyophilized materials. However, it is equally applicable to liquid preparations as exemplified below. A monoclonal antibody to Rhesus D antigen was studied for its stability in the liquid state when stored at 4, 25 and 37 °C compared with a reference temperature of −40 °C (data kindly supplied by the Research & Development Department, Bio Products Laboratory, Elstree, UK, and used with permission). Two functional assays were compared: antibody-dependent cellular cytotoxicity (ADCC) against target ^{51}Cr-labelled red blood cells, and the autoanalyser agglutination test with enzyme-treated red blood cells. Both assays generated data that indicated that decay at elevated temperatures could be fitted using the Arrhenius model to predict stability at 4 °C. Both sets of data fitted the model well, the actual decay data correlating well to that predicted by the Arrhenius model. However, whereas ADCC predicted a decay rate of 1.8 % per annum at 4 °C (Table 27.10) the decay by autoanalyser was 0.11 % per annum at the same temperature (Table 27.11).

This not only shows the value of the Arrhenius method for predicting stability in liquid preparations but also shows that the method used for the determination of activity may be important in determining the level of deterioration observed, therefore the method used should be selected carefully so as to most closely correspond to the *in vivo* potency.

Table 27.10 Degradation in monoclonal anti-D based on lymphocyte (NK cell) antibody-dependent cellular cytotoxicity (ADCC) method.

Temperature (°C)	Degradation rate $K_{\{t\}}$ (per month)[a]	Standard error of $K_{\{t\}}$	Potency loss(% per month)[a]	95 % upper confidence limit of potency loss[b] (% per month)
−20	0.00003	0.00034	0.003	0.175
4	0.00154	0.01043	0.154	5.229
20	0.01613	0.05091	1.600	23.715
37	0.14962	0.16671	18.896	62.587

[a] *Relative to samples stored at the reference temperature (−40°C).*
[b] *Based on five times the standard error of* $K_{\{t\}}$.

Table 27.11 Degradation in monoclonal anti-D based on autoanalyser method.

Temperature (°C)	Degradation rate $K_{\{t\}}$ (per month)[a]	Standard error of $K_{\{t\}}$	Potency loss (% per month)[a]	95 % upper confidence limit of potency loss[b] (% per month)
−20	0.00000	0.00002	0.000	0.011
4	0.00009	0.00322	0.009	1.603
20	0.00219	0.03936	0.219	18.043
37	0.04743	0.16690	4.633	58.601

[a]*Relative to samples stored at the reference temperature (−40°C).*
[a]*Based on five times the standard error of* $K_{\{t\}}$.

27.6 CONCLUSIONS

Stability studies are a key element in the development of a biological product. For biological medicines, only when such studies have shown that the candidate product is adequately stable at the chosen storage conditions does the development process usually proceed via clinical trials to the market place. Inevitably, stability data will at first be limited and it is at this point that accelerated degradation can indicate potential stability pitfalls and assist the development process. Such accelerated degradation studies, when backed up with powerful molecular analytical techniques, can establish the pathways of degradation that are likely to occur in the biological under study, and which will have impact on biological efficacy.

Data from such studies can be analysed mathematically using Arrhenius and similar model equations. Accelerated degradation studies, though valuable, may however be misleading in terms of the rate of deterioration that will ultimately occur at the chosen storage conditions. When determining the stability of biological materials, there is no substitute for real time stability data under the intended storage conditions.

ACKNOWLEDGEMENTS

The authors thank Tony Hubbard, NIBSC, for supplying data on the accelerated stability studies of Factor VIII, and J. Saldanha, H. Holmes and R. E. Gaines Das for use of their published studies. They also thank Bio Products Laboratory for kindly supplying accelerated stability data on a monoclonal antibody in solution.

REFERENCES

Adams GDJ. Ramsay JR (1996) *J. Pharm. Sci.*; 86: 1301–1305.

Andya JD, Maa Y-F, Costantino HR, *et al.* (1999) *Pharm. Res.*; 16: 350–358.

Becker GW, Tackitt PM, Bromer WW, Lefeber DS, Riggin RM (1988) *Biotechnol. Appl. Biochem.*; 10: 326–337.

Bell LN, Hageman MJ, Muraoka LM (1995) *J. Pharm. Sci.*; 84: 707–712.

Bennett PS, Maigetter RZ, Olson MG, Provost PJ, Scattergood EM, Scholfield TL (1992) *Dev. Biol. Stand.*; 74: 213–221.

Breen ED, Curley JG, Overcashier DE, Hsu CC, Shire SJ (2001) *Pharm. Res.*; 18: 1345–1353.

Chang BS, Beauvais RM, Dong AC, Carpenter JF (1996) *Arch. Biochem. Biophys.*; 331: 249–258.

Clarke S, Stephenson RC, Lowenson JD (1992) In *Stability of Protein Pharmaceuticals Part A: Chemical and Physical Pathways of Protein Degradation.* Eds Ahern TJ, Manning MC. Plenum, New York; 1–29.

Cleland JL, Powell MF, Shire SJ (1993) *Crit. Rev. Therap. Drug Carrier Sys.*; 10: 307–377.
Costantino HR, Andya JD, Nguyen PA, *et al.* (1998a) *J. Pharm. Sci.*; 87: 1406–1411.
Costantino HR,Carrasquillo KG, Cordero RA, Mumenthaler M, Hsu CC, Griebenow K (1998b) *J. Pharm. Sci.*; 87: 1412–1420.
Costantino HR, Schwendeman SP, Langer R, Klibanov AM (1998c) *Biochemistry (Moscow)*; 63: 357–363.
Duddu SP, Zhang G, Dal Monte PR (1997) *Pharm. Res.*; 14: 596–600.
Farber JH, Levine RL (1986) *J. Biol.chem*; 261:4574–8.
Fitzpatrick S, Saklatvala R (2003) *J. Pharm. Sci.*; 92: 2495–2501.
Ford AW, Dawson PJ (1994) *Biologicals*; 22: 191–197.
Franks F (1990) *CryoLetters*; 11: 93–110.
Franks F (1994) *Trends Biotechnol.*; 12: 114–117.
Hageman MJ (1992) In *Stability of Protein Pharmaceuticals Part A: Chemical and Physical Pathways of Protein Degradation*. Eds Ahern TJ, Manning MC. Plenum, New York; 273–309.
Harris RJ, Kabakoff B, Macchi FD *et al.* (2001) *J. Chromatogr. B Biomed. Sci. Appl.*; 752: 233–245.
Hoitink MA, Beijnen JH, Bult A, van der Houwen OA, Nijholt J, Underberg WJ (1996) *J. Pharm. Sci.*; 85: 1053–1059.
Holmes H, Davis C, Heath A, Hewlett I, Lelie N (2001) *J. Virol. Methods*; 92: 141–150.
Hsu CC, Ward CA, Pearlman R, Nguyen HM, Yeung DA, Curley JG (1992) *Dev. Biol. Stand.*; 74: 255–71.
ICH Guideline Q1A (R2): *Stability Testing of New Drugs and Products.* (revised). Available at www.ich.org
ICH Guideline Q2(R1) *Validation of Analytical Procedures: Text and Methodology.* Available at www.ich.org
ICH Guideline Q1B (1996) *Stability Testing, Photostability Testing of New Drug Substances and Products.* Available at www.ich.org
ICH Guidleine Q5C (1996) *Stability Testing of Biotechnological/Biological Products.* Available at www.ich.org
ICH Guideline Q1D (2002) *Bracketing and Matrixing Designs for Stability Testing of New Drug Substances and Products.* Available at www.ich.org
ICH Guideline Q1E (2003) *Evaluation of Stability Data.* Available at www.ich.org
Izutsu K, Yoshioka S, Kojima S (1994) *Pharm. Res.*; 11: 995–999.
Jang SW, Woo BH, Lee TJ, Moon SC, Lee KC, DeLuca PP (1997) *Pharm. Dev. Technol.*; 2: 409–414.
Joshi AB, Kirsch LE (2004) *Int. J. Pharm.*; 273: 213–219.
King SY, Kung MS, Fung HL (1984) *J. Pharm. Sci.*; 73: 2332–2344.
Kirkwood TB (1977) *Biometrics*; 33: 736–742.
Kirkwood TBL (1984) *J. Biol. Stand.*; 12: 215–224.
Kirkwood TBL, Tydeman MS (1984) *J. Biol. Stand.*; 12: 207–214.
Kroon DJ, Baldwin-Ferro A, Lalan P (1992) *Pharm. Res.*; 9: 1386–1393.
Lai MC, Topp EM (1999) *J. Pharm. Sci.*; 88: 489–500.
Li S, Patapoff TW, Overcashier D, Hsu C, Nguyen TH, Borchardt RT (1996) *J. Pharm. Sci.*; 85: 873–877.
Liu WR, Langer R, Klibanov AM (1991) *Biotech. Bioeng.*; 37: 77–84.
Magari RT (2002) *J. Pharm. Sci.*; 91: 893–899.
Manning MC, Patel K, Borchardt RT (1989) *Pharm. Res.*; 6: 903–918.
Matejtschuk P, Baker RM, Chapman GE (1998) In *Bioseparation and Bioprocessing.* Volume II: *Processing, Quality and Characterization, Economics, Safety and Hygiene.* Ed. Subramanian G., Wiley VCH. Weinheim; 223–252.
Meli M, Frey J, Perier C (2003) *J. Nutr. Health Aging*; 7: 263–266.
McNally E J (2000) In *Protein Formulation and Delivery.* Marcel Dekker, New York; 111–200.
Moss DW, Maire I, Calam DH, *et al.* (1994) *Ann. Biol. Clin.*; 52: 189–198.
Nei T, Souzu H, Araki T (1966) *Cryobiology*; 2: 276–279.
Nelson WB (2004) *Accelerated Testing Statistical Models, Test Plans and Data Analysis.* Wiley, New York; second edition.
Nguyen TH (1994) *ACS Symposium Series*; 567: 59–71.
O'Brien J (1996) *J. Food Sci.*; 61: 679–682.
Page C, Dawson P, Woollacott D, Thorpe R, Mire-Sluis A. (2000) *J. Pharm. Pharmacol.*; 52: 19–26.
Perez C, Castellanos IJ, Costantino HR, Al-Azzam W, Griebenow K (2002) *J. Pharm. Pharmacol.*; 54: 301–313.

Pikal MJ, Dellerman K, Roy ML (1992) *Dev. Biol. Stand.*; 74: 21–28.
Poxon SW, Hughes JA (2000) *Pharm. Dev. Tech.*; 5: 115–122.
Roberts GD, Johnson WP, Burman S, Anumula KR, Carr SA (1995) *Anal. Chem.*; 67: 3613–3625.
Saldanha J, Lelie N, Heath A (1999) *Vox Sang.*; 76: 149–158.
Sane SU, Wong R, Hsu CC (2004) *J. Pharm. Sci.*; 93: 1005–1018.
Sun WQ, Davidson P (1998) *Biochim. Biophys. Acta*; 1425: 235–244.
Sun WQ, Davidson P (2001) *CryoLetters*; 22(5): 285–292.
Tydeman MS, Kirkwood TBL (1984) *J. Biol. Stand.*; 12: 195–206.
Wang DQ, Hey JM, Nail SL (2004) *J. Pharm. Sci.*; 93: 1253–1263.
Wang W (2000) *Int. J. Pharm.*; 203: 1–60.
Wang Y-CJ, Hanson MA (1988) *J. Parenter. Sci. Technol.*; 42(suppl): S3–S26.
Williams ML, Landel RF, Ferry JD (1955) *J. Am. Chem. Soc.*; 77: 3701–3707.
Yoshioka S, Aso Y, Nakai Y, Kojima S (1998) *J. Pharm. Sci.*; 87: 147–151.
Zhang W, Czupryn MJ (2003) *J. Pharm. Biomed. Anal.*; 30: 1479–1490.

Useful Web Site

International Conference on Harmonisation http://www.ich.org

Cells as Products

28 Cell Culture in Tissue Engineering

TE Hardingham, CM Kielty, AE Canfield, SR Tew, SG Ball, NJ Turner and KE Ratcliffe

28.1 INTRODUCTION

Tissue engineering and regenerative medicine are concerned with ways in which we can harness biological processes to achieve healing and functional repair in patients. Tissue engineering aims to develop our scientific understanding of how living cells function, to enable us to gain control and direct their activity to promote the repair of damaged and diseased tissues. A patient with a chronic, persistent leg ulcer may not lack the inherent capacity to heal a skin wound, but rather may lack at the wound site the biological signals, the chemical messengers and physical cues, that initiate the events of cell migration, blood vessel formation and tissue assembly, that characterize normal wound healing. Strategies in tissue engineering are aimed at providing physical tissue replacement combined with the biological cues to initiate a repair process that the patient's own tissues can then go on to complete.

The precise form of a tissue-engineered therapy will vary with the medical application for which it is designed, but the typical elements are one or more types of living cells with particular tissue functions, and a material support that forms a structure for culturing the cells in the laboratory and for delivering the tissue equivalent to the damaged or diseased site in the patient at surgery. It might be in the form of a cartilage repair, a blood vessel, or a more complex organ structure, depending on the clinical application. The 'package' therefore contains several important and quite different material components and it frequently involves the culture of living cells combined with fabrication techniques for three-dimensional structures. This requires the coordination of a broad range of different disciplines. The future development of tissue engineering in research groups thus depends very much on bringing together broadly based research teams to form interdisciplinary collaborations, with input from cell biologists, molecular biologists, biomaterial scientists, bioengineers and health care clinicians.

Much work has focused initially on skin, to develop products for burns and non-healing dermal ulcers, and on musculoskeletal tissues including bone, cartilage, meniscus, tendon and ligament for fractures, joint repair and craniofacial reconstruction. There is also great interest in small-diameter blood vessels and other vascular repair including cardiac muscle for the treatment of ischaemic heart disease. For organ function, greatest progress has been with liver-assist devices, where liver function has been transduced into a two dimensional array of cells and channels that can act as temporary extra-corporeal life support for patients lacking liver function, preceding and immediately after transplantation (Tilles *et al.* 2002). In neural tissues there is much research on promoting peripheral nerve regeneration using biomaterial guides enhanced by

Medicines from Animal Cell Culture Edited by G. Stacey and J. Davis

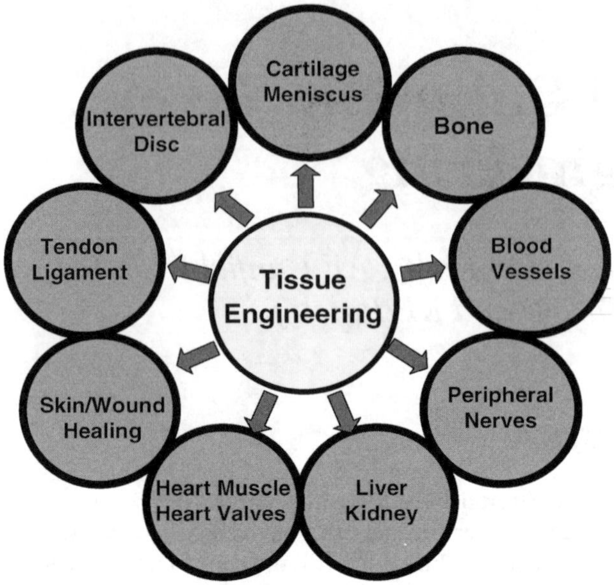

Figure 28.1 Tissue engineering approaches are being made to replace or repair many tissue and organ functions. Initial successes have mainly been in skin, cartilage and bone, but many other applications are under study and the broad range of those is illustrated.

the delivery of factors for chemotaxis and neurite growth (Mohanna *et al.* 2003). This parallels work on cell-based treatment for brain injury, including stroke and neurodegenerative diseases such as Parkinson's disease, although these generally involve no scaffold and may not formally be regarded as tissue engineering. This research field is still young and the approaches and the range of potential applications is expanding fast (Figure 28.1). There are still major questions on the sourcing of primary cells for tissue engineering applications (Williams, 2004). Initial products for skin tissue engineering used primary fibroblasts, and it has been interesting to observe that even allogeneic cells are well tolerated in recipient patients (Mansbridge 2002). With other primary cells, such as endothelial cells, there are immediate problems with allogeneic donors. It remains to be seen if tolerance to allogeneic cells extends to other cells of use in tissue engineering. This is also an issue for stem cell sources and although embryonic stem (ES) cells may provide almost unlimited progeny, it remains to be established how many ES cell lines might be needed to provide sufficiently close tissue typing to ensure compatibility in delivery within tissue-engineered applications.

28.2 CELL BIOLOGY

A key to many of the new developments in tissue engineering lies in the current progress in research in cell biology. Sequencing of the human genome is leading to a rapid increase in the understanding of the biological signals and cellular interactions that govern natural repair processes. Many past developments in cell biology have used monolayer cell culture, on the assumption that it modeled *in vivo* behaviour. For many investigations of intracellular function this has been adequate, but it has long been recognized that many differentiated cell phenotypes are not sustained in monolayer culture (Levintow & Eagle 1961). For tissue engineering applications where differentiated function is crucially important, the culture of cells and control of their phenotype has become one of the most important issues. Thus there is great interest in conditions of monolayer and multilayer culture, the delivery of specific growth factors, the response to physical forces, such as mechanical loading, and how to organize these factors to achieve cell expansion or extracellular matrix assembly. The sources of cells are also most important, and the issue of whether primary cells from patients can be co-opted for repair or if adult stem cells or ES cells offer greater potential has yet to be resolved.

The extracellular matrix (ECM) provides the shape and form of all tissues. In those applications where some immediate mechanical properties are required, such as in cartilage, skin, blood vessel, or bone, there is a need for a structural matrix. Although natural or artificial biomaterials may be used to provide temporary support, the ability of cultured cells to drive the assembly of organized ECM is a very important issue in generating a tissue with properties suitable for engraftment in a patient. There is currently much research and discovery in matrix biology that is leading to a better understanding of ECM assembly. In the following three sections we illustrate strategies for culturing chondrocytes, vascular smooth muscle cells and endothelial cells, in all of which there are important requirements for ECM assembly, and which illustrate the use of gene transfer, delivery of growth factors, effects of physical forces, use of stem cells and the need for establishing criteria for identifying phenotype and for monitoring it in culture.

28.3 ARTICULAR CHONDROCYTES

28.3.1 Introduction

Articular cartilage is found covering the ends of long bones, providing a smooth, mechanically resilient surface that allows efficient articulation. The articular chondrocyte is responsible for the construction and maintenance of the extracellular matrix of articular cartilage. Its use in tissue engineering has been well explored, not only due to the need for effective techniques for resurfacing damaged articular surfaces, but because it is the only cell type found in the tissue. This could deceive one into thinking that cartilage tissue engineering is a simple matter of growing chondrocytes on a biodegradable scaffold until a functional piece of tissue is formed. However, the reality is very different and cartilage tissue engineering has remained a challenge, mainly due to difficulties in the maintenance of cell phenotype and in the formation of a correctly organized and resilient extracellular matrix.

28.3.2 Cartilage Extracellular Matrix

Chondrocytes characteristically have a spherical morphology and are encased within a collagenous pericellular capsule. This provides cell adhesion interactions as well as mechanical protection and is further surrounded by a region of proteoglycan-rich matrix termed the territorial matrix. This unit of cell, pericellular capsule and territorial matrix is called the chondron (Figure 28.2). The intervening ECM is termed the inter-territorial matrix and also contains large amounts of proteoglycans, mainly aggrecan. It has a highly organized framework of collagen fibrils of which collagen type II is the major constituent (Buckwalter & Mankin 1997). Both collagen type II and aggrecan have limited expression in other tissues and so make useful markers for the chondrocyte phenotype. The inter-territorial matrix is highly hydrated due to the large negative charge present on glycosaminoglycans (GAG) that are covalently attached to aggrecan. The rigid collagen framework within the tissue coupled to the swelling pressure created by this hydration is essential for the correct mechanical function of cartilage. Combinations of cell shape, interactions with specific extracellular molecules and the influence of mechanical forces all contribute to the development of a fully differentiated chondrocyte (Buckwalter & Mankin 1997), the primary role of which is to maintain and organize this ECM.

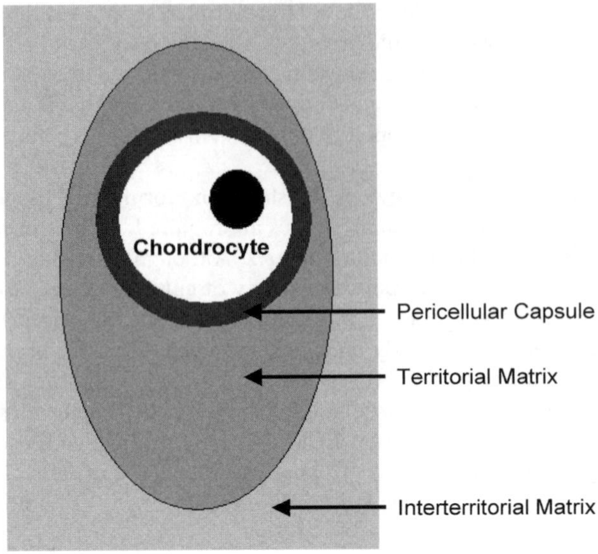

Figure 28.2 Schematic diagram illustrating the chondrocyte with its surrounding pericellular capsule and territorial matrix. This cell matrix unit, which has been termed the chondron, is embedded within the cartilage interterritorial matrix.

28.3.3 Development and Disease of Cartilage

To understand the difficulties that can be involved in interpreting experiments with cultured chondrocytes, it is useful to appreciate some aspects of developing cartilage. The best-characterized model of cartilage development is that found during long bone growth. Long bones grow by the calcification of an initial cartilage template that is itself laid down within developing limb buds by the condensation of mesenchymal cells. The process involves rapid proliferation, hypertrophy and programmed cell death (deCrombrugghe *et al.* 2001). Articular cartilage grows on a second front of hypertrophy found at the secondary ossification centre at the epiphysis, and may involve a specialized precursor cell residing at the articular surface (Hayes *et al.* 2001). What is clear is that chondrocyte proliferation only occurs at an appreciable rate during these developmental processes. In adult articular cartilage the cells are effectively quiescent and those that are present at skeletal maturity are probably retained for life. This lack of proliferation obviously presents a challenge to potential cartilage repair therapies. Intrinsic injuries to cartilage, which are contained within the tissue itself, are not repaired, as no cells are able to migrate and proliferate to fill the wound (Hunziker & Rosenberg 1996). Some cell clustering is seen in artificially induced wounds (Mankin 1962) and in cartilage degenerative diseases such as osteoarthritis (Mankin 1974). However, this occurs slowly and seems to be a generalized response to mechanical insult. Tissue engineering approaches to cartilage resurfacing have been based on the isolation of cells and their expansion *in vitro* and attempts to encourage them to form a new functional extracellular matrix.

28.3.4 Establishing Chondrocyte Cultures

Like many primary cells derived from connective tissues, chondrocytes have to be released from their extracellular matrix in order to expand their number in culture. However, when chondrocytes are removed from the tissue and cultured as monolayers, their differentiated phenotype is rapidly lost. Within a few passages the cells stop expressing collagen type II and instead produce collagen

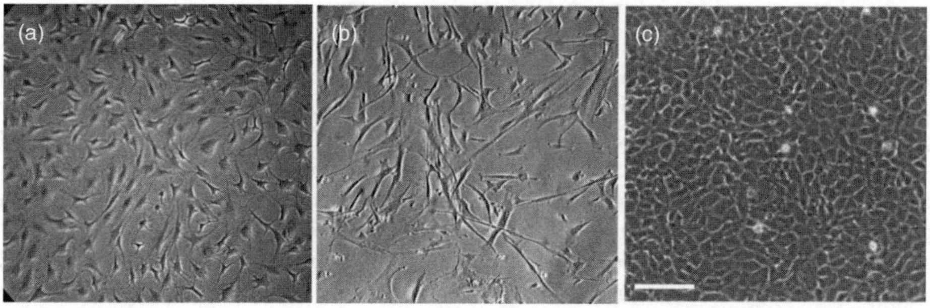

Figure 28.3 Phase contrast images demonstrating (a) the flattened morphology of serially passaged human chondrocytes, (b) the classic spindle shape of human aortic smooth muscle cells, and (c) the cobblestone morphology of confluent endothelial cells. All of the cells were grown as monolayers on tissue culture plastic. Scale bar = 100 μm.

type I, which is characteristic of fibroblasts. The standard release protocol consists of a 1-hour digestion with trypsin, hyaluronidase or pronase, followed by a longer (typically overnight) digestion in bacterial collagenase (Kuettner *et al.* 1982). The cells initially adhere to the tissue culture plastic and retain their round shape but over the course of a week will spread out, forming focal contacts (Figure 28.3a). The change in the shape of the cells is also accompanied by induction of cell division. Cell division in serum-supplemented medium such as Dulbecco's modified Eagle's medium (DMEM) is steady but not exceptionally rapid. A confluent monolayer split at a 1:2 ratio will typically be confluent again around 7 days later. If a high rate of proliferation is required, then recent evidence with human chondrocytes has shown that addition of a combination of platelet derived growth factor BB (PDGF-BB), transforming growth factor β-1 (TGFβ-1) and fibroblast growth factor-2 (FGF-2) to serum-containing culture medium markedly upregulates the proliferative rate of articular chondrocytes in monolayer culture (Babero *et al.* 2003). We have used this high proliferation rate to increase retroviral transduction efficiencies in chondrocytes, thus improving the potential for gene therapy in these cells (Li *et al.* 2004). Transduction efficiency was increased up to fivefold when carried out in conjunction with this growth factor treatment. Use of this growth medium caused the cells to produce even less collagen type II, but increased their capacity to re-differentiate into functional chondrocytes as well as other cell types, such as adipocytes or osteoblasts (Barbero *et al.* 2003).

28.3.5 Chondrogenic Cell Culture Systems

Following the amplification of cell number in monolayer culture, conditions for chondrocyte re-differentiation usually attempt to induce either the rounded shape displayed by the chondrocyte *in vivo,* or the densely packed cellular condensations seen during development. It seems likely that chondrocyte phenotype is linked closely with cell shape, which is modulated by the cytoskeleton. For instance, chondrocyte monolayer cultures produce increased cartilage-like collagen proteins when cultured with cytochalasin D, an actin stress fibre antagonist (Benya *et al.* 1988). To force the cells to adopt a round shape they can be held in suspension above a non-adhesive substratum such as agarose or poly (2-hydroxyethyl methacrylate) (HEMA). Alternatively they can be encased within a gel such as alginate. Alginate is a polysaccharide composed of D-mannuronic and L-guluronic acid derived from kelp, which has a calcium-dependent gelling property (Guo *et al.* 1989). Cells are suspended in a solution of 1.2 % sodium alginate in 150 mM sodium chloride, typically at around 2×10^6 cells/ml. This suspension is dropped from a syringe through a fine needle into a 100 mM calcium chloride solution and upon contact the alginate begins to polymerize, forming 'beads'. The beads should be kept in the calcium chloride solution for 10 minutes for full polymerization to occur and then rinsed in 150 mM sodium chloride solution before they are cultured in growth medium in 50 ml centrifuge tubes or non-tissue culture treated Petri dishes. To recreate cellular condensations, chondrocytes and chondrocyte precursor cells can be cultured as cell pellets or micromasses in an attempt to improve direct cell-to-cell communication. Pellet cultures are very simple and involve counting an appropriate number of cells and centrifuging them at 160 *g*. The pellets should be cultured individually in the tube that they were spun in. After 24 hours the pellets are generally coherent enough to withstand some gentle agitation and media can be changed. To form robust pellets from chondrocytes, addition of 25 μg/ml ascorbate 2-phosphate, 10^{-7}M dexamethasone and 10 ng/ml TGFβ-1 to the medium is advisable. Chondrocytes in pellets are well suited to serum-free culture as this slows proliferation rates and encourages extracellular matrix formation. Our studies have indicated that culturing monolayer-expanded human articular chondrocytes in alginate gives rise to a more immediate shift in phenotype than culturing in pellet culture. Levels of collagen II gene expression as assessed by real time PCR can be upregulated as much as threefold after a 14-day culture period in alginate beads. Under the same conditions the expression of SOX9, a transcription factor that regulates chondrocyte phenotype, is increased tenfold in only 3 days. Similar cells grown as pellets show no increase, or even a decrease, in collagen II gene expression. The exact reaction of the cells to these culture conditions seems to depend on the number of passages that the cells have undergone. Earlier-passage cells retain some re-differentiation capacity, whilst later-passage cells, typically after five or six doublings, show poor re-differentiation capacity over 14 days.

28.3.6 Stem Cells as Chondroprogenitors

Research on stem cells as a source of chondrocytes for tissue repair applications has increased rapidly in the last 10 years, and much of this focused on the use of bone marrow-derived mesenchymal stem cells. These cells are an obvious choice as they are simple to isolate and have been shown to form part of the marrow-based cartilage repair response that occurs following full thickness cartilage injuries (those that penetrate from the bone marrow through the subchondral bone) (Shapiro *et al.* 1993). They can be differentiated into chondrocytes using the pellet culture system described above (Mackay *et al.* 1998), and can also be driven down a number of lineages including bone, fat and neural tissue (Pittenger *et al.* 1999). More recently, methods for inducing chondrogenesis in ES cells have been examined (Kramer *et al.* 2000; Nakayama *et al.* 2003). Whilst stem cells are an exciting potential cell source for use in cartilage tissue engineering, there are many ethical and practical concerns with their use. Differentiation in pellet culture is not always homogeneous throughout the tissue and it can progress to a hypertrophic state very easily (Barry *et al.* 2001; Nakayama *et al.* 2003). This would be very undesirable in cartilage repair. Avoiding these aspects of the developmental pathways is therefore a major challenge in the use of stem cells in cartilage repair.

28.3.7 Molecular Mechanisms Controlling Chondrocyte Phenotype During Culture

The molecular mechanisms that result in the loss of chondrocyte phenotype and their potential to re-differentiate are beginning to be elucidated. During monolayer culture, as collagen II gene expression levels are falling, the transcription factor SOX9, which has been shown to promote the expression of collagen II (deCrombrugghe *et al.* 2001) as well as other cartilage matrix genes (Bridgewater *et al.* 1998; Zhang *et al.* 2003), is down regulated (Stokes *et al.* 2001, 2002; Hardingham *et al.* 2002). With this in mind we have transfected the SOX9 gene into passaged human chondrocytes using retroviruses (Li *et al.* 2004) and then subjected them to alginate or pellet culture to examine their responses. We found that the cells that were expressing SOX9 at elevated levels were better able to re-express the chondrocytic phenotype. They synthesized nine times more type II collagen RNA in monolayer culture and this was far more readily upregulated by alginate or pellet culture conditions, compared with cells transduced with control vector, even though the cells had undergone six or more passages (Li *et al.* 2004; Tew *et al.* 2003). Furthermore the SOX9-transduced articular chondrocytes accumulated far more cartilaginous extracellular matrix in the three-dimensional cultures. It therefore seems likely that the loss of SOX9 expression is a critical factor in the loss of chondrocytic potential in these cells during monolayer expansion.

28.3.8 Tissue Engineering Applications

For over 10 years, chondrocytes have been used in surgical applications as dense cellular suspensions that are injected into wound sites on the articular surface (Brittberg *et al.* 1994). Genzyme has successfully marketed this technique as the Carticel procedure. However, many new tissue-engineering approaches seed the chondrocytes onto fabricated matrices formed from a range of substances such as poly(glycolic acid) (PGA), poly(caprolactone) and collagen, which are relatively simple to fabricate. Many polymer systems have been used in research but few have been developed for commercial application. These include the esterified hyaluronan-based Hyaff product line from Fidia Advanced Biopolymers, and self-assembling peptide matrices marketed as Puramatrix by 3DM, Inc. To allow seeding, the cells are grown in suspension along with the scaffold. A great deal of work has been carried out in attempting to find the best conditions to seed the cells (Vunjak-Novakovic *et al.* 1998). Generally, a static culture is not conducive to good seeding. Bioreactors have been designed that keep the cells suspended and provide better fluid flow into the centre of the scaffolds. Better results are generally seen in these systems (Vunjak-Novakovic *et al.* 1999). Nevertheless, significant problems remain with these processes, and problems such as effective population of the interior of a construct, and cell death due to poor nutrient penetration of the construct still remain. Improvements in our understanding of the chondrocytes and their interactions with these materials, as well as better bioreactor designs, are needed to address these problems.

28.4 VASCULAR SMOOTH MUSCLE CELLS

28.4.1 Introduction

Vascular smooth muscle cells (SMC) present within the medial layer of vessel walls provide essential contractility and are responsible for secretion and deposition of critical ECM structural elements of vessel walls, including elastic fibres and fibrillar collagens. These SMC are axially associated and form continuous lamellar contractile layers around the vessel wall that are interspersed with the fibres that endow elasticity (Figure 28.4). In contrast to mature cardiac or skeletal muscle cells that are considered to undergo irreversible differentiation, SMC exhibit a broad range of reversible phenotypic states in response to different stimuli. These range from a synthetic phenotype that is metabolically active, synthesizing and secreting ECM and proliferating rapidly, to a fully differentiated contractile phenotype that proliferates very slowly and is rich in contractile proteins.

28.4.2 Current Strategies for the Repair/Reconstruction of Vascular Tissues

A challenging field in tissue engineering research is the generation of small diameter vascular grafts with long-term potency. The gold standard for a prosthetic blood vessel is that it should possess all the properties of a normal vessel. Therefore, to fulfil this objective, SMC are required as a critical component of any tissue-engineered vascular graft.

Initial attempts at constructing functional blood vessels involved culturing SMC within a collagen gel surrounding a tubular mould, with additional mechanical support provided by a Dacron

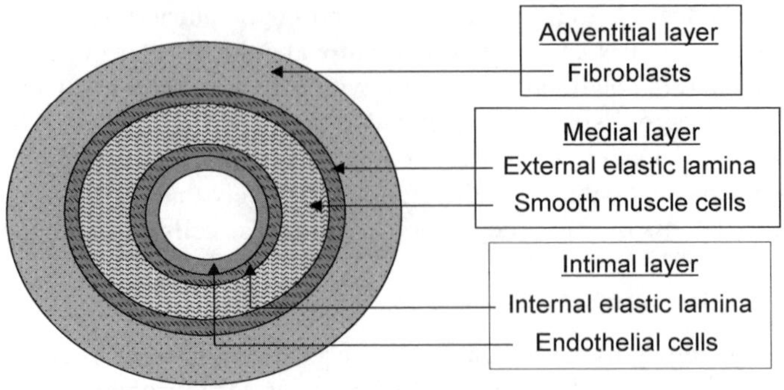

Figure 28.4 Schematic diagram illustrating the cellular structure of an arterial blood vessel wall. Endothelial cells surrounding the lumen are separated from medial layer smooth muscle cells by a subendothelial matrix and internal elastic lamina. Medial smooth muscle cells are arranged in concentric lamellar layers that are separated by an elastic fibre-rich matrix. Fibroblasts, the predominant cellular component of the outer collagen-rich adventitial layer, are separated from medial smooth muscle cells by an external elastic lamina.

mesh (Weinberg & Bell 1986). However, these structures were not suitable for *in vivo* grafting, since they were unable to withstand physiological pressures. Subsequently it was found that culturing SMC in a bioreactor under pulsatile flow defined SMC orientation and increased the synthesis of ECM molecules and elastic fibre deposition, which enhanced the mechanical strength of vascular grafts (Niklason *et al.* 2001). The importance of the correct ECM to provide the appropriate mechanical properties was demonstrated when a vascular graft incorporating sheets of cultured SMC placed around a temporary tubular support, resulted in a burst strength equivalent to a native vessel during short term implantation in a canine model (L'Heureux *et al.* 1993). In our laboratory, we are using bone marrow-derived mesenchymal stem cells (MSC) as SMC progenitors to seed small diameter vascular grafts. These MSC are seeded onto electrostatically spun porous polyurethane, coated with vascular ECM molecules to enhance cell attachment and regulate their phenotype. With this approach we aim to improve vascular graft properties by emulating events occurring during blood vessel development.

Primary SMC can be isolated from tissues and cultured *in vitro*. However, the phenotype of cultured SMC must be carefully monitored since these cells are subject to marked phenotypic variations between synthetic proliferative, and quiescent contractile, phenotypes, depending on culture conditions (medium, presence of foetal calf serum or specific growth factors, tissue source, ECM substratum). No single SMC marker exists, but a well-defined panel of cytoskeletal and ECM molecules can be used to define the early or late differentiation status of cultured SMC.

28.4.3 SMC in Vascular Development

SMC in the large vessels close to the heart are derived from the neural crest and from local mesenchymal stem cells that, early in gestation, begin to elongate and synthesize smooth muscle proteins. Changes in cell shape to an elongated form are known to play an essential role in regulating smooth muscle myogenesis (Beqaj *et al.* 2002). RhoA and other GTPases play a critical role in regulating the actin cytoskeleton. During vessel wall development, SMC generally have a proliferative synthetic phenotype, with rapid cell division and concurrent deposition of abundant matrix. However, in mature vessels SMC exhibit a contractile quiescent phenotype with limited extracellular matrix synthesis.

28.4.4 Vascular SMC Markers

Because no single marker has been identified that alone distinguishes the smooth muscle lineage, analysis of the simultaneous expression of multiple SMC proteins is critical in defining the smooth muscle phenotype (Owens 1995). Smooth muscle (SM) α-actin is one of the most useful markers for early smooth muscle differentiation, although other cells such as astrocytes and myofibroblasts may also express SM α-actin in culture. A number of other cytoskeletal markers reflect intermediate to late SMC differentiation. Mid-differentiation markers include calponin and caldesmon, and late markers include SM myosin heavy chain (MHC) and desmin. Some cytoskeletal markers, including smoothelin, SM22α and SM MHC are almost exclusively expressed by SMC. Other cytoskeletal markers are also expressed by other cell types but are useful when assessed together with SMC-specific markers, because they reflect a given differentiated SMC state; these include vimentin and vinculin. In addition to cytoskeletal markers, SMC phenotype can also be monitored on the basis of extent and type of vascular ECM deposited. Expression of fibrillar collagens and elastic fibre molecules reflects a synthetic proliferative SMC phenotype, whereas deposition of basement membrane molecules such as laminin is characteristic of a quiescent contractile state.

28.4.5 Establishing Vascular SMC Cultures

When isolated from blood vessels and grown in culture, contractile SMC generally 'de-differentiate' to the synthetic phenotype, the rate of change being a function of the culture conditions (Chamley-Campbell *et al.* 1979). They display a classic spindle-shaped morphology (Figure 28.3b). Culture parameters such as initial seeding density, cell passage, use of serum-free medium and specific batches of serum, have a pronounced effect on modulating the SMC phenotype. In a representative study, virtually all freshly isolated aortic SMC were initially positive for SM α-actin and SM myosin proteins, but after proliferating in culture less than 5 % subsequently expressed and retained these markers (Shanahan *et al.* 1993). When confluent after seven days, approximately 40 % of cells re-expressed the contractile proteins, whilst subculturing for more than five passages resulted in less than 1 % of these cells expressing SM α-actin and SM myosin. Other reports have described the culture of SMC expressing specific contractile proteins for many passages (Murray *et al.* 1990; Rothman *et al.* 1992; Tagami *et al.* 1986).

28.4.6 Stem Cell Differentiation into SMC

Cultured cells displaying some of the morphological and phenotypic characteristics of SMC have been reported following *in vitro* differentiation of different stem cells, including ES cells, mesenchymal stem cells and endothelial precursor cells in circulating blood. Murine ES cells, which were positive for the vascular endothelial growth factor (VEGF) receptor Flk-1, were reported to generate both endothelial and SM cells (Yamashita *et al.* 2000), suggesting a common progenitor for both cell types. The Flk-1 stem cells, cultured on a type IV collagen-coated surface with medium containing 10 % foetal calf serum (FCS), were differentiated to SMC by addition of PDGF-BB and to endothelial cells by administration of VEGF. When the Flk-1 stem cells were seeded in a type I collagen gel, together with both growth factors, they differentiated and interacted to form *in vitro* assembled blood-vessel-like structures. Phenotypic similarities have been reported between mesenchymal stem cells and immature vascular SMC (Galmiche *et al.* 1993; Ball *et al.* 2004). Bone marrow mesenchymal stem cells are a source of smooth-muscle-like cells found in aortic transplant neointimal lesions (a vascular injury resulting in SMC accumulation in the intimal layer of a blood vessel wall) (Shimuzu *et al.* 2001; Sata *et al.* 2002). Recently, it has been demonstrated that there is a cell population among bone marrow stromal cells, which are committed to a SMC lineage (Kashiwakura *et al.* 2003). A recent report suggested that SMC can be cultured from smooth muscle progenitor cells in human peripheral blood (Simper *et al.* 2002).

These cells were positive for SM α-actin, SM MHC and calponin by immunofluorescence and Western blotting, and were also positive for CD34, Flt1, and Flk1 receptor, but negative for Tie-2 receptor expression, suggesting a bone marrow angioblastic origin.

28.4.7 Culture Conditions that Influence SMC Differentiation

28.4.7.1 Passage and medium

In order to retain contractile SMC status in culture it is important to passage cells when subconfluent in order to avoid selecting proliferating cells lacking contact inhibition. Culture of SMC in medium that contains little or no foetal calf serum (a rich source of cytokines and growth factors) induces a quiescent contractile phenotype.

28.4.7.2 Effects of ECM

Growth of cultured SMC on different ECM substrates profoundly regulates the SMC phenotype. Fibronectin promotes change from a contractile to a synthetic phenotype, whereas laminin retards the transition (Hedin *et al.* 1988). In response to cellular adhesion and stretching on ECM, upregulation of SMC contractile proteins have been shown to be controlled by cell shape (Yang *et al.* 1999). ECM has been found to direct SMC differentiation by regulating GTPase activity and cell shape (Beqaj *et al.* 2002). Thus culture of cells on laminin-2 induces SMC differentiation by decreasing levels of RhoA and allowing cell elongation. These ECM-induced effects reflect specific integrin receptor interactions that lead to focal adhesion formation and activation of intracellular signalling responses. Elastic fibre molecules also exert a profound influence on SMC phenotype. It has been reported that elastin binds SMC through the elastin binding protein receptor and influences SMC proliferation (Mochizuki *et al.* 2002). We have shown that SMC bind fibrillin-1 molecules and microfibrils through $\alpha 5\beta 1$ and $\alpha v\beta 3$ integrin receptors, and induce SMC spreading (Bax *et al.* 2003). These microfibrils form the outer mantle of elastic fibres within the medial layer of the vessel wall, and interact directly with SMC through focal contacts.

28.4.7.3 Effects of mechanical and haemodynamic forces

Pulsatile flow and shear stress are critical determinants of the contractile SMC phenotype (Hu & Clark 1989). Such mechanical forces stimulate the production of vascular ECM molecules such as elastic fibre molecules, thus providing altered cell-matrix contacts that are sensed by cells through receptor-induced mechanotransduction. The cells respond to altered cell-matrix communications by signalling and cytoskeletal changes.

28.4.7.4 Tissue source and method of isolation

The phenotype of SMC in culture is strongly influenced by source tissue location, cellular populations isolated, and age of donor (Lemire *et al.* 1994). In general, the younger the individual, the more proliferative the SMC. The method of SMC isolation from tissues is also critical. SMC isolated by proteolytic enzyme (collagenase and trypsin) tissue dissociation tend to retain a contractile phenotype to a greater extent than those isolated by explant culture, a method that essentially selects the synthetic proliferative SMC that migrate out of the tissue explant and establish proliferating colonies (Murray *et al.* 1990).

28.4.7.5 Co-culture with endothelial cells

Contact and communication between endothelial and SM precursor cells is an important determinant of SMC differentiation. Co-culturing bovine aortic endothelial cells and murine embryonic fibroblasts (10T1/2 cells) with medium containing 10% FCS, induced 10T1/2 cells to develop a

SMC-like morphology and express three SMC markers (SM myosin, SM22α and calponin) that were undetectable in 10T1/2 cells cultured alone (Hirschi *et al.* 1998). This effect was mediated by TGFβ-1, which is activated in endothelial-mural cell co-contact (Antonelli-Orlidge *et al.* 1989). We have shown that indirect co-culture of SMC-like mesenchymal stem cells with endothelial cells enhances SM α-actin cytoskeletal organization, whereas direct co-culture downregulates cytoskeletal organization (Ball & Kielty 2004). Thus direct cellular communication is critical for regulating the contractile cytoskeleton.

28.4.7.6 Vascular growth factors

Supplementation of SMC cultures with vascular growth factors directly and profoundly influences their phenotype, and may also influence other vascular cells to transdifferentiate. Under the influence of TGFβ-1, various endothelial cells in culture have been reported to be transdifferentiated to a SMC-like phenotype, with reversible expression of SM α-actin (Amberger *et al.* 1991) and the display of both synthetic and contractile ultrastructures (Arciniegas *et al.* 1992). Transitional cells expressing both endothelial and SMC markers were observed after a brief incubation with TGFβ-1. Culturing endothelial cells on denatured collagen attenuated transdifferentiation, which was reversed by adding TGFβ-1. Culturing them on plastic facilitated the transdifferentiation process, which was inhibited by TGFβ-1 antibodies (Frid *et al.* 2002). However, within primary endothelial cultures the percentage of cells capable of transdifferentiation to a SMC phenotype was estimated at only 0.01 to 0.03 %. A more significant fraction of endothelial cells, which could be induced to differentiate to a SMC-like phenotype that displayed contractility, was observed when the cultures were deprived of FGF-1 (Ishisaki *et al.* 2003). We have found that direct supplementation of SMC culture medium with TGFβ-1 rapidly leads to the formation of a well organized SM α-actin cytoskeleton, whilst transient transfection with the TGFβ-1 gene using an adenoviral vector induces both early (SM α-actin) and late SMC markers (including SM MHC). This distinction may reflect differences in cellular access to the active form of this potent growth factor.

Human mononuclear cells isolated from peripheral blood and cultured on denatured collagen in endothelial growth medium differentiated into endothelial cells, while addition of PDGF-BB to the medium resulted in enhanced proliferation of smooth muscle-like cells (Simper *et al.* 2002). Adult human mesenchymal stem cells cultured in medium containing either 10% or 2% FCS, expressed SM α-actin at approximately similar levels (Kinner *et al.* 2002). Addition of TGFβ-1 significantly increased and PDGF-BB decreased SM α-actin expression and also contractility of the cells as determined on a cell-seeded type I collagen-GAG gel. Similarly, PDGF-BB downregulates SM α-actin expression in SMC cultures (Holycross *et al.* 1992), which is in contrast to its SMC-differentiating effects on ES cells (Yamashita *et al.* 2000). Thus, the actions of PDGF-BB on SMC differentiation depend on the cell type.

28.5 ENDOTHELIAL CELLS

28.5.1 Introduction

Blood vessels were originally thought to be mere tunnels through tissue until Friedrich Daniel von Recklinghausen discovered a cellular lining, the endothelium. The endothelium, and the endothelial cells of which it is comprised, play a vital role in blood vessel function and are the primary deterrent to thrombosis. In the capillaries, endothelial cells control the transport of solutes into the surrounding tissues. In larger vessels these cells help regulate blood pressure and control haemostasis and inflammation through the release of chemicals such as nitric oxide and vonWillebrand factor. Endothelial cells also play a significant role in wound healing through the formation of new blood vessels by angiogenesis. The complex interactions between endothelial cells and neighbouring cells or extracellular matrix also means that endothelial cells are intimately involved in regulating numerous disease processes, including thrombosis, diabetic angiopathy, hyperlipidaemia, atherosclerosis and peripheral vascular and heart disease.

Peripheral vascular and heart disease are an increasing health and socio-economic burden in most developed countries, with surgical bypass with vein the major form of treatment. In the United States alone over 1.4 million patients a year will undergo arterial replacement surgery (Vacanti & Langer 1999). Unfortunately, a number of these patients will require bypass of small diameter vessels (<6 mm) but have no useable autologous vessels for grafting. In these cases synthetic substitutes are used. Expanded polytetrafluorethylene (ePTFE) and polyethylene terephthalate fibre (Dacron) are the most widely used materials for synthetic grafts. Although successful in large diameter vessels (>6 mm), in small diameter vessels they are compromised by their thrombogenicity. Furthermore, these materials do not fully match the elastic properties of the native vessels. This compliance mismatch results in cell proliferation at the anastamoses of the graft and a narrowing of the lumen, termed intimal hyperplasia. Tissue engineering is now being used to develop endothelialized synthetic grafts with the long-term goal of producing an autologous 'living' graft using biodegradable materials.

28.5.2 Endothelial Cell Sourcing

There is increasing evidence to show that endothelial cells from different tissues and vessel types are morphologically, biochemically and functionally diverse. This finding of endothelial heterogeneity has led researchers to question which type of endothelial cells they utilize for *in vitro* studies, and is a major consideration for tissue engineering; as a result numerous different sources have been investigated. These include microvascular endothelial cells from adipose tissue, circulating endothelial progenitor cells from peripheral or cord blood, human umbilical vein, and peripheral vessels such as the saphenous vein.

Adipose tissue is highly vascularized, and based on total cells present per volume of fat is composed of approximately 85 % endothelial cells (Williams *et al.* 1994). Methods for isolating endothelial cells from adipose tissue were first described in 1984 by Jarrell and Williams (Jarrell *et al.* 1984) who used a crude clostridial collagenase to digest the tissue and release the cells. The endothelial cells could then be separated from the adipocytes by density gradient centrifugation. Adipose tissue can produce high yields of endothelial cells ($>1.0 \times 10^6$ cells/gram fat). Despite this, it is not commonly used in tissue engineering due to possible contamination with mesothelial cells from the serosal tissue that surrounds adipose tissue.

Recently, there has been a significant shift towards the use of stem cells in tissue engineering, particularly endothelial progenitor cells. These cells are derived from the bone marrow and many researchers have shown that it is possible to extract large numbers of these cells from the blood using magnetic particles coated with anti-endothelial antibodies, or using density gradient centrifugation (Asahara *et al.* 1997; Boyer *et al.* 2000). It has also been demonstrated that in numerous vascular diseases levels of progenitor cells are significantly elevated, suggesting that these cells may be ideal for use in developing an autologous vascular graft. However, the behaviour of these cells in culture is not fully understood and currently it is difficult to stimulate these progenitors to convert in large numbers to mature endothelial cells, and cell yields are variable (Hernandez *et al.* 2000; Rafii & Lyden 2003). To be successful in tissue engineering applications the cells used must act together in unison. Therefore, while these cells have potential for use in tissue engineering, a great deal of research is needed before they can be used clinically.

In our laboratory we isolate human endothelial cells from umbilical vein and non-essential blood vessels, particularly sapheneous vein. Jaffe *et al.* (1973) first described the isolation of human umbilical vein endothelial cells (HUVECs), and over the past three decades they have been one of the favoured endothelial cell types for *in vitro* studies investigating angiogenesis and vascular function. HUVECs have the benefit of being easy to isolate and culture from umbilical veins, and supply of tissue is not usually a limiting factor. In order to obtain healthy, viable cells, HUVECs must be isolated from fresh human umbilical cords within 4 hours of delivery. Firstly, the cords are inspected for damage as small puncture wounds or tears are sometimes found due to handling during delivery. To allow the vein to be distended the cord has to be cut at these points to produce a leak-free vessel. Veins are canulated at one end, and the canula held in place by a double-grip umbilical cord clamp. Using a syringe, the veins are flushed with Hank's balanced salt solution (HBSS) to remove any blood or fluid. The free end of the cord is then clamped and the vein is gently perfused with 0.1 % (w/v) collagenase type 1A. Collagenase is added until the vein is as distended as possible without rupture. The canula and syringe are left in the vein to prevent leakage of the collagenase and the cord incubated at 37 °C for 10–15 minutes. Longer incubation times generally result in contamination of cultures with fibroblasts. The effluent is flushed through with 20 ml of HBSS and centrifuged at 1000 rpm for 10 minutes. Endothelial coverage of a segment of vein is 1×10^5 cells/cm^2 (Williams 1995). However, 100 % recovery is rarely achieved and is typically around 80–90 %.

While HUVECs are useful for *in vitro* research, their use in tissue engineering is limited due to their immunogenicity. In order to make the transfer to clinical practice, adult human endothelial cells are required. Source tissue can usually be obtained during vein grafting or coronary arterial bypass operations, as excess tissue is always taken to ensure graft success. In this case, the vein is washed briefly to remove traces of blood and any side-vessels closed with sutures or liga-clips. Isolation of endothelial cells can then be performed using the same method as for umbilical vein endothelial cells. In our laboratory we commonly use tissue obtained during varicose vein surgery. Here, vein strippers are used, which results in the evertion of the vein. However, this procedure can result in damage to the endothelium and yields are therefore not as high as using traditionally isolated vein. In order to isolate the endothelial cells, the lumen must be closed to prevent contamination by fibroblasts and smooth muscle cells from the medial and adventitial layers. To achieve this, the ends of the vein are tied with sutures or clamped with liga-clips. The whole vein is then bathed in 0.2 % collagenase (type 1A) for 30 minutes at 37 °C with constant agitation. The supernatant is then decanted and centrifuged at 200 *g* for 10 minutes.

28.5.3 Endothelial Cell Culture

In vivo endothelial cells are 99 % quiescent and have an estimated turnover rate of between 47 and 23 000 days. Therefore special culture conditions are required for rapid growth *in vitro*. Growth of endothelial cells is attachment-dependent, and cells in suspension will rapidly undergo apoptosis. As a result, tissue culture dishes are generally coated with a 0.1–1 % gelatin solution, which greatly improves endothelial attachment. An enriched culture medium is also required. In our laboratory we use medium 199 supplemented with foetal calf serum. For HUVECs, 10 % serum can be used. However, for human saphenous vein endothelial cells (HSVECs), 20 % serum is required due to the increased age of the cells. This is further supplemented with endothelial cell growth supplement (ECGS). This contains a cocktail of growth factors that stimulate proliferation and prevent the cells from becoming quiescent. Despite the addition of these growth supplements the lifespan of endothelial cells is limited, surviving for only 12–14 passages at most. In our laboratory we discard the cells after passage 7 to ensure that the quality of the cells is maintained.

28.5.4 Endothelial Cell Characterization

Endothelial cells in culture have a very distinctive, well described, cobblestone appearance at confluence. Typically, they appear as flat, 1–2 μm thick, cells about 10–20 μm in diameter with a central nucleus (Figure 28.3c). Endothelial cells also exhibit contact inhibition and can be seen growing in small patches or clusters in sparsely seeded cultures.

Immunocytochemistry is one of the most widely used tools for characterizing cell types by using antibodies specific to the cell of interest. However, one significant difficulty of this technique has been the lack of a dependable marker for endothelial cells. As a result, a range of markers must be used and there are a number of antibodies with high specificity that are useful in characterizing endothelial cells. Two of the most commonly used antibodies are directed against vonWillebrand factor (vWF) and platelet endothelial cell adhesion molecule-1 (CD31). Venous-derived endothelial cells contain cytoplasmic inclusions called Weibel-palade bodies that contain high concentrations of vonWillebrand factor resulting in a characteristic punctate cytoplasmic staining (Figure 28.5a). However, not all endothelial cells contain Weibel-palade bodies. Microvascular endothelium for example contains relatively few Weibel-palade bodies and can therefore produce a negative reaction when stained for vWF. CD31 is expressed at the cell/cell junction of confluent endothelium resulting in positive staining around the periphery of the cell (Figure 28.5b). Again, although characteristic of endothelial cells, positive staining is not always guaranteed. For tissue engineering applications where a clear identification is required, other antibodies such as E-Selectin (CD62E) and the lectin *Ulex europaeus* agglutinin-1 can be used for confirmation. Finally, reverse

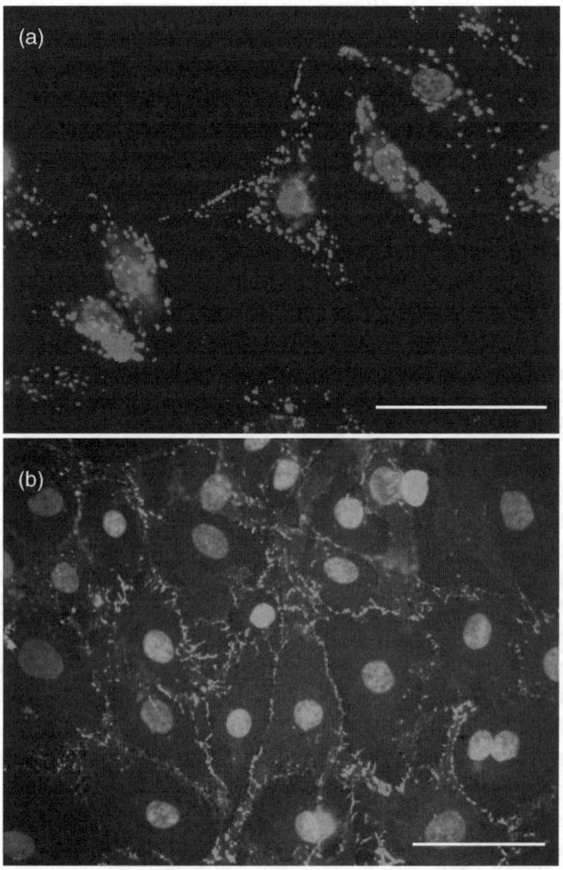

Figure 28.5 (a) Human saphenous vein endothelial cells, grown as monolayers, stained with antibodies against (a) von Willebrand factor or (b) CD31. Cell nuclei stained with DAPI. Scale bar = 20 μm.

transcription-polymerase chain reaction (RT-PCR) can be used. By designing primers to known specific endothelial cell messenger RNAs, cells can be confirmed to be endothelial. Commonly used markers include endothelial nitric oxide synthase, *fms*-like tyrosine kinase receptor (Flt-1), and kinase insert domain-containing receptor (Ratcliffe *et al.* 1999).

28.5.5 Use of Endothelial Cells in Tissue Engineering

Vascular surgery is considered to be one of the pioneering disciplines in the field of tissue engineering, with research into developing tissue-engineered vascular grafts beginning in the mid-1960s. The initial concept, first proposed by Malcolm Herring and colleagues, was that of seeding autologous endothelial cells onto the lumen of a synthetic graft since prosthetic vascular grafts in humans do not naturally develop an endothelial cell lining. Numerous groups over the next two decades investigated methods for endothelializing grafts, developing a variety of systems using both venous-derived endothelial cells and microvascular endothelial cells from adipose tissue. Peter Zilla and colleagues developed one of the most successful of these 'biolized' grafts, comprising an endothelialized fibrin-coated ePTFE graft (Meinhart *et al.* 2001). This group used a two-stage procedure where autologous endothelial cells from sub-dermal vein were cultured *in vitro* prior to seeding the fibrin-coated graft. To date, such grafts have been in use clinically for over 10 years for peripheral bypass grafting of small diameter arteries with results comparable to native vein, the current gold standard for vascular bypass grafting.

In addition to being used on their own to line synthetic grafts, endothelial cells have also been used in conjunction with smooth muscle cells and fibroblasts to produce completely tissue-engineered vascular grafts. One of the first attempts was by Weinberg and Bell (1986). This model was based on a collagen gel seeded with bovine vascular cells, but still used a synthetic Dacron mesh for support. This technique initially proved successful *in vitro* and was used as the basis for research by many other groups. The development of a scaffold-free graft has also been investigated, most notably by L'Heureux *et al.* (1993), who cultured sheets of cells and rolled them around a central spindle to produce a complete graft that mimicked the structure of a normal artery. However, while both these methods were successful *in vitro*, *in vivo* results were poor and neither has progressed into clinical use.

Our group is investigating the use of biodegradable polymers that aim to provide the initial strength required to resist blood flow and gradually to degrade, leaving a graft composed solely of host tissue. Biodegradable polymers, such as PGA, have been used for decades as suture materials and are now being used extensively for tissue engineering purposes. One of the more successful of these grafts was developed by Nicklason *et al.* (2001), utilizing a PGA scaffold. The appearance of this graft was similar to a native artery, with a complete layer of tissue formed around and through the PGA material resulting in the formation of a smooth luminal surface.

In our laboratory we are investigating the use of Hyaff-11, a hyaluronan-based biomaterial, for developing tissue-engineered vascular grafts. We have shown that it is possible to achieve 94 % attachment of endothelial cells when seeded at 1.0×10^6 cells/cm^2, resulting in the formation of an intact monolayer after 20 days in culture. We have also shown that endothelial cells grown on Hyaff-11 scaffolds secrete an organized sub-endothelial matrix comprising fibronectin, laminin and collagen type IV and VIII. As complete monolayer formation and extracellular matrix deposition are essential factors for the success of any vascular prosthesis, Hyaff-11 seems to be a very promising scaffold for small vessel reconstruction (Turner *et al.* 2004).

28.5 PRODUCTS FOR HEALTHCARE

When will tissue-engineering products be available in healthcare? Progress so far has been slow (Hardingham 2003). The first products for chronic skin wounds have been effective, but not great commercial successes. However, several countries, including the UK, have recognized that with a potential for rapid advance, this area justifies greater research investment. There is an expanding need worldwide for better treatments for many chronic clinical problems, particularly those common in the elderly. It remains a challenge for tissue engineering that the products need to meet real medical needs. Also, for them to be funded from healthcare resources they need to be cost-effective, and this will be more easily achieved in some applications than others (Hardingham 2003). However, despite these many hurdles, the potential advantages are large and it seems clear that over the next 5–15 years, there will be a steady introduction of tissue-engineered products that will offer novel and radical solutions to some important medical problems.

REFERENCES

Amberger A, Bauer H, Tontsch U, Gabbiani G, Kocher O, Bauer HC (1991) *FEBS Lett.*; 287: 223–225.
Antonelli-Orlidge A, Saunders KB, Smith SR, D'Amore PA (1989) *Proc. Natl. Acad. Sci. USA*; 86: 4544–4548.
Arciniegas E, Sutton AB, Allen TD, Schor AM (1992) *J. Cell Sci.*; 103: 521–529.
Asahara T, Murohara T, Sullivan A *et al.* (1997) *Science*; 275: 964–967.
Babero A, Ploegert S, Heberer M, Martin I (2003) *Arthritis Rheumatism*; 48: 1315–1325.
Ball SG, Kielty CM (2004) *J. Biochem. Cell Biol.*; 37: 714–727.
Barry F, Boynton RE, Liu B, Murphy JM (2001) *Exp. Cell Res.*; 268: 189–200.
Bax DV, Bernard SE, Lomas A *et al.* (2003) *J. Biol. Chem.*; 278: 34605–34616.
Benya PD, Brown PD, Padilla SR (1988) *J. Cell Biol.*; 10: 161–170.
Beqaj S, Jakkaraju S, Mattingley MR, Pan D, Schuger L (2002) *J. Cell Biol.*; 156: 893–903.
Boyer M, Townsend LE, Vogel LM *et al.* (2000) *J. Vasc. Surg.*; 31: 181–189.
Bridgewater LC, Lefebvre V, de Crombrugghe B (1998) *J. Biol. Chem.*; 273: 14998–15006.
Brittberg M, Lindahl A, Nilsson A, Ohlsson C, Isaksson O, Peterson L (1994) *New Engl. J. Med.*; 331: 889–895.
Buckwalter JA, Mankin HJ (1997) *J. Bone Jt. Surg.*; 79A: 600–611.
Chamley-Campbell JH, Campbell GR, Ross R (1979) *Physiol. Rev.*; 59: 1–6.
de Crombrugghe B, Lefebvre V, Nakashima K (2001) *Curr. Opin. Cell Biol.*; 13: 721–727.
Farndale RW, Sayers CA, Barrett AJ (1982) *Connect. Tiss. Res.*; 9: 247–248.
Frid MG, Kale VA, Stenmark KR (2002) *Circ. Res.*; 90: 1189–1196.
Galmiche MC, Koteliarsky VE, Briere J, Herve P, Charbord P (1993) *Blood*; 82: 66–76.
Guo J, Jourdian GW, MacCallum DK (1989) *Connect. Tiss. Res.*; 19: 277–297.
Hardingham T (2003) *Tiss. Eng.*; 9: 1063–1064.
Hardingham T, Tew S, Murdoch A (2002) *Arthritis Res.*; 4: S63–68.
Hayes AJ, MacPherson S, Morrison H, Dowthwaite G, Archer CW (2001) *Anat. Embryol.*; 203: 469–479.
Hedin U, Bottger BA, Forsberg E, Johansson S, Thyberg J (1988) *J. Cell Biol.*; 107: 307–319.
Heid CA, Stevens J, Livak KJ, Williams PM (1996) *Genome Res.*; 6: 986–994.
Hernandez DA, Townsend LE, Uzieblo MR *et al.* (2000) *Am. Surg.*; 66: 355–359.
Hirschi KK, Rohovsky SA, D'Amore PA (1998) *J. Cell Biol.*; 141: 805–814.
Holycross BJ, Blank RS, Thompson MM, Peach MJ, Owens GK (1992) *Circ. Res.*; 71: 1525–1532.
Hu N, Clark EB (1989) *Circ. Res.*; 65: 1665–1670.
Hunziker EB, Rosenberg LC (1996) *J. Bone Jt. Surg.*; 78A: 721–733.
Ishisaki A, Hayashi H, Li AJ, Imamura T (2003) *J. Biol. Chem.*; 278: 1303–1309.
Jaffe EA, Nachman RL, Becker CG, Minick CR (1973) *J. Clin. Invest.*; 52: 2745–2756.
Jakob M, Démarteau O, Schäfer D *et al.* (2001) *J. Cell Biochem.*; 81: 368–377.
Jarrell B, Levine E, Shapiro S *et al.* (1984) *J. Vasc. Surg.*; 1: 757–764.
Kashiwakura Y, Katoh Y, Tamayose T *et al.* (2003) *Circulation*; 107: 2078–2081.
Kinner B, Zaleskas JM, Spector M (2002) *Exp. Cell Res.*; 278: 72–83.
Kramer J, Hegert C, Guan K, Wobus AM, Muller PK, Rohwedel J (2000) *Mech. Dev.*; 92: 193–205.
Kuettner KE, Pauli BU, Gall G, Memoli VA, Schenk RK (1982) *J. Cell Biol.*; 93: 743–750.
Lemire JM, Covin CW, White S, Giachelli CM, Schwartz SW (1994) *Am. J. Pathol.*; 144: 1068–1081.
Levintow L, Eagle H (1961) *Ann. Rev. Biochem.*; 30: 605–640.

L'Heureux N, Germain L, Labbe R, Auger FA (1993) *J. Vasc. Surg.*; 17: 499–509.
Li Y, Tew SR, Russell AM, Gonzalez K, Hardingham TE, Hawkins RE (2004) *Tiss. Eng.*; 10: 575–584.
Mackay AM, Beck SC, Murphy JM, Barry FP, Chichester CO, Pittenger MF (1998) *Tiss. Eng.*; 4: 415–428.
Mankin HJ (1962) *J. Bone Jt. Surg.*; 44A: 688–698.
Mankin HJ (1974) *New Eng. J. Med.*; 291: 1285–1292.
Mansbridge J (2002) *Expert Opin. Biol. Ther.*; 2: 25–34.
Meinhart JG, Deutsch M, Fischlein T, Howanietz N, Froschl A, Zilla P (2001) *Ann. Thorac. Surg.*; 71: S327–331.
Mochizuki S, Brassart B, Hinek A (2002) *J. Biol. Chem.*; 277: 44854–44863.
Mohanna PN, Young RC, Wiberg M, Terenghi G (2003) *J. Anat.*; 203: 553–565.
Murray TR, Marshall BE, Macarak EJ (1990) *J. Cell Biol.*; 143: 26–38.
Nakayama N, Duryea D, Manoukian R, Chow G, Han CY (2003) *J. Cell Sci.*; 116: 2015–2028.
Niklason LE, Abbott W, Gao J *et al.* (2001) *J. Vasc. Surg.*; 33: 628–638.
Pittenger MF, Mackay AM, Beck SC *et al.* (1999) *Science*; 284: 143–147.
Owens GK (1995) *Physiol. Rev.*; 75: 487–517.
Rafii S, Lyden D (2003) *Nat. Med.*; 9: 702–712.
Ratcliffe KE, Anthony FW, Richardson MC, Stones RW (1999) *Hum. Reprod.*; 14: 1549–1554.
Reno C, Marchuk L, Sciore P, Frank CB, Hart DA (1997) *Biotechniques*; 22: 1082.
Rothman A, Kulik TJ, Taubman MB, Berk BC, Smith CW, Nadal Ginard B (1992) *Circulation*; 86: 1977–1986.
Sata M, Saiura A, Kunisato A *et al.* (2002) *Nat. Med.*; 8: 403–409.
Shanahan CM, Weissberg PL, Metcalfe JC (1993) *Circ. Res.*; 73: 193–204.
Shapiro F, Koide S, Glimcher MJ (1993) *J. Bone Jt. Surg.*; 75A: 532–553.
Shimuzu K, Sugiyama S, Aikawa M *et al.* (2001) *Nat. Med.*; 7: 738–741.
Simper D, Stalboerger PG, Panetta CJ, Wang S, Caplice NM (2002) *Circulation*; 106: 1199–1204.
Stokes DG, Liu G, Dharmavaram R, Hawkins D, Piera-Velazquez S, Jimenez SA (2001) *Biochem. J.*; 360: 461–470.
Stokes DG, Liu G, Coimbra IB, Piera-Velazquez S, Crowl RM, Jimenez SA (2002) *Arthr. Rheum.*; 46: 404–419.
Tagami M, Nara Y, Kubota A *et al.* (1986) *Cell Tiss. Res.*; 245: 261–266.
Tew SR, Li Y, Russell AM, Prince JM, Hawkins RE, Hardingham TE (2003) *Trans. Orthop. Res. Soc.*; 49: 226.
Tilles AW, Berthiaume F, Yarmush ML, Tompkins RG, Toner M (2002) *J. Hepatobiliary Pancreat. Surg.*; 9: 686–696.
Turner NJ, Kielty CM, Walker MG, Canfield AE (2004) *Biomaterials*; 25: 5955–5964.
Vacanti JP, Langer R (1999) *Lancet*; 354: SI32–4.
Vunjak-NovakovicG, Obradovic B, Martin I, Bursac PM, Langer R, Freed LE (1998) *Biotech. Prog.*; 14: 193–202.
Vunjak-Novakovic G, Martin I, Obradovic B *el al.* (1999) *J. Orthop. Res.*; 17: 130–138.
Weinberg CB, Bell E (1986) *Science*; 231: 397–400.
Williams D (2004) *Mater. Today*; 7: 24–29.
Williams SK (1995) *Cell Transpl.*; 4: 401–410.
Williams SK, Wang TF, Castrillo R, Jarrell BE (1994) *J. Vasc. Surg.*; 19: 916–923.
Yamashita J, Itoh H, Hirashima M *et al.* (2000) *Nature*; 408: 92–96.
Yang Y, Relan NK, Przywara DA, Schuger L (1999) *Development*; 126: 3027–3033.
Zhang P, Jimenez SA, Stokes DG (2003) *J. Biol. Chem.*; 278: 117–123.

29 The Use of Stem Cells in Cell Therapy

F Martín, J Jones, P Vaca, G Berná and B Soria

29.1 INTRODUCTION

More than 20 years have passed since the first isolation of mouse pluripotent embryonic stem cells (ES) (Martin 1981; Evans & Kaufman 1981). At that time, these cells were of interest for their use in the generation of transgenic mice. For the past 5 years, though, they have been of great interest due to their potential in tissue engineering and subsequent transplantation therapies.

Tissue engineering makes possible the generation of live tissues and organs for transplantation. Unfortunately, this technique needs a significant number of tissue-specific cells, so much attention has been paid to the possibility of using multipotential progenitor cells. As of yet, little is known about the intrinsic molecular programmes defining self-renewal and differentiation, despite a great deal of work having been done in this field. In this review, we will describe the most recent advances in stem cell biology and *in vitro* differentiation techniques, as well as discussing the most important applications of stem cell technology.

29.2 THE STEM CELL CONCEPT

As of yet, there is no universally accepted definition for the term 'stem cell', despite our growing knowledge of these cells. Nevertheless, they possess certain characteristics that serve to distinguish them from other cell types. The most obvious characteristics are: (i) an unlimited or prolonged self-renewal capacity, i.e. the ability to divide continuously and create new stem cells, and (ii) multi-potentiality, which is the ability to differentiate into a number of cell types using specific molecular pathways. There are also a number of properties that are frequently ascribed to stem cells, such as the ability to undergo asymmetric cell division, to exist in a mitotically quiescent form, and to regenerate clonally. These properties are not displayed by all the types of stem cell (embryonic and adult), so it is difficult to give a universally applicable definition. For this reason, at the moment, a certain tolerance of ambiguity in the definition of stem cell is necessary, even though new data from transcriptional profiling of purified stem cell populations are beginning to help to establish the concept of 'stemness'.

29.3 TYPES AND PROPERTIES OF STEM CELLS

Stem cells can be classified according to: (i) their developmental potency, and (ii) their origin.

When a stem cell divides, the daughter cells may give rise to a variety of cell types, and the potency of a stem cell refers to this variety. In terms of potentiality, stem cells may be classified

Medicines from Animal Cell Culture Edited by G. Stacey and J. Davis

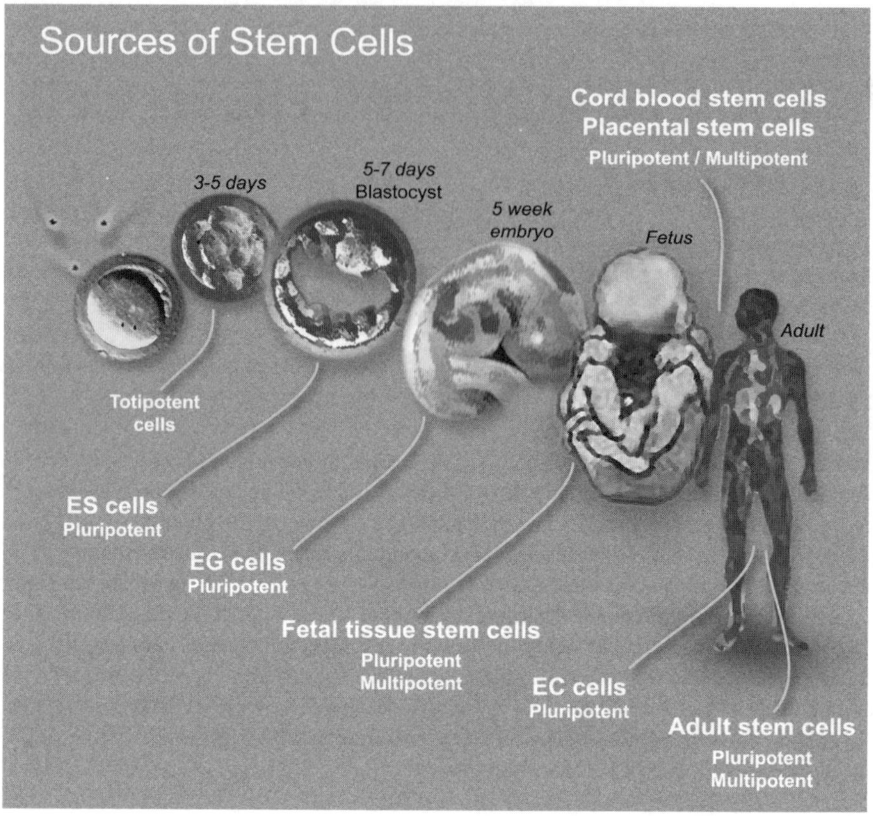

Figure 29.1 Principal sources of stem cells. The illustration indicates the main sources of stem cells and correlates them with their developmental potential. ES: embryonic stem cells; EG: embryonic germ cells; EC: embryonal carcinoma cells.

as totipotential, pluripotential, multipotential, or unipotential. Totipotential stem cells are capable of differentiating into any cell type, including the trophoblast of the placenta; pluripotential cells can give rise to any cell type of the embryo, including the germ line; multipotential cells can only give rise to a limited number of cell types of a certain tissue, normally from the same germ layer; and unipotential stem cells are cell lineage progenitors committed to differentiate into a particular cell type (Berna *et al.* 2001) (Figure 29.1).

Depending on their origin, stem cells may be: (i) embryonic stem (ES) cells; (ii) embryonic germ (EG) cells; (iii) embryonal carcinoma (EC) cells; (iv) trophoblast stem (TS) cells, or (v) foetal or adult stem cells, of various sources.

The remainder of this section will examine the properties of the different sources of stem cells, focusing mainly on ES cells.

29.3.1 Embryonic Stem Cells

Pluripotent ES cells were originally isolated from *in vitro* outgrowths of mouse blastocysts cultured on mitotically inactivated fibroblasts in the presence of conditioned media (Martin 1981; Evans & Kaufman 1981). The ES cells derive from the inner cell mass (ICM), which at an early developmental stage constitutes the entire internal cell component of the blastocyst, and later segregates into a sub-compartment named the epiblast prior to implantation. ES cells can: (i) propagate as homogeneous stem cell cultures with sustained symmetrical self-renewal; (ii) expand without apparent limit, maintaining a stable euploid karyotype; (iii) differentiate into the three germ layers; (iv) give rise to teratocarcinomas; (v) integrate into the developing embryo, and (vi) colonize the germ line. ES cells are not capable of generating a blastocyst and should therefore not be considered totipotent (Smith 2001). In addition, ES cells have several characteristic cell surface molecules (including stage-specific embryonic antigens–SSEA), express several specific enzymes such as alkaline phosphatase and telomerase, and express several molecular markers, which are rapidly down-regulated upon differentiation (Eigess & Benvenisty 2002).

Unfortunately, none of these markers is exclusive to ES cells and can be found in other cell types. ES cell lines have also been isolated from chickens, mink, hamsters, marmosets, rabbits, pigs, monkeys and humans, although none of them possess all the six characteristics mentioned above. This seems to indicate that the situation in the mouse is the exception rather than the rule, or that the isolation techniques of ES cells in the other species must be improved. In fact, even in the mouse, there are considerable difficulties in deriving ES cells from inbred mouse strains other than C57BL/6 and 129 (Rathjen & Rathjen 2001).

In 1998, Thomson *et al.* derived human ES cells from the inner cell mass (ICM) of normal human blastocysts (Thompson *et al.* 1998). Two years later, Pera *et al.* (2000) and Amit *et al.* (2000) confirmed that possibility. Human ES cells meet the first four criteria, but obviously, for ethical reasons, the critical functional tests of chimaera contribution and gamete production (the last two criteria) have not been performed. In addition, human ES cells do not respond to leukaemia inhibitory factor (LIF) and are capable of differentiating into trophoblasts (Reubinoff *et al.* 2000), suggesting that they might not be in exactly the same developmental stage as mouse ES cells.

29.3.2 Embryonic Germ Cells

These stem cells derive from primordial germ cells (PGC), which are the precursors of the mature gametes. The first time PGCs can be clearly distinguished from somatic cells in the mouse embryo is around 7.5 days post-coitum (dpc). The first mouse EG cell lines were established in 1992 by Matsui *et al.* (1992) and Resnick *et al.* (1992). These cells express many of the same cellular markers, and possess the same properties, as ES cells.

Shamblott *et al.* derived human EG-like cells from the gonadal ridge and mesenchyma of 5- to 9-week old foetal tissue (Shamblott *et al.* 1998). These cells differ in a number of properties with respect to human ES cells: they display different morphology, surface markers and growth requirements, do not form teratomas when xenografted, and proliferate less (Shamblott *et al.* 2001).

29.3.3 Embryonal Carcinoma Cells

EC cells are the undifferentiated cells of a teratocarcinoma, which is a tumour from a germinal tissue. EC cells were first isolated in mice by Kahn and Ephrussi (Kahn & Ephrussi 1970), and a few years later in humans by Hogan *et al.* (Hogan *et al.* 1977). Studies performed with EC cells laid the experimental background for the later expansion of ES cell studies. In general, EC cells have similar properties to ES cells, only with several differences: (i) their differentiation potential is lower; (ii) they are unlikely to be transmitted through the germ line, and (iii) very often have karyotypic abnormalities (Eigess & Benvenisty 2002).

29.3.4 Trophoblast Stem Cells

Mouse TS cells can be obtained from several stages of development and from various places, such as blastocysts at 3.5 dpc, from extraembryonic ectoderm in 6.5 dpc embryos, or from chorionic ectoderm of 7.5 dpc embryos. TS cell lines obtained at different stages of development have similar properties, such as cell culture conditions, gene expression and developmental potential. These cells need to be cultured in the presence of fibroblast growth factor-4 (FGF-4) and heparin, as well as in primary embryo fibroblast-conditioned medium (Tanaka *et al.* 1998). Like ES cells, TS cells can continue the embryonic development when reintroduced into early embryos, although these cells only contribute to the trophoblast lineages and never to embryonic lineages or extraembryonic tissues derived from mesodermal or endodermal lineages (Tanaka *et al.* 1998).

The embryonic and foetal environment also seems to be a source of stem cells. Two recent reports from the same group (Prusa & Hengstschläger 2002; Prusa *et al.* 2003) describe the presence in the amniotic fluid of an Oct-3/4 (a POU-V related DNA-binding transcription factor) positive cell line, which also expresses stem cell factor and alkaline phosphatase. It has been suggested that these cells, which have a high proliferation rate, are pluripotent stem cells.

29.3.5 Adult Stem Cells and Amnion Stem Cells

Adult stem cells have a limited differentiation potential when compared with embryonic stem cells. Most adult stem cells, theoretically, possess the same properties as other stem cells: (i) self-renewal; (ii) multi-lineage differentiation of a single cell, and (iii) *in vivo* functional reconstitution of a given tissue. It has long been known that in adult stem cells, the degree of self-renewal and differentiation potential is less than that of ES cells. However, a recent report suggests the possible existence, in mice and humans, of a small population of mesenchymal stem cells with pluripotent characteristics (Jiang *et al.* 2002). These cells might acquire the characteristics of cells located outside the limb-bud mesoderm, including endothelium, neuroectoderm and endoderm (Jiang *et al.* 2002).

The existence of new sources for pluripotent adult stem cells is still a matter of debate (Wagers 2002). Whilst some reports clearly indicate that some adult stem cells retain a potency broader than expected (Tosh & Slack 2002), the possibility of stem cells fusing with somatic cells (Ying *et al.* 2002) has cast doubt upon it. On the other hand, cell fusion may induce nuclear reprogramming of such differentiated cells. Since the cloning of 'Dolly' the sheep (Campbell *et al.* 1996), many investigations have shown that adult cells from many mammalian species can be reprogrammed to acquire pluripotentiality. Table 29.1 lists both adult stem cells committed to differentiate into tissue-specific cell types, and adult stem cells that display broader potentiality. Bone marrow represents the most promising source. Some reports even indicate that bone marrow cells may participate in the healing of tissues other than those of the haematopoietic lineage, opening new avenues of investigation in tissue regeneration.

Table 29.1 Sources of adult stem cells.

Committed Adult Stem Cells	Pluripotent Adult Stem Cells
Intestinal crypt cells	Bone marrow
Haematopoietic stem cells	mesenchymal, MAPC
Muscle satellite cells (myoblasts)	monocyte derived
Liver stem cells	Umbilical cord
Spermatogonia	Mesangioblast
Oogonia	Peripheral blood stem cells
	Deciduous teeth
	Neural stem cells
	Muscle stem cells
	Fat cells

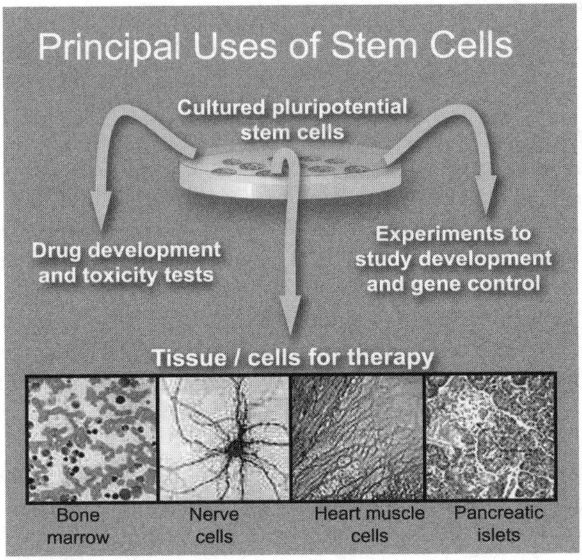

Figure 29.2 Potential applications of stem cells. The illustration shows some of the more important stem cell applications.

29.4. STEM CELL POTENTIAL

Stem cell research has an enormous potential in the understanding of fundamental human biology. In this regard, stem cells offer unprecedented opportunities to develop new treatments for numerous diseases. Moreover, they also present a new alternative for exploring fundamental questions of developmental biology, such as determining the basic mechanisms of tissue development or tissue specialization. This knowledge is necessary in developing effective therapies. Also, stem cells can be used to investigate the effects of drugs and environmental factors in embryotoxicity and pharmacology. As discussed in the previous section, there are different types of stem cell, of distinct origin and developmental potential. This suggests that different types of stem cell could have different applications. In this section we will discuss the most important possible stem cell applications, which are regenerative medicine and research applications (Figure 29.2).

29.4.1 Stem Cells and Regenerative Medicine

The ability of stem cells to differentiate *in vitro* into a wide variety of cell types makes them ideal for the generation of cells for tissue engineering and transplantation therapies. The possible contributions of stem cells to regenerative medicine are:

- Cell therapy for the treatment of diseases such as diabetes, diseases of the nervous system (for example multiple sclerosis, Parkinson's and Alzheimer's diseases), stroke, cardiovascular diseases, skin replacement for burns and grafts, as well as some respiratory diseases;
- Gene therapy for cancers, cystic fibrosis, Huntington's chorea, blood diseases (thalassaemia and haemophilia) and primary immune-deficiency diseases;
- Bioengineering of organs and tissues (such as nerve bundles for spinal cord repair, bone and cartilage).

Although some of these medical applications have proved useful in animal models (Bjorkland *et al.* 2002; McDonald *et al.* 1999; Soria *et al.* 2000), we need to answer some fundamental questions before they are used in human regenerative medicine. These questions can be summarised as below:

What causes stem cells to maintain themselves in an undifferentiated state? Up until now, the mechanisms by which stem cells remain undifferentiated are unknown. Nevertheless, research in mouse ES cells has elucidated two unrelated pathways, LIF (Niwa 2000) and Oct-3/4 (Smith *et al.* 1988), which play a role in maintaining pluripotency and proliferation. Even though both LIF and Oct-3/4 pathways may be involved in the control of these characteristics, the molecular mechanisms and the genes that govern them are unknown. Also, as has been indicated above, the mechanisms and molecules that play a role in human ES pluripotency are probably different (Reubinoff *et al.* 2000). Identification of those genes that drive pluripotent SC differentiation into specific pathways, both in early differentiation (Figure 29.3), and in late differentiation to more specific cell types (Soria 2001) will be instrumental in order to implement *in vitro* differentiation protocols.

What genetic and environmental signals affect differentiation? One of the major goals in stem cell research is to understand the decision-making processes in lineage commitment and cell type differentiation of pluripotent cells. At present, two basic techniques of *in vitro* differentiation are used: (i) differentiation through the formation of embryoid bodies (EBs), and (ii) differentiation in monolayer culture. The formation of EBs is achieved by cultivating stem cells in bacterial Petri

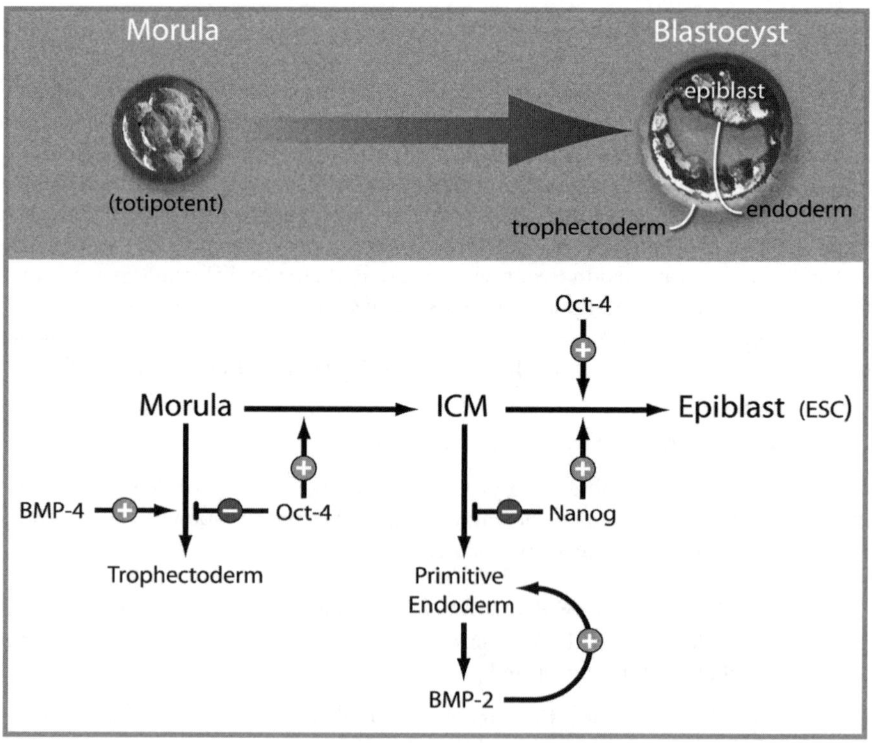

Figure 29.3 Early differentiation of totipotent stem cells (morula) into blastocyst. The illustration summarizes the main factors that govern the early differentiation process. ICM: inner cell mass; ESC: embryonic stem cell; BMP: bone morphogenic protein; Oct-4: POU-V-related DNA-binding transcription factor.

dishes or in hanging drops. The EB formation is usually enhanced by the withdrawal of LIF from the media. ES cells also differentiate in monolayer culture when deprived of LIF, feeder support, or are cultured in methyl cellulose. These procedures, together with the proper combination of growth and differentiation factors plus extracellular matrix factors, help to control the differentiation process. Until now, a number of exogenous factors have been used to direct stem cell differentiation. In mice, soluble factors are known to direct differentiation toward certain cell types. For example, interleukin-1 (IL-1) directs stem cells to become macrophages or neutrophils (Wiles & Keller 1991), retinoic acid induces neuron formation (Bain *et al.* 1995), transforming growth factor-β (TGF-β) induces myogenesis (Slager *et al.* 1993), and nicotinamide induces the formation of pancreatic islet-cells (Soria et al. 2000). The effects of soluble factors on human ES cells are only beginning to be studied. Schuldiner *et al.* (2000) studied the effects of eight growth factors on human ES differentiation and found that each growth factor directed the cells to a unique set of differentiated cells. Moreover, it has been demonstrated that nerve growth factor-β (NGF-β) can be a potent enhancer of human ES neural differentiation (Schuldiner *et al.* 2001). Another way to differentiate stem cells is by genetic approaches, such as the over-expression of certain genes in stem cells ('gain-of-function') (Rohwedel *et al.* 1995) or inducing homozygous mutations ('loss-of-function') (Rohwedel *et al.* 1998). To select a pure population of differentiated cell types, a genetic selection has to be performed. A good approximation could be the one performed by Klug *et al.* (1996) and by Soria *et al.* (2000, 2001). In both studies, the authors used an 'intelligent' transgene that coupled the promoter of a specific gene (α-cardiac myosin heavy chain and insulin, respectively) to a gene that encoded resistance to an antibiotic, such as neomycin. This strategy allows selection of the desired cells based on their antibiotic resistance.

What cues do stem cells use to initiate or terminate cell divisions? This is an important question that needs to be resolved. Although it has been demonstrated that well-differentiated ES cells do not form tumours, undifferentiated early embryonic cells may give rise to teratomas or teratocarcinomas when transplanted into extrauterine sites (Smith 2001).

What physiological properties guide the functional integration of newly generated tissues into existing organs? Studies on mouse EG (Labosky *et al.* 1994) and ES cells (Dean *et al.* 1998) have shown that stem-cell-derived tissues or cells often fail to control the expression of imprinted genes properly. Given the variety of imprinting-related developmental abnormalities observed in humans and experimental animals (Reik & Walter 2001), the possibility that imprinted gene expression might be unregulated in stem-cell-derived tissues raises a potentially serious problem for human stem cell transplantation therapy. On the other hand, a study published by Onyango *et al.* (2002) strongly suggests that failure to regulate imprinted gene expression is not characteristic of differentiated cells derived from human embryonic stem cell lines.

Can stem cells be rejected on transplantation? The answer to this question seems to be yes. Drukker *et al.* (2002) demonstrated that although human ES cells present low expression levels of Class I MHC (MHC-1) molecules, this expression is dramatically elevated in the presence of cytokines and thus may lead to rejection on transplantation. To overcome this problem, new strategies must be developed, such as: (i) the formation of a 'histocompatibility bank' in which all human ES cell lines are HLA isotyped; (ii) the production of specific human ES cell lines for each patient, and (iii) HLA manipulation of the donor ES cell, in order to make the cell suitable for all patients.

Can adult stem cells be used for cell therapy? One of the major advantages of adult stem cells is that they can be derived from the actual patient and can avoid rejection when transplanted. During the last three decades, haematopoietic stem cells from bone marrow, blood or umbilical cord blood have been used clinically to re-establish the haematopoietic system. Although there are still some controversies about adult stem cell plasticity, the data published (Jiang *et al.* 2002) seem to prove that some types of adult stem cell might be more pluripotent than others. However, for adult stem cell plasticity to be clinically useful, a ready supply of well-characterized pluripotent adult stem cells will be required. This implies a better knowledge of the phenotype of such cells, their environment, and their processes of proliferation and differentiation. On the other hand, stem cell fusion with somatic cells (Ying *et al.* 2002) has introduced new ways to interpret some of the data previously accepted as the best examples of transdifferentiation ever published (Lagasse *et al.* 2000), even though they may represent a pathway to induce reprogramming of differentiated cell types.

29.4.2 Research Applications of Stem Cells

Pluripotent stem cells reproduce cellular developmental processes and gene expression patterns of early embryogenesis during *in vitro* differentiation. These properties are important for their use as tools in understanding some of the larger problems in basic and clinical biology. These include:

- *Human development biological studies.* This aspect is constrained by practical and ethical limitations. Human ES cells may allow scientists to investigate how cells are committed to the major lineages of the body. The knowledge gained will help many fields of work, such as cancer biology and the understanding of the causes of birth defects.
- *Development of genetic modification techniques to create culture models for human diseases.* Some models of human diseases are actually limited by the current animal and cell culture models. For example, a number of pathogenic viruses (like HIV and hepatitis C) grow only on human or some non-human primate cells.
- *Gene therapy.* For gene-based therapies, it is critical for the desired gene to be introduced into organ stem cells in order to achieve long-term expression and therapeutic effect.
- *Drug discovery and toxicology.* The mutagenic, cytotoxic and embryotoxic effects of chemical substances could be tested using *in vitro* systems of pluripotent embryonic stem cells.

29.5 CURRENT STATUS AND POTENTIAL

Transplantation of cells, tissues and organs to restore adequate physiological responses and/or anatomical structures is an old idea in medicine that became a reality with the discovery of modern immunosuppression techniques. Despite the success in surgical procedures and the discovery of new immunosuppressants, this approach will always be limited by the supply of donors. Thus, stem cells, with their high proliferation and differentiation potential, hold great promise for future use in regenerative medicine. Some diseases such as diabetes, Parkinson's disease and others in which the damage is localized to a specific cell type, seem to be good candidates for cell therapy. Clinical experience in islet transplantation from dead donors (diabetes) and foetal tissues (Parkinson's disease) established the 'proof of concept' that transplantation may work. The scarcity of donated tissues compared with demand has driven the search for new sources of differentiated tissue that may be used in cell therapy.

29.5.1 Diabetes

There are several studies where blood glucose levels have been normalized in STZ-diabetic mice after transplantation of embryonic-stem-cell-derived insulin-containing cells (Soria *et al.* 2000; Hori *et al.* 2002; Vaca *et al.* 2006). Table 29.2 summarizes the different approaches used so far to generate insulin-producing cells from stem cells. Cell trapping methods are very efficient in selecting cell lineages in which the selected gene is expressed (Soria *et al.* 2000), whilst coaxial methods are more suitable for obtaining a higher proportion of insulin-containing cells. Unpublished results from our group indicate that a combination of coaxial and cell-trapping methods

Table 29.2 *In vitro* generation of insulin-producing cells from stem cells.

Embryonic stem cells	Adult stem cells
Cell trapping	Cell trapping
Soria *et al.* (2000)	
Soria *et al.* (2001)	
Moritoh *et al.* (2003)	
Voca *et al.* (2006)	
Coaxial	Coaxial
Lumelsky *et al.* (2001)	Bonner-Weir *et al.* (2000)
Assady *et al.* (2001)	Ramiya *et al.* (2000)
Hori *et al.* (2002)	Hori *et al.* (2005)
D'Amour *et al.* (2006)	Ruhnke *et al.* (2005)
	Hao *et al.* (2006)
Gene expression	Gene expression
Blyszczuk *et al.* (2003)	Ferber *et al.* (2000)
Blyszczuk & Wobus (2006)	Kojima *et al.* (2002)
	Zalzman *et al.* (2003)

decreases that variability and increases the proportion of islet cells and their progenitors. Different strategies to induce maturation of the exocytotic machinery are particularly relevant in order to obtain functional beta-cells that synthesize, store and release insulin in a regulated form. Adult progenitor cells of the pancreas (duct cells) have been induced to form islet cell clusters (Bonner-Wier *et al.* 2000). Also, the expression of genes that direct the differentiation of a cell into a particular fate has been explored in ES cells (Blyszczuk *et al.* 2003; Blyszczuk & Wobus 2006).

Bone marrow harbours cells that have the ability to differentiate into non-haematopoietic tissues such as neurons, endothelia, epithelia and muscle. Two recent reports show that bone marrow cells may repopulate pancreatic islets of Langerhans (Ianus *et al.* 2003; Hess *et al.* 2003). *In vivo* tissue regeneration will be discussed below.

29.5.2 Neurological Diseases

Neurological diseases, including neurodegenerative diseases (Parkinson's disease, Alzheimer's disease, Huntington's chorea, multiple sclerosis, etc.), stroke and spinal lesions are also potential targets for stem-cell-based therapies. Table 29.3 summarizes the possible applications of stem cell therapies for neurological diseases. Although only two areas of the mammalian brain (the hippocampus and the olfactory bulb) generate significant numbers of new nerve cells, stem cells are widely distributed throughout the brain. This unexpected plasticity of the nervous system has inspired researchers to devise neuro-replacement therapies (Ostenfeld & Svendsen 2003; Holden 2002). Most clinical investigations have used cells derived from foetal tissues as a source of transplantable cells, demonstrating a 'proof of principle' for cell transplants in the treatment of neurological diseases. Intrastriatal transplants of human foetal mesencephalic tissue in Parkinson's disease patients show that grafted dopaminergic neurons re-innervate the striatum, and restore regulated dopamine release and movement-related frontal cortical activation, with significant symptomatic relief (Lindvall 2003). Ethical concerns and the lack of sufficient amounts of tissue impede large-scale clinical trials. New sources include ES cells, EG cells, adult stem cells or even foetal cells from pigs and other animals, or cells cultivated from a certain type of tumour. Efforts are being addressed to the generation of dopaminergic

Table 29.3 Stem cell therapy for neurological diseases.

Disease	Neural damage	Tentative treatment	References
Spinal lesions and traumatic injury of CNS	Loss of neural connections	Yes	Karimi-Abddrezaee *et al.* (2006)
Stroke	Loss of neural connections	Yes	Willing *et al.* (2003)
Neurodegenerative diseases			
Parkinson's	Dopaminergic cells (basal ganglia)	Yes	Lindvall (2003)
			Arenas (2002)
			Isacson (2003)
			Lindvall and McKay (2003)
Alzheimer's	Cholinergic cells (forebrain)	Yes	Wan *et al.* (2006)
Amyotrophic lateral sclerosis	Cholinergic cells (spinal motorneurons)	Yes	Silani and Leigh (2003)
Huntington's chorea	GABAergic cells (striatum)	Yes	Vazey *et al.* (2006)
			Peschanski and Dunnett (2002)
Retinitis pigmentosa	Photoreceptors (retina)	Yes	Smith (2004)
			Tropepe *et al.* (2000)
Demyelinating diseases multiple sclerosis	Autoimmune attack	Yes	Pluchino *et al.* (2003)
			Su *et al.* (2006)

neurons from these sources. Animal studies have shown successful *in vitro* differentiation from ES cells to dopaminergic neurons that, when transplanted into animal models, improve their condition (Arenas 2002). An alternative approach is to engineer stem cells to release glial-derived neurotrophic factor (GDNF), a potent growth factor for dopaminergic neurons, which may rescue these cells from degeneration.

Like Parkinson's, Huntington's disease entails degeneration of one cell type, the medium spiny neurons of the striatum that release GABA, although they establish more diffuse connections than the dopaminergic neurons in Parkinson's disease. Even though stem cell-derived GABAergic neurons have still not been produced, a promising clinical trial with foetal tissue has shown that many neurons integrate in the brain.

Alzheimer's disease involves a wholesale attack on forebrain cholinergic neurons. The disease is so widespread that one could hardly replace them. Most scientists are sceptical about replacement therapies for Alzheimer's disease. However, activation of SC regenerates neural tissue in the rat hippocampus, an important area affected in Alzheimer's disease in humans. Cholinergic neurons are also affected in Lou Gehrig's disease (amyotrophic lateral sclerosis (ALS)), in which spinal motoneurons are lost. Since ALS causes damage throughout the spinal cord, stem cell therapy would require implanting neural precursors at a fairly primitive stage of development when they are able to migrate. Kerr *et al.* (2003) have shown that human embryonic germ cell derivatives facilitate motor recovery of rats with diffuse motor neuron injury.

In multiple sclerosis there is a widespread de-myelination and subsequent axonal loss. Two therapeutic interventions have been suggested: (i) cell replacement with neural precursor cells promoting multifocal re-myelination, and (ii) autologous blood stem cell transplantation combined with high-dose immunosuppressive therapy (Pluchino *et al.* 2003). Clinical assays using CD34+

autologous peripheral blood cells are in progress. Bone marrow stem cell transplantation for multiple sclerosis has shown positive early results in multicentre trials. However, a significant mortality risk requires further study. More recently, a multiple sclerosis patient has received a transplant of autologous Schwann cells (Stangel & Hartung 2002). Whilst other cell sources are being studied in experimental models, it seems reasonable to think that stem-cell-based therapies will provide better treatments for multiple sclerosis.

Neural cell loss after ischaemic brain insults and/or loss of connections in spinal lesions are also candidates for stem cell therapy. Olfactory ensheathing cells have been successfully used in spinal lesions and it is expected that a combination of growth factors and SC, from different origins, may be used in stroke recovery.

29.5.3 Therapeutic use of Haematopoietic Stem Cells

Bone marrow transplantation and haematopoietic stem (HS) cell-based therapies may become a powerful strategy for the treatment of haematologic disorders (leukaemia, aplastic anaemia, etc.), congenital immunodeficiencies, metabolic and autoimmune disorders, gene therapy, cancer therapy and induction of transplantation tolerance.

Several primary immunodeficiencies have been treated with transplantation of allogeneic haematopoietic stem cells. The European Registry for Stem Cell Transplantation in severe combined immunodeficiencies (SCID) and in other immunodeficiency disorders (non-SCID) reported the improvement of survival and more effective prevention of infections and graft-versus-host disease (Scherer & Schoenfeld 1998). New strategies are being devised for the effective treatment of adenosine deaminase deficiency or HIV infection.

HS cells are also an excellent vehicle for gene therapy of inherited and acquired disorders (van Damme *et al.* 2002). Self-renewal, proliferation and multilineage differentiation of HS cells, together with efficient expression of the transgene transferred, may provide an efficient tool for gene therapy. Marrow-derived cells are good vehicles for gene therapy of the pulmonary epithelium, repair of cartilage, inflammed glomeruli or beta-thalassaemia.

Animal experimentation has demonstrated that allogeneic bone marrow transplantation can be used to treat autoimmune diseases such as rheumatoid arthritis, type 1 diabetes mellitus, lupus erythematosus, immune thrombocytic purpura, chronic glomerulonephritis, etc. In contrast, transplantation of purified HS cells from autoimmune-prone mice leads to the appearance of autoimmune diseases in the recipients.

Some preliminary data from clinical trials show that allogeneic bone marrow transplantation may resolve human autoimmune diseases such as Crohn's disease, multiple sclerosis and rheumatoid arthritis. Moreover, the adoptive transfer of autoimmune disease (Grave's disease, myasthenia gravis) after bone marrow transplantation has been reported (Antoine *et al.* 2003). Based on these results, the term 'stem cell diseases' has been coined for autoimmune diseases.

Since HS cells play a major role in the generation of immunity against tumour cells, they may be used to generate vaccines for cancer therapy. Dendritic cells are the best-equipped antigen-presenting cells to overcome tolerance to self antigens presented by cancer cells. Moreover, since bone marrow contributes to different cell types in the tumour stroma, including vascular endothelial cells, HS cells may be engineered to deliver a 'suicide gene', achieving substantial inhibition of angiogenesis and slower tumour growth without systemic toxicity.

Finally, HS cells may be used to promote chimaerism and enhance tolerance in organ transplantation.

29.5.4 Myocardial and Skeletal Muscle Regeneration

Adult human myocardium does not regenerate because muscle cells cannot re-enter the cell cycle. Foetal and neonatal cardiomyocytes, skeletal myoblasts (satellite cells), bone marrow stem cells and embryonic stem-cell-derived cardiomyocytes have been transplanted in experimental settings to replace lost myocardial tissue (Anversa & Nadal-Ginard 2002). Clinical trials with autologous skeletal myoblasts or bone marrow stem cell transplantation have been initiated. Bone marrow stem cells differentiate into cardiomyocytes when transplanted into the heart, as well as into vascular cells and scar tissue, participating in the remodelling process of the myocardium. Early human trials have so far proved the process to be safe, with improvement of myocardial flow and ventricular function. In contrast, satellite cells integrate into the tissue but do not establish gap junctions with myocardial fibres, which results in the appearance of arhythmias.

Mesangioblasts, a form of vessel-associated stem cells, correct the dystrophic phenotype (morphologically and functionally) in adult immunocompetent alfa-sarcoglycan null mice, a model for muscular dystrophy (Jensen & Drapeau 2002).

29.6 AN 'ASTONISHING' HYPOTHESIS

Animal studies and preliminary observations suggest that a sub-population of HS cells may have the ability to differentiate into diverse cell types such as hepatocytes, myocytes, neuronal cells, etc., especially after tissue damage. In contrast to the dogma that adult stem cells are developmentally committed to a restricted number of cell types, this developmental plasticity may be of clinical use both in the healing of tissues other than blood and in inducing tolerance in solid organ transplantation (Terai *et al.* 2002).

Mobilization of stem cells from the bone marrow and their migration to various tissues may be a normal process of tissue regeneration and repair (Imai & Ito 2002) (Figure 29.4). Evidence of bone marrow contribution derives from experimental studies and from the observation of chimaerism in the transplanted organ (Quaini *et al.* 2002).

In this sense, identification of the mechanisms and molecules involved in the mobilization and tissue re-colonization may be crucial for the potential treatment of various degenerative diseases (Figure 29.2). Given that bone marrow stem cells may repopulate other tissues, it is tempting to speculate that mobilization of progenitors from the bone marrow will reactivate damaged tissue and organ regeneration. Nevertheless, there are several questions that have to be addressed: (i) which are the competent bone marrow progenitors?; (ii) in peripheral blood, which are the bone marrow-derived progenitor cells?; (iii) are they the same for all tissues?; (iv) which are the local signals that promote or impede tissue regeneration?; and (v) do bone marrow stem cells colonize endothelial tissues and promote vasculogenesis?. This last point is particularly important. Provided that bone marrow stem cells (CD34+ cells?) contribute to endothelialization of the transplanted (solid) organ, the strategy may include bone marrow mobilization and apheresis of autologous bone marrow stem cells, together with organ transplantation. Experimental work in Rhesus monkeys (Contreras *et al.* 1999), using allogeneic skin grafts and pretreatment of the graft recipient with the cytokine granuloctye-colony-stimulating factor (G-CSF), which mobilizes bone marrow progenitor cells, has shown facilitated graft acceptance in the absence of immunosuppressive therapy. Levels of peripheral CD34+ cells were enhanced in the recipient, replacing the endothelium of the graft, aborting the allogeneic response and promoting graft acceptance.

29.7 CONCLUDING REMARKS

Research in developmental biology and haematology has led to the discovery of stem cells. Recently, several techniques have been developed for the *in vitro* culture and differentiation of stem cells. This knowledge provides a great opportunity for the study and understanding of many aspects of human biology that will improve therapies and help to provide future cures for diseases.

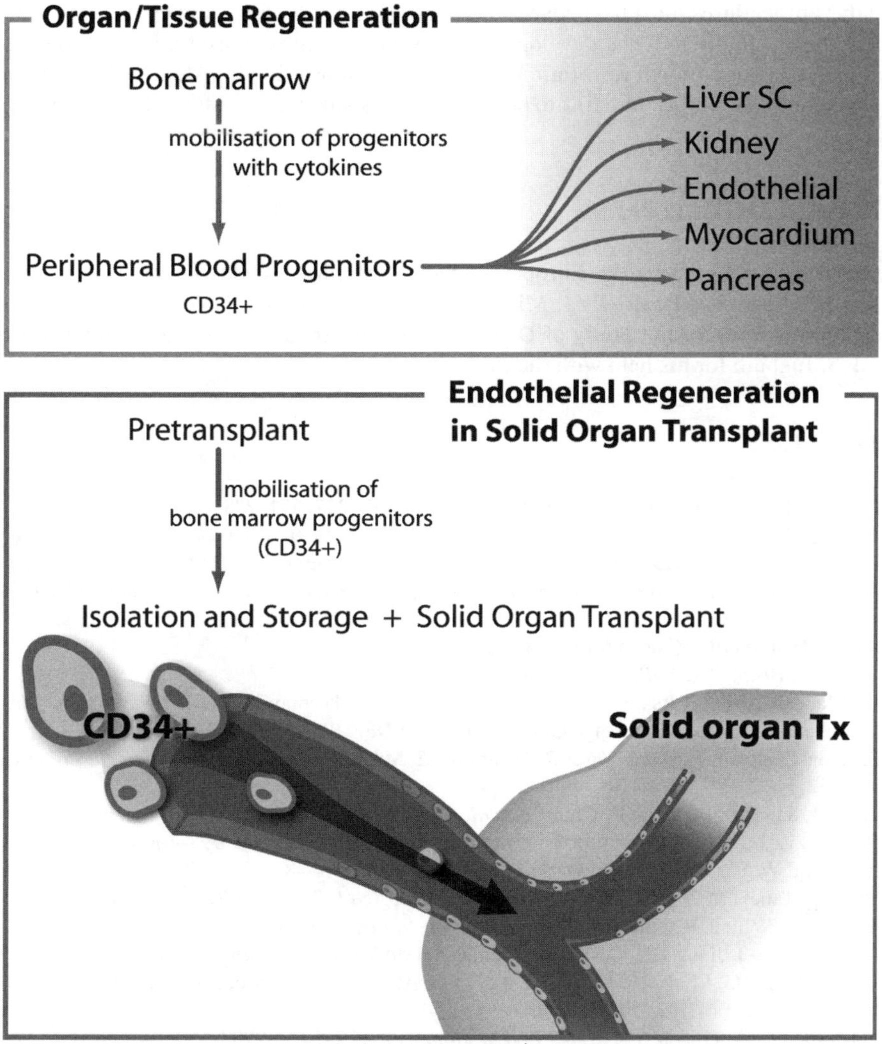

Figure 29.4 Importance of bone marrow stem cells for tissue regeneration after transplant. The illustration suggests that mobilization of stem cells from the bone marrow and their migration to various tissues may be a normal process of tissue and endothelial regeneration. SC: stem cell.

Studies of both embryonic and adult stem cells, of animal and human origin, will be required to improve the scientific and therapeutic potential of regenerative medicine. Thus, research on both adult and embryonic stem cells should be pursued, since embryonic stem cells will be important to improve the research in adult stem cells, and *vice versa*. Although stem cell research is at the cutting edge of biological science, it is still in its infancy. Before the enormous potential and capabilities of stem cells can be developed, several goals need to be achieved, namely: (i) markers must be identified that characterize specific types of stem cell; (ii) similarly, markers must be identified that distinguish the stages of a stem cell committed to differentiate into a particular cell lineage; (iii) profiles of gene expression in stem cells and their progeny must be described;

(iv) standard procedures must be devised for isolating and culturing stem cells; (v) techniques to propagate them reliably must be developed; and (vi) a consensus must be reached on the physiological criteria that confirm restoration of tissue function following stem cell transplantation. Finally although much work remains to be done, there is sufficient evidence to warrant continued efforts on stem cell research.

ACKNOWLEDGEMENTS

Our research is supported by the European Union, Juvenile Diabetes Foundation International, Cardion-AG, Fundació Marató TV3, Ministry of Science and Technology, Ministry of Health, European Foundation for the Study of Diabetes Jonta de Andalucia and Generalitat Valenciana. We thank S. Ingham for his help with the illustrations.

REFERENCES

Amit M, Carpenter MK, Inokuma MS *et al.* (2000) *Dev. Biol.*; 227: 271–278.

Antoine C, Muller S, Cant A, Cavazzana-Calvo M, Veys P, Vossen J, Fasth A, Heilmann C, Wulffraat N, Seger R, Blanche S, Friedrich W, Abinun M, Davies G, Bredius R, Schulz A, Landais P, Fischer A; *European Group for Blood and Marrow Transplantation; European Society for Immunodeficiency.* (2003) Lancet; 361: 541–542.

Anversa P, Nadal-Ginard B (2002) *Nature*; 415: 240–243.

Arenas E (2002) *Brain Res. Bull.*; 57: 795–808.

Assady S, Maor G, Amit M, Itskovitz-Eldor J, Skorecki KL, Tzukerman M (2001) *Diabetes*; 50: 1691–1697.

Bain G, Kitchens D, Yao M, Huettner J, Gottlieb D (1995) *Dev. Biol.*; 168: 342–357.

Berna G, León-Quinto T, Enseñat-Waser R, Montanya E, Martín F, Soria B (2001) *Biomed. Pharmacother.*; 55: 206–212.

Bjorklund LM, Sanchez-Pernaute R, Chung S *et al.* (2002) *Proc. Natl. Acad. Sci. USA*; 99: 1755–1757.

Blyszczuk P, Czyz J, Kania G *et al.* (2003) *Proc. Natl. Acad. Sci. USA*; 100: 998–1003.

Blyszczuk P, and Wobus AM. (2006). *Methods Mol Biol;* 330: 373–385

Bonner-Weir S, Taneja M, Weir G *et al.* (2000) *Proc. Natl. Acad. Sci. USA*; 97: 7999–8004.

Campbell KH, McWhir J, Ritchie WA, Wilmut I (1996) *Nature*; 380: 64–66.

Contreras J, Navarro J, Jiang XL, Curien D, Thomas J (1999) *Transplantation*; 68: S9.

DAmour KA, Bang AG, Eliazer S, Kelly OG, Agulnick AG, Smart NG, Moorman MA, Kroon E, Carpenter MK, Baetge EE. (2006). *Nat. Biotechnol;* 24: 1392–1401.

Dean W, Bowden L, Aitchison A *et al.* (1998) *Development*; 125: 2273–2282.

Drukker M, Katz G, Urbach A *et al.* (2002) *Proc. Natl. Acad. Sci. USA*; 99: 9864–9869.

Eigess R, Benvenisty N (2002) *FEBS Lett.*; 529: 135–141.

Evans MJ, Kaufman MH (1981) *Nature*; 292: 154.

Ferber S, Halkin A, Cohen H *et al.* (2000) *Nat. Med.*; 6: 568–572.

Hao E, Tyrberg B, Itkin-Ansari-P, Lakey JRT, Geron I., Monosov EZ, Barcova M, Merkola M. and Levine F. (2006). *Nature Medicine;* 12: 310–310.

Hess D, Li L, Martin M *et al.* (2003) *Nature Biotechnol.*; 21: 763–770.

Hogan B, Fellows M, Avner P, Jacob F (1977) *Nature*; 270: 515–518.

Holden C (2002) *Science*; 297: 500–502.

Hori Y, Gu X, Xie X, Kim SK. (2005). *PlosMed;* 2: e103.

Hori Y, Rulifson I, Tsai B, Heit J, Cahoy J, Kim S (2002) *Proc. Natl. Acad. Sci USA*; 99: 16105–16110.

Ianus A, Holz GG, Theise ND, Hussain MA (2003) *Clin. Invest.*; 111(6): 843–850.

Imai E, Ito F (2002) *Pediatr. Nephrol.*; 17: 790–794.

Isacson O (2003) *Lancet Neurol.*; 2: 417–424.

Jiang Y, Jahagirdar BN, Reinhardt RL *et al.* (2002) *Nature*; 418: 41–49.

Jensen GS, Drapeau C (2002) *Med. Hypotheses*; 59: 422–428.

Kahn BW, Ephrussi B (1970) *J. Natl. Cancer Inst.*; 44: 1015–1029.

Karimi-Abdolrezaee S, Eftekharpour E, Wang J, Morshead CM, Fehlings MG (2006). *J Neurosci*; 26: 3777–3789.
Kerr DA, Llado J, Shamblott MJ *et al.* (2003) *J. Neurosci.*; 23: 5131–5140.
Klug MG, Soonpaa MH, Koh GY, Field LJ (1996) *J. Clin. Invest.*; 98: 216–224.
Kojima H, Nakamura T, Fujita Y *et al.* (2002) *Diabetes*; 51: 1398–1408.
Labosky PA, Barlow DP, Hogan BL (1994) *Development*; 120: 3197–3204.
Lagasse E, Connors H, Al-Dhalimy M *et al.* (2000) *Nat. Med.*; 6: 1229–1234.
Lindvall O (2003) *Pharmacol. Res.*; 47: 279–287.
Lindvall O, McKay R (2003) *Proc. Natl. Acad. Sci. USA*; 100: 7430–7431.
Lumelsky N. Blondel O, Laeng P, Velasco I, Ravin R, McKay R (2001) *Science*; 292: 1389–1394.
Martin GR (1981) *Proc. Natl. Acad. Sci. USA*; 78: 7634–7638.
Matsui Y, Zsebo K, Hogan BLM (1992) *Cell*; 70: 841–847.
McDonald JW, Liu XZ, Qu Y *et al.* (1999) *Nat. Med.*; 5: 1410–1412.
Moritoh Y, Yamato E, Yasui Y, Miyazaki S, Miyazaki J (2003) *Diabetes*; 52: 1163–1168.
Niwa H, Miyazaki J, Smith AG (2000) *Nat. Genet.*; 24: 372–376.
Onyango P, Jiang S, Uejima H *et al.* (2002) *Proc. Natl. Acad. Sci. USA*; 99: 10599–10604.
Ostenfeld T, Svendsen CN (2003) *Adv. Tech. Stand. Neurosurg.*; 28: 3–89.
Peschanski M, Dunnett SB (2002) *Lancet Neurol.*; 1(1): 81.
Pera MF, Reubinoff B, Trounson A (2000) *J. Cell Sci.*; 113: 5–10.
Pluchino S, Quattrini A, Brambilla E *et al.* (2003) *Nature*; 422: 688–694.
Prusa AR, Hengstschläger M (2002) *Med. Sci. Monit.*; 8(11): 253–257.
Prusa AR, Marton E, Rosner M, Barnaschek G, Hengstschläger M (2003) *Hum. Reprod.*; 18: 1489–1493.
Quaini F Urbanek K, Beltrami AP *et al.* (2002) *N. Engl. J. Med.*; 346: 5–15.
Ramiya VK, Maraist M, Arfors KE, Schatz DA, Peck AB, Cornelius JG (2000) *Nat. Med.*; 6(3): 278–282.
Rathjen J, Rathjen PD (2001) *Curr. Opin. Genet. Dev.*; 11: 587–594.
Reik W, Walter J (2001) *Nat. Rev. Genet.*; 2: 21–32.
Resnick JL, Bixler LS, Cheng L, Donovan PJ (1992) *Nature*; 359: 550–551.
Reubinoff BE, Pera MF, Fong CY, Trouson A, Bongso A (2000) *Nat. Biotechnol.*; 18: 399–404.
Rohwedel J, Horak V, Hebrok M, Füchtbauer EM, Wobus A. (1995) *Exp. Cell Res.*; 239: 214–225.
Rohwedel J, Guan K, Zuschratter W *et al.* (1998) *Dev. Biol.*; 201: 167–184.
Ruhnke M, Ungefroren H, Nussler A, Klapper W, Martin F, Ulrichs K, Hutchinson JA, Soria B, Sehle G, Parwaresch RM, Heeckt P, Kremer B and Fändrich F (2005). Gastroenterology; 128: 1774–1786.
Schuldiner M, Yanuka O, Iskovitz-Eldor J, Melton DA, Benvenisty N (2000) *Proc. Natl. Acad. Sci. USA*; 97: 11307–11312.
Schuldiner M, Eiges R, Eden A *et al.* (2001) *Brain Res.*; 913: 201–205.
Shamblott MJ, Axelman J, Wang S *et al.* (1998) *Proc. Natl. Acad. Sci. USA*; 95: 13726–13731.
Shamblott MJ, Axelman J, Littlefield JW *et al.* (2001) *Proc. Natl. Acad. Sci. USA*; 98: 113–118.
Sherer Y and Shoenfeld Y. (1998). Bone Marrow Transplant; 22: 873–881.
Silani V, Leigh N (2003) *Amyotroph. Lateral Scler. Other Motor Neuron Disord.*; 4(1): 8–10.
Slager HG, Van Inzen W, Freund E, Van den Eijnden-Van Raaij AJ, Mummery CL (1993) *Dev. Genet.*; 14: 212–224.
Smith A (2001) In *Stem Cell Biology.* Eds Marshak DR, Gardner RL, Gottlieb D. Cold Spring Harbor Laboratory Press, New York; 205–230.
Smith AG, Heath JK, Donaldson DD *et al.* (1988) *Nature*; 336: 688–690.
Smith LE (2004). J. Clin Invest; 114: 755–757.
Soria B (2001) *Differentiation*; 68(4–5): 205–219.
Soria B, Roche E, Berná G, León-Quinto T, Reig JA, Martín F (2000) *Diabetes*; 49: 157–162.
Soria B, Real FX, Martín F (2001) *Diabetologia*; 44: 407–415.
Stangel M and Hartung HP. (2002). Progress in Neurobiology; 68: 361–376.
Su L, Xu J, Ji Bx, Wang SG, Lu CY, Dong HQ, Yu YY, Lu DP (2006). *Int J Hematol*; 84: 272–275.
Tanaka S, Kunath T, Hadjantonaskis AK, Nagy A, Rossant J (1998) *Science*; 282: 2072–2075.
Terai S, Yamamoto N, Omori K, Sakaida I, Okita K (2002) *J. Gastroenterol.*; 37 (Suppl.14): 162–163.
Thomson JA, Itskovitz-Eldor J, Shapiro SS *et al.* (1998) *Science*; 282: 1145–1147.
Tosh D, Slack JM (2002) *Nat. Rev. Mol. Cell. Biol.*; 3: 187–194.

Tropepe V, Coles BL, Chiasson BJ *et al.* (2000) *Science*; 287: 2032–2036.
Vaca P., Martin F, Vegara-Meseguer JM, Rovira JM, Berna G and Soria B (2006). Stem Cells; 24: 258–265.
Van Damme A, Vanden-Driessche T, Collen D, Chuah MK (2002) *Curr. Gene Ther.*; 2: 195–209.
Vazey EM, Chen K, Hughes SM, Connor B (2006). *Exp Neurol*; 199: 384–396.
Wagers AJ, Sherwood RI, Christensen JL, Weissman IL (2002) *Science*; 297: 2256–2259.
Wan Q, Matsumoto Y, Shindo T, Miyake K, Shindo A, Kawanishi M, Kawai N, Tamiya T, Nagao S (2006). *J Med Invest*; 53: 61–69.
Wiles MV, Keller G (1991) *Development*; 111: 259–267.
Willing AE, Lixian J, Milliken M *et al.* (2003) *J. Neurosci. Res.*; 73: 296–307.
Ying QL, Nichols J, Evans EP, Smith AG (2002) *Nature*; 416: 545–548.
Zalzman M, Gupta S, Giri RK *et al.* (2003). *Proc. Natl. Acad. Sci. USA*; 100(12): 7253–7258.

Useful Web Sites

NIH Task Force/General Information/Grants/ DNA libraries	http://stemcells.nih.gov/fedPolicy/ NIHTaskForce.asp
American Association for the Advancement of Science	http://www.aaas.org/spp/sfrl/projects/stem/ main.htm
SC Research News	http://www.stemcellresearchnews.com/
NIH Information	http://stemcells.nih.gov/registry/
Time Magazine Debate	http://www.time.com/time/2001/stemcells/
SC Research Foundation	http://www.stemcellresearchfoundation.org/
SC Network in Canada	http://www.stemcellnetwork.ca/
Institute for Stem Cell Research in Edinburgh	http://www.iscr.ed.ac.uk/
Medline Database web page	http://www.nlm.nih.gov/medlineplus/ stemcellsstemcelltransplantation.html
Washington Post SC web page	http://www.washingtonpost.com/wp-dyn/ politics/specials/stemcells/
International Society for Stem Cell Research	http://www.isscr.org/
House of Lords Report on SC	http://www.parliament.the-stationery-office. co.uk/pa/ld200102/ldselect/ldstem/83/8301. htm
National Institute of General Medical Sciences	http://www.nigms.nih.gov/funding/stemcells. html
Nuffield Council in Bioethics	http://www.nuffieldbioethics.org/home/
ReNeuron	http://www.reneuron.com/reneuron/reports/ stemcell/stemcell.pdf
The position of the Catholic Church	http://www.americancatholic.org/News/ StemCell/default.asp
UK SC Bank	http://www.nibsc.ac.uk/divisions/cbi/stemcell. html
Karolinska Institute SC Bank	http://info.ki.se/news/items/stemcell_lines_ en.html
University of California in San Francisco Training program	http://escells.ucsf.edu/Training/Trng.asp
American Society of Cell Biology	http://www.ascb.org

30 Cells as Vaccines

AG Dalgleish and MA Whelan

30.1 INTRODUCTION

The use of human cells as vaccines for cancer therapy arose from the study of tumour immunogenicity in animal models in the early 1960s (Burnet 1967, 1971). Tumour cell lines were established that were used to explore the transplantability of the tumour in different hosts. This led directly to the discovery that resected immunogenic tumours rendered the host resistant to re-challenge. Cell lines have now been used as vaccines in the treatment of a number of human cancers. More recently they have been used as vehicles for transfected genes, such as cytokines and co-stimulatory molecules that enhance the immunogenicity of the cell lines, and this forms the backbone of early forays into gene therapy for cancer treatment.

Cell-based cancer vaccines have also been developed in another form, namely as treatments based on dendritic cells (DC). DC are the essential professional antigen-presenting cells of the immune system and are able both to induce immunological tolerance and break it (Shortman 2000). Expanding dendritic cells *in vitro* with the use of GM-CSF and either IL-4 or TNFα, allows a culture to be prepared that can then be pulsed with tumour antigens in a number of forms, including highly specific peptides, whole cell lines or tumour lysates. Optimization of this process can induce strong anti-tumour responses, and the clinical development of this technique and selection of relevant antigens in specific cancers has made this probably the single most popular type of cancer vaccination at the time of writing.

Human cells may also be used as vaccines for infectious diseases, in which the virus buds through the cell wall incorporating many cell components into its own envelope. Early studies in the SIV model of AIDS showed that this approach could provide an extremely effective prophylactic vaccine (Chan *et al.* 1992). However, later experiments suggested that the vaccine was only effective for a specific virus strain grown through the same cell as the vaccine (Almond *et al.* 1997). A more recent report, however, suggests that this approach may yet lead to more effective infectious disease vaccines (Lu *et al.* 2003).

Classical vaccinology uses a derivative of the disease-causing agent as a prophylactic. Recent advances in the field have shown that this logic may even be applied to cancers, in which the tumour itself is inactivated and then used to generate protective immunity. In this chapter the authors explore this concept, ranging from the original concept to some of the exciting new therapies based on transfected cells and dendritic cell-based vaccines. The various options available for immunotherapy strategies are given in Figure 30.1.

30.2 TUMOUR CELL LINES AS VACCINES FOR CANCER

In the first half of the Twentieth century a number of experimental systems showed the rejection of tumours following transplantation within outbred populations or between different species

Medicines from Animal Cell Culture Edited by G. Stacey and J. Davis

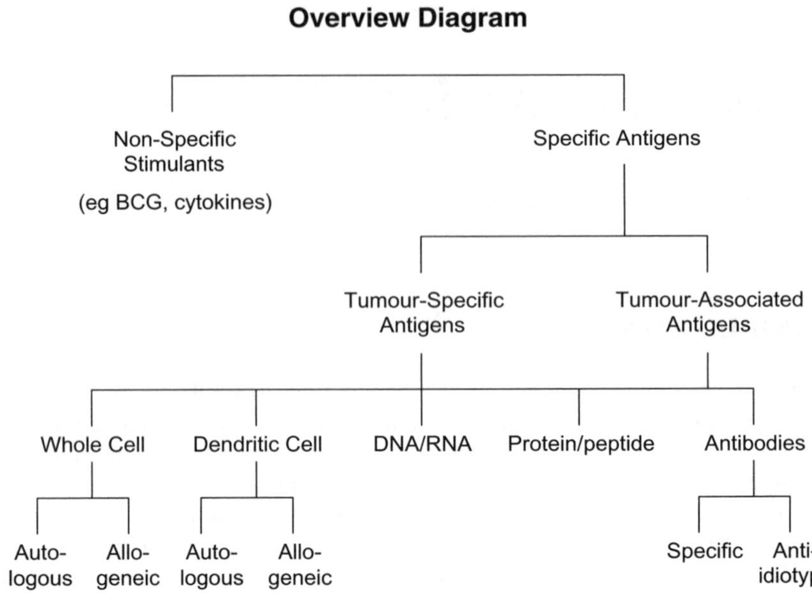

Figure 30.1 An overview of immunotherapy strategies. Many conventional therapies employ non-specific stimulation, e.g. BCG in bladder cancer. Current research tends to focus on non-specific antigen therapies that can be delivered in a variety of modalities.

(Baldwin 1973; Globerson & Feldman 1964). The introduction of inbred laboratory animal strains allowed similar experiments to be carried out using spontaneously arising transplantable tumours in syngeneic recipients. The fact that tumours could be induced to grow in immunocompetent animals suggested that they were non-immunogenic, until it was demonstrated that animals rendered disease-free by surgical excision were then resistant to re-challenge with the same tumour (Hewitt *et al.* 1976; Prehn 1994). These experiments showed that the tumours induced a specific immune response with memory.

In the early 1960s, using a variety of cell lines in animal model systems, it was demonstrated that the immunogenicity of different cell lines was very variable, with virally induced cell lines being the most immunogenic (Hewitt *et al.* 1976). The fact that the immunogenicity was highly tumour-specific led to the conclusion that there had to be specific antigenic determinants, which were initially called tumour-associated transplantation antigens. This gave rise to the concept of tumour antigens, of which hundreds have now been identified, divided up into several specific families (see Table 30.1). The last few years has seen an explosion in the identification of new tumour antigens, leading to the realization that tumours may have specific antigens that can be recognized by the immune system, but which do not induce an effective immune response *in vivo*, meaning that the tumours are often not immunogenic. The major goal of cancer immunotherapy is to induce responses to these 'silent' antigens.

The problem is that, with all these specific antigens, no one really knows which is the best candidate to choose for effective vaccination. There are many single antigen vaccine approaches, although very few have demonstrated a sustained clinical response. This is likely to be due to immunological escape; for example, loss of CTL activity to the tumour antigens MART-1 and MAGE-3 has been shown to be associated with disease progression (Andersen *et al.* 2001), suggesting deletion of these antigens leads to loss of important antigenic targets. In spite of all the elegant techniques identifying specific tumour antigens and targeting them

Table 30.1 Cancer antigens. Cancers are essentially 'self' cells that have by-passed normal homeostatic regulation mechanisms. Consequently, the immune system has difficulty differentiating them from non-malignant cells. Fortunately, there are a number of markers that may be employed specifically to identify, and thus attack, tumour cells.

Nature of antigen	Description
Tumour-specific antigens (TSA)	This is a relatively small group of antigens and can be exemplified by the cancer-testis antigens. These genes are completely silent in normal tissue, but are expressed by cancerous cells, thus making them highly specific markers of disease. Examples include tyrosinase and the MAGE antigens, both from melanoma.
Tumour-associated antigens (TAA)	These are largely differentiation antigens, expressed by normal cells, but massively over-expressed in cancerous tissue. Many targets initially thought to be tumour-specific are indeed widely spread over many tumours such as the gangliosides and mucin antigens.
Mutational antigens	Point mutations are common in many cancers, and indeed these are often in a similar location. An example is the common mutation of the P53 oncogene.
Viral antigens	Some viruses have been shown to be oncogeneic and examples are the E6 and E7 oncogenes from human papilloma virus.

with elegant technologies (i.e. DNA vaccines, viral vectors or peptide-pulsed dendritic cells), it is becoming increasingly recognized that more than one tumour antigen will need to be targeted to induce a broad and effective immune response. This is why cell lines are potentially so valuable, as they contain a wide variety of tumour-associated, if not completely tumour-specific, antigens.

The most detailed and thorough development of cell vaccines has taken place in melanoma, due to the relative resistance of the disease to more conventional therapies and to the occurrence of occasional spontaneous remissions and responses to immunologic manipulation (Morton *et al.* 2002). Most melanoma vaccines are based on whole, autologous or allogeneic tumour cells or cell extracts/lysates, with or without an immunostimulant such as *Bacille Calmette Guerrin* (BCG), Freund's complete adjuvant, Newcastle disease virus, *Vaccinia* virus, PPD (from *Mycobacterium tuberculosis*), alum or combinations thereof. Given the number of different antigens available, it is logical to develop autologous cell lines from the patient's own tumour and enhance the immunogenicity with powerful adjuvants. Unfortunately, raising cell lines from tumours is far from simple, requiring the right facilities, considerable manpower, and luck. Melanoma is one of the easiest tumours from which to raise cell lines in culture, having a success rate, in the best hands, of approximately 50% (Stamps *et al.* 1992). Unfortunately, the disease often progresses faster than cell lines can be established in culture, which further limits the strategy. Moreover, early research using autologous irradiated tumour cells met with only limited success (Currie *et al.* 1971; Morton 1972).

Allogeneic cell-based vaccines, despite being non-HLA matched with the patients, have the advantage that they can be selected for high immunogenicity and antigenicity. Production of such vaccines can be carried out well ahead of time, thus removing the critical time constraints on patient vaccination. The lines used are stable and provide a good source of cells for multiple vaccinations. Although melanoma has a wide diversity of physical manifestations in clinical presentation, there are numerous shared antigens present on many melanoma cells, including gangliosides, MAGE, MART, GP100 and tyrosinase, all of which have been shown to be effective targets for immunotherapy against melanoma (reviewed in Romero *et al.* 2002). The main challenge foreseen with allogeneic cell lines is that the antigens must be HLA (human leukocyte antigen) restricted. Morton and his colleagues initially used one, subsequently two, and finally chose a combination of three cell lines that expressed six melanoma-associated antigens, on the cell surface. The HLA background of the cells was taken into consideration so that they would cover the majority of the population with melanoma, using direct presentation of antigens by the tumour to T-cells.

Morton and colleagues went on to show that a reliable supply of these cells could be made available for use in the clinic by irradiating and then cryopreserving them. The vaccine was injected intradermally in the axilla and inguinal regions at intervals of 2 weeks for 4 weeks, then monthly for a year with the first two treatments being administered with BCG as an adjuvant. The survival of these patients correlated significantly with delayed-type hypersensitivity (DTH) and antibody responses to the cell surface antigens (Barth *et al.* 1994; Habal *et al.* 2001). Data from a large Phase II study on patients with both stage III and stage IV disease confirmed the survival benefit of the vaccine. Chung *et al.*(2003), showed that the vaccine was most effective when accompanied by complete resection. Five-year survival of 39 % for vaccine patients was observed, as opposed to similarly treated non-vaccinated patients showing a 19 % 5-year survival. Unfortunately a large randomised study in both resected stage III and IV patients failed to show a significant advantage over the control group containing BCG alone.

Analysis of the immune response in this study showed a highly statistically significant correlation with survival. Of interest, it would appear that high IgM titres to cell surface antigens such as the gangliosides, but not IgG antibodies, correlated with survival (Takahashi *et al.* 1999). Delayed-type hypersensitivity (DTH) responses to vaccine cells were also a very strong indicator of increased survival. The DTH response correlated with strong mixed lymphocyte tumour reactions, whilst complement-dependent cytotoxicity correlated with increased survival (Hsueh *et al.* 1998). One possible interpretation of these data is that cytotoxity, either of cellular or humoral origin, is the main mechanism of tumour eradication in these subjects.

The excellent immunological results seen to date with a combination of three cells, differing in HLA types, raises the question as to whether the allogeneic response itself may be beneficial. Using B16 and K1735 murine models of melanoma, the authors have been able to show that allogeneic murine melanoma cell vaccines are often superior to autologous (Knight *et al.* 1996). We have subsequently looked at a number of tumour models and found that in several cases allogeneic vaccination appears to out-perform autologous (Hrouda *et al.* 2000). This is not the predicted response based on the MHC/HLA restriction dogma. However, not all allogeneic vaccine models induce better protection than autologous, and it is perhaps reasonable to consider that a shared antigen is lacking in these systems. It is therefore likely that the allogeneic character of this cell-line based vaccine is of benefit and is in itself a 'danger signal' (Matzinger 2002) to the immune system and thus can act like an adjuvant.

More recently, cell lines have been fused with antigen-presenting cells to produce hybridomas that could be the perfect cell-based vaccine (Jantscheff *et al.* 2002). Although these studies are not as impressive as the early data from the Morton vaccine trial, either in survival or numbers, they do show similar trends; those recipients who do make a specific immune response tend to live longer than those who do not.

Another tumour type that has been treated with cell lines is colorectal cancer, where Hanna and colleagues developed autologous tumour cells from surgical resections in stage II and III colon cancer (Hoover *et al.* 1993). They were able to obtain enough material for three injections of 10^7 irradiated autologous tumour cells administered with BCG following surgical resection. Encouraging data in the original study led the oncology group at University Hospital, Antwerp, Belgium, to conduct a double blind study on stage II and stage III colorectal patients. In this study they were able to generate four vaccine doses per sample, with the fourth being given 6 months after surgery. In this 254-patient randomized trial, all patients receiving the vaccine remained free of tumour recurrence for a significantly longer period, with a 42 % reduction in risk of recurrence or death and with a trend towards overall improved survival (Vermorken *et al.* 1999). This study confirms the potential benefit of cell-based vaccines in treating colorectal cancer. Its main limitation was the small amount of vaccine available due to the small volume of the autologous tumour, which raises the question of a potential advantage of using allogeneic cell lines in a larger study. A number of groups are actively pursuing this approach.

Also under investigation with cell-line based vaccines is prostate cancer. Although this tumour bears none of the hallmarks of being potentially immunologically responsive (it is not a tumour that increases in size with immunosuppression or is subject to spontaneous remission) the identification of prostate-specific antigens, of which there are now several, makes it an attractive target. Hrouda *et al.* confirmed that an allogeneic approach was practical using the allogeneic cell lines in the Copenhagen and Lobund Wistar models (Hrouda *et al.* 2000). Eaton *et al.* (2002) tested four different vaccines, each utilizing three cell lines in a Phase I/II study on sixty patients and found evidence of a PSA response corresponding with cell-mediated Th-1 associated cytokines in some patients (Eaton *et al.* 2002). This was a group of patients with very advanced disease, some of whom did not make any significant immune response. Our laboratories have selected an optimal cell combination and are currently performing Phase II trials in hormone-resistant patients who do not have bulky or terminal disease. A study of hormone-relapsed, but metastases-free patients showed a 40% reduction in the rate of rise of PSA and an increased progression-free survival time (Michael *et al.* 2005).

Cell-based vaccines make good candidates for a number of diseases, including lung cancer, although increasingly they are being used as vehicles for gene therapy where the transduced genes enhance the immunogenicity of the cell line. Such genes include the cytokines, IL-2, IL-4, IL-12 and GM-CSF, as well as co-stimulatory molecules such as CD80 or CD86 (reviewed in Dalgleish & Browning 1995). Indeed, it may even be beneficial to insert both a cytokine and a co-stimulatory molecule. GM-CSF-secreting cell lines are currently being trialled in lung, prostate and pancreatic cancer.

30.3 DENDRITIC CELLS AS VACCINES

Dendritic cells (DCs) are the most potent antigen-presenting cells known and, consequently, are very rare. Their ability to stimulate T-cells is at least an order of magnitude better than any other known cell type, and they possess a phenotype such that their primary function is antigen presentation (Banchereau & Steinman 1998). They are characterized by having large amounts of both MHC class I and class II molecules, as well as an abundance of molecules that assist in cell adhesion, for example CD80, CD86 and integrins (Shortman 2000). There are few specific markers for DCs, although CD11c and CD83 can be useful (Banchereau & Steinman 1998). DCs also possess the unique ability to 'cross-prime', in which exogenous antigen is processed and presented via the MHC class I pathway, rather than by the more normal class II route (Carbone & Bevan 1990; Heath & Carbone 2001). It is this latter property that makes them attractive for vaccination purposes, since conventional dogma dictates that a CD8 CTL (cytotoxic T-lymphocyte) response would be most beneficial in a tumour environment, and CTL responses are restricted by MHC class I. However, most vaccines tend to be administered such that the antigen lies outside the cell. Without the DC's ability to cross prime, it would be impossible to generate CTL activity.

Given the potency of dendritic cells, most clinical trials tend to use relatively small numbers with only a few vaccinations. This is fortunate given that one of the enduring problems of using DCs clinically is that large amounts of blood are needed to generate them, due to their rarity, and that it is difficult to store them since most freezing protocols lead to their maturation, which is a primary function of these cells. Some work has been carried out on the latter problem and it is now possible to freeze immature DCs, which may be revived and matured in the presence of antigen prior to immunization (Feuerstein *et al.* 2000; John *et al.* 2003). The problem of rarity remains, although some work has been done using Flt3 ligand to stimulate large numbers of DCs *in vivo*

(Rini *et al.* 2002). Unfortunately, initial clinical studies in renal carcinoma have shown that whilst DC numbers did increase, no patient benefit was observed (Rini *et al.* 2002).

As understanding of the specific molecular interactions that occur during an immune response has increased, so has the use of targeted DC cancer vaccination protocols. Initial experiments in this field focused on the ability of DCs to generate specific CTL responses when loaded with individual peptides. Examples of this include human papilloma virus (HPV) (Schoell *et al.* 1999), Muc-1 (Brossart *et al.* 2000) and prostate specific membrane antigen (PSMA) (Horiguchi *et al.* 2002). In all of these cases specific peptides, binding to a particular HLA molecule (usually HLA-A2) were loaded onto syngeneic DCs and then used to stimulate autologous T-cells from healthy volunteers. These data clearly show that this methodology can indeed generate CTLs, hence the obvious extension is to examine if there is any clinical benefit to be gained.

This may still be regarded as a new area of research and, consequently, no major clinical breakthroughs have yet been made. However, there are certainly signs that this approach may prove efficacious as the technology matures. DCs loaded with CEA peptide showed only a minor response (Morse *et al.* 1999), although disease stabilization was seen in prostate cancer using a PSMA peptide (Simmons *et al.* 1999) and in melanoma using tyrosinase or gp100 peptides (Lau *et al.* 2001). Interestingly, although many workers have demonstrated specific cellular immunity, very few see clinical regression of tumours (Lodge *et al.* 2000; Mackensen *et al.* 2000; Sadanaga *et al.* 2001). It has recently been suggested that this might be a simplistic point of view, since although tumour regression is rare and CTLs are common, patient survival is undoubtedly enhanced (Andersen *et al.* 2001). Despite this observation, it is almost certain that current vaccination strategies can be enhanced. Yang has suggested that many CTL clones are ineffective due to their low affinity for MHC and it is only by increasing this that tumour regression will be seen (Yang *et al.* 2002). Similarly, it has been postulated that peptides could be delivered in more effective ways to DCs, for example using chaperone proteins (Noessner *et al.* 2002).

Whilst the elegance of these systems is undoubtedly attractive, they are fundamentally limited by their reliance on using HLA-matched patients. Peptides used to mature DCs will only bind to specific motifs on individual MHC molecules; if the patient does not carry these alleles then the therapy will fail. An alternative approach is to allow the DC itself to make the necessary selection of peptides. This is particularly attractive given that DCs possess an armoury of antigen processing machinery and excel at producing MHC binding peptides. Furthermore, the source of antigen can be standardized and used in all patients. The most obvious pool of antigens can be found in tumour cells themselves. If DCs are fed lysates of tumour cells, then peptide processing and selection can take place leading to the presentation of antigen on all HLA types. Initial clinical data from DCs loaded with autologous tumour lysates are encouraging (Geiger *et al.* 2001; Marten *et al.* 2002), as is the finding that allogeneic tumour lysates can also be employed (Holtl *et al.* 2002).

Another possible route for enhancing DC vaccination is to transduce them with either cytokines or co-stimulatory molecules. This is undoubtedly a useful approach, but it is complicated by the fact that DCs are a dynamic population and, therefore, stable transfections are not usually possible. Few DC cell lines exist, and those that do are not always capable of maturing in response to antigen. Hence, another approach for gene transduction is necessary. As vectors for carrying exogenous genes, viruses can rarely be bettered by artificial constructs. They are particularly adept at infecting target cells and at avoiding unwanted immune responses. Consequently, it has become common to infect DC populations with viruses carrying exogenous genes. Again, these studies are at an early stage and most work has only demonstrated proof of principle. A variety of virus-gene constructs have been employed including vaccinia carrying *LacZ* (Di Nicola *et al.* 1998), canarypox carrying MAGE-A1 (Chaux *et al.* 1999) and avipox loaded with CD80 (Tsang *et al.* 2001). In all cases, there is evidence of successful transduction of DCs. Dendreon have reported significant clinical benefit from a DC based vaccine in prostate cancer and are applying for FDA approval. Many other vaccine studies using dendritic cells are in progress (Small *et al.* 2006).

30.4 CELL-BASED VACCINES FOR INFECTIOUS DISEASES

The previous section has shown that as knowledge of DC growth conditions and methodologies improves there is enormous potential for employing them in many different roles. Another major use for them may be pulsing them with viral antigens and using this to immunize against the whole virus. Such an approach has already been claimed to be effective in an SIV model (Lu *et al.* 2003).

The intriguing protection seen using allogeneic cell lines in SIV monkey models clearly shows protection correlating with antibodies to HLA (Almond *et al.* 1997; Chan & Morton 1998). Using cell lines as hosts for the growth of infectious agents that are difficult to immunize against, and then, after inactivation, using the infected cell lines as vaccines, may be a useful new mode for the presentation of epitopes to the immune system.

ACKNOWLEDGEMENT

Angus Dalgleish's work is supported by the Cancer Vaccine Institute.

REFERENCES

Almond N, Corcoran T, Hull R *et al.* (1997) *J. Med. Primatol.*; 26: 34–43.
Andersen MH, Keikavoussi P, Brocker EB *et al.* (2001) *Int. J. Cancer*; 94: 820–824.
Baldwin RW (1973) *Adv. Cancer Res.*; 18: 1–75.
Banchereau J, Steinman RM (1998) *Nature*; 392: 245–252.
Barth A, Hoon DS, Foshag LJ *et al.* (1994) *Cancer Res.*; 54: 3342–3345.
Brossart P, Wirths S, Stuhler G (2000) *Blood*; 96: 3102–3108.
Burnet FM (1967) *Lancet*; 1: 1171–1174.
Burnet FM (1971) *Transplant. Rev.*; 7: 3–25.
Carbone FR, Bevan MJ (1990) *J. Exp. Med.*; 171: 377–387.
Chan AD, Morton DL (1998) *Semin. Oncol.*; 25: 611–622.
Chan WL, Rodgers A, Hancock RD *et al.* (1992) *J. Exp. Med.*; 176: 1203–1207.
Chaux P, Luiten R, Demotte N *et al.* (1999) *J. Immunol.*; 163: 2928–2936.
Chung MH, Gupta RK, Hsueh E *et al.* (2003) *J. Clin. Oncol.*; 21: 313–319.
Currie GA, Lejeune F, Fairley GH (1971) *Br. Med. J.*; 2: 305–310.
Dalgleish A, Browning M (1995) *Tumour Immunology–Immunotherapy and Cancer Vaccines.* Cambridge University Press Cambridge, UK.
Di Nicola M, Siena S, Bregni M *et al.* (1998) *Cancer Gene Ther.*; 5: 350–356.
Eaton JD, Perry MJ, Nicholson S *et al.* (2002) *BJU. Int.*; 89: 19–26.
Feuerstein B, Berger TG, Maczek C *et al.* (2000) *J. Immunol. Methods*; 245: 15–29.
Geiger JD, Hutchinson RJ, Hohenkirk LF *et al.* (2001) *Cancer Res.*; 61: 8513–8519.
Globerson A, Feldman M (1964) *J. Natl. Cancer Inst.*; 32: 1229–1243.
Habal N, Gupta RK, Bilchik AJ *et al.* (2001) *Ann. Surg. Oncol.*; 8: 389–401.
Heath WR, Carbone FR (2001) *Ann. Rev. Immunol.*; 19: 47–64.
Hewitt HB, Blake ER, Walder AS (1976) *Br. J. Cancer*; 33: 241–259.
Holtl L, Zelle-Rieser C, Gander H *et al.* (2002) *Clin. Cancer Res.*; 8: 3369–3376.
Hoover HC Jr, Brandhorst JS, Peters LC *et al.* (1993) *J. Clin. Oncol.*; 11: 390–399.
Horiguchi Y, Nukaya I, Okazawa K *et al.* (2002) *Clin. Cancer Res.*; 8: 3885–3892.
Hrouda D, Todryk SM, Perry MJ *et al.* (2000) *BJU Int.*; 86: 742–748.

Hsueh EC, Famatiga E, Gupta RK, Qi K, Morton DL (1998) *Ann. Surg. Oncol.*; 5: 595–602.
Jantscheff P, Spagnoli G, Zajac P, Rochlitz CF (2002) *Cancer Immunol. Immunother.*; 51: 367–375.
John J, Hutchinson J, Dalgleish A, Pandha H (2003) *J. Immunol. Methods*; 272: 35–48.
Knight BC, Souberbielle BE, Rizzardi GP, Ball SE, Dalgleish AG (1996) *Melanoma Res.*; 6: 299–306.
Lau R, Wang F, Jeffery G, Marty V *et al.* (2001) *J. Immunother.*; 24: 66–78.
Lodge PA, Jones LA, Bader RA, Murphy GP, Salgaller ML (2000) *Cancer Res.*; 60: 829–833.
Lu W, Wu X, Lu Y, Guo W, Andrieu JM (2003) *Nat. Med.*; 9: 27–32.
Mackensen A, Herbst B, Chen JL *et al.* (2000) *Int. J. Cancer*; 86: 385–392.
Marten A, Flieger D, Renoth S *et al.* (2002) *Cancer Immunol. Immunother.*; 51: 637–644.
Matzinger P (2002) *Science*; 296: 301–305.
Michael A, Ball G, Quatan N *et al.* (2005) *Clin. Cancer Res.*; 11: 4469–78.
Morse MA, Deng Y, Coleman D *et al.* (1999) *Clin.Cancer Res.*; 5: 1331–1338.
Morton DL (1972) *Natl. Cancer Inst. Monogr.*; 35: 375–378.
Morton DL, Hsueh EC, Essner R *et al.* (2002) *Ann. Surg.*; 236: 438–448; discussion: 448–49.
Noessner E, Gastpar R, Milani V *et al.* (2002) *J. Immunol.*; 169: 5424–5432.
Prehn RT (1994) *Cancer Res.*; 54: 908–914.
Rini BI, Paintal A, Vogelzang NJ, Gajewski TF, Stadler WM (2002) *J. Immunother.*; 25: 269–77.
Romero P, Valmori D, Pittet MJ, *et al.* (2002) *Immunol. Rev.*; 188: 81–96.
Sadanaga N, Nagashima H, Mashino K *et al.* (2001) *Clin. Cancer Res.*; 7: 2277–2284.
Schoell WM, Mirhashemi R, Liu B (1999) *Gynecol. Oncol.*; 74: 448–455.
Shortman K (2000) *Immunol. Cell Biol.*; 78: 161–165.
Simmons SJ, Tjoa BA, Rogers M *et al.* (1999) *Prostate*; 39: 291–297.
Small EJ, Schellhammer PF, Higano CS *et al.* (2006) *J. Clin. Oncol.*; 24: 3089–94
Stamps AC, Gusterson BA, O'Hare MJ (1992) *Eur. J. Cancer*; 28A: 1495–1500.
Takahashi T, Johnson TD, Nishinaka Y, Morton DL, Irie RF (1999) *J. Invest. Dermatol.*; 112: 205–209.
Tsang KY, Zhu M, Even J, Gulley J, Arlen P, Schlom J (2001) *Cancer Res.*; 61: 7568–7576.
Vermorken JB, Claessen AM, van Tinteren H *et al.* (1999) *Lancet*; 353: 345–50.
Yang S, Linette GP, Longerich S, Haluska FG (2002) *J. Immunol.*; 169: 531–539.

Useful Web Sites

www.onyvax.com
www.sgul.ac.uk
www.jwci.org
www.nci.nih.gov/clinicaltrials
www.cancerresearchuk.org

Risk Assessment and Regulatory Aspects

31 Risk Assessment of Cell Culture Procedures

G Stacey

31.1 INTRODUCTION

Risk management in its broadest form is an increasingly important part of life that is required at high levels of operation to promote confidence in plans in industry and government. Within individual organizations it is used as an important management tool to enable issues of liability to be addressed and may help to shape strategic development. In the field of biological medicines, risk assessment and risk management have a particularly important role to play in assuring the safety and quality of products for clinical use. At the level of individual research scientists, production managers, etc., risk management is largely a scientific and technical activity to ensure operational safety for any personnel working in, or affected by, the laboratory environment and its products and waste materials. However, at this level it can also be used to optimize laboratory operation by reducing down-time and enhancing outputs and economies. This chapter will focus on those aspects of risk assessment most relevant to the cell and tissue culture laboratory and the patients receiving cell culture products.

31.2 THE PROCESS OF RISK MANAGEMENT AND RISK ASSESSMENT

31.2.1 Nomenclature and Components of Risk Management

The process of risk management revolves around the concepts of 'hazard' and 'risk', and our ability to quantify these factors and prioritize our response to them provide the fundamental elements of risk management. A hazard is a condition that could result in an adverse event. Risk is the likelihood that the adverse event will occur. Typically, risk management can be broken down into a sequence of actions:

- identification of hazards;
- evaluation of risks;
- assessment of controls on these risks;
- consideration of residual risks;
- reviewing risks.

Medicines from Animal Cell Culture Edited by G. Stacey and J. Davis

This basic process can be applied to any operation from the commercial development of a new therapy to the operation of a routine laboratory procedure.

31.2.2 Types of Risk Assessment

Risk assessment can be approached in a number of different ways. A common methodology is to list all hazards in a 'risk table' and then score each risk according to the severity of the consequences of the associated adverse event and the likelihood that this event will occur. Typically these values are used to calculate a 'risk score' by allocating each risk a score of relative severity, and a separate score for likelihood and using the product of these two values to derive the risk score. This risk score can be used to focus attention on the highest level risks. Controls on each risk can then be assigned a relative value to enable calculation of residual risk. Those responsible for risk management must assess whether the scores are realistic, and whether the residual risk is acceptable, otherwise further remedial action will be required to bring the risk within acceptable limits. It is also important to remember that risks can change, either in a gradual and imperceptible way (e.g. change in cohort of patients donating tissue, alterations to protocols) or by sudden and dramatic shifts (e.g. application of new technology, discovery of a new pathogen). Both possibilities indicate the need for regular and careful re-evaluation of risk, which may also identify completely new risks to be addressed.

A number of different methods of assessing risk can be applied in cell and tissue culture including HAZOP (IEC 2001), 'fish bone' analysis and fault tree analysis (Sheeley 1998) and it is important to ensure that whatever method is selected is the most appropriate for the intended purpose. Hazard Analysis Critical Control Point (HACCP) is another form of risk analysis that has been used in a wide range of different fields (e.g. World Health Organization 2003) including clean-room cell culture processes (Reinmuller & Ljungqvist 1999). For medicines the key elements of risk management, including the approaches indicated above, have now been captured in a guidance of the International Conference on Harmonization (2005). All of these techniques rely on accurate identification of hazards and assessment of risk to the cultures, laboratory operators and recipients of products, which are the topics of the following sections.

31.3 GENERAL BIOLOGICAL HAZARDS

Clearly there is a very broad range of chemicals that could cause cytotoxic effects in biological systems, including high levels of some cell culture reagents(e.g. antibiotics, HEPES buffering agent). Chemical toxicants may cause acute toxic effects, sensitizing reactions in individuals/animals (e.g. formaldehyde) or more long-term or cumulative effects as seen with carcinogens. Biological hazards include the toxins produced by microorganisms and plants, allergens (animal, plant and microbial) and infectious agents. Numerous examples of these chemical and biological agents are either known to be used in cell culture processes for specific technical purposes or may be introduced inadvertently by contamination of reagents or cells. Where a hazardous material is well characterized it can be readily addressed in terms of risk. In the case of infectious agents the appropriate action can be taken according to the risk category for the agent involved and national guidance on containment (Table 31.1). For chemical agents similar guidance is available for risk assessment, containment and accident procedures, including general risk assessment under national regulations (COSHH 2002; World Health Organization 1993) or specific guidance on individual agents or chemical toxicants (e.g. ACDP 1995; CHIP 2002).

Inevitably, some risks are uncertain or poorly characterized. The former group includes the risk that certain materials of biological origin may be contaminated with microorganisms, and the latter group include the risks from some microorganisms and putative carcinogens. The risk posed

Table 31.1 Risk groups for microorganisms (WHO 1993).

Risk group (category)	Description	Examples from cell culture
1	No or very low individual risk, low community risk – a microorganism that is unlikely to cause human disease.	Non-pathogenic environmental bacteria and mycoplasma, BVDV in bovine serum
2	Moderate individual risk, low community risk – a pathogen that can cause human or animal disease, but is unlikely to be a serious hazard to laboratory workers, the community or livestock.	Epstein Barr virus in B-lymphoblastoid cells such as B95-8
3	High individual risk, low community risk – pathogen that usually causes serious human or animal disease, but does not ordinarily spread from one infected individual to another. Effective treatment and preventative measures are available.	Hepatitis B in primary human hepatocytes
4	High individual and community risk – pathogen that usually causes serious human or animal disease and which may be readily transmitted from one individual to another, directly or indirectly. No available treatment.	Examples usually include Ebola virus but may also include novel pathogenic strains of avian influenza isolated *in vitro.*

by such inadvertent or adventitious agents should be addressed on the basis of the generic handled types of material to be in the cell and tissue culture laboratory. These may include:

- manufactured products of biological origin, including those of environmental, plant, animal or human origin;
- cell cultures supplied by a research collaborator or culture collection;
- clinical or veterinary samples whereby different agents are likely to be present in urine, faeces, tissue samples, etc.;
- environmental samples;
- biopsy and autopsy samples;
- forensic and archeological samples.

Materials of human or animal origin will be associated with a risk of exposure to certain pathogens and this risk can be assessed by considering the following features of the causative agent: route of transmission, severity of disease caused, person-to-person communicability, environmental survival, susceptibility to inactivation by chemical and physical treatments (ACDP 2005).

Cell culture laboratories are often multi-user environments accommodating multiple procedures/agents and diverse risks. In the management of risk under such circumstances, communication, coordination and cooperation are vital and may be a requirement under national law [e.g. Regulation 11 of the Management of Health and Safety at Work Regulations (MHSWR 2000)]. Also of particular importance in such environments is the need for a high degree of staff awareness of hazards, and adequate training in their containment and handling (see Coecke *et al.* 2005, and below).

31.4 EVALUATION OF HAZARDS FROM CELLS, AND RISK REDUCTION

31.4.1 Microbiological Hazards

Since cell cultures are derived directly from animal or human tissue, the most obvious risk is microbiological contamination arising in the tissue of origin. The effect of such contamination might be passive transfer to, and infection of, laboratory workers (e.g. Hummeler 1959; Lloyd & Jones 1984; Mahy *et al.* 1991) or even recipients of clinical products derived from the cells (Sweet *et al.* 1960; Tedder *et al.* 1995).

The key risk factor, other than contamination of the original tissue, is the potential of any *in vitro* cell culture to be contaminated with, and possibly support the growth of, pathogenic microorganisms (e.g. viruses, prions, mycoplasma). As already discussed, there are a variety of factors that can influence infectious risk and, in particular, the species and tissue of origin will influence the level of risk associated with a particular cell culture. The relative risks from different species/tissues has already been reviewed (Frommer *et al.* 1993; Pocchiari 1991) but will need to be updated in the light of developing microbiological knowledge. Where tissues and primary cells are handled, the level of risk can be related to the number of samples handled from different individuals, the risk associated with any particular cohort of donors, and also the geographical origin of the cells or tissue (Cobo *et al.* 2005 and references therein).

For most routine cell culture laboratories, the major concerns will relate to the potential for cell contamination with serious human pathogens that would put laboratory workers at risk (see Table 31.1). However, the products of cell culture may be used in a wide range of clinical applications: from inoculation of highly purified recombinant vaccines into healthy vaccinees, to transplantation of cultured cells (e.g. stem cells) into immuno-compromised or terminally ill patients (Figure 31.1). A careful assessment of the risk/benefit balance for patients, combined with careful post-market surveillance, is therefore vital in the application of new animal cell-derived products(DoH 2002).

High-risk infectious organisms that may be present in blood or brain of human or primate origin include haemorrhagic fever viruses, T-lymphotrophic viruses, and hepatitis viruses (Frommer *et al.* 1993). However, blood and tissues from a broad range of other mammalian species may also pose a direct risk of zoonotic infections that cause human disease and even death(for a general review see Krauss 2003). Such species include non-human primates (e.g. B virus, hepatitis A, Marburq virus, simian immunodeficiency virus, *Mycobacterium tuberculosis*) (Fleming 2006), goats (e.g. *Coxiella burnetti* the causative agent of Q-fever) (Frommer *et al.* 1993) and rodents (e.g. lymphocytic choriomeningitis virus, hantaan virus, reovirus 3) (Kraft & Meyer 1990; Lloyd & Jones 1984; Mahy *et al.* 1991). In addition, even cell lines from insect cells may harbour organisms of concern (Stacey & Possee 1996; Vaughn 1991). In the past, contamination of laboratory rodent colonies was also found to cause contamination of cell lines and antibody preparations (Niklas *et al.* 1993). It is important to remember that a number of zoonotic infections in animals may be asymptomatic in a proportion of animals, and some may become persistent secretors of virus. Zoonotic infections in humans may also be asymptomatic in a proportion of individuals, or difficult to diagnose initially (e.g. *Chlamydophila pscittaci*, lymphocytic choriomeninigitis virus).

A major event in the very early days of the development of vaccines from animal cell cultures was the contamination of live polio vaccine with simian virus 40 (SV40) from the primary monkey kidney cells used as the substrate for virus production in the 1950s and 1960s (see Chapter 1). More recently, detection of SV40 gene sequences in some cases of human cancer have prompted concerns that these arose from vaccine contamination. However, investigations of the potential role of SV40 have proved inconclusive (Rollinson *et al.* 2004; Magnani 2005) and individuals exposed to the SV40-contaminated vaccines in the 1950s and 1960s have not shown any evidence of ill effects (Elmishad *et al.* 2006). Whatever conclusion is finally drawn on this issue, it did

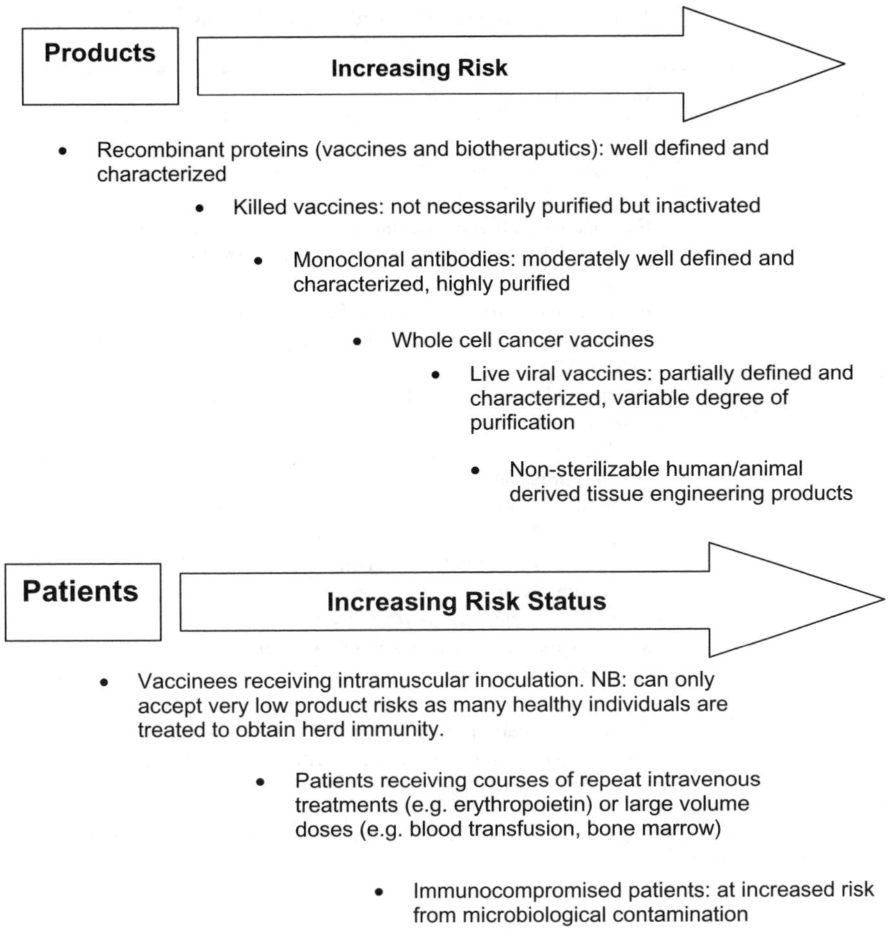

Figure 31.1 Increasing risk of contamination transmission.

promote the use of alternative cell cultures, such as diploid fibroblasts and continuous cell lines, that could be cryopreserved as cell banks and subjected to safety testing prior to use in production (see Table 31.2 and Chapter 32).

For the purposes of biomedical products derived from rodent cell cultures, the European Medicines Agency has identified a range of viruses to be considered in products derived from such cells (Committee for Proprietary Medicinal Products 1997), which are outlined in Table 31.3. A group of these viruses is known to have the potential to infect human or primates or their cells *in vitro* (Table 31.3) and there may be other viruses that need to be added to this list (such as parainfluenza-3, Miyata *et al.* 2005), which could represent a hazard to human cell therapy products in which rodent-derived materials are used. In general it may be assumed that any viral or other microbial contamination in a product for clinical use is undesirable, particularly in the case of cell therapies, even if the organism is of little direct risk to the health of patients.

Persistent infection with non-cytopathic strains of bovine viral diarrhoea virus is commonly observed in bovine cell lines (and also the RK13 rabbit cell line) and, not surprisingly, can affect subsequent infection studies using other viruses (Nakamura *et al.* 1995). Infection of human cell lines with BVDV has been reported (Harasawa & Mizusawa 1995), but in such cases it is possible

Table 31.2 Some of the typical analytical tests applied to banks of cell lines used in the manufacture of biologicals.[a]

Characteristic of cells	Investigation
Identity	Karyology Isoenzyme profiles DNA profiling
Purity	For rodent cell lines: inoculation of laboratory animals and detection of immunological responses to viral agents (MAP, RAP and HAP, see Table 31.3) For human cell lines: PCR-based tests for DNA/RNA of key serious human viral pathogens including HepB, HepC, HIV 1 and 2, HTLV I and II, HCMV Bovine and porcine viral tests for cells exposed to bovine serum and trypsin. Product-enhanced reverse transcriptase (PERT) test for retroviral contamination Inoculation of animal and human cell lines (observed for cytopathic effects and viral haemadsorption of red blood cells) Adventitious agent detection using inoculation of laboratory animals, e.g. suckling mice, rabbits, guinea pigs, fertile hens eggs Transmission electron microscopy of cell sections
Sterility	Mycoplasma tests (culture and DNA stain)
Tumorigenicity	Inoculation of suckling immunocompromised rats and observation for tumours
Oncogenicity	3T3 transformation assay for oncogenicity of host cell genomic DNA Inoculation of laboratory animals with lysed production cells

[a]*NB: The tests described for adventitious agents may not all be required for the different cell banks established, namely master cell bank, working cell bank and cells beyond the passage limit for production purposes (see Committee for Proprietary Medicinal Products 1997).*

that BVDV in the serum used to culture the cells could have introduced apparent infection with BVDV, since BVDV can bind effectively to cells of non-bovine origin including primate cells without causing infection (Xue & Minocha 1996). Another bovine virus known to have the potential to infect cells from a broad range of species is bovine polyoma virus and, unlike BVDV, it also has the potential to transform bovine and non-bovine cells, thus presenting an oncogenic hazard (Schuurman *et al.* 1992).

Contamination and persistent infection of cell lines with organisms, which could not have been predicted by assessment of species and tissue of origin, have also arisen. Examples include HIV infection in rat epithelial cell lines (Canivet *et al.* 1990), lymphocytic choriomenigitis virus in HeLa cells (Simon *et al.* 1982) and epizootic haemorrhagic disease virus in hamster cells (Rabenau *et al.* 1993). In addition, type D retroviral proviruses have been found in some human cell lines (Uckert *et al.* 1986), and some of these viruses are capable of infecting human cells *in vitro*, notably some EBV-transformed cell lines (Sun *et al.* 1995). Such persistent infections are also found that are due to organisms that are not necessarily pathogenic for humans but that could influence the performance or acceptability of cell culture systems.

All mammalian tissues and cell lines contain endogenous retrovirus sequences and retrovirus-like sequences, some of which are known to be expressed under certain circumstances (Harris 1998). Expression of these viral sequences is a common feature in cancer cell lines (Patzke *et al.* 2001) and viral particles are found in eggs used for vaccine production (Weiss 2001), as well

Table 31.3 Risk groups for viruses likely to arise from mouse and rat materials (adapted from EMEA 1997).

Murine virus species known to be detected in antibody production tests for adventitious agents	Known to cause infections in humans or primates	Known to be capable of infecting human or primate cells	No evidence for the capacity to cause infection in humans	Antibody Production test[a] in which virus can be detected
Hantaan virus	+	+		MAP, RAP
Lymphocytic choriomeningitis virus	+	+		MAP, HAP
Reovirus type 3	+	+		MAP, HAP, RAP
Sendai virus[b]	+	+		MAP, HAP, RAP
Lactate dehydrogenase virus	+	+		MAP
Ectromelia virus		+	+	MAP
Minute virus of mice		+	+	MAP
Mouse adenovirus		+	+	MAP
Mouse cytomegalovirus		+	+	MAP
Mouse rotavirus (EDIM)		+	+	MAP
Pneumonia virus of mice		+	+	MAP, HAP, RAP
Kilham rat virus		+	+	RAP
Toolan virus (HI)		+	+	RAP
K virus			+	MAP
Mouse encephalomyelitis virus (Theiler's, GDV III)			+	MAP, RAP
Mouse hepatitis virus			+	MAP
Polyoma virus			+	MAP
Thymic virus			+	MAP
Rat coronavirus			+	RAP
Sialoacryoadenitis virus			+	RAP

[a]*MAP, mouse antibody production test; HAP, hamster antibody production test; RAP, rat antibody production test.*
[b]*NB: Due to an immunological cross-reaction some infections in animal colonies identified as due to Sendai virus may in fact be due to parainfluenzavirus type 3 (Miyata et al., 2005).*

as in immortalized mouse and hamster cells used in the manufacture of monoclonal antibodies and recombinant glycoproteins (Adamson 1998). These agents do not appear to have presented a hazard to laboratory workers. However, some diseases, such as multiple sclerosis and rheumatoid arthritis, have been associated with the expression of endogenous retroviruses (Perl 2003). Whilst it is apparent that endogenous retroviruses have probably been present in a variety of cellular transplanted materials used for decades, such as blood and bone marrow, there have been no reports of adverse effects associated with their presence (Weiss 2001; Lower 1999). Nevertheless, potentially infectious endogenous retrovirus particles are released in high titres from some murine cells. This has led regulatory bodies to require products made using murine cell cultures to be adequately purified and inactivated to ensure that viral particles are not present in the final product (see Chapter 19) even though they do not infect human cells.

31.4.2. Reduction of Microbiological Risks from Cells

The reduction of microbiological risk associated with cells used in production of biological medicines must be dealt with both from the perspective of the risk to laboratory workers preparing the cells and also from the perspective of the patients receiving cell-derived products.

It is clear that cell cultures provide the ideal laboratory substrate for growth of microorganisms such as viruses. Accordingly, it is commonly recommended that all cell cultures should be treated as if they were potentially infectious. Risk can be reduced through a careful process of donor selection and testing of cell lines. Human cell lines used routinely for laboratory research and development work or testing should ideally be derived from donors that have been screened for serious human pathogens such as HIV-1 and -2, HTLV I/II and hepatitis B and C (Coecke *et al.* 2005). However, since infected donors may provide samples, when the markers of infection are not apparent (e.g. detectable levels of viral RNA, virus-specific antibody), it may be more appropriate to test the cell lines once they have been established. Obviously, anti-viral antibody tests are not relevant to testing of established cell lines and other nucleic-acid-based tests will need to be qualified specifically for use in cell lines used for therapy. Indeed it is possible that in some cases of cell line infection, the level of virus could fluctuate over time providing the potential for intermittent positive results or false negative tests.

Tissues and primary cells for transplantation are generally required to have been tested, and found negative, for the key blood-borne viral pathogens, and to have fulfilled other requirements of donor selection from populations in which additional risk factors due to changing social norms, including increasing foreign travel, must also be addressed (Advisory Committee on the Microbiological Safety of Blood, Tissues and Cells 2000; Cobo *et al.* 2005; European Union 2004; Food and Drugs Administration 2004; Stacey & Hartung, in press, and see web sites at the end of this chapter). These requirements will also apply to cells and tissues used in other biomedical technologies, although additional product-specific safety testing may be required in the areas of somatic cell therapy (Committee for Proprietary Medicinal Products 2001a), xenogeneic cell therapy (Committee for Proprietary Medicinal Products 1999) and gene therapy (Committee for Proprietary Medicinal Products 2000). The use of cell lines, such as stem cell lines, will also need special evaluation and will probably need the application of a combination of the approaches described above for transplantation and the production of biotherapeutics (see Chapter 32).

Cell lines used as the substrates for manufacture of vaccines and biotherapeutics are prepared as cell banks that are subjected to exhaustive testing before use. This testing is outlined in guidance from regulatory authorities (Food and Drugs Administration 1993; International Conference on Harmonization 1997; World Health Organization 1998) but final testing regimes must include tests indicated by specific risk assessments on the production cell line, the production process and the final product. Key issues that must be addressed for the quality and safety of each production cell line, and for any cell line used in the laboratory, include purity (absence of microorganisms), authenticity (correct cell line and absence of other cells), and stability (maintenance of genotypic and phenotypic characteristics on passage *in vitro*) (Stacey 2002). Examples of the types of test typically performed for cell lines used in production processes are included in Table 31.2.

Specific guidance has also been produced for the preparation of monoclonal antibodies, which include requirements for the host cells and the product (Committee for Proprietary Medicinal Products 1995; Food and Drugs Administration 1997). In the production of antibodies for therapy, production and testing of antibody preparations requires careful attention. The risks associated include:

- production cells: endogenous adventitious agents (in the cells and other biological materials), identity/source of cells, genetic stability and any agents used if there is a cell immortalization process involved (e.g. EBV transformation);
- antibody product: purity, consistency, immune reactivity, cytokine contamination, presence of fragments and aggregates of the product;
- methods used for production (e.g. production of ascites *in vivo*, batch culture or continuous bioreactor culture of antibody-producing cells) (see Chapter 10) and purification (see Chapter 18).

31.4.3 Non-microbiological Cell-associated Hazards

Cells can also present a source of non-infectious hazards relating to oncogenicity of the host cell DNA and the presence of biologically active proteins from the host cell, which may contaminate cell-derived medical products. In addition the tumorigenic potential of viable cells is an issue of concern not only in terms of the DNA and protein contamination risk already mentioned, but also in terms of the risk of viable cells giving rise to tumours, especially from preparations of cells for implantation into humans.

The risks associated with genomic DNA from different host cells can be assessed using the 3T3 transformation assay (Copeland & Cooper, 1979), although this may provide data biased towards the activity of certain types of oncogene (e.g. *ras*). Risk assessment relating to host cell DNA has been addressed by the establishment of limits for maximum quantities of host cell DNA per final dose of product. The maximum levels for culture products have been an ongoing topic of discussion led by the World Health Organization. Highly purified recombinant DNA products are expected to contain low levels of host DNA (e.g. less than 10 pg/dose). For other types of product, less stringent limits have been set but there is no generally acceptable limit and this requires careful evaluation, particularly in the case of products from human tumorigenic cells (Committee for Proprietary Medicinal Products 2001b). The permitted maximum is therefore set according to product-specific risk assessment that takes into account factors including:

- the characteristics of the host cell line;
- the nature of any viral sequences present in the host cell;
- molecular size and nature of any residual genomic DNA in the product;
- efficiency of the product purification process for removal of host cell DNA;
- sensitivity and specificity of the assays used to evaluate residual DNA in the product;
- the dose size and regimen (e.g. requirement for multiple doses).

In addition to residual cell substrate DNA, tissue or cell culture-derived products may also contain biologically active molecules such as growth factors and cytokines that may elicit immunological reactions and will need to be taken into consideration in risk assessment and safety testing for products (see Chapter 25). Cell lines used for manufacturing processes are known to express a range of cytokines (Gearing *et al.* 1989) and this will also be a more serious consideration for new cell therapies where there may be little opportunity in processing to remove contaminating proteins, and expression profiles of such molecules could vary between batches of product.

The use of tumorigenic cell lines for the manufacture of viral vaccines has been avoided (Committee for Proprietary Medicinal Products 2001b) and it is accepted that tumour cell line DNA is potentially hazardous (Krontiris & Cooper 1981). The 'tumorigenic' hazard of certain cell types is generally related their ability to cause tumours in immuno-compromised animals, although the results from specific protocols for these *in vivo* tumorigenicity tests remain to be standardized (see below). The observed tumours are generally assumed to arise from survival and proliferation of the original cells. Such tests are controlled by the use of highly tumorigenic cells such as HeLa cells and cells that do not cause tumours such as the MRC-5 human diploid cells (World Health Organization 1998). Factors that should be considered when carrying out risk assessment of tumour-cell-derived products are the same as those described above (Committee for Proprietary Medicinal Products 2001b), although the nature/origin of the cells and the nature of any viral sequences or oncogenes present may receive special attention in the case of tumorigenic cells. A defined risk assessment process for products derived from tumorigenic cells has also been published by the Food and Drugs Administration (Lewis *et al.* 2001). Some other cell substrates used for manufacturing, such as Vero cells, may become tumorigenic at higher passage number or under altered culture conditions. Such issues relate to the characteristics of the production cell line and do not relate directly to the safety of cell-derived products. However, cells exhibiting a highly tumorigenic nature may need to be subjected to more detailed investigations such as enhanced oncogenicity testing.

The area of tumorigenicity testing of cell substrates used for manufacturing is complex. It is used to characterize a fundamental biological property of the cells in question, but this characteristic (the ability to form tumours) is not well understood. Methods employed for the testing of production cells usually involve the inoculation of a range of different animals (including immunodeficient mice or rats, rabbits, guinea-pigs and fertile hen eggs) and inspection for the development of tumours. Such testing may be affected by a range of experimental variables such as length of observation period, age and strain of animals used, mechanism of inducing immunodeficiency in animals, mode and site of inoculation, and the presence of infectious agents in the test sample or the inoculated animal. The value of such tumorigenicity testing is questionable, not only because of the numbers of animals sacrificed, but also because of the lack of standardization for this assay both in its performance and interpretation, and its relevance to modern cell-derived products (Noguchi 1992). Very similar tests are performed on lysed cells and these are called 'oncogenicity' tests that are intended to answer a different question relating to the detection of adventitious agents and other agents that might cause tumours in animals. Oncogenicity tests are much more directly related to product safety than are tests for tumorigencity, as the various components of the cells including DNA and bioactive molecules may pass into the product. However, as in the case of tumorigenicity tests, they remain to be standardized.

The matters of tumorigenicity and oncogenicity are of concern for highly purified products and may therefore be of even more pressing concern for cell therapy, where purification steps may not exist, and there may be elevated levels of extracellular DNA (see above) arising in products that are cryopreserved prior to use. Stem cell cultures have tremendous therapeutic potential as outlined in Chapter by 29. However, chromosome instability (e.g. Draper *et al.* 2004) and cell fusion (Chen *et al.* 2004; Ishikawa *et al.* 2006), may arise during the culture of stem cells, and these could lead to the development of populations of aberrant cells that could present a threat to transplanted patients. Methods for identifying shifts to a tumorigenic nature and detection of low levels of aberrant cells will be required to validate the safety of new stem cell and other cell therapies. Such new assays may initially involve improvements in current oncogenicity assays (see above) and *in vivo* assays using transgenic animals (e.g. Onions 2001; Sistare 2001). However, *in vitro* cell-based assays for tumour cell properties, such as metastasis, and molecular expression profiling may also be required.

31.4.4 Other Hazards for Cell-based Products

Whilst risk assessment for the production cells, or cells for transplantation themselves, tends to be focused on the issues already covered, there are other factors that can also have significant impact on the quality, safety and commercial viability of a cell-derived product. These factors include ethical status, genetic and phenotypic stability during *in vitro* culture, and response to altered culture conditions in the final stages of production of the biological medicine.

31.4.4.1 Ethical issues

Ethical status is not necessarily the first issue that comes to mind in relation to risk, but could be critical to the ability of a manufacturer to deliver a product. Failure to ensure adequate ethical governance of the source of cellular materials and cell lines could prevent the manufacturer from being able to continue to make and sell the product, and might expose the manufacturer to litigation. Ethical and legal regulation of human-derived cells and tissue vary considerably between countries. Thus, it is wise to ensure that procedures used to obtain original human tissue, and particularly issues of donor consent, meet the requirements of the countries in which the products will be made available (Coecke *et al.* 2005). Failure to adhere to certain standards for animal husbandry and *in vivo* testing clearly could also expose a company to criticism. Ethical and legal issues also arise with non-human cells, tissues and cell lines derived from certain endangered species. Information on the international regulation of such materials under the CITES treaty can obtained at www.cites.org.

31.4.4.2 Variability of cell state

Testing cell banks exhaustively for likely pathogens and for cell culture stability and performance will not necessarily give information on the overall safety and efficacy of the cells at the production stage. Even where cells are tested at or beyond the passage limits used in production, the culture of cells at the production stage under conditions different from the standard cell culture expansion protocols may cause changes in the cells that could affect their efficacy and safety. This is beginning to be addressed in guidance documents (Coecke *et al.* 2005) and is a critical issue for standardization of cell cultures, which is dealt with in Chapter 32.

Delivery of standardized cell preparations will be especially important for use in the developing area of cell therapy and tissue engineering, in order to control the characteristics of cells prepared for transplantation into patients. This will be vital to assure the reproducibility, reliability, safety and efficacy of a particular cell therapy, and each new therapy and application will need characterization parameters and techniques to be carefully validated for use. A range of analytical tests may be required to do this, including assessment of protein expression (e.g. cell surface markers, enzyme activities, components of key biochemical pathways, RNA expression profiles). More complex functional assays may also be needed to demonstrate potential efficacy or 'potency' for each batch of cells prepared for clinical use, but such assays will take some time to develop. In addition tests would be required to detect the presence of cells that might generate aberrant or uncontrolled cell growth, or that could present a tumorigenic hazard to patients, e.g. residual undifferentiated cells in cells derived from human embryonic stem cells, cell transformation, cell fusion. Where cell products have a short shelf life once prepared, any critical quality control and safety tests that must be applied will have to give results rapidly. The areas of rapid batch release testing and functional assays for product efficacy represent challenging areas for the successful application of cell therapies, but will be important to assist in risk reduction.

31.5 EVALUATION OF HAZARDS FROM GROWTH MEDIA

Traditionally the growth of cells *in vitro* has depended on animal serum, primarily of bovine origin, which is a largely undefined product that is also highly variable from batch to batch and would be damaged by many physical sterilization methods (see Chapter 4). The subculture of adherent cell lines has also been largely dependent on the use of porcine trypsin, which has been identified as a source of infection of cells in culture due to porcine viruses, and the porcine parvovirus in particular (Hallauer 1971). Until recently, there have been no preparations of these key culture reagents of animal origin that had received any kind of viral evaluation or inactivation. Thus, unless stringent efforts have been made to avoid it, any cell line in regular use has probably been exposed to bovine, porcine and possibly murine viruses from other cell culture reagents including antibodies (Nicklas *et al.* 1993). Inevitably, the longer a cell line's history of *in vitro* passage, the more likely it is to have had multiple exposures to viruses and possibly prions. Thus any cell line received into the laboratory should be considered at risk of such contamination.

The presence of viral contamination in serum has become a serious issue for manufacturers of biomedical products since the 1970s, when an outbreak of foot-and-mouth disease in the UK focused attention on seeking sources of serum that present a low risk of viral contamination. More recently, the outbreak of BSE in cattle and subsequently new variant CJD in humans, both in the UK, again focused attention on the sourcing and traceability of serum, although the risk associated with products derived from cell culture is considered to be extremely low (Minor *et al.* 2001).

Commercial supplies of calf serum are known to be potentially contaminated with a wide range of bovine viruses (Erickson *et al.* 1991; Rabenau *et al.* 1993) and BVDV and bovine polyoma virus are known to be especially prevalent, with BVDV contamination being present in at least 50% of commercially available serum batches tested (Bolin *et al.* 1991; Erickson *et al.* 1991;) and bovine polyomavirus in at least 66% (Kappeler *et al.* 1996; Schuurman *et al.* 1991). Contamination with polyomavirus is increasing in incidence and is of particular concern due the tumourigenic potential of this virus in non-bovine cells (Schuurman *et al.* 1992).

Risk assessment protocols for such reagents have been established, and certification systems have also been introduced for specific products (Committee for Proprietary Medicinal Products 2003; European Pharmacopoeia 2005). Whilst the usual sources of serum used in cell culture are either human or bovine, other species may also be used in biological medicines (Eloit 1999) including rodent serum for complement and horse serum for anti-venoms (Burnouf *et al.* 2004) and cell culture, and specific risk assessment will be required for products made using such materials. Serum manufacturers now offer irradiated serum and trypsin, and this is a further effective measure to reduce the risk of viral infection of cell cultures (see Chapter 4) although careful qualification is required to assure homogeneous irradiation.

In an attempt to eliminate risks from growth media components, many manufacturing processes using animal cell lines for the production of antibodies, therapeutic proteins and viral vaccines have switched to the use of serum-free and even animal protein-free media (Merten 1999; Froude 1999). In addition, as recombinant DNA technology has progressed it has provided a growing range of recombinant versions of natural cell culture reagents such as fibronectin (now a common proprietary reagent), collagen (available from Prospec-Tany TechnoGene Ltd and BD Biosciences) and trypsin (available from Prospec-Tany Technogene Ltd) and a range of growth factors and cytokines. A trypsin-like replacement of microbial origin is also now available(Tryp LE, Invitrogen).

The actual infectious risk represented by different cell growth media products should be evaluated in the light of a number of factors including the following:

- species and tissue of origin and location of the originating organisms. The basic factors influencing likelihood of contamination have been outlined in Section 31.4.2. Controlling these risks can be achieved through good practice in animal husbandry (including sourcing good quality breeding stock, screening colonies for infection, and maintaining specific pathogen-free status), avoiding use of wild or semi-wild animals, and aseptic collection and processing of harvested materials;
- nature of the processing, purification and formulation methods, including assessment of any added materials and the capacity of different steps in this process for reducing the level of any potential contaminants;
- nature, effectiveness and reliability of any sterilization steps. It is unlikely that any proteins such as growth factors would survive standard physical methods of sterilization, such as autoclaving or standard sterilizing doses of irradiation. Nevertheless, processes such as pasteurization, filtration, chemical treatment and low level irradiation (see Chapter 4) can be quite effective at

reducing the viable microbial load in cell culture products. Such procedures however, need to be carefully controlled to ensure equivalent treatments within a batch and between batches;

- testing performed on the product. Viral testing of animal-derived products will be most appropriate when no sterilization process can be used. Care should be taken to assess viral testing to ensure that appropriate methods are employed to give defined and acceptable levels of sensitivity and specificity, and that they are performed by a qualified or accredited laboratory with appropriate experience in the tests performed.

It should be also be borne in mind that numerous products used in cell culture are derived from natural bacterial strains (e.g. collagenase, pronase) and recombinant microorganisms (e.g. various cytokines, growth factors, collagen). In these cases it is obvious that the product processing and purification methods must remove any residual organisms, spores and endotoxin, but other processing intermediates, including the bacteriological growth media, will also need to be considered in risk assessment for medical products as they may contain raw materials of animal origin.

31.6 HAZARD IDENTIFICATION, EVALUATION AND CONTROL IN CELL CULTURE PROCESSES

31.6.1 Hazard Identification in Cell Culture Processes

In order to ensure the safety of a process for the manufacture of a biological medicine it is vital that each step in the process is evaluated for risks associated with exposure to starting materials, open processes, or other risk-bearing steps that could affect the contamination status or performance of the product. The study of risk associated with processes is well developed in manufacturing industry for a variety of biological medicines and is now being applied routinely in tissue banking. Such risk assessment will be highly relevant for new cell therapies, from the derivation of new cell lines, through the cell banking process to ultimate application of the cells in humans. A diagram of all elements of a process can be prepared as a process map for the manufacture of a product from a cell line and an example is shown in Figure 31.2. This mapping process should be detailed and each element scrutinized to identify those critical steps where there could be contamination or other effects on the cell phenotype that could affect the quality and safety of the final product. Potential effects of individual steps in the process map may be difficult to quantify and may benefit from evaluation of the degree of uncertainty, or variability, introduced by the different variables involved (British Standards Institute 2004).

The toxic and allergenic properties of all processing materials should be carefully considered, including materials involved in their preparation and any that might be leached from vessels that come in intimate contact with cells. In addition, any culture component or raw material/reagent of animal origin will obviously need to be assessed for the risk it represents in terms of viral or other biological contamination. According to the general model for risk management (Section 31.2), these contamination risks can then be modified in the light of various kinds of risk control that have been applied. These include the purification/inactivation properties of the reagent preparation and the purification process and the safety testing programme applied to the final reagent. Inactivation procedures need to be assessed carefully to determine whether they are sterilizing (achieving exponential killing rates that result in very low probabilities for viable contaminants to remain) or disinfection (achieving a reduction of any viable potential microbial load) processes, and both kinds of process will need to be validated for the type of reagent in question. It is important to remember that sterilizing processes may fail due to poor maintenance of equipment (e.g. uncalibrated autoclaves) or challenge by unusual microorganisms [e.g. passage of

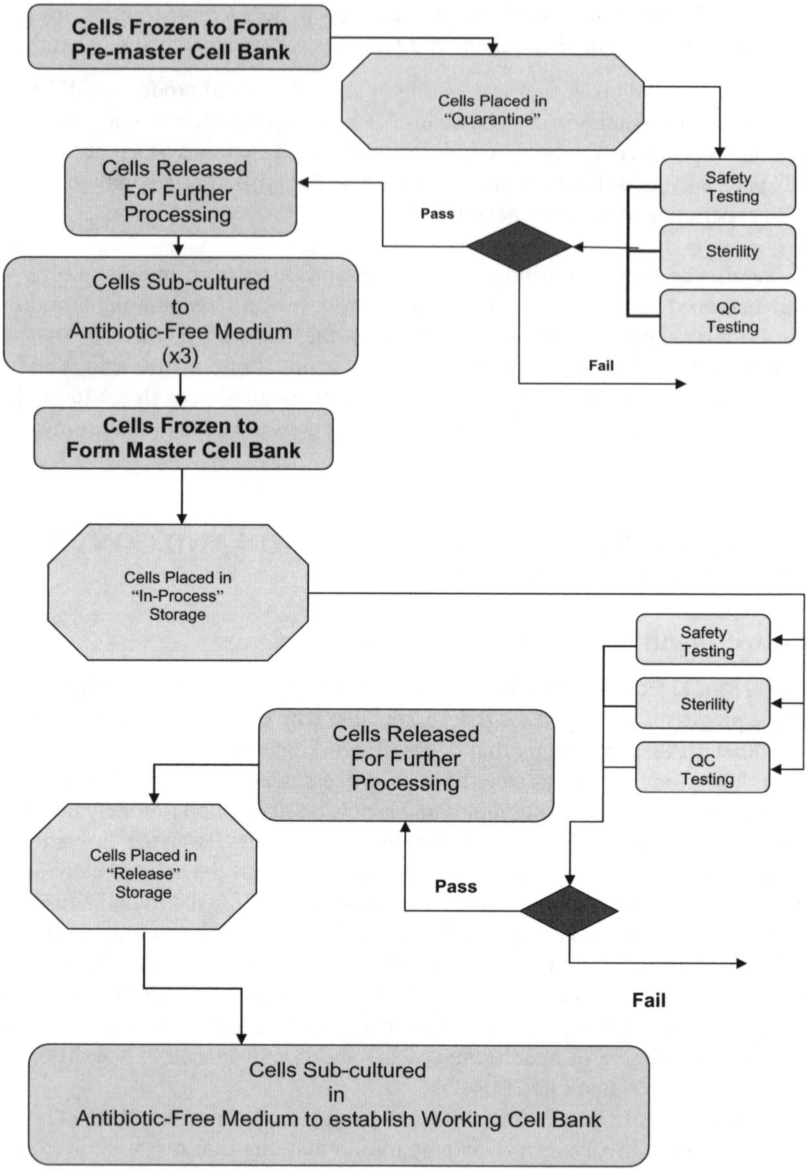

Figure 31.2 Example of a cell banking process map. (Adapted from Healy *et al.* 2005)

exceptionally small bacterial cells through 0.22 μm filters (Oie *et al.* 1998) or irradiation-resistant viral particles (see Chapter 4)]. Other elements that may be considered in this risk assessment process are the history of safe use of a reagent together with the resistance of any cell culture process to viral replication as already described. When such risk assessment has been performed, manufacturers may be reluctant to supply materials for clinical use if the reagent was not originally intended for this purpose. However, in such cases the manufacturer may be willing to supply a material to a specification established and validated by customers for cell culture procedures for therapy (Stacey 2004).

31.6.2 Evaluating Risk and Risk Reduction in Cell Culture Processes

Hazards must be evaluated and controlled in cell culture processes for the safety of both laboratory workers and recipients of cell-culture-derived medicines. In the setting of a research or testing laboratory, where safety of laboratory workers is the primary safety issue, adoption of good aseptic technique and containment of cell culture work in a Class II microbiological safety cabinet will provide general protection from contaminants (Coecke *et al.* 2005). Screening of the source of cells or the cells themselves may enhance safety, but is not a perfect solution as discussed in Section 31.4.1.

Many cell-derived products are highly purified using different sequential treatments that provide effective multiple viral removal steps, each of which contribute to reducing risk of viable viral contamination in the final product. Unfortunately, in cell therapy applications where the final product comprises viable cells, the types of purification step used rarely assist in removal or inactivation of viruses. Table 31.4 shows a comparison of key elements of production processes for

Table 31.4 Differences in the processing of products for transplantation and manufactured biological medicines.

Stage in processing	Biotherapeutics and bioprophylactics[a]	Blood products	Tissues, cells and organs for transplantation
Starting material	Intensively characterized cell banks	Bulk serum and plasma preparations pooled from multiple donors subjected to donor selection[b]	Single 'units', tissue or cells subjected to donor selection[b]
Contamination status of material during processing	Cells maintained free of contamination: antibiotic-free culture to expose any contamination	Aseptic collection and preparation means low bioburden (bacteria and fungi). Contamination with certain viruses at low levels (e.g. B19, TTV) may not contra-indicate use	May well carry some bioburden (bacterial contamination, viruses not detected by viral marker screens on donor blood samples). However, processing must be aseptic and not introduce contaminants or facilitate their growth.
Downstream processing: purification, etc.	Product is usually highly purified which significantly reduces potential viral contamination of product (see Chapter 19	Filtration may be used with inactivation by chemical or physical (e.g. pH) methods. Fractionation also may contribute to removal of adventitious agents	Where purification occurs it is generally not a viral removal step
Qualification of product	Bulk product and batch release used to demonstrate each batch is fit for purpose (i.e. potency and safety)	Bulk product and batch release testing demonstrate each batch of product is fit for purpose.	Validation of process, and environmental monitoring, are used to assure reproducible and safe product. Potency testing is often not feasible and microbiological testing may not give useful data

[a]*Vaccines, recombinant proteins, gene therapy vectors, etc.*
[b]*Acceptable ethical sourcing, medical history and viral screening.*

biotherapeutics, blood products and transplantation/cell therapies. It illustrates the key differences in critical process points in each group of products that must be addressed to assure product safety. In products incorporating donor cells or cell components, the emphasis for demonstrating risk reduction for the cell component lies in donor screening procedures and cell bank safety testing.

Reducing risks for biomedical products through qualification of cell culture reagents and other materials used in cell culture processing is vital for both purified products and cell therapies. Areas of obvious and immediate concern are those products of animal origin and, as described above (Section 31.5), a combination of careful and traceable sourcing, testing for animal pathogens and use of qualified viral reduction procedures, will be required. For some starting materials of animal origin the manufacturing process may be adequate for removing or inactivating likely potential viral contaminants, and detailed information on the manufacturing process and any viral challenge studies will be helpful (see above and Chapter 19). Testing for viruses likely to be present in the original tissue may also be carried out, and lists of viruses for certain animal products have been published (Committee for Proprietary Medicines 1997). In order to ensure reliable and reproducible supplies of such reagents, it is important to put service level agreements in place with suppliers, and any new batches of raw materials should be reviewed to ensure they have the correct composition and have received appropriate testing by a qualified laboratory. With some biological materials this may include product-specific pre-use quality control not performed by the supplier of the medium or reagent (Stacey 2004). For any cell culture product, it is important to evaluate all elements of the growth medium to identify all possible sources of biological contamination. Most manufacturers are attempting to switch cell culture production processes to serum- and protein-free media. Whilst such media may support growth of cells and high product generation rates, there can still be complications, and careful investigation should be carried out to exclude adverse effects of serum-free culture. The cell stress involved in adaptation to such media could include endogenous virus reactivation, increased cellular pinocytosis, increased cell susceptibility to potential contaminants, altered cell performance, and increased rates of cell death releasing cell contents, including enzymes and DNA, that may contaminate or possibly damage the product.

Specialized tools may be used for assessing risk in production processes. These include 'failure mode and effect analysis' (Burgmeier 2002) and 'uncertainty analysis' (Lim *et al.* 2005; Biwer *et al.* 2005). International standards have also been established for risk assessment applying to pharmaceuticals (World Health Organization 2003).

31.7 CONCLUSION

Identification and evaluation of hazards for cells, raw materials and processes is a fundamental first step in risk analysis of cell-derived products. It should also be remembered that this process should encompass hazards to the health of laboratory workers, to the quality of the product and to the recipients of the products. Risks arising from use of biological reagents are rarely well defined and it is important to assess all potential contaminants. Control of risks can be achieved through traceability of raw materials, containment of hazardous processes, validation of viral removal steps, etc. Whilst general principles have been laid down for risk assessment of some new developments in the use of cells to produce new cell therapies, assessment of some of the specific risks involved are yet to be tackled, particularly for those products that have very short shelf lives.

It must also be borne in mind that the specification for release of each product must address the intended use (e.g. route of injection, patient group details) and a risk benefit analysis must be performed for each product/application to ensure that any risks are in balance with patient benefit.

REFERENCES

ACDP (1995) *Categorisation of Biological Agents According to Hazard and Categories of Containment.* HSE Books, Sudbury, UK (www.tso.co.uk); fourth edition.

ACDP (2005) *Biological Agents: Managing the Risks in Laboratories and Healthcare Premises.* Advisory Committee on Dangerous Pathogens, Health and Safety Executive, TSO, Sudbury, UK (www.hse.gov.uk).

Adamson SR (1998) *Dev. Biol. Stand.*; 93: 89–96.

Advisory Committee on Microbiological Safety of Blood, Tissues and Cells (MSBT) (2000) *Guidance on the Microbiological Safety of Human Organs, Tissues and Cells Used in Transplantation.* Medical and Healthcare products Regulatory Agency, London, UK.

Biwer A, Griffith S, Cooney, C (2005) *Biotech. Bioeng.*; 20: 167–179.

Bolin SR, Matthews PJ, Ridpath JF (1991) *J. Vet. Diagn. Invest.*; 3: 199–203.

British Standards Institute (2004) *General Metrology. Part 4: Practical Guide to Measurement Uncertainty.* British Standards Institute, London.

Burnouf T, Griffiths E, Padilla A, Seddik S, Stephano MA, Guttierrez J-M (2004) *Biologicals*; 32: 115–128.

Burgmeier J (2002) *Jt. Comm. J. Qual. Improv.*; 28: 331–339.

Canivet M, Hofman AD, Hardy D *et al.* (1990) *Virol.*; 178: 543–541.

Chen KA, Laywell ED, Marshall G, Walton N, Zheng T, Steindler DA (2006) *Exp. Neurol.*; 198: 129–35.

CHIP (2002) *Chemicals (Hazard Information and Packaging for Supply) Regulations 2002* (CHIP). Health and Safety Executive, UK (see web list for H&SE Books and also the CHIP (Amendment) Regulations 2005 (CHIP 3.1)).

Cobo F, Stacey GN, Hunt C *et al.* (2005) *Appl. Micro. Biotechnol.*; 68: 456–466.

Coecke S, Balls M, Bowe G *et al.* (2005) *ATLA*; 33: 261–287.

Committee for Proprietary Medicinal Products (1995) *Production and Quality Control of Monoclonal Antibodies* (3AB4a) European Agency for the Evaluation of Medicinal Products, London, UK, October 1997. http://www.emea.europa.eu

Committee for Proprietary Medicinal Products (1997) *Note for Guidance on Quality of Biotechnological Products: Viral Safety Evaluation of Biotechnology Products Derived from Cell Lines of Human or Animal Origin.* (CPMP/ICH/295/95).

Committee for Proprietary Medicinal Products (1999) *Concept Paper on Xenogeneic Cell Therapy (draft 1), 8.2 CPMP Guidelines.* European Agency for the Evaluation of Medicinal Products, London, UK.

Committee for Proprietary Medicinal Products (2000) *Note for Guidance on the Quality, Preclinical and Clinical Aspects of Gene Transfer Medicinal Products.* European Agency for the Evaluation of Medicinal Products, London, UK.

Committee for Proprietary Medicinal Products (2001a) *Points to Consider on Human Somatic Cell Therapy* (CPMP/BWP/41450/98). The European Agency for the Evaluation of Medicinal Products, London, UK.

Committee for Proprietary Medicinal Products (2001b) *Position Statement on the use of Tumourigenic Cells of Human Origin for the Production of Biological and Biotechnological Medicinal Products* (CPMP/BWP/1143/00). The European Agency for the Evaluation of Medicinal Products, London, UK.

Committee for Proprietary Medicinal Products (2003) *Note for Guidance on the use of Bovine Serum in the Manufacture of Human Biological Medicinal Products* (CPMP/BWP/1793/02). The European Agency for the Evaluation of Medicinal Products, London, UK.

Copeland NG, Cooper GM (1979) *Cell*; 16: 347–356.

COSHH (2002) *Control of Substances Hazardous to Health* (fifth edition). The Control of Substances Hazardous to Health Regulations 2002 (as amended). Approved code of practice and guidance. Health and Safety Executive, UK, ISBN 0717629813. (see also the COSHH general guidance document in website list).

DoH (2002) Code of Practice for Production of Human-derived Therapeutic Products, UK Department of Health, TSO Sudbury, UK. www.mhra.gov.uk/home (see "How We Regulate – Tissue Engineered Products")

Draper JS, Smith K, Gokhale P *et al.* (2004) *Nat. Biotechnol.*; 22: 53–54.

Elmishad AG, Bocchetta M, Pass HI, Carbone M (2006) *Dev. Biol. (Basel)*; 123: 109–117.

Eloit M (1999) *Dev. Biol. Stand.*; 99: 9–16.

Erickson GA, Bolin SR, Landgraf JG (1991) *Dev. Biol. Stand.*; 75: 173–175.

European Pharmacopoeia (2005) Minimising the risk of transmitting animal spongiform encephalopathy agents via human and veterinary medicinal products (5.2.8). *European Pharmacopoeia* 5.0, 01/2005:50208, 463–471.

European Union (2004) *Official J. Europ. Union;* L102: 48–58.

Fleming DO (2006) Chapter 2, in: Biological Safety: Principles and Practice (4th edn) eds Fleming DO, Hunt DC, ASM Press, pp 19–33.

Food and Drug Administration (1993) *Points to Consider in the Characterization of Cell Lines used to Produce Biologicals.* Center for Biologics Evaluation and Research, Food and Drug Administration, Bethesda, USA (http://www.fda.gov/cber/).

Food and Drug Administration (1997) *Points to Consider in the Manufacture and Testing of Monoclonal Antibody Products for Human Use.* Food and Drugs Administration, Rockville, MD20857, USA.

Food and Drug Administration (2004). 21 CFR Parts 16, 1270 and 1271. *Current Good Tissue Practice for Human Cell, Tissue, and Cellular and Tissue-based Products Establishments; Inspection and Enforcement; Final Rule.* Department of Health and Human Services, November 24.

Frommer W, Archer L, Boon B, Brunuis G, Collins CH, Crooy P *et al.* (1993) *Appl. Microbiol. Biotechnol.*; 39: 141–147.

Froude SJ (1999) *Dev. Biol. Stand.*; 99: 157–166.

Gearing AJ, Cartwright JE, Bird CR, Wadhwa M, Priest R, Thorpe R (1989) *Lancet*; 2(8670): 1011–1012.

Hallauer C, Kronauer G, Seigl G (1971) *Arch. Gesamte Virusforsch.*; 35: 80–90.

Harasawa R, Mizusawa H (1995) *Microbiol. Immunol.*; 39: 979–985.

Harris R (1998) *Bioessays*; 20: 307–316.

Healy L, Hunt C, Young L, Stacey G (2005) *Adv. Drug Del. Rev.*; 57: 1981–1988.

Hummeler K, Davidson WL, Henle W, Laboccetta AC, Ruch HG (1959) *N. Eng. J. Med.*; 261: 64–68.

IEC (2001) *Hazard and Operability Studies (HAZOP Studies) Application Guide.* International Eletrotechnical Commission, Geneva, Switzerland.

International Conference on Harmonization (1997) ICH Topic Q 5 D *Quality of Biotechnological Products: Derivation and Characterisation of Cell Substrates Used for Production of Biotechnological/Biological Products* (CPMP/ICH/294/95). ICH Technical Coordination, European Medicines Evaluation Agency, London.

International Conference on Harmonization (2005) ICH Topic Q9 *Quality Risk Management (Current Step 4 Version).* ICH, ICH Technical Coordination, European Medicines Evaluation Agency, London. See also website for ICH Q9 Quality Risk Management.

Ishikawa F, Shimazu H, Shultz LD *et al. FASEB J.*; 20: 950–952.

Ishikawa F, Shimazu H, Shultz LD *et al.* (in press). *FASEB J.* (E-publication ahead of print April 2006).

Kappeler A, Lutz-Wallace C, Sapp T, Sidhu M (1996) *Biologicals*; 24: 131–135.

Kraft V, Meyer B (1990) *Z. Versuchstierkd.*; 33: 29–35.

Kraus H, Weber A, Appel M *et al.* (2003) Zoonoses: Infectious Diseases Transmissble from Animals to Humans (3rd edn) ASM Press; Washington DC.

Krontiris TG, Cooper GM (1981) *Proc. Natl. Acad. Sci. USA*; 78: 1181–1184.

Lewis AM, Krause P, Peden K (2001) *Dev. Biol. Stand.*; 106: 513–535.

Lim AC, Zhou Y, Washbrook J (2005) *Biotechnol. Prog.*; 21: 1231–1242.

Lloyd G, Jones N (1984) *J. Inf.*; 12: 117–125.

Lower R (1999) *Trends Microbiol.*; 7: 350–356.

Magnani C (2005) *Med. Law*; 96: 347–353.

Mahy BW, Dykewicz C, Fischer-Hoch S, Ostroff S, Tipple M, Sanchez A (1991) *Dev. Biol. Stand.*; 75: 183–189.

Merten O-W (1999) *Dev. Biol. Stand.*; 99: 167–180.

MHSWR (2000) *Management of Health and Safety at Work Regulations* (1999) Approved Code of Practice L21. HSE books.

Minor PD, Will RG, Salisbury D (2001) *Vaccine*; 19: 3607.

Miyata H, Kanazawa T, Shibuya K, Hino S (2005) *J. Gen. Virol.*; 86: 733–741.

Nakamura S, Shimazaki T, Sakamoto K, Fukusho Y, Ogawa N (1995) *J. Vet. Med. Sci.*; 57: 677–681.

Nicklas W, Kraft V, Meyer B (1993) *Lab. Anim. Sci.*; 43: 296–300.
Noguchi PD (1992) *Dev. Biol. Stand.;* 76: 47–50.
Oie S, Oomaki M, Yorioka K *et al.* (1998) *J. Hosp. Infection*; 38: 61–65.
Onions D (2001) In *Evolving Scientific and Regulatory Perspectives on Cell Substrates for Vaccine Development.* Eds Brown F, Lewis A, Peden K, Krause P. Karger, Basel; Vol. 106: 135–142.
Patzke S, Lindeskog M, Munthe E, Aasheim H-C (2002) *Virology*; 303: 164–173.
Perl A (2003) *Rheum. Dis. Clin. N. Am.*; 29: 123–143.
Pocchiari M (1991) *Dev. Biol. Stand.*; 75: 87–95.
Rabenau H, Ohlinger V, Anderson J *et al.* (1993) *Biologicals*; 21: 207–214.
Reinmuller B, Ljungqvist B (1999) *Nordic J. Contam. Cont. Cleanroom Technol.*; 28: 9–12.
Rollinson DE, Page WF, Crawford H *et al.* (2004) *Am. J. Epidemiol.*; 160: 317–324.
Schuurman R, van Steenis B, van Strien A, van der Noorda J, Sol C (1991) *J. Gen. Virol.*; 72: 2739–2745.
Sheeley H (1998) In *Safety in Cell and Tissue Culture.* Eds Stacey GN, Doyle A, Hambleton P. Kluwer Academic Publishers, Dordrecht, Netherlands: 173–188.
Simon M, Domok I, Pinter A (1982) *Acta Microbiol. Acad. Sci. Hung.*; 29: 201–208.
Sistare FD (2001) In *Evolving Scientific and Regulatory Perspectives on Cell Substrates for Vaccine Development.* Eds Brown F, Lewis A, Peden K, Krause P. *Dev. Biol. Stand.* Karger, Basel; Vol. 106: 123–132.
Stacey GN (2002) *Dev. Biologicals*; 111: 259–272.
Stacey GN (2004) *Human Fertility*; 7: 113–118.
Stacey G (2005) *Eur. Biopharm. Rev. winter edn*: 112–115.
Stacey GN, Hartung T (in press) In *Drug Testing in vitro: Breakthroughs and Trends in Cell Culture Technology.* Eds Marx U, Sandig V. Wiley-VCH.
Stacey GN, Possee R (1996) *Cytotechnology*; 20: 299–304.
Sweet BH, Hilleman MR, (1960) *Proc. Soc. Exp. Med.*; 105: 420–427.
Sun R, Grogan E, Shedd D *et al.* (1995) *Virology*; 209: 374–383.
Tedder RS, Zuckerman MA, Goldstone AH *et al.* (1995) *Lancet*; 346.
Uckert W, Fleischhacker M, Kettmann R (1986) *Virology*; 155: 742–746.
Vaughn JL (1991) *Dev. Biol. Stand.*; 76: 319–214.
World Health Organization (1993) *Laboratory Biosafety Manual.* World Health Organization, Geneva; second edition.
World Health Organization (1998) *Requirements for the Use of Animal Cells as in vitro Substrates for the Production of Biologicals.* Requirements for Biological Substances No. 50. WHO Technical Report Series No. 878. World Health Organization, Geneva.
World Health Organization (2003) Technical Report Series No. 908, Appendix 7: *Application of Hazard Analysis and Critical Control Point (HACCP) Methodology to Pharmaceuticals*, WHO, Geneva.
Weiss RA (2001) *Emerging Infect. Dis.*; 7: 153–154.
Xue W, Minocha HC (1996) *Vet. Microbiol.*; 49: 67–79.

Useful Web Sites

COSHH: a brief guide	http://ptcl.chem.ox.uk/MSDS/simplecoshh.pdf
EMEA	http://www.emea.europa.eu
European Pharmacopoeia	http://www.pheur.org
ICH Quality Risk Management	http://www.ich.org/LOB/media/MEDIA1957.pdf
FDA	http://www.fda.gov
FDA CBER Tissue Action Plan	http://www.fda.gov/tissue
Health and Safety Executive Books (UK)	http://www.hsebooks.com/books
WHO	http://www.who.int

32 Standardization of Cell Culture Procedures

G Stacey

32.1 INTRODUCTION

This chapter deals with the generic issues involved in the standardization of cell culture procedures over a very broad range of activities, from derivation of the original cell cultures to their use in biological assays and manufacturing processes. Cell cultures are inherently prone to variation and can be difficult to control, thus making standardization a critical activity to ensure that reliable data and products are obtained from cell-based systems. Appropriate use of reference materials, and preparation of qualified reference cell banks, are vital parts of this standardization process and will be addressed in the following sections.

32.2 INHERENT VARIABILITY OF CELL CULTURES

32.2.1 Instability

During the process of isolating new continuous cell lines the original cultures of primary cells and subsequent early passages are often observed to be extremely unstable, as exemplified by loss of chromosomes in hybridoma cultures or the karyological changes and loss of differentiated cell types in tumour cell cultures. However, following this initial period a stable cell line may arise that can be passaged many times without apparently changing its characteristics (for examples see Kuechler *et al.* 2002). Whilst genetic profiling of cell lines has, in many cases, demonstrated genetic stability over extended periods of *in vitro* culture, some cell lines have proved to be more prone to genetic changes than others (Stacey *et al.* 1992; Hampe *et al.* 1992). Furthermore, whilst each continuous cell line is known to have a typical consensus karyotype (i.e. modal chromosome complement), the frequency of particular karyotypes observed within a culture is known to fluctuate with passage (Table 32.1) around this modal number, and may change over time during *in vitro* culture (Cailleau *et al.* 1978).

Cell line studies for the purposes of establishing suitability for production processes will often follow key characteristics of the cells over many passages, and changes that may occur include change in DNA profile (Hampe *et al.* 1992; Racher *et al.* 1994; Stacey 1992) and increasing levels of *in vitro* colony-forming capability of cells in agar, which may be associated with cellular tumorigenicity (Figure 32.1). Finite cell lines, with a limited replicative capacity (e.g. MRC-5, WI-38), whilst considered highly stable genetically and phenotypically, undergo a progressive ageing process, associated with loss of telomere length amongst other changes (Swim & Parker 1957; Allsopp *et al.* 1992). Ultimately, this process results in cessation of growth where the culture is in a viable

Medicines from Animal Cell Culture Edited by G. Stacey and J. Davis

Table 32.1 Typical variation in karyology but not modal number with passage *in vitro*. Unpublished data courtesy of L. Young, NIBSC.

Cell line/type	Modal karyotype	Variation in range of chromosome number per cell determined at different time points
MRC-5/human diploid fibroblasts	46	43–50
HeLa/human cervical carcinoma	60	50–73
RK13/rabbit epithelial-like cells	66	57–73
L20B/recombinant mouse cell line expressing the human polio receptor	45	37–51
Vero	58	50–65

yet non-replicative state, commonly called senescence. This ageing process can be accelerated by *in vitro* conditions including suboptimal subculture procedures (Ryan *et al.* 1975), and the presence of certain antibiotics and analogues of amino acids and nucleotides (Rattan & Stacey 1994).

Microbial contamination will also have dramatic effects on cell cultures. Whilst in many cases a bacterial or fungal contamination will cause catastrophic loss of the cell culture, contamination with mycoplasma and viruses can become established as persistent infections of the cells, and these agents may replicate with the cells over long periods of *in vitro* culture (see Chapter 31). These persistent infections may substantially alter the characteristics of the cell line. In the case of mycoplasma there may be irreversible changes including genetic effects, physiological changes and transformation (Rottem & Naot 1998). Effects can also be quite subtle, including increased chromosomal abnormality (e.g. Polyanskaia & Samokisch 1999), induction of cytokines in the infected cells (Zurita-Salinas *et al.* 1996; Fabisiak *et al.* 1993) and interference with selection of hybrids (e.g. Boyle *et al.* 1981).

32.2.2 Irreproducibility

Reversible variation in characteristics of cell lines can also arise as a result of technical variation in the handling and culture of cells. If seeding densities in monolayer cultures for use in assays are not enumerated, and the assay is performed with different cell numbers at different times,

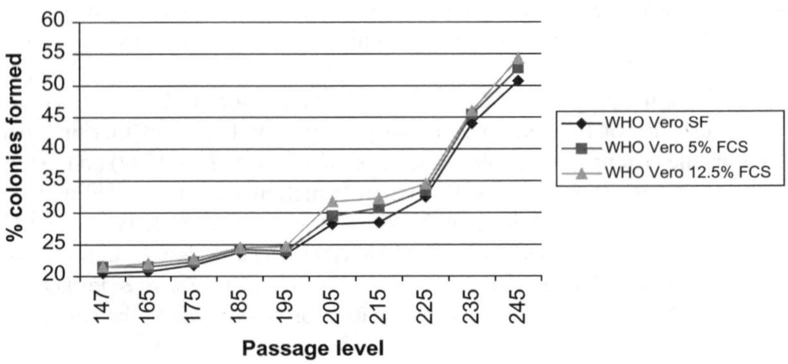

Figure 32.1 Increasing levels of *in vitro* colony forming capability in 0.3 % agarose as demonstrated for Vero cells. Unpublished data provided by L Young, NIBSC.

it is likely that there will be increased variability in data and possibly failure to provide reliable cultures for routine use. Also, if cells are not harvested at a reproducible point in the growth phase of the culture, they may again yield highly variable assay data. This may be particularly acute if cultures are used some time after confluency is reached, as in certain cell lines (such as MDCK and Caco-2) cell differentiation is initiated following contact inhibition of the confluent monolayer, and the function and structure of the cells will become progressively altered after the culture achieves confluency.

Variation in the maintenance of cell cultures can also cause apparent differences in the levels of protein expression in recombinant cell lines, and certain cells at high passage level are known to have altered characteristics (e.g. Caco-2) (Yu *et al.* 1997; Briske-Anderson *et al.* 1997) as may cells grown under altered culture conditions (Riley *et al.* 1991).

32.3 CRITICAL PARAMETERS OF CELL CULTURES

32.3.1 Viability

Many cell-based assays have end-point measurements based on cell 'viability'. This can be measured using diverse techniques that will all give subtly different assessments of cell growth and/or function (Table 32.2). Since different viability assays will measure different aspects of the cell's biology, (e.g. membrane integrity, enzyme activity, metabolic activity) it is important to ensure that an appropriate viability assay has been selected for the intended purpose. Not only is it possible for different viability assays to give different results from a single cell sample, it is also possible for a single bioassay to deliver different values for viability if the time for the endpoint measurement is varied within an otherwise standardized viability test protocol. A significant example is the neutral red assay where the end point measurement may need to be optimized for each cell type used (Babich & Borenfreund 1990; Borenfreund & Babich 1995). Also, in assays measuring biochemical activity such as the MTT [3-(4,5-dimethylthiazol-2-yl)-2,5-diphenyltetrazolium) assay, where the effect of a particular treatment is inhibitory rather than lethal to the activity, 'viability' may appear to drop significantly but subsequently recover to control levels as the cells survive a non-lethal insult. Cell-based bioassays may therefore require significant work to optimize the time point after application of test samples to the cells, and any incubation step within the viability assay that may be affected by the cell substrate performance. For a general reference on cytotoxicity assays see Freshney (2001).

32.3.2 Population Doublings

The *in vitro* age of a culture may not be known if the culture has been passed from one laboratory to another, although most laboratories would keep a record of the number of passages a particular culture has undergone in their hands, in recognition of the effects that can be observed at high passage levels of some cell lines. Each passage will increase the replicative age (cumulative number of cell divisions) of the culture, depending on a range of factors that may be deliberately adjusted, such as split ratio (i.e. a split ratio of 1:3 implies a passage of all cells from one flask into three flasks of the same size) and seeding density (cells/ml or cells/cm^2), or on factors intrinsic to the growth medium, reagents and characteristics of the cell line, such as viability after passage, and plating efficiency of viable passaged cells. The plating efficiency for a particular cell line under particular culture conditions can be established experimentally (Freshney 1994). A much more accurate measure of replicative age of a culture is provided by population doubling level (Freshney 1994). This may be estimated [e.g. $PDL = \{(\log_{10}$ cell count at harvest$) - (\log_{10}$ cell count at inoculation$)\}/0.301$] (Doyle & Griffiths 2000) through a general knowledge of the

Table 32.2 Comparison of different assays of cell viability.

Assay	Scientific basis	Reference	Complications	Applications
Trypan blue dye exclusion	Ability of functionally integral membranes to exclude dye from the cells	Patterson (1979)	Cells in early stage apoptosis retain integrity of cell membrane and will appear viable	General applications in research and biomedicine
Neutral red assay	Neutral red dye accumulates in lysosomes of live cells and can be measured spectrophotometrically	Babich and Borenfreund (1990)	Protocol may need to be optimized for particular cell types and toxicity assay May be affected by the presence of serum	Toxicology assays
3-(4,5-Dimethyl-thiazol-2-yl)-2,5-diphenyl-tetrazolium bromide (MTT)	MTT is reduced by viable cells to an insoluble formazan product	Mossman (1983)	Inhibition of biochemical activity may be misinterpreted as loss of viability	This assay is widely used as a convenient multi-sample assay by dissolving the formazan in dimethylsulphoxide and measuring using an ELISA 'plate reader'.
Lactate dehydrogenase assay	Spectrophotometric measurement of LDH enzyme released from cells	Wagner (1992) Racher (1990)	LDH activity and stability can vary with cell type. Culture conditions influence cellular LDH content Incomplete cell lysis will give low values.	Used in a wide range of applications including analysis of cell cultures in bioreactors
Fluorescein diacetate	Activity of membrane esterases in cells with complete membranes	Widholm (1972)	In some instances dying cells can give positive results	General applications in cell culture work

plating efficiency and seeding density, but is best calculated following enumeration of cells at each passage and an accurate determination of the plating efficiency for the cell line under specific culture conditions. Calculation of PDLs for some cell lines may be challenging, as cells that grow as colonies (some tumour cell lines and human embryonic stem cell lines), spheroids or natural aggregates, may be difficult to disaggregate in order to allow accurate cell counts, and careful evaluation may be required (Moreira *et al.* 1994) to avoid loss of cell viability that would give rise to false PDL values.

32.3.3 Culture Scale-up and Scale-down

Expanding cells on a large scale is most commonly achieved by passage of cells into more and more culture flasks. However, the requirement for larger single batches of product and the labour-intensive nature of manual subculture led to the use of chemical engineering technology in the form of stirred tank bioreactors (see Chapter 1). In these systems cells either had to be adapted to grow in suspension or on 'microcarriers' (see Chapter 10). The new mode of cell culture, the significant effects of 'mass action' in large volumes of medium and the need for culture agitation requires careful optimization to provide a standardized product. The challenges of controlling and standardizing such systems are dealt with in more detail in Chapters 13 and 14.

However, whilst these effects are well known in the development of large-scale production processes, the reverse process of scaling down can be equally problematic. In a scaled-down micro-well culture, such as those used for high throughput testing, the environmental factors may be altered but more significantly, certain cell cultures that have complex subpopulations or that may undergo differentiation can show marked well-to-well variation for smaller well sizes (i.e. below 24-well plate format). This might be anticipated for certain tumour cell lines that grow as colonies (e.g. embryonal carcinoma cells, human embryonic stem cells). However, it also occurs in some continuous cell lines such as Caco-2, in which the cells differentiate to produce large zones of specialized cells (Vachon *et al.* 1992). In the case of such cells used for testing and particularly in high throughput analytical systems, there will be a natural lower limit below which standardization may become impossible.

The utility of new technologies intended to analyse responses in single cells, such as fibre optics (Vo-Dinh & Kasili 2005), could clearly be affected by the kind of intra-culture variation described above and other epigenetic variations that may occur between cells even in apparently homogeneous cell populations. In such cases the individual cells studied will need to be appropriately characterized and controlled or sufficient cells analysed to give representative results.

32.4 CONTROLLING VARIATION IN CELL CULTURES

32.4.1 Controlling the Source of Cells

The source of any cell preparation or cell line used should be well characterized and qualified for its particular purpose. For cells and tissues taken directly from animals or humans this may be achieved through careful documentation and tracking of the material. For cultures that can be cryopreserved, it is possible to quality control the cells prior to use. When obtaining cell lines, it is vital to obtain cells from a laboratory that has carried out appropriate quality controls to ensure the fundamental characteristics required for all cell culture work:

- identity, i.e. the cells are what they purport to be;
- purity, i.e. freedom from microbiological contamination;
- stability on growth or passage *in vitro*.

Sourcing cells from unqualified sources can potentially lead to invalid data and wasted time and resources if the cells have been inadvertently switched or cross-contaminated (MacLeod *et al.* 1999; Masters *et al.* 2001; Stacey *et al.* 2000). The laboratory that originally isolated the cell line would seem the most desirable source for ensuring that the correct cell line is obtained. However, some research laboratories perform little basic quality control and have even been reported to provide the wrong cells inadvertently (MacLeod *et al.* 1999). It is generally a good principle to obtain cell lines from culture collections or other *bona fide* biological resource centres that endeavour to ensure that the cells have received at least the basic quality control discussed above. However, it will be extremely difficult or impossible to demonstrate that a particular cell line is authentic unless samples of tissue from the original donor or animal are available for comparison. Where new cell lines are to be established it is therefore wise to arrange for a sample of original cells or tissue to be retained for genotyping.

In the case of identity testing, cell lines have traditionally been tested using karyology and isoenzyme analysis, which provide confirmation of the species of origin of the cells and in some cases may also specifically identify a particular cell line or strain of origin respectively (Doyle & Stacey 2000). More recently DNA fingerprinting and DNA profiling have been taken up, as they can provide individual-specific identification in a single test (Gilbert *et al.*1991; Stacey *et al.* 1992, 2006). Reliable and specific profiling methods, developed for paternity testing and forensic work, are readily available for human cells and numerous companies now provide relatively cheap human DNA profiling services. However, there are relatively few other species where equivalently validated methods are readily available, and reference methods that can provide cell line-specific profiles over a broad range of species remain valuable tools (Stacey *et al.* 1992, 2006). It is also important to recognize that a DNA profile simply provides a DNA 'bar code' of allele sizes that can confirm identity when compared with a sample of the original tissue, but cannot assure that key phenotypic characteristics have remained. In this respect, proteomic analysis may prove useful in screening for phenotypic consistency (Wagg & Lee 2005).

Freedom from microbiological contamination of cell lines is important since contaminated cultures may: (i) be lost due to cell death; (ii) have altered characteristics that affect the performance of the cells and the reliability of resulting data; (iii) cross-contaminate other cells in the laboratory, and (iv) represent a hazard to laboratory workers. Detection of common contaminants is discussed below in relation to cell bank testing, and issues relating to worker and patient safety are explored in Chapter 31.

The stability of a cell line's features *in vitro* may relate to its susceptibility to change under altered culture conditions or the potential for the cells to undergo genetic drift on serial passage (see Figure 32.1). Each of these possibilities must be investigated experimentally and this is discussed further below in relation to cell banking.

32.4.2 Standardizing the Cell Culture Environment and Procedures

When proceeding to standardize a cell culture process, it is important to evaluate the potential sources of variability in all starting materials including all elements in direct contact with cells (e.g. media, culture supplements, culture surfaces). It is also important to consider those additional factors that will have a direct impact on the growth and response of the cells (e.g. temperature, gas atmosphere). For more unusual culture systems, additional features such as pressure, gravity, shear stress and magnetic fields may also need to be controlled, as the range of applications of *in vitro* cell cultures grows. In most types of cell culture the level of carbon dioxide can have significant effects on cells (Swim & Parker 1958; deZengotita *et al.* 2002; Chakrabarti & Chakrabarti 2001) as will the incubation temperature, and in all laboratory work with cells, some control on the calibration and monitoring of carbon dioxide and temperature for cell culture should be made. As cell-based systems for manufacturing, testing and therapy become increasingly complex, it will be vitally important to document and control cell culture reagents and materials that influence the growth and response of cells. Critical reagents (as defined above) should be carefully documented and specified. Different manufacturing methods for reagents and their purification may affect cells in different ways, and the activity of some key biological molecules, such as growth factors, can vary subtly in different preparations (e.g. post-translational modification, aggregation, dissociation, degradation and contamination), which may well affect the cell response (see Chapters 22–24).

A further source of variability amongst cell cultures is the method of subculture. Separate workers passaging the same cell line even in the same laboratory may produce different results (e.g. apparent product expression levels) and this effect will also be observed between different laboratories. In addition some continuous cell lines, such as Caco-2, can undergo permanent loss of key functions if subjected to suboptimal passaging.

Over the many years that microorganisms and cell cultures have been used in research and industry, a fundamental principle that has emerged is the establishment of a well characterized cryopreserved master seed-stock from which to develop all future cultures. The ability to cryopreserve a batch of vials each containing identical preparations of cells (a cell bank) that can be characterized, safety tested and made available for use over many decades, is a vital tool for reliable research and development of safe and standardized cell-derived products and cell therapies.

An approach developed over many years has been to establish the master bank, subject it to appropriate characterization and quality control and then to use individual vials of this stock to generate large 'working' cell banks that are retested for critical characteristics prior to use. This tiered system is central to assuring long term provision of good quality cells (both prokaryotic and eukaryotic) and should be considered best practice for any cell culture laboratory (Figure 32.2).

The quality control applied to each cell bank should include, as a minimum, some means of determining viability, mycoplasma testing and a sterility test (Stacey 2002). The suitability of the viability test method used should be established for each cell line, and the specificity and sensitivity of the mycoplasma and sterility tests should be carefully considered, as should the use of appropriate controls and reference strains and materials. It should be borne in mind that some fastidious organisms may not be detected in standard protocols, and sterility tests may need to be extended to include microbiological growth media that will support the growth of organisms dependent on carbon dioxide, anaerobic conditions and serum.

The preparation of a master cell bank provides an important reference point for any project in which cell cultures are a critical component, and reference cell banks have been formally adopted by the WHO for large international collaborative projects (WHO 2003a,b). However, it is also important to recognize that cells are prone to variation when individual bank vials are recovered, and careful specification of the culture media, conditions and subculture regime are important to assist the standardization of results in different laboratories and over time. In addition, for critical applications it is good practice to passage cells beyond the expected limit of use, establishing extended cell banks at intervals of several passages and then comparing the characteristics of these banks in parallel to determine any drift that may occur in the culture over time.

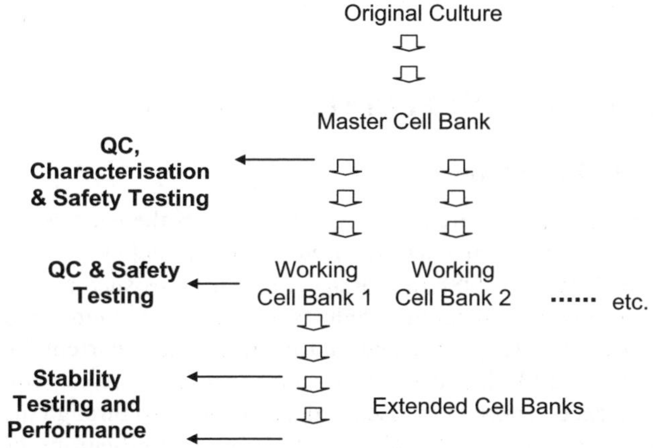

Figure 32.2 Master and working cell bank principle.

32.4.3 Reference Materials

It is quite unexceptional for laboratories on a national or worldwide basis to use a variety of different bioassay techniques for the same biological assay. Even where the same technique or kit is used, local conditions and raw materials may well vary significantly. In order to achieve a measure of how bioassay precision and accuracy varies between laboratories, a valuable approach has been to establish reference materials of a particular analyte comprising large batches of samples prepared from a homogeneous batch of material. The batch is assigned a nominal biological activity (in 'international units') and this is used periodically to qualify other local working standards used to control experiments. This process enables assessment of assay accuracy and any drift of the assay over time. Such reference materials are fully qualified through extensive collaborative studies using the reference preparation in a broad range of laboratories (Seagroatt & Kirkwood 1986; Saldanha 1999). Many such international reference materials (IRMs) have been established through the World Health Organization's Expert Committee for Biological Standardization, which scrutinizes and approves such IRMs and their validation studies through an established protocol (WHO 1995). A wide variety of different IRMs have been produced for different cell-based bioassays, including virus standards (e.g. The WHO Polio Laboratory Manual, see www.who.int/), antibodies and hormones. Very recently, cells themselves were used to provide genomic DNA reference material for genetic testing (Barton *et al.* 2004). Already some IRMs of cytokines are being used to help standardize cell therapies and there will no doubt be more candidate reference materials to assist in new stem cell therapies and applications of stem cells for *in vitro* diagnosis and in tissue engineering.

32.4.4 Best Practice Guidelines for Cell Culture

For research and development work there are numerous books written from the experience of individuals on good cell culture practice (Davis 2002; Doblhof-Dier & Stacey 2000; Freshney 1994; Masters 2000). Guidelines providing consensus advice on cell culture have been developed by expert groups (e.g. Masters *et al.* 2000; Hartung *et al.* 2002; Coecke *et al.* 2005). The most recent consensus guidance, coordinated by the European Centre for the Validation of Alternative Methods (ECVAM) (Coecke *et al.* 2005), identified six core 'principles' (Table 32.3) for cell and tissue culture that address the broad range of issues of importance to laboratory workers. In addition, in order to standardize the preparation of cells for use in an assay, the degree of reliability of performance of the cells delivered for such use should be determined and optimized (Stacey, 2002).

32.5 CONSIDERATIONS IN SPECIALIST FIELDS

32.5.1 Vaccines and Production

For cell lines used for the manufacture of biological medicines there are regulatory guidelines to ensure that the cells meet the fundamental requirements for acceptance for use in manufacturing (CBER 1993; ICH 1998; WHO 1998) and for specific cell types such as tumorigenic cell lines (ICH 2001). Standardization of cell culture products is achieved through the detailed documentation and validation of procedures, equipment and facilities under current Good Manufacturing Practice (see Chapters 34,12,13,14, and 15). Cell culture products are subsequently standardized through the use of reference materials as discussed above, but also through the definition and use of reference methods (www.jrc.irmm.be; Emons *et al.* 2004). Cell-culture-derived products for therapeutic or prophylactic application are also subject to testing requirements on a batch-by-batch basis (www.who.int/biologicals).

Table 32.3 Principles of good cell culture practice (Coecke *et al.* 2005).[a]

GCCP principle	Key components	Relevance
(1) Establishment and maintenance of a sufficient understanding of the *in vitro* system and of the relevant factors that could affect it.	Authenticity, including identity, provenance and confirmation of genotypic and/or phenotypic characteristics Ensuring freedom from microbiological contamination Stability and functional integrity of the system in relation to its intended use.	Ensuring that data from cell culture systems are not invalidated due to fundamental lack of understanding about the effects of cross-contamination, microbial contamination or the culture environment and its effects on cells.
(2) Assurance of the quality of all materials and methods, and of their use and application, in order to maintain the integrity, validity, and reproducibility of any work conducted.	Cells and tissues: authentication and monitoring Culture conditions: specification and monitoring of critical reagents/culture-ware Specification, calibration and monitoring of critical equipment that may influence cell growth/ or product.	Ensuring that work is not invalidated due to lack of control on cells and the culture environment
(3) Documentation of the information necessary to track the materials and methods used, to permit the repetition of the work, and to enable the target audience to understand and evaluate the work	Origins of cells and tissues Handling, maintenance and storage Reporting including public communication	Providing traceability which is an essential scientific requirement to enable experimental work to be repeated. The ability to report data accurately is also fundamental, including the ability to communicate with non-scientific audiences
(4) Establishment and maintenance of adequate measures to protect individuals and the environment from any potential hazards.	Physical: including proper handling of liquid nitrogen during cryopreservation of cells and tissues and retrieval of vials from frozen storage Biological: including proper use of laminar-flow cabinets	Providing for safe working and living environments
(5) Compliance with relevant laws and regulations, and with ethical principles	Reference to key areas of law and ethics that may well impinge on the work of scientists working with cell cultures of human and animal origin	Ensuring that scientists are aware of the legal and ethical issues that may affect their work in an environment where they are increasingly accountable for the work they perform
(6) Provision of relevant and adequate education and training for all personnel, to promote high quality work and safety.	Approach to training and monitoring Core culture techniques General training of key importance to quality of cell culture work	Appropriately trained staff are the greatest resource in any science laboratory, and staff training is critical to efficient, accurate and valid data.

[a]*For a PDF version of the GCCP 2005 document see www.ecvam.jrc.it*

32.5.2 Toxicology Testing

Toxicology testing of products has traditionally been performed using *in vivo* tests on animals but there has been a long history of sustained pressure to eliminate the use of animals for this purpose, and this has been enshrined in what is universally known as the 3Rs principle (for reduction, refinement and replacement of the use of animals for testing) (Russell & Burch 1959). The European Centre for the Validation of Alternative Methods (Ispra, Italy) has been the official European laboratory coordinating the validation of replacement cell-based toxicological test methods. ECVAM has published formal procedures for such validation (Balls & Fentem 1997), and at an international level the OECD has coordinated the development of Good Laboratory Practice (GLP) as the overarching guidance on product safety testing in general (OECD 2004; see also Chapter 33). In addition, specific test protocols have been standardized and published as pharmacopoeial standards (e.g. European Pharmacopoeia, US Pharmacopoeia).

32.5.3 Diagnosis

Cell cultures have been used for many years for the detection of viruses, and standardization of cell lines used for these purposes has been organized on an international basis by the World Health Organization. Today cell lines are being developed for an increasingly wide variety of other diagnostic tests including the detection of functional antibody in the serum of vaccinees (Fleck *et al.* 2003) and cytokine assays (Wadhwa & Thorpe, 1998; Meager *et al.* 2001). In addition it should be remembered that monoclonal antibodies, cell culture products that are hugely important for a diverse range of biological assays, are themselves standardized through the use of reference materials (RMs). One further rapidly developing area is use of cells to provide validated RMs in the form of cell line genomic DNA for HLA typing and diagnosis of genetic disorders (Barton *et al.* 2004).

32.5.4 Therapeutic Applications

The guidance required for new therapeutic products derived from stem cell and other human and animal cell lines is a combination of those already described and guidelines for human tissues for transplantation. In some countries within Europe, codes of practice are in use (e.g. DH 2001, DH 2002, JACIE 2000) and the European Human Tissues Directive will provide the overarching regulatory framework in Europe with two technical annexes providing guidance. Similar guidelines have also been established for the regulation of human tissues and cellular products in the USA (www.fda.gov/cber/tissue/docs.htm).

As already discussed, reference materials are vitally important in the control of biological assays and batch release for procurement, processing, labelling and storage (EU 2004). In the US, tests for biological products, such as cytokines and growth factors, are controlled using pharmacopoeial protocols and international reference materials (Meager *et al.* 2001; Wadhwa & Thorpe 1998). Tissue-engineered products will also require some elements of standardization and in some countries guidance is already in place (e.g. DH 2002). In Europe an advisory group has been established for the evaluation of the quality and safety of 'advanced medical therapies' (EU 2003). The rapidly developing field of stem cell research and the establishment of human embryonic stem cell lines also provides great promise for future therapies. However, there are only a few examples of specific guidance available for the preparation and use of such cell lines (UK Steering Committee for the UK Stem Cell Bank and the Use of Stem Cell Lines 2003, 2005) and international standardized approaches for the characterization of stem cell lines are just beginning to be developed (Andrews *et al.* 2005).

32.6 CONCLUSIONS

Cells themselves can be very difficult to standardize. However, control of the cell culture environment, including physical conditions as well as culture reagents, currently provides the best means to provide reproducible cell cultures for use in research, industry and medicine. It is highly likely that proteomics tools will play an important role in the analysis of cell culture systems for various applications, and it will be important to provide appropriate guidance and reference materials to promote standardization of results.

The laboratory operator probably will remain a significant variable along with the cells themselves and, accordingly, control of the culture environment and documented and quality controlled cell banking procedures should be accompanied by a careful staff training programme to ensure competence in laboratory procedures and awareness of the factors that could affect the quality of the cell cultures and their *in vitro* response.

REFERENCES

Allsopp, RC, Vaziri H, Patterson C. *et al.* (1992) *Proc. Natl Acad. Sci. USA*; 89: 10114–10118.

Andrews PA, Benvenisty N, McKay R *et al.* (2005) *Nature Biotechnol.*; 23: 795–797.

Babich H, Borenfreund E (1990) *ATLA*; 18: 129–144.

Balls M, Fentem JH (1997) *ATLA*; 25: 33–43.

Barton D, Klein C, Stacey GN (2004). In *Encyclopaedia of Diagnostic Genomics and Proteomics.* Eds Fuchs J, Podda M. Marcel Dekker, New York.

Borenfreund E, Babich H (1995). In *Cell and Tissue Culture Laboratory Procedures.* Eds Griffiths JB and Doyle A. John Wiley & Sons, Chichester; 4B 7.1–7.7.

Boyle JM, Hopkins J, Fox M, Allen TD, Leach RH (1981) *Exp. Cell Res.*; 132(1):67–72.

Briske-Anderson MJ, Finley JW, Newman SM (1997) *Proc. Soc. Exp. Biol. Med.*; 4: 248–257.

Cailleau R, Olive M, Cruciger QVJ (1978) *In Vitro*; 14: 911–915.

CBER (Center for Biologics Evaluation and Research) (1993) *Points to Consider in the Characterization of Cell Lines Used to Produce Biologicals.* Food and Drugs Administration, Bethesda, USA (http://www.fda.gov/cber/).

Chakrabarti R. Chakrabarti R (2001) *J. Cell Biochem.*; 83: 200–203.

Coecke S, Balls M, Bowe G *et al.* (2005), *Guidance on Good Cell Culture Practice.* A report of the second ECVAM Task Force on Good Cell Culture Practice, *ATLA*; 33: 1–27.

Davis JM (2002) *Animal Cell Culture: A Practical Approach.* Oxford, University Press, oxford, UK; second edition.

DeZengotita VM, Schelzer AE, Miller WM (2002) Biotechnol. Bioeng.; 77: 369–380.

DH (2001) *Code of Practice for Tissue Banking.* UK Department of Health, TSO, Sudbury, UK. www.dh.gov.uk/Publications

DH (2002) *Code of Practice for Production of Human-derived Therapeutic Products.* UK Department of Health, TSO, Sudbury, UK. www.mhra.gov.uk/home (see "How We Regulate – Tissue Engineered Products")

Doblhoff-Dier O, Stacey G (2000). *Cell Lines: Applications and Biosafety. Biological Safety–Principles and Practices*, third edition. Eds Fleming DO Hunt DL. ASM Press Washington; 221–241.

Doyle A, Griffiths JB (2000) In *Cell and Tissue Culture for Medical Research.* Eds Doyle A, Griffiths JB. John Wiley & Sons, New York.

Doyle A, Stacey GN (2000). In *Encyclopedia of Cell Technology.* Ed Spier R. John Wiley & Sons, New York; 285–293.

Emons H, Linsinger T, Gawlik B (2004) *Trends Anal. Chem.*; 23, 442–449.

EU (2003) Directive 2003/63/EC Part IV: *Advanced Medical Therapy Products.* http://pharmacos.eudra.org/F2/advtherapies/docs/COM_2005_567_EN

EU (2004) Directive 2004/23/EC of the European Parliament and the Council of 31 March 2004 on setting standards of quality and safety for the donation, processing, preservation, storage and distribution of human tissues and cells. *Official Journal of the European Union*; L102: 48–58.

Fabisiak JP, Weiss RD, Powell, Dauber JH (1993) *Am. J. Respir. Cell Mol. Biol.*; 8:358–364.

Fleck RA, Athwal H, Bygraves JA, Hockley DJ, Feavers IM, Stacey GN (2003) *In Vitro Cell Dev. Biol. Anim.*; 39(6):235–242.

Freshney IR (1994) *Culture of Animal Cells.* Wiley Liss Inc. New York; third edition: 282.

Freshney IR (2001) In *Cell Culture Methods for in vitro Toxicology.* Eds Stacey GN, Doyle A, Ferro M. 213–230.

Gilbert DA, Reid YA, Gail MH *et al.* (1991) *Am. J. Hum. Genet.*; 47: 499–514.

Hampe J, Nurnberg P, Epplen C, Jahn S, Grunow R, Epplen JT (1992) *Human Antib. Hybridomas*; 3: 186–190.

Hartung T, Balls M, Barduoille C, Blanck O, Coecke S, (2002) *ATLA*; 30: 407–414.

ICH (1998) Topic Q 5 D *Quality of Biotechnological Products: Derivation and Characterisation of Cell Substrates Used for Production of Biotechnological/Biological Products* (CPMP/ICH/294/95), ICH Technical Coordination, European Medicines Evaluation Agency, London.

ICH (2001) *Position Statement on the Use of Tumourigenic Cells of Human Origin for the Production of Biological and Biotechnological Medicinal Products* (CPMP/BWP/1143/00), ICH Technical Coordination, European Medicines Evaluation Agency, London.

JACIE (Joint Accreditation Committee of ISHAGE-Europe and EBMT) (2000) *Cytotherapy*; 21: 225–246 (for updates see www.jacie.org).

Kuechler A, Weise A, Michel S, Schaeferhenrich A, Pool-Zobel BL, Claussen U, Liehr T (2002) *Genes Chromosomes Cancer*; 34:1–8.

MacLeod RAF, Dirks WG, Matsuo Y, Kaufman M, Milch H, Drexler HG (1999) *Int. J. Cancer*; 83: 555–563.

Masters JRW (2000) *Animal Cell Culture: A Practical Approach.* Oxford University Press, Oxford, UK; third edition.

Masters JRW, Thompson J, Daly-Burns B *et al.* (2001) *Proc. Natl. Acad. Sci. USA*; 98: 8012–8017.

Masters JRW, Twentyman P, Arlett C *et al.* (2000) *Bnt. J. Cancer*; 82: 1495–1509.

Meager A, Gaines-Das R, Zoon K, Mire-Sluis A (2001) *J. Immunol. Methods*; 257: 17–33.

Moreira, JL, Alves PM, Aunins JG, Carrondo JM (1994) *App. Micr. Biotechnol.*; 41: 203–209.

Mossman T (1983) *J.Immunol. Methods*; 65: 55–59.

Nelson-Rees WA, Daniels DW, Flandermeyer RR (1981) *Science*; 212: 446–452.

OECD (2004) *Draft Advisory Document of the OECD Working Group on the Application of GLP Principles to* in vitro *Studies.* OECD, Paris.

Patterson MK (1979) *Methods Enzymol.*; 58: 141–152.

Polyanskaia GG, Samokisch VA (1999) *Tsitologiia*; 41: 752–757.

Racher AJ, Looby D, Griffiths JB. (1990) *J. Biotechnol.*; 15: 129–146.

Racher AJ, Stacey GN, Bolton BJ, Doyle A, Griffiths JB (1994) In *Animal Cell Technology: Products of Today, Prospects for Tomorrow.* Eds. Spier RE, Griffiths JB, Berthold W. Butterworth-Heinemann, Oxford; 69–75.

Rattan SIS, Stacey GN (1994) In *Cell and Tissue Culture Procedures.* Eds Doyle A, Griffiths JB, Newell DG, John Wiley & Sons, Chichester; 6D: 2.1–2.12.

Riley SA, Warhurst G, Crowe PT, Turnberg LA (1991) *Biochim. Biophys. Acta*; 1066: 175–182.

Rottem S, Naot Y (1998) *Trends Microbiol.*; 6: 436–440.

Russell WMS, Burch RL (1959) *The Principles of Humane Experimental Techniques.* Methuen, London UK.

Ryan JM, Sharf BB, Cristofalo VJ (1975) *Exp. Cell Res.*; 91: 389–392.

Saldanha J (1999) *Biologicals*; 27: 285–289.

Seagroatt V, Kirkwood TB (1986) *Lymphokine Res.*; 5 (suppl.): S7–11.

Stacey GN (2002) *Dev. Biologicals*; 111: 259–272.

Stacey GN, Bolton BJ, Morgan D, Clark SA, Doyle A (1992) *Cytotechnology*; 8: 13–20.

Stacey GN, Hoelzl H, Stephenson JR, Doyle A (1997) *Biologicals*; 25: 75–83.

Stacey GN, Masters JRW, Hay R, Drexler HG, MacLeod RAF, Freshney RI (2000) *Nature*; 403: 356.

Stacey GN, Byrne E, Hawkins R (2006) *Animal Cell Biotechnology Methods and Protocols.* Humana, in press.

Steering Committee for the UKSCB and for the Use of Stem Cell Lines (2003) *Code of Practice for the UK Stem Cell Bank.* Secretariat for the UK Steering Committee for the UK Stem Cell Bank and for the Use of Stem Cell Lines, Medical Research Council, 20 Park Crescent, London. (www.mrc.ac.uk)

Steering Committee for the UKSCB and for the Use of Stem Cell Lines (2005) *Code of Practice for the Use of Stem Cell Lines*. Secretariat for the UK Steering Committee for the UK Stem Cell Bank and for the Use of Stem Cell Lines, Medical Research Council, 20 Park Crescent, London. (www.mrc.ac.uk)

Swim HE, Parker RF (1957) *Am. J. Hygiene*; 66: 235–243.

Swim HE, Parker RF (1958) *J. Biophys. Biochem, Cytol.*; 4: 525–528.

Vachon PH, Bealieu J-F (1992). *Gastroenterology*; 103: 414–423.

Vo-Dinh T, Kasili P (2005) *Anal. Bioanal. Chem.*; 382(4); 918–25.

Wadhwa M, Thorpe R (1998) *J. Immunol. Methods*; 219: 1–5.

Wagg SK, Lee LE. (2005) *Proteomics*; 5: 4236–4244.

Wagner A, Marc A, Engasser JM, Einsele A, (1992) *Biotech. Bioeng.*; 39: 320–325.

WHO (1995) *Recommendations for the Preparation, Characterisation and Establishment of International and other Biological Reference Standards.* WHO/BS/04.

WHO (Expert Committee on Biological Standardization and Executive Board) (1998) *Requirements for the Use of Animal Cells as* in vitro *Substrates for the Production of Biologicals* (Requirements for Biological Substances No. 50). WHO Technical Report Series No. 878. Geneva: World Health Organization.

WHO (2001) *Requirements for the Use of Animal Cells as* in vitro *Substrates for the Production of Biologicals,* WHO, Geneva, Switzerland.

WHO (2003a) *WHO Tech. Rep. Ser.*; 926.

WHO (2003b). *Requirements for the Use of Animal Cells as* in vitro *substrates for the production of biologicals*: Addendum 2003. WHO, Geneva, Switzerland.

Widholm JM (1972) *Stain Technol.* 47: 189.

Yu H, Cook TJ, Sinko PS (1997) *Pharm. Res.*; 14: 757–762.

Zurita-Salinas CS, Palacios-Boix A, Yanez A, Gonzalez F, Alcocer-Varela J (1996) *FEMS Immunol. Med. Microbiol.*; 15:123–128.

Web Sites

Department of Health (UK) code of practice for the production of human tissue-derived products (2002) (see *How We Regulate – Tissue Engineered Products)*	www.mhra.gov.uk/home
Department of Health (UK) code of practice for tissue banks (2001)	www.dh.gov.uk/Publications
European Centre for the Validation of Alternative Methods	www.ecvam.jrc.it
European Pharmacopoeia	www.pheur.org/
Food and Drug Administration, Center for Biologics and Evaluation and Research	www.fda.gov/cber/guidance
Food and Drug Administration, Center for Drugs Evaluation and Research	www.fda.gov/cder/guidance
International Conference on Harmonization	www.ich.org/
Joint Accreditation Committee for ISHAGE Europe	www.jacie.org
Institute for Research Methods and Materials	www.jrc.irmm.be
National Institute for Biological Standards and Control	www.nibsc.ac.uk/aboutus/research.html
Organization for Economic Cooperation and Development	www.oecd.org/
United States Pharmacopeia	www.usp.org/
World Health Organization	www.who.int/biologicals

33 Good Laboratory Practice for Cell Culture Processing

B Orton

33.1 OVERVIEW AND HISTORY

33.1.1 What Does Good Laboratory Practice Mean to You?

For most researchers good laboratory practice means working methodically, and making appropriate and accurate records in any laboratory experiment. So when you subculture your cells you would turn on the safety cabinet, collect fresh medium and the flask of cells and 'split' the cell suspension between new flasks according to the level of confluency or cell concentration and add new medium as normal. Afterwards you would clear up, returning the fresh culture to the incubator and spare medium to storage, cleaning the cabinet and disposing of the used solutions. Then if you just write a general record of the subculture in your laboratory notebook you have finished. After all, everyone in the laboratory knows these standard techniques.

However there is more to a formal Good Laboratory Practice (GLP) system than this. 'GLP is concerned with the organizational processes and the conditions under which certain laboratory studies are planned, performed, monitored, recorded, archived and reported. Adherence by test facilities to the principles of GLP ensures the proper planning of studies and the provision of adequate means to carry them out. It facilitates the proper conduct of studies, promotes their full and accurate reporting, and provides a means whereby the validity and integrity of the studies can be verified' (UK GLPMA 2000).

In the mid-1970s the US Food and Drug Administration (FDA) investigated some of the laboratories submitting data for animal toxicity studies in support of applications for licences of new medicines. This review identified careless and, in some cases, fraudulent practices. Inspectors found that some data was inaccurate and some reports submitted were completely fabricated, with no laboratory records of the experiments being performed. Following this review, in 1978 the US became the first country to introduce regulations for GLP. The first UK scheme was a voluntary code of practice published in 1982, which was based on the first OECD *Principles of Good Laboratory Practice*.

GLP is now the standard required by law in many countries for studies performed to determine the safety of materials with respect to human or animal health. In the UK and other EU states the regulations are equivalent to the revised OECD *Principles of GLP*. These were agreed by the OECD group of nations with the intention of harmonizing standards, so that data would be mutually acceptable in different countries. However some parties, such as the USA and Japan, who signed up to the OECD document have also issued their own standards. Although there are some differences the intent is still the same, because regulatory authorities need to be able to rely on the safety data submitted in support of applications to manufacture new medicines.

Medicines from Animal Cell Culture Edited by G. Stacey and J. Davis

GLP can be applied to many kinds of safety studies, for example:

- development of new medicines from cells;
- investigating the safety of pesticides and herbicides, e.g. residue studies;
- determining the environmental effects of industrial chemicals;
- toxicology of new pharmaceutical products (animal or human) to laboratory animals.

In the UK the requirements of GLP are laid down in *The Good Laboratory Practice Regulations* (Statutory Instrument 1999), applicable to all these studies. In the USA the FDA published *Good Laboratory Practice for Nonclinical Laboratory Studies*, but other government agencies produced separate regulations for other types of safety studies.

A specific GLP guide for toxicity studies that often involve cell cultures has also been published (and their 2004 amendment (statutory Instrument 2004)) as well as ISO standards for GLP in general (OECD 2004/ISO/IEC 2005).

To comply with the GLP requirements, senior laboratory managers need to set up quality systems to support the actual cell culture work involved in testing procedures. These will contribute to assuring the quality and integrity of the experiments. This is the purpose of GLP standards, and surely this aim is also worthwhile even when GLP is not legally required. If the work is compliant with GLP, it should be possible to determine exactly what was done and how, who did it and when, i.e. to reconstruct the experiment after the event from the documents and materials retained. The key people who ensure these aims are met in a GLP laboratory are *management*, the *study director* and *quality assurance* (auditing) personnel. Their responsibilities as described in the *Principles of GLP* will be explained in this chapter, together with general systems and the GLP study lifecycle.

33.2 ORGANIZATION

GLP allocates specific responsibilities to certain job roles in the laboratory, and therefore management should nominate staff to these positions. As well as staff who work directly on the studies, archive and auditing duties must be performed by independent personnel. A simple organization structure is illustrated in Figure 33.1.

The study director is a key person who has responsibilities for both the science and the GLP compliance of the study. He or she should be a senior scientist with expertise in the majority of the techniques of the study, e.g. an experienced cell scientist. However, good knowledge of GLP is also very important.

All personnel working in the laboratory need to be appropriately qualified and experienced for the work they are expected to perform. This should be demonstrable from their personal files of CVs and training records, which should be periodically reviewed and updated when training in

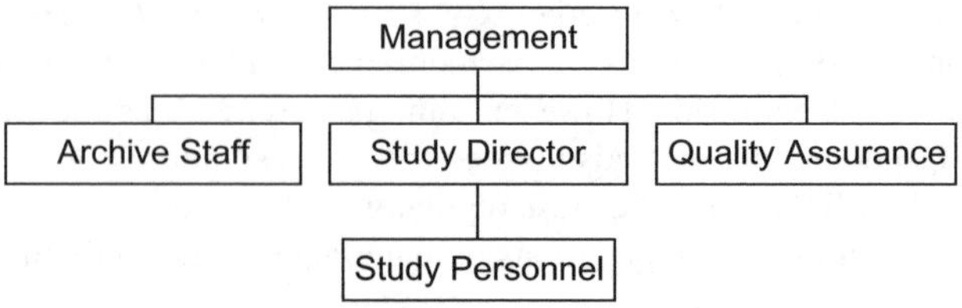

Figure 33.1 GLP Facility Organogram.

Table 33.1 Types of document in a GLP laboratory.

Document type	Examples
Facility documentation	Laboratory plans
	Organization chart
Standard operating procedures (SOPs)[a]	Control of SOPs
	Routine laboratory methods
	Labelling convention for cell cultures
	Equipment maintenance and calibration
	Media preparation and sterilization
	Archiving procedures
	Audit procedures
Study plans	To include study-specific methods and required measurements
Records and forms	Staff training records
	Job descriptions and CVs
	Equipment maintenance and calibration
	Media preparation and sterilization
	Temperature monitoring
	Archive index
	Archive transfer forms
	Audit records
Study reports	Study experimental methods, results and conclusions

[a]*See Section 33.3.3*

new procedures is received. Competency statements are often used to summarize training, especially for study directors and management.

33.3 GENERAL SYSTEMS

In order to operate to GLP, a laboratory needs to establish some general systems to support the conduct of the studies. These will include general documentation, facilities and equipment, standard operating procedures, archiving and auditing.

33.3.1 Documentation

Many different kinds of documents and records will be needed in a GLP laboratory. Examples are listed in Table 33.1.

Some laboratories may also choose to use higher-level documents such as quality manuals and policies in addition to SOPs.

33.3.2 Facilities, Equipment and Reagents

To perform any experiment it is essential to have adequate laboratory facilities. This applies to any work, whatever the quality standard required, or the results will not be reliable. Hence the rooms and areas, storage and waste disposal facilities must be appropriate for the studies. It is the responsibility of laboratory management to provide suitable facilities and apparatus for the experiments.

All equipment used to prepare reagents or maintain cultures needs to have acceptable capacity and accuracy, to be properly maintained, and calibrated where appropriate.

Critical equipment for a GLP cell culture laboratory usually includes:

- incubators (CO_2 supply, controlled humidity, monitored and with alarms);
- temperature-monitored fridges and freezers with alarms;
- Class II microbiological safety cabinets;
- centrifuges and waterbaths (controlled temperature);
- autoclave and filter sterilization equipment for media and other reagents.

For example, incubators for cell cultures need regular maintenance and temperature monitoring. A programme of scheduled tasks should be established to ensure that the equipment is fit for purpose. The precise details are for laboratory management to decide, e.g. the frequency and range of calibration of temperature probes, but once documented the programme should be followed. Usually a calibration label is attached to the item to allow staff to verify readily when the last calibration was performed and when the next is due, before they start work. The actual record of data values obtained during calibration must be retained and stored in an equipment log along with records of maintenance and repair.

In addition, any reagents used need to be made carefully or the cell cultures may be affected. Thus during the preparation of culture media the various components should be measured appropriately, and records of the preparation, including sterilization where necessary, kept to provide traceability. If there should be a problem with one of the cultures, an investigation could confirm that the media had been prepared correctly.

33.3.3 Are SOPs Any Use, or Are They Just Piles of Paper?

Standard operating procedures (SOPs) document the required methods for regular tasks. Describing clearly how to perform the routine operations means that all staff can be trained to perform them in the same manner, to ensure consistency and quality. Involving the users as far as practicable in the preparation of these documents helps to produce accurate procedures with the appropriate level of detail.

The procedures should be controlled so that they are available in the latest authorized format in any area where staff will need to use them. However if procedures are revised, all old versions in laboratory files must be withdrawn to ensure that the correct new procedure is followed.

Increasingly laboratories are implementing electronic document control, which has many advantages over paper systems; nevertheless the degree of control needs to be the same whichever method is used. Relevant electronic SOPs should be available to each laboratory area, which will probably mean making arrangements for printing copies. Any paper copies printed must not be stored long-term or they could become invalid because the electronic SOP version has changed. One way to achieve this could be to ensure that all printed copies are marked with a short expiry date, e.g. one week. There also need to be mechanisms to show authorization of the electronic procedures by senior management, and for the retention of the superseded documents.

Procedures that should be formally documented in this way in the cell culture laboratory are likely to include media preparation, temperature monitoring (either manual or describing the continuous monitoring system), routine laboratory cleaning, disinfection and discard techniques. Written SOPs are also needed for auditing and administrative-type tasks such as report preparation, archiving and maintaining training records.

33.3.4 Why Should so Much Paper be Archived?

After the completion of any study submitted to regulatory authorities there may be questions that can only be resolved by 'reconstructing' the work using the records. In addition, reference to the detailed experimental records could be useful to laboratory staff working on related experiments at a later date.

Management must designate a person to be responsible for managing the facility archive. As well as study-specific records, all equipment maintenance and calibration records, a copy of every superseded SOP, personnel training records and other general records need to be retained.

If data is captured electronically, it is of course possible to store it long term on, for example, CDs. In the USA the electronic records rule 21 CFR Part 11 describes the controls necessary when computerized systems are used. These controls include security aspects, data trails and archiving that are designed to ensure that electronically recorded data cannot be altered by unauthorised personnel or without any trace. If the requirements of this rule are met, the FDA will consider that electronic records may be used instead of paper. Subsequently guidance was published to clarify the FDA's intentions concerning the application of the Part 11 rule.

33.3.5 If Everyone is Experienced Why Are Audits Necessary?

Anyone can make mistakes or get into bad habits, and it has been found that an objective assessment by a person not closely involved in the work is the best mechanism to detect them. In addition, an impartial third party can sometimes point out that the description in a procedure or report is not quite clear to someone who has not worked with the study throughout its lifetime. The QA personnel are also reviewing the compliance status of the work with the GLP regulations, since they may have greater knowledge of the requirements than do study staff. Audits are required under GLP, but the auditors should also be able to make suggestions to improve the laboratory's quality systems for the benefit of all.

QA personnel must set up and perform a planned programme of quality audits, following the audit SOPs. The main purpose of these audits is to ensure that the GLP principles are being complied with. However audits could focus on different aspects such as verification of practical compliance with written SOPs, the adequacy and management of the quality systems, or traceability of data through the records. Any combination of facility, process and study-specific audits may be used, in accordance with the company's documented policy on audits. Facilities audits cover topics such as archiving, general equipment systems and training. Process audits can monitor certain procedures that are performed frequently in the laboratory, but are not specific to any one study, and there could also be study-specific audits of the practical work.

33.4 THE LIFECYCLE OF A GLP SAFETY STUDY

Each individual study performed has a well-defined process or 'life cycle', from initiation by the client (who may belong to the same organization) to completion and acceptance by QA and the client. The various stages of a GLP safety study are illustrated in Figure 33.2.

33.4.1 Study Initiation

Once it is agreed that the safety study is needed, laboratory management must appoint a study director. This person will have overall scientific responsibility for the study and its compliance with GLP. The first task is to prepare the study plan, which describes the objective, proposed starting and completion dates and detailed experimental procedures, including all materials and measurements required. Some studies may follow fixed protocols such as EP or USP methods, whereas others will be bespoke. Any routine procedures described in SOPs may be included by referring

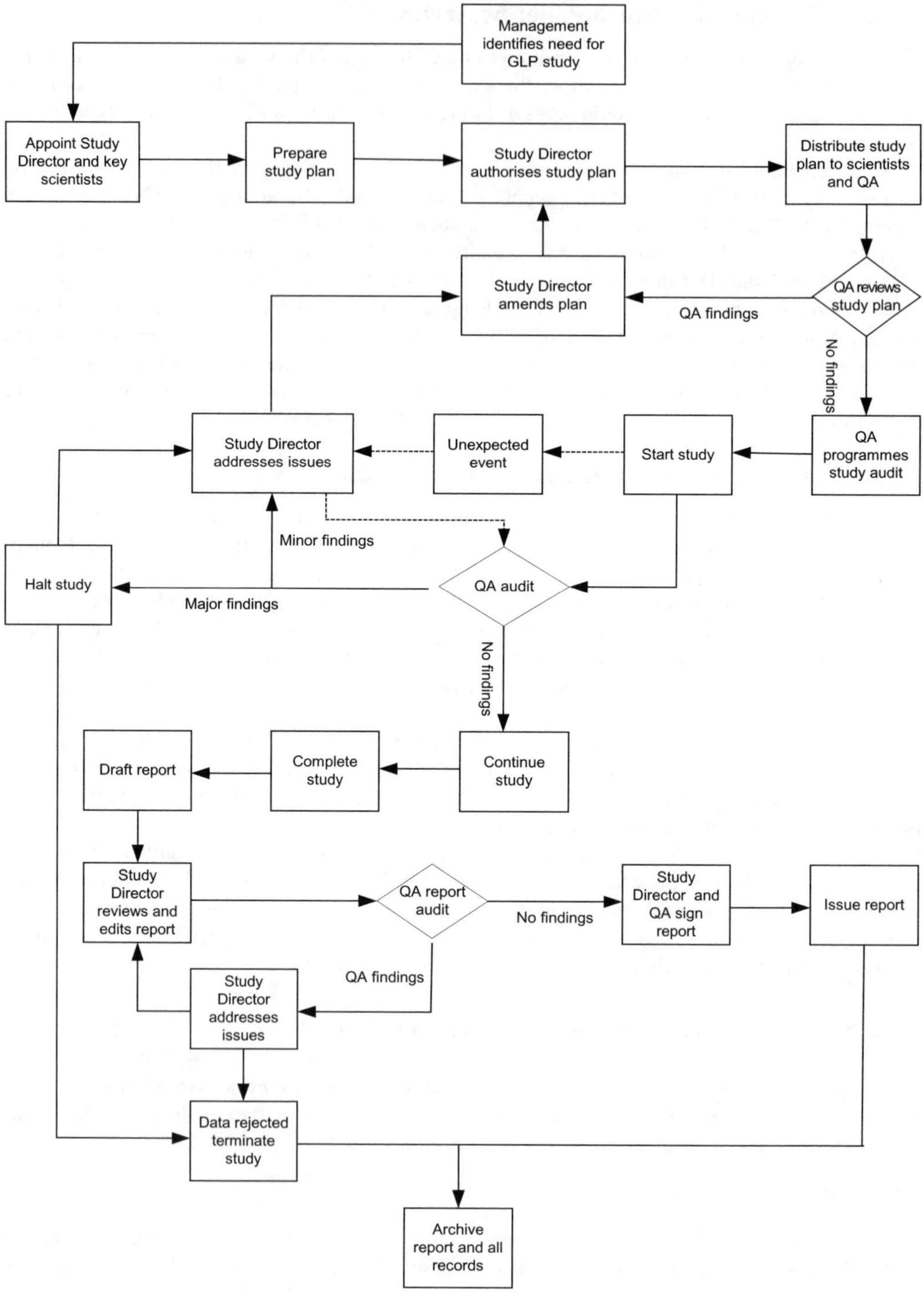

Figure 33.2 GLP study lifecycle.

to the relevant documents. The study director (in conjunction with the client) decides the design of the study, e.g. whether a piece of work is one larger study or several smaller ones. The study plan controls all the study activities, and must be authorized by the study director. Before it is signed, management must ensure that all the necessary resources are available to the study director – laboratory equipment, support services, and a sufficient number of scientists to allow the study to be performed according to the proposed timescale.

Under the UK *GLP Regulations*, the study plan must be reviewed by QA, to ensure it is compliant. This review informs QA of the impending study and allows a study audit to be scheduled.

33.4.2 Performing the Study

During the experimental stage of the study, all personnel must ensure that the procedures described in the study plan and referenced SOPs are followed and all necessary details recorded. If there should be any problems, unexpected events or results, the study director must be informed promptly in order that he or she can consider the implications. It is possible that there may be consequences for other parts of the study, and in some circumstances an amendment to the study plan might be needed. The study director must assess and document the impact of any deviations from the study plan on the quality and integrity of the study.

It is the responsibility of all personnel to record promptly, accurately and legibly all experimental methods and results obtained. These records will be important to show that the correct methods were used. Writing details on scraps of paper or informal pocket notepads is not acceptable and formal methods of documentation must be established. In addition, any changes made in the records must provide a full audit trail showing who made the change, when and why. Where data is captured automatically, such as for temperature monitoring, it is still essential to have audit trails and retain the data long term, whether electronically or in paper format.

In order to ensure that all the appropriate experimental details are recorded it is often helpful to design standard forms for experimental stages that are common to several studies. These can incorporate method instructions as well as confirmation of performing the steps, identification and quantities of reagents used, and any measurements and observations.

During all experimental work the test material must be handled appropriately to protect against contamination and mix-ups. It is also a GLP requirement to record the amounts used in the study, in order to be able to reconcile the quantity supplied with what remains. For most studies (short-term studies are exempt in the UK) it is also necessary to retain samples of all batches used, for analytical purposes.

33.4.3 Study Monitoring

Many organizations have a policy that every GLP study is audited by QA during the practical stage. This is not a requirement in the UK, but it is for companies working in the USA and inspected by the FDA. A study audit is conducted at a time agreed with the study director, depending on the study stage selected for audit, and is intended to provide independent confirmation that procedures specified in the study plan are being correctly followed and appropriate records made. The QA auditor reports any observations to the study director for corrective actions to be initiated.

33.4.4 Study Reporting

Once all experimental data is obtained, the study scientists will draft the report, describing the methods, results and conclusions, following the reporting SOP. After review by the study director all GLP study reports must be audited by QA, who will record any errors or discrepancies noted. However it is the responsibility of the study scientists and study director to prepare an accurate report, and they should not assume that any and every error will be found by QA.

The purpose of the QA report audit is to ensure that the laboratory's procedures for report preparation reliably produce a true and accurate record of experiments conducted. The study director must address any comments made by QA before the report can be issued. The study director then signs a statement concerning the GLP compliance status of the study, and if there have been significant GLP deviations these must be described. All audits performed by QA on the study must be listed in a statement in the report that is signed by QA, and also confirms that the report reflects the raw data.

33.4.5 Study Archiving

Once the study has been reported it is the responsibility of the study director to ensure that all study-specific documents are transferred formally to the laboratory archive according to the documented procedures. The archive staff take responsibility for their long-term retention, and must ensure that no one has unrestricted access to the documents. If papers are temporarily removed from the archive store, records must be kept of the borrower, items concerned and date of retrieval. Such items must then be returned within a reasonable period and checked back into the archive.

Electronic archiving systems need to ensure that similar security is achieved, but also future readability, which may mean the medium and software used, should be considered carefully. Using a dedicated area of the networked system and allowing read access to general users is one way to archive electronic data, if the network capacity is sufficient.

33.5 SUMMARY

GLP is a formal laboratory quality system intended to ensure the integrity and validity of data generated from safety studies. The regulatory authorities aim to issue medicine manufacturing licences based on the data submitted so it is essential that the data is reliable and accurate. All safety evaluation tests specified by the authorities need to be done under GLP. For products from animal cell cultures these would include showing the safety of the cell lines to be used in manufacture, e.g. by viral testing, isoenzyme analysis, karyology, DNA fingerprinting and tumorigenicity testing.

33.5.1 GLP v GMP

Good Laboratory Practice (GLP) and Good Manufacturing Practice (GMP, see Chapter 34) are both regulatory standards intended to ensure quality, integrity and reconstruction of the work after the event. GLP will apply to laboratory studies performed to establish the safety of cell cultures with respect to human or animal health. Data from these experiments is then submitted to the regulatory authority, and GLP is intended to ensure that such results are valid and reliable. The results of trials to determine the safety and efficacy of the new medicine when given to patients will also be required. These trials need to be performed to Good Clinical Practice (GCP) standards. If the new drug application is approved, licences will be issued and a facility can then manufacture the medicines, provided that the requirements of GMP are followed during the manufacturing process.

The good practice standards require continuous official endorsement. Routine biennial inspections are performed in the UK by the Inspection and Enforcement Division of the Medicines and Healthcare Products Regulatory Agency (MHRA). Any findings observed by the inspectors must be addressed or approval may be withdrawn. The quality systems introduced by the facility should guarantee the safety, quality and efficacy of medicines and the adequacy of laboratory data for hazard assessment. As the purpose of both GLP and GMP is similar there are no fundamental differences between them, although there are some parts of the regulations tailored for each, and GMP also includes guidance for special kinds of manufacturing processes such as those for sterile and biological products.

REFERENCES

Food and Drug Administration (1978) *Good Laboratory Practice for Nonclinical Laboratory Studies*, Code of Federal Regulations Title 21, Part 58, and revisions.

Food and Drug Administration (1997) *Electronic Records; Electronic Signatures; Final Rule*, Code of Federal Regulations Title 21 Part 11.

Food and Drug Administration (2003) *Guidance for Industry Part 11, Electronic Records; Electronic Signatures – Scope and Application.*

ISO/IEC (2005) *General Requirements for the Competence of Testing and Calibration Laboratories*, 17025, Edition Z. International Standards Organisation, Geneva.

OECD (1998) *Principles of Good Laboratory Practice.* OECD Environmental Health and Safety Publications.

OECD (2004) *Draft Advisory Document Document of the OECD Working Group on the Application of GLP Principles to* in vitro *Studies.* OECD, Paris.

Statutory Instrument (1999) *The Good Laboratory Practice Regulations 1999, No. 3106.* HMSO, London.

Statutory Instrument (2004) *The Good Laboratory Practice (Codification Amendments etc.) Regulations 2004, No. 994.* HMSO, London.

The UK Good Laboratory Practice Monitoring Authority (UK GLPMA) (2000) *Guide to UK GLP Regulations 1999.*

34 Good Manufacturing Practice for Cell Culture Processing

A Green, G Sharpe

34.1 SIGNIFICANT HISTORICAL EVENTS

The regulations covering the manufacture, storage and distribution of pharmaceutical products in the UK originate from The Medicines Act of 1968. This was introduced following probably the most publicized adverse event to hit the pharmaceutical industry in the twentieth century: the thalidomide tragedy. It was described by the medical historian Hans Ruesch (1991) as follows:

> 'an estimated 10 000 children – but probably many more – born throughout the world as phocomelics, deformed, some with fin-like hands grown directly on the shoulders; with stunted or missing limbs; deformed eyes and ears; ingrown genitals; absence of a lung; a great many of them stillborn or dying shortly after birth; parents under shock, mothers gone insane, some driven to infanticide.'

The cause of such an horrific event was found to be insufficient testing and research into the side effects this prescription drug may have on certain groups within the population, namely pregnant women.

In the US, a brand of thalidomide, 'Kevadon', was refused a licence by medical officer Frances Kelsey on the grounds of insufficient safety data in 1959, a foresighted view of what would become part of Good Manufacturing Practice (GMP) working practices. Unfortunately two million tablets were distributed for 'investigational use' throughout the US. The full scope of the tragedy was realized by 1962, as many deformed infants were born due to the use of the investigational material. For her work in preventing the greater catastrophe, which would have occurred had she permitted the use of Kevadon throughout the US, Frances Kelsey received the President's Distinguished Federal Civilian Service Award in 1962, the highest civilian honour available to a government employee in the US.

A second significant incident occurred in the UK and became known as the Devonport incident. This was not a problem with the actual pharmaceutical product but was caused by serious errors in the steam sterilization process and absence of Good Manufacturing Practice (GMP) in the sampling of a batch of sterile injectable solution in 1972. The bags of dextrose infusion fluids were sent to the Devonport section of Plymouth General Hospital, where some of the insufficiently sterilized bags of product were administered to patients. The problem only came to light when several of the patients died after receiving the microbially contaminated infusion solution.

These are just two of the high profile cases that have triggered the need for a framework of rules, regulations and guidelines for pharmaceutical manufacturers to work to. The EU, US and other members of the global community have taken the knowledge gained after such tragedies and

Medicines from Animal Cell Culture Edited by G. Stacey and J. Davis

created guidelines to address every aspect of manufacture, sampling, testing, storage, packaging and distribution of pharmaceutical materials, and this is commonly referred to as GMP – Good Manufacturing Practice.

34.2 PROGRESS: MEETING THE NEEDS OF A CHANGING WORLD

As more knowledge is gained every year, the pharmaceutical industry and its regulatory bodies are constantly learning about the potential pitfalls in this complex business sector. It is important therefore to note a small but significant addition to GMP, the letter 'c' to create cGMP. This now refers to *current* Good Manufacturing Practice. The constant progress in the technology of manufacture and control of medicines demands that the legislation designed to manage these areas evolves, to provide adequate protection for patients.

34.3 INSPECTIONS AND ENFORCEMENT

To manage, update and enforce cGMP, many regulatory authorities were set up across the world. The agencies most relevant to manufacturing operations within the UK are the EMEA— European Medicines Agency and the UK's MHRA – Medicines & Healthcare products Regulatory Agency. The EMEA provides guidance on interpretation of the legislation surrounding pharmaceutical manufacture and provides a centralized system that all members of the EU must satisfy to gain authorization for marketing biotechnology products. The MHRA acts as the national agency to enforce the guidelines laid down by the EMEA and interpret them for the UK. The MHRA's guidelines are currently the *Rules and Guidance for Pharmaceutical Manufacturers and Distributors 2002* (MCA 2002) more commonly known as 'The Orange Guide' due to the distinctive livery of the publication.

The largest pharmaceutical market in the world is the USA, with drug sales of $229.5 billion in 2003 (Reuters 2004). This makes it a very important area for many UK-based pharmaceutical manufacturers. In order to market products in the US a similar system of authorizations to that outlined above must be satisfied. These are administered by the FDA (Food and Drug Administration) and the guidelines are set out in the *Code of Federal Regulations* or CFRs. In general, the UK and US systems are similar, but with some differences in detail and interpretation.

These standards are set by representatives from the regulatory bodies, industry and professional quality associations within the respective geographical regions. This allows the free flow of relevant information to all parties concerned, with the aim of creating suitable guidelines that are effective and, just as importantly, practical.

Inspection of pharmaceutical manufacturers by audit is the fundamental method used by regulatory bodies to monitor their activities. If the audits show inadequate controls or deficient practices within the organizations, the regulators may issue warning letters describing these in the public domain. A further step would be sanctioning the company by preventing approval and sale of products, and ultimately legal proceedings against companies and individuals that could result in criminal proceedings and custodial sentences.

In May 2004, the regulations within the EU governing the manufacture of licensed pharmaceutical products extended to the manufacture of clinical trials material, according to the EU Directive on Good Clinical Practice in Clinical Trials – 2001/20/EC. This requires any site manufacturing material to be used in a clinical trial to hold a licence from the national regulatory authority (MHRA for UK operations). Once again, this is an evolution of the regulations to protect the end user of the drug, whether it is a licensed or experimental product. Chapter 35 gives a broader overview of international regulations.

34.4 SPECIFICS OF cGMP

As described earlier, the potential risks involved in the pharmaceutical industry are high, with little room for error. The purpose of GMP in this environment is to ensure that the drug product is *fit for its intended use*. To be more specific, the drug must have the following qualities:

- it has been manufactured with the appropriate level of control;
- it is the correct product;
- it is the correct strength;
- it is free from contamination;
- it is suitably stable;
- it is correctly packaged;
- it is correctly labelled;
- it is properly sealed;
- it is suitably stored and distributed.

These indicators of GMP cannot be retrospectively built into a product. It must be thought of as a philosophy of work from the acquisition and use of raw materials, through the design and development of the manufacturing process, the testing and certifying that the material meets specification and ultimately through to the appropriate handling and distribution of the product. If GMP is not built in from day one, the product cannot be considered to be GMP compliant.

34.4.1 Quality Assurance

The Orange Guide (MCA 2002) describes quality assurance (QA) as '..the sum total of the organized arrangements made with the object of ensuring that the medicinal products are of the quality required for their intended use. QA therefore incorporates GMP plus other factors outside the scope of this Guide.' A simple representation is given in Figure 34.1.

The management of an organization has a responsibility to ensure that a drug product is manufactured in accordance with the current GMP guidance and it is from this definition that a quality system is designed. Most often the system consists of written procedures that clearly describe all aspects of the design, manufacturing, storage and distribution process.

In order to ensure that the necessary attributes are present in a drug product, QA personnel are employed to check that all factors that may influence the quality of the product are suitably controlled. The documenting of these controls and data acquired using these systems is the major part of the record to demonstrate compliance with GMP. A batch of product must have all the relevant information required to trace each critical step within the process and all details of materials, personnel, dates, times and many other process parameters. The task for QA is to verify that this information is complete, true and accurate, in order that the material can be approved by a 'qualified person' for use. The main duty of a qualified person is to verify that the batch satisfies the provisions set out in the manufacturing licence and in particular the GMP aspects. An example of the QA and Production framework, showing the staff relationships is given in Figure 34.2.

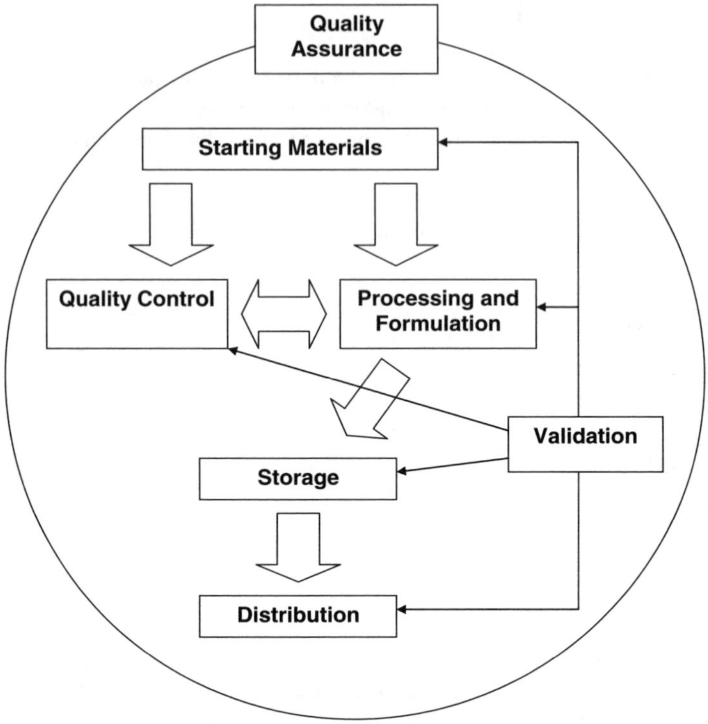

Figure 34.1 Simplified example of how some of the most important elements of a quality system might interrelate for a cell culture product, showing the need for validation at each stage (Stacey, 2004).

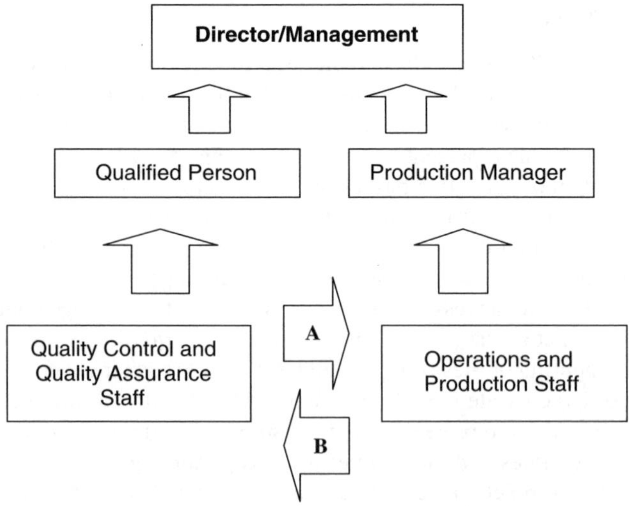

Figure 34.2 Outline of interrelationships of staff involved in manufacture and quality assurance. It is important to note the independence of QA and production/operational activities in terms of line management. However, it is vital that the management should promote close interaction and cooperation between the two areas. Formal links between the two groups include; inspections, audits, reviews and authorization (A) and materials for quality control and reports (B). (Figure provided by G Stacey, NIBSC).

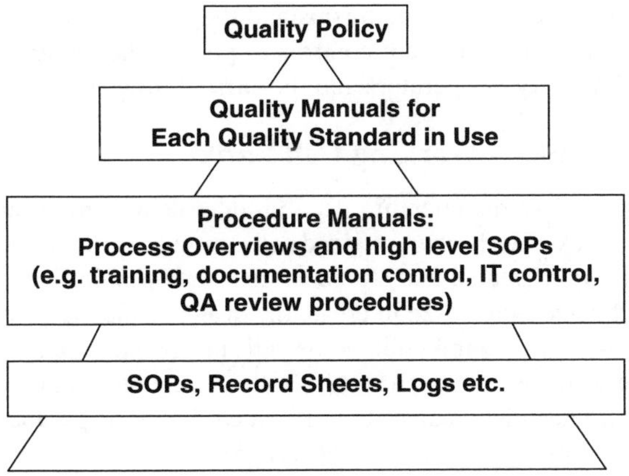

Figure 34.3 Hierarchies of documentation in a GMP quality system.

34.4.2 Documentation

The documentation of actions performed during GMP manufacture, and records of the design, construction, performance and maintenance of building or equipment is the primary means by which assurance is provided that the drug product is of the desired quality. The documentation in a GMP system is structured in several levels, as illustrated in Figure 34.3.

There is a phrase widely used in GMP-regulated businesses to emphasize this point, 'If it ain't written down, it didn't happen'. To this end, strict data recording procedures and control of both paper and electronic documentation is required.

All documentation relating to GMP procedures, materials or equipment must be written, reviewed, approved and distributed according to written procedures. The procedures are commonly referred to as standard operating procedures (SOPs).

The main constituent of documentation directly relating to the manufacturing operations for a drug product is contained in the batch manufacturing record (BMR). This is the means by which manufacturing personnel directly record the real-time events and observations during the production process.

A drug product will finally require a certificate of analysis showing that it was suitably tested and the results obtained were within the specifications laid down prior to manufacture.

34.5 GMP IN CELL CULTURE

34.5.1 Principal Differences from Chemical Processing

The 'Orange Guide' (MCA 2000) and CFRs describe in great detail the requirements a manufacturing organization must meet in order to comply with cGMP. Much of the general information regarding conventional or classical manufacture of pharmaceuticals applies to cell culture manufacturing but extra guidance is considered necessary for this highly specialized and relatively new discipline in manufacturing.

Annex 2 of the 'Guide' (MCA 2002) provides specific considerations, directly linked to the mode of production and the use of live organisms, in order to highlight the specific precautions required when dealing with processes that are not simply chemical in nature. Some of the cells commonly used in cell culture are: mammalian, insect, bacterial, and yeast.

The variability in all of these living organisms and the potential for contamination with other organisms necessitates extremely stringent controls to protect the integrity of the product and in some cases, to ensure the safety of operators and the surrounding environment.

34.5.2 Master Cell Banks – Working Cell Banks

All production cultures used to manufacture drug product using cell culture techniques must be initiated from a reserve stock of well-characterized cells, more commonly referred to as a cell bank. A master cell bank (MCB) is created and stored as the starting point for manufacturer's working cell banks (MWCB), which are then used to create the drug product. If such cells are subsequently used to produce virus products then infection of the cells must be from a corresponding reservoir of virus known as a master virus seed stock (MVSS). Both of these stocks of material can be pivotal in the production of biopharmaceutical materials and as such should be produced to comply fully with GMP, even in early stage development, to provide assurance of their manufacture, storage and use.

Clearly for such important material as cell banks and seed stocks, full characterization and safety testing is paramount. Most manufacturers choose to use a contract service organization (CSO) to perform this work, in order to separate the testing procedures from the manufacturing areas. This is to remove the risk of contamination from the positive control organisms.

The information for cell banks forms a significant part of the licence application since their behaviour and purity are of great importance for the later stages of manufacture. It should be stressed once more that the cell banks should be produced to comply fully with GMP.

It is critical that there is no significant contamination or instability in this material, as this would have a dramatic effect on the performance of the cells and the quality of material produced. This requires close monitoring of the procedures in manipulating cell bank materials and in the analysis of their characteristics between successive lots of product. Microbial and mammalian cell cultures should be shown to be pure with respect to their clonal make-up, and sterile or monoseptic. Virus material must be highly purified during processing to produce master and working virus seed stocks.

Due to the importance (and therefore monetary value) of master materials, the GMP guidelines advise that the stored seed stock is split and maintained at different locations to minimize the chances of total loss. The control of access and storage of these master materials is also paramount, with the need for strict procedures and the presence of suitably responsible personnel when handling. In order to maintain adequate documentation of the identity and integrity of materials, they should be stored with suitable labelling and auditable evidence of their movement, storage conditions and access records.

34.5.3 Personnel

This is possibly the greatest variable in manufacturing and one that deserves careful consideration. The standard of training for staff in the biological manufacturing environment must be of the highest order and general principles are given in the Orange Guide (MCA 2002). It is expected that the people working with cell culture processes receive training additional to that required for 'normal' pharmaceutical manufacture, and have an appreciation of the impact their actions may have on success or failure in the creation of biologically derived material.

By its very nature, cell culture technique provides conditions suitable for rapid cell growth. This is vital for manufacturing the target organism but the process is vulnerable to microbial contamination derived from the manufacturing personnel. Contamination from people is potentially the main source, but consideration must also be given to minimizing any environmental factors that may create a risk, such as airborne particles and other adulterating material. This gives great weight to the need for procedural controls and personal hygiene in this environment, and special consideration should also be given to the health of manufacturing personnel (MCA 2002).

34.5.4 Premises and Equipment

The design and control of premises suitable for cell culture manufacture of pharmaceuticals is a massive subject, and is more fully covered in Chapter 12. The main reason pharmaceutical manufacturing facilities require specific and exacting standards is to provide effective protection for the product. The environment in which a biopharmaceutical is generated has great potential to impact on the finished product quality.

Some major GMP factors to be considered within a biologics facility are:

- Consideration should be given to the design and construction of the building, facilities and equipment. This is to ensure ease of cleaning and prevention of microbiological contamination, particulate contamination and product-to-product cross contamination. If fumigation procedures are to be employed in the facility, these should be validated to show that they are effective.
- All utilities that could impact on product quality such as water, compressed air, steam, heating, ventilation and air-conditioning (HVAC), should be qualified and monitored with limits set for their supply, and corrective actions implemented when their analysis nears alert levels, prior to breaching the set specifications.
- Both equipment and facilities should have specific maintenance and cleaning schedules outlined as part of an ongoing programme to maintain their required performance and suitability for use. For cell culture processes, consideration should also be given to the supplies of medicinal gases (carbon dioxide, carbon dioxide/air mixtures, nitrogen, oxygen) and compressed air.
- There should be sufficient space within the facility, and the process flow should be designed in such a way, as to avoid mix-ups, and contact between product or starting materials and waste.

From a cGMP viewpoint, thorough documentation of all aspects relating to the validation, control and maintenance of a manufacturing facility and its equipment is critical. These are the documents that would be presented at audit to regulatory bodies or customers in order to verify the level of design and control a manufacturer has over the environment in which biological products are prepared.

34.6 QUALITY CONTROL

There would be little point in manufacturing a drug product to cGMP requirements if there was no way of verifying that it met specifications set according to regulatory requirements. This is where quality control (QC) is employed in the GMP framework. Using a broad array of analytical techniques, the QA team will identify and quantify all the factors deemed critical for any specific drug product.

Biological products require extensive analysis to characterize fully the drug substance. The product specifications should address the following as a minimum:

- identity (confirms correct material);
- quantity (confirms dose parameter);
- purity (confirms material has correct purity);
- impurity (confirms product safety);
- potency (confirms activity);
- Sterility including adventitious agents (confirms material is sterile).

The tests employed in final QC analysis of drug products should be validated to show that they are effective in producing accurate and reproducible results. For a limited number of established cell-based products, pharmacopoeial methods may be available but many products require development of specific analytical tests. Once the material has undergone a full suite of testing, and been shown to comply with all specifications, a certificate of analysis would be produced and approved by relevant personnel within the organization. This document should accompany the material to demonstrate that it complies with all the analytical specifications for that product. When confirmation is available that the material is satisfactory, the status labelling of the product can be updated to reflect its 'approved' status. Conversely the material would be labelled as 'failed' if compliance with the associated specifications were not achieved, and the material dealt with accordingly. The certificate of analysis is a significant part of the documentation required before a product can be released for use, but all aspects of the documentation must show that the material is suitable for its intended use before it can be finally approved for release.

Many products that are made from or contain cells require very careful handling and storage. It is very unlikely that cell culture materials would be able to exist at ambient conditions for any practical length of time, and therefore appropriate measures must be used to maintain their integrity. Freeze drying, freezing and cryopreservation are some of the techniques employed to ensure that the material has as long an effective shelf life as possible. This provides time to perform the quality testing and release of the product, and ultimately allows the product to be stored for a practical length of time before its activity is lost.

34.7 SUMMARY

GMP in cell culture processing is a huge subject, with many areas of expertise required to create and maintain the systems that underpin the manufacture of biologically derived pharmaceuticals. The main point that should not be lost in all this detail is the central premise that material must be *fit for its intended use*. The importance of this statement cannot be emphasized too greatly when the intended use is for administration to people. Every assurance must be made to ensure that these end users are protected from harm by the diligence and professionalism of today's biotechnology manufacturers and regulatory bodies.

REFERENCES

Ruesch H (1991) *Slaughter of the Innocent*, CIVITAS Publications, NY.

Medicines Control Agency (MCA) (2002) *Rules and Guidance for Pharmaceutical Manufacturers and Distributors*. HMSO, London.

Reuters (15 March 2004) *UPDATE 1 – Global drug sales up 9 pct to almost $500 blm – IMS*, [Online].

Stacey G (2004) *Hum. Fertil.* (Camb.); 7(2): 113–118.

Useful Web Sites

Medicines & Healthcare products Regulatory Agency	http://www.mhra.gov.uk/
Food and Drug Administration	http://www.fda.gov/http://pharmacos.eudra.org/
International Conference on Harmonisation	http://www.ich.org/
European Medicines Agency	http://www.emea.europa.eu

35 International Regulatory Framework

R Guenther

35.1 INTRODUCTION

The registration of new pharmaceuticals is controlled by a multitude of laws, guidance documents and regulations. Health authorities have a responsibility to protect the public from the potential risk of unsafe drugs, but they should also support the rapid introduction of new drugs onto the market in order to enhance the health of patients. Their responsibility covers not only marketed drugs, but also new medical products in development, starting with the first trials in humans. Based on a risk-benefit evaluation, taking into account all available data on quality, safety, and efficacy, the health authority will decide whether a new pharmaceutical drug will get marketing authorization or, in the case of a product in development, an authorization for clinical trials.

Because the regulation of pharmaceuticals, and especially of biological medicines, is a rapidly evolving area, this chapter gives only a current picture, which will certainly change with time.

35.2 HISTORY OF PHARMACEUTICAL REGULATION

35.2.1 The Start of Pharmacopoeial Standards

The control of pharmaceutical products can be traced back to approximately 3000 years ago in Egypt with the inspection of syrup manufacturers who produced medicinal products (Penn 1979). In the Middle Ages, the control, preparation, and use of drugs was described for the first time in so called pharmacopoeias. The first pharmacopoeia in the UK was the *London Pharmacopoeia*, published in 1618 and followed by several revised editions until the publication of the first *British Pharmacopoeia* in 1864. The first US pharmacopoeia was established in 1820 by eleven physicians in Washington. More recently, the *European Pharmacopoeia* was inaugurated in 1964, and its production is currently overseen by the European Department for the Quality of Medicines (EDQM).

In the twentieth century, the UK *Therapeutic Substances Act* from 1925 was an important step towards modern registration and control procedures for drugs. This legislation regulated processes and documentation, for example the inspection of manufacturing sites, batch records, and labeling. At the same time in the USA, the Food and Drug and Insecticide Administration was created (1927), working under the 1906 US Pure Foods and Drugs Act against misbranding. In 1930 the name of the agency was changed to its current name, the Food and Drug Administration (FDA).

Medicines from Animal Cell Culture Edited by G. Stacey and J. Davis

35.2.2 Key Adverse Events that Helped Shape the Regulatory Environment

Specific regulations for biologics started to appear in 1902, when the Biological Controls Act was born as a consequence of the St Louis tragedy, in which 13 children died of tetanus owing to the administration of diphtheria antitoxin derived from blood serum of horses that had been transported in a tetanus-infected retired milk wagon.

In 1937 in the USA 107 persons died (mostly children) after taking a sulfanilamide elixir containing diethylene glycol. This incident provided the impetus for significantly expanding the regulation of drug control, which was done with the Food, Drug and Cosmetic Act that came into force in 1938 and which has been the legal basis for the work of the FDA ever since.

The most important incident of the twentieth century was the thalidomide tragedy of the late 1950s early 1960s (see also Chapter 34). This had a lasting influence on the further development of regulatory requirements. In 1957, the German pharmaceutical company Grünenthal brought sleeping pills containing the new drug thalidomide onto the market. Two years later, the first observations of peripheral neuropathies were recorded in patients treated with thalidomide. In 1960 and 1961 several side effects, mainly involving the central nervous system and developmental defects in foetuses were reported in the literature and presented at medical congresses. Thousands of babies with limb deformities were born to women who had been treated with thalidomide during pregnancy. As a consequence, in November 1961, Grünenthal withdrew their drug from the market.

After the experience with thalidomide, further legislation regulating the evaluation and approval process for new drugs was created with high priority. This was accomplished in the USA with the Kefauver–Harris Drug Amendments to the Food, Drug and Cosmetic Act in 1962. These included several new requirements, such as manufacturing according to GMP (Good Manufacturing Practice) and the notification of adverse drug effects to the FDA. In the UK, the Committee on Safety of Drugs (CSD) was set up in 1963 to evaluate safety and efficacy. In the European Union (EU), the first European directive to control medicines was introduced in 1965, Directive 65/65EEC, which is still the basis for the European legislation of pharmaceuticals.

35.3 REGIONAL REGULATIONS

35.3.1 European Union

For all the current 27 member states of the EU there are uniform procedures for registration of new pharmaceuticals. A summary of these regulations (for example pharmaceutical legislation, Notice to Applicants, guidelines for medicinal and veterinary medicinal products, and information on Good Manufacturing Practice) is published in EudraLex, *The Rules Governing Medicinal Products in the European Union*. The electronic version of EudraLex is available on the website of the Pharmaceutical Unit (see List of Useful Web Sites).

The European bodies involved in regulatory issues are:

- *The European Commission*: This is the administrative body within the EU that proposes new policies and legislation, and monitors the implementation of European legislation by the national authorities. The European Commission is structured into several directorates general (DGs). Human and veterinary pharmaceutical legislation is the responsibility of the Pharmaceutical Unit, which is a part of the Enterprise and Industry DG.
- *The Council of the European Union* (Council of Ministers): This body wields the legislative power of the EU, is comprised of representatives from all member states, and decides whether the proposals of the European Commission should become European law. These laws can be:

Regulations: Laws that apply directly to all EU member states.

Directives: These define objectives, but must be implemented into national laws before becoming effective. Directives normally leave some room for interpretation and flexibility by the individual countries.

- *The Pharmaceutical Committee*: Provides expert advice to the European Commission.
- *The European Medicines Agency* (EMEA): The EMEA was established in 1993 by Council Regulation 2309/93, and became active in 1995. The EMEA has the central role of coordinating the scientific evaluation of safety, quality, and efficacy of human and veterinary pharmaceuticals in the EU.
- *The Committee for Medicinal Products for Human use* (CHMP): This committee, which is a scientific body within the EMEA, has the role of giving an opinion as to whether a new drug for human use should receive marketing approval. The corresponding committee for veterinary pharmaceuticals is the Committee for Medicinal Products for Veterinary use (CVMP). Each Member State appoints one scientific expert for these committees.

35.3.1.1 The clinical trial authorization procedure

Authorization to perform clinical trials with new biologicals in development falls under the responsibility of the individual member states. However, since 1 May 2004 the format and content of a request for a clinical trial authorization (CTA) should follow the detailed guidance of the European Commission based on Directive 2001/20EC in any EU member state. The information about quality, manufacture and control, and clinical and non-clinical data, are summarized in an investigational medicinal product dossier (IMPD). The request for a CTA has to be submitted to the national health authority and the ethics committee(s). Both the health authority and the ethics committee(s) will review the documentation. The clinical trial can be started, if the ethics committee has issued a favourable opinion and if the health authority did not voice its non-acceptance of the proposed trial. The time period for examination of the request is fixed to a maximum of 60 days according to the European Directive.

35.3.1.2 The marketing authorization procedure

To gain marketing authorization for a new biological/biotechnological product for all EU countries, the so-called 'centralised procedure' must be followed (see Figure 35.1). This procedure is the compulsory licensing procedure in the EU for all biotechnology products (and for any new medicinal products for certain diseases, such as for example cancer or diabetes). The detailed process for getting marketing authorisation is laid down in the *Notice to Applicants*, Volume 2A.

35.3.1.2a Submission

After submitting the registration dossier to the EMEA, the agency performs an administrative check regarding completeness and correctness of the format and content of the dossier and then passes it on to the CHMP. Responsible for the coordination of the evaluation of the dossier are the rapporteur and co-rapporteur, assigned experts of the CHMP. The rapporteur and co-rapporteur are appointed by the CHMP before submission, based on the applicant's suggestions and the preferences of the CHMP members.

35.3.1.2b Assessment

The (co)-rapporteurs evaluate the dossier based on the information provided on clinical safety and efficacy, toxicology and chemistry, manufacturing, and control of the product. The results

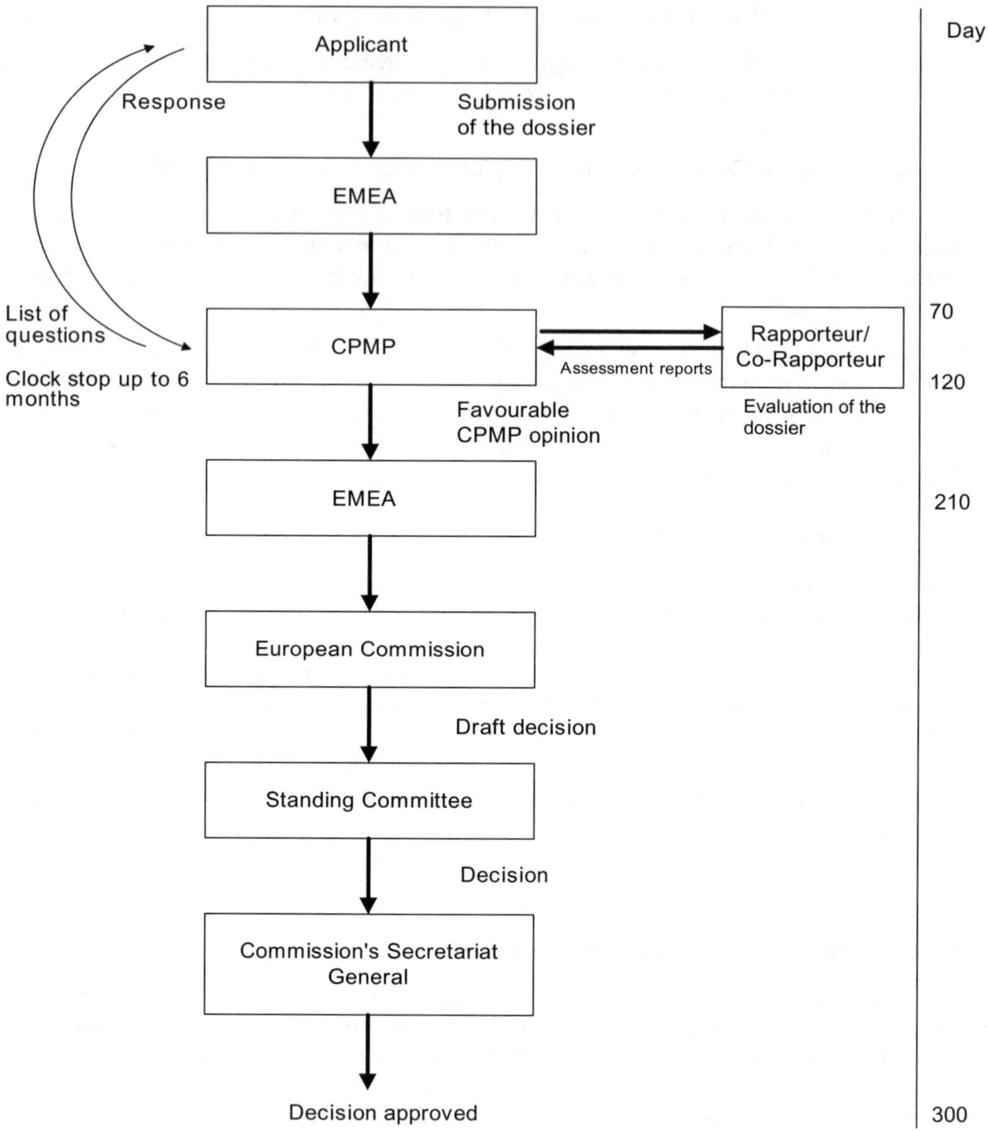

Figure 35.1 Centralized Procedure for achieving marketing authorization in the EU.

of this evaluation, including the (co)-rapporteurs' concerns and questions to the applicant, are summarized in the assessment report. The initial assessment reports from rapporteur and co-rapporteur are released on day 70 after submission, with a view to giving other CHMP members the opportunity to comment on these reports. In the majority of cases, questions and concerns about the dossier are raised, and consequently the final CHMP list of questions should be available not later than 120 days after submission of the dossier. The applicant then has a maximum of 6 months in which to respond to open questions and concerns (which causes a 'clock stop' of the review process). Based on the assessment report and the applicant's response, the CHMP will give a favourable or unfavourable opinion after an additional 90 days (day 210).

35.3.1.2c Decision-making process

In the case of a favourable opinion, the EMEA will forward the decision to the Pharmaceutical Unit of the DG of the European Commission. The European Commission will then make a final decision to grant a marketing authorization, taking into account all member states' opinions by consulting the Standing Committee on Medicinal Products for Human Use. In some cases, for example where the product is considered important from the public health point of view, an accelerated assessment is possible and the CHMP opinion can be adopted as soon as day 120, with commitments to conduct follow-up studies. The final decision of the European Commission is binding for all EU member states and as such is valid in the entire EU, with the exception of price and reimbursement negotiations for which national law in each member state applies. The marketing authorization is valid for 5 years and can be extended on request by the applicant 3 months before the marketing authorization expires.

35.3.2 USA

The regulatory body responsible for approval of new drugs in the USA is the Food and Drug Administration (FDA), which was preceded by the Agriculture Department, founded in 1862 by President Lincoln. The FDA is headed by a commissioner, who is appointed by the President of the United States. The FDA has extensive responsibility, not only for human medicines, but also for veterinary medicines, devices, and foods. The agency is organized into several specialized departments, so called program centres. The program centres listed below are the most important ones for biological products:

- *Center for Drug Evaluation and Research (CDER)* CDER is the branch of the FDA that is responsible for the review and approval, for market distribution, of all new and generic synthetically derived drug products, monoclonal antibodies and most of the therapeutic biological products (Table 35.1).
- *Center for Biologics Evaluation and Research (CBER)* CBER is the branch of the FDA that has similar responsibilities for products classified as 'biologics', which are not covered by CDER.

A detailed list of the various types of non-synthetic drugs and the program centre of the FDA responsible for their review and approval is given in Table 35.1.

- *Office of Regulatory Affairs (ORA)* ORA is the unit responsible for the FDA field offices who prepare for, and conduct, GMP and pre-approval inspections of pharmaceutical manufacturers.

Table 35.1 Assignment of different groups of biological/biotechnological products to the corresponding agency.

CDER	CBER
• Monoclonal antibodies • Cytokines • Growth factors • Enzymes • Other proteins intended for therapeutic use that are extracted from animals or microorganisms, including their recombinant versions	• Viral vectors (e.g. used for gene therapy) • Products composed of human or animal cells or part of those cells • Vaccines, including therapeutic vaccines • Clotting factors • Blood, blood components and related products • *In vitro* diagnostics

- *Center for Veterinary Medicine (CVM)* CVM is responsible for ensuring the safety and efficacy of veterinary drugs.

A compilation of all effective government regulations for food and drugs can be found in title 21 of the Code of Federal Regulations (21CFR), published by the US Printing Office. FDA guidelines for a sponsor are available in specific documents called *Guidance for Industry*, which are published in the *Federal Register*, the official News Bulletin of the Federal Government. These documents can be downloaded from the FDA website.

35.3.2.1 Application to perform clinical trials

In the development phase of a new drug before the start of any clinical trial, an investigational new drug (IND) application has to be on file. The IND application must contain information on preclinical data (animal pharmacology and toxicology), clinical protocols, and information on chemistry, manufacturing, and control (CMC) of the drug substance and drug product. The FDA will assess the acceptability of the IND application based on the information provided by the sponsor about the planned clinical trial, and about the quality and safety of the drug. In most cases a pre-IND meeting will be held before submission of the IND package. An IND does not need formal approval from the FDA. It automatically becomes effective 30 days after the FDA receives the IND, unless the FDA has any concerns and issues a 'clinical hold'.

35.3.2.2 The licensing procedure

35.3.2.2a Submission

Based on a positive outcome of the clinical trials performed with a new drug, which show safety and efficacy in its proposed use, the applicant can submit a biologics license application (BLA) to the FDA to gain an approval to market a new biological drug for human use. The BLA must include all information about animal studies (safety) and human clinical trials (safety and efficacy) obtained during development, and information on chemistry, manufacturing and controls (quality). Before submitting the dossier to the FDA, there is usually a pre-BLA meeting that takes place between the sponsor and the health authority. During this meeting, the applicant has the opportunity to present a summary of all relevant data referring to safety, efficacy and quality of the product. General issues can be discussed directly with the reviewers of the dossier. After submission of the BLA, the FDA performs an administrative dossier check and has to inform the applicant within 60 days about the acceptability or rejection (refusal-to-file letter) of the dossier for assessment. Reasons for rejection may be, for example, incomplete data or incorrect format of the dossier.

35.3.2.2b Assessment

After the file has been accepted, the FDA will start reviewing the various sections of the dossier. Minor issues that may come up during the review process can be discussed directly between the reviewer and the applicant. After the review process is completed, the FDA may send a complete response letter to the applicant, containing a list of all the outstanding questions and concerns related to the dossier. This will give the applicant the opportunity to answer the questions and to resolve all open issues with the FDA, prior to the official action date. The action date is 10 months from the date of submission (6 months if an expedited review is granted for extraordinary circumstances). The sponsor is expected to make every effort to respond as quickly as possible to FDA questions during the review of the application. Whether or not the FDA is able to review completely responses to their questions is dependent upon the size of the response and the amount of time left on the review clock. During the review process there is the possibility of making changes to, or submitting additional data to, the file. This can be initiated by the sponsor, or upon request by the FDA. It should be noted that changes should not be submitted to the application without up-front discussions with the FDA. In addition they should not be submitted when there is less than 3 months left on the review clock (unless agreed up-front by the FDA). The corresponding data and information are submitted to the agency as an 'amendment to a pending BLA'.

In addition to the internal review process, the FDA can organize an advisory committee meeting, a meeting at which experts on the relevant medical indication can give their opinion on the product and thus influence the final decision.

When the overall assessment of the BLA leads to a positive decision, the FDA will send an 'approval letter' to the applicant. This allows the applicant to market the drug immediately in the USA. If the agency comes to the conclusion that, based on the available data on safety, efficacy and quality, the product cannot be approved, they will send a 'not approvable letter' to the applicant. The applicant can either withdraw the application or amend the application or notify the agency of an intent to file an amendment. Additional clinical studies may be performed, or changes may be made to the labelling, which can change the result of the assessment to an 'approval' decision. As a third option there is also the possibility of issuing an 'approvable letter', which means that the drug is about to be approved, but some deficiencies have been observed, and the definitive approval cannot be given unless these issues are resolved or the specific information is provided. The sponsor will then have the opportunity to provide more data or make the suggested changes to the dossier such that an approval may be issued at a future date.

It should be mentioned that, as a prerequisite for a positive decision, i.e. an approval, a satisfactory pre-approval inspection (PAI) of the production and testing site is required to guarantee the manufacturer's compliance with current GMP guidelines and all of the provisions set forth in the application.

35.3.3 Japan

The competent Japanese health authority is the Ministry of Health, Labour, and Welfare (MHLW) and the approval and licensing system in Japan is regulated by the Pharmaceutical Affairs Law (PAL). A major revision of the PAL took place in recent years, which significantly affected both the organization of the authority and the approval process. The last pieces of the revised PAL came into force in April 2005. The Pharmaceutical and Medical Device Agency (PMDA) was created in April 2004 by merging the Pharmaceuticals and Medical Devices Evaluation Centre (PMDEC) and the Organization of Pharmaceutical Safety and Research (OPSR, 'KIKO'). PMDA is responsible for data review and GLP/GCP inspections, and plays a key role in the approval process.

The dossier has to be submitted to PMDA, which will organize a review meeting (MENDAN), to give the applicant the opportunity to present the new drug to the agency. Following the MENDAN, a review team of PMDA will assess the dossier and prepare a review report. A meeting will take place between the review team and experts from the Pharmaceutical Affairs and Food Sanitation Council (PAFSC), who have an advisory function for the PMDA, to discuss the evaluation of the data submitted. Any questions that may arise will be sent to the applicant, who then has the opportunity to resolve the issues and discuss any instructions of the agency during an interview review meeting. After a second expert review between the PMDA team and PAFSC experts, PMDA will announce whether or not the new drug is approvable or not approvable. The final decision on drug approval will be made by the MHLW. The overall review time for a registration dossier is approximately 12–18 months. For serious, life-threatening diseases or in cases where no alternative therapy exists, a priority consultation and review is possible with a reduced review time of 6 months.

35.4 INTERNATIONAL CONFERENCE ON HARMONIZATION

In the context of globalization, country-specific requirements for national guidelines and laws became an obstacle for the multinational registration of new drugs. To achieve harmonization between the three regions, Europe, USA, and Japan, the International Conference on Harmonization of Technical Requirements for Registration of Pharmaceuticals for Human use (ICH) was inaugurated in 1990, based on activities that were initiated at the WHO Conference of Drug Regulatory Authorities in 1989. From today's point of view, the ICH process can be seen as a successful initiative to reach a significant degree of international harmonization toward ensuring that good, safe, and effective medicines are developed and registered in the most efficient and economical manner. As a joint initiative of regulatory bodies and the pharmaceutical industry, each region is represented by their health authority and their corresponding industry association (see Table 35.2).

In addition to the official ICH members, the WHO, Canada, and the European Free Trade Area (EFTA), represented by Switzerland, have observer status in the ICH.

The International Federation of Pharmaceutical Manufacturers Association (IFPMA), of which many industry associations from outside the ICH area are also members, is closely associated with the ICH and ensures the connection of non-ICH countries to the ICH process.

The topics addressed by ICH are divided into three sub-groups: Safety, Efficacy, and Quality. Additional subjects that do not fit with these three categories, are dealt with under a 'multidisciplinary' topic. Most of the relevant technical guidelines dealing with biotechnological/biological products can be found in ICH Q5, *Quality of Biotechnological Products.* Although most of the currently valid guidelines were finalized between 1995 and 1999, international harmonization is still an ongoing process. In 2003, the ICH steering committee decided to develop a harmonized pharmaceutical quality system, applicable across the life cycle of the product emphasizing an integrated approach to quality risk management and science. This resulted in establishing ICH Q8, *Pharmaceutical Development ("Quality by Design"),* and ICH Q9, *Quality Risk Management.* In May 2005, ICH Q10, *Pharmaceutical Quality Systems*, was adopted as the topic for a new tripartite guideline.

The ICH process shows that there is a common interest between regulators and the pharmaceutical industry in standardising requirements for assessment of safety, efficacy and quality of new medicines, and that it is possible to overcome historical and cultural differences in the regulation of pharmaceuticals in different regions.

35.5 WORLD HEALTH ORGANIZATION

Whilst WHO does not issue marketing authorizations, it has a long history of supporting the development of the quality of medicines in key areas, primarily vaccines. WHO now provides regulatory advice on a range of biological products through its Expert Committee on Standardization. In addition, WHO authorizes a wide range of international reference materials used

Table 35.2 Health authorities and industry associations of ICH.

Region	Regulatory body	Industry association
EU	European Commission/EMEA	European Federation of Pharmaceutical Industries and Associations (EFPIA)
USA	FDA	Pharmaceutical Research and Manufacturers of America (PhRMA)
Japan	Ministry of Health, Labour and Welfare (MHLW)	Japan Pharmaceutical Manufacturers Association (JPMA)

to provide standardization for assays for the biological activity of biologicals. In relation to the use of animal cells for the manufacture of medicinal products, WHO has organized a series of meetings over the years to deal with specific products and issues, and published its *Guidelines for Cell Substrates Used in the Manufacture of Biological Medicines* in 1997. In addition, they organized the production of reference cell banks, including a WHO Vero 'master bank', which qualifies as a reference bank for Vero cells used in the manufacture of vaccines, and a bank of CHO-K1 cells for use in genotoxicity studies. The development of such banks is monitored by the WHO Cell Bank Monitoring Group, which engages a range of experts in the field periodically to review the available reference cell banks and propose new areas where they might be utilised.

35.6 THE GUIDELINES

Regulatory guidelines provide general guidance and recommendations on many issues specific to biotechnological products. The following section provides an overview of important information that is expected by the regulators. The section focuses mainly on animal cell products, including recombinant proteins and monoclonal antibodies, covering the whole manufacturing process from the expression vector to the final drug substance. Most of the principles outlined here also apply to vaccines and non-recombinant proteins produced by cell culture expression systems.

Information on regulatory guidelines for some of the advanced technology medicinal products, like products used for cell therapy, xenotransplantation, and products manufactured by transgenic organisms, are provided in Chapter 36.

A non-exhaustive summary of relevant guidelines for biotechnology products is provided in Appendix 1.

35.6.1 Expression Construct

Detailed information that must be given on any recombinant DNA constructs:

- source of the cell line from which the protein coding sequence was originally obtained;
- methods used to assemble the expression construct;
- regulatory parts of the expression construct (physical map), e.g. promoters, enhancers, sequences for antibiotic resistance and their location on the DNA;
- confirmation of DNA sequence of the gene of interest by sequence analysis;
- the method of introduction of the expression construct into the host cell.

35.6.2 Cell Bank System

The host cell into which the expression construct is introduced to produce the entire molecule must be described in detail, including the source and history of the parental cell line.

Furthermore, all procedures used to generate the cell bank should be described:

- transfection, amplification, and selection of the final clone;
- reagents used during cell culture;
- methods used to adapt from, for example, serum-containing to serum-free media.

Based on the selected clone, a cell bank, usually a master cell bank (MCB), and from that a working cell bank (WCB), is established. Information on the cell banking system should be provided. A crucial component of the quality assessment of the biotechnological product is the characterization and testing of the cell bank. As part of the characterization exercise, the correct nucleotide sequence of the expression construct after incorporation into the cell should be verified. The copy number and number of integration sites analysed by, for example, restriction endonuclease mapping, should also be reported. Testing of the cell bank also includes identity testing by, for example, phenotypic or genotypic characteristics, and purity testing. Special emphasis has to be given to testing for adventitious agents, in particular testing for endogenous and exogenous viruses (See Chapter 19). Normally, the full characterization of the cell bank is only done once on either the MCB or the WCB.

35.6.3 Genetic Stability

Because of the inherent potential for mutations to occur in all living organisms, genetic stability of the cell line under production-scale conditions must be shown, and the *in vitro* cell age of the cell line to be used in production has to be defined. To demonstrate that the cell line is stable at maximum *in vitro* cell age (or beyond) and that no alteration of the protein structure occurs, analysis of either the final protein product and/or the nucleotide sequence must be performed using appropriate analytical methods, and is compared with the coding sequence of the expression construct verified at the MCB level.

35.6.4 Production and Purification

The production (cell culture process) and purification (downstream processing) must be described in detail (for example media components, chromatographic methods, in-process controls, etc.). To show that the process is well controlled, appropriate validation must be performed, to ensure reproducible yield and quality of the product (see Chapter 15). The ability of the process to clear contaminants like media components and cell-derived substances is of particular importance (see Chapter 18).

Material must be manufactured according to current GMP guidelines, a prerequisite for all material to be used in humans.

35.6.5 Characterization

Characterization of cell culture-derived biological products has to include the physico-chemical and biological properties. Physico-chemical characterization covers the detailed structural determination of the primary sequence, including post-translational modifications and, if possible, higher order structures of the protein (see Chapters 23 and 24). An essential part of the characterization of a cell culture-derived product is the determination of biological activity. Depending on the nature of the product, this can be a bioassay, a binding assay, or an immunoassay. If possible the result of the biological assay should reflect the intended biological effect of the drug in therapeutic use.

In many cases, biological products cannot be assigned a single defined molecular structure, but show a certain degree of heterogeneity. If the different molecules derive from the biosynthetic process itself, as in the case of varied post-translational modifications, and have similar biological properties in relation to safety and efficacy, they can be defined as part of the product. Variants originating from the manufacturing process or during storage, but without any impact on biological activity, safety, or efficacy, are defined as *product-related substances*. There are no limits defined in the guidelines for the permitted amount of product-related substance, but it is highly important that the manufacturing process can yield a product of consistent quality, i.e. the amount of the individual variants has to stay within a specified range.

Molecular variants that have different characteristics to the desired molecule with respect to activity, safety, and efficacy, are considered to be *product-related impurities*. If the impurities come from the cell culture process (such as media components), or the purification process (for example 'ligands' from an affinity column), or from the production cell itself (host cell proteins, DNA) they are considered to be *process-related impurities*. Manufacturers have to ensure that the amount of such impurities in the drug substance is fixed at the lowest possible level. Clearance studies together with appropriate process controls should demonstrate the ability of the purification steps consistently to remove process-related impurities. Any material that is introduced adventitiously (= adventitious agents) is considered a *contaminant*. Critical for all cell culture products is microbiological contamination by bacteria, fungi, or yeast, the absence of which must be strictly controlled. Of significant importance in the overall assessment of the quality of a cell line-derived product is the virus safety evaluation (see Chapter 31). Viruses can be adventitiously introduced in the production process by, for example, contaminated raw materials, but they can also be an integral part of the cell line used and therefore already present in the MCB. Murine cell lines infected with murine leukaemia virus are an example of such an endogenous contamination. To avoid viral contamination of the product and the production facility, the manufacturer has to perform an extensive viral testing and validation program. The concept of performing a viral safety evaluation is based on a three-tiered, complementary approach:

(i) Testing of cell banks and raw materials for absence of viruses.

(ii) Determination of the capacity of the production process to remove or inactivate viruses. In many cases specific virus removal steps (e.g. filtration) or virus inactivation steps (e.g. low pH treatment) are part of the purification process.

(iii) In-process testing for absence of viruses at appropriate production steps.

For more detailed information on virus clearance and inactivation, please refer to Chapter 19.

35.6.6 Product Control

To ensure the quality of the cell culture-derived product, release analysis of each production batch must be performed. The analytical methods used to evaluate identity, purity, potency, and quality of the product by routine testing are typically a subset of methods applied for characterization of the molecule. These methods have to be appropriately validated and should relate to the individual properties of the molecule, which are relevant to ensure consistent quality. In addition, the quality attributes tested should, whenever possible, be based on the existing clinical and non-clinical experience with the product. It should be considered that product control is not restricted to lot-release testing of the final drug substance only. Process validation and in-process testing are essential parts of the control system for confirming quality and consistency of a medicinal product. Testing of process-related impurities is often more meaningful when analysed at the relevant purification stages rather than performing these tests on the final drug substance. For some impurities it is quite sufficient to demonstrate removal by process validation. Examples of this validation approach are media components or residual host cell DNA, for which testing of every manufacturing batch may not be necessary, if the capability of the purification process to eliminate these impurities can be demonstrated.

35.6.7 Stability

A detailed description of the requirements for stability testing is provided in the ICH guidelines on stability testing. In general, the principles, as outlined in ICH Q1A (R2), *Stability Testing of New Drug Substances and Products*, apply for all pharmaceutical products. However, the specific biotechnology-related issues are addressed in an annex to the main guideline, ICH Q5C, *Stability Testing of Biotechnological/Biological Products*. Guidance is given on the batches and samples to be tested, frequency of testing, and test methods to be used. Although, in principle, only well characterized or well specified proteins fall into the scope of this guideline, the guidance provided may in many cases also be applicable to other products, such as conventional vaccines.

35.6.8 Comparability

Changes in the production and purification process are frequently made during development. For products already on the market it is sometimes necessary to introduce process changes, for technical or economic reasons. To ensure that the change does not have any effect on quality, safety, and efficacy of the product, the manufacturer must show comparability between the products made before and after the change. Initially, the comparability exercise is mainly based on the evaluation of quality control analyses, physicochemical and biological characterization data, and stability studies. When these data indicate that a manufacturing change has an impact on the quality of the product, it may be necessary to conduct additional non-clinical and clinical trials to assess the impact of the change on safety and efficacy.

35.7 FORMAT AND CONTENT OF A REGISTRATION DOSSIER

One of the initiatives of ICH (ICH M4) was the harmonization of the format of registration dossiers for new medicinal products, resulting in the 'Common Technical Document' (CTD). The CTD includes all technical, non-clinical (toxicological and pharmacological) and clinical information on a new drug, and any new registration dossier must be submitted in CTD format in the European Union, Japan, and Switzerland since 1 July 2003. In USA and Canada, although not a requirement, the CTD format is strongly recommended. The dossier is structured into five different modules:

- *Module 1*: Regional administrative information (not considered part of the CTD).
- *Module 2*: Introduction, quality overall summary (QOS), and non-clinical and clinical overviews and summaries.
- *Module 3*: Quality part.
- *Module 4*: Non-clinical study reports.
- *Module 5*: Clinical study reports.

The required structure for format of content for the drug substance part is presented in Table 35.3. The quality module of the CTD contains all relevant information about the chemistry, manufacturing and control of the drug substance and drug product.

Table 35.3 Headings of a drug substance part of a registration dossier.

S.1 General information	
	S.1.1 Nomenclature
	S.1.2 Structure
	S.1.3 General properties
S.2 Manufacture	
	S.2.1 Name of manufacturer
	S.2.2 Description of manufacturing process
	S.2.3 Control of materials
	S.2.4 Controls of critical steps and intermediates
	S.2.5 Process validation
	S.2.6 Manufacturing process development
S.3 Characterization	
	S.3.1 Elucidation of structure and other characteristics
	S.3.2 Impurities
S.4 Control of drug substance	
	S.4.1 Specifications
	S.4.2 Analytical procedures
	S.4.3 Validation of analytical procedures
	S.4.4 Batch analyses
	S.4.5 Justification of specification
S.5 Reference standards or materials	
S.6 Container closure system	
S.7 Stability	
	S.7.1 Stability summary and conclusions
	S.7.2 Post-approval stability protocol and stability commitment
	S.7.3 Stability data
A.1 Facilities and equipment	
A.2 Adventitious agents safety evaluation	
A.3 Excipients	

35.8 SUMMARY AND FUTURE PROSPECTS

Application of medicinal products is strictly controlled by the responsible health authorities of the different countries and regions. Despite national and regional differences in pharmaceutical legislation, the regulations all have the common goal of providing patients with safe and efficacious products of high and consistent quality. With the aim of overcoming the issue of different national guidelines with different requirements that have to be followed by the pharmaceutical industry, ICH started to work on the harmonization of regulatory guidance documents in Europe, the USA, and Japan more than 10 years ago. This process of harmonization will certainly continue and with ICH Q8, Q9 and Q10 the discussion on a new harmonized approach to implementing quality in the manufacturing process ('quality by design') has been initiated.

Several guidelines are in effect that provide instructions on the necessary information concerning manufacturing and control for the registration of animal cell products. These guidance documents will, however, be re-examined from time to time in order to take into account new experience and scientific and technological advances.

An example of a new approach to ensure the quality of the pharmaceutical product is the PAT (process analytical technology) initiative of the FDA and EMEA. One important approach of PAT is to include on-line monitoring of critical steps with the possibility of applying 'real time' corrective actions to the process parameters, if necessary (see Chapter 13). The goal of PAT is to include the control of quality in the manufacturing process, instead of performing quality control testing with individual samples of the final product.

Another field of change is the area of 'biogenerics' or 'biosimilar' products. For several recombinant biopharmaceuticals the patent protection has expired or will expire within the next few years. Because of the importance of these products, it can be expected that further legislation covering biogenerics will come into force in the near future.

Although the regulatory framework is changing rapidly, the main goal will still be to ensure that safe and efficacious pharmaceuticals can be provided to the public.

REFERENCE

Penn RG (1979) *Brit. J. Clin. Pharmac.*; 8: 293–305.

List of Useful Web Sites

Pharmaceutical Unit of the European Commission	http://pharmacos.eudra.org
Europa, the web site of the European Commission	http://ec.europa.eu
EMEA	http://www.emea.europa.eu
EU Health Authorities	http://www.hma.eu
FDA	http://www.fda.org
WHO	http://www.who.int
ICH	http://www.ich.org
European Directorate for the Quality of Medicines, EDQM	http://www.edqm.eu

APPENDIX 1

Non-exhaustive summary of relevant guidelines for biotechnology products

ICH Guidelines

ICH code	Title
Q5A (R1)	Viral safety evaluation of biotechnology products from cell lines of human or animal origin
Q5B	Quality of biotechnological products: analysis of the expression construct in cells used for production of rDNA derived protein products
Q5C	Quality of biotechnological products: stability testing of biotechnological/biological products
Q5D	Derivation and characterization of cell substrates used for production of biotechnological/biological products
Q5E	Comparability of biotechnological/biological products subject to changes in their manufacturing process
Q6B	Specifications: test procedures and acceptance criteria for biotechnological/ biological products
Q7	Good manufacturing practice guide for active pharmaceutical ingredients
S6	Preclinical safety evaluation of biotechnology-derived pharmaceuticals

WHO Guidelines

Code	Title
Technical Report Series, No. 800, 1990, Annex 4 and Addendum 1999, Technical Report Series, No. 904, 2002	Guidelines for the preparation, characterization and establishment of international and other standards and reference reagents for biological substances
Technical Report Series, No. 814, 1991	Guidelines for assuring the quality of pharmaceutical and biological products prepared by recombinant DNA technology
Technical Report Series, No. 878, 1998, Annex 1	Use of animal cells as *in vitro* substrates for the production of biologicals
Technical Report Series, No. 878, 1998, Annex 3	Guideline for assuring the quality of DNA vaccines
Technical Report Series, No. 908, 2003, Annex 4	Good manufacturing practices for pharmaceutical products: main principles

EMEA Guidelines

Year or Code	Title
1989	Preclinical biological safety testing on medicinal products derived from biotechnology
1990	Production and quality control of cytokine products derived by biotechnological processes
1995	Production and quality control of medicinal products derived by recombinant DNA technology
1995	Production and quality control of monoclonal antibodies
CPMP/BWP/268/95	Note for guidance on virus validation studies: The design, contribution and interpretation of studies validating the inactivation and removal of viruses
CPMP/BWP/382/97	CPMP position statement on DNA and host cell proteins (HCP) impurities, routine testing versus validation studies
CPMP/BWP/2490/00	Cell culture inactivated influenza vaccines – Annex to note for guidance on harmonization of requirements for influenza vaccines (CPMP/BWP/214/96)
CPMP/BWP/3088/99	Note for guidance on the quality, preclinical, and clinical aspects of gene transfer medicinal products
CPMP/BWP/41450/98	Points to consider on the manufacture and quality control of human somatic cell therapy medicinal products
CPMP/BWP/1143/00	Position statement on the use of tumourigenic cells of human origin for the production of biological and biotechnological medicinal products
CPMP/BWP/764/02-Draft	Points to consider on quality aspects of medicinal products containing active substances produced by stable transgene expression in higher plants
CPMP/BWP/1793/02	Note for guidance on the use of bovine serum in the manufacture of human biological medicinal products
EMEA/410/01 Rev.2	Note for guidance on minimizing the risk of transmitting animal spongiform encephalopathy agents via human and veterinary medicinal products
EMEA/CPMP/BWP/3207/00/Rev1	Guideline on comparability of medicinal products containing biotechnology-derived proteins as drug substance: Quality issues

(*continued*)

EMEA/ CPMP/3097/02	Guideline on comparability of medicinal products containing biotechnology-derived proteins as drug substance: Non-clinical and clinical issues
CPMP/1199/02	Points to consider on xenogeneic cell therapy medical products
EMEA/CPMP/ VEG/134716/2004	Guideline on adjuvants in vaccines
CHMP/ QWP/185401/2004	Guideline on the requirements to the chemical and pharmaceutical quality documentation concerning investigational medicinal products in clinical trials

FDA Guidelines

Year	Title
1985	Points to consider in the production and testing of new drugs and biologicals produced by recombinant DNA technology
1992	Supplement: Nucleic acid characterization and genetic stability
1993	Points to consider in the characterization of cell lines used to produce biologicals
1995	Points to consider in the manufacture and testing of therapeutic products for human use derived from transgenic animals
1996	FDA guidance concerning demonstration of comparability of human biological products, including therapeutic biotechnology-derived products
1996	Guidance on applications for products comprised of living autologous cells manipulated *ex vivo* and intended for structural repair or reconstruction
1997	Points to consider in the manufacture and testing of monoclonal antibody products for human use
1997	Proposed approach to regulation of cellular and tissue-based products
1997	Guidance for industry: changes to an approved application for specified biotechnology and synthetic biological products
1997	Guidance for industry: changes to an approved application: biological products
1998	Guidance for industry: guidance for human somatic cell therapy and gene therapy
2001	PHS guideline on infectious disease issues in xenotransplantation
2001	Guidance for industry: monoclonal antibodies used as reagents in drug manufacturing
2001	Guidance for industry: biological product deviation reporting for licensed manufacturers of biological products other than blood and blood components
2002	Guidance for industry: preventive measures to reduce the possible risk of transmission of Creutzfeld–Jakob Disease (CJD) and variant Creutzfeld–Jakob Disease (vCJD) by human cells, tissues, and cellular and tissue-based products (HCT/Ps), draft guidance
2002	Guidance for industry: drugs, biologics, and medical devices derived from bioengineered plants for use in humans and animals, draft guidance
2003	Guidance for industry: source animal, product, preclinical, and clinical issues concerning the use of xenotransplantation products in humans
2003	Guidance for industry: comparability protocols – protein drug products and biological products – chemistry, manufacturing, and controls information
2004	Guidance for Industry: PAT - a framework for innovative pharmaceutical development, manufacturing and quality assurance
2006	Guidance for Industry: characterization and qualification of cell substrates and other biological starting materials used in the production of viral vaccines for the prevention and treatment of infectious diseases (draft guidance)

Additional Guidance Documents for Good Manufacturing Practice

- Eudralex, The rules Governing Medicinal Products in Europe, Volume 4 – *Good Manufacturing Practice.*
- PIC/S – Pharmaceutical Inspection Convention/Pharmaceutical Inspection Co-operation Scheme: *Guide to Good Manufacturing Practice for Medicinal Products.*

36 New Areas: Cell Therapy and Tissue Engineering Products – Technical, Legal and Regulatory Considerations

L Tsang

36.1 INTRODUCTION

Medicines arising from the application of recombinant DNA and hybridoma technologies are already in clinical use, and the search for innovative developments in new areas of biotechnology is still ongoing. Progress in the understanding of biochemistry, cell and molecular biology, genetics, material science and biomedical engineering have stimulated interest in the clinical development of cell- or tissue-based therapeutic products for treating individuals suffering from life-threatening or seriously debilitating conditions, which may not be amenable to conventional clinical interventions. All these changes have continued to present technical and regulatory challenges to industry and regulatory authorities to ensure product quality, safety and efficacy.

There has been a long history of using mammalian cells as substrates for the production of therapeutic and prophylactic medicinal products. Human diploid fibroblast cells such as MRC-5, a cell line that was developed in the 1960s, have been used in the development of live attenuated viral vaccines. Established animal cell lines such as BHK and Chinese hamster ovary cells (CHO) have been widely used in producing a range of glycosylated recombinant proteins for treating, for example, infertility and haemophilia. The principles for the quality and safety evaluation of cell substrates are well established and have been published in various international or region-specific regulatory guidelines (see www.emea.europa.eu for EMEA web site; www.fda.gov. for US FDA web site). For example,

- *Note for Guidance on Quality of Biotechnological Products: Derivation and Characterization of Cell Substrates used for Production of Biotechnological/Biological Products* (Q5D) CPMP/ICH/294/95.
- *Note for Guidance on Quality of Biotechnological Products: Analysis of the Expression Construct in Cell Lines used for Production of r-DNA-derived Protein Products* (Q5B) CPMP/ICH/139/95.
- *CPMP Guideline on Production and Quality Control of Medicinal Products Derived by Recombinant DNA Technology.*

Medicines from Animal Cell Culture Edited by G. Stacey and J. Davis

This chapter, by contrast, reviews current scientific developments and highlights the related regulatory, legal and bioethical issues where the actual cells and tissues are the active components of the final products.

36.2 CELL-BASED IMMUNOTHERAPY

The use of cells for therapeutic purposes, such as adoptive immunotherapy, is an attractive proposition for several reasons. The technology established for manufacturing recombinant proteins can be adapted for the production of therapeutic proteins *in situ* using cells as the delivery vehicles following transplantation. Furthermore, appropriately engineered cells or irradiated cells can be used as therapeutics for treating diseases such as cancers and immune deficiencies by stimulating the patient's own immune system.

It is theoretically and experimentally possible to train a patient's own immune system to destroy a tumour or beat off an offending chronic viral infection. For the development of cancer therapeutic vaccines, it has been widely accepted that tumours express antigens that can be specifically recognized by the host immune system and these constitute the targets for cancer vaccine therapy (see Chapter 30). Advances in basic immunobiology such as the isolation and characterization of certain antigen-presenting cells, e.g. dendritic cells, have renewed interest in developing cell-based immunotherapeutics as potential clinical modalities for treating diseases like cancer. Two broad classes of tumour antigen have been identified that can be exploited for cancer vaccine development:

- tumour-specific antigens (TSA), which are mutated proteins or glycoproteins that are expressed by neoplastic tissue and not by non-malignant cells;
- tumour-associated antigens (TAA), which are antigens expressed by normal tissues, but aberrantly expressed by tumours in terms of levels and/or sites.

The principles underlying cancer therapeutic vaccines are to induce an immune response to TSAs and TAAs by the use of vaccine adjuvants. It has been argued that multivalent antigenic vaccines are preferable to a single component vaccine in order to avoid the phenomenon of 'antigen escape' where mutating cancer cells lose expression of a particular antigen, rendering it refractory to the particular single antigen vaccine. Lysed or whole cancer cells that have been inactivated by, for example, irradiation, contain a biological milieu with a broad spectrum of TAAs and TSAs. For such products, it is vitally important:

- to define the history of the derived cell-lines including the reagents used, and the procedures involved including the passage numbers;
- to define the antigen profiles, including their genotypic and phenotypic fidelity following passage;
- to control the reproducibility of the process to ensure batch consistency.

For cell-based vaccines derived from tumour cells, one area that deserves particular attention from a patient safety perspective is validation of the capability of the process to inactivate proliferating tumour cells. The availability of reliable, quantitative and sensitive assays is an essential prerequisite for the determination of the efficacy of the inactivation step prior to cell administration to patients.

Antigen-presenting cells can be used as delivery vehicles. For example, vaccination with dendritic cells pulsed with antigenic peptides derived from various tumour antigens has great and as yet unrealized potential in cancer treatment. Mechanistically, this seems to enhance the antitumour immunity by prolonging MHC class-restricted antigen presentation on dendritic cells (Wang & Wang 2002).

The strategy of bypassing the normal immune regulatory mechanism by generating an immune response *ex vivo*, and then returning the fully activated immune cells to the patient has also shown promise. Artificially engineered antigen-presenting cells can provide a defined system for the expansion of cytotoxic T cells *ex vivo* (Maus *et al.* 2002). This approach can augment current methods of T-cell stimulation [the prime candidate is cytotoxic T lymphocytes (CTL)], and provide a springboard for the generation of effector T cells for adoptive immunotherapy for tumour rejection or virus elimination (Dudley 2002).

Further, in the oncology arena, allogeneic, autologous or xenogeneic cells have been explored as candidates for *ex vivo* gene transfer before injection into the patient, where they localize specifically to tumour foci. The most obvious choice of the cell type as gene carriers would be certain immune cells, including macrophages, T cells, natural killer (NK) cells, and eosinophils (Rosenberg, 1993; Di Carlo *et al.* 2001). However, non-immune cells can be recruited as potential candidates, including stem cells. Such clinical approaches can augment anti-tumour immune responses offered by allogeneic cell-based cancer vaccines (Morton *et al.* 1993; Dalgleish 1996; Hsueh *et al.* 1998; Fabre 2001).

Despite this, further research is still required in a number of areas to realize the full potential of cell-based immunotherapies. These include:

- identification of immunogenic determinants responsible for mediating the relevant T cell immune response;
- optimization of strategies to diminish self-tolerance and to generate strong, long lasting, immunity through manipulation of both the antigen and the delivery system.

36.3 REGENERATIVE MEDICINES AND TISSUE ENGINEERING PRODUCTS

There is also an increasing interest in developing therapeutic products for restructuring or regenerating damaged tissues or organs. Growth factors have been or are being investigated for the regeneration of tissues such as bone, skin, cartilage, neurones and blood vessels, and the differentiation of stem cells into specific cell lineages. Some of these products have already been approved or are in clinical development. One example is bone morphogenetic proteins, such as osteogenic protein 1, which has been approved in the European Union under the European Centralized Procedure as a medicinal product for healing leg fractures. Another example is keratinocyte growth factor 2 that has been investigated for the healing of chronic leg ulcers. For such products, the active ingredients are embedded in a biocompatible matrix, which regulates the release of the growth factors.

In addition to the scientific endeavours in developing growth-factor-based regenerative products, there is also an increasing interest in using cells for repairing damaged or injured tissues or organs. There are a number of clinical targets including cartilage replacement using autologous chondrocytes (see Chapter 28), and skin replacements for diabetic ulcers using fibroblasts and/or keratinocytes. Indeed, the launch of a cartilage replacement therapy product, Carticel, has infused new interest into the development of tissue engineering products (Brittberg *et al.* 1994). Skin is a difficult organ to transplant because of its inherently strong immune defence system. Nevertheless, it is a useful testing ground for the potential clinical application of cell-based products. However, Apligraf, an artificial skin replacement product, has been approved in the United States as a device.

Notwithstanding the foregoing, the most pressing area for clinical development of cell-based therapeutics is to regenerate tissues that have a limited capacity for self-repair, and those that are readily damaged by injury or disease. The most obvious target is the brain and there are very few effective treatments currently available for devastating neurological conditions such as stroke, Alzheimer's and Parkinson's diseases. There have been considerable research efforts to develop new sources of therapeutic cells based on neuroprogenitor cells or neural stem cells; these are undifferentiated cells capable of self-renewal, proliferation, and differentiation into neurones or glia (Renfranz *et al.* 1991; Snyder *et al.* 1995; Gage 2000; McKay 2000). In experimental animals, these progenitor cells can differentiate into neural cells bearing specific characteristics to form integrated neuronal network and architecture (Shin *et al.* 2000). Such cells, with the capability of migrating extensively in the developing brain and to areas of injury (Snyder *et al.* 1995; Alvarez-Buylla *et al.* 2000) or tumours (Aboody *et al.* 2000), can therefore be exploited to enlarge the potential of gene transfer (see below).

36.4 STEM CELLS

The study of stem cells, begun 40 years ago within the haematopoietic system, has demonstrated that a single precursor cell present in the bone marrow (BM) of adult animals is capable of both extensive self-renewal and multi-lineage differentiation (Till & McCulloch, 1961; Wu *et al.* 1967, 1968). There is now a body of evidence to demonstrate that pluripotential stem cells from a variety of sources can be induced to differentiate into any of several lineages of choice (Pereira 1995; Goodell *et al.* 1996; Asahara *et al.* 1997; Prockop, 1997; Ferrari *et al.* 1998; Shamblott *et al.* 1998; Shi *et al.* 1998; Bittner *et al.* 1999; Bjornson *et al.* 1999; Brustles *et al.* 1999; Gussoni *et al.* 1999; Reyes *et al.* 2001). All these scientific endeavours have opened up an enormous vista of opportunity for all types of cell-based therapeutics. The plasticity of stem cells could therefore be exploited to create, under defined cultivation conditions, any cell or tissue type that is required for any particular clinical application, and may obviate the need to harvest specific cell types. Approaches include the use of adult, foetal or embryonic stem cells. An example of the former is the differentiation of bone marrow-derived stem cells into endothelial cells (Asahara *et al.* 1997; Shi *et al.* 1998).

Despite the on-going debate on the bioethical issues regarding the potential use of embryonic stem cells (ES), such cells have the dual ability to proliferate and to differentiate into all the cells and tissues of the body and hence are a very attractive proposition for further preclinical and clinical development (Orkin & Morrison 2002; see below on legal aspects). ES cells can also be produced by a process called nuclear replacement, which was used to create Dolly the sheep in 1996. All mammals are diploid genetically and receive a haploid set of genes from each parent at fertilization. In order to bypass the step of fertilization, an unfertilized ovum recovered after ovulation is used as the starting material for the initiation of embryonic development. The haploid set of chromosomes from the oocyte is removed and the donor somatic cell containing a diploid nucleus is fused with the oocyte. An external stimulus such as an electrical pulse is artificially applied to activate the ovum and to mediate the fusion process (Campbell *et al.* 1996; Schneicke, 1997; Seidel 1983; Wilmut *et al.* 1997).

Whilst adult stem cells are an ethically preferable alternative, scientists have not been able to show that they have the same versatility as ES cells. In experimental models, animal ES cells have shown their ability to differentiate into various cell types including insulin-secreting cells and neuronal cells (Reubinoff *et al.* 2000; Park *et al.* 2002).

It may be possible to isolate stem cell populations and engineer them under defined cultivation conditions or by genetic modification, to deliver therapeutic genes or to repopulate specific cell types in the injured areas. The true plasticity of stem cells at present is still being investigated

(Lemischka 2002). Indeed, there is currently a paucity of information on selective gene expression in ES cells, and in organ- or tissue-associated adult stem cells in different stages of their maturation into specific cell lineages. Such information is necessary to further our understanding of the genomic plasticity of human somatic cells, and the isolation and characterization of stem cells including cell-specific gene expression and transcriptional control. It will also enhance our approach to developing markers, which are necessary for the quality control of such products to ensure their genotypic and phenotypic fidelity with their cellular counterparts *in situ*.

36.5 CELL-BASED GENE TRANSFER

Cell-based gene transfer has been exploited for the potential treatment of neurological diseases and of cancer (Hsich 2002; Harrington 2002). It has been recognized that cells can be utilized as delivery vehicles for subsequent gene transfer into target cells. Macrophages, T cells and tumour cells or other non-immune differentiated cells together with stem cells have all been explored as potential candidates for gene transfer. The current technology for the manufacture of viral vectors for gene transfer has not achieved sufficient titres for therapeutic targeting because of one or a combination of the following reasons: low initial viral titres, immune inactivation, and non-specific adhesion and loss of viral particles. There are strategies to circumvent these biological and/or technical difficulties. One of them is to protect and chaperone the vectors until they have reached target sites. This is of particular interest in the field of cancer gene transfer (Wagers *et al.* 2002; Harrington *et al.* 2002). The vectors can be designed to incorporate payload mechanisms to achieve site-inducible expression of a tissue- or cell-specific gene expression cassette. Cells can also be engineered to contain a suicide gene, such as HSV-TK, which can be used to activate a pro-drug such as ganciclovir for cancer treatment (Garin *et al.* 2001).

For clinical applications, cells should ideally be normal and derived from the patient (autologous) in order to reduce immune rejection. Skin fibroblasts and endothelial cells are considered to be good initial candidates as they are usually readily grown in culture. Currently, research is underway to engineer universal donor cells by suppressing immune recognition, hence minimizing their rejection following transplantation (Schmouder 2000).

36.6 BIOMATERIALS FOR CELL-BASED THERAPY

Collaborative research between material scientists and stem cell biologists has resulted in novel approaches to tissue engineering for a variety of clinical indications. Polymer scaffold has been created to allow the growth of artificial skin and cartilage for use in hospitals to treat burns and damaged joints. Recently, using immortalized neural precursor cells seeded in a scaffold consisting of biodegradable polymer such as polyglycolic acid (PGA), it has been demonstrated that this combination approach can be utilized for potential treatment of extensive injuries sustained in the central nervous system (CNS) (Park *et al.* 2002; Ourednik *et al.* 2002).

Continued research and development of new biomaterials is, therefore, pivotal to the development of cell-based products particularly in relation to tissue engineering (Omstead *et al.* 1998; Mahoney & Saltzman 2001). The biomaterial component is important in the following respects:

- to provide the transplanted cells or tissues with a three-dimensional lattice or scaffold for the creation of an extracellular matrix;
- to sustain the viability and functionality of the transplanted cells or tissues, e.g. by bypassing the immune surveillance system of the recipient;
- to act as an adjuvant in the case of adoptive immunotherapy.

There are a number of key biochemical or physico-chemical properties that determine the suitability of a biomaterial for *in vivo* use:

- biocompatibility;
- sterilizability;
- processability and porosity;
- biodegradability;
- mechanical strength;
- surface properties that are conducive to site- or tissue-specific delivery of the transplant.

All of these properties will be significant factors in the development of new biomaterials for clinical use.

36.7 REGULATORY AND LEGAL ASPECTS

Although this chapter focuses on the regulatory system in operation within the European Union, references are made to other jurisdictions particularly in relation to the United States for the purpose of comparison.

In order to harness the on-going scientific endeavours in the development of tissue- and cell-based therapeutics, it is vitally important to establish a clear regulatory framework governing the control of such products. Although regulatory guidance has been developed by the Committee for Human Medicinal Products (CHMP) (previously known as Committee for Proprietary Medicinal Products (CPMP)) that sets out the broad principles and standards for the regulation of cell-based or tissue-derived medicinal products, there is still a lack of clarity and certainty about their classification and assessment on a pan-European basis, and this has rendered the commercialization of cell- and tissue-based products extremely difficult (see Wilson 2002 and below). This possibly stems from the fact that the legal definition of 'medicinal product' developed in the early 1960s is inadequate to accommodate the technologies emerging since the 1990s. Although the directive governing medical devices (Directive 93/42/EC) makes clear that certain combination products should be regulated as medical devices incorporating a medicinal substance with an ancillary action, there have been many instances where the European competent authorities have misclassified such products as medicinal. In policy terms, there is a need to stimulate debate as to whether or not the harmonized rules for medicinal products would be the most appropriate regulatory regime given that European Medical Device Directive 93/42/EEC explicitly excludes viable cells from its scope. If it is not, then, as proposed by the European Commission in 2002, there is a need to consider a Community legal framework for the regulation of human tissue engineering products. If this is the case, then there is a need to define what human tissue engineering products are. There is a need to consider the evolution of biotechnology-derived products that has taken place since the 1980s and the underlying scientific principles for safeguarding public health and patient safety. Irrespective of the regulatory regime to be applied, the approval of such products should be based on:

- an adequate quality system in relation to the control of the design, processing and supply of the product; this includes the control of raw materials including cells and culture media used in the process;
- an evaluation of the product performance which determines the risk/benefit or effectiveness of a given product taking into account the target disease and the patient population for which the product is intended.

Interests in the development of embryonic stem cell research and its relationship with cloning have also sparked off great controversies, some of which have precipitated challenges before courts. In the UK, the primary act of Parliament is The Human Fertilization and Embryology Act (HFEA) 1990, which purported to outlaw the reproductive cloning of human beings by virtue of Section 3. This provision prohibits 'replacing a nucleus of a cell of an embryo with a nucleus taken from a cell of any person, embryo or subsequent development of an embryo'. Section 1 defines an embryo as: 'a live human embryo where fertilization is complete, and ...references to an embryo include an egg in the process of fertilization and, for this purpose, fertilization is not complete until the appearance of a two cell zygote.'

The UK Government had held the position that the primary legislation was sufficient in scope to encompass any proposed development of human cloning whilst acknowledging the fact that the technique used in the creation of Dolly (see above) was different from the procedure as set out in Section 3. The validity of this assumption was tested in the legal challenge made by the Pro-Life Alliance in November 2001 in the case of *The Queen on the application of Bruno Quintavalle on behalf of the Pro-Life Alliance and the Secretary of State for Health*. This case was not about the cloning techniques themselves, but addressed the question as to whether or not the legislation, as currently worded, would apply to an organism (e.g. an embryo) that was created as a result of cloning, using a technique called cell nuclear replacement (CNR). The argument was run as follows: if CNR did not involve fertilization, then the corollary of this was that the embryo created by this procedure would fall outside the scope of the legislation and therefore the embryo would be unregulated by the Act. This case also challenged the Government's inclusion of stem cell research in the list of permissible research on human embryos by virtue of Human Fertilization and Embryology (Research Purposes) Regulation 2001 (24 January 2001). When this case was first considered in the High Court before Mr Justice Crane, the judge declared that, with some reluctance, organisms produced by CNR would fall outside the statutory and licensing framework. In the subsequent appeal before the Court of Appeal on 16 and 18 January 2002, the Court held that by reliance on other case law, it was in fact appropriate to give the relevant section a purposive construction so that an embryo created by CNR would fall within the regulatory framework of the Act. It concluded:

> '...a regulatory regime that excludes from its ambit embryos created by cell nuclear replacement is contrary to the intention of Parliament in introducing the 1990 Act. The prospect of such a regime is both startling and alarming. These considerations provide the most cogent reason to reach an interpretation of the Act which embraces embryos produced by cell nuclear replacement, subject to consideration of any countervailing considerations or incoherence'.

In December 2001, the UK enacted the Human Reproductive Cloning Act 2001, which '*prohibits the placing in a woman of a human embryo which has been created otherwise than by fertilization*'.

36.7.1 European Definition of a Cell Therapy Medicinal Product

Despite legal constraints in the definition of a medicinal product, the Committee for Proprietary Medicinal Products (CPMP) of the European Medicines Agency has promulgated a 'points-to-consider' regulatory document to assist with the definition of a cell therapy medicinal product. This working definition takes into account the Commission Communication issued in 1998 (98/C 229/03) regarding the scope of the harmonized rules governing the marketing authorization of medicinal products within the European Union, and the legal definition of a medicinal product pursuant to Article 1 of Directive 2001/83/EC. The working definition lays down the criteria for defining a cell therapy medicinal product for the purpose of regulation under the European harmonized rules for medicinal products.

In Annex I to 2001/83/EC (as amended by Directive 2003/63/EC), a somatic cell therapy medicinal product is defined as:

'For the purpose of this Annex, somatic cell therapy medicinal products shall mean the use in humans of autologous (emanating from the patient himself), allogeneic (coming from another human being) or xenogeneic (coming from animals) somatic living cells, the biological characteristics of which have been substantially altered as a result of their manipulation to obtain a therapeutic, diagnostic or preventive effect through metabolic, pharmacological and immunological means. The manipulation includes the expansion or activation of autologous cell populations *ex vivo* (e.g. adoptive immunotherapy) the use of allogeneic and xenogeneic cells associated with medical devices used *ex vivo* or *in vivo* (e.g. microcapsules, intrinsic matrix scaffolds, bio-degradable or not)'.

The CHMP has already adopted its regulatory guidance on xenogeneic cell therapy medicinal products following a period of public consultation.

36.7.2 United States FDA Definition for a Cell Therapy Product

The FDA, it appears, has a broader jurisdiction to regulate tissue- and cell-based products compared with its counterpart in the European Union. On 14 October 1993, the United States FDA articulated its statutory authority regarding human somatic cell therapy and gene therapy products (58 Fed Reg 53,248). Somatic cell therapy product is defined as:

'the prevention, treatment, cure, diagnosis, or mitigation of disease or injuries in humans by the administration of autologous, allogeneic, or xenogeneic cells that have been manipulated or altered *ex vivo*'.

In 1997, the agency proposed a framework for the regulatory oversight of cell- and tissue-based products within which it has laid down the regulatory approach based on the level of public health concern the products may pose. In parallel with this proposal, the agency has developed a number of guidelines concerning, amongst others, xenotransplantation, the use of xenotransplantation products in humans and good tissue practice.

According to the FDA rules, manufacturers are required to demonstrate that their product is qualitatively and quantitatively defined as safe. Similar to the European 'points-to-consider' document for cell therapy medicinal products, cells that are more than minimally manipulated should be characterized in terms of their biological properties. On 19 January 2001, the FDA issued its final rule, which requires establishments for human cells, tissues, and cellular and tissue-based products to be registered with the agency and to require such products to be listed.

36.7.3 Regulatory Standards and Requirements

In the light of the above and analogous to the regulation of blood products, donor testing for various viral markers is required. For certain cell-based products, such as those derived from animals other than humans (xenogeneic), additional safety measures such as bio-security in animal husbandry and testing of the animal tissues and cells are required. There is also a need to develop a strategy for patient surveillance and registry for the purpose of wider public health protection such as the monitoring of possible transmission of zoonotic agents. Details regarding the mechanism for this regulatory oversight have not yet been elaborated. No doubt this will stimulate legal and ethical debate on matters, for example, of privacy.

The regulatory guidance for human somatic cell therapy products issued by the US FDA and the EU CPMP share certain common features:

- the materials and safety parameters employed in the product manufacture including control of the starting materials in terms of their quality and safety, e.g. microbiological and virological purity;
- control and definition of the cell culture procedures;

- validation of handling procedures;
- characterization of the cells;
- specification and control of the final cell therapy product;
- control of the cell culture procedures to ensure product and process consistency;
- batch identification;
- lot release testing and quality control.

The United Kingdom Medical Devices Agency (MDA), currently part of the United Kingdom Medicines and Healthcare products Regulatory Agency, has also issued a *Code of Practice for the Production of Human-derived Therapeutic Products* (see www.medical-devices.gov.uk or www.mhra.gov.uk) that sets out the general principles, similarly addressed by the FDA and the CPMP, underpinning:

- product safety;
- quality systems and standards in the selection of donors, retrieval of tissues, testing, processing, storage and delivery;
- post-marketing surveillance.

A new directive (Directive 2004/23/EC) has been adopted in the European Union following a proposal made by the European Commission (under the Treaty provision for public health protection) to establish a pan-European framework for ensuring harmonized standards for the quality and safety of tissues and cells of human origin used for application in the human body. This directive must be transposed into the domestic laws of the member states in 2006 to give effect to the legal requirements. This regulatory framework will strengthen requirements related to the suitability of donors of tissues and cells and the screening of donated substances of human origin in the European Union, and to harmonize the requirements for establishments involved in procurement, testing, processing, storage, and distribution of tissues and cells of human origin, as well as national accreditation and monitoring structures. The directive is particularly wide-ranging as its scope covers human tissues and cells used for 'human applications', which according to the recitals should be interpreted broadly. In the UK, the Human Tissue Authority, which has been established under the Human Tissue Act 2004, is the competent authority for the regulations required by the EC directive.

As mentioned before, in July 2002, the European Commission Enterprise Directorate issued a public consultation document on the need for a legislative framework for human tissue engineering and tissue-engineered products. It has been said that such products should be regulated under the 'third pillar' regulation as they do not necessarily have the same characteristics as a medicinal product or a medical device. At the time of writing, the debate is still ongoing in the European Union as to the best way to regulate such products. "In November 2005, the European Commission adopted a legislative proposal for regulating tissue engineered products as an advanced therapy medicinal product. Under this proposal, tissue engineered product is defined to mean a product. Under this proposal, tissue engineered product is defined to mean a product that (1) contains or consists of engineered cells or tissues and (2) is presented as having properties for, or is used in or administered to human beings with a view to regenerating, repairing or replacing a human tissue. A cell or tissue is said to be engineered if it has been subject to substantial manipulation so that the original biological characteristics, physiological functions or structural properties relevant for the intended regeneration, repair or replacement are altered. The legislative proposal is, at the time writing, being considered by the European Parliament.

36.8 BIOETHICAL ISSUES

This section focuses on, in general terms, bioethical issues in relation to informed consent. Informed consent is needed in the following situations:

(a) before starting treatment or physical investigation or providing personal care for a patient;

(b) in the context of clinical investigations such as clinical trials (see e.g. Directive 2001/20/EC);

(c) procurement of tissues and cells from donors.

In general, above-mentioned activities should only be commenced if the individuals concerned have had the opportunity to give their valid consent. A valid consent means that:

- it is given voluntarily and freely, without pressure or undue influence being exerted on the individual concerned, based on information provided in a manner that he/she can understand;
- he/she is given the opportunity to ask questions;
- the individual has the capacity to give consent;
- the individual is able either to register his/her objection or to give explicit consent to, for example, particular tissue removal, storage or use.

In order to give valid consent, an individual must understand in broad terms the nature and purpose of the procedure. In considering what information to provide, the health professional should try to ensure that the patient or donor is able to make a balanced judgement on whether to give or withhold consent (see also the Department of Health *Consultation Report on Human Bodies, Human Choices: the law on human organs and tissue in England and Wales*).

Article 5 of the *Convention on Human Rights and Dignity of the Human Being with regard to the Application of Biology and Medicine: Convention on Human Rights and Biomedicine (Oviedo, 4.IV.1997)* provides:

> 'An intervention in the health field may only be carried out after the person concerned has given free and informed consent to it. This person shall beforehand be given appropriate information as to the purpose and nature of the intervention as well as on its consequences and risks. The person concerned may freely withdraw consent at any time'.

The convention also states that the broad principle of informed consent enshrined in Article 5 is applicable to activities regarding scientific (clinical) research and organ and tissue removal.

In the context of clinical trials, both the European Union and the United States FDA require informed consent, the rules of which are promulgated respectively under Directive 2001/20/EC and 21 Code of Federal Regulation (21 CFR 50.20). The European Directive as well as the International Conference on Harmonization Guideline for Good Clinical Practice make specific reference to the Declaration of Helsinki; the latest version was adopted following the General Assembly in Edinburgh in 2000. The key principles are set out below:

- Research investigators should be aware of the ethical, legal and regulatory requirements for research on human subjects in their countries as well as applicable international requirements.
- It is the duty of the physician in medical research to protect the life, health, privacy and dignity of the human subject.

- Medical research involving human subjects must conform to generally accepted scientific principles, and be based on a thorough knowledge of the scientific literature.
- Predictable risks and burdens should be assessed in comparison with foreseeable benefits.
- Informed consent should be obtained.

In 1998, the European Commission's Group on Ethics (EGE) reported on the following aspects:

- the ethical imperative to protect public health to minimize risk of disease transmission;
- the integrity of the human body;
- the importance of seeking the prior, informed and valid consent of donors; the donor's consent must be given on the basis of information provided in as clear and precise terms as possible;
- the protection of identity of the donor and recipient to prevent possible discrimination.

The main issues raised by the EGE report have been included in the latest draft of the proposed directive for regulating the procurement and storage of human tissues and cells. Articles 13 and 14 respectively concern matters regarding consent, data protection and confidentiality.

36.9 CONCLUSIONS

The realization of the full potential of tissue- and cell-based products in a clinical setting demands robust scientific evidence, underpinned by legitimate regulatory requirements to ensure their safety and efficacy. The arena of tissue- and cell-based products cuts across the realms of biology, physics, chemistry, materials sciences and clinical practice. There is, therefore, a need to adopt a multidisciplinary approach to addressing key issues pertaining to the design and functional assessment of such products. The accrual of such knowledge, together with equally important information on the mechanisms of cell differentiation and tissue repair, will form a rational basis for future clinical developments of these products. In parallel with such scientific endeavours, a clear and transparent regulatory system is needed, having regard to accepted bioethical principles. There is an urgent need to consider how such products should be optimally and appropriately regulated in the interest of public health protection, without causing unnecessary hindrance to patients wishing to access innovative treatments. In order to achieve this, it is imperative for the regulatory authorities to work in partnership with the wider scientific community as well as the patient groups.

REFERENCES

Aboody KS, Brown A, Rainov NG *et al.* (2000) *Proc. Natl. Acad. Sci. USA*; 97: 12846–12851.
Alvarez-Buylla A, Herrera DG, Wichterle H (2000) *Prog. Brain Res.*; 127: 1–11.
Asahara T, Murohara T, Sullivan A *et al.* (1997) *Science*; 275: 964–967.
Bittner RE, Schofer C, Weipoltshammer K *et al.* (1999) *Anat. Embryol. (Berl)*; 199: 391–396.
Bjornson CR, Rietze RI, Reynolds BA, Magli MC, Vescovi AL (1999) *Science*; 283: 534–537.
Blackburn CC, Manley NR, Palmer DB, Boyd RL, Anderson G, Ritter MA (2002) *Trends Immunol.*; 23(8): 391–395.
Brittberg M, Lindahl A, Nilsson A, Ohlsson C, Isaksson O, Peterson L (1994) *New Eng. J. Med.*; 331: 889–895.
Brustles O, Jones KN, Learish RD *et al.* (1999) *Science*; 285: 754–756.
Campbell KHS, McWhir J, Ritchie WA, Wilmut I (1996) *Nature*; 380: 64–66.

Committee for Proprietary Medicinal Products (1999) *Note for Guidance on the Quality, Preclinical and Clinical Aspects of Gene Transfer Medicinal Products* (CPMP/BWP/3088/99).

Committee for Proprietary Medicinal Products (1998) *Points to Consider on the Manufacture and Quality Control of Human Somatic Cell Therapy Medicinal Products* (CPMP/BWP/41450/98).

Committee for Proprietary Medicinal Products (2002) *Points to Consider on Xenogeneic Cell Therapy Medicinal Products* (Draft) (CPMP/1199/02).

Dalgleish A (1996) *Melanoma Res.*; 6: 5–10.

Di Carlo E, Forni G, Lollini P, Colombo MP, Modesti A, Musiani P (2001) *Blood*; 97: 339–345.

Department of Health (2002) *Human Bodies, Human Choices- The Law on Human Organs and Tissue in England and Wales* (A Summary Consultation Report) July 2002.

Dudley ME (2002) *Nature Biotechnol.*; 20: 125–126.

European Group on Ethics in Science and New Technologies (1998) *Report on Ethical Aspects of Human Tissue Banking: Opinion to the European Commission*, No. 11, 21 July 1998.

The European Parliament and Council (2001) *Directive* 2001/20/EC *on the Approximation of the Laws, Regulations and Administrative Provisions of the Member States Relating to the Implementation of Good Clinical Practice in the Conduct of Clinical Trials on Medicinal Products for Human Use.* OJ 2001 L121: 34–44.

The European Parliament and Council (2002) *Directive Setting Standards of Quality and Safety for the Donation, Procurement, Testing, Processing Storage and Distribution of Human Tissues and Cells* (draft) COM (2002) 319 (final) 19 June 2002.

Fabre JW (2001) *Nat. Med.*; 7: 649–652.

Ferrari G, Cussella-De Angelis G, Coletta M *et al.* (1998) *Science*; 279:1528–1530.

Food and Drug Administration (1991) *Points to Consider in Human Somatic Cell Therapy and Gene Therapy* as updated.

Gage FH (2002) *Science*; 287: 1433–1438.

Garin MI, Garrett E, Tiberghien P *et al.* (2001) *Blood*; 97: 122–129.

Goodell MA, Brose K, Paradis G, Conner AS, Mulligan RC (1996) *J. Exp. Med.*; 183: 1797–1806.

Gussoni E, Soneoka Y, Strickland CD *et al.* (1999) *Nature*; 401: 390–394.

Harrington K, Alvarez-Vallina L, Crittenden M *et al.* (2002) *Human Gene Ther.*; 13: 1263–1280.

Hsich G, Sena-Esteves M, Breakefield XO (2002) *Human Gene Ther.*; 13: 579–604.

Hsueh EC, Gupta RK, Qi K, Morton DL (1998) *J. Clin. Oncol.*; 16: 2913–2920.

Kim J-H, Auerbach JM, Rodriguez-Gomez JA *et al.* (2002) *Nature*; 418: 50–56.

Lemischka I (2002) *Nature Biotechnol.*; 20: 425.

Lincoln Tsang, Partner, Arnold & Porter LLP (1998) (C98/C 229/03) on the Community marketing authorization procedures for medicinal products, OJ 22 July 1998.

Mahoney MJ, Saltzman WM (2001) *Nature Biotechnol.*; 19: 934–939.

Maus MV, Thomas AK, Leonard D *et al.* (2002) *Nature Biotechnol.*; 20: 143–148.

McKay R (2000) *Nature*; 406: 361–364.

Medical Devices Agency (2002) *A Code of Practice for the Production of Human-derived Therapeutic Products.* HMSO, London.

Morton DL, Foshag LJ, Hoon DSB *et al.* (1993) *Ann. Surg.*; 216: 463–482.

Omstead DR, Baird LG, Christensen L *et al.* (1998) *Tiss. Eng.*; 4: 239–266.

Orkin SH, Morrison SJ (2002) *Nature*; 418: 25–27.

Ourednik J, Ourednik V, Lynch WP, Schachner M, Snyder EY (2002) *Nature Biotechnol.*; 20: 1103–1110.

Park KI, Teng YD, Snyder E (2002) *Nature Biotechnol.*; 20: 1111–1117.

Pereira RF (1995) *Proc. Natl. Acad. Sci. USA*; 92: 4857–4861.

Prockop DJ (1997) *Science*; 276: 71–74.

Renfranz PJ, Cunningham MG, McKay RD (1991) *Cell*; 66: 713–729.

Reubinoff BE, Pera MF, Fong C-Y, Trounson A, Bongso A (2000) *Nature Biotechnol.*; 18: 399–404.

Reyes M, Lund T, Lenvik T, Aguiar D, Kodie L, Verfaillie CM (2001) *Blood*; 98: 2615–2625.

Rosenberg SA, Lotze MT, Yang JC *et al.* (1993) *J. Natl. Cancer Inst.*; 85: 622–632.

Shamblott MJ, Axelman J, Wang S *et al.* (1998) *Proc. Natl. Acad. Sci. USA*; 95: 13726–13731.

Schneicke AE, Kind AJ, Richie WA *et al.* (1997) *Science*; 278: 2130–2133.

Schmouder RL (2000) *Transplant Proc.*; 32: 1463–1467.

Seidel GE (1983) *J. Expt. Zool.*; 228: 347–354.
Shi Q, Rafii S, Hong-De Wu M *et al.* (1998) *Blood*; 92: 362–367.
Shin JJ, Fricker-Gates RA, Perez FA, Leavitt BR, Zurakowski D, Macklis JD (2000) *J. Neurosci.*; 20: 7404–7416.
Snyder EY, Taylor RM, Wolfe JH (1995) *Nature*; 374: 367–370.
Till J, McCulloch E (1961) *Radiat. Res.*; 14: 1419–1430.
Wagers AJ, Christensen JL, Weissman IL (2002) *Gene Ther.*; 9: 606–612.
Wang R-F, Wang HY (2002) *Nature Biotechnol.*; 20: 149–154.
Wilmut I, Schnieke AE, McWhir J, Kind AJ, Campbell KHS (1997) *Nature*; 385: 810–813.
Wilson A (2002) *The Regulatory Affairs Journal (Devices)*; third quarter: 207–210.
Wu A, Till J, Siminovitch L, McCulloch EA (1967) *J. Cell Physiol.*; 69: 177–184.
Wu A, Till J, Siminovitch L, McCulloch E (1968) *J. Exp. Med.*; 127: 455–467.

Index

Medicines from Animal Cell Culture Edited by G. Stacey and J. Davis

This index was prepared by Neil Manley.